Lecture Notes in Computer Science   **13451**

More information about this series at https://link.springer.com/bookseries/558

Alexander Gelbukh (Ed.)

Computational Linguistics and Intelligent Text Processing

20th International Conference, CICLing 2019
La Rochelle, France, April 7–13, 2019
Revised Selected Papers, Part I

 Springer

Editor
Alexander Gelbukh
Instituto Politécnico Nacional
Mexico City, Mexico

ISSN 0302-9743   ISSN 1611-3349 (electronic)
Lecture Notes in Computer Science
ISBN 978-3-031-24336-3   ISBN 978-3-031-24337-0 (eBook)
https://doi.org/10.1007/978-3-031-24337-0

This Springer imprint is published by the registered company Springer Nature Switzerland AG
The registered company address is: Gewerbestrasse 11, 6330 Cham, Switzerland

Preface

CICLing 2019 was the 20th International Conference on Computational Linguistics and Intelligent Text Processing. The CICLing conferences provide a wide-scope forum for discussion of the art and craft of natural language processing research, as well as the best practices in its applications.

This set of two books contains three invited papers and a selection of regular papers accepted for presentation at the conference. Since 2001, the proceedings of the CICLing conferences have been published in Springer's Lecture Notes in Computer Science series as volumes 2004, 2276, 2588, 2945, 3406, 3878, 4394, 4919, 5449, 6008, 6608, 6609, 7181, 7182, 7816, 7817, 8403, 8404, 9041, 9042, 9623, 9624, 10761, 10762, 13396, and 13397.

The set has been structured into 14 sections representative of the current trends in research and applications of natural language processing: General; Information Extraction; Information Retrieval; Language Modeling; Lexical Resources; Machine Translation; Morphology, Syntax, Parsing; Name Entity Recognition; Semantics and Text Similarity; Sentiment Analysis; Speech Processing; Text Categorization; Text Generation; and Text Mining.

In 2019 our invited speakers were Preslav Nakov (Qatar Computing Research Institute, Qatar), Paolo Rosso (Universidad Politécnica de Valencia, Spain), Lucia Specia (University of Sheffield, UK), and Carlo Strapparava (Foundazione Bruno Kessler, Italy). They delivered excellent extended lectures and organized lively discussions. Full contributions of these invited talks are included in this book set.

After a double-blind peer review process, the Program Committee selected 95 papers for presentation, out of 335 submissions from 60 countries.

To encourage authors to provide algorithms and data along with the published papers, we selected three winners of our Verifiability, Reproducibility, and Working Description Award. The main factors in choosing the awarded submission were technical correctness and completeness, readability of the code and documentation, simplicity of installation and use, and exact correspondence to the claims of the paper. Unnecessary sophistication of the user interface was discouraged; novelty and usefulness of the results were not evaluated, instead they were evaluated for the paper itself and not for the data.

The following papers received the Best Paper Awards, the Best Student Paper Award, as well as the Verifiability, Reproducibility, and Working Description Awards, respectively:

Best Verifiability, Reproducibility, and Working Description Award: "Text Analysis of Resumes and Lexical Choice as an Indicator of Creativity", Alexander Rybalov.

Best Student Paper Award: "Look Who's Talking: Inferring Speaker Attributes from Personal Longitudinal Dialog", Charles Welch, Veronica Perez-Rosas, Jonathan Kummerfeld, Rada Mihalcea.

Best Presentation Award: "A Framework to Build Quality into Non-expert Translations", Christopher G. Harris.

Best Poster Award, Winner (Shared): "Sentiment Analysis Through Finite State Automata", Serena Pelosi, Alessandro Maisto, Lorenza Melillo, and Annibale Elia. And "Toponym Identification in Epidemiology Articles: A Deep Learning Approach", Mohammad Reza Davari, Leila Kosseim, Tien D. Bui.

Best Inquisitive Mind Award: Given to the attendee who asked the most (good) questions to the presenters during the conference, Natwar Modani.

Best Paper Award, First Place: "Contrastive Reasons Detection and Clustering from Online Polarized Debates", Amine Trabelsi, Osmar Zaiane.

Best Paper Award, Second Place: "Adversarial Training based Cross-lingual Emotion Cause Extraction", Hongyu Yan, Qinghong Gao, Jiachen Du, Binyang Li, Ruifeng Xu.

Best Paper Award, Third Place (Shared): "EAGLE: An Enhanced Attention-Based Strategy by Generating Answers from Learning Questions to a Remote Sensing Image", Yeyang Zhou, Yixin Chen, Yimin Chen, Shunlong Ye, Mingxin Guo, Ziqi Sha, Heyu Wei, Yanhui Gu, Junsheng Zhou, Weiguang Qu.

Best Paper Award, Third Place (Shared): "dpUGC: Learn Differentially Private Representation for User Generated Contents", Xuan-Son Vu, Son Tran, Lili Jiang.

A conference is the result of the work of many people. First of all, I would like to thank the members of the Program Committee for the time and effort they devoted to the reviewing of the submitted articles and to the selection process. Obviously, I thank the authors for their patience in the preparation of the papers, not to mention the development of the scientific results that form this book. I also express my most cordial thanks to the members of the local Organizing Committee for their considerable contribution to making this conference become a reality.

November 2022   Alexander Gelbukh

Organization

CICLing 2019 (20th International Conference on Computational Linguistics and Intelligent Text Processing) was hosted by the University of La Rochelle (ULR), France, and organized by the L3i laboratory of the University of La Rochelle (ULR), France, in collaboration with the Natural Language and Text Processing Laboratory of the CIC, IPN, the Mexican Society of Artificial Intelligence (SMIA), and the NewsEye project. The NewsEye project received funding from the European Union's Horizon 2020 research and innovation programme under grant agreement No 770299.

The conference aims to encourage the exchange of opinions between the scientists working in different areas of the growing field of computational linguistics and intelligent text and speech processing.

Program Chair

Alexander Gelbukh	Instituto Politécnico Nacional, Mexico

Organizing Committee

Antoine Doucet (Chair)	University of La Rochelle, France
Nicolas Sidère (Co-chair)	University of La Rochelle, France
Cyrille Suire (Co-chair)	University of La Rochelle, France

Members

Karell Bertet	L3i Laboratory, University of La Rochelle, France
Mickaël Coustaty	L3i Laboratory, University of La Rochelle, France
Salah Eddine	L3i Laboratory, University of La Rochelle, France
Christophe Rigaud	L3i Laboratory, University of La Rochelle, France

Additional Support

Viviana Beltran	L3i Laboratory, University of La Rochelle, France
Jean-Loup Guillaume	L3i Laboratory, University of La Rochelle, France
Marwa Hamdi	L3i Laboratory, University of La Rochelle, France
Ahmed Hamdi	L3i Laboratory, University of La Rochelle, France
Nam Le	L3i Laboratory, University of La Rochelle, France
Elvys Linhares Pontes	L3i Laboratory, University of La Rochelle, France
Muzzamil Luqman	L3i Laboratory, University of La Rochelle, France
Zuheng Ming	L3i Laboratory, University of La Rochelle, France
Hai Nguyen	L3i Laboratory, University of La Rochelle, France

| Armelle Prigent | L3i Laboratory, University of La Rochelle, France |
| Mourad Rabah | L3i Laboratory, University of La Rochelle, France |

Program Committee

Alexander Gelbukh	Instituto Politécnico Nacional, Mexico
Leslie Barrett	Bloomberg, USA
Leila Kosseim	Concordia University, Canada
Aladdin Ayesh	De Montfort University, UK
Srinivas Bangalore	Interactions, USA
Ivandre Paraboni	University of São Paulo, Brazil
Hermann Moisl	Newcastle University, UK
Kais Haddar	MIRACL Laboratory, Faculté des Sciences de Sfax, Tunisia
Cerstin Mahlow	ZHAW Zurich University of Applied Sciences, Switzerland
Alma Kharrat	Microsoft, USA
Dafydd Gibbon	Bielefeld University, Germany
Evangelos Milios	Dalhousie University, Canada
Kjetil Nørvåg	Norwegian University of Science and Technology, Norway
Grigori Sidorov	CIC-IPN, Mexico
Hiram Calvo	Nara Institute of Science and Technology, Japan
Piotr W. Fuglewicz	TiP, Poland
Aminul Islam	University of Louisiana at Lafayette, USA
Michael Carl	Kent State University, USA
Guillaume Jacquet	Joint Research Centre, EU
Suresh Manandhar	University of York, UK
Bente Maegaard	University of Copenhagen, Denmark
Tarık Kişla	Ege University, Turkey
Nick Campbell	Trinity College Dublin, Ireland
Yasunari Harada	Waseda University, Japan
Samhaa El-Beltagy	Newgiza University, Egypt
Anselmo Peñas	NLP & IR Group, UNED, Spain
Paolo Rosso	Universitat Politècnica de València, Spain
Horacio Rodriguez	Universitat Politècnica de Catalunya, Spain
Yannis Haralambous	IMT Atlantique & UMR CNRS 6285 Lab-STICC, France
Niladri Chatterjee	IIT Delhi, India
Manuel Vilares Ferro	University of Vigo, Spain
Eva Hajicova	Charles University, Prague, Czech Republic
Preslav Nakov	Qatar Computing Research Institute, HBKU, Qatar

Software Reviewing Committee

Ted Pedersen
Florian Holz
Miloš Jakubíček
Sergio Jiménez Vargas
Miikka Silfverberg
Ronald Winnemöller

Best Paper Award Selection Committee

Alexander Gelbukh
Eduard Hovy
Rada Mihalcea
Ted Pedersen
Yorick Wilks

Contents – Part I

Information Retrieval

Language Modeling

Lexical Resources

Machine Translation

Morphology, Syntax, Parsing

Contents – Part II

Sentiment Analysis

Speech Processing

Text Categorization

Text Generation

Text Mining

General

Visual Aids to the Rescue: Predicting Creativity in Multimodal Artwork

Carlo Strapparava$^{(\boxtimes)}$, Serra Sinem Tekiroglu, and Gözde Özbal

FBK-irst, Trento, Italy
strappa@fbk.eu

Abstract. Creativity is the key factor in successful advertising where catchy and memorable media is produced to persuade the audience. Considering not only advertising slogans but also the visual design of the same advertisements would provide a perceptual grounding for the overall creativity, consequently the overall message of the advertisement. In this study, we propose the exploitation of visual modality in creativity assessment of naturally multimodal design. To the best of our knowledge, this is the first study focusing on the computational detection of multimodal creative work. To achieve our goal, we employ several linguistic creativity detection features in combination with bag of visual words model and observable artistic visual features. The results of the creativity detection experiment show that combining linguistic and visual features significantly improves the unimodal creativity detection performances.

Keywords: NLP for Creative Language · Images and language · Multimodality

1 Introduction

Making an advertisement catching and memorable is the core task of creative people behind any original and effective campaign. Especially analysing award-winning ads, it is possible to appreciate a range of approaches including ways of visualizing concepts, the use of rhetorical devices, such as exaggeration, paradox, metaphor and analogy, and taking advantage of shock tactics and humour [22]. In any case, visual and text modalities are carefully thought to have a complementary and coordinated effect. While a computational treatment of the most subtle techniques is still very challenging, we think it is worthwhile to start exploring this topic in order to have a better understanding of contributing factors that can be utilized in computational creativity.

Accordingly, automatically quantifying the creativity level of multimodal design might be beneficial in various purposes such as choosing the potential successful advertisements, creative language and image generation for educational material or even a computational assessment of artistic value of the multimodal artwork.

© Springer Nature Switzerland AG 2023
A. Gelbukh (Ed.): CICLing 2019, LNCS 13451, pp. 3–16, 2023.
https://doi.org/10.1007/978-3-031-24337-0_1

As a topic being on the rise in computational linguistics, multimodality is mostly exploited on top of the linguistic models to perceptually ground the current tasks. For instance, semantic representations benefit from the reinforcement of linguistic modality with visual [4,5,9] and auditory [13] modalities. In the same manner, we propose devising visual modality in collaboration with linguistic modality in creativity detection task. To the best of our knowledge, this is the first study aiming to identify multimodal creativity in a computational manner. Moreover, the multimodality type of the dataset stands out amongst the others since the linguistic channel of an advertisement is complementary to the visual channel instead of being a scene description or an image label.

We used a set of naturally creative images; an advertising dataset which is composed of 500 images and corresponding slogans. As the counterpart of the creative data class, we investigated WordNet synset definitions and corresponding images from ImageNet. Figure 1 exemplifies the creative, non-creative tuples in the final dataset. Although these images seem to be very similar at the first glance with a child in the middle of the frame, the subtle and creative details in the left picture of Fig. 1, such as the male figure held by the child, immediately draw audience's attention. In this study, our focus will be to capture these creative properties both in visual and linguistic modalities to generate a multimodal creativity detection model.

We conjecture that in order to get meaningful information from a small dataset, we need to use the features that are as generalized as possible. To this end, we employ a Bag-of-Visual-Words (BoVW) model to determine if creativity in images display common visual characteristics and if these characteristics have a positive effect on overall creativity of a multimodal advertisement.

The rest of the paper is organized as follows. We first give a brief summary of the relevant multimodality and computational creativity studies in Sect. 2. In Sect. 3, we give the basis of our work in terms of the dimensions of advertising creativity. We present the creativity dataset that we collected in Sect. 4. Section 5 and 6 include the visual and linguistic creativity detection models, while Sect. 7 summarizes the experiments that we conducted.

2   Related Work

Considering that the essential focus of this study is computational creativity and multimodality, we summarize the most relevant studies conducted on these topics.

[2] proposes a categorization declaring two kinds of creativity: historical and psychological. Psychological creativity correlates with a surprising, valuable idea which might only be new to the thinker, whereas historical creativity dictates a chronologically novel idea within the human history. [8] quantifies the creativity in paintings within the context of historical creativity where creative paintings adequately differ from the antecedent paintings and influence the subsequent. They present a computational framework that is based on a creativity implication network. The proposed framework is exploited in creativity quantification task in paintings.

Regarding the linguistic creativity, [18] present a creative sentence generation framework, BRAINSUP, on which several semantic aspects of the output sentence can be calibrated. The syntactic information and a huge solution space are utilized to produce catchy, memorable and successful sentences. In another study, [20] explore common and latent methods of brand and product naming and produce a gold standard for creative naming. [14] focus on identifying creativity in lexical compositions. They consider two computational strategies, first investigating the information theoretic measures and the connotation of words to find the correlates of perceived creativity and then employing supervised learning with distributional semantic vectors.

Concerning the multimodality, [3] analyze the affect of different types of visual features such as SIFT and LAB on semantic relatedness task, and present a comparison of unimodal and multimodal models. [25] propose a model that employs stacked autoencoders to learn joint embeddings from both modalitie encoded as vectors of attributes. They performed similarity judgment and concept categorization tasks to evaluate the model. [23] experiment on a complementary multimodal dataset similar to ours. They explore the influence of the metadata (i.e., titles, description and artist's statement) of an abstract painting for the computational sentiment detection task. For the combination of modalities, they propose a novel joint flexible Schatten p-norm model exploiting the common patterns shared across visual and textual information. [24] exploit visual modality to improve the metaphor detection performance.

3   Creativity in Advertising

It would be misleading to define creativity, which is a highly broad term, from a single point of view. For this study, we only try to find a subset of properties that would be considered as the clues of creativity in the products of creative thinking and production process in line with the objective creativity proposed by [12]. In this categorization, subjective creativity refers to the creative mental activities, while objective creativity refers to the creativity in the resultant product.

The creativity elements and dimensions in advertising have been investigated thoroughly. [1] explore the influence of dimensions of creativity such as novelty (expectancy), meaningfulness (relevancy), and emotion (valence of feelings) to the effectiveness of the advertisement. While novelty could be identified as the unexpectedness and out-of-box degree of an advertisement, meaningfulness is the relevancy of the advertisement to the message aimed to be conveyed. The third dimension, emotional content, focuses on the feelings awakened in the audience. These three dimensions should manifest themselves in a creative advertising media.

[26], on the other hand, elaborate on the divergence, which is the encapsulation of novel, different, or unusual elements, in ads proposing that the most significant characteristic of creative ads is their divergence. They empirically assemble five factors of divergence as the dimensions of advertising creativity: i) Originality: An advertisement should contain surprising and uncommon elements to exhibit the dimension of originality; ii) Flexibility: An advertisement

should include different ideas or the capability of switching between the ideas; iii) Elaboration: Unexpected details, stressing and extending the subtle properties form the basis of the elaboration dimension; iv) Synthesis: Blending or putting together unrelated objects or ideas should be included in a creative advertisement; v) Artistic Value: Advertising creativity implies artistic creativity where the creative item contains aesthetically appealing verbal, visual, or sound elements. When considered as a piece of art, ad images can be described with the artistic concepts such as space, texture, form, shape, color, tone, line, variety, balance, brush strokes and many others [8, 10].

Considering the dimensions of creativity described by [1, 26], we design our features in order to cover as many dimensions as possible in linguistic and visual representations of advertising. To be more precise, we intend to capture surprisal, novelty, emotional and unusual properties in an advertising slogan. Yet another creativity infusion strategy in advertisement production is using sensory words especially generating linguistic synaesthesia as an imagination boosting tool [22]. To this end, we also try to quantify the human sense relations in linguistic modality. Moreover, we aim to extract artistic components in the visual elements in addition to the latent visual descriptions.

Fig. 1. Creative sample (Left) with slogan 'Adopt. You will receive more than you can ever give.', Non-creative sample (Right) with definition 'a young person of either sex'

4 Multimodal Creativity Dataset

To investigate the creativity in a multimodal configuration, we first identify two sets of data each of which reflects a distinctive level of creativity. Advertisements, as our first dataset, are manufactured at the end of a well-designed creation process. We collected 500 award winning advertising images and their slogans as the creative class of our dataset from Ads of the World Awards[1], Art Directors Club Awards[2], and The Cannes Lions International Advertising Festival[3]. We chose the mentioned websites respecting their leadership in advertising and design community. To obtain the non-creative counterparts for multimodal advertisements, similar to [14] we explored the dictionary glosses in WordNet, which aim to convey the meaning of a word in the least figurative way. Therefore, glosses are one of the best candidates for a non-creative dataset as an opposite ending of the creativity spectrum.

Considering the visual description of WordNet glosses and as the non-creative counterparts of advertising images, we exploit ImageNet in which we can find images associated with the WordNet synsets. For each advertising slogan, we extracted a list of keywords from the product type, product category and slogan itself. Since ImageNet covers a limited number of synsets, we tried to extract as many keywords as possible to guarantee a conjugate non-creative example for each advertising image. Then, these keywords are queried in WordNet to generate a candidate synset list as a query for ImageNet search. The resultant multimodal dataset consists of 500 advertising slogan-image tuples and 500 synset image-gloss tuples.

Since the creative and non-creative images might differ from each other in terms of the image quality and the image components, we pre-processed the images in order to be able to make the task as fair as possible. To this respect, after we find a counterpart for each ad image from ImageNet, we embed the definition part of the gloss of the associated synset into each image. By doing so, we aim to have a convention between the classes since all the ad images have slogan text embedded on them. We randomize the color, the font type, the font size and the position of the embedded text. In addition to preparing the non-creative images with embedded text, we also standardize the image sizes by resizing the images so that the longest edge of an image can have a length of maximum 300 px.

5 Creativity Detection in Visual Modality

Creativity on the images is attempted to be quantified by utilizing 2 visual models: i) Bag of Visual Words (BOVW) and ii) Observable Visual Features (OVF) including lines, circles and dominant colors.

---

[1] http://adsoftheworld.com/.

[2] http://adcglobal.org/awards/.

[3] http://www.canneslions.com/.

5.1   Bag of Visual Words

In regard to the intuition that creative images display common latent characteristics, we intend to find a representation that allows us to easily capture these commonalities. To this respect, we adopt the bag of visual words (BOVW) approach that is considered as a discrete representation for images [3] with a common vocabulary. Therefore we construct the first visual model to identify the creative images by constructing a Bag of Visual Words (BOVW) pipeline which is highly embraced by the community of multimodal studies [3,23,29].

The first step of the BOVW pipeline is to obtain a visual vocabulary from an adequately large image set. Inspired by [5] for the whole pipeline, we exploit ESP Game Dataset[4] as the training corpus to create the visual vocabulary. We extract large number of local features, i.e. descriptors, from 10K images. We employ one of the most appreciated descriptor extraction methods, the Scale Invariant Feature Transform (SIFT) [15], which is invariant to orientation, scaling and partly invariant to illumination change, affine distortion and noise. To identify the SIFT features we devise *vl_phow* wrapper that extracts the descriptors for densely sampled SIFT keypoints at multiple scales, which is included in the the VLFeat toolbox [28]. Rather than detecting particular keypoints using Difference of Gaussian Filtering as SIFT does, Dense Sift computes the descriptors in densely sampled locations on the image, in our setup, on every five pixels as a regular grid at the scales (10, 15, 20, 25 pixel radii). Instead of default greyscale computation of descriptors, we utilize the HSV color space. As a result, we obtain descriptors for each H, S and V channels. After the SIFT feature extraction, the descriptors are quantized by k-means clustering so that we ultimately obtain 5000 visual words, which is the optimal number stated by [5].

The next step after the vocabulary generation is to represent the images in our dataset in terms of the visual words. In this phase, we use hard quantization to associate the descriptors extracted from an image to the corresponding visual words. We also apply 4×4 spatial binning resulting in the 80K components for an image vector. After generating the image vectors, we map the spatial histograms into KCHI2 kernels and train a linear SVM solver for the two classes: i) Creative, ii) NonCreative by employing Stochastic Dual Coordinate Ascent implementation of VLFeat and exploit 800 training images from Creativity Dataset.

We should especially point to the fact that the parametric and algorithmic decisions in our setup are mostly originated from the previous and well-accepted studies on multimodality. We sincerely encourage others to experiment with the alternative decisions and improve the creativity detection performances on images.

5.2   Observable Visual Features

Together with the implicit properties belonging to creative and non-creative classes, we also utilize the explicit objects such as lines and circles found in the

---

[4] http://www.cs.cmu.edu/~biglou/resources/.

images. In addition, we seek the impact of the color information per se in the creativity detection by encoding the dominant colors as another feature. These features are mostly related to the artistic dimension of the creativity.

Dominant Colors. We employ 5 dominant colors in the images encoded as normalized R, G, B values.

Lines. We extract lines in an image through Hough Transform [7]. This feature set contains number of lines, length of the longest line and average line length.

Circles. In addition to the lines, we also extract the circles by Hough Transform and generate three features as the number of circles, max radius length, and average radius length.

Figure 2 displays detected lines on an example creative image and a set of extracted circles from the example non-creative image.

Fig. 2. Detected lines on a creative and circles on a non-creative sample.

6 Creativity Detecting Features in Linguistic Modality

To attain the features that can disclose the creativity in advertising slogans, we first need a corpus that can depict the natural order of the language. To this end, we used a subset of English GigaWord 5th Edition released by Linguistic Data Consortium (LDC)[5]. For our study, we worked on a randomly chosen file which we will call the *Corpus* from now on.

Sentence Self Information. Self-information, which might also be defined as surprisal, can be interpreted as the information load of a specific outcome of an

---

[5] http://www.ldc.upenn.edu/Catalog/catalogEntry.jsp?catalogId=LDC2011T07.

event. We calculate the self-information s of a bigram B by $s(B) = -log(p(B))$ exploiting the conditional probability distribution of bigram model trained on the corpus. We obtain the sentence self information as the average s of the bigrams extracted from the sentence.

Emotion and Human Sense Associations. Emotional score features are designed to quantify the feelings generated by a sentence. As a dimension of creativity, emotions have an impact on the effectiveness of an advertisement [1]. To this respect, we attempt to capture the emotional connotation of a given sentence through calculating the scores for each emotion, i.e. anger, fear, anticipation, trust, surprise, sadness, joy, and disgust, through the Equation $S(e) = (\sum_{s_i} sim(s_i, e)) - (\sum_{e_j \neq e} sim(s_i, e_j))$ [19]. For a given emotion e in the set of emotions, s_i is the ith word in the sentence and sim is the association function. The emotion associations of the words are obtained from the resource created by [17]. For sensorial associations of the sentences, we apply the same equation and obtain the word-human sense associations from Sensicon[6] generated by [27]. Both features are normalized with respect to the sentence length.

Unusual Word Score. Another surprising effect might be provided by introducing unusual words that increase the originality of a sentence. Therefore, we would like to employ a feature where we capture the contribution of uncommon words to the creativity in a sentence. We adopt the equation to calculate the unusual word score, $f(s) = (1/|s|)(\sum_{s_i} 1/c_i)$ introduced by [19]. For a given sentence s, c_i is defined as the frequency of the ith word, s_i, observed in the corpus.

Variety Score. We employ this feature to detect whether creative language displays a particularly different word variety then a non-creative, descriptive language. As calculated by [19], we obtain variety score as the number of distinct words in the sentence over the sentence length.

Phonetic Score. The exploitation of phonetic features in creative and persuasive sentence analysis has been deeply explored in the works of [11,19]. Following these studies, we explore the alliteration, homogeneity, rhyme and plosive scores generated by using the HLT Phonetic Scorer[7] [19].

7   Creativity Prediction Experiment

We organized the creativity prediction experiment similar to [6,11], where the classifier identifies the more creative item for a given item tuple as an instance. To be more precise, the feature vectors of two items are concatenated and each feature is renamed as *feature_name_left* or *feature_name_right*. The prediction task, then, is transformed into two class classifier for *left* and *right* classes. Moreover, we reversed the positions of items in an instance and thus we obtain a balanced dataset with 500 instances for *left* and 500 instances for *right* classes.

---

[6] https://hlt-nlp.fbk.eu/technologies/sensicon.
[7] https://hlt-nlp.fbk.eu/technologies/hlt-phonetic-scorer.

For the experimental setup, we split the dataset into training, testing and development sets as 800, 100 and 100 instances, respectively. Since we collected the non-creative samples by querying the automatically collected keywords on WordNet and ImageNet, there might be some creative instances matching with the same synset, causing the same linguistic non-creative definition for more than one item. Although we retrieved distinct images for each instance and guarantee that all image-sentence tuples are unique, the same sentences occurring both in train and test sets might affect the performance of linguistic experiment. Therefore, we pre-split the training data into 10 folds ensuring that all non-creative instances associated to the same synset are included in the same fold.

7.1 Multimodal Creativity Score

Multimodal fusion approaches vary depending on the requirement of the task on which it is applied or the characteristics of the multimodal data. For instance, on a word similarity task, while early fusion technique has an objective of jointly learning the multimodal representations, middle fusion operates on the concatenation of independent representations of the modalities and obtains the similarity score. On the other hand, late fusion first calculates the scores for each modality and combine them as a single score [13].

The early(joint) fusion approach is designed to identify an object or a notion via various modalities. However, multimodal advertising instances may represent quite different but complementary properties of the same message instead of referring to the same concept. Moreover, while we have a sparse feature set for the visual modality, the linguistic modality features are very dense, hence, the early fusion of these two feature sets might cause a poor performance [24]. Due to these concerns, we embraced the late fusion (score level) strategy to obtain the multimodal creativity score. To combine the scores from each modality, we employed Eq. 1 where sl_i and sv_i denote the linguistic and the visual creativity scores, respectively.

$$mscore(ad_i) = \alpha \times sl_i + (1 - \alpha) \times sv_i \tag{1}$$

We obtained the peak point for alpha as around 0.5 by tuning it on the development dataset. After calculating the multimodal creativity scores, namely *mscore*, for the left and the right parts of an instance, we labeled the creative part by finding the maximum value among the scores.

7.2 Unimodal Experiments

We first investigated the performances of linguistic features with a classification task employing Support Vector Machine implemented within the *scikit-learn* package [21].

The first row labeled 'L' in Table 1 shows the cross validation accuracy and test accuracies of all the linguistic features. Each row in the rest of the table shows the ablation of the indicated feature. We also marked the results that are

Table 1. The linguistic modality experiment results, ** denotes $p - val < 0.01$

Model	Training		Testing
	# Features	CV	F1
L	40	.85	.89
L-Phonetic	32	.74**	.82
L-SelfInfo	38	.83	.85
L-Emotion	24	.84	.90
L-Unusual	38	.90	.90
L-Variety	38	.85	.90
L-Senses	30	.81**	.85

statistically significant in terms of the drop of the performance in comparison to all features according to McNemar's test [16]. In addition, we listed the number of features included in each model in the second column of the table.

In this initial experiment, we surprisingly found out that *Emotion, Unusual Words* and *Variety* scores affect the creativity detection performance adversely in spite of being theoretically invaluable for the advertising creativity. On the other hand, phonetic scores have the most significant role in the detection performance ($p - val \approx 0.0000$). Considering the fact that phonetic properties are frequently devised in advertising creativity, this result is in line with our expectations. Furthermore, removal of the human sense association features also decreases the cross validation performance significantly ($p - val = 0.0003$). We can discuss that the test set that we sampled might be undersized since we cannot observe any significance of the test performance change when the cross validation has changed dramatically, especially for the phonetic features. In Table 4, we present example instances that are resolved by the associated feature. For the test sample shown in the first line, phonetic features empower the classifier especially with the help of the high alliteration score of the creative slogan. In the second example, the sentence self-information in the non-creative definition is much greater than the slogan, of which creative load mostly emerges when combined with its ad image[8]. Sentence self-information score has a low but negative correlation (Pearson Correlation r-value: -0.304) with the creativity on our dataset.

Regarding the visual modality, we exploited SVM solver of VLFeat toolbox as we discussed in Sect. 5.1. We trained the BoVW model on 800 balanced image dataset. After constructing the BoVW creativity model, we performed the test phase in 100 test images that are associated with the linguistic test set. At the end of the test phase, we obtained creativity and non-creativity scores for each test image. We employed a straightforward decision process where the labeling is conducted by finding the higher value among the creative and non-creative scores. The creativity prediction performance of the BoVW model is shown in

---

[8] https://adsoftheworld.com/media/print/eco_lamp?size=original.

Table 2. The BoVW model visual modality experiment results

Visual Model	#Words	Testing
BoVW Model	5000	.84

Table 3. The OVF model visual modality experiment results

Model	Training		Testing
	# Features	CV	F1
OVF	42	.66	.69
OVF-Lines	36	.63	.67
OVF-Circles	36	.65	.72
OVF-Colors	12	.67	.72

Table 2. Although not as strong as linguistic model, BoVW model is still feasible to assess the creativity in multimodal design.

For the second visual model, which exploits the observable visual features, we performed the same training strategy that we employed in the linguistic experiment. In comparison with the performance of BoVW model, OVF model yields obviously poor results as they are shown in Table 3. However, it still adds a value to the overall experimental results since we can imply that the visual creativity in ad images has more subtle properties than the basic artistic concepts, solely.

7.3   Multimodal Fusion Results

Considering the unimodal results of linguistic and BoVW models, we can safely say that they are already successful to resolve the creativity test for our dataset by themselves with 0.89 and 0.84 F1 values respectively. However, as shown in Table 5, the multimodal model (L+BoWV) significantly outperforms both models implying that each modality captures different aspects of creativity. More specifically, late fusion of linguistic and BoVW models shows an improvement with a strong significance ($p-val = 0.008$) with respect to the linguistic model and ($p-val \approx 0.0000$) with respect to BoVW model. For the only unresolved tuple, shown in Fig. 3, the BoVW model misclassified the creative image with a strong non-creativity score which is as high as its non-creative conjugate shown in the figure, leading to the misclassification via multimodal score.

We especially stay away from the claim that we resolved 'the problem' with the performance with an F1 value of 0.98. Instead, our outcome remarks that we could make use of the difference of the creativity aspects from the linguistic and visual modality feature sets. In fact, we designed our dataset to be able to leverage the distance between the creativity level of the data classes so that the creativity capturing features become obvious.

Table 4. Example sentences resolved with the associated linguistic features.

Creative Ad Slogan	Non-Creative Definition	Feature
Creatives, keep fighting the good fight	Photographs or other visual representations in a printed publication	Phonetic
Renewable energy makes the future bright	A barrier constructed to contain the flow of water or to keep out the sea	Self-Information

Table 5. Multimodal fusion results and comparisons to unimodal experiments. ** denotes $p-val < 0.01$ and * denotes $p-val < 0.05$.

Model	F1	α
L+BoVW+OVF	.98	0.33
L+OVF	.92*	0.50
OVF+BoVW	.92	0.50
L+BoVW	.98	0.50
BoVW	.84**	1.00
L	.89**	1.00

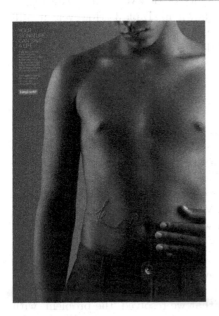

Fig. 3. An unresolved instance with multimodal fusion. Creative ad (Left) with slogan 'Your signature can save a life', Non-creative gloss-image (Right) with definition 'a living person'

8   Conclusion

In this study, we propose computational exploitation of multimodality in creativity detection. In particular, we suggest that visual modality is an essential part of the effectiveness and creativity of an advertisement. Our results show that the combination of visual and linguistic models significantly outperforms the performance of unimodal experiments. As an overall argument that we can raise, the seclusion of the other modalities and focusing only on the textual analysis for multimodal advertising prints or for any other multimodal data would cripple the competence of the findings.

We believe that the multimodal fusion of creativity quantifying features would be beneficial for various artistic endeavours such as for copywriters while choosing the potential successful advertisements. The creativity dataset is also publicly available upon request to the authors so that it can be utilized for research purposes.

Although the negative and positive samples in our dataset are overly distant in terms of the level of creativity, we specifically focus on this divergence in order to form a set of linguistic and visual features that can be exploited and modified in the more challenging tasks such as the assessment of advertising creativity.

References

1. Ang, S.H., Low, S.Y.: Exploring the dimensions of ad creativity. Psychol. Mark. **17**(10), 835–854 (2000)
2. Boden, M.A.: Dimensions of Creativity. MIT Press, Cambridge (1996)
3. Bruni, E., Boleda, G., Baroni, M., Tran, N.K.: Distributional semantics in technicolor. In: Proceedings of the 50th Annual Meeting of the Association for Computational Linguistics: Long Papers-Volume 1, pp. 136–145. Association for Computational Linguistics (2012)
4. Bruni, E., Tran, G.B., Baroni, M.: Distributional semantics from text and images. In: Proceedings of the GEMS 2011 Workshop on Geometrical Models of Natural Language Semantics, pp. 22–32. Association for Computational Linguistics (2011)
5. Bruni, E., Tran, N.K., Baroni, M.: Multimodal distributional semantics. J. Artif. Intell. Res. (JAIR) **49**, 1–47 (2014)
6. Danescu-Niculescu-Mizil, C., Cheng, J., Kleinberg, J., Lee, L.: You had me at hello: how phrasing affects memorability. In: Proceedings of the 50th Annual Meeting of the Association for Computational Linguistics: Long Papers-Volume 1, pp. 892–901. Association for Computational Linguistics (2012)
7. Duda, R.O., Hart, P.E.: Use of the hough transformation to detect lines and curves in pictures. Commun. ACM **15**(1), 11–15 (1972)
8. Elgammal, A., Saleh, B.: Quantifying creativity in art networks. In: Proceedings of the 6th International Conference on Computational Creativity, p. 39 (2015)
9. Feng, Y., Lapata, M.: Visual information in semantic representation. In: Human Language Technologies: The 2010 Annual Conference of the North American Chapter of the Association for Computational Linguistics, pp. 91–99. Association for Computational Linguistics (2010)
10. Fichner-Rathus, L.: Foundations of Art and Design: An Enhanced, Media Cengage Learning, Boston (2011)

11. Guerini, M., Özbal, G., Strapparava, C.: Echoes of persuasion: the effect of euphony in persuasive communication. In: Proceedings of the 2015 Conference of the North American Chapter of the Association for Computational Linguistics (NAACL-2015), pp. 1483–1493. Denver, Colorado (2015)
12. Jarvie, I.: The rationality of creativity. In: Dutton, D., Krausz, M. (eds.) Thinking About Society: Theory and Practice, pp. 282–301. Springer, Dordrecht (1981)
13. Kiela, D., Clark, S.: Multi-and cross-modal semantics beyond vision: grounding in auditory perception. In: Proceedings of the EMNLP (2015)
14. Kuznetsova, P., Chen, J., Choi, Y.: Understanding and quantifying creativity in lexical composition. In: EMNLP, pp. 1246–1258 (2013)
15. Lowe, D.G.: Distinctive image features from scale-invariant keypoints. Int. J. Comput. Vision **60**(2), 91–110 (2004). https://doi.org/10.1023/B:VISI.0000029664.99615.94
16. McNemar, Q.: Note on the sampling error of the difference between correlated proportions or percentages. Psychometrika **12**(2), 153–157 (1947). https://doi.org/10.1007/BF02295996
17. Mohammad, S.M., Turney, P.D.: Emotions evoked by common words and phrases: Using mechanical turk to create an emotion lexicon. In: Proceedings of the NAACL HLT 2010 Workshop on Computational Approaches to Analysis and Generation of Emotion in Text, pp. 26–34. Association for Computational Linguistics (2010)
18. Özbal, G., Pighin, D., Strapparava, C.: Brainsup: Brainstorming support for creative sentence generation. In: ACL, vol. 1, pp. 1446–1455 (2013)
19. Özbal, G., Pighin, D., Strapparava, C.: BrainSup: Brainstorming support for creative sentence generation. In: Proceedings of ACL'13 (Volume 1: Long Papers), pp. 1446–1455. Association for Computational Linguistics (2013). http://aclweb.org/anthology/P13-1142
20. Özbal, G., Strapparava, C., Guerini, M.: Brand pitt: a corpus to explore the art of naming. In: Proceedings of the LREC, pp. 1822–1828. Citeseer (2012)
21. Pedregosa, F., et al.: Scikit-learn: machine learning in Python. J. Mach. Learn. Res. **12**, 2825–2830 (2011)
22. Pricken, M.: Creative Advertising Ideas and Techniques from the World's Best Campaigns. Thames & Hudson, 2nd edn. (2008)
23. Sartori, A., Yan, Y., Ozbal, G., Salah, A., Salah, A., Sebe, N.: Looking at mondrian's victory boogie-woogie: what do i feel? In: International Joint Conferences on Artificial Intelligence (IJCAI) (2015)
24. Shutova, E., Kiela, D., Maillard, J.: Black holes and white rabbits: metaphor identification with visual features (2016)
25. Silberer, C., Lapata, M.: Learning grounded meaning representations with autoencoders. In: ACL, vol. 1, pp. 721–732 (2014)
26. Smith, R.E., MacKenzie, S.B., Yang, X., Buchholz, L.M., Darley, W.K.: Modeling the determinants and effects of creativity in advertising. Mark. Sci. **26**(6), 819–833 (2007)
27. Tekiroglu, S.S., Özbal, G., Strapparava, C.: Sensicon: an automatically constructed sensorial lexicon. In: Proceedings of the 2014 Conference on Empirical Methods in Natural Language Processing (EMNLP), pp. 1511–1521. Association for Computational Linguistics, Doha, Qatar (2014). http://www.aclweb.org/anthology/D14-1160
28. Vedaldi, A., Fulkerson, B.: VLFeat: an open and portable library of computer vision algorithms (2008). http://www.vlfeat.org/
29. Yang, J., Jiang, Y.G., Hauptmann, A.G., Ngo, C.W.: Evaluating bag-of-visual-words representations in scene classification. In: Proceedings of the International Workshop on Workshop on Multimedia Information Retrieval, pp. 197–206. ACM (2007)

Knowledge-Based Techniques for Document Fraud Detection: A Comprehensive Study

Beatriz Martínez Tornés[1], Emanuela Boros[1], Antoine Doucet[1]([⊠]),
Petra Gomez-Krämer[1], Jean-Marc Ogier[1], and Vincent Poulain d'Andecy[2]

[1] University of La Rochelle, L3i, F-17000, La Rochelle, France
`{beatriz.martinez_tornes,emanuela.boros,antoine.doucet,`
`petra.gomez-Kramer,jean-marc.ogier}@univ-lr.fr`
[2] Yooz, 1 Rue Fleming, 17000 La Rochelle, France
`vincent.dandecy@getyooz.com`

Abstract. Due to the availability of cost-effective scanners, printers, and image processing software, document fraud detection is, unfortunately, quite common nowadays. The main challenges of this task are the lack of freely available annotated data and the overflow of mainly computer vision approaches. We consider that relying on the textual content of forged documents could provide a different view on their detection by exploring semantic inconsistencies with the aid of specialized knowledge bases. We, thus, perform an exhaustive study of existing state-of-the-art methods based on knowledge-graph embeddings (KGE) using a synthetically forged, yet realistic, receipt dataset. We also explore additional knowledge base incremental data enrichments, in order to analyze the impact of the richness of the knowledge base on each KGE method. The reported results prove that the performance of the methods varies considerably depending on the type of approach. Also, as expected, the size of the data enrichment is directly proportional to the rise in performance. Finally, we conclude that, while exploring the semantics of documents is promising, document forgery detection still poses a challenge for KGE methods.

Keywords: Fraud detection · Knowledge base · Knowledge graph

1 Introduction

Document forgery is quite common nowadays due to the availability of cost-effective scanners, printers, and image processing software. Most administrative documents exchanged daily by companies and public administrations lack the technical securing to authenticate them, such as watermarks, digital signatures or other active protection techniques. Document forgery is a gateway to other

This work was supported by the French defense innovation agency (AID) and the VERINDOC project funded by the Nouvelle-Aquitaine Region.

A. Gelbukh (Ed.): CICLing 2019, LNCS 13451, pp. 17–33, 2023.
https://doi.org/10.1007/978-3-031-24337-0_2

types of fraud, as it can produce tampered supporting documents (invoices, birth certificates, receipts, payslips, etc.) that can lead to identity or tax fraud. The amount of fraud detected in tax and social matters reached €5.26 billion in 2019, according to the CODAF[1] (French Departmental anti-fraud operational committees). According to Euler Hermes [24], European fraud insurance company, and the French Association of Financial Directors and Management Controllers (DFCG) on their annual fraud trend analysis the most common fraud attempts suffered by companies in France in 2019 are: fake supplier fraud (by 48% of respondents), fake president fraud (38%), other identity theft, e.g., banks, lawyers, auditors etc., (31%) and fake customer fraud (24%).

One of the main challenges of document fraud detection is the lack of freely available tagged data, as many studies around fraud do not consider the actual documents and focus on the transactions (such as credit card fraud, insurance fraud or even financial fraud) [10,36,43]. Collecting real forged documents is rather difficult [39,48,54], because real fraudsters would not share their work, and companies or administrations are reluctant to reveal their security breaches and cannot share sensitive information. Taking an interest in real documents actually exchanged by companies or administrations is important for the fraud detection methods developed to be usable in real contexts and for the consistency of authentic documents to be ensured.

Most of the recent research in document forensics is focused on the analysis of images of documents as a tampering detection task [13,17,19,25]. Likewise, most of the existing corpora for fraud detection are based on the creation of synthetic content by introducing variations allowing a particular approach to be tested [10,39,43]. For example, documents were automatically generated [12], as well as the noise and change in size of some characters, their inclination or their position. A corpus of payslips was artificially created by randomly completing the various fields required for this type of document [48]. Another corpus consists of the same documents scanned by different scanners in order to evaluate source scanner identification [42]. These corpora examples, suitable for fraud detection by image-based approaches [18,20,21], are not appropriate for content analysis, because they do not include realistic information, nor frauds that are semantically more inconsistent or implausible than the authentic documents.

However, we consider that the textual content of the document could provide a different vision towards detection fraud that would not rely on graphical imperfections, but on semantic inconsistencies. Existing fraud detection approaches mainly focus on supervised machine learning (e.g., neural networks, bagging ensemble methods, support vector machine, and random forests) based on hand-crafted feature engineering [10,35–37,39]. However, these approaches do not consider documents as their only input. Knowledge graph representation methods [32,50,56] are used to predict the plausibility of the facts or *fact-checking* using external knowledge bases (YAGO [49], DBpedia [6] or Freebase [14]). Despite the popularity of knowledge graph-based approaches, these are still underex-

---

[1] Comités opérationnels départementaux anti-fraude https://www.economie.gouv.fr/codaf-comites-operationnels-departementaux-anti-fraude.

plored in analyzing document coherence and detecting fraud in semi-structured administrative documents (receipts, payslips). Semi-structured data is data that presents some regularity, but not as much as relational data [1]. Most studies focused on coherence analysis and approached it from a discursive point of view [9], with tasks such as sentence intrusion detection [46] or text ordering: this approach is not compatible with administrative documents. We, thus, propose to perform an exhaustive study on knowledge-based representation learning and its applicability to document forgery detection, in a realistic dataset regarding receipt forgery [3, 4].

The remainder of the paper is organized as follows. First, Sect. 2 presents this dataset in detail along with the ontology based on its topical particularity. Section 3 defines the notions of fact-checking and document fraud detection with knowledge bases. Section 4 provides an exhaustive list of the knowledge-based different state-of-the-art methods that further are explored in the experimental setup in Sect. 5. The results are presented and visualized in Sect. 6. Finally, results are discussed and conclusions drawn in Sect. 7.

2   Receipt Dataset for Fraud Detection

The dataset [3, 4] is composed of 999 images of receipts and their associated optical character recognition (OCR) results, of which 6% were synthetically forged. It was collected to provide a parallel corpus (images and texts) and a benchmark to evaluate image- and text-based methods for fraud detection.

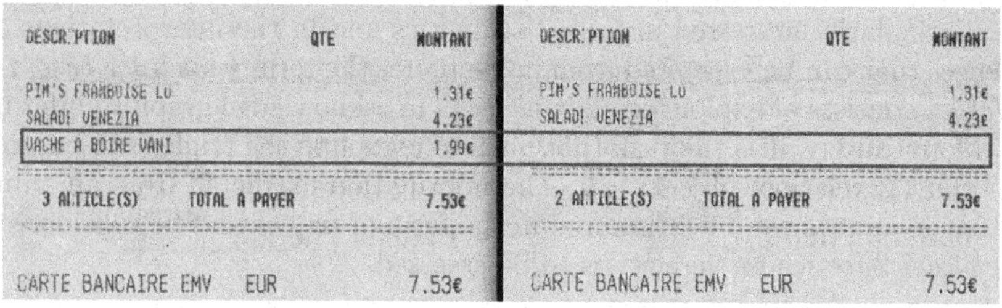

Fig. 1. A scan of a normal receipt (left) and a forged receipt (left) from the dataset [4]. The red box reveals the removal of a grocery item. (Color figure online)

The forged receipts are the result of forgery workshops, in which participants were given a standard computer with several image editing software to manually alter both images and associated OCR results of the receipts. The workshop fraudsters were free to chose the image modification method. The corpus contains copy-move forgeries (inside the document), splicing (copying from a different document), imitation of font with a textbox tool, etc. Not only are the forgeries diverse, but they are also realistic, consistent with real-world situations such as a fraudulent refund, an undue extension of warranty or false mission expense reports, as shown in Fig. 1. The dataset also provides an ontology [5]. The ontology accounts for all the information present in a receipt: the classes that define *Company* entities, through their contact information (telephone numbers, website, address, etc.) or information enabling them to be identified (SIREN[2], SIRET[3]). Other classes concerns purchases (purchased products, by which companies, means of payment, etc.).

3   Knowledge-Based Techniques for Document Fraud Detection

Document fraud detection is closely related to *fact-checking* [28]. *Fact-checking* is the NLP task that refers to the evaluation of the veracity of a claim in a given context [51]. We consider that the objectives between fact-checking and document authentication are similar enough to focus on their commonalities and their intersection for the detection of document fraud.

Although tampering, or fraud, can be mentioned as part of a verification or authentication task, this positions tampering as a barrier impacting data quality and not as a characteristic of the data to be detected. Data falsification can be identified as an obstacle to information verification [11] that is resolved by assessing the reliability of sources or by relying on information redundancy (majority voting heuristic).

Recent *fact-checking* methods utilize knowledge bases to asses the veracity of a claim. A *knowledge base* (KB) is a structured representation of facts made up of entities and relations. Entities can represent concrete or abstract objects and relations represent the links maintained between them. The terms *knowledge graph* and *knowledge base* are often used interchangeably [33]. As we are more particularly interested in formal semantics and in the interpretations and inferences that can be extracted from it, we prefer the term *knowledge base*. *Link prediction* consists of exploiting existing facts in a knowledge graph to infer new ones. More exactly, it is the task that aims to complete the triple (subject, relation,?) Or (?, relation, object) [44]. The information extracted from an administrative document can be structured in the form of triples and be considered as a *knowledge base* whose veracity is to be assessed.

4   Knowledge-Based Fact-Checking Methods

Many knowledge base fact-checking methods have been proposed, as well as many variations and improvements thereof. First considered as a logic task, the goal was to find and explicitly state the rules and constraints. Those rules were either manually crafted or obtained with rule mining methods [26]. More recently, with the rise of machine learning techniques, knowledge base methods with graph embeddings (KGE) methods have been proposed [31,33,44,55]. These methods encode the entities and the relations between them of the knowledge base in a

---

[2] https://en.wikipedia.org/wiki/SIREN_code.
[3] https://en.wikipedia.org/wiki/SIRET_code.

vector space of low dimensionality. These vectors aim to capture latent features of the entities (and relations) of the graph.

Next, while we provide an exhaustive list of the different methods, we focus on detailing the KGE-based approaches that we further consider in the experimental setup in Sect. 5. We separated the methods according to the following taxonomy [44]: *matrix factorization models, geometric models,* and *deep learning models.*

4.1   Matrix Factorization Models

RESCAL [41] is a approach that represents entities as vectors (h and t) and relations r as matrices.

DistMult [57] is a simplification of RESCAL for which the matrices of relations W_r have the constraint of being diagonal. This constraint lightens the model, because it considerably reduces the space of the parameters to be learned, however, becoming less expressive. For DistMult, all the relations are represented by a diagonal matrix, and are therefore considered as symmetrical.

ComplEx [52] is an extension of DistMult which represents relations and entities in a complex space ($h, r, t \in \mathbb{C}^d$). Thus, despite the same diagonal constraint as DistMult, it is possible to represent asymmetric relations thanks to the non-commutativity of the Hadamard product in the complex space.

QuatE [58] is an extension from ComplEx that goes beyond the complex-space to represent entities using quaternion embeddings and relations as rotations in the quaternion space. This models aims to offer better geometrical interpretations. QuatE is able to model symmetry, anti-symmetry and inversion.

SimplE [34] takes a 1927 approach [30] to tensor factorization (canonical polyadic decomposition) and adapts it to link prediction. This approach learns two independent vectors for each entity, one for the subject role and the other for the object role. They propose to make the representation of the subject and object vectors of one entity by relying on any relation r and its inverse r'.

TuckER [8] is a linear model for which a tensor of order 3 $T \in \mathbb{R}^{I \times J \times K}$ can be decomposed into a set of three matrices A, B and C and a core (a tensor of lower rank).

HolE [40] is an approach based on holographic embeddings. These make use of the circular correlation operator to compute interactions between latent features of entities and relations. That allows for a compositional representation of the relational data of the knowledge base.

4.2   Geometric Models

Structured Embedding [16] approach represents each relation by two matrices $M_r^h, M_r^t \in \mathbb{R}^{d \times d}$ that allow to perform projections specific to each relation of subject t and object t entities. This model thus makes it possible to distinguish the different types of relations, as in the role of subject and object of an entity.

TransE [15] utilized word embeddings and their capacity to account for the relations between words through translation operations between their vectors: $h + r \approx t$. Thus, $f(h, r, t) = -||h + r - t||_p$ where $p \in \{1, 2\}$ is a hyper-parameter. TransE is limited regarding symmetrical or transitive relations, as well as for high cardinalities $(1 \ldots N$ to 1, or 1 to $1 \ldots N$ relations). This method has become popular because of its calculation efficiency.

TransH [56] addresses the expressivity limits of TransE for relations with cardinalities greater than 1. Each relation is represented as a hyperplane.

TransR [38] is an extension of the previously presented geometric models, which explicitly considers entities and relations as different objects, representing them in different vector spaces.

TransD [32] is another extension of TransR. Entities and relations are also represented in different vector spaces. The difference concerns the projection matrix which, unlike TransR, is not the same for all entities and only depends on the relation.

CrossE [59] extends the traditionnal models as it explicitly tackles crossover interactions, the bi-directional effects between entities and relations.

RotatE [50] represents relations as rotations between subject and object in complex space.

MurE [7] is the Euclidean counterpart of MuRP, a hyperbolic interaction model developed to effectively model hierarchies in KG.

KG2E [29] aims to model the uncertainties linked to entities and relations, compared to the number of observed triples containing these entities and relations. Thus, entities and relations are represented by distributions, more particularly Gaussian multivariate distributions.

4.3 Deep Learning Models

ConvE [22] is a multi-layer convolutional network model for link prediction, yielding the same performance as DistMult. This approach is particularly effective at modelling nodes with high in degree – which are common in highly-connected, complex knowledge graphs such as YAGO [49], DBpedia [6] or Free-base [14].

ERMLP [23] is a multi-layer perceptron based approach that uses a single hidden layer and represents entities and relations as vectors.

ProjE [47] is a neural network-based approach with a combination and a projection layer for candidate-entities.

5   Experimental Setup

The KGE methods presented in Sect. 4 allow more efficient use of knowledge bases by transforming them into vector space while maintaining their latent semantic properties. We present in the following section our experimental setup regarding the exploitation of these methods in the context of our dataset and additional data enrichment.

5.1   Evaluation

The evaluation is carried out by comparing the score of the actual triples against the score of all the other triples that could be predicted. Ideally, the score of the original triple is expected to have a better score against the others. This classification can be done according to two configurations: *raw* and *filtered*. In the *raw* configuration, all triples count for ranking, even valid triples belonging to the graph, while in the *filtered* configuration, these are not taken into account. We perform all the evaluations in a *filtered* configuration. From these ranks, the global metrics commonly used [44] are the following:

- **Mean Rank (MR)**: the average of these ranks. The lower this is, the more the model is able to predict the correct triples. Since this is an average, this measurement is very sensitive to outliers.
- **Mean Reciprocal Rank (MRR)**: the average of the inverse of the ranks. This metric is less sensitive to extreme values and is more common.
- **Hits@K**: the rate of predictions for which the rank is less than or equal to threshold K.

5.2   Data Pre-processing

The dataset and the ontology [3, 5] presented in Sect. 2 serve as the starting point for our study. First, we are interested in the instances that populate the ontology. Table 1 presents the object properties present in the ontology, along with their domain and their image (the classes between which they express a relation). For creating the sets for training, testing and validation, we eliminated the reverse object properties of the ontology to ensure the elimination of the risk of *data leakage*: the inverse relations could serve as information for the model during the testing stage. For example, determining if the triple (*Carrefour City*, has_address, "48 impasse du Ramier des Catalans") is true can be assisted by the realisation that has_address and is_address_of are inverse relations and by the triple (*"48 impasse du Ramier des Catalans", is_address_of, Carrefour City*).

5.3   Data Enrichment: External Verification

In order to allow an efficient verification of the information coming from the receipts, we are also interested in a data enrichment. We have about 15,000

Table 1. Receipt ontology object properties.

Domain	Object property	Reverse property	Image
City	has_zipCode	is_zipCode_of	ZipCode
Company	has_contactDetail	is_contactDetail_of	ContactDetail
Company	has_adress	is_address_of	Address
Company	has_email_address	is_email_address_of	EmailAdress
Company	has_fax	is_fax_of	FaxNumber
Company	has_website	is_website_of	Website
Company	has_phone_number	is_phone_number_of	PhoneNumber
Company	issued	is_issued_by	Receipt
Product	has_expansion	is_expansion_of	Expansion
Company	has_registration	is_registration_of	Registration
Receipt	has_intermediate_payment		IntermediatePayment
Receipt	concerns_purchase		Product
Receipt	contains	is_written_on	Product, Registration, ContactDetail
SIREN	includes	is_component_of	SIRET, RCS, TVA IntraCommunity
City, ZipCode	part_of		Address
Company	is_located_at		City
Company	sells	is_sold_by	Product

triples (subject, predicate, object). This knowledge base, built solely from information from sales receipts, can be enriched using external resources. The reference base is intended to be easy to build and set up, based on existing databases or resources as structured as possible, such as company catalogs, in order to extract the price of products, the contact information of companies or even company registration information. To increase knowledge on companies, we utilized data from French national institute of statistics and economic studies (INSEE[4]). The INSEE provides freely available resources about companies, referred to as the SIRENE database.

Enrichment of Data from the SIRENE Database. The methodology used, which aims to be fast and to build on existing resources, can be extended to other types of documents that contain the same relations (such as *has_address*, *has_SIRET*, etc.). Thus, we wish to incorporate data from the SIRENE database, which provides a database of data on French companies[5]. Given that the SIRENE registry has entries for 31 million French companies and that in the receipts corpus there are 387 different companies, we propose to measure the impact of the increase in data on the performance of fact-checking incrementally. To the 15,000 triples from the receipts, we added 1,000 to 9,000 triples from the SIRENE database, in steps of 1,000. This leaves us with ten different data sets, which size is presented in Table 2. The relations of the receipt ontology available in the SIRENE database are *has_zip_code*, *has_address*, *has_registration*,

---

[4] https://www.insee.fr/en/accueil.
[5] http://sirene.fr/siren/public/home.

is_component_of, *part_of* and *is_located_at*. All of them relate to either registration information about the companies, or their addresses. In order to match the information from the receipt to the external reference database, certain approximations have been made. First, we consider that *City* class of ontology is equivalent to *Municipality*, field of the SIRENE database. We consider that in the receipt corpus we are dealing mainly with cities, a case in which this equivalence is respected. In addition, we chose to keep the relation *has_registration* as well as its hyponymous relations (*has_siren*, *has_siret* and *has_nic*). The relations that were added despite not being defined in the ontology are the following: *has_nic* was added in the same way as *has_siren* and *has_siret*; *has_main_activity* and *is_headquarter*. Table 2 compares the size of the obtained knowledge bases.

Table 2. Size of the knowledge base extracted from sales receipts. The receipts_k with $k \in [1000, \ldots, 9000]$ represent the datasets resulting from the incremental data enrichment using the SIRENE database 5.3.

Dataset	No. Entity	No. Object Property	No. Triple
receipts	7,108	16	14,379
receipts_1000	7,613	21	15,379
receipts_2000	8,081	21	16,379
receipts_3000	8,542	21	17,379
receipts_4000	8,990	21	18,379
receipts_5000	9,406	21	19,379
receipts_6000	9,808	21	20,379
receipts_7000	10,196	21	21,379
receipts_8000	10,500	21	22,379
receipts_9000	10,918	21	23,379

Address Alignment. One of the limits of considering the textual content of documents as a knowledge graph is that lexical variation is not dealt with the extracted text from the receipts becomes the label of the entities. In the ontology, the addresses are broken down into their components (ZipCode and City). However, one address expressed in two different ways would not be recognized as the same. We therefore used the application programming interface (API) of the National Address Database (BAN[6]), the only database of addresses officially recognized by the French administration. We added a query allowing to associate each address with its id in the BAN. The id associated with each address depends on its specificity: 17300_7593_00033 is the id of *33 Rue de la Scierie 17000 La Rochelle*; 17300_7593 is the id of *Rue de la Scierie 17000 La Rochelle*.

---

[6] https://api.gouv.fr/les-api/base-adresse-nationale.

This approach makes it possible to reconcile identical addresses, but it also constitutes an implicit verification step. Addresses that were not associated with an id were kept with their full text.

6   Results

We trained the methods presented in Sect. 3 on the receipts corpus with and without the different levels of data enrichment presented in Table 2.[7]

Results without Data Enrichment. As we can see in Fig. 3, the best results come from matrix factorization models, particularly QuatE. DistMult, ComplEx and QuatE, are both based on the RESCAL method. However, ComplEx shows very poor results, even lower than RESCAL. QuatE represents entites in the quaternion space, which extends the expressivity of the model compared to DistMult, which represents every relation with a diagonal matrix and imposes symmetry. Second best results come from geometric models, especially MuRE, TransD and RotatE. Lastly, deep learning models show very low performance: the small size of the receipt dataset can explain this low result. We can observe that the methods struggle to accurately represent the entities and relations found in the receipts.

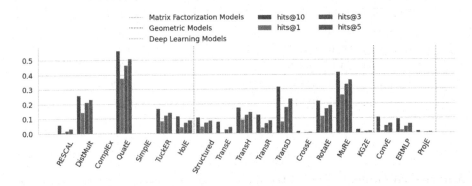

Fig. 2. Hits@K without data enrichment.

Figure 4 shows the receipt entities embeddings as represented by the different models. We can observe that most methods do not allow for an efficient receipt classification: all the documents are represented in a unique cluster. This is in accordance with the poor results the methods show in regards to the mean

---

[7] The dataset has been split into a training and test set (80% and 20% respectively) thanks to the PyKEEN library https://github.com/pykeen/pykeen [2], to avoid redundant triples being found both in training and test. The previously presented methods are implemented by PyKEEN, library that we chose to use for its completeness, flexibility and ease of use.

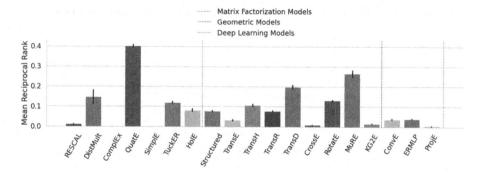

Fig. 3. MRR scores without data enrichment.

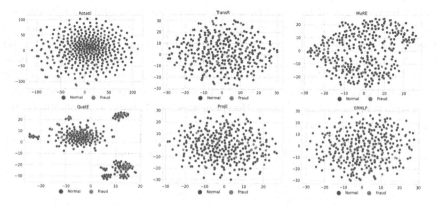

Fig. 4. Representation of the entities without data enrichment with T-SNE [53]. We only chose the models that obtained a rather higher performance in Fig. 2.

rank scores in Fig. 2 and mean reciprocal rank scores in Fig. 3. The only method that yields document classification results is QuatE, as it is able to distinguish receipts from others according to their issuing company. However, the classification is not enough to detect forgery, as we can see with the distribution of forged receipts (in orange in Fig. 4). When analyzing the MR results, the lower they are, the better, as it means the correct triples have a lower rank. The opposite goes for MRR, which is the inverse of the MR. The lowest results are for the SimplE and ComplEx methods (Fig. 3), both matrix factorization methods. One of our data pre-processing stages was the elimination of inverse triples in order to avoid the risk of test leakage. However, SimplE relies on inverse triples to accurately represent the entities and relations of the KG: our implementation choice explains the rather weak results of the SimplE method.

We performed an additional evaluation by filtering the relations in the test set in order to understand what relations were learned better by QuatE (as seen in Table 3). We also provide the number of triples containing each relation in order to be cautious when interpreting those results.

Table 3. Hits@10 results of the QuatE model for the six best learned relations.

Relation	Hits@10	Triple count
has_zipCode	66%	343
has_address	66%	243
has_phone_number	61%	315
sells	43%	4,555
contains	41%	1,906
concerns_purchase	39%	4,555

It can be noted in Table 3 that the best-learned relations are related to the contact information of the companies (zip code, address, and phone number). However, relations expressing the structure of the receipts (contains and concerns_purchase) have weaker results. Learning the semantic structure of the KB built from the receipt dataset proves to be a difficult task. Most of the information available is related to the purchases made and which companies they were made from. That induces a bias, and as we are approaching the problem as a prediction task, this implies that we may be dealing with an underlying consumer behavior study that is not the object of our research, and for which we have no ground truth. Indeed, we are interested in learning the document structure and the ability of KGE to assess document coherence.

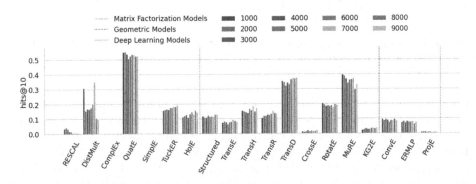

Fig. 5. Hits@10 scores variation across the different levels of data enrichment.

Results with Data Enrichment. We then evaluated the models across the different data enrichment steps. As we trained the different models on different size datasets, we do not show the MR results, as they cannot be compared: with more triples to sort, an equivalent performance would rank the correct triples lower.

Figure 5 shows the variation of the hits@K scores for all the models with triples from the SIRENE database. QuatE shows the best performance across all enrichment steps. However, we can observe how the enrichment process improves the results of several methods (DistMult, MuRE, RotatE, TransD, HolE, and TransH). That improvement is however not linear but could be explained by the nature of the receipt dataset. Indeed, the collected receipts mainly come from the same city, and thus contain very local information, while the enrichment comes from a national database and was done randomly.

7   Discussion and Conclusions

Document forgery detection poses a challenge for KGE methods aiming to represent the information in a document to authenticate it. However, data augmentation can improve the results. A perspective of our work was to approach the data augmentation in a specialized manner. Instead of adding triples with a relation semantic matching criteria, we could focus more on the documents to authenticate and add reference information about the companies, addresses, products, and other entities that are more prevalent in the data.

Another perspective would be to turn to methods that take into account literal information [27]. Most KGE methods focus on the entities and the relations between them, without taking into account the additional information available. In the particular case of forgery detection, a lot of relevant information is ignored (prices, dates, quantities, etc.).

A limitation concerning the data used to train and evaluate KGEs [45] is their imbalanced nature: 15% of the entities of FB15K is contained in 80% of the triples. For example, the entity "United States" is contained in almost all nationality relations. It is more profitable in terms of performance to predict an American nationality regardless of the subject entity than to learn this relationship with its underlying structure. Our dataset suffers from this bias.

The evaluation metrics are holistic as they do not distinguish the relations by type [45]. This holistic evaluation presents several shortcomings in the context of fraud detection, as the different relations play different roles in the semantic coherence structure of the document. Some relations are harder to predict because of their structural features, in particular, the high cardinality of the relations related to the products sold in the receipts - having a higher number of target peers for a triple makes link prediction harder [44]. Those relations also play a less important role in document authentication, as they are more related to user behavior than information we have the means to verify.

Our study, thus, confronted the scientific obstacles linked to the opacity of fact-checking methods for document fraud detection based on knowledge bases. Experimenting with these different methods in the concrete and complex application case of fraud detection however made it possible to set up an extrinsic evaluation of these methods.

References

1. Abiteboul, S.: Semistructured data: from practice to theory. In: Proceedings 16th Annual IEEE Symposium on Logic in Computer Science. IEEE (2001)

2. Ali, M., Berrendorf, M., Hoyt, C.T., Vermue, L., Sharifzadeh, S., Tresp, V., Lehmann, J.: Pykeen 1.0: a python library for training and evaluating knowledge graph emebddings (2020)
3. Artaud, C., Doucet, A., Ogier, J.M., d'Andecy, V.P.: Receipt dataset for fraud detection. In: First International Workshop on Computational Document Forensics (2017)
4. Artaud, C., Sidère, N., Doucet, A., Ogier, J.M., Yooz, V.P.D.: Find it! fraud detection contest report. In: 2018 24th International Conference on Pattern Recognition (ICPR). IEEE (2018)
5. Artaud, C.: Détection des fraudes : de l'image á la sémantique du contenu : application á la vérification des informations extraites d'un corpus de tickets de caisse. Ph.D. thesis (2019)
6. Auer, S., Bizer, C., Kobilarov, G., Lehmann, J., Cyganiak, R., Ives, Z.: DBpedia: a nucleus for a web of open data. In: Aberer, K., et al. (eds.) ASWC/ISWC -2007. LNCS, vol. 4825, pp. 722–735. Springer, Heidelberg (2007). https://doi.org/10.1007/978-3-540-76298-0_52
7. Balazevic, I., Allen, C., Hospedales, T.: Multi-relational poincaré graph embeddings. In: Advances in Neural Information Processing Systems, vol. 32 (2019)
8. Balažević, I., Allen, C., Hospedales, T.M.: Tucker: tensor factorization for knowledge graph completion (2019)
9. Barzilay, R., Lapata, M.: Modeling local coherence: an entity-based approach. Comput. Linguist. **34**(1), 1–34 (2008)
10. Behera, T.K., Panigrahi, S.: Credit card fraud detection: a hybrid approach using fuzzy clustering & neural network. In: 2015 2nd International Conference on Advances in Computing and Communication Engineering. IEEE (2015)
11. Berti-Équille, L., Borge-Holthoefer, J.: Veracity of data: from truth discovery computation algorithms to models of misinformation dynamics. Synth. Lect. Data Manag. **7**(3), 1–155 (2015)
12. Bertrand, R., Gomez-Kramer, P., Terrades, O.R., Franco, P., Ogier, J.M.: A system based on intrinsic features for fraudulent document detection. In: 2013 12th International Conference on Document Analysis and Recognition, pp. 106–110. IEEE, Washington, DC, USA (2013)
13. Bertrand, R., Terrades, O.R., Gomez-Krämer, P., Franco, P., Ogier, J.M.: A conditional random field model for font forgery detection. In: 2015 13th International Conference on Document Analysis and Recognition (ICDAR). IEEE (2015)
14. Bollacker, K., Evans, C., Paritosh, P., Sturge, T., Taylor, J.: Freebase: A collaboratively created graph database for structuring human knowledge. In: Proceedings of the 2008 ACM SIGMOD International Conference on Management of Data, pp. 1247–1250. SIGMOD'08, Association for Computing Machinery, New York, NY, USA (2008)
15. Bordes, A., Usunier, N., Garcia-Durán, A., Weston, J., Yakhnenko, O.: Translating embeddings for modeling multi-relational data. In: Proceedings of the 26th International Conference on Neural Information Processing Systems, NIPS'13, Lake Tahoe, Nevada, vol. 2, pp. 2787–2795. Curran Associates Inc., Red Hook, NY, USA (2013)
16. Bordes, A., Weston, J., Collobert, R., Bengio, Y.: Learning structured embeddings of knowledge bases. In: Proceedings of the 25th AAAI Conference on Artificial Intelligence, AAAI'11, San Francisco, California, pp. 301–306. AAAI Press (2011)
17. Cozzolino, D., Gragnaniello, D., Verdoliva, L.: Image forgery detection through residual-based local descriptors and block-matching. In: 2014 IEEE International Conference on Image Processing (ICIP). IEEE (2014)

18. Cozzolino, D., Poggi, G., Verdoliva, L.: Efficient dense-field copy-move forgery detection. IEEE Trans. Inf. Forensics Secur. **10**(11), 2284–2297 (2015)
19. Cozzolino, D., Verdoliva, L.: Camera-based image forgery localization using convolutional neural networks. In: 2018 26th European Signal Processing Conference (EUSIPCO). IEEE (2018)
20. Cozzolino, D., Verdoliva, L.: Noiseprint: a CNN-based camera model fingerprint (2018)
21. Cruz, F., Sidere, N., Coustaty, M., d'Andecy, V.P., Ogier, J.M.: Local binary patterns for document forgery detection. In: 2017 14th IAPR International Conference on Document Analysis and Recognition (ICDAR), vol. 1. IEEE (2017)
22. Dettmers, T., Minervini, P., Stenetorp, P., Riedel, S.: Convolutional 2d knowledge graph embeddings (2018)
23. Dong, X., et al.: Knowledge vault: a web-scale approach to probabilistic knowledge fusion. In: Proceedings of the 20th ACM SIGKDD International Conference on Knowledge Discovery and Data Mining, KDD'14, New York, New York, USA, pp. 601–610. Association for Computing Machinery, New York, NY, USA (2014)
24. EulerHermes-DFCG: Plus de 7 entreprises sur 10 ont subi au moins une tentative de fraude cette annye. https://www.eulerhermes.fr/actualites/etude-fraude-2020.html
25. Fridrich, J., Kodovsky, J.: Rich models for steganalysis of digital images. IEEE Trans. Inf. Forensics Secur. **7**(3), 868–882 (2012)
26. Galárraga, L., Teflioudi, C., Hose, K., Suchanek, F.M.: Fast rule mining in ontological knowledge bases with AMIE+. VLDB J. **24**(6), 707–730 (2015). https://doi.org/10.1007/s00778-015-0394-1
27. Gesese, G.A., Biswas, R., Alam, M., Sack, H.: A survey on knowledge graph embeddings with literals: which model links better literal-ly? (2020)
28. Goyal, N., Sachdeva, N., Kumaraguru, P.: Spy the lie: fraudulent jobs detection in recruitment domain using knowledge graphs. In: Qiu, H., Zhang, C., Fei, Z., Qiu, M., Kung, S.-Y. (eds.) KSEM 2021. LNCS (LNAI), vol. 12816, pp. 612–623. Springer, Cham (2021). https://doi.org/10.1007/978-3-030-82147-0_50
29. He, S., Liu, K., Ji, G., Zhao, J.: Learning to represent knowledge graphs with gaussian embedding. In: Proceedings of the 24th ACM International on Conference on Information and Knowledge Management (2015)
30. Hitchcock, F.L.: The expression of a tensor or a polyadic as a sum of products. J. Math. Phys. **6**, 1–4 (1927)
31. Huynh, V.P., Papotti, P.: A benchmark for fact checking algorithms built on knowledge bases. In: 28th ACM International Conference on Information and Knowledge Management, CIKM'19, 3rd-7th November 2019, Beijing, China (2019)
32. Ji, G., He, S., Xu, L., Liu, K., Zhao, J.: Knowledge graph embedding via dynamic mapping matrix. In: Proceedings of the 53rd Annual Meeting of the Association for Computational Linguistics and the 7th International Joint Conference on Natural Language Processing (volume 1: Long papers) (2015)
33. Ji, S., Pan, S., Cambria, E., Marttinen, P., Yu, P.S.: A survey on knowledge graphs: representation, acquisition and applications (2020)
34. Kazemi, S.M., Poole, D.: Simple embedding for link prediction in knowledge graphs (2018)
35. Kim, J., Kim, H.-J., Kim, H.: Fraud detection for job placement using hierarchical clusters-based deep neural networks. Appl. Intell. **49**(8), 2842–2861 (2019). https://doi.org/10.1007/s10489-019-01419-2

36. Kowshalya, G., Nandhini, M.: Predicting fraudulent claims in automobile insurance. In: 2018 2nd International Conference on Inventive Communication and Computational Technologies (ICICCT). IEEE (2018)

37. Li, Y., Yan, C., Liu, W., Li, M.: Research and application of random forest model in mining automobile insurance fraud. In: 2016 12th International Conference on Natural Computation, Fuzzy Systems and Knowledge Discovery (ICNC-FSKD). IEEE (2016)

38. Lin, Y., Liu, Z., Sun, M., Liu, Y., Zhu, X.: Learning entity and relation embeddings for knowledge graph completion. In: Proceedings of the AAAI Conference on Artificial Intelligence, vol. 29 (2015)

39. Mishra, A., Ghorpade, C.: Credit card fraud detection on the skewed data using various classification and ensemble techniques. In: 2018 IEEE International Students' Conference on Electrical, Electronics and Computer Science (SCEECS). IEEE (2018)

40. Nickel, M., Rosasco, L., Poggio, T.: Holographic embeddings of knowledge graphs. In: Proceedings of the AAAI Conference on Artificial Intelligence, vol. 30 (2016)

41. Nickel, M., Tresp, V., Kriegel, H.P.: A three-way model for collective learning on multi-relational data. In: Proceedings of the 28th International Conference on Machine Learning, ICML'11, pp. 809–816 (2011)

42. Rabah, C.B., Coatrieux, G., Abdelfattah, R.: The supatlantique scanned documents database for digital image forensics purposes. In: 2020 IEEE International Conference on Image Processing (ICIP). IEEE (2020)

43. Rizki, A.A., Surjandari, I., Wayasti, R.A.: Data mining application to detect financial fraud in indonesia's public companies. In: 2017 3rd International Conference on Science in Information Technology (ICSITech). IEEE (2017)

44. Rossi, A., Firmani, D., Matinata, A., Merialdo, P., Barbosa, D.: Knowledge graph embedding for link prediction: a comparative analysis (2020)

45. Rossi, A., Matinata, A.: Knowledge graph embeddings: are relation-learning models learning relations? In: EDBT/ICDT Workshops (2020)

46. Shen, A., Mistica, M., Salehi, B., Li, H., Baldwin, T., Qi, J.: Evaluating document coherence modeling. Trans. Assoc. Comput. Linguist. **9**, 621–640 (2021)

47. Shi, B., Weninger, T.: Proje: Embedding projection for knowledge graph completion. In: Proceedings of the AAAI Conference on Artificial Intelligence, vol. 31 (2017)

48. Sidere, N., Cruz, F., Coustaty, M., Ogier, J.M.: A dataset for forgery detection and spotting in document images. In: 2017 7th International Conference on Emerging Security Technologies (EST). IEEE (2017)

49. Suchanek, F.M., Kasneci, G., Weikum, G.: Yago: a core of semantic knowledge. In: Proceedings of the 16th International Conference on World Wide Web, WWW'07, pp. 697–706. Association for Computing Machinery, New York, NY, USA (2007)

50. Sun, Z., Deng, Z.H., Nie, J.Y., Tang, J.: Rotate: knowledge graph embedding by relational rotation in complex space (2019)

51. Thorne, J., Vlachos, A.: Automated Fact Checking: task formulations, methods and future directions. CoRR (2018)

52. Trouillon, T., Welbl, J., Riedel, S., Éric Gaussier, Bouchard, G.: Complex embeddings for simple link prediction (2016)

53. Van Der Maaten, L.: Accelerating t-SNE using tree-based algorithms. J. Mach. Learn. Res. **15**(1), 3221–3245 (2014)

54. Vidros, S., Kolias, C., Kambourakis, G., Akoglu, L.: Automatic detection of online recruitment frauds: characteristics, methods, and a public dataset. Future Internet **9**(1), 6 (2017)

55. Wang, Q., Mao, Z., Wang, B., Guo, L.: Knowledge graph embedding: a survey of approaches and applications. IEEE Trans. Knowl. Data Eng. **29**(12), 2724–2743 (2017)
56. Wang, Z., Zhang, J., Feng, J., Chen, Z.: Knowledge graph embedding by translating on hyperplanes. In: Proceedings of the AAAI Conference on Artificial Intelligence, vol. 28 (2014)
57. Yang, B., tau Yih, W., He, X., Gao, J., Deng, L.: Embedding entities and relations for learning and inference in knowledge bases (2015)
58. Zhang, S., Tay, Y., Yao, L., Liu, Q.: Quaternion knowledge graph embeddings (2019)
59. Zhang, W., Paudel, B., Zhang, W., Bernstein, A., Chen, H.: Interaction embeddings for prediction and explanation in knowledge graphs. In: Proceedings of the 12th ACM International Conference on Web Search and Data Mining (2019)

Exploiting Metonymy from Available Knowledge Resources

Itziar Gonzalez-Dios[✉], Javier Álvez, and German Rigau

IXA and LoRea Groups, University of the Basque Country UPV/EHU,
Donostia/San Sebastian, Spain
{itziar.gonzalezd,javier.alvez,german.rigau}@ehu.eus

Abstract. Metomymy is challenging for advanced natural language processing applications because one word or expression is used to refer to another implicit related concept. So far, metonymy has been mainly explored in the context of its recognition and resolution in texts. In this work we focus on exploiting metonymy from existing lexical knowledge resources. In particular, we analyse how metomynic relations are implicitly encoded in WordNet and SUMO. By using an existing automated reasoning framework to test the new modelling acquired from WordNet and SUMO, we propose a practical way to deal with figurative expressions.

Keywords: Metonymy · WordNet · SUMO · Knowledge representation · Commonsense reasoning

1 Introduction

Metonymy is a cognitive and linguistic process where one thing is used to refer another [1]. Together with the metaphor, metonymy is an imaginative device (a figure of speech) that is an object of prime interest for cognitive linguistics. In the conceptual projection of metonymy, the target, which belongs to an experiential domain, is understood in terms of the source, which belongs to other domain [2]. Common metonymic cases are expressing parts as wholes, continents for contents, places for products... For example, in (1) the word *Bordeaux* is not referring to the French city, but to the wine produced in its region. That is, a place is used for a product.

(1)   I never drink a young <u>Bordeaux</u>.

The use of figurative language is crucial for natural language understanding and, therefore, for Natural Language Processing (NLP). In the case of metonymy, many data-driven works have dealt with its resolution. For example, Markert et al. [3] present the results of the SemEval-2007 shared task on that topic and Gritta et al. [4] focus on locations. There are also some works such as [5,6] which investigate logical metonymy, an elliptical construction that occurs when an event-subcategorising verb is combined with an entity-denoting direct object.

© Springer Nature Switzerland AG 2023
A. Gelbukh (Ed.): CICLing 2019, LNCS 13451, pp. 34–43, 2023.
https://doi.org/10.1007/978-3-031-24337-0_3

In this work we analyse the characterisation of metomymy in existing Lexical Knowledge Resources (LKRs). The LKRs we use in this paper are the semantic network WordNet [7] and the SUMO ontology [8]. WordNet [7] has implicit metonymic relations that need to be extracted and SUMO has no formal representation of metonymy. That is why, we explore how we can reuse the information in WordNet to model it in SUMO, and start a discussion towards making these relations explicit and systematised in WordNet. We concentrate on the membership relation of locations and organisations that are coded in WordNet. For instance, the words *Andorra* and *Hanseatic League* denote respectively in (2) and (3) a government that is formed by group of people and a group of cities formed by group of people, and not as regions or organisations on their own, but we use them with metonymic reading to get a more efficient and economic communication.

(2)   Andorra does not close border with Catalonia.

(3)   The Hanseatic League carried out an active campaign against pirates.

If we look for Andorra and Hanseatic League in WordNet we do find that Andorra (nation) has Andorran (inhabitant) as member and Hanseatic League (political organisation) has Bremen (city) as its member. That is, Andorra and Hanseatic League are represented also as group of people in WordNet and not only as locations or organisations. That is why, we think that the member relation in WordNet can be a good clue to detect metonymic readings.

So, the aim of this paper is to detect metonymic candidates via the member relation in WordNet and propose an approach to show that it is possible to deal with figurative expressions. Exactly, we explore how we can reuse the information in WordNet to model it in SUMO, and, moreover, we propose to start a discussion towards making these relations explicit and systematised in WordNet. Another aim of this work is to ease the burden and the cost of creating new resources by exploring the interoperability of the ones that are available in order to create a future framework for commonsense reasoning.

This paper is structured as follows: in Sect. 2 we present the analysis of metonymy in the LKRs, in Sect. 3 we explain our approach to treat metonymy that we evaluate in Sect. 4, we discuss the results in Sect. 5 and we conclude and outline the future work in Section 6.

2   Analysing Metonymy in WordNet and SUMO

Metonymic readings in WordNet are not specifically labelled, but some are coded within the *member* relation. Based on the annotation framework of [3] where they distinguished metonymic patterns for locations (place-for-people, place-for-event...), organisations (org-for-members, org-for-event...) and class-independent categories (object-for-name and object-for-representation), we have decided to extract the synsets denoting locations and organisations, which, in our opinion, are candidates for being metonymic.

Table 1. Selected BLCs to find metonymic candidates

Candidate type	Selected BLCs
Locations	$area_n^1$, $capital_n^3$, $district_n^1$, $country_n^2$, $state_n^1$, $American\_State_n^1$, $geographical\_area_n^1$, $desert_n^1$, $city_n^1$, $national\_capital_n^1$, $state\_capital_n^1$, $national\_park_n^1$, $town_n^1$, $region_n^1$, $boundary_n^1$, end_n^1, $port_n^1$
Organisations	$organization_n^1$, $gathering_n^1$, $lineage_n^1$, $terrorist\_organization_n^1$, $artistic\_movement_n^1$

In order to extract the candidates, we have selected the Basic Level Concepts (BLCs) [9] presented in Table 1 relating locations and organisations. In this paper, we will use the format $word_n^s$, where s is the sense number of the $word$ and n means that it is a noun. For brevity, we will only show a representative variant of each selected BLC synset in the table. Based on those BLCs we have also gathered their hyponyms as long as their semantic file was *location* for locations and *group* for organisations. In total we have selected 156 candidate synsets.

Following, among those candidates we have extracted the synset pairs that are linked via the *member* relation. This way, we have obtained a dataset of 672 metonymic pairs. This way, we have detected some of the implicit metonymic readings in WordNet.

The SUMO counterpart of the WordNet *member* relation is the predicate $member_r$ which relates an individual object (i.e. an instance of the SUMO class $Object_c$, the class for ordinary objects) as part of a collection (i.e. an instance of $Collection_c$, the class where its member can be altered without changing the nature of the collection itself). To denote SUMO concepts in this paper we will use the symbol c for SUMO classes and r for SUMO relations.

Moreover, WordNet and SUMO are linked via a semantic mapping [10]. This mapping links a synset to a SUMO concept by means of the relations *equivalence* (both mean the same) or *subsumption/instantiation* (the semantics of the SUMO concept is more general than the semantics of the synset in WordNet). The semantic mapping relation is denoted, in this paper, by appending as suffix the symbols '=' (equivalence), '+' (subsumption) and '@' (instantiation) to the corresponding SUMO concept. This way, the synset $Hardy_n^1$ presented in (4) is mapped to Man_c+ and $Laurel\_and\_Hardy_n^1$ is to $Group_c+$. It is important to mention that no metonymic readings are coded in the original mapping, only literal.

So, if we cross-check the knowledge in WordNet (the pairs in the dataset we have extracted) and SUMO (see Sect. 4), we see that example (4) is validated: from the knowledge in SUMO it is possible to infer that some instances of Man_c and $Group_c$ are related be $member_r$. In other words, humans can be members of groups. On the contrary, Example (5) is unvalidated since $France_n^1$ is connected to $Nation_c+$ and $Nation_c$ is not instance of $Collection_c$. That is, $France_n^1$ is not understood as a group of people in the mapping to SUMO, but as a mere region (an object) and, therefore, cannot have members.

(4)   a. $Hardy_n^1$ United States slapstick comedian who played (...) (Man_c+)
        —MEMBER OF—

        b. $Laurel\_and\_Hardy_n^1$ United States slapstick comedy duo (...) $(Group_c+)$

(5)   a. $French\_person_n^1$ a person of French nationality. $(EthnicGroup_c+)$
        —MEMBER OF—

        b. $France_n^1$ a republic in western Europe; (...) $(Nation_c+)$

Once we have extracted the metonymic candidates, we pursue to model metomymic readings, but where should it be modelled? In the mapping? In the ontology? In both? In the Sect. 3 we present our practical approach to deal with this figure of speech.

3   Modelling Metonymy

In order to model metonymy based on the WordNet dataset we have analysed both the mapping and the ontology.

3.1   Exploring the Mapping

By inspecting the mapping manually by class frequency, we have realised that there is lack of systematicity above all with demonyms: for example (6) and (7) are mapped to $FreshWaterArea_c+$ (the class for rivers and lakes) and $NaturalLanguage_c+$ (the class for languages) respectively although each one is an individual, (8) is mapped to $EthnicGroup_c+$ (the class for groups of people that share a country, a language...) despite being also a person. The latest is the most frequent case in the dataset.

(6) $Sabine_n^2$ a member of an ancient Oscan-speaking people (...)

(7) $Sotho_n^1$ a member of the Bantu people who (...)

(8) $Sherpa_n^1$ a member of the Himalayan people (...)

So, in order to correct automatically the mapping of synsets like (8), we have created the following heuristic based on the semantic files (SF), the glosses of WordNet the BLCs:

– Change the mapping of a synset from $EthnicGroup_c$ to $Human_c$ (the class for modern man and woman) if its BLC is $person_n^1$; its SF is *person*; and there is one of the following expressions *a member of, a native or inhabitant of, an inhabitant of, a person of, a speaker of, a German inhabitant of, a native or resident or, a resident of, a native of, a Greek inhabitant of, a native or resident of, a person of, inhabitant of, a Polynesian native or inhabitant of* or *an American who lives in* in its gloss.

This way, we have corrected the mapping of 272 synsets.

We have also manually analysed the mapping of the synsets that are connected to a SUMO concept different from $EthnicGroup_c$ but that fulfil the remaining above conditions. Out of 75 synsets we have corrected 30, most of them corrected to the SUMO class $Human_c$, namely cases such as (6) and (7).

For the other kind of errors (not relating demonyns), we have extracted the synsets containing the expressions *an island territory of, an island in* and *a glacial lake*. Out of 6 synsets we have corrected 3 manually.

In total, we have corrected the mapping 308 synsets: 33 manually and 272 automatically.

After having corrected the mapping, we have modelled the metonymy. To that end, we have added the $Group_c+$ class to the synsets in the b part of the pair. This way, we have created a corrected and modelled mapping (CMM).

3.2 Exploring the Ontology

In order to start exploring the ontology we have performed an error analysis on the most frequent metonymic pairs that remained unresolved in [11]. The first inspection revealed us the respective members of cities, countries... and groups or organisations as members of other groups or organisations were not validated. As mentioned before, the predicate $member_r$ in SUMO relates instances of $Object_c$ and $Collection_c$, but regions, for example, are not connected to (subconcepts of) $Collection_c$ and groups are not connected to (subconcepts of) $Object_c$. So, according to SUMO, they cannot be related as part or whole in member pairs.

After this analysis, we have included 22 new general axioms in the ontology for organisations and regions that modelled metonymy and added missing related knowledge. Next, we describe two of the introduced axioms:

- Relating the regions, any SUMO concept related to some instance of $LandArea_c$ (the general class for areas where predominates solid ground and includes all the nations, states, provinces...) by $member_r$ as whole is restricted to be instance of $Agent_c$ (the class for something or someone that can act on its own and produce changes), which is superclass of both $Organization_c$ (the class for corporations or similar institutions) and $Human_c$. That means that countries can have humans and organisations as members. This way, we will validate example (2): $Andorra_n^1$ ($Nation_c+$) can have $Andorran_n^1$ ($Human_c+$) as member.

```
(forall (?LAND ?MEMBER)
    (=>
        (and
            (instance ?LAND LandArea)
            (member ?MEMBER ?LAND))
        (instance ?MEMBER ?Agent)))
```

– Relating the organisations, any instance of $PoliticalOrganization_c$ (the class for organisation that attempt some sort of political change) is restricted to be related by $member_r$ as whole to some instance of $GeographicArea_c$ (the general class for geographic locations with definite boundaries). In other words, political organisations can have countries, cities... as members. This way, we will validate example (3): $Hanseatic\_League_n^1$ ($PoliticalOrganization_c+$) can have $Bremen_n^1$ ($City_c+$) as member.

```
(forall (?PORG)
    (=>
        (instance ?PORG PoliticalOrganization)
        (exists (?GAREA)
            (and
                (instance ?GAREA GeographicArea)
                (member ?GAREA ?PORG)))))
```

With this kind of axioms we look also for the validation of candidates such as $Breton_n^1$ as member of $Bretagne_n^1$ ($StateOrProvince_c+$) and $Belgique_n^1$ ($Nation_c+$) as member of $Benelux_n^1$ ($PoliticalOrganization_c+$).

4   Evaluation Framework and Results

In order to evaluate our interventions, we have used the framework presented by [11,12], where competency questions (CQ) based on predefined question patterns (QP) are automatically created taking the knowledge of the WordNet dataset into account. In Fig. 1 we present an example of the CQs that are created.

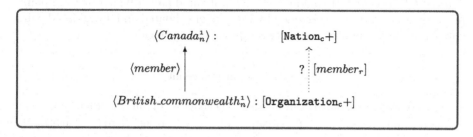

Fig. 1. Example of competency question about metonymy relating countries and organizations

In the present work, we have used the QPs presented in the above mentioned work in order to create the CQs for the *member* relation with the original mapping (OM), the original ontology (OO), our corrected and modelled mapping

(CMM) and our modelled ontology (OM). In order to see which intervention has more impact, we have considered the following cases:

1. OM+OO: The starting point with the original mapping and original ontology
2. CMM+OO: Measuring the impact of the mapping with the corrected and modelled mapping and original ontology
3. OOM+MO: Measuring the impact of the ontology with the original mapping and modelled ontology
4. CMM+MO: Measuring the impact of both the mapping and the ontology with the corrected and modelled mapping and modelled ontology

In the first and third cases, 89 CQs are created, while 124 CQs are created in the second and fourth cases. Then, the resulting sets of CQs are experimentally evaluated by using the ATP Vampire v4.2.2 [13] in a Intel® Xeon® CPU E5-2640v3@2.60GHz with 2GB of RAM memory per processor. For each test, we have set an execution-time limit of 300 s and a memory limit of 2GB.[1] Depending on the outcome provided by the ATP, each CQ can be classified as i) *passing* (the ATP proves that the knowledge encoded in the CQ is entailed by SUMO); ii) *non-passing* (the knowlegde encoded in the CQ is incompatible with SUMO since the ATP proves that its negation is entailed); or iii) *unknown* (if the ATP finds no proof).

The results of these experiments are presented in Table 2. More concretely, we provide the number of metonymic pairs that are validated (Validated column), unvalidated (Unvalidated column) and unclassified (Unclassified column). Roughly speaking, a metonymic pair is classified as validated/unvalidated/unclassified if the corresponding CQ is decided to be passing/non-passing/unknown respectively. Further, a metonymic pair is also classified as unvalidated if the mapping information is incompatible with the SUMO predicate $member_r$, since no CQ can be obtained. We provide the number of metonymic pair with incompatible mapping information in the column Unvalidated between brackets. On this basis, we calculate the recall (Recall column), precision (Precision column) and the $F1$ measure (F1 column): recall is validated pairs divided by total pairs, precision is validated pairs divided by the sum of validated and unvalidated pairs and $F1$ is the harmonic mean between precision and recall.

Table 2. Experimental results

Experiment	Validated	Unvalidated		Unclassified	Recall	Precision	$F1$
OM+OO	4	357	(356)	311	0.006	0.011	0.008
CMM+OO	5	12	(11)	655	0.007	0.294	0.015
OM+MO	197	359	(356)	116	0.293	0.354	0.321
CMM+MO	483	12	(11)	177	0.719	0.976	0.828

[1] Parameters: `--proof tptp --output_axiom_names on --mode casc -t 300 -m 2048`.

Looking at the experimental results (Table 2), we see that in the first case (OM+OO) only 4 pairs are validated. Thus, the $F1$ is really low (0.008). By correcting and modelling the mapping (CMM+OO), only one more pair is validated, and although the $F1$ increases the (0.015), the results are still very poor. The important comment on this experimental phase is that more pairs are unclassified (345 more) and that they are no longer unvalidated. That is, this time the mapping information is not incompatible with $member_r$, but the knowledge encoded in the corresponding CQ is not yet entailed by the ontology. When adding metonymic knowledge in the ontology, but still using the original mapping (OM+MO), we see that there is a big improvement in the number of validated pairs: 193 more pairs are validated (total 197) and, thus, $F1$ is increased if compared to the two previous cases. Further, if we compare the results in the cases CMM+OO and OM+MO, we see that modelling the metonymic knowledge in the ontology has more impact than modelling it in the mapping. Finally, by combining both mapping and ontology modellings (CMM+MO), we see that the results are undoubtedly better: 483 pairs are validated (more than the two thirds of the pairs) and $F1$ is 0.828, i.e. 0.820 more than in the initial case. However, there is room for improvement: 177 pairs are unclassified and in the case of 12 pairs the mapping information is not compatible with $member_r$.

In order to look at the results qualitatively, we present two validated examples: (9) and (10). Like (2), (9) is validated now because we added axioms that allow countries, regions... to have people and regions as members. Similar to (3), (10) is validated now because we added axioms to relate countries, nations... to organisations.

(9) a. $Mexican_n^1$ a native or inhabitant of Mexico (...) ($Human_c+$)
        —MEMBER OF—

     b. $Mexico_n^1$ a republic in southern North America(...) ($Nation_c+$)

(10) a. $France_n^1$ a republic in western Europe (...) ($Nation_c+$)
        —MEMBER OF—

     b. $NATO_n^1$ an international organization (...) ($Organization_c+$)

5   Discussion

As mentioned in the previous sections, we have a big improvement, but there still room to work with. That is why we have also performed a detailed analysis of the unvalidated and unclassified pairs.

Regarding the unvalidated pairs, only one pair with compatible mapping information is unvalidated: $crew_n^2$ (connected to $GroupOfPeople_c+$) as member of $workforce_n^1$ (connected to $SocialRole_A+$). The reason is that the class of attributes $SocialRole_A$ can be applied to only individuals, thus not to groups. Being the only one, this case was part of the long tail that we did not analyse during the inspection of the mapping. The other unvalidated pairs have incompatible mapping information: one of the synsets is mapped to a process and therefore

CQs cannot be created e.g. $conferee_n^2$ (connected to $Human_c+$) as member of $conferee_n^1$ (connected to $FormalMeeting_c+$, which is a SUMO process).

Regarding the unclassified pairs, we need to keep on modelling specific data, as we only modelled general metonymic relations in the ontology. We have also detected mapping errors in 4 synsets whose expressions did not cover such as *an Oscan-speaking member of*. Moreover, we have identified at least 38 pairs could be validated if we gave the ATP more time.

On the other hand, while modelling the information in the ontology we have realised that the *member* relation is not always clear in WordNet. Is an organisation *member of* another organisation? Or is it *part of* it? Does it depend? We think that sometimes the distinction is not very clear. For example, the synset $workforce_n^1$ has a member which is $crew_n^2$ (an organized group of workmen) and a part which is $shift_n^7$ (a crew of workers who work for a specific period of time).

That is why we consider that a discussion on the WordNet relations can be necessary. In our opinion, there is a lack of formality in the relations: some of them are ambiguous. So, we propose to start thinking about systematising them, axiomatising them. Moreover, we think that making explicit i.d. modelling implicit or hidden relations such as the case of metonymy we have presented here can be advantageous and profitable for NLP applications.

6   Conclusion and Future Work

In this paper we have presented a practical approach to show that it is feasible to work with figurative language in well-known knowledge resources by reusing their information and correcting their discrepancies. We have detected implicit metonymic relations regarding organisations and locations in WordNet and we have modelled this knowledge in SUMO. We have explored two general formalisations: modelling in the mapping and modelling in the ontology, and we have seen the changes in the ontology have more impact than the changes in the mapping. Moreover, we have corrected discrepancies in the mapping between them. As for the results, we see that there is considerable improvement (from $F1$ 0.008 to $F1$ 0.828) when merging both formalisations, but there is still room for improvement by given the ATP more resources or continuing adding more specific information. But, we have also seen that relations are not well-defined.

Therefore, future work involves also examining more precisely the meaning and the coherence of the relations both in WordNet and SUMO in order to systematise and axiomatise them. Moreover, we would like to study other kind of metonymic relations, for instance, relating processes and events. Evaluating the correctness of the implicit metonymic knowledge in WordNet is also one of our future goals.

Acknowledgments. This work has been partially funded by the Spanish Projects TUNER (TIN2015-65308-C5-1-R) and GRAMM (TIN2017-86727-C2-2-R) and the Basque Projects BERBAOLA (KK-2017/00043) and LoRea (GIU18/182).

References

1. Littlemore, J.: Metonymy. Cambridge University Press, Cambridge (2015)
2. Barcelona, A.: Metaphor and Metonymy at the Crossroads: A Cognitive Perspective. Walter de Gruyter, Berlin (2012)
3. Markert, K., Nissim, M.: Data and models for metonymy resolution. Lang. Resour. Eval. **43**(2), 123–138 (2009). https://doi.org/10.1007/s10579-009-9087-y
4. Gritta, M., Pilehvar, M.T., Limsopatham, N., Collier, N.: Vancouver welcomes you! minimalist location metonymy resolution. In: Proceedings of the 55th Annual Meeting of the Association for Computational Linguistics (Volume 1: Long Papers), pp. 1248–1259 (2017)
5. Shutova, E., Kaplan, J., Teufel, S., Korhonen, A.: A computational model of logical metonymy. ACM Trans. Speech Lang. Process. (TSLP) **10**(3), 1–28 (2013)
6. Chersoni, E., Lenci, A., Blache, P.: Logical metonymy in a distributional model of sentence comprehension. In: 6th Joint Conference on Lexical and Computational Semantics (* SEM 2017), pp. 168–177 (2017)
7. Fellbaum, C. (ed.): WordNet: An Electronic Lexical Database. MIT Press, Cambridge (1998)
8. Niles, I., Pease, A.: Towards a standard upper ontology. In: Guarino, N., et al., (ed.): Proceedings of the 2nd International Conference on Formal Ontology in Information Systems (FOIS'01). ACM, pp. 2–9 (2001)
9. Izquierdo, R., Suárez, A., Rigau, G.: Exploring the automatic selection of basic level concepts. In: Proceedings of the International Conference on Recent Advances on Natural Language Processing (RANLP'07), vol. 7 (2007)
10. Niles, I., Pease, A.: Linking lexicons and ontologies: mapping wordnet to the suggested upper merged ontology. In: Arabnia, H.R., (ed.): Proceedings of the IEEE International Conference on Information and Knowledge Engineering (IKE'03). vol. 2, pp. 412–416. CSREA Press (2003)
11. Álvez, J., Gonzalez-Dios, I., Rigau, G.: Cross-checking using meronymy. In: Calzolari, N., et al., (eds.): Proceedings of the 11th International Conference on Language Resources and Evaluation (LREC'18). European Language Resources Association (ELRA) (2018)
12. Álvez, J., Lucio, P., Rigau, G.: A framework for the evaluation of -based ontologies using. IEEE Access **7**, 1–19 (2019)
13. Kovács, L., Voronkov, A.: First-order theorem proving and VAMPIRE. In: Sharygina, N., Veith, H. (eds.) CAV 2013. LNCS, vol. 8044, pp. 1–35. Springer, Heidelberg (2013). https://doi.org/10.1007/978-3-642-39799-8_1

Robust Evaluation of Language–Brain Encoding Experiments

Lisa Beinborn[1]([✉]), Samira Abnar[2], and Rochelle Choenni[2]

[1] Vrije Universiteit Amsterdam, Amsterdam, Netherlands
l.beinborn@vu.nl
[2] University of Amsterdam, Amsterdam, Netherlands
{s.abnar,r.m.v.k.choenni}@uva.nl

Abstract. Language–brain encoding experiments evaluate the ability of language models to predict brain responses elicited by language stimuli. The evaluation scenarios for this task have not yet been standardized which makes it difficult to compare and interpret results. We perform a series of evaluation experiments with a consistent encoding setup and compute the results for multiple fMRI datasets. In addition, we test the sensitivity of the evaluation measures to randomized data and analyze the effect of voxel selection methods. Our experimental framework is publicly available to make modelling decisions more transparent and support reproducibility for future comparisons.

Keywords: Evaluation of language · fMRI datasets · Language–brain encoding

1 Introduction

Representing language in a computationally usable format has been a research goal since the beginning of computational linguistics. In the last decade, distributional representations which interpret words, phrases, sentences, and even full stories as a high-dimensional vector in semantic space have become the most common standard. These representations are obtained by training language models on large corpora to optimally encode contextual information.

The quality of language representations is commonly evaluated on a set of downstream tasks. These tasks are either driven by engineering adequacy (e.g. the effect of the language representations on the performance of systems such as machine translation) or by the ability to reproduce human decisions (e.g. the performance of the representations on semantic similarity or entailment tasks).

The experiments were conducted in 2018 when all three authors were employed at the Institute of Logic, Language and Computation at the University of Amsterdam. The paper was presented in 2019. Since then, language modeling has progressed immensely. Experimental standards for robust, comparable, and reproducible evaluation for interpreting language–brain encoding experiments with respect to reasonable random permutation baselines need to be further developed and more widely adopted.

A. Gelbukh (Ed.): CICLing 2019, LNCS 13451, pp. 44–61, 2023.
https://doi.org/10.1007/978-3-031-24337-0_4

Many language researchers, however, are driven by the urge to better understand the underlying principles of human language processing.

With the increasing availability of brain imaging data, it has become popular to evaluate computational models by their ability to simulate brain signals related to human language processing [18, 19, 27]. If we can develop models that encode linguistic information in a way that is comparable to the activity in human brains, we will get one step closer to cognitively plausible models of human language understanding. While experimenting with human brains is evidently strictly constrained and regulated due to ethical reasons, we can easily query, adapt, constrain, degrade, and manipulate the computational model and analyze the effect on its language processing capabilities.

Although working with brain imaging data is highly promising from a cognitive perspective, it comes with many practical limitations. Brain datasets are usually too small for powerful machine learning models, the imaging technology produces noisy output that needs to be adjusted by statistical correction methods, and most importantly, only very few datasets are publicly available. Experiments in previous work are usually performed on a single dataset, so that it is unclear whether the observed effects are generalizable. In addition, the applied evaluation procedures have not yet been standardized. Understanding the subtle differences in the experimental setup to interpret the results can be particularly difficult because it has not yet become a common practice to publish the experimental code along with the results.

To the best of our knowledge, this paper provides the first analysis of language–brain encoding experiments which applies a consistent evaluation scenario across multiple fMRI datasets. We examine whether different evaluation measures provide different interpretations of the predictive power of the encoding model. Our experimental framework is publicly available to make modelling decisions more transparent and facilitate reproducibility for future comparisons. Due to its modular architecture, the pipeline can easily be extended to experiment with other datasets and language models.[1]

Table 1. 4 fMRI datasets for language–brain encoding. In WORDS and STORIES, stimuli have been isolated by averaging over the brain responses. The ALICE and HARRY datasets contain continuous stimuli.

Name	Stimuli	Presentation mode	Subj.	Scans	Voxel size	Reference
WORDS	60 words	Word + image	9	360	3x3x6	Mitchell et al. [24]
STORIES	40 stories	Read sentences	30	40	3x3x3	Dehghani et al. [13]
ALICE	1 chapter	Listen to audio book	27	362	3x3x3	Brennan et al. [10]
HARRY	1 chapter	Read word by word	8	1351	3x3x3	Wehbe et al. [32]

---

[1] The code is available at https://github.com/beinborn/brain-lang.

2 Human-Centered Evaluation of Computational Models

As computational language models are trained on human-generated text, their performance is inherently optimized to simulate human behavior. Although novel architectural solutions attract notable interest in the research community, the ultimate benchmark for a model is the ability to approximate human language processing abilities. Models are supposed to reach a gold standard of human annotation decisions [29] and the difficulty of a task is often estimated by the inter-annotator agreement [5] or by error rates of human participants [8]. While these product-oriented evaluations focus on a final outcome, procedural measures of response times [25] or eye movements [7] are analyzed to provide deeper insights on sequential phenomena like attention or processing complexity. As neural network models are inspired by neuronal activities in the human brain, it is particularly interesting to analyze similarities and differences between distributed computational representations and low-level brain responses.

Electroencephalography (EEG) measures can be used to study specific semantic or syntactic phenomena [15,18,31] and compare the processing complexity of computational models to brain responses, for example, with respect to the N400 and P600 effects [14]. Signals with higher spatial resolution like magnetoencephalography (MEG) and functional magnetic resonance imaging (fMRI) are often used for experiments which are known as brain decoding and brain encoding. In the decoding setup, a computational model learns to identify differences in the signal and to discriminate between the responses for abstract and concrete words [4], for different syntactic classes [9,22], for levels of syntactic complexity [10], and many other linguistic categories. Mitchell et al. [24] have shown that it is not only possible to distinguish between semantic categories but that a model can even learn to distinguish which word a participant is reading. The reverse direction of predicting the brain response that would most likely be observed for a novel linguistic stimulus is commonly called encoding. The encoding task requires a strong computational representation of the stimulus that reflects the shared properties of different stimuli and the relations between stimuli. For the remainder of this paper, we will focus on the language–brain encoding task and on fMRI datasets.

Many word representations have been tested on the Mitchell et al. [24] data including information from lexical resources, distributional, and multimodal representations [1,4,11,34]. It has also been proposed to directly feed the brain signal into the language model as an additional source of information [6,16]. Recently, new approaches for encoding and decoding of datasets using longer linguistic stimuli such as sentences [27] and even full stories [10,13,19,32] are emerging. In some experiments, it has been shown that contextualized representations obtained from recurrent neural networks [19,33] seem to represent the continuous stimuli slightly better than models that represent sentences as a conglomerate of context-independent word representations [13,27]. However, these results are hard to generalize because they have been tested only on a single dataset. Gauthier and Ivanova [17] raise doubts about the informativeness of encoding results because differences between models are not reflected. Our robust evaluation experiments can serve as a comparative testbed for future analyses.

3   Datasets

We use four fMRI datasets that have been collected by different researchers (see Table 1). All datasets use English language stimuli and the participants are native speakers. Standard fMRI preprocessing methods such as motion correction, slice timing correction and co-registration to an MNI template had already been applied.

3.1   Isolated Stimuli

We use two datasets that work with isolated stimuli. The stimuli are not related and can be presented in varying order to the participants. Each stimulus is represented with only a single brain activation vector by averaging over several scans obtained during the presentation of the stimulus.

Words. For the WORDS dataset, 9 participants were shown a word paired with a line drawing of the object denoted by the word and were instructed to think about the properties of the object [24]. Six scans were taken during the presentation of each word. The scans were temporally detrended and smoothed. The activation values were normalized by computing the percent signal change relative to the fixation condition. Scans and stimuli were aligned with an offset of 4 s to account for the haemodynamic delay. The brain activation for each word is calculated by taking the mean over the six scans.

Stories. For the STORIES dataset, 30 participants were reading 40 short personal stories that had been collected from weblogs [13]. The stories consisted of 11 sentences on average and were presented in three consecutive batches on a screen. The dataset also contains data for Farsi and Chinese stories but for the sake of comparison, we focus on the English subset here. The scans were preprocessed with detrending, temporal smoothing and spatial smoothing. The activation values were normalized by calculating z-scores with respect to the fixation condition. The authors then discretized the continuous story stimulus by calculating the mean over all story scans. We exclude subject 30 from the data because the voxel values are all zero.

3.2   Continuous Stimuli

Humans process language incrementally and in context. In order to simulate a more naturalistic language setting, recent approaches to brain encoding use continuous stimuli and analyze the fMRI scans as a sequence of responses.

Harry. For the HARRY dataset by Wehbe et al. [32], 8 participants read chapter 9 of *Harry Potter and the Sorcerer's stone* [30]. The story was split into four blocks and presented word by word on a screen. Each word was displayed for 0.5 s and an fMRI scan was taken every 2 s. We follow their protocol and apply detrending and temporal smoothing but do not smooth spatially because it did not have an effect on the results in pilot experiments.

Alice. For the ALICE dataset by Brennan et al. [10], 27 participants were listening to an audio recording of the first chapter of *Alice in Wonderland* [12]. The published data contains the preprocessed signal averaged for 6 regions of interests defined using functional and anatomical criteria. The raw signal is not available.

4   Encoding Model

The fMRI data is obtained by measuring the so-called blood-oxygenation level dependent (BOLD) response. This signal indicates the level of oxygen in the blood (approximated by its magnetic susceptibility) and an increased BOLD response in an area of the brain is interpreted as increased neuronal activity in this region. In order to analyze the response, the brain is fragmented into stacked voxels which are cubes of constant size (e.g. $3 \times 3 \times 3$ mm). The response thus consists of a three-dimensional matrix with activation values for each voxel. This matrix is flattened into a one-dimensional vector $\mathbf{v}$. In the brain encoding approach, the goal is to predict $\mathbf{v}$ given the stimulus $\mathbf{s}$ that was presented when measuring the response.

Mapping Model. A multiple linear ridge regression model is usually applied as encoding model to learn the response pattern $\mathbf{v_n} \in \mathbb{R}^m$ for stimulus $\mathbf{s_n} \in \mathbb{R}^d$ on a training set $V \in \mathbb{R}^{m \times n}$ of responses to n other stimuli.[2] It requires a strong computational representation of the stimulus that reflects the relations between stimuli. The predictive power of this mapping model is evaluated on a set of held-out stimuli $S \in \mathbb{R}^{d \times n}$. The mapping model learns a separate regression equation for every voxel v_i which is fitted by learning a weight w_d for each dimension s_d of the stimulus representations and the weights are regularized by the L2 norm. The cost function f for learning the weight vector $\mathbf{w}$ for a voxel vector $\mathbf{v_i}$ is:

$$f(\mathbf{v_i}) = \sum_{n=1}^{N}(v_{i_n} - \sum_{d=1}^{D} w_d \cdot s_{d_n})^2 + \lambda \sum_{d=1}^{D} w_d{}^2$$

4.1   Language Model

The linguistic stimuli are represented using vectors obtained from a language model. Previous work has compared the performance of different language models for brain encoding tasks showing that contextual models like long short-term

---

[2] Whether a linear model is a plausible choice is debatable. We use it here for comparison with previous work.

memory networks perform better than standard word-based representations [19]. For a more robust comparison, we keep the language model constant for all datasets. We choose the language model *ELMO* because it produces contextualized representations on the sentence level and performs very well on semantic tasks [28]. *ELMO* is based on a bi-directional long short-term memory network and it uses character-based representations of the input which makes it perform very well on out-of-vocabulary words. This is an important property for modeling fictional texts. We use a pre-trained pytorch version of *ELMO* available on github.[3]

For WORDS, we use the representations from the token layer. For all other datasets, we obtain contextualized representations from the first layer. We restrict the representation to the forward language model to simulate incremental processing and obtain a 512-dimensional vector. We take the representation of the last token of each sentence and average over all sentences for each story in STORIES. For the continuous stimuli, we feed the language model the whole chapter and extract the representation of the last token of the sequence which had been presented between the previous and the current scan.

Haemodynamic Delay. The fMRI signal measures a brain response to a stimulus with a delay of up to ten seconds [23]. This delay needs to be considered when aligning stimuli with responses. Similarly to Mitchell et al. [24], we align scans to stimuli with a fixed offset of 4 s. The haemodynamic response decays slowly over a duration of several seconds. For continuous stimuli, this means that the response to previous stimuli will have an influence on the current signal. Wehbe et al. [32] use a feature-based representation and learn different weights for stimuli occurring at previous time steps. In this approach, the number of features increases linearly with the number of time steps considered. In contextual language models, a representation is build up incrementally using recurrent connections. The representation of a word thus implicitly contains information from the previous context. As *ELMO* processes language sentence by sentence, our context window comprises the current sentence up to the current word but the number of dimensions remains constant.

4.2   Voxel Selection

The number of voxels in a brain varies with respect to the voxel size and the shape of the subject's brain. In the datasets used here, the number of voxels ranges from 20,000 to more than 40,000. The activity measured in many of these voxels is most likely not related to language processing but might change due to physical processes like the noise perception in the scanner. In these cases, learning a mapping model from the stimulus representation to the voxel activation will not succeed because the stimulus has no influence on the variance of the voxel signal. Whole-brain evaluations of mapping models thus only have limited informative value. In previous work, different voxel selection models have been applied to

---

[3] https://github.com/allenai/allennlp/blob/master/tutorials/how_to/elmo.md.

analyze only a subset of interesting voxels. Wehbe et al. [32] and Brennan et al. [10] reduced the voxels by using previous knowledge about regions of interests. Restricting the brain response to voxels that fall within a pre-selected set of regions of interests can be considered as a theory-driven analysis.

Information-Driven Voxel Selection. In contrast to the theory-driven region of interest analysis, Kriegeskorte et al. [20] propose a more information-driven approach. So-called searchlight analyses move a sphere through the brain to select voxels (comparable to sliding a context window over text) and analyze the predictive power of the voxel signal within the sphere. Dehghani et al. [13] and Wehbe et al. [32] use this searchlight approach for the decoding task. In brain encoding, the predictive direction is reversed. The ability to predict voxel activation based on the stimulus is carefully interpreted as an indicator that processing the stimulus influences the activity in this particular voxel. For WORDS, Mitchell et al. [24] analyze all six brain responses for the same stimulus and select 500 voxels that exhibit a consistent variation in activity across all stimuli. Jain and Huth [19] calculate the model performance for a single voxel as the Pearson correlation between real and predicted responses on the test set and analyze voxels with a correlation above a threshold. Gauthier and Ivanova [17] recommend to evaluate voxels based on explained variance. We select the 500 most predictive voxels on the training set for WORDS by four selection methods: stability, Pearson correlation, explained variance, and random.

Table 2. The effect of voxel selection on the pairwise accuracy on WORDS. Accuracy and stable voxels are calculated as described in [24].

Metric	None	Stable	by EV	by R	Random
Cosine	.57	.65	.67	.56	.57
Euclidean	.57	.66	.67	.56	.57
Pearson	.58	.67	.68	.57	.58

Results of Voxel Selection. Table 2 shows the results for different voxel selection methods. It can be seen that voxel selection by explained variance performs on par with the selection of stable voxels. We had speculated that simply reducing the number of voxels might already lead to improvements because similarity measures tend to perform better in lower-dimensional spaces [2] but a random selection of voxels has no effect. For the remainder of the paper, we report results on the 500 voxels that obtained the highest explained variance results on the training set unless indicated otherwise because the option of selecting stable voxels is not available for the other datasets.

5   Evaluation Experiments

The voxel selection results show that a small experimental parameter can have a strong effect. We thus perform three experiments using different evaluation procedures: pairwise accuracy, voxel-wise evaluation, and representational similarity analysis. We repeat each experiment with a language model that assigns a random (but fixed) vector to each word to analyze the sensitivity of the evaluation metric. Random story representations are obtained by averaging over words.

5.1   Pairwise Evaluation

As the fMRI datasets are very small for machine learning purposes, Mitchell et al. [24] introduced an evaluation procedure that maximizes the training data. Given a set of n samples, a mapping model is trained on $n-2$ samples and tested on the two remaining samples. Mitchell et al. [24] call this procedure leave-two-out cross-validation but it differs from standard cross-validation setups because each sample occurs n times in the test set leading to $\binom{n}{2}$ different models. The performance is evaluated by calculating the pairwise accuracy over all models.

A pair of two test samples (s_1, s_2) is considered to be classified correctly if the model prediction p_1 is more similar to the true target s_1 than to s_2, and p_2 is more similar to s_2. This general idea of pairwise accuracy has been implemented in different ways. The applied similarity metrics f are cosine similarity [24], euclidean similarity [32], and Pearson correlation [11,27]. The prediction for a pair can be considered to be correct by comparing the summed similarity of the correct alignments with the false alignments [11,13,24]. Wehbe et al. [32] and Wehbe et al. [33] calculate the accuracy by comparing the predictions only for the first sample. A stricter interpretation of the pairwise accuracy would only consider the prediction to be correct, if both samples are correctly matched to their prediction. We refer to the different interpretations as *sum match* (1), *single match* (2), and *strict match* (3):

$$f(s_1, p_1) + f(s_2, p_2) > f(s_1, p_2) + f(s_2, p_1) \tag{1}$$

$$f(s_1, p_1) > f(s_1, p_2) \tag{2}$$

$$f(s_1, p_1) > f(s_1, p_2) \land f(s_2, p_2) > f(s_2, p_1) \tag{3}$$

Experimental Setup. We calculate the pairwise accuracy for all four datasets, for the two similarity metrics cosine and euclidean and for the three match definitions sum, single, and strict. The leave-two-out evaluation only works well for isolated stimuli as in WORDS and STORIES. For the continuous stimuli, we perform standard cross-validation. The HARRY data can be split into four folds according to the experimental blocks. For the ALICE data, we determined six folds. The predictions for each fold are then paired with a randomly selected sample. We set a distance constraint between the two samples of at least 20 time steps to avoid overlapping response patterns. For each sample, we average the result over 1,000 random pairs as in Wehbe et al. [32].

Table 3. Pairwise accuracy results measured with cosine similarity, Euclidean similarity, and Pearson correlation and different match definitions averaged over all subjects. The results for the random language model are indicated in parentheses.

	Match	\small Encoding Model (Random LM) WORDS	STORIES	ALICE	HARRY
	Sum	.67 (.54)	.57 (.53)	.54 (.53)	.50 (.49)
Cosine	Single	.60 (.53)	.53 (.53)	.53 (.51)	.49 (.49)
	Strict	.26 (.13)	.14 (.02)	.28 (.27)	.25 (.24)
	Sum	.67 (.53)	.56 (.53)	.53 (.53)	.50 (.49)
Euclidean	Single	.59 (.50)	.51 (.50)	.52 (.51)	.50 (.49)
	Strict	.24 (.08)	.11 (.02)	.17 (.11)	.12 (.07)
	Sum	.68 (.53)	.56 (.54)	.53 (.53)	.50 (.50)
Pearson's R	Single	.61 (.53)	.52 (.52)	.52 (.52)	.50 (.49)
	Strict	.26 (.10)	.11 (.02)	.27 (.27)	.25 (.24)

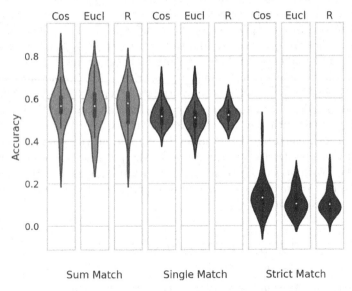

Fig. 1. Violin plot for the pairwise accuracy results for all subjects in STORIES for each evaluation metric.

Results. The results in Table 3 are averaged over all subjects. It can be seen that the differences between the three similarity metrics and the sum and the single match are very small. The strict match is consistently more rigorous than the other match types. This indicates that both predictions would often be matched to the same stimulus when ignoring the pairwise exclusivity constraint. We conclude that the other two match types tend to slightly overestimate the discriminability of the stimulus. We also note that the difference to the random language model is more pronounced for the strict match for WORDS and STORIES. For these two datasets, the results vary strongly across subjects. Subjects 1,3 and 4 in WORDS yield high accuracy results (0.87, 0.87, 0.76 for the cosine sum match) whereas the prediction for subject 6 is below chance level. We provide violin plots in Fig. 1 for a better impression of the variance across subjects in STORIES. Although the results are worse than for WORDS, the accuracy is quite high for some subjects (0.80, 0.78, 0.7). The results obtained for the isolated stimuli are comparable to those reported previously by Mitchell et al. [24] and Dehghani et al. [13]. For the continuous stimuli, the encoding model is not able to learn a robust signal. Wehbe et al. [32] reported better results for the HARRY data but they performed the decoding task. Brennan et al. [10] did not report encoding or decoding results but focused on correlating the fMRI signal with computational models for surprisal.

5.2 Voxel-Wise Evaluation

The pair-wise distance measures are an abstraction over all voxels. A model that mostly predicts constant values and only varies a few indicative voxels could perform well. As the mapping model independently predicts each voxel, we can take a closer look at the predictability of each voxel. This procedure accounts for the assumption that not every voxel in our brain will be influenced by the stimulus. In previous work, prediction results have often been reported only over significant voxels.

Table 4. Voxel-wise results for cross-validation when taking the **average** over voxels. The results are averaged over all folds and all subjects. The results for the random language model are given in parentheses.

Voxels	Dataset	EV	Average R^2	$r^2 simple$
Whole brain	Words	-.21 (-.09)	-.41 (-.35)	.01 (.01)
	Stories	-.05 (.00)	-.26 (-.20)	.02 (.01)
	Harry	-.34 (-.05)	-.27 (-.05)	.00 (.00)
Top 500 on train	Words	-.14 (-.08)	-.33 (-.26)	.07 (.11)
	Stories	-.07 (.00)	-.27 (-.19)	.04 (.02)
	Harry	-.43 (.01)	-.44 (-.07)	.00 (.00)
Top 500 on test	Words	.42 (.21)	.34 (.05)	.51 (.37)
	Stories	.41 (.11)	.34 (.08)	.68 (.67)
	Harry	-.12 (.01)	-.12 (.01)	.02 (.02)

Experimental Setup. The explained variance (EV) and the coefficient of determination (R^2) are the most common metrics for evaluating linear regression. They measure the proportion of the variance in the dependent variable that is predictable by the model. The two metrics are closely related but explained variance also accounts for the mean error. We use the implementation of these scores in the python library *scikit-learn* [26]. Jain and Huth [19] calculate a different r^2 value: they multiply the Pearson correlation between the predictions and the observed activations for voxel v_i with the absolute correlation $(r^2(v_i) = r_{v_i} \times |r_{v_i}|)$. We refer to this measure as $r^2 simple$. We calculate all three metrics and compare the results for the whole brain with a selection of the 500 best-performing voxels on the training and on the testing set respectively. Selection on the test set is not recommended but added to compare previous work.

Results. Tables 4 shows the results for the voxel-wise evaluation averaged over all subjects and over all voxels. It can be seen that the models are highly overfitted as we get much better results when voxels are directly selected on the test results than when they are pre-selected on the training data. In the conditions which control for overfitting, the explained variance and the R^2 are always negative. A value of zero for explained variance is obtained for a model that constantly predicts the mean. It is almost impossible to identify which one of two very negative models performs less bad based on this value alone. The prediction quality should generally be interpreted with caution as the number is averaged over all voxels, all folds and all subjects. Both, the inter-subject variance and the variance in voxel predictability are very high, so that positive and negative results cancel each other out. The $r^2 simple$ metric almost always returns a positive score. This might be a more satisfying result when evaluating the encoding quality; however, the metric also returns high positive scores for the random language model in some cases.

Accumulation Method. Instead of averaging the encoding quality over voxels, Jain and Huth [19] report the sum. For comparison, the summed results are provided in the appendix in Table 6. Sum metrics depend on the number of voxels over which they are calculated. For the whole brain analysis, averaged sum metrics are thus not interpretable in absolute terms because the number of voxels in the brain varies between subjects (see Fig. 2 for an illustration). When accumulating the sum over a fixed set of selected voxel, we see that the results for the $r^2 simple$ metric are consistently better (3) but the extreme change on the x-axis in the two figures indicates that sum scores should be interpreted with caution.

Model-Driven Voxel Selection. We additionally determine the voxels with the highest explained variance on the test set when training on 80% of the data. We set a threshold (0.3 for STORIES and WORDS, 0 for ALICE) and plot predictive voxels for the subjects for which we obtained highest accuracy in the pairwise comparison in Fig. 4. The results are rather inconclusive. There is almost no

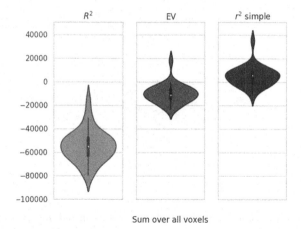

Fig. 2. Violin plots of the voxel-wise results (summed over all voxels) for all subjects in STORIES. It can be seen that the sum score conceals very high inter-subject variance.

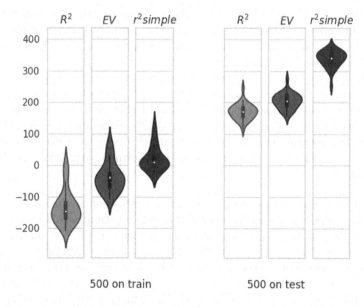

Fig. 3. Violin plots of the voxel-wise results (summed over all voxels) for all subjects in STORIES for all voxels. Note the extreme change in the scale of the y-axis compared to Fig. 2 due to the number of voxels. If the number of voxels over which the sum is calculated is unknown, the result cannot be interpreted.

overlap in the voxels and they are spread over several brain regions. This indicates that model-driven voxel information should only be interpreted on larger datasets.

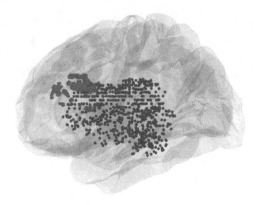

Fig. 4. Predictive voxels for WORDS in blue, STORIES in red and HARRY in yellow.

5.3   Representational Similarity Analysis

The previous methods indicate that the continuous stimuli cannot be well encoded. In order to be able to attribute this flaw more directly to the language model, we perform representational similarity analysis [21] to compare the relations between brain activation vectors to the relations between stimulus representations without the intermediate mapping model. The approach assumes that similar brain activation patterns are caused by strongly related stimuli. The quality of the computational representation of the stimuli can then be assessed by its ability to model these relations [3,11,34]. As commonly performed in previous work, we measure the relations between vectors by the cosine distance and compare brain scans and representations by Spearman correlation and Pearson correlation.

Results. At first glance, the results in Table 5 seem to confirm the impression that the encoding model performs better for the isolated stimuli. However, the same results can be obtained with the random language model. The random model could capture word identity (recall that the same random vector is assigned to different occurrences of the same word) which might serve as a relevant signal for the story stimuli but this would not explain the results for the WORDS dataset with 60 different words. It can be seen that generally the more conservative rank-based Spearman correlation is much lower than the Pearson correlation. For the current setup, the representational similarity analysis results are unsatisfactory. However, the methodology largely reduces the number of parameters and facilitates the comparison of different computational models. We thus think that it could be a promising analysis method for future experiments.

6   Discussion

The setup of encoding experiments requires many modelling decisions for the stimulus representation, the stimulus–response alignment, the mapping model

Table 5. Results for representational similarity analysis calculated for the whole brain using pearson correlation and spearman correlation. The results for the random language model are indicated in parentheses.

Metric	WORDS	STORIES	ALICE	HARRY
Spearman	0.09 (0.05)	0.08 (0.09)	0.03 (0.01)	0.00 (0.01)
Pearson	0.41 (0.44)	0.19 (0.22)	0.06 (0.02)	0.06 (0.03)

and its learning parameters, the noise reduction techniques for the brain responses, the voxel selection, and the evaluation metric. Experimenting with a single dataset bears the danger of overfitting the experimental setup. We have seen that different evaluation metrics can interpret the predictive power of an encoding model very differently. Encoding results should thus always be compared to a reasonable baseline and hypotheses should be tested over several datasets. In this comparison, we intentionally restricted the experimental setup by choosing the same language model for all datasets. At this point, it remains unclear, whether the close to random results in many settings result from an unfortunate choice of the language model parameters or from a noisy signal. Our experimental pipeline is modular and provides a useful testbed for future experiments with alternative stimuli representations.

More sophisticated context models might increase the number of dimensions. From a machine learning perspective, most encoding experiments are problematic because the number of features is often higher than the number of samples. In addition, similarity metrics are known to sometimes behave unexpectedly when applied on high-dimensional data [2]. One could apply dimensionality reduction on the language representations but these methods change the structure of the representation and make it difficult to derive cognitive insights for the original model. For future data collections, it would be important to obtain more data points from fewer subjects to facilitate more powerful pattern analyses.

FMRI encoding is an intriguing but also very challenging task because of the noisy signal. Within the current state of the art, even a tiny signal that is significantly different from chance, can be seen as a success. The pairwise estimation measures can present the results in a more pronounced way. However, as our analysis with the strict match have shown, the other match definitions tend to give an overly optimistic impression of the discriminability of the stimuli. A similar problem occurs, when summing the $r^2 simple$ value only over predictive voxels. We are convinced that in the long run, the field benefits from a more conservative estimate of the predictive power of the developed models.

7   Conclusions

We have performed a robust comparison for language–brain encoding experiments and receive very diverse results for different evaluation metrics. It is our hope that our experimental framework can pave the way for future experiments to gradually determine the optimal encoding parameters. We plan to extend our experiments to the datasets by Pereira et al. [27] and to other languages. We can already provide a set of practical recommendations for evaluation: 1. For the pairwise evaluation, it is helpful to additionally report the strict match to put the results in perspective. 2. Averaging over subjects is not very informative, violin plots can give a better impression of the variance. 3. For sum metrics, it is important to clearly specify the number of voxels that are taken into consideration. 4. Voxel selection methods should only be performed on the training set and should be transparently documented because they have a strong effect on the results.

Acknowledgements. The work presented here was funded by the Netherlands Organisation for Scientific Research (NWO), through a Gravitation Grant 024.001.006 to the Language in Interaction Consortium.

Appendix

Table 6. Voxel-wise results for cross-validation when taking the **sum** over voxels. The results are averaged over all folds and all subjects. The results for the random language model are given in parentheses. The results in this table are hard to interpret. We discourage the use of the sum method as accumulation method.

Voxels	Data	EV	Sum R^2	$r^2 simple$
Whole	Words	-4,3k (-1,9k)	-8,4k (-5,6k)	250.37 (184.33)
	Stories	-10,2k (-47.56)	-54,6k (-42,2k)	4,9k (2,8k)
	Harry	-10,7k (-1,5k)	-10,8k (-1,4k)	-6.29 (-3.10)
500 train	Words	-68.82 (-39.84)	-164.67 (-129.35)	33.34 (-0.38)
	Stories	0.00 (-0.55)	-134.31 (-96.88)	21.36 (9.43)
	Harry	-215.45 (-37.28)	-218.14 (-37.63)	0.11 (-0.09)
500 test	Words	209.98 (104.76)	253.98 (25.66)	171.56 (187.20)
	Stories	204.90 (56.30)	339.33 (39.52)	171.08 (334.99)
	Harry	-58.66 (7.40)	-59.70 (7.23)	10.41 (9.86)

References

1. Abnar, S., Ahmed, R., Mijnheer, M., Zuidema, W.: Experiential, distributional and dependency-based word embeddings have complementary roles in decoding brain activity. In: Proceedings of the 8th Workshop on Cognitive Modeling and Computational Linguistics (CMCL'18), pp. 57–66. Association for Computational Linguistics (2018). http://aclweb.org/anthology/W18-0107

2. Aggarwal, C.C., Hinneburg, A., Keim, D.A.: On the surprising behavior of distance metrics in high dimensional space. In: Van den Bussche, J., Vianu, V. (eds.) Database Theory – ICDT 2001, pp. 420–434. Springer, Berlin Heidelberg, Berlin, Heidelberg (2001). http://kops.uni-konstanz.de/bitstream/handle/123456789/5715/On_the_Surprising_Behavior_of_Distance_Metric_in_High_Dimensional_Space.pdf?sequence=1

3. Anderson, A.J., Bruni, E., Bordignon, U., Poesio, M., Baroni, M.: Of words, eyes and brains: Correlating image-based distributional semantic models with neural representations of concepts. In: Proceedings of the 2013 Conference on Empirical Methods in Natural Language Processing, pp. 1960–1970. Association for Computational Linguistics (2013). http://aclweb.org/anthology/D13-1202

4. Anderson, A.J., Kiela, D., Clark, S., Poesio, M.: Visually grounded and textual semantic models differentially decode brain activity associated with concrete and abstract nouns. Trans. Assoc. Comput. Linguist. **5**, 17–30 (2017). http://aclweb.org/anthology/Q17-1002

5. Artstein, R., Poesio, M.: Inter-coder agreement for computational linguistics. Comput. Linguist. **34**(4), 555–596 (2008). https://www.mitpressjournals.org/doi/pdfplus/10.1162/coli.07-034-R2

6. Athanasiou, N., Iosif, E., Potamianos, A.: Neural activation semantic models: computational lexical semantic models of localized neural activations. In: Proceedings of the 27th International Conference on Computational Linguistics, pp. 2867–2878 (2018). http://www.aclweb.org/anthology/C18-1243

7. Barrett, M., Bingel, J., Hollenstein, N., Rei, M., Søgaard, A.: Sequence classification with human attention. In: Proceedings of the 22nd Conference on Computational Natural Language Learning, pp. 302–312 (2018). http://www.aclweb.org/anthology/K18-1030

8. Beinborn, L., Zesch, T., Gurevych, I.: Predicting the difficulty of language proficiency tests. Trans. Assoc. Comput. Linguist. **2**(1), 517–529 (2014). http://www.aclweb.org/anthology/Q14-1040

9. Bingel, J., Barrett, M., Søgaard, A.: Extracting token-level signals of syntactic processing from fMRI - with an application to pos induction. In: Proceedings of the 54th Annual Meeting of the Association for Computational Linguistics (Volume 1: Long Papers), 1, pp. 747–755 (2016). http://www.aclweb.org/anthology/P16-1071

10. Brennan, J.R., Stabler, E.P., Van Wagenen, S.E., Luh, W.M., Hale, J.T.: Abstract linguistic structure correlates with temporal activity during naturalistic comprehension. Brain Lang. **157**, 81–94 (2016). https://www.sciencedirect.com/science/article/pii/S0093934X1530068

11. Bulat, L., Clark, S., Shutova, E.: Speaking, seeing, understanding: correlating semantic models with conceptual representation in the brain. In: Proceedings of the 2017 Conference on Empirical Methods in Natural Language Processing, pp. 1081–1091. Association for Computational Linguistics (2017). http://aclweb.org/anthology/D17-1113

12. Carroll, L.: Alice's Adventures in Wonderland. Macmillan, London (1865)

13. Dehghani, M., et al.: Decoding the neural representation of story meanings across languages. Human Brain Mapp. **38**(12), 6096–6106 (2017). https://www.ncbi.nlm.nih.gov/pubmed/28940969

14. Frank, S.L., Otten, L.J., Galli, G., Vigliocco, G.: Word surprisal predicts n400 amplitude during reading. In: Proceedings of the 51st Annual Meeting of the Association for Computational Linguistics (Volume 2: Short Papers), 2, pp. 878–883

(2013). https://www.semanticscholar.org/paper/Word-surprisal-predicts-N400-amplitude-during-Frank-Otten/0998e0763328764935e74db7c124ee4ee277c360

15. Fyshe, A., Sudre, G., Wehbe, L., Rafidi, N., Mitchell, T.M.: The semantics of adjective noun phrases in the human brain. bioRxiv (2016). https://www.biorxiv.org/content/biorxiv/early/2016/11/25/089615.full.pdf

16. Fyshe, A., Talukdar, P.P., Murphy, B., Mitchell, T.M.: Interpretable semantic vectors from a joint model of brain-and text-based meaning. In: Proceedings of the Conference. Association for Computational Linguistics. Meeting, vol. 2014, p. 489. NIH Public Access (2014). http://aclweb.org/anthology/P14-1046

17. Gauthier, J., Ivanova, A.: Does the brain represent words? An evaluation of brain decoding studies of language understanding. arXiv:1806.00591 (2018). https://arxiv.org/pdf/1806.00591.pdf

18. Hale, J., Dyer, C., Kuncoro, A., Brennan, J.R.: Finding syntax in human encephalography with beam search. In: Proceedings of the 56th Annual Meeting of the Association for Computational Linguistics, Volume 1 (Long Papers), pp. 2727–2736. Association for Computational Linguistics (2018). http://aclweb.org/anthology/P18-1254

19. Jain, S., Huth, A.: Incorporating context into language encoding models for fMRI. bioRxiv (2018). https://www.biorxiv.org/content/early/2018/05/21/327601

20. Kriegeskorte, N., Goebel, R., Bandettini, P.: Information-based functional brain mapping. Proc. National Acad. Sci. **103**(10), 3863–3868 (2006). http://www.pnas.org/content/103/10/3863.full

21. Kriegeskorte, N., Mur, M., Bandettini, P.A.: Representational similarity analysis-connecting the branches of systems neuroscience. Front. Syst. Neurosci. vol. 2, p. 4 (2008). https://www.ncbi.nlm.nih.gov/pmc/articles/PMC2605405/

22. Li, J., Fabre, M., Luh, W.M., Hale, J.: The role of syntax during pronoun resolution: evidence from fMRI. In: Proceedings of the 8th Workshop on Cognitive Aspects of Computational Language Learning and Processing, pp. 56–64. Association for Computational Linguistics (2018). http://aclweb.org/anthology/W18-2808

23. Miezin, F.M., Maccotta, L., Ollinger, J., Petersen, S., Buckner, R.: Characterizing the hemodynamic response: effects of presentation rate, sampling procedure, and the possibility of ordering brain activity based on relative timing. Neuroimage **11**(6), 735–759 (2000). https://doi.org/10.1006/nimg.2000.0568

24. Mitchell, T.M., et al.: Predicting human brain activity associated with the meanings of nouns. science **320**(5880), 1191–1195 (2008). https://www.cs.cmu.edu/tom/pubs/science2008.pdf

25. Monsalve, I.F., Frank, S.L., Vigliocco, G.: Lexical surprisal as a general predictor of reading time. In: Proceedings of the 13th Conference of the European Chapter of the Association for Computational Linguistics, pp. 398–408. Association for Computational Linguistics (2012). https://aclanthology.info/pdf/E/E12/E12-1041.pdf

26. Pedregosa, F., et al.: Scikit-learn: machine learning in Python. J. Mach. Learn. Res. **12**, 2825–2830 (2011). http://www.jmlr.org/papers/volume12/pedregosa11a/pedregosa11a.pdf

27. Pereira, F., et al.: Toward a universal decoder of linguistic meaning from brain activation. Nat. Commun. **9**(1), 1–13 (2018). https://doi.org/10.1038/s41467-018-03068-4

28. Peters, M., et al.: Deep contextualized word representations. In: Proceedings of the 2018 Conference of the North American Chapter of the Association for Computational Linguistics: Human Language Technologies, Volume 1 (Long Papers), 1, pp. 2227–2237 (2018). http://www.aclweb.org/anthology/N18-1202

29. Resnik, P., Lin, J.: Evaluation of NLP systems. The Handbook of Computational Linguistics and Natural Language Processing, vol. 57, pp. 271–295 (2010). https://pdfs.semanticscholar.org/41ef/e3fb47032d609bbb13b7c850bb8b1dbd544d.pdf

30. Rowling, J.K.: Harry Potter and the Sorcerer's Stone. Levine Books, Arthur A (1998)

31. Sudre, G., et al.: Tracking neural coding of perceptual and semantic features of concrete nouns. NeuroImage **62**(1), 451–463 (2012). https://www.ncbi.nlm.nih.gov/pmc/articles/PMC4465409/

32. Wehbe, L., Murphy, B., Talukdar, P., Fyshe, A., Ramdas, A., Mitchell, T.: Simultaneously uncovering the patterns of brain regions involved in different story reading subprocesses. PLoS One **9**(11), e112575 (2014). https://doi.org/10.1371/journal.pone.0112575

33. Wehbe, L., Vaswani, A., Knight, K., Mitchell, T.: Aligning context-based statistical models of language with brain activity during reading. In: Proceedings of the 2014 Conference on Empirical Methods in Natural Language Processing (EMNLP). Association for Computational Linguistics (2014). https://doi.org/10.3115/v1/d14-1030

34. Xu, H., Murphy, B., Fyshe, A.: Brainbench: a brain-image test suite for distributional semantic models. In: Proceedings of the 2016 Conference on Empirical Methods in Natural Language Processing, pp. 2017–2021 (2016). http://www.aclweb.org/anthology/D16-1213

Connectives with Both Arguments External: A Survey on Czech

Lucie Poláková and Jiří Mírovský[(⊠)]

Faculty of Mathematics and Physics, Institute of Formal and Applied Linguistics,
Charles University, Prague, Czech Republic
{polakova,mirovsky}@ufal.mff.cuni.cz

Abstract. Determining a relative position of a discourse connective and the two arguments (text segments) it connects is an important part of a full discourse parsing task. This paper investigates discourse connectives whose position in a text deviates from the usual setting – namely connectives that occur in neither of the two arguments – and as such present a challenge for discourse parsers. We find syntactic patterns for this phenomenon and describe it linguistically on the basis of Czech discourse-annotated corpus material, with the aim to facilitate an automatic detection of such connectives and a correct localization of their arguments.

Keywords: Discourse connectives · Text coherence · Attribution · Syntactic structure · Text mining

1 Introduction

Any task concerned with identifying discourse relations presupposes a solid segmentation of a text into discourse units. Syntax plays a crucial helpful role in this, but it is also specific syntactic constructions that cause trouble in the segmentation. It has been argued previously that the mismatch in alignment of syntactic and discourse units is caused largely by attribution, i.e. ascription of contents to the agents who uttered them [2]. In this paper, we arrive at a similar observation from a very different perspective: we investigate the position of discourse connectives, the most apparent markers of discourse relations, with regard to the two discourse segments they connect (discourse arguments). In particular, we study connective tokens that have both arguments external, that means, the connective is not included in either of the arguments, compare the basic graphical scheme in Fig. 2 and Examples 1 and 2 below.[1] The existence of such cases is documented at least in two manually annotated discourse corpora for two languages, but since searching for this type of connective-argument configuration is quite complex, there may be more in other discourse-annotated corpora.

---

[1] In all examples in the paper, the left-sided argument of a discourse relation is highlighted in italics, the other one in bold, the connective is underlined.

© Springer Nature Switzerland AG 2023
A. Gelbukh (Ed.): CICLing 2019, LNCS 13451, pp. 62–72, 2023.
https://doi.org/10.1007/978-3-031-24337-0_5

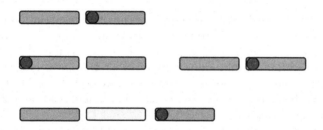

Fig. 1. Typical syntactic positions of connectives with regard to their two arguments: Line 1: coordinating conjunctions, line 2: subordinating conjunctions, line 3: adverbs

The study is based on one of the most prominent approaches to discourse coherence in the recent decade, the analysis of local coherence in the Penn Discourse Treebank (henceforth PDTB, version 2.0 [9] and lately version 3.0 [11] and [13]). This approach was also adopted for annotation of discourse relations in Czech in the Prague Dependency Treebank (henceforth PDT, [1]), the primary data source for our survey.

(1)   *Compaq*, which said it discovered the bugs, *still plans to announce new 486 products on Nov. 6.* Because of the glitch, <u>however</u>, the company said **it doesn't know when its machine will be commercially available.** (PDTB)

Our study first discusses basic syntactic configurations of discourse connectives with regard to their PoS characteristics (Sect. 1.1) and then focuses on the non-typical positions of connectives (Sect. 1.2). Section 2 describes the frameworks, data a tools used and the way of detection of the targeted connectives and relations. The linguistic analysis itself follows in Sect. 3, with introduction of testing criteria for connective "scopes" (Sect. 3.2) and corpus evidence (3.3). The implications of our findings are summed up in the concluding Sect. 4.

1.1   Part of Speech and The Position of Connectives

Discourse connectives, the core of which includes coordinating and subordinating conjunctions (e.g. *and, but, because*) and adverbs (e.g. *moreover, then, otherwise*), are typically located within one of the two discourse segments (arguments) they connect.[2] Coordinating conjunctions are typically placed at the beginning of the second argument in the linear order, compare the first line of Fig. 1, the connective is represented by a dark dot. Subordinating conjunctions are to be found within the argument represented by a dependent clause, either preceding the main clause (the other argument) or following it, compare line two of Fig. 1. Sentence adverbs occur within the second argument in the linear order; the first argument can be non-adjacent, they are also referred to as anaphoric connectives [14], see line three in Fig. 1 – the colorless box represents a (faculta-

---

[2] In this study, we do not address secondary discourse connectives like *that is why* or prepositional connectives with nominalized arguments like *after his arrival*.

tive) text segment that does not take part in the discourse relation in question. Claiming this, we refer to the properties of discourse connectives in Czech and also draw on our experience with English and German connectives. Intuitively, these syntactic settings would apply also for other European languages, with a possible minor variation. However, in this paper, we do not want to make claims about syntactic behaviour of connectives in other languages than Czech, this is yet a topic to be researched.

Except for intra-sentential connectives in dependent clauses, these regularities make the task of determining the arguments of a connective a search for the external (left-sided) argument [8], or Arg 1 in the terminology of the authors. The non-adjacency of the left-sided argument of anaphoric connectives has been a known issue in discourse analysis and parsing and it is being dealt with [4,14][3].

1.2   Connectives with Both Arguments External

The aim of our study, though, are connectives with **both arguments external**, that means, connectives that are located outside both the arguments they connect, in a third clause, compare the scheme in Fig. 2. Such cases are not very frequent, but they still could be documented in Czech in the annotated data of the Prague Dependency Treebank 3.0 (Example 2) and also in the Penn Discourse Treebank 2.0 for English, as Example 1 above shows.

(2)   *Ministerstvo vnitra považuje před dnešním jednáním o charakteristice rozpočtu na příští rok za předčasné poskytovat detailní informace*, sdělila tisková mluvčí ministerstva. Řekla <u>nicméně</u> také, **že i ministerstvo vnitra počítá v porovnání s letošním rokem s posílením investic.**

Before the negotiations on the budget profile, the ministry of the interior considers it premature to provide detailed information, the ministry spokeswoman announced. She also said, <u>however</u>, **that even the ministry of the interior expects strengthening of investments.**

[Lit: She said however also that...][4]

---

[3] Some frameworks for discourse analysis, e.g. [6], though, do not allow for a non-adjacent interpretation, such a long-distance relation is non-existent.

[4] Note that in Czech, the *nicméně*–connective (*however*) is undoubtedly located within the main clause. The English translations of Czech examples are the best possible approximations to the original sentences given the more relaxed word order in Czech, even for the *ale*–connective (*but*). Where needed, we use literal translations.

Fig. 2. A connective with both arguments external – general scheme

2   Method, Data and Tools

Our method for analysis of discourse connectives follows (i) the Penn Discourse Treebank 2.0 [9] style of annotation – a discourse connective is defined as a predicate of a binary relation between abstract objects (events, states, actions...) called discourse arguments [10], and (ii) dependency syntax of the Prague Functional Generative Description [12]. For this study, we especially use the framework for looking at inclusion/exclusion of a connective in syntactically different types of clauses. The primary source for our survey was the data of the Prague Dependency Treebank 3.0 [1] which comprises approx. 50 thousand sentences of newspaper text with manual syntactic annotation, manual annotation of discourse relations and their semantic types, connectives and arguments. The treebank provided interesting evidence but it proved too small for our study. That is why we subsequently also used the SYN V3 collection of the Czech National Corpus (CNC [3]), a reference corpus of approx. 178.5 million sentences of written contemporary Czech, automatically lemmatized and tagged, and queried it as described in Sect. 3.3 using the KonText query engine [5]. Evidence for the connectives with external arguments in the PDT was collected via search engine PML-Tree Query [7].[5]

3   Analysis

The distant connective position is a setting that clearly deviates from the default connective positions, compare the general scheme in Fig. 2 to the previous typical settings in Fig. 1. According to the corpus figures, it is a rare setting, yet the evidence texts make full sense and are not in any respect ungrammatical or stylistically incorrect. We were able to document 72 such connectives in the data of PDT which had to be manually sorted out for irrelevant material, double hits etc. with the final 48 relevant connective tokens. In majority of the detected

---

[5] The best approximation to finding the relevant connectives was achieved with the following PML-TQ query:

```
t-node $t :=
  [ !descendant $n4, !sibling $n3, !parent $n4,
    member discourse
    [ t-connectors.rf t-node $n4 :=
      [ functor != "RHEM",
        0x parent t-node
        [ nodetype = "coap" ] ],
      target_node.rf t-node $n3 :=
      [ !descendant $n4, !parent $n4 ] ] ];
```

Fig. 3. Argument scheme for sentences in Example 2

cases, a connective placed outside both its arguments syntactically belongs to the governing clause but, from the semantic viewpoint, it is interpreted in the dependent clause. We illustrate the situation on Example 2 from above: in the main clauses plan, there is the *také*–connective (*also*) anchoring a relation of *conjunction* between the two main clauses (here: attribution clauses with verbs of saying). The contrastive meaning expressed by the *nicméně*–connective (*however/but*) can only be inferred from the meanings of reported contents (lower syntactic level), not from the attribution spans. The argument configuration is schematized in Fig. 3 and the sentences can be paraphrased as follows:

- Main clauses plan:

   The spokeswoman announced A. **She <u>also</u> said B.**

   *She announced A. <u>However/but</u> she said B.

- Subordinate clauses plan (reported contents):

   The ministry does not want to reveal any information too early. **<u>However</u>, it reveals expectations on strengthening of investments.**

Over 70% of the relevant connective tokens appear in structures with verbs of saying and thinking (i.e. with governing attribution clauses) or with verbs of existence and general meanings. A closer exploration revealed that structures with attribution clauses are more likely to involve connectives interpreted lower at the level of content clauses, as just demonstrated, but also that some of the cases are difficult to judge as the phenomenon is quite complex, see Table 1. That is why we introduced a set of transformation tests that could partially sort out the different interpretations. In the rest of the paper, we focus solely on structures with verbs of saying and connectives with meanings of comparison/contrast, as they represent the most numerous and distinctive patterns of the studied phenomenon.

Table 1. Corpus evidence for connectives with both arguments external (PDT)

	Structures with verbs of saying and thinking	Structures with verbs of general meaning, verbs of existence etc.
Connective interpreted lower	18	11
Possibly both interpretations	8	11

3.1 Wide and Narrow Connective "scopes"

If a connective can be interpreted in a distant clause, lower in the syntactic structure, it is similar to the transfer of negation known from sentences like 3, which is a typical form of hedging:

(3) I don't think Mary will come. = I think Mary will not come.

In the first sentence here, the negation is located in the main clause but it scopes over the dependent clause only, as the paraphrased second sentence shows. This is what we call the narrow scope. Analogous structures with narrow connective "scope" were just described on Example 2 above. These cases are not to be confused with other cases where the connective in fact syntactically and semantically relates two clauses of attribution (wide "scope", including their respective content clauses). This was demonstrated by [2] with the English example from PDTB (4) and is also visible on the Czech Example 5 from the PDT.

(4) *Advocates said the 90-cent-an-hour rise, to $4.25 an hour by April 1991, is too small for the working poor,* <u>while</u> **opponents argued that the increase will still hurt small business and cost many thousands of jobs.** (PDTB)

(5) *Prezident Havel při odchodu z jednání vlády uvedl, že v diskusi zvítězil názor, aby na oslavy byli přizváni i zástupci Německa.* **Premiér Václav Klaus** <u>však</u> **po skončení jednání LN řekl, že vláda o této věci včera ještě nerozhodla.**

When leaving the government meeting President Havel stated that the view prevailed in the debate that the representatives from Germany should also be invited to the celebrations. **Prime Minister Václav Klaus,** <u>however</u>, **said to the LN after the meeting that yesterday the government had not yet decided on the matter.**

3.2 Testing Criteria

Determining whether the connective actually semantically relates the attribution clauses or whether it operates at a lower level between the attributed contents can be a problematic issue. We introduce two transformation tests (that can be used independently and should provide the same results) based on substitution and eliding for determining the connective scope in attribution clauses:

It is (i) **moving the connective lower**, to the content clause and (ii) **omitting the attribution clause** in the second sentence in the surface order while preserving the connective in question. After the application of any of the two tests, if the connection of content clauses related with an (originally distant) connective preserves the original meaning, we can speak about the narrow connective scope. On the other hand, if such a connection gets semantically weird or incomplete, the connective will relate the governing, attribution clauses and take the wide scope. Compare the situations under (i1-ii1) and (i2-ii2), where the tests are applied for the original corpus Examples 2 and 5 from above:

(i1) Lowering of the connective: the connective operates at the level of attributed contents

(6) *Ministerstvo vnitra považuje před dnešním jednáním o charakteristice rozpočtu na příští rok za předčasné poskytovat detailní informace*, sdělila tisková mluvčí ministerstva. Řekla také, **že** <u>nicméně</u> **i ministerstvo vnitra počítá v porovnání s letošním rokem s posílením investic.**

Before the negotiations on the budget profile, the ministry considers it premature to provide detailed information, the ministry spokeswoman announced. She also said that, <u>however</u>, **even the ministry expects strengthening of investments.**

[Lit: She said also, that however even the ministry...]

(ii1) Omitting the attribution clause: the connective operates at the level of attributed contents

(7) *Ministerstvo vnitra považuje před dnešním jednáním o charakteristice rozpočtu na příští rok za předčasné poskytovat detailní informace*, sdělila tisková mluvčí ministerstva. **Nicméně i ministerstvo vnitra počítá v porovnání s letošním rokem s posílením investic.**

Before the negotiations on the budget profile, the ministry considers it premature to provide detailed information, the ministry spokeswoman announced. **However, even the ministry expects strengthening of investments.**

[Lit: However even the ministry...]

(i2) Lowering of the connective: awkward. The connective operates most likely at the level of main clauses.

(8) *?Prezident Havel při odchodu z jednání vlády uvedl, že v diskusi zvítězil názor, aby na oslavy byli přizváni i zástupci Německa.* **Premiér Václav Klaus po skončení jednání LN řekl, že vláda** <u>však</u> **o této věci včera ještě nerozhodla.**

?When leaving the government meeting President Havel stated that the view prevailed in the debate that the representatives from Germany should also be invited to the celebrations. **Prime Minister Václav Klaus said to the LN after the meeting that yesterday the government,** <u>however</u>, **had not yet decided on the matter.**

The placement of the connective *však* (*however*) in Example 8 appears disruptive in the dependent clause. This is most likely because we expected it to come earlier in the sentence. The contrast here apparently relates to the person of Prime

Minister claiming something else than the President, the contrastive connective tends to stand in close proximity of both these agents.

(ii2) Omitting the attribution clause: the connective operates at the level of attributed contents, but with a loss of important information. This transformation is thus not possible.

(9)   *Prezident Havel při odchodu z jednání vlády uvedl, že *v diskusi zvítězil názor, aby na oslavy byli přizváni i zástupci Německa.* **Vláda však o této věci včera ještě nerozhodla.**

*When leaving the government meeting President Havel stated that *the view prevailed in the debate that the representatives from Germany should also be invited to the celebrations.* **However,** **the government had not yet decided on the matter yesterday.**

Example 9 makes perfect sense even after the transformation but the meaning has shifted: the omitted information that the second argument is a (contradictory) statement of the Prime Minister, not the President, is an important one. In this case, the connective *však* (*however*) operates between the attribution clauses, which means, the syntactic and the discourse interpretations go hand in hand.

As far as we could observe, a contrastive connective syntactically connecting two main attribution clauses takes the wide scope, that means the syntactic and the discourse interpretations match, in structures with non-identical sources of attribution. In other words, the connective must take a wide scope, if the two contents in dependent clauses are expressed by two different agents (as in 4 and 5 above).

If we accept the fact that a (contrastive) connective scopes over a clause dependent to the one in which it appears, it can be treated as "raised". But the connective does not always "climb" in the syntactic structure like the negation in Example 3 above. If we switched the order of the clauses, so that the attribution clause is sentence-last, the connective would be most likely located in the content clause (in this case the first clause in linear order). Disregarding the ordering of the attribution clause and the content clause, a contrastive connective **tends to stand as far to the left as possible,** i.e. as close as possible to its first, left-sided argument. If we now consider the previous setting – the attribution clause containing the connective at the sentence-initial position – we can say that the connective has moved left.[6]

---

[6] In an earlier phase of this research, we called the phenomenon *connective movement.* A colleague later pointed out the possible confusing connection of this term with generative grammar in sense of N. Chomsky, which we do not want to make here, so we abandoned the use of this term.

3.3   Corpus Evidence

To confirm or disprove these assumptions about preferential positions of connectives in various settings with attribution, we used two different corpora of Czech. The results from the PDT are displayed above in Table 1. In the much larger CNC, we searched for the two following patterns and their variation. In the first pattern, the contrastive connective appears in the main clause, whereas in the second one, it appears in the dependent (content) clause introduced in Czech obligatorily by *že (that)*:

1. [The beginning of the sentence - verb[7] - (0-3 positions) - connective *však/ale* - comma - subordinating conjunction *že*]

Example (lit.): *He_said but/however(...), that*

2. [The beginning of the sentence - verb - (0-3 positions) - comma - subordinating conjunction *že* - connective *však/ale*]:

Example (lit.): *He_said (...), that but/however*

The results are summarized in Table 2.

Table 2. Connectives in main clauses and dependent *"that"* clauses (CNC)

Pattern No	Pattern variation	Total occurrences	Instances per million tokens
1	Verb (...) but, that	41,842	15.58
1	Verb (...) however, that	50,162	18.68
2	Verb (...), that but	606	0.23
2	Verb (...), that however	294	0.11
2	Verb (...), that (...) but	3,594	1.34
2	Verb (...), that (...) however	1,570	0.58
1	Verb but, that	24,477	9.12
1	Verb however, that	31,822	11.85
2	Verb, that but	415	0.15
2	Verb, that however	235	0.10

Structures with *ale/však*–connectives in the governing clause are much more frequent in Czech. We have divided the results to structures with one to three arbitrary positions between the verb and the connective/comma, which allows for inclusion of most multi-word verb forms: first 6 rows of Table 2; and to fixed single-verb structures: rows 7–10. Rows 5 and 6 represent structures where both in the main clause and in the dependent clause one to three arbitrary positions

---

[7] As Czech is a pro-drop language, we did not need to search for a pronominal subject in this case.

are allowed – we added these two rows to the table in order to reflect the possible positions of Czech clitics (= in front of the connective). For both patterns, it is clearly visible that structures with connectives in the governing clause are more frequent than structures with lower placement of the connective. This supports the claim that (contrastive) connectives tend to be located close to their first argument in the linear order. A secondary test is represented by the fact that although the semantics of the verb in the query was not restricted in any way, the query returned almost exclusively verbs of saying.

4   Conclusion

In the annotation of discourse relations and connectives in Czech written texts, externally placed connectives were found (connectives located outside both their arguments). These cases occur in vast majority in connection with attribution (reporting). According to our findings, this pattern usually involves a contrastive connective located in an attribution clause, which is at the same time the governing clause. The connective expresses contrast between two reported contents, but only so, if a single speaker utters them both; involvement of more speakers suggests a "wide connective scope", in other words, a regular interpretation (contrast between the main clauses).

In these structures, a connective expressing contrast between two reported contents **moves left** to the sentence-initial attribution clause in order to stand closer to its left (first in the linear order) argument. In opposite cases, i.e. where the attribution clause is sentence-last, the connective is unlikely to appear that far right. Testing criteria for Czech were partly helpful, although not always univocally decisive. With the meaning of addition (conjunction), it is difficult to decide about the scope of the connective, for other meaning classes, there was no sufficient material found. We provided corpus evidence from a small manually annotated treebank with syntactic and discourse information (PDT): the phenomenon is rare but traceable; results from a large reference corpus of Czech (CNC) show that contrastive connectives *ale/však* are much more frequent in attribution (main) clauses than in the dependent content clauses.

Connectives with both arguments external and discourse relations expressed by them present a challenge for discourse parsers. In our study, we attempted to define conditions under which connective movement occurs. Our results and testing criteria may help devise rules for recognition of such cases and may contribute to improving the specification of argument extent and the precision of the automatic discourse parsing.

Acknowledgments. This work has been supported by the Grant Agency of the Czech Republic (projects 20-09853S and 17-03461S). The research reported in the present contribution has been using language resources developed, stored and distributed by the LINDAT/CLARIAH-CZ project of the Ministry of Education, Youth and Sports of the Czech Republic (project no. LM2018101).

References

1. Bejček, E., et al.: Prague Dependency Treebank 3.0. Data/Software. Charles University, Faculty of Mathematics and Physics, Institute of Formal and Applied Linguistics, Prague (2013). https://www.lindat.cz/
2. Dinesh, N., Lee, A., Miltsakaki, E., Prasad, R., Joshi, A., Webber, B.: Attribution and the (non-) alignment of syntactic and discourse arguments of connectives. In: Proceedings of the Workshop on Frontiers in Corpus Annotations II: Pie in the Sky, pp. 29–36. Association for Computational Linguistics (2005)
3. Křen, M., et al.: Czech National Corpus - SYN, version 3. Data/Software. Institute of the Czech National Corpus, Charles University, Faculty of Arts, Prague (2014). https://www.korpus.cz/
4. Lee, A., Prasad, R., Joshi, A., Dinesh, N., Webber, B.: Complexity of dependencies in discourse: are dependencies in discourse more complex than in syntax? In: Proceedings of the 5th International Workshop on Treebanks and Linguistic Theories, Prague, Czech Republic, pp. 79–90 (2006)
5. Machálek, T., Křen, M.: Query interface for diverse corpus types. Natural Language Processing, Corpus Linguistics, E-learning, pp. 166–173 (2013)
6. Mann, W.C., Thompson, S.A.: Rhetorical structure theory: toward a functional theory of text organization. Text-Interdisc. J. Study Discourse **8**, 243–281 (1988)
7. Pajas, P., Štěpánek, J.: System for querying syntactically annotated corpora. In: Lee, G., im Walde, S.S. (eds.) Proceedings of the ACL-IJCNLP 2009 Software Demonstrations, pp. 33–36. Association for Computational Linguistics, Suntec (2009)
8. Prasad, R., Joshi, A., Webber, B.: Exploiting scope for shallow discourse parsing. In: Chair, N.C.C., et al. (eds.) Proceedings of the Seventh International Conference on Language Resources and Evaluation (LREC 2010), Valletta, Malta. European Language Resources Association (ELRA) (2010)
9. Prasad, R., et al.: Penn Discourse Treebank Version 2.0. Data/Software. University of Pennsylvania, Linguistic Data Consortium, Philadelphia. LDC2008T05 (2008)
10. Prasad, R., et al.: The Penn Discourse Treebank 2.0 Annotation Manual. Technical report, University of Pennsylvania, Philadelphia (2007)
11. Prasad, R., Webber, B., Lee, A., Joshi, A.: Penn Discourse Treebank Version 3.0. Data/Software, Linguistic Data Consortium. University of Pennsylvania, Philadelphia. LDC2019T05 (2019)
12. Sgall, P., Nebeský, L., Goralčíková, A., Hajičová, E.: A Functional Approach to Syntax in Generative Description of Language. American Elsevier Pub. Co., New York (1969)
13. Webber, B., Prasad, R., Lee, A., Joshi, A.: The Penn Discourse Treebank 3.0 Annotation Manual. Technical report (2018)
14. Webber, B., Stone, M., Joshi, A., Knott, A.: Anaphora and discourse structure. Comput. Linguist. **29**(4), 545–587 (2003)

Recognizing Weak Signals in News Corpora

Daniela Gifu[1,2](✉) (iD)

[1] Faculty of Computer Science, "Alexandru Ioan Cuza" University of Iasi, 700483 Iasi, Romania
daniela.gifu@info.uaic.ro
[2] Institute of Computer Science, Romanian Academy, 700481 Iasi Branch, Romania

Abstract. This paper presents a suit of experiments carried out with different machine learning systems developed to learn how to find weak signals in news corpora. The task is particularly challenging as the weak signals are not marked at word or sentence level, but rather at document/paragraph level and while there is no explicit definition that strictly applies to them, people are very good at recognizing them. Purposely lacking strict annotation guidelines, a large news corpus was annotated via tacit knowledge and then a supervised learning technique to reproduce the weak signal label was used. A large improvement in weak signals recognition using deep learning over other approaches is reported, useful for investors, entrepreneurs, economists, and normal users, to give them a clue on how to invest.

Keywords: Deep learning · Weak signals · News corpora · Statistics

1 Introduction

How would be it like going back in time and reading scientific predictions from 10 years ago? Some of them would sound very strange, which supports once again the Old Danish proverb that "making predictions is hard, especially about the future". It is a process that can be measured by the degree in which he/she is able to recognize different signals, which heralds a new wave. For instance, the case of elections, when we have to predict which candidate is better [1]. However, sometimes, some predictions may prove quite accurate [2, 3]. A subset of these is the result of corroborating small evidence existing in separate places for which the necessity to put them together was found by some visionary capable minds. Concisely, this paper is about building NLP (Natural Language Processing) systems to improve the probability for this type of predictions to be true.

The need for detecting weak signals, defined as imprecise and early indicators of impending important trends or events, has been under focus in the last years. It becomes critical to realize what the next trends in economy will be. In science, making predictions almost equates to having a bright idea on how apparently disparate small achievements may be brought together to make something that is very promising. There is presented a methodology able to learn to discern between what may be vs. what is not a weak signal, and to retrieve and corroborate document that seems to amplify it.

© Springer Nature Switzerland AG 2023
A. Gelbukh (Ed.): CICLing 2019, LNCS 13451, pp. 73–85, 2023.
https://doi.org/10.1007/978-3-031-24337-0_6

Initially, this problem was treated as a supervised task. A relatively large corpus of news from scientific journals was compiled, and a group of annotators was initiated to mark paragraphs in news from scientific journals, which may contain weak signals according to the annotators' judgment. Then, a series of learning algorithms was run to see the extent to which it would automatically be identified what paragraph may contain weak signals. As it turned out that accuracy can be improved, the next step is to see how we could learn recognizing weak signals by reducing dependency on annotated data. To reach this aim, a possible way was to use Google Ngram Viewer to analyze the time distribution over certain topics that were identified automatically from the corpus. Some of the topics described by an acceding ratio were selected, following a technology similar to the one presented in [4] or [5]. The assumption of this survey is that the description of these topics at some point in time, before they had a large impact, was made by using weak signals. A corpus was automatically compiled from the time just before these topics became clearly important. Then, there the word embedding approach has been used for what appeared to be the common vocabulary in describing these topics. A deep learning algorithm was trained based on gradient descent, then a LSTM (Long Short-Term Memory) neural network was implemented. Next, these systems were tested on a set of documents for which it was known to contain weak signals or not, thanks to the previous supervised experiment. The results showed that we could learn to make predictions in an unsupervised manner, based on unknown weak signals, obtaining an accuracy, which is close to the supervised one. This is the fundamental result reported in this paper.

2   Related Work

The literature on weak signals is not very large, as this field is about to emerge. A groundbreaking paper [6, 7] was looking mainly to weak signals for detecting deviational behavior to efficiently provide preemptive counter measures. The probabilistic model presented here is similar to the one used in language modelling (an estimation of posterior probability of certain class via chain formula).

In [7] an automatic detection of crime using tweets is presented. They use LDA (Latent Dirichlet Allocation) to predict classes of similar words for topics related to violence.

While we can gain a valuable insight from these papers, their scope is limited because there is a direct connection between the overt information existing in text and speaker's intention.

Another emerging field, diachronicity (i.e., the evolution of certain topics in mass media over time) is linked to detection of weak signals. The statistical tests presented for epoch detection in [8], or temporal dynamics in [10–12] proves to be useful.

In [13, 14] we find very useful insights from dealing with discriminative analysis and SVM (Support Vector Machine) respectively, can be found. In order to improve the results of this study, for a start we need to understand how we could restrict further the objective function. In addition [14], the principle of an attentive neural network is presented.

3 Materials and Methods

Let's find out information about the data set and different machine learning systems used in to predict how to find weak signals in news.

3.1 Dataset

The study started by compiling a corpus of scientific papers and articles (42,916) from online freely available repositories published between 1960 and 2017.

Our assumption is that weak signals represent a form of tacit knowledge. As such, it may be counterproductive to define a formal guideline aiming to identify the weak signal. Rather, the annotator has to be the liberty to mark a whole document as containing weak signals or not. In a second round of annotations, it was restricted the scope to paragraph rather than whole document. Most of the annotated paragraphs have between 100 to 250 words. Therefore, two annotated corpora were obtained, which for convenience they will be referred to short and long respectively. The long corpora, LC, refer to full documents as training/test corpora. The short corpora, SC, refer to paragraphs.

There is not a perfect overlap between these two corpora; approximatively 15% of paragraphs come from different documents that the ones considered on LC corpora.

The annotation is binary, YES or NO signaling the existence of weak signals or not respectively. In case of SC, all paragraphs, not explicitly classified as "YES" from the analyzed documents, were be considered "NO". However, in order to make sure that there are as little as possible misclassifications, the SC "NO", for some of these paragraphs, were double-checked. Table 1 shows their distribution.

Table 1. Size of hand annotated corpora.

Corpora type	YES	NO
LC	4,100	14,020
SC	3,700	14,500

To ease a fair comparison of the performances for these two corpora, it is need to have a similar ratio of weak *vs.* non weak in both corpora. The team consists of 18 undergrads volunteers. On a given 300 documents the annotators were encouraged to discuss their doubts and to defend their position in case of disagreement.

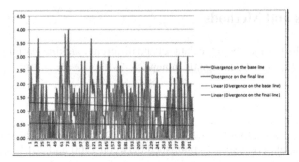

Fig. 1. Towards reaching a stable shared tacit knowledge.

In Fig. 1, the evolution of the average number of documents on which there was a strong disagreement, for samples of 10 documents out of the chosen 300, is plotted. The average disagreement lowered from 1.4 to 1.1 and the divergence decreased from .55 to .38.

It seems that 1.1 is a hard threshold for this task. After 1,200 of documents were annotated as carriers of weak signals, the experiment was repeated. The average of disagreement for samples of ten documents, was still 1.1. For making a decision decreased for time between these two experiments, the average time was measured (Fig. 2).

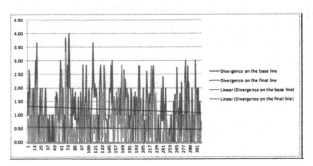

Fig. 2. Average time for making a decision.

These results suggest that this task, in spite of being one driven by tacit knowledge, is learnable by algorithmic probabilistic hypothesis space search. The annotators developed patterns – they seem to filter out a lot of the content, otherwise the time to reach a decision would not have decreased that dramatically, and there is gray zone where experience does not help. This behavior tends to help an automatic classifier, as it does not have to be very precise in order to obtain a human similar performance.

After a preliminary round of trial annotation of several hundred of documents, a taxonomy was created that sprung naturally from this experiment based on the a several components: *technology*; *innovation in services*; trend shift; *behavioural change*; *major actor move*; *breakthrough discovery*; *top research*; *wild card* (Fig. 3).

Categories	Technology	Others								total votes	Total votes/ total events
		Innovation in Services	Trend shift	Behavioral Change	Major actor move	Discovery	Studies	Wildcard	NS		
Votes	606	126	176	80	184	104	132	13	401	1802	1.24
Unique classification	367	26	53	26	70	46	100	11	401	1096	

Fig. 3. Weak signal taxonomy distribution.

The intention in using these labels was to try to capture the annotator intuition on why a certain document/paragraph is considered as carrier of weak signals. This taxonomy helps us to see if there are indeed any subjective differences that may affect the learning process. It came as a surprise that the annotators did not want to use often the wild card taxonomy. The number of documents that received just one category is relatively high and quasi constant (50%). The number of documents that received more than three categories is non-significant, less than 3%.

In Fig. 4, the dynamics of reaching consensus among annotators is drawn (Fig. 5).

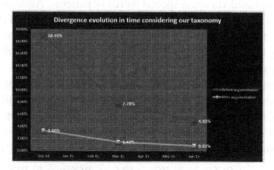

Fig. 4. Reaching consensus over taxonomies.

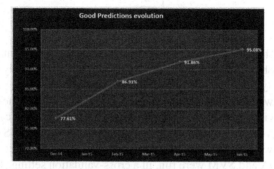

Fig. 5. Control group judgment.

There was checked whether this consensus was reached due to an increasingly strong and commonly shared tacit knowledge, i.e., due to acquiring an expertise, or due to accepting a dominant view. A control group checked the validity of the agreement. The results strongly suggest the first alternative – acquiring an expertise.

Finally, all these experiments strongly suggest that there has a tacit knowledge about weak signals that is shared at least 80% of the time. However, there will be a 10% individual hard kernel that constitutes a potential disagreement area. Maybe this is exactly the prediction on the immediate future.

3.2 Machine Learning

Below, a series of learning approaches are presented, being tried systematically. The experience and insight gained from previous steps guided our decision in designing the next step.

3.2.1 Baseline

In a supervised approach, finding the pieces of news containing weak signals is a binary classification task. A first approach is to use tf-idf weights to compute the similarity between a document and the documents in one of the two classes. This provides us with a weak baseline. However, it is an informative one. It tells how much of the weak signals are judged to be expressed via some special words or patterns. Anticipating, it turns out that this is not the case at all. This baseline has negligible accuracy, far distanced from the best results obtained eventually. This preliminary finding confirmed that the task is not trivial at all and that many clues based on which the human judges the correct answer are not necessarily expressed by clearly defined overt phrases.

As such, a couple of off the shelf approaches were used that will provide a set of baselines for this task. Two libraries were considered, which implement quadratic discriminative analysis, QDA (Quadratic Discriminant Analysis) from SCIKIT library, and SVM, linear SVM form WEKA (Waikato Environment for Knowledge Analysis) library, respectively. See also the Eqs. 1 and 2.

$$P(y = k|X) = \frac{P(X|y = k)\,P(y = k)}{P(X)} - \frac{P(X|y = k)\,P(y = k)}{\sum_l P(X|y = l) \cdot P(y = l)} \tag{1}$$

$$\widehat{R}_n(f) = \frac{1}{n} \sum_{i=1}^{n} \pi(f(X_i) \neq Y_i) \tag{2}$$

The reasons behind this choice have to do with the type of data employed here. The fact that the *tf-idf* obtained a very low score does not immediately imply that maximizing the prior probability P(*word|weak signal*) is inefficient.

At this point, we have to understand whether the projection of the data into a bidimensional space will lead to conelike structures, that is, that the data can separated by a quadratic function. On the other hand, if the difference between the SVM and QDA is large enough this will show that QDA suffers from the masking effect.

Here, both QDA and SVM were run, in a cross-validation setting, 10 folds 1/8 ratio for train/test and 1/8 ratio for development/train.

Table 2. QDA, SVM for SC.

Method	Weak signal	No signal
QDAcr	0.412	0.877
SVMcr	0.663	0.913
QDAts	0.403	0.865
SVMts	0.610	0.905

That is a tenth of the corpus for test and development, respectively was used. For test, 500 weak-signals and 500 no-signals were used. In Table 2, the results for QDA and SVM for SC, and in Table 3 the results for LC for cross validation are presented.

The *tf-idf* scored 0.18 for and 0.12 respectively. For the moment, both QDA and SVM scored significantly better than that. In addition, indeed there is a non-random difference between QDA and SVM result.

Table 3. QDA, SVM for LC.

Method	Weak signal	No signal
QDAcr	0.38	0.901
SVMcr	0.472	0.946
QDAts	0.365	0.890
SVMts	0.455	0.930

To understand better the nature of this difference a series of experiments alternating the ratio of weak signals in the training corpus was ran. Note that no significant differences from Table 2 and 3! This shows that probably we cannot improve these results by adding more training. Given that SVM is a constraint over a large boundary for ‖Ein-Eout‖ and that the differences from QDA are large, Eq. 1 follows that it is possible to search for a better model even further. That is, particularly for this task, it has to be found a better estimation, as the worst-case scenario seems not to characterize this corpus. Because, the number of dichotomies cannot be directly computed, and therefore, the exact VC (*Vapnik-Chervonenkis*) dimension is unknown, based on the Tables 2, 3 it is intuitively tempting to consider that the VC bound is indeed too loose for this task. That is, we can do better in estimating the posterior probability. The right question is whether we have enough data to train a more detailed classifier. Indeed, deep learning methods may be up to the task.

3.2.2 Deep Learning Approach

Here, two experiments carried out, using deep learning methods, are described. The first method uses a simple gradient descent model with CR (Cross Entropy Loss) or log loss, function. The second is a LSTM neural network. The cross entropy is described by Eqs. (3).

$$H_{y'}(y) = -\sum_i y_i' log(y_i)$$ (3)

LSTM

$$H = [X_t \cdot h_{t-1}] \quad \widehat{i}t = \sigma(W^i H + b_i) \quad ft = (W^f H + b_i) \quad ot = \sigma(W^0 H + b_0)$$

The experiments were organised in the following manner: a test corpus of 1,500 documents, 1000 non signals and 500 signals, was made. The rest of the corpus was sent to the deep learning algorithm. Each document was represented as a vector via word embedding. A batch of 100 vectors, selected randomly, at one training cycle over a couple of thousands of cycles was used. In order to both (i) have a better assessment of the accuracy and (ii) analyze the influence of signal *vs.* non signal ratio in training the algorithm was ran three times. The number of test non-signal was 1,000 and the number of test weak-signal was 500. The ration weak-signal/no-signal in one batch of 100 random input example to train was varied. CR_n/m, represents a system with cross entropy loss function that has 2,000 cycles with a batch of 100 containing *m* random weak-signal and *n* random no-signal from training corpus. There was tested for n/m in {5/5, 3/7} (rows in tables below). For example, LS_5/5 represents the system of LSTM with 100 batches made of 50 weak and 50 no signals, randomly chosen from training set. Table 4 and 5 show the results. The columns represent the accuracy for each of the three runs for weak and no signals respectively.

Table 4. CR, LS for SC.

System_n/m	1W	1N	2W	2N	3W	3N
CR_5/5	0.75	0.85	0.74	0.86	0.79	0.86
LS_5/5	0.75	0.89	0.73	0.92	0.75	0.87
CR_3/7	0.88	0.93	0.86	0.93	0.85	0.92
LS_3/7	0.86	0.94	0.86	0.95	0.88	0.95

Table 5. CR, LS for LC.

System_n/m	1W	1N	2W	2N	3W	3N
CR_5/5	0.75	0.85	0.74	0.86	0.79	0.86
LS_5/5	0.75	0.89	0.73	0.92	0.75	0.87
CR_3/7	0.88	0.93	0.86	0.93	0.85	0.92
LS_3/7	0.86	0.94	0.86	0.95	0.88	0.95

As expected, the results are usually better on SC than on LC. However, the most important thing was that both approaches produces consistent results for each run, there were no big jumps/drops in accuracy at any particular experiment.

Here, a great improvement, over the SVM performances, can be seen, which prove that the data was enough to compute accurate Softmax estimation. It also shows that for this particular problem, the VC bound is too loose indeed.

To evaluate the dependency of the deep learning algorithm on the number of examples, a second round of experiments considering for training only q half, two thirds, 3 quarts and 7/8 of the initial training data was ran. There is no statistically significant difference once we feed more than 3 quarts of the data. This means that it is unlikely that these results will be further improved by providing more data alone.

3.3 Unsupervised Learning

In this section, a methodology of learning weak signals in an unsupervised way is presented. In the previous section, very good results via training examples were obtained. The annotated data, in spite of coming by a long-time consuming process, may still contain errors, may still be subjectively biased by guidelines. A better way is to induce the training data and then fed it to the deep learning algorithm. Concisely, a set of topics that were introduced by weak signals before they become very important topics that everybody talks about are going to consider (Fig. 6).

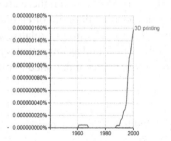

Fig. 6. 3Dprinting N-gram distribution.

For example, self-driving cars, 3D printing, Higgs boson, etc. This set of topics as a side effect of the annotation process explained in Sect. 3. The Google N-gram was used to observe the diachronic evolution of a particular topic. As expected, these topics have a boom, which has a very particular steeply ascending plot starting after a certain year.

This booming trend can be captured via statistical tests. It is very easy to detect the booming period due to its clearly defined shape, but, in fact, the pre boom period is actually matters the most. In order to detect the pre boom statistically correlated period, the Welch and ratio tests were used. In this period both the Welch and ratio test, indicate that something is going on, but there is no boom yet. That is, a non-random variation into the distribution of previous years is caught, but there is no confirmed boom yet. This is the period when that particular topic was rather described using weak signals. In a few years, after the boom, the description is far from "weak signal", it is a confirmed trend. In Fig. 7, the pre boom period is automatically marked, being found for *Higgs boson.*

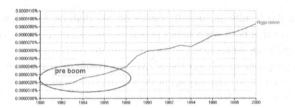

Fig. 7. Pre boom period for Higgs boson.

The pre boom period is characterized by a positive ratio test followed by a period of relatively stability, thus negative Welch test (Fig. 8).

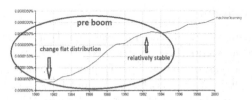

Fig. 8. Machine learning pre boom.

A human interpretation of this behavior is that there are initial excitements about a certain achievement, which make people talk more often that before about it, so the old flat distribution changes. Then there is a period of relatively silence, and then there is the boom. Also, machine learning was found as important topic, with a clear definite pre boom and boom period. For the four topics mentioned above, in Table 6 the pre boom periods are plotted.

Table 6. Pre boom period.

Topics	Period
Machine learning	1978–1990
3D printing	1960–1995
Higgs boson	1981–1988
Self-driving	1960–1998

Once the pre boom period is known, only articles from that period were considered. There, should be something in these articles that a human annotator would call weak signals. In this way, the signal corpus can be compiled. To compile the non-signal corpus, which corresponds to the negative learning, the documents from pre boom period and any other articles from the pre boom period that do not exhibit any unusual distribution were taken. There was kept the same ratio between the number of signal vs. non signal documents as the one for the corpus described in Sect. 3. By feeding into the deep learning algorithm the vectors representing these documents, we get a higher accuracy if the data set contains more topics and more texts from the pre-boom period. There is going to test the accuracy of (i) the unsupervised learning and (ii) the dependency of accuracy on those parameters: number of topics, the detection accuracy of the pre boom period and the number of documents considered as weak signals. Then, then follows the testing stage of 1,500 documents, randomly chosen, out of which 500 are signals, and none of them contains any references to the topics used to collect the training corpus. Table 7 contains the results, following this process (Fig. 9).

Table 7. Unsupervised weak signal prediction.

System	Weak signal	No signal
CR	0.45	0.89
LSTM	0.44	0.93

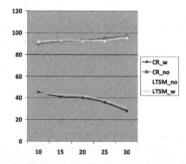

Fig. 9. Dependency on pre boom accuracy.

The results considering 10 topics are very good. They are well above the *tf-idf* baseline and they are close to the performances of the SVM. In order to include more topics from the pre boom period, by increasing the boundaries by 10% to 30%, the pre boom period has been artificially modified. It can be noticed a major deterioration in results; the accuracy was dropped to 0.3. By including more documents, the results did not go up in a statistical significant way. These findings suggest that pre boom epoch accurate detection plays a very important role, and that probably few thousands of documents, cumulatively; represent a sufficient statistics for this task. Little is gained by adding any other information, either in a supervised way or not.

Finally, the following experiment to complete the picture was considered. First, there was tried to predict whether there will be a boom only in the distribution of the individual topics in the pre boom. That is, by just looking at the pre boom period it is tried to predict whether that respective topic will materialize in a fully-fledged trend. Here, was considered the mean difference between the pre boom period and the pre boom period with an empirical threshold. The intuition reveals that a topic exhibiting a large difference; it is likely to witness a boom. The accuracy was less than .08, showing that the analysis of individual topics will tell us something about trends only when it is too late.

4 Conclusions

This study is an experiment on weak signals prediction. The possibility of trend prediction based on weak signals is very exciting and it has many applications. This study shows that even when it is not known what the weak signals are, there is still the ability to be used them in predicting future trends via deep learning methods with unsupervised learning. A carefully analysis of the experiments, performed systematically from tf-idf to LSTM, provide us with insights on choosing one type of learning over other possible candidates. Next, other deep learning algorithms are to be experimented. A starting point is to understand better how it could be narrowed down the search for weak signals. The results suggest that a major improvement of several points could be possible, if a paragraph could be indicated instead of a document as a source of weak signals. Therefore, our next effort is to narrow down the search for pre boom period at the paragraph level, rather than document level.

Acknowledgments. This project is funded by the Ministry of Research and Innovation within Program 1 – Development of the national RD system, Subprogram 1.2 – Institutional Performance – RDI excellence funding projects, Contract no. 34PFE/19.10.2018".

References

1. Delmonte, R., Tripodi, R., Gîfu, D. Opinion and factivity analysis of italian political discourse. In: Proceedings of the 4th edition of the Italian Information Retrieval Workshop (IIR 2013), pp. 88–99. Pisa, Italy, CEUR-WS on-line Proceedings Series (2013)
2. Popescu, O., Strapparava, C.: Behind the times: detecting epoch changes using large corpora. In: International Joint Conference on Natural Language Processing, pp. 347–355. Nagoya, Japan (2013)
3. Gifu, D.: Contrastive diachronic study on romanian language. In: Proceedings FOI-2015, S. Cojocaru, C. Gaindric (eds.), Institute of Mathematics and Computer Science, pp. 296–310. Academy of Sciences of Moldova (2015)
4. Popescu, O., Strapparava, C.: Time corpora: epochs, opinions and changes. Knowl. Based Syst. **69**, 3–13 (2014)
5. Amarandei, S., Fleşcan, A., Ioniţă, G., Turcu, R., Trandabăţ, D., Gîfu, D.: Key Biomedical Concepts Extraction. In: Gîfu, D., Trandabăţ, D. (Eds.). Proceedings of the International Workshop on Curative Power of Medical Data, D. Gîfu and D. Trandabăţ (Eds.). "Alexandru Ioan Cuza," pp. 21–26 University Publishing House, Iaşi (2017).
6. Brynielsson, J., Horndahl, A., Johansson, F., Kaati, L., Martenson, C., Svenson, P.: Harvesting and analysis of weak signals for detecting lone-wolf terrorists. Secur. Inf. **2**(1), 1–15 (2013)
7. Cohen, K., Johansson, F., Kaati, L., Mork, J.: Detecting linguistic markers for radical violence in social media. In: Terrorism and Political Violence, vol. 26, no. 1, pp. 246–256 (2014)
8. Wang, X., Gerber, M.S., Brown, D.E.: Automatic crime prediction using events extracted from twitter posts. In: Yang, S.J., Greenberg, A.M., Endsley, M. (eds.) SBP 2012. LNCS, vol. 7227, pp. 231–238. Springer, Heidelberg (2012). https://doi.org/10.1007/978-3-642-29047-3_28

9. Mihalcea, R., Nastase, V.: Word epoch disambiguation: finding how words change over time. In: Proceedings of the 50th Annual Meeting of the Association for Computational Linguistics, vol. 2, pp. 259–263 (2012)
10. Wang, X., McCallum, A.: Topics over time: a non-markov continuous-time model of topical trends. In: KDD 2006, USA (2006)
11. Wang, C., Blei, D.M., Heckerman, D.: Continuous time dynamic topic models. proceedings of uncertainty in artificial intelligence. In: Zunino, L., Bariviera, A.F., Guercio, M.B., Martinez, L.B., Rosso, O.A. (eds.): On the efficiency of sovereign bond markets. Phys. A Stat. Mech. Appl, vol. 391, pp. 4342–4349 (2012)
12. Gerrish, S.M. Blei, D.M.: A language-based approach to measuring scholarly impact. In: Proceedings of International Conference of Machine Learning (2010)
13. Hastie, T., Tibshirani, R., Friedman, J.: The Elements of Statistical Learning. Data Mining, Inference, and Prediction, 2nd ed. Springer, Cham (2008)
14. Abu-Mostafa, Y., Magdon-Ismail, M., Lin, H.-T.: Learning from Data (2013). amlbook.com
15. Augenstein, I., Rocktaeschel, T., Vlachos, A., et al.: Stance detection with bidirectional conditional encoding. In: Proceedings of the Conference on Empirical Methods in Natural Language Processing. Empirical Methods in Natural Language Processing, Austin, Texas, USA (2016)

Low-Rank Approximation of Matrices for PMI-Based Word Embeddings

Alena Sorokina, Aidana Karipbayeva[✉], and Zhenisbek Assylbekov

Department of Mathematics, Nazarbayev University, Astana, Kazakhstan
{alena.sorokina,aidana.karipbayeva,zhassylbekov}@nu.edu.kz

Abstract. We perform an empirical evaluation of several methods of low-rank approximation in the problem of obtaining PMI-based word embeddings. All word vectors were trained on parts of a large corpus extracted from English Wikipedia (enwik9) which was divided into two equal-sized datasets, from which PMI matrices were obtained. A repeated measures design was used in assigning a method of low-rank approximation (SVD, NMF, QR) and a dimensionality of the vectors (250, 500) to each of the PMI matrix replicates. Our experiments show that word vectors obtained from the truncated SVD achieve the best performance on two downstream tasks, similarity and analogy, compare to the other two low-rank approximation methods.

Keywords: Natural language processing · Pointwise mutual information · Matrix factorization · Low-rank approximation · Word vectors

1 Introduction

Today word embeddings play an important role in many natural language processing tasks, from predictive language models and machine translation to image annotation and question answering, where they are usually 'plugged in' to a larger model. An understanding of their properties is of interest as it may allow the development of better performing embeddings and improved interpretability of models using them. One of the widely-used word embedding models is the Skip-gram with negative sampling (SGNS) of Mikolov et al. (2013). Levy and Goldberg (2014) showed that the SGNS is implicitly factorizing a pointwise mutual information (PMI) matrix shifted by a global constant. They also showed that a low-rank approximation of the PMI matrix by truncated singular-value decomposition (SVD) can produce word vectors that are comparable to those of SGNS. However, truncated SVD is not the only way of finding a low-rank approximation of a matrix. It is optimal in the sense that it minimizes the approximation error in the Frobenius and the 2-norms, but this does not mean that it produces optimal word embeddings, which are usually evaluated in downstream NLP tasks. The question is: Is there any other method of low-rank matrix approximation that produces word embeddings better than the truncated SVD factorization? Our experiments show that the truncated SVD is actually

© Springer Nature Switzerland AG 2023
A. Gelbukh (Ed.): CICLing 2019, LNCS 13451, pp. 86–94, 2023.
https://doi.org/10.1007/978-3-031-24337-0_7

a strong baseline which we failed to beat by another two widely-used low-rank approximation methods.

2   Low-Rank Approximations of the PMI-matrix

The simplest version of a PMI matrix is a symmetric matrix with each row and column indexed by words[1], and with elements defined as

$$\text{PMI}(i,j) = \log \frac{p(i,j)}{p(i)p(j)}, \tag{1}$$

where $p(i,j)$ is the probability that the words i, j appear within a window of a certain size in a large corpus, and $p(i)$ is the unigram probability for the word i. For computational purposes, Levy and Goldberg (2014) suggest using a positive PMI (PPMI), defined as

$$\text{PPMI}(i,j) = \max(\text{PMI}(i,j), 0). \tag{2}$$

They also show empirically that the low-rank SVD of the PPMI produces word vectors which are comparable in quality to those of the SGNS.

The low-rank matrix approximation is approximating a matrix by one whose rank is less than that of the original matrix. The goal of this is to obtain a more compact representation of the data with a limited loss of information. In what follows we give a brief overview of the low-rank approximation methods used in our work. Since both PMI (1) and PPMI (2) are square matrices, we will consider approximation of square matrices. For a thorough and up-to-date review of low-rank approximation methods see the paper by Kishore Kumar and Schneider (2017).

Singular Value Decomposition (SVD) factorizes $\mathbf{A} \in \mathbb{R}^{n \times n}$, into the matrices $\mathbf{U} \in \mathbb{R}^{n \times n}$, $\mathbf{S} \in \mathbb{R}^{n \times n}$ and $\mathbf{V}^\top \in \mathbb{R}^{n \times n}$:

$$\mathbf{A} = \mathbf{U}\mathbf{S}\mathbf{V}^\top,$$

where $\mathbf{U}$ and $\mathbf{V}$ are orthogonal matrices, and $\mathbf{S}$ is a rectangular diagonal matrix whose entries are in descending order, $\sigma_1 \geq \sigma_2 \geq \cdots \geq \sigma_n \geq 0$, along the main diagonal, and are known as the singular values of $\mathbf{A}$. The rank d approximation (also called *truncated* or *partial SVD*) of $\mathbf{A}$, $\mathbf{A}_d$ where $d < \text{rank}\,\mathbf{A}$, is given by zeroing out the $n - d$ trailing singular values of $\mathbf{A}$, that is[2]

$$\mathbf{A}_d = \mathbf{U}_{1:n,1:d}\mathbf{S}_{1:d,1:d}\mathbf{V}^\top_{1:d,1:n}.$$

By the Eckart-Young theorem (Eckart and Young 1936), A_d is the closest rank-d matrix to A in Frobenius norm, i.e. $\|\mathbf{A} - \mathbf{A}_d\|_F \leq \|\mathbf{A} - \mathbf{B}\|_F, \forall \mathbf{B} \in \mathbb{R}^{n \times n}$: $\text{rank}(\mathbf{B}) = d$. Levy and Goldberg (2014) suggest factorizing the PPMI matrix with truncated SVD, and then taking the rows of $\mathbf{U}_{1:n,1:d}\mathbf{S}^{1/2}_{1:d,1:d}$ as word vectors, and we follow their approach.

---

[1] Assume that words have already been converted into integer indices.

[2] $\mathbf{A}_{a:b,c:d}$ is a submatrix located at the intersection of rows $a, a+1, \ldots, b$ and columns $c, c+1, \ldots, d$ of a matrix $\mathbf{A}$.

QR Decomposition with column pivoting of $\mathbf{A} \in \mathbb{R}^{n \times n}$ has the form $\mathbf{AP} = \mathbf{QR}$, where $\mathbf{Q} \in \mathbb{R}^{n \times n}$ is orthogonal, $\mathbf{R} \in \mathbb{R}^{n \times n}$ is upper triangular and $\mathbf{P} \in \mathbb{R}^{n \times n}$ is a permutation matrix. The rank d approximation to $\mathbf{A}$ is then

$$\mathbf{A}_d = \mathbf{Q}_{1:n,1:d}[\mathbf{RP}^\top]_{1:d,1:n}$$

which is called *truncated QR decomposition* of $\mathbf{A}$. After factorizing the PPMI matrix with this method we suggest taking the rows of $\mathbf{Q}_{1:n,1:d}$ as word vectors.

However, we suspect that a valuable information could be left in the $\mathbf{R}$ matrix. A promising alternative to SVD is a Rank Reveling QR decomposition (RRQR). Assume the QR factorization of the matrix $\mathbf{A}$:

$$\mathbf{AP} = \mathbf{Q} \begin{bmatrix} \mathbf{R}_{11} & \mathbf{R}_{12} \\ 0 & \mathbf{R}_{22} \end{bmatrix}$$

where $\mathbf{R}_{11} \in \mathbb{R}^{d \times d}$, $\mathbf{R}_{12} \in \mathbb{R}^{d \times (n-d)}$, $\mathbf{R}_{22} \in \mathbb{R}^{(n-d) \times (n-d)}$. For RRQR factorization, the following condition should be satisfied:

$$\sigma_{\min}(\mathbf{R}_{11}) = \Theta(\sigma_d(\mathbf{A}))$$
$$\sigma_{\max}(\mathbf{R}_{22}) = \Theta(\sigma_{d+1}(\mathbf{A}))$$

which suggests that the most significant entries are in $\mathbf{R}_{11}$, and the least important are in $\mathbf{R}_{22}$. Thus, we also suggest taking the columns of $[\mathbf{RP}^\top]_{1:d,1:n}$ as word vectors.

Non Negative Matrix Factorization (NMF). Given a non negative matrix $\mathbf{A} \in \mathbb{R}^{n \times n}$ and a positive integer $d < n$, NMF finds non negative matrices $\mathbf{W} \in \mathbb{R}^{n \times d}$ and $\mathbf{H} \in \mathbb{R}^{d \times n}$ which minimize (locally) the functional $f(\mathbf{W}, \mathbf{H}) = \|\mathbf{A} - \mathbf{WH}\|_F^2$. The rank d approximation of $\mathbf{A}$ is simply

$$\mathbf{A}_d = \mathbf{WH}.$$

When factorizing the PPMI matrix with NMF, we suggest taking the rows of $\mathbf{W}$ as word vectors.

3   Experimental Setup

3.1   Corpus

All word vectors were trained on the `enwik9` dataset[3] which was divided into two equal-sized splits. The PMI matrices on these splits were obtained using the `hypewords` tool of Levy et al. (2015). All corpora were pre-processed by removing non-textual elements, sentence splitting, and tokenization. PMI matrices were derived using a window of two tokens to each side of the focus word, ignoring words that appeared less than 300 times in the corpus, resulting in vocabulary sizes of roughly 13000 for both words and contexts. A repeated measures design was used for assigning the method of factorization (SVD, QR, NMF) and dimensionality of the vectors (250, 500) to each PMI matrix replicate. We used two replicates per each level combination.

---

[3] http://mattmahoney.net/dc/textdata.html.

3.2 Training

Low-rank approximations were performed using the following open-source implementations:

- Sparse SVD from SciPy (Jones et al. 2014),
- Sparse RRQR from SuiteSparse (Davis and Hu, 2011), and
- NMF from scikit-learn (Pedregosa et al. 2011).

For NMF we used the nonnegative double SVD initialization. We trained 250 and 500 dimensional word vectors with each method.

3.3 Evaluation

We evaluate word vectors on two tasks: similarity and analogy. A similarity is tested using the WordSim353 dataset of Finkelstein et al. (2002), containing word pairs with human-assigned similarity scores. Each word pair is ranked by cosine similarity and the evaluation is the Spearman correlation between those rankings and human ratings. Analogies are tested using Mixed dataset of 19544 questions such as "a is to b as c is to d", where d is hidden and must be guessed from the entire vocabulary. We filter questions with out of vocabulary words, as standard. Accuracy is computed by comparing $\arg\min_d \|\mathbf{b} - \mathbf{a} + \mathbf{c} - \mathbf{d}\|$ to the labelled answer.

4 Results

The results of evaluation are provided in Table 1, which we analyze using the two-factor ANOVA with factors being (1) low-rank approximation method and (2) dimensinality of word vectors, and response being the performance in similarity or analogy task. We analyze the tasks separately (Fig. 1).

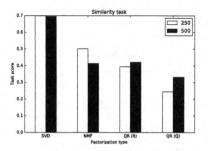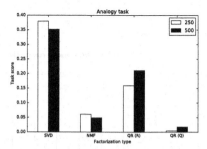

Fig. 1. Test scores for different factorization methods on Similarity and Analogy tasks.

Table 1. Results

Method of low-rank approximation	Dimensionality of vectors	Replicate #	Similarity task	Analogy task
SVD	250	1	0.7010	0.3778
SVD	250	2	0.6969	0.3817
SVD	500	1	0.6989	0.3568
SVD	500	2	0.6914	0.3458
NMF	250	1	0.5265	0.0660
NMF	250	2	0.4780	0.0563
NMF	500	1	0.4499	0.0486
NMF	500	2	0.3769	0.0487
QR (R)	250	1	0.4077	0.1644
QR (R)	250	2	0.3822	0.1533
QR (R)	500	1	0.4717	0.2284
QR (R)	500	2	0.3719	0.1925
QR (Q)	250	1	0.2870	0.0034
QR (Q)	250	2	0.2009	0.0059
QR (Q)	500	1	0.3573	0.0165
QR (Q)	500	2	0.3048	0.0186

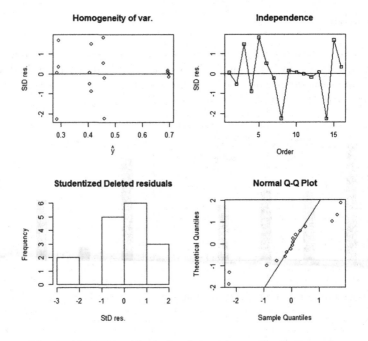

Fig. 2. ANOVA residuals for the results on Similarity task.

4.1   Similarity Task

The standard residual analysis is used to check whether the ANOVA assumptions are satisfied. From Fig. 2 we see that the residuals have constant variability around zero, are independent and normally distributed. The normality is confirmed using Shapiro-Wilk test, p-value = 0.7923.

Table 2. ANOVA table for the similarity task results

Source	DF	Sum of squares	Mean square	F Value	Pr > F
Model	7	0.37055017	0.05293574	29.68	<.0001
Error	8	0.01426728	0.00178341		
Corrected total	15	0.38481745			
	R-Square	Coeff Var	Root MSE	Score Mean	
	0.962925	9.126819	0.042230	0.462707	

Table 3. Main and Interaction Effects in the Similarity task

Source	DF	Sum of squares	Mean square	F Value	Pr > F
Factorization	3	0.35433596	0.11811199	66.23	<.0001
Dimension	1	0.00011159	0.00011159	0.06	0.8088
Interaction	3	0.01610263	0.00536754	3.01	0.0945

Table 4. ANOVA Table for the Analogy task results

Source	DF	Sum of squares	Mean square	F Value	Pr > F
Model	7	0.30745304	0.04392186	424.61	<.0001
Error	8	0.00082753	0.00010344		
Corrected total	15	0.30828057			
	R-Square	Coeff Var	Root MSE	Score Mean	
	0.997316	6.602449	0.010171	0.154043	

The SAS package was used to obtain ANOVA table (Table 2), which shows the effects of the factors on the similarity score. F-test for equality of the factor level means was conducted, $F = 29.68$ and p-value < 0.0001. Hence, it can be concluded that at least one factor level mean is different from the others. $R^2 = 0.962925$ shows that more than 96% of variation in the similarity score is explained by the factors considered.

Proceeding with analysis of main and interaction effects, one can conduct F-test for each of the factors and the interaction between them. From Table 3, we

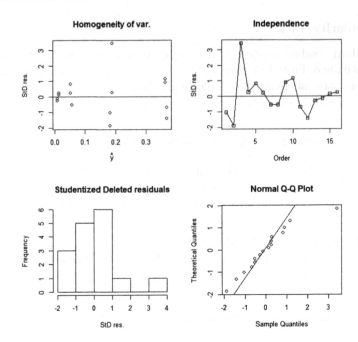

Fig. 3. ANOVA Residuals for the Analogy task results

Table 5. Main and Interaction Effects in the Analogy task

Source	DF	Sum of squares	Mean square	F Value	Pr > F
Factorization	3	0.30365768	0.10121923	978.52	<.0001
Dimension	1	0.00013820	0.00013820	1.34	0.2811
Interaction	3	0.00365715	0.00121905	11.78	0.0026

see that the method of low-rank approximation affects the performance of words vectors in the similarity task, $F = 66.23$, p-value < 0.0001. The dimensionality of word vectors has no effect on the performance in the similarity task, $F = 0.06$ with p-value > 0.8. Also, there is no interaction between the method of factorization and the dimensionality of word vectors, $F = 3.01$ with p-value 0.0945. Thus, SVD significantly outperforms the other factorization methods.

4.2   Analogy Task

Again, we first need to check whether the ANOVA assumptions are satisfied. From Fig. 3 we see that the residuals have constant variability around zero, are independent and normally distributed. The normality is confirmed using Shapiro-Wilk test, p-value $= 0.112$. The ANOVA Table (Table 4) shows that at least one level mean is different from the others. R^2 is 0.997316, thus, 99% of variation in the analogy score is explained by the considered factors (Table 5).

We proceed to the analysis of main and interaction effects. The method of low-rank approximation affects the performance of word vectors in the analogy task, $F = 978.52$ with p-value < 0.0001. The dimensionality of word vectors has no effect on the performance in the analogy task, $F = 1.34$ with p-value > 0.2. Unlike the similarity task, there is an interaction effect between the two factors, $F = 11.78$ with p-value $= 0.0026$.

5 Discussion

Why Dimensionality is Critical in Similarity Task for NMF? We obtained the highest results in the similarity task using the SVD-based low-rank approximation, for which the dimensionality of word vectors did not influence the performance much. On the contrary, the performance in similarity task using the NMF method of factorization is significantly affected by the dimension of the word vector: 250-dimensional word vectors give significantly better results than 500-dimensional ones. This can be explained by the specific characteristics of the NMF method of factorization. When we look at the word vectors produced by NMF, we can see that they contain many zeros. Hence, an increase in the dimensionality makes them even sparser. Similarity task is based on finding the cosine of the angle between two word vectors. Therefore, when the vectors become sparser, the result of element-wise multiplication, which is necessary for obtaining cosine, becomes smaller. Thus, there is a much higher possibility that the cosine similarity score between two vectors, containing many zeros, will give a number closer to zero than to 1. This, as a result, leads to the worse performance in the similarity task. Our suggestion is to decrease the dimensionality of the NMF method to 100. We expect that this may give better results.

Why NMF Performs Poorly in the Analogy Task? We provide a theoretical analysis of the poor performance of the NMF in the analogy task. We model word vectors produced by the NMF as independent and identically distributed random vectors from an isotropic multivariate Gaussian distribution $\mathcal{N}(\mathbf{4.5}, \mathbf{I})$[4], since for a 500-dimensional $\mathbf{v} \sim \mathcal{N}(\mathbf{4.5}, \mathbf{I})$ there is a big chance that it is nonnegative:

$$\Pr(\mathbf{v} \in [0, +\infty)^{500}) = [\Pr(4.5 + Z > 0)]^{500} \approx 0.9983,$$

where $Z \sim \mathcal{N}(0, 1)$ is a standard normal random variable. For a triplet of word vectors $\mathbf{a}$, $\mathbf{b}$ and $\mathbf{c}$ we have $\mathbf{b} - \mathbf{a} + \mathbf{c} \sim \mathcal{N}(\mathbf{4.5}, \mathbf{3I})$, and therefore

$$\Pr(\mathbf{b} - \mathbf{a} + \mathbf{c} \in [0, +\infty)^d) = [\Pr(3 + \sqrt{3}Z \geq 0)]^d$$
$$= [\Pr(Z \geq -4.5/\sqrt{3})]^d < [0.9953]^d.$$

When $d = 500$, this probability is ≈ 0.1, i.e. there is a small chance that $\mathbf{b} - \mathbf{a} + \mathbf{c}$ is non negative, and thus we will likely not find a non-negative $\mathbf{d}$ when we minimize

---

[4] The isotropy is motivated by the work of Arora et al. (2016); $\mathbf{4.5}$ is a vector with all elements equal to 4.5.

$\|\mathbf{b} - \mathbf{a} + \mathbf{c} - \mathbf{d}\|$. This is confirmed empirically: for *all* word triplets (a, b, c) from the analogy task, the vector $\mathbf{b} - \mathbf{a} + \mathbf{c}$ has at least one negative component.

Why Using R is Better than Using Q in the QR Decomposition? The $\mathbf{Q}$ matrix from QR factorization gives the worst results in the similarity task, and it does not depend on the dimensionality of the vector. The reason is that the necessary information is left in the $\mathbf{R}$ matrix. Truncation of $\mathbf{RP}^\top$ gives better approximation to the original matrix than the truncated $\mathbf{Q}$, because the most significant entries of $\mathbf{RP}^\top$ are in the top left quarter and remain after truncation.

6   Conclusion

We analyzed the performance of the word vectors obtained from a word-word PMI matrix by different low-rank approximation methods. As it was expected, the truncated SVD provides a far better solution than the NMF and the truncated QR in both similarity and analogy tasks. While the performance of the NMF is relatively good in the similarity task, it is significantly worse in the analogy task. NMF produces only non-negative sparse vectors and we showed how this deteriorates the performance in both tasks. $\mathbf{RP}^\top$ matrix from QR factorization with column pivoting gives better word embedding than $\mathbf{Q}$ matrix in both tasks.

Acknowledgement. The work of Zhenisbek Assylbekov has been funded by the Committee of Science of the Ministry of Education and Science of the Republic of Kazakhstan, contract # 346/018-2018/33-28, IRN AP05133700.

References

Arora, S., Li, Y., Liang, Y., Ma, T., Risteski, A.: A latent variable model approach to PMI-based word embeddings. Trans. Assoc. Comput. Linguist. **4**, 385–399 (2016)

Davis, T.A., Hu, Y.: The university of Florida sparse matrix collection. ACM Trans. Math. Softw. (TOMS) **38**(1), 1 (2011)

Eckart, C., Young, G.: The approximation of one matrix by another of lower rank. Psychometrika **1**(3), 211–218 (1936)

Finkelstein, L., et al.: Placing search in context: the concept revisited. ACM Trans. Inf. Syst. **20**(1), 116–131 (2002)

Jones, E., Oliphant, T., Peterson, P.: {SciPy}: open source scientific tools for {Python} (2014)

Kishore Kumar, N., Schneider, J.: Literature survey on low rank approximation of matrices. Linear Multilinear Algebra **65**(11), 2212–2244 (2017)

Levy, O., Goldberg, Y.: Neural word embedding as implicit matrix factorization. In: Advances in Neural Information Processing Systems, pp. 2177–2185 (2014)

Levy, O., Goldberg, Y., Dagan, I.: Improving distributional similarity with lessons learned from word embeddings. TACL **3**, 211–225 (2015)

Mikolov, T., Sutskever, I., Chen, K., Corrado, G.S., Dean, J.: Distributed representations of words and phrases and their compositionality. In: Advances in Neural Information Processing Systems, pp. 3111–3119 (2013)

Pedregosa, F., et al.: Scikit-learn: machine learning in python. J. Mach. Learn. Res. **12**(Oct), 2825–2830 (2011)

Text Preprocessing for Shrinkage Regression and Topic Modeling to Analyse EU Public Consultation Data

Nada Mimouni[✉] and Timothy Yu-Cheong Yeung

Governance Analytics, University Paris-Dauphine, PSL Research University,
Place du Maréchal de Lattre de Tassigny, Paris 75016, France
{nada.mimouni,yu-cheong.yeung}@dauphine.fr

Abstract. Most text categorization methods use a common representation based on the bag of words model. Use this representation for learning involve a preprocessing step including many tasks such as stopwords removal and stemming. The output of this step has a direct influence on the quality of the learning task. This work aims at comparing different methods of preprocessing of textual inputs for LASSO logistic regression and LDA topic modeling in terms of mean squared error (MSE). Logistic regression and topic modeling are used to predict a binary position, or stance, with the textual data extracted from two public consultations of the European Commission. Texts are preprocessed and then input into LASSO and topic modeling to explain or cluster the documents' positions. For LASSO, stemming with POS-tag is on average a better method than lemmatization and stemming without POS-tag. Besides, tf-idf on average performs better than counts of distinct terms, and deleting terms that appear only once reduces the prediction errors. For LDA topic modeling, stemming gives a slightly lower MSE in most cases but no significant difference between stemming and lemmatization was found.

Keywords: Text representation · Feature selection · Topic modeling

1 Introduction

Textual analysis has recently become popular in social sciences, thanks to the advances of computer sciences. However, the quality of the analysis depends on the preparation of the inputs. "Garbage in, Garbage out" often happens if the practitioners are unaware of the importance of the preprocessing of the texts. This work attempts to compare different preprocessing methods in terms of the prediction performance of LASSO (least absolute shrinkage and selection operator) logistic regression and interpretability of topics and clusters for Latent Dirichlet Allocation (LDA) topic modeling. The variable to predict is a binary position, or a stance on an issue. For LASSO, we contrast methods in three dimensions combining lemmatization, stemming, term counts and frequencies and the size of the corpus (whether to consider or not most frequent and infrequent terms). For topic modeling, we only compare lemmatization and stemming while deleting terms appeared fewer than or equal to five times.

© Springer Nature Switzerland AG 2023
A. Gelbukh (Ed.): CICLing 2019, LNCS 13451, pp. 95–109, 2023.
https://doi.org/10.1007/978-3-031-24337-0_8

LASSO regression identifies the most relevant terms, which give the highest prediction power, associated with some (mainly binary) predefined sources, types or positions. It produces a list of terms that allow us to predict the position of a document on which we are unable or too difficult to classify by human efforts. Similarly, topic modelling produces lists of terms that are more likely be associated with some topics, which are recognized ex-post. It helps researchers identify firstly topics and secondly important terms for each topic. It also produces clusters of documents, defined by how likely documents are talking about the same topic or co-occurrence of terms. In principle, they serve similar purposes as they classify documents. Nevertheless, no matter what is the purpose, we are concerned with the quality of inputs, and thus the preprocessing step is crucial to the quality of the products of the two methods.

Besides, comparing the two methods is interesting itself. Suppose that different types of documents are talking about different topics, or different positions are associated with different topics, or different authors have different habits of writing, we should find consistent results from topic modelling. However, we have not seen any research work trying to compare them side-by-side. At the very least, a consistent result would be evidence that the same type of documents use the language similarly. An inconsistent result would suggest that co-occurrence of terms is not sufficient for identifying the type or position of a document. The main difference of the two methods can be summarized as follows: co-occurrence of terms define topics but distinctive terms identify positions. For example, "company" and "price" are more likely to appear together in a document talking about "private sector", while "government" and "politics" are more likely to co-occur in a document talking about "public sector". Both types of documents may talk about "good governance" in which "accountability" is a distinctive term, while both types may discuss "bad governance" in which "corruption" is a distinctive term. In most cases, the two dimensions do not align exactly.

For LASSO, we have the observed choices as the reference of comparing the quality of different preprocessing methods by measuring the distance between the reference and the prediction. For topic modelling, a natural choice of reference is absent. We proceed as the following. If topics and positions (the numbers of them are assumed to be equal) are perfectly aligned, the observed choices are also the reference for measuring the quality of inputs for topic modelling. Even if topics and positions are not exactly aligned, by comparing preprocessing methods to a single hypothetical reference is still meaningful. For example, if the true correlation between topics and positions is only 0.8, we should expect the better inputs would give us a smaller errors. An illustration is given in Sect. 6.

The data are collected from two public consultations launched by the European Commission on the roaming wholesale market and audiovisual market. Public consultations are designed for the stakeholders and the public to participate in the decision-making process. Their positions and any supporting arguments could be taken by the Commission and written into the directive issued to all member states. It is arguably an important channel for stakeholders to exert influence at the European level and thus most of them take it seriously. Besides, public consultations well-suit our purpose because they are systematically designed, vary in topics, and contain useful information for research in other disciplines. We consider each response to a question in the consultation as a document. We select six questions which contain a check-box and an open text-box. The information input in the check-box helps us construct the binary position of the document on a certain matter. The open text-box allows us to extract additional textual information that supports the stance.

The constructed binary position is the reference to which we compare the predictions coming from LASSO and topic modeling. It is also taken as the dependent variable of the LASSO regression, while not needed for topic modeling. To the contrary, topic modeling clusters documents based not on predetermined topics but on the probability of co-occurrence of terms. However, to assess the quality of the inputs, we use the same binary position of each question as the reference.

Our work sheds light on the search of the best method of text preprocessing in text classification for LASSO and LDA topic modeling. For social sciences research purposes, our work helps researchers fill in missing values as some responses do not give a clear answer. Second, LASSO helps us construct a bag of related terms for each position along a dimension of a topic or policy, by which we can classify any related documents along the same dimension. LDA helps discovering patterns within a set of documents without any prior idea about their content, by topics generation and document classification.

2 Related Work

Feature selection methods have been widely researched in different contexts [1,2]. For text mining, using feature selection targets applying machine learning and statistical methods on text. Document clustering and classification are often the focus [3,4]. To improve the classification accuracy, the choice of appropriate preprocessing tasks is decisive [5]. Text preprocessing is the first step of any text mining process [6]. It is mainly based on the use of natural language processing (NLP) techniques (e.g. syntactic parsing, part of speech tagging (POS)). Common preprocessing tasks, such as tokenization, filtering, stemming and lemmatization, prepare the input to generate a vector space representation. In most cases, transforming inputs from text to data involves a choice between lemmatization and stemming [7]. The development of support vector machines calls for the avoidance of lemmatization and stemming [8,9]. In [10], authors show that removing stop words is generally accepted as helpful but both lemmatization and stemming can worsen the performance. We assume that tuning preprocessing steps is a key factor for optimal efficiency regarding the data and the targeted machine learning method.

Logistic regression has been used as a classifier for binary text categorization in several research works [11,12]. LASSO logistic regression is the recommended regression method for text classification [13]. It stands out among other text classification methods especially in social sciences because of its simple implementation and interpretation.

We employ LDA topic modeling as a text clustering algorithm to create groups of homogeneous texts according to the topic they are dealing with, Which reflect the positions of their authors. A topic model is a probabilistic generative model that is widely used in the field of text mining for unsupervised topic discovery. It allows to have insight of what a corpus is talking about by transforming the word space of documents into "topic" space, much more smaller and easily interpretable by humans. We focus here on Latent Dirichlet Allocation (LDA) [14] as it is the most popular model used in social sciences [15]. This model considers each document as a mixture of various topics, and that each word in the document belongs to one of the document's topics. The results from the two methods will be compared, both on the basis of the preprocessing pipelines and the final categorization output.

A quite related approach to our work is described in [16] where shrinkage regression and LDA topic modeling are combined to analyze public opinion (supporting or opposing a specific candidate) about an election process. Another close approach presented in [17] explores the possibility of analyzing open-ended survey responses by topic modeling.

3   The Data

The policy-making of the European Union is chosen as the target domain. In a regular manner, the European Commission involves stakeholders in the legislation process via open public consultations. A consultation is presented as a list of questions each attached with a list of pre-defined answers (check-boxes) and a free text area where users can explain their choices. All consultations from the Commission are open and are gathered in one list accessible in their web site[1].

In this study, we choose the *Public consultation on the review of national wholesale roaming markets, fair use policy and the sustainability mechanism referred to in the Roaming Regulation 531/2012 as amended by Regulation 2015/2120* and *Consultation on Directive 2010/13/EU on audiovisual media services (AVMSD): A media framework for the 21st century*. In total, we collected 43 and 187 responses respectively from the two consultations, who are mainly companies and trade unions. We manually studied the proposed directives that were drafted after the consultation in order to extract the final decisions with respect to the questions of the consultation. We finally pick 6 questions from the consultation for which we can identify the exact final position of the Commission. Next, we extracted the answers from the check-boxes manually to identify the positions or stances of the stakeholders on these 6 questions, and from them we construct a binary variable that corresponds to whether the stakeholder's position is aligned with the Commission's (value 1) or not (value 0). Information given in the open-boxes is then extracted by automated techniques.

4   Methods to Prepare the Inputs

The collected data consists of 43 and 187 responses to the roaming and audiovisual consultations respectively. We manually selected 6 questions and thus in total the dataset contains 548 documents. For technical aspects of this study we used the python libraries

---

[1] Consultations of the European Commission: https://ec.europa.eu/info/consultations_en.

spaCy[2] and nltk[3] for NLP, gensim[4] to generate bi-grams and topic models and sklearn[5] for features extraction (to create the vector space model) (Table 1).

Table 1. Input dataset example - the question Q20

Question text	Do you consider that the functioning of the national wholesale roaming markets absent regulation would be capable of delivering RLAH at the retail level in accordance with the domestic charging model?
Check-box options	"Yes"; "No"; "It depends on the Member State"; "Don't know"
Free text-box	Please explain your response and provide examples

4.1 Cleaning, Filtering and Tokenization

Cleaning consists of removing all special characters from the original text (delimiters, etc.). For filtering, we used a pre-compiled stop-words list for English defined by the spaCy NLP environment. The first ten elements of the list out of 305 are: 'five', 'further', 'regarding', 'whereas', 'and', 'anywhere', 'below', 'him', 'due' and 'already'. We also defined a new list of domain-specific terms that could be removed (e.g. 'answer', 'question', 'response', 'explain'). This is a good practice to avoid those frequent terms but not very important for defining the semantics of a given domain. The stream of text is then broken into pieces to generate a list of tokens. We generated bi-grams (sequences of two words that appear more then a fixed threshold) to better capture the semantics of text classes.

4.2 Lemmatization and Stemming

Both lemmatization and stemming aim at reducing inflectional forms of a word to a common base form [18]. Stemming is used to reduce words to their root by deleting prefixes and suffixes. In this study, we used the Porter's stemming algorithm [19]. Lemmatization uses vocabulary and morphological analysis of words to return the base or the dictionary form of a word. We used the WordNet lemmatizer [20].

Lemmatization and stemming prepare the inputs differently. For example, A and B wrote "regulation" and "regulatory" respectively. Stemming groups two words together, and therefore "regul" as an input appears two times. But lemmatization does not consider them as equal and each word is a distinct input that appears only once. When predicting a position that is associated with regulation in general, we may question whether lemmatization really values the concept sufficiently for LASSO and whether using lemma or stem makes a big change for LDA.

---

[2] https://spacy.io/.

[3] https://pypi.org/project/nltk/.

[4] https://pypi.org/project/gensim/.

[5] https://scikit-learn.org/.

4.3   Test Corpora Generation

To test the effect of using one or the other technique, we used combinations with terms frequency and POS-tags of words to generate different test corpora. First, we compare lemmatization and stemming (denoted in the following by "lemma" and "stem" respectively). Second, we prepare a sample that includes all terms ("min-freq = 1") and a sub-sample that excludes the terms that appear only once in the corpus ("min-freq = 2"). Then, we either use the counts of distinct terms as inputs or compute the tf-idf (term frequency - inverse document frequency) weight for each term[6]. In addition, based on part of speech tagging of words, we work with a trimmed sample that includes only nouns (denoted by "POS-tags"). Combining these different text preprocessing methods, we generated 9 test corpora for LASSO regression as follows : (1) lemma + counts + min-freq = 1, (2) lemma + tf-idf + min-freq = 1, (3) lemma + tf-idf + min-freq = 2, (4) stem + counts + min-freq = 1, (5) stem + tf-idf + min-freq = 1, (6) stem + tf-idf + min-freq = 2, (7) stem + postag + counts + min-freq = 1, (8) stem + postag + tf-idf + min-freq = 1, (9) stem + postag + tf-idf + min-freq = 2. For LDA, we compared the use of lemmatization and stemming on the entire sample and the trimmed sample. These methods are discussed and evaluated in the following.

5   LASSO Regression

5.1   Methodology

LASSO regression solves the problem of dimensionality that complicates textual analysis because inputs (tokens) are too numerous compared to a standard size of sample. It penalizes those variables with large coefficients by adding a penalty term in the process of maximizing the log-likelihood. With a proper weight (often mentioned as λ or λ_1 in the literature) chosen, LASSO assigns zeros to some coefficients and helps reducing the dimensions significantly. Precisely, we write the penalized linear model as follows:

$$min\left\{l(\alpha, \beta) + n\lambda \sum_{j=1}^{p} \omega_j |\beta_j|\right\} \tag{1}$$

where $l(\alpha, \beta)$ is the negative log likelihood function in the binomial logistic regression, n is the total number of documents, and λ is the weight given to each penalty term, which is $\omega_j |\beta_j|$ for term $j = 1, 2..., p$. We standardized each coefficient by the standard deviation of that covariate. In effect, the standardization will give a larger weights to rare terms and is recommended in the literature [21].

Note that we do not divide the data into train-set and test-set because we do not have a sufficiently big dataset. But we are less concerned about this because our aim is to develop a method to fill missing values and to establish a dictionary that helps us map different stakeholders concerning the same issues.

---

[6] Tf-idf is a statistical measure that evaluates how a term is important to a document in a collection, computed as: $tf\text{-}idf_{t,d} = tf_{t,d} \times idf_t$ where $idf_t = log(\frac{n_{documents}}{df_t})$ and $df_t =$ number of documents containing t.

Apart from basic stemming, we employ POS-tag to keep only nouns in the sample before stemming. To further reduce the size of inputs, we consider also a subsample that excludes those terms that appear only once. Apart from reducing the dimensionality, we consider both counts of distinct terms and tf-idf values. In effect, deleting those words that appear only once does not make a difference when working with counts because LASSO will consider their unimportance and assign zeros to them. Therefore, we compare in total 9 different methods (see Table 2).

As mentioned above, each selected question has a check-box for the contributors to precisely state their position and is followed by an open-box for any additional comments on the issue. We can thus identify where the stakeholders actually stand, either aligned with the Commission or not, which is the dependent variable of our LASSO regression, while we capture the textual inputs in the open-box and use them as the explanatory variables in the regression. As our objective is to predict a contributor's position, the best set of inputs should give the smallest errors. We thus compute the mean squared errors of the 9 types of inputs for the 6 questions. The mean squared errors are computed as follows:

$$MSE = \frac{1}{n} \sum_{i=1}^{n} (y_i - \hat{y}_i)^2 \qquad (2)$$

The actual value y_i is either 1 (aligned with the Commission ex-post) and 0 (not aligned). The predicted value $\hat{y}_i$ is the prediction of the LASSO regression, which ranges from 0 to 1. A smaller MSE value says that the prediction is more accurate. Therefore, we can compare the values and may be able to conclude if one feature selection method is superior than others.

Besides, LASSO is usually done with frequencies as the inputs. The tf-idf is an alternative, which increases a term's weight proportionally to the number of times it appears in the document, but offset by the frequency of it in the corpus. Values after the transformation are smaller than 1, and LASSO is unable to work with too small values. We therefore adjust the values by multiplying them by 10 before inputing in the LASSO regression. Since we standardized the coefficients of LASSO and transformed the inputs into tf-idf, rare terms are weighted up twice. Our results will also tell if it overweights rare terms and worsens the prediction.

The choice of λ is subject to debate. Some recommend to select the value that minimizes the average error, while some others propose to use a high-dimension-adjusted Akaike Information Criteria [22]. Due to the relatively low dimensions of our corpus, we cannot afford losing too many information. Therefore, we simply picked the smallest value of λ that gives a number of estimates which is smaller than the number of unique terms. Consequently, λ ranges from 1 to 7. The number of non-zero coefficients ranges from 5 to 29. Table 2 reports the λ chosen and the number of non-zero coefficients of the LASSO regression.

5.2   Result Comparison

Table 3 reports the comparison of results of the 9 methods. We bold the lowest MSE value for each question. In short, we do not find an absolutely superior method of

Table 2. Summary of λ values and the corresponding number of non-zero coefficients

	Roaming				AVMSD	
	Q20	Q25	Q26	Q27	Q2.1	Q3.2
Method 1: Lemma+Counts min freq = 1						
λ	2	3	2	2	3	4
Non-zero	9	6	7	5	22	7
Method 2: Lemma+tfidf min freq = 1						
λ	2	2	2	3	3	7
Non-zero	13	9	8	8	26	4
Method 3: Lemma+tfidf min freq = 2						
λ	3	3	3	3	3	5
Non-zero	10	12	13	10	25	6
Method 4: Stem+Counts min freq = 1						
λ	2	2	2	2	4	2
Non-zero	10	9	8	6	19	18
Method 5: Stem+tfidf min freq = 1						
λ	2	3	2	3	3	5
Non-zero	6	9	11	8	20	5
Method 6: Stem+tfidf min freq = 2						
λ	2	3	2	3	3	5
Non-zero	7	10	11	11	22	6
Method 7: Stem+POStag+Counts min freq = 1						
λ	2	2	1	2	3	1
Non-zero	10	8	11	5	22	29
Method 8: Stem+POStag+tfidf min freq = 1						
λ	3	3	2	4	3	2
Non-zero	10	12	11	5	23	18
Method 9: Stem+POStag+tfidf min freq = 2						
λ	3	2	3	3	3	2
Non-zero	19	13	16	13	25	18
Sample	34	41	39	35	175	98

Table 3. Result comparison

MSE	Roaming				Audiovisual	
	Q20	Q25	Q26	Q27	Q2.1	Q3.2
Lemmatization						
Method 1	0.16900	0.16279	0.12581	0.19358	0.0988	0.1145
Method 2	**0.16342**	0.26995	0.10801	0.18665	0.1007	0.1131
Method 3	0.19548	0.12005	0.09230	0.16459	0.0952	0.0982
Stemming						
Method 4	0.16394	0.26207	0.11248	0.20035	0.1156	0.0927
Method 5	0.20257	0.13818	0.09744	0.19463	0.0957	0.1027
Method 6	0.19925	0.12412	0.08930	0.17784	0.0933	0.1019
Stemming+POStag						
Method 7	0.18635	0.11629	**0.08315**	0.19324	0.1077	0.0712
Method 8	0.19893	0.12005	0.09060	0.19012	0.0858	**0.0674**
Method 9	0.17489	**0.08115**	0.15997	**0.14560**	**0.0833**	0.0676

preparing the inputs for LASSO. It shows that the research question indeed does not have an obvious answer and needs further investigation. Tf-idf (min frequency = 1) beats counts 11 out of 18 times and tf-idf (min frequency = 2) beats counts 14 out of 18 times. Tf-idf (min frequency = 2) performs better than tf-idf (min frequency = 1) 15 out of 18 times.

If we only consider counts, stemming with POS-tag beats others 4 out of 6 times. If we only consider tf-idf (min frequency = 2), stemming with POS-tag performs better than others 4 out of 6 times. On average, stemming plus POS-tag with tf-idf (min frequency = 2) gives the lowest average MSE. It is also the method that gives the highest number of non-zero coefficients. It suggests that LASSO works better if the inputs have been significantly trimmed.

Stemming seems to be a superior choice. Lemmatization may allow too many forms of the same concept to appear separately, that in effect diminishes their importance relative to some terms which do not have many different forms. For example, the root "regul" is in certain context more important than "market", but the former could be expressed in "regulation", "regulatory", etc. Tf-idf performs on average better than counts because of its advantage to value more the rare terms. Besides, we find no evidence that tf-idf transformation should not be employed together with standardizing coefficients in LASSO regression. Weighting up rare terms twice performs better compared to the baseline result using counts. POS-tag that keeps only nouns tends to work well in our context.

6   LDA Topic Modeling

6.1   Process

Documents of the collection are transformed into a bag of words (BOW) described as a word-document matrix which values w_{ij} represent the frequency of word i in document j. Then the model is trained using the vocabulary matrix as input. The LDA model outputs two matrices, one presents topic probability distribution over words (per-topic word distribution), the other presents document probability distribution over the topics (per-document topic distribution).

In order to get accurate topics estimate that guarantees topics interpretability, training parameters should be carefully set. In our case, we considered 20 passes through the entire corpus after each of which the model is updated and used small chunks (sub-corpora) for updates, so that the model estimation converges faster. The model is trained until the topics converge or the maximum number of iterations (we fixed at 400) is reached. The number of topics to be generated by the model is another parameter that should be fixed. Before we start the experiments with LDA, we manually analyzed our documents to get their positions. As described above, positions are either aligned, 1, or not, 0, with the position of the commission and we used LASSO to predict this binary position. We had than the intuition that each group will use different arguments to explain its position and thus, for topic modeling, we should expect to have two topics emerging from the data. To confirm this intuition, we calculated the coherence score, a measure used to evaluate the quality of topics according to their clearness and interpretability by a human [23]. In the state of the art, two types of measures are usually used: the intrinsic measure (we use in this study), which uses the corpus trained to generate the topics, and the extrinsic measure, which uses external corpus to calculate these probabilities. Both measures compute the coherence as the sum of scores on top frequent terms describing a topic $(t_1, \ldots, t_n)$ by pairs $(score(w_i, w_j) =$ number of times they co-occur). We implemented the c_v measure, we chose among others defined in the literature, as it was proved to be the best performing measure (achieving the highest average correlation to human ratings) [24]. It uses a pairwise score function to calculate the probability of a term w_i to appear in a random document and of terms w_i and w_j to co-occur in a given document.

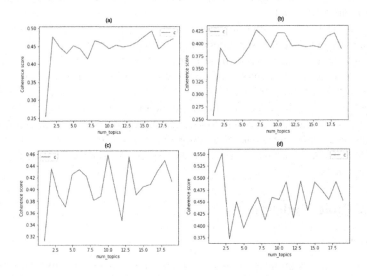

Fig. 1. Coherence graphs of questions Q20 (a) and Q25 (b) for lemma method, and Q26 (c) and Q27 (d) for stem method.

Figure 1 shows the coherence values by the number of topics for the roaming input dataset (questions 20, 25, 26 and 27) that we preprocessed with lemmatization and stemming. We consider the first peak in the graph as corresponding to the optimal number of topics that gives the most human-interpretable ones. In graphs (a), (b), (c) and (d) this peak is found for a number of topics which equals 2, which corresponds to the intuition we had.

Figure 2 shows the distribution of topics over the vocabulary of the dataset. We use a web-based interactive topic model visualization to help us interpret the topics in the fitted model [25]. The left part of the figure displays topics as circles (two topics are generated), the right part shows the top 30 most relevant terms for topic one. The blue and red bars give the overall term frequency and the estimated term frequency within the selected topic respectively (topic one is selected in the figure). The second type of output is the document-topic distribution. This distribution gives for each document the probability of each topic represented in the document. Classes of documents can be calculated by assigning a document to the highest probability topic. Table 4 gives the

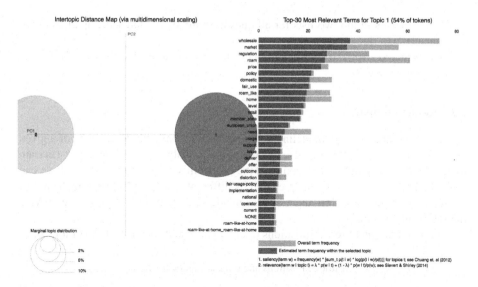

Fig. 2. Topic-term distribution for topic 1 of question Q20, preprocessed with lemma method.

distribution of topics for seven randomly selected companies' answers (which correspond to our initial documents) prepared with lemma and stem methods. Companies are assigned to class 1 (aligned with the commission) or class 2 (not aligned) according to how much their answers are representative of topic 1 or topic 2 as given by the topic-distribution values.

Table 4. Document-topic distribution with lemma (left) and stem (right)

Company name	Topic	Topic distribution	Company name	Topic	Topic distribution
Melitaplc	1	0.941384375095	Melitaplc	1	0.938471198082
MVNOEurope	1	0.769798099995	MVNOEurope	1	0.692744493484
UnitedInternetAG	1	0.938454866409	UnitedInternetAG	1	0.979364931583
CETIN	2	0.747453808784	CETIN	1	0.748222112656
eircom	2	0.747464954853	eircom	1	0.748222589493
QSCAG	1	0.751199483871	QSCAG	1	0.737836956977
SciamuSKft	2	0.765129625797	SciamuSKft	2	0.798504829406

The terms displayed in Fig. 2 were preprocessed with the lemma method described above. At this point of experiments we didn't know whether using lemmatization or stemming to preprocess the input text gives better result with LDA topic model. We run experiments with both methods and defined two criteria upon to which we can evaluate the output:

– Human interpretability for the per-word topic distribution. We compared the lists of stemmed and lemmatized top relevant terms and we didn't find any significant

change. Terms in the two lists for all questions are almost the same apart from the truncated part due to stemming.
- Mean squared error and accuracy for the per-document topic distribution. The results are given in Sect. 6.2.

6.2 Result Comparison

As mentioned above, topic modeling is mainly used to extract hidden topics of a given text collection. As LDA model also calculates the probability of belonging to a certain topic of the document, we can compare the probability to a reference so that we can evaluate the quality of the inputs prepared by different methods. To motivate the use of the observed binary position taken by stakeholders as the reference point, we provide the following example.

Suppose that the observed positions of eight documents are as follows (Table 5):

Table 5. Positions and topics of 8 documents

Position or topic	P1	P2	T1	T2
Document	a	e	a	e
	b	f	b	f
	c	g	c	g
	d	h	h	d

We also assume that the probability of belonging to a certain topic is either 0 or 1. The correlation between topics and positions is thus 0.75. Suppose that we take the observed positions as the reference to evaluate the quality of the textual inputs. We compare the following four outcomes (Table 6):

Table 6. Comparison of 4 methods of preprocessing

Method	(1)		(2)		(3)		(4)	
Topic	T1	T2	T1	T2	T1	T2	T1	T2
Document	a	e	a	e	a	e	a	b
	b	f	b	f	b	f	h	c
	c	g	c	g	g	c		d
	d	h	h	d		d		e
						h		f
								g
If true topics are known								
MSE	0.25		0		0.375		0.75	
If positions are used as reference								
MSE	0		0.25		0.375		0.50	

Method 1 produces clusters of documents exactly the same as the clusters of positions. If true topics are known, the MSE is 0.25 but it is zero if the positions are taken as the reference. Method 2 produces clusters of documents exactly the same the true clusters of topics. The MSE is 0 if true topics are known but is 0.25 if positions are used as reference. Method 2 is in fact the best preprocessing method but we wrongly rank it below Method 1. However, as long as the MSEs obtained using positions as the reference are lower than the 0.25, i.e. 1 - true correlation, the ranking of methods is correct. For example, Method 3 gives 3 wrong predictions while Method 4 gives 4 wrong predictions when positions are the reference, and we conclude correctly that Method 3 is better than Method 4. In short, as long as MSEs are smaller than 1 minus the true correlation, the higher the MSE the poorer the performance of the inputs. Therefore, using observed positions as the reference is acceptable as long as the alignment between positions and topics is close, and the closer the better.

We report the comparison result in Table 7. MSE refers to the mean squared errors defined in Eq. (2) where the reference point is the manually coded position. We also report the accuracy rate that is the proportion of correct classifications of documents. We do not find a significant difference between lemmatization and stemming in terms of performance, but stemming gives a lower MSE in four out of six cases.

Table 7. Result comparison

	Roaming				Audiovisual	
	Q20	Q25	Q26	Q27	Q2.1	Q3.2
Lemmatization						
MSE	0.3007	0.3264	0.2214	0.2438	0.3245	0.2718
Accuracy	0.6060	0.5366	0.5641	0.5758	0.6171	0.7041
Stemming						
MSE	0.3021	0.3110	0.1441	0.3056	0.3169	0.2621
Accuracy	0.5758	0.5122	0.8462	0.5278	0.6264	0.7041

7   Discussion

To predict a binary position, LASSO shrinkage regression and topic modeling employ different approaches. LASSO regression relies on predetermined positions or topics of a subset of documents, either by manual classification or through some additional information, that help train the regression to classify terms. Topic modeling instead does not require any predetermined topics but extracts hidden patterns by identifying co-occurring terms. The advantage of the latter is that it does not require manual classification or additional information for training propose. However, using topic modeling in predicting positions of a certain matter does not give satisfactory result, at least in the context we explore in this paper. To be fair, LASSO regression benefits from the additional information given in the check-box, and thus its superiority is well expected. The poor performance of topic modeling may be due to the fact that both sides are using very similar terms in presenting their arguments, i.e. the correlation between topics and positions is low. Even though both LASSO and topic modeling can be used for text classification, LASSO is a more suitable tool if topics or positions are well defined and predetermined.

8   Conclusion

This work aims at comparing different ways of preprocessing of texts for predicting a binary position or stance by LASSO logistic regression and LDA topic modeling. We collect textual data from a public consultation launched by the European Commission and manually identify alignments and misalignments between stakeholders and the Commission, which are explained by the textual inputs given in the open-box. For LASSO, stemming with POS-tag is on average a better method than lemmatization and basic stemming, tf-idf on average performs better than counts, and deleting terms that appear only once tends to reduce the prediction errors. In short, stemming plus POS-tag with tf-idf and minimum term frequency equal to 2 generates the lowest average mean squared errors. For topic modeling, stemming and lemmatization are not significantly different.

References

1. Jović, A., Brkić, K., Bogunović, N.: A review of feature selection methods with applications. In: 2015 38th International Convention on Information and Communication Technology, Electronics and Microelectronics (MIPRO), pp. 1200–1205. IEEE (2015)
2. Gao, W., Hu, L., Zhang, P., Wang, F.: Feature selection by integrating two groups of feature evaluation criteria. Expert Syst. Appl. **110**, 11–19 (2018)
3. Labani, M., Moradi, P., Ahmadizar, F., Jalili, M.: A novel multivariate filter method for feature selection in text classification problems. Eng. Appl. Artif. Intell. **70**, 25–37 (2018)
4. Uysal, A.K., Gunal, S.: A novel probabilistic feature selection method for text classification. Knowl.-Based Syst. **36**, 226–235 (2012)
5. Uysal, A.K., Gunal, S.: The impact of preprocessing on text classification. Inf. Process. Manag. **50**, 104–112 (2014)
6. Vijayarani, S., Ilamathi, M.J., Nithya, M.: Preprocessing techniques for text mining-an overview. Int. J. Comput. Sci. Commun. Netw. **5**, 7–16 (2015)
7. Korenius, T., Laurikkala, J., Järvelin, K., Juhola, M.: Stemming and lemmatization in the clustering of finnish text documents. In: Proceedings of the Thirteenth ACM International Conference on Information and Knowledge Management, pp. 625–633. ACM (2004)
8. Leopold, E., Kindermann, J.: Text categorization with support vector machines. How to represent texts in input space? Mach. Learn. **46**, 423–444 (2002)
9. Méndez, J.R., Iglesias, E.L., Fdez-Riverola, F., Díaz, F., Corchado, J.M.: Tokenising, stemming and stopword removal on anti-spam filtering domain. In: Marín, R., Onaindía, E., Bugarín, A., Santos, J. (eds.) CAEPIA 2005. LNCS (LNAI), vol. 4177, pp. 449–458. Springer, Heidelberg (2006). https://doi.org/10.1007/11881216_47
10. Toman, M., Tesar, R., Jezek, K.: Influence of word normalization on text classification. Proc. InSciT **4**, 354–358 (2006)
11. Genkin, A., Lewis, D.D., Madigan, D.: Large-scale Bayesian logistic regression for text categorization. Technometrics **49**, 291–304 (2007)

12. Onan, A., Korukoğlu, S., Bulut, H.: Ensemble of keyword extraction methods and classifiers in text classification. Expert Syst. Appl. **57**, 232–247 (2016)
13. Gentzkow, M., Kelly, B.T., Taddy, M.: Text as data. Technical report, National Bureau of Economic Research (2017)
14. Blei, D.M., Ng, A.Y., Jordan, M.I.: Latent dirichlet allocation. J. Mach. Learn. Res. **3**, 993–1022 (2003)
15. Sukhija, N., Tatineni, M., Brown, N., Moer, M.V., Rodriguez, P., Callicott, S.: Topic modeling and visualization for big data in social sciences. In: 2016 Intl IEEE Conferences on Ubiquitous Intelligence Computing, Advanced and Trusted Computing, Scalable Computing and Communications, Cloud and Big Data Computing, Internet of People, and Smart World Congress (UIC/ATC/ScalCom/CBDCom/IoP/SmartWorld), pp. 1198–1205 (2016)
16. Yoon, H.G., Kim, H., Kim, C.O., Song, M.: Opinion polarity detection in twitter data combining shrinkage regression and topic modeling. J. Informet. **10**, 634–644 (2016)
17. Roberts, M.E., Stewart, B.M., Tingley, D., Lucas, C., Leder-Luis, J., Gadarian, S.K., Albertson, B., Rand, D.G.: Structural topic models for open-ended survey responses. Am. J. Polit. Sci. **58**, 1064–1082 (2014)
18. Manning, C.D., Raghavan, P., Schutze, H.: Stemming and lemmatization. In: Introduction to Information Retrieval. Cambridge University Press, Cambridge (2008)
19. Porter, M.F.: An algorithm for suffix stripping. Program **14**, 130–137 (1980)
20. Fellbaum, C.: Wordnet. Wiley Online Library (1998)
21. Manning, C.D., Raghavan, P., Schütze, H.: Introduction to Information Retrieval. Cambridge University Press, Cambridge (2008)
22. Flynn, C.J., Hurvich, C.M., Simonoff, J.S.: Efficiency for regularization parameter selection in penalized likelihood estimation of misspecified models. J. Am. Stat. Assoc. **108**, 1031–1043 (2013)
23. Newman, D., Lau, J.H., Grieser, K., Baldwin, T.: Automatic evaluation of topic coherence. In: Human Language Technologies: The 2010 Annual Conference of the North American Chapter of the Association for Computational Linguistics, HLT 2010, Stroudsburg, PA, USA, pp. 100–108. Association for Computational Linguistics (2010)
24. Röder, M., Both, A., Hinneburg, A.: Exploring the space of topic coherence measures. In: Proceedings of the Eighth ACM International Conference on Web Search and Data Mining, WSDM 2015, pp. 399–408. ACM, New York (2015)
25. Sievert, C., Shirley, K.E.: LDAvis: a method for visualizing and interpreting topics. In: Proceedings of the Workshop on Interactive Language Learning, Visualization, and Interfaces (2014)

Intelligibility of Highly Predictable Polish Target Words in Sentences Presented to Czech Readers

Klára Jágrová and Tania Avgustinova[✉]

CRC 1102: Information Density and Linguistic Encoding, Department of Language Science and Technology, Saarland University, Saarbrücken, Germany
{kjagrova,avgustinova}@coli.uni-saarland.de

Abstract. This contribution analyses the role of sentential context in reading intercomprehension both from an information-theoretic and an error-analytical perspective. The assumption is that not only cross-lingual similarity can influence the successful word disambiguation in an unknown but related foreign language, but also that predictability in context contributes to better intelligibility of the target items. Experimental data were gathered for 149 Polish sentences [1] with highly predictable target words in sentence final position presented to Czech readers in a web-based cloze translation task. Psycholinguistic research showed that predictably of words in context correlates with cognitive effort to process the information provided by the word and its surprisal [3]. Our hypothesis is that intelligibility of highly predictable words in sentential context of a related language also correlates with surprisal values obtained from statistical trigram language models. In order to establish a baseline, the individual words were also presented to Czech readers in a context-free translation experiment [4]. For the majority of the target words, an increase in correct translations is observable in context, as opposed to the results obtained without context. The overall correlations with surprisal are low, the highest being the joint surprisal of the Polish stimulus sentence. The error-analysis shows systematic patterns that are at least equally important intercomprehension factors, such as linguistic distance or morphological mismatches.

Keywords: Slavic receptive multilingualism · Czech · Polish · Statistical language modeling · Context in intercomprehension · Reading · Surprisal · Linguistic distance

1 Introduction

In previous research in cross-lingual intelligibility of written text, the role of linguistic distance (lexical, orthographic, morphological, syntactic, phonetic) was investigated as a predictor for human performance [cf., for instance, 5, 6, 9, 10, 13, 15]. Thus, linguistic distance is supposed to reflect the (dis)similarity of two related codes on the different linguistic levels: the lower the linguistic distance, the more similar and mutually intelligible the two codes should be. Lexical distance is determined as the percentage of non-cognates in a language pair, while orthographic and morphological distances are usually measured as string similarity by means of the Levenshtein distance (LD) [18].

© Springer Nature Switzerland AG 2023
A. Gelbukh (Ed.): CICLing 2019, LNCS 13451, pp. 110–125, 2023.
https://doi.org/10.1007/978-3-031-24337-0_9

As for the linguistic distance and intelligibility of Polish (PL) sentence material for Czech readers, findings from the literature are summarized in Table 1. Heeringa et al. found that PL is an outlier in terms of orthography among the other Slavic languages spoken in the EU [9]. Jágrová et al. [15] found that in relation to the low lexical distance (10%) between PL and Czech (CS), their orthographic distance (34%) is extraordinarily high when compared to Bulgarian and Russian (RU) that have similar levels of both orthographic (13.5%) and lexical distance (10.5%).

Table 1. PL for Czech readers: comparison of distances and intelligibility (in %) in related research.

Distance	Heeringa et al. [9]	Golubović [5][a]	Jágrová et al. [15]	Jágrová et al. [14]
Lexical	23	17.7	10	12
Orthographic	31	31.7	34	38
Morphological	–	31.4	–	–
Intelligibility	64.29[b]	41.01	–	–

[a]Data for the written cloze test [5]
[b]Data for the written translation task of the most frequent Ns from the British National Corpus as published in [5, p. 77] on the material of [9]

The role of sentential context for the understanding of a particular language Lx, however, was subject to relatively few studies in this research field [14]. Muikku-Werner [19] qualitatively analysed the role of co-text in a study where Finnish students were asked to translate Estonian sentences. She found that the role of neighbourhood density – the number of available similar word forms – changes with words in context, as potential other options have to fit the restricted syntactic frame or be collocated [19, p. 105]. She states that "when recognizing one word, it is sometimes simple to guess the unfamiliar word frequently occurring with it, that is, its collocate. If there are very few alternatives for combination, this limitedness can facilitate an inference of the collocate" [ibid.].

In a study on the disambiguation of cross-Slavic false friends in divergent sentential contexts, Heinz [11] confronted students of different Slavic L2 backgrounds with spoken sentence samples in other Slavic Lx. He points out that the amount of perceived context is decisive for a successful comprehension of Lx stimuli. He also speaks of a negative role that context could play, namely if respondents attempt to formulate a reasonable utterance, they might revise their lexical decision [11], meaning that the target word might be misinterpreted due to misleading or misinterpreted context.

Another concept that is therefore likely to play a role in the intercomprehension of sentences is that of semantic priming [8]. Gulan and Valerjev [7] provide an overview of the types of priming that are identified in psycholinguistic literature (semantic, mediated, form-based, and repetition). The relevant type of priming for the present study appears to be semantic priming with both sub-types – associative and non-associative priming [7, p. 54]. During associative priming, a certain word causes associations of other words with the reader that might, but do not have to be related in meaning. Typical associations can be *engine–car* or *tree–wood*. A reader then might expect such a target word fitting a prime to occur in the sentence, for instance, at the position of an unfamiliar, unidentifiable word in the Lx. Cases of non-associative priming are words that are usually not mentioned together in such association tasks, but that are "clearly associated in meaning" [ibid.], for instance *to play – to have fun*. Semantic priming, of course, can only work if the prime is correctly recognized as such.

2   Hypothesis

Successful disambiguation of target words in a closely related foreign language relies on both cross-lingual similarity (measurable as linguistic distance) and predictability in sentential context (in terms of surprisal obtained from 3-gram LMs).

In a monolingual setup, it was shown that the more predictable a word is in context, the lower is the cognitive effort to process the information provided by the word – this corresponds to a low surprisal value [3]. On the contrary, words that are unpredictable in context and thus cause greater cognitive effort have higher surprisal values (see Sect. 3.2 for details). In the current multilingual setup, target words that have low linguistic distance to the reader's language and are predictable in context are expected to be translated correctly more often than words that are less similar and unpredictable. Since (dis-) similarity is measured by LD and predictability in context is captured by surprisal, the correct answers per target word should correlate with LD and surprisal better than only with LD.

Of course, the amount of correctly perceived sentential context plays a crucial role, too. If the context was not intelligible enough for the reader, then the supportive power of the context in terms of predictability might lose its effect. With a context that is helpful enough, it should be possible to recognize even non-cognates and maybe even false friends in sentences. The effects of semantic priming are not expected to be predictable by the trigram language models (LMs) applied here.

Consequently, the research questions can be formulated as follows:

1. Are PL target words more comprehensible for Czech readers when they are presented in a highly predictive sentential context?
2. If so, do surprisal values obtained from trigram LMs correlate with the intelligibility scores of the target words?

3   Method

3.1   Design of the Web-Based Cloze Translation Experiment

The cloze translation experiments were conducted over an experiment website [4]. After having completed the registration process including sociodemographic data, participants were introduced to the experimental task by a short tutorial video on the website. They were asked to confirm to have understood the task and to set their keyboard to CS. With each stimulus sentence, they would initially see only the first word of the sentence. They were prompted to click on the word in order to let the next word appear. They were asked to follow this procedure until the end of the sentence. Only after they have clicked on the last word in the sentence, the cloze gap with the target word for translation was displayed. This method ensures that participants read each sentence word by word. There are two separate time limits: one for clicking and reading through the sentence and one for entering the translation of the target word. The latter was automatically set to 20–30 s, depending on the length of the sentence. For each target word, data from at least 30 respondents were collected in both conditions (Fig. 1).

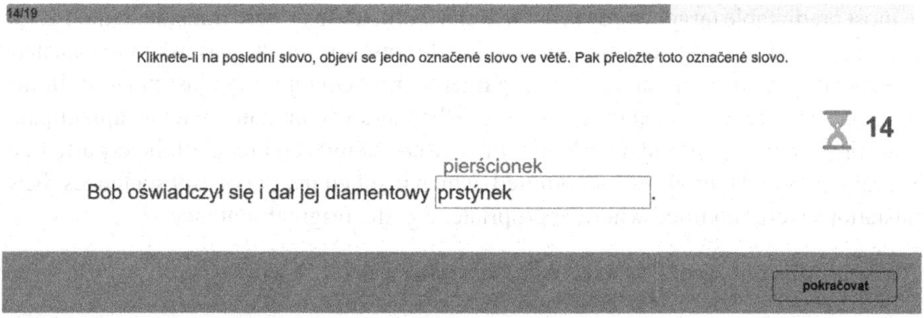

Fig. 1. Experimental screen in cloze translation experiments as seen by Czech respondents. The instruction on top says: 'When you click on the last word, a marked word will appear. Then translate this marked word'. The target word is displayed on top of the frame, the correct CS translation is inside the frame.

As a baseline, the target word forms from the sentences were also presented without context and in their base forms to other Czech respondents over the same experimental website – see Fig. 2. Target words with identical base forms in both Ls were not tested without context. The respondents were asked to translate each presented word with a time constraint of 10 s.

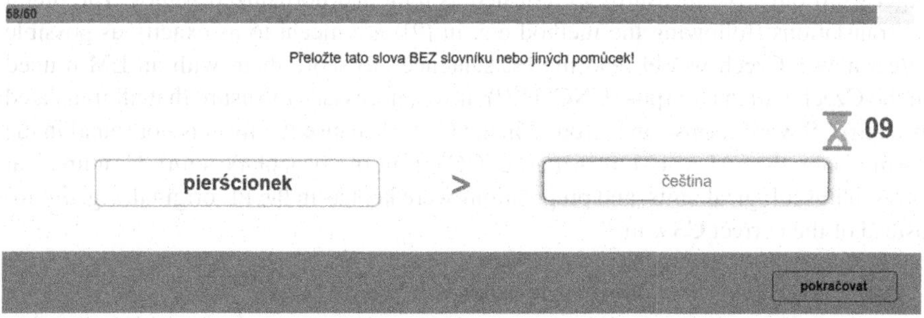

Fig. 2. Experimental screen in context-free experiments as seen by Czech respondents. The instruction on top says: 'Translate these words without a dictionary or other aids.' Respondents have 10 s time to enter their CS translation.

3.2 Stimuli

In order to use stimuli with predictive context systematically, sentences from a monolingual cloze probability study by Block and Baldwin [1] were adapted. They tested a set of 500 sentences in a cloze completion task where the completion gap was always placed on the last position in each sentence. In addition to the cloze experiments, they validated the sentences in psycholinguistic ERP experiments. Their study resulted in a dataset of 400 high-constraint, high cloze probability sentences. From these sentences, those with the most predictable target words (90–99% cloze probability) were translated into PL for the present study. A colleague and professional translator for PL was asked to translate the sentences in such a manner that the target words remain on the last position in the sentences. The translated sentences are published as a resource in the data supplement.

In the original (American) EN set, there were sentences that contained particular cultural topics and therefore were omitted, which resulted in a set of 149 sentences. Few translations were modified where appropriate, e.g. the original sentence

When Colin saw smoke he called 911 to report a fire. [1].

was modified into

Gdy Colin zobaczył dym, zadzwonił do straży pożarnej i zgłosił pożar.
'When Collin saw the smoke he called the fire department and reported a fire.' [1]

The respondents were not informed that the sentential context presented is a helpful, high-constraint context or that the target words should be highly predictable.

Literal Translation for Measuring Linguistic Distance and Surprisal. Linguistic distance and surprisal as predictors of intelligibility were measured for the literal CS translations and for the original PL stimuli. These two measures were applied (i) to the whole sentences, (ii) to the final trigram, (iii) to the final bigram, and (iv) to the target word only. All measures were tested as total and normalized values. The literal CS translations (following the method e.g. in [9]) are meant to as exactly as possible reflect how a Czech would read the PL sentence. To score them with an LM trained on the Czech national corpus (CNC, [17]), it was necessary to ensure that all translated (pseudo) CS word forms can be found in the CNC, because if a form is not found in the training data, the LM would treat it as an OOV (out of vocabulary item). Grammatical forms, phraseological units, and prepositions were kept as in the PL original, e.g. *do* 'to' instead of the correct CS *k* in

Poszła do fryzjera, żeby ufarbować włosy.
'She went to the salon to color her hair.' [1].

which was transformed into

*Zašla do kadeřníka, žeby obarvit vlasy.

for the calculation. Another example would be *genealogiczne drzewo* 'family tree' that was transformed into *genealogický strom* 'genealogical tree' instead of *rodokmen* 'family tree'. PL words existing in colloquial CS or in CS dialects and reflected in the CNC were also preserved in the literal translations, for instance the conjunction *bo* 'as, since' in

*Nie mogła kupić koszulki, **bo** nie pasowała.*
'She could not buy the shirt because it did not fit.' [1],

which would be *protože* 'because' in a written standard CS translation. PL negations and verb forms in the past tense or in the conditional mood required for their CS correspondences an explicit division of negation particles, verb forms, and auxiliaries. For instance, the negation particle *ne* was separated from CS verbs, and the PL example above was consequently transformed into

*Ne mohla koupit košilku, bo ne pasovala.

instead of keeping the correct CS negated verb forms *nemohla* '(she) could not' and *nepasovala* '(it) did not fit'. Other examples are verb forms that are reflexive in only one of the languages, for instance, *dołączyła do zespołu* 'she joined the band' is not reflexive in PL, while in CS equivalent *přidala se do kapely* is reflexive. The reflexive pronoun was therefore omitted in the literal CS translation: **přidala do kapely*.

Non-cognates and false friends were replaced by their correct CS translations. The literal translations and their surprisal values and distance measures can be found in the data supplement.

Surprisal is an information-theoretic measure of unpredictability [3]. Statistical LMs inform us about the probability that a certain word w_2 follows a certain other word w_1. For a given word, the surprisal is the negative log-likelihood of encountering this word in its preceding context [3, 14]. It is defined as:

$$Surprisal(unit|context) = -\log_2 P(unit|context) \qquad (1)$$

Thus, surprisal reflects frequency and predictability effects in the corpus on which the LM was trained. Figure 3 illustrates the principle of 3-gram LM counts.

```
Bob oświadczył się
    oświadczył się i
        się i dał
            i dał jej
                dał jej diamentowy
                    jej diamentowy pierścionek
                        diamentowy pierścionek .
```

Fig. 3. Example for 3-gram as they could occur in a PL corpus.

The PL stimuli sentences were scored by an LM trained on the PL part of InterCorp [2] and the CS literal translations were scored by an LM trained on the CNC [15]. Both were LMs with Kneser-Ney smoothing [16]. As the 3-gram LMs applied here cannot capture links between items further apart from each other than in a window of three words, the surprisal is expected to predict only such relations that are in direct successive position. Schematic implications such as

Farmer spędził ranek dojąc swoje krowy.

'The **farmer** spend the morning milking his **cows**.' [1]

or hyponymy such as in

Ellen lubi poezję, malarstwo i inne formy sztuki.

'Ellen enjoys **poetry**, **painting**, and other forms of **art**.' [1]

are not expected to be predictable with surprisal obtained from the 3-gram LMs.

Linguistic Distance. Orthographic distance was calculated as the Czechoslovak to PL pronunciation-based LD, i.e. always towards the closest CS or Slovak (SK) translation equivalent under the assumption that the Czech readers have receptive skills in SK (cf. method of Vanhove with Germanic distance [20, p. 139]). No costs were charged for the alignment of *w:v, ł:l, i:y, y:i, ż:ž* since their pronunciation is transparent to the readers [12]. If a target word is a non-cognate, its distance is automatically set to 1. Lexical distance is determined by the number of non-cognates per sentence in the language pair. A separate variable for the category false friends has been added, as false friends can be both cognates and non-cognates (see Sect. 4.2).

4 Results

4.1 Comparison: Target Words With vs. Without Context

The mean intelligibility of target words improved significantly from 49.7% without context to 68% in highly predictive contexts (t(298) = 4.39, p < .001). This means that the hypothesis that predictive sentential context contributes to a better intelligibility of highly predictable words in an unknown related language can be confirmed for the scenario PL read by Czech respondents. Figure 4 contains a trend line at $f(x) = 1x$ which divides the data points into those for which intelligibility improved in context (above the line) and those for which intelligibility decreased with the provided context (beneath the line). The points on the line are those for which no difference between the conditions with or without context could be discovered.

In the condition with context, a correctness rate of 100% could be observed for 26 target words, and 18 other target words were correctly translated by 96.7% of the respondents. In the condition without context, there were only 19 target words with a correctness rate of 100%, and 11 with $\geq$96.7%.

Cases of context-driven decisions are frequently observed in the responses, e.g.

Bob oświadczył się i dał jej diamentowy pierścionek.

'Bob proposed and gave her a diamond ring' [1].

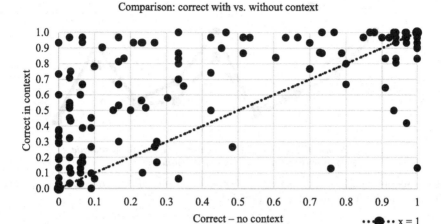

Fig. 4. Comparison of results for target words with vs. without context.

When presented in this sentence, 90% translated the PL target *pierścionek* 'ring' correctly, while in the condition without context only 45.5% gave the correct CS cognate *prstýnek*. The trigram LM confirms that the target *pierścionek* 'ring' is highly predictable after *diamentowy* 'diamond [adjective]' [1], which is indicated by the dropping surprisal curve after *diamentowy* in Fig. 5. The surprisal values are provided in the unit Hart (Hartley).

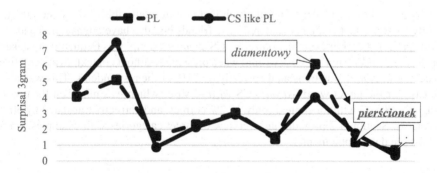

PL: *Bob oświadczył się i dał jej diamentowy **pierścionek**.*
CS: *Bob se zasnoubil a dal jí diamantový **prstýnek**.*

Fig. 5. Surprisal graph for the PL sentence *Bob oświadczył się i dał jej diamentowy pierścionek.* 'Bob proposed and gave her a diamond ring' [1].

In contrast to the sentence in Fig. 5, there is an increase in surprisal in Fig. 6 at the target *siłownię* 'gym [accusative]' for the sentence

Sportowiec lubi chodzić na podnoszenie ciężarów na siłownię.

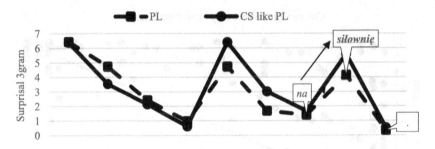

PL: **Sportowiec** *lubi chodzić na podnoszenie ciężarów na* **siłownię**.
CS: **Sportovec** *rád chodí na vzpírání do* **posilovny**.

Fig. 6. Surprisal graph of *Sportowiec lubi chodzić na podnoszenie ciężarów na siłownię.* 'The athlete is enjoying lifting weights at the gym' [1].

'The sportsman likes to do weightlifting at the gym.'[1]

'The athlete is enjoying lifting weights at the gym.'[2]

In the monolingual cloze completion task [1], 95% of English native speakers provided the response (*gym*), which suggests that the word *athlete* or *sportowiec* 'athlete' functions here as a semantic prime. So the higher rate of correct translations in context (58.1% vs. 30.3% without context) might be explained by the thematic association of the target word *siłownię* 'gym [accusative]' with the sentence-initial *sportowiec* 'athlete, sportsman' rather than with its directly preceding words *ciężarów na* 'weights [genitive pl.] at'.

Figure 4 shows an extraordinarily high increase in intelligibility for some targets, mostly for those that can be considered false friends but also have cognate translations (FF-C in Sect. 4.2). For example, *znaczek* – CS *známka* 'stamp' was frequently mistaken for *znak* or *značka* 'sign' (93.9% wrong) when presented without context. In a predictive context, however, it was translated correctly by 71% of the respondents. This was also the case for the target word *wazon* – CS *váza* – 'vase' which was mistaken for *vagon* 'wagon' (48.5%) without context (only 15.2% correct) and correctly translated by 50% in context. Section 4.2 provides an overview of target categories with examples; for a full list see data supplement.

4.2 Different Categories of Target Words

The intelligibility scores vary with different categories of target words in both conditions, i.e. with and without context – cf. Table 2.

Cognates Identical in Base Form. (C-IB, n = 11) This sub-category of cognates includes target words with a base form that is identical in both languages, but not in the inflected forms as presented in the context. For instance, *ryba* 'fish' is identical in

---

[1] The PL translator was instructed to keep the target word at the last position in the sentences. Therefore, some translations might vary slightly from their original EN versions (cf. [1]).

[2] Original version of the sentence as of [1].

its base forms in both PL and CS but the PL target *rybę* in accusative differs from its CS correspondence *rybu*.

Real or True Cognates. (C, n = 89) differ only in orthography and/or in morphological features. We can observe a ceiling effect (maximum scores in both conditions) of target words with very low orthographic distance, such as PL *mokry* and CS *mokrý* 'wet' that differ only in diacritics and were translated correctly by all respondents. The same applies to target words with easily identifiable pronunciation, for instance, in *czasu* – CS *času* – 'time [genitive]' that was translated correctly by 96.8%. Interestingly, there are target words with a relatively high LD, e.g., PL *obiad* 'lunch' with an LD of 40% to the CS *oběd* 'lunch', but an intelligibility score of 100% in context (cf. the sentence below) and 93.3% without context.

> *Zrobiła sobie kanapkę i frytki na obiad.*
> 'She made herself a sandwich and chips for lunch.' [1]

The intelligibility of true cognates correlates significantly with linguistic distance of the target word (r = .549, p < .001), but not with surprisal (r = .043, p < 1).

Cognates in Other Contexts. (C-OC, n = 3): The items are cognates not in the presented sentence but in other contexts. For instance, PL *szczotka* 'brush, broom' can correspond to CS *štětka* 'brush' only in come contexts, e.g. as a brush for shaving, but not as a broom (correct CS *smeták*) for sweeping the floor.

Non-cognates with Correct Associations. (NC-A, n = 7): CS translations are not cognates of the PL targets, but they do share some common features. Thus, PL *latawiec* 'kite' might be associated with the CS verb *létat* 'to fly'. Respondents are likely to associate the stimulus with a concept in their language and then come up with the correct CS translation *drak*.

Real Non-cognates. (NC, n = 5): Unrelated lexical items that are not expected to be intelligible for the reader without context, e.g., PL *atrament* vs. CS *inkoust* 'ink'.

False Friends as Cognates: (FF-C, n = 15) These items are cognates frequently mistaken for another more similar CS word in at least one of the conditions: with or without context. This is one of four categories of false friends. As a threshold for false friends, the percentage of the particular wrong type of response must have been higher than the sum of no responses and correct responses. In addition, the particular wrong response must have been more frequent than the sum of all other wrong responses.

False Friends that are Cognates in Other Contexts: (FF-OC, n = 9) These frequently misinterpreted items are cognates in another context than in which they were presented. For example, PL *przebrać* 'to change clothes' is frequently mistaken for CS *přebrat* 'to pick over', while the correct translation would be *převléct se*.

False Friends with Correct Associations. (FF-A, n = 5) are words that are frequently mistaken for other more similar CS words which have some common semantic features with a correct cognate translation. Respondents might associate the stimulus with a concept in their language and then come up with the correct translation in context. For instance, PL *drzewo* 'tree' is frequently mistaken for CS *dřevo* 'wood', which at the same time can lead to a correct association with CS *strom* 'tree' in the respective context.

Real False Friends. (FF, n = 5) are frequently mistaken for another more similar CS word. For instance, PL *gwóźdź* 'nail' is frequently mistaken for CS *hvozd* 'forest', while the correct translation would be *hřebík*.

Table 2. Intelligibility of target words with vs. without context in the different categories.

	C-IB	C	C-OC	NC-A	NC	FF-C	FF-OC	FF-A	FF
No context	94.5%	65.9%	4.0%	8.7%	6.3%	18.4%	3.83%	3.33%	4.9%
Context	81.4%	80.1%	16.6%	49.8%	31.1%	68.6%	19.33%	42.5%	26.3%
t-test	ns	t = 3.05	Ns	t = 5.07	t = 1.90	t = 5.28	t = 2.45	t = 2.72	ns
Significance		p < .01		p < .001	p < .05	p < .001	p < .05	p < .05	

The differences between the intelligibility of target words with vs. without context are significant for all categories except for cognates identical in their base form, cognates in other context and real false friends. The greatest and highly significant difference between the two conditions was found for target words that are false friends but have cognate translations.

4.3 Analysis of Wrong Responses

The error analysis of responses reveals some features of target words that linguistic distance and surprisal can account for only to a limited extent, if at all:

Differences in Government Pattern. In some sentences, the target words seem to have been more difficult, probably because of differences in government patterns. For instance, the target word *dzień* 'day' was translated more often correctly without context (80%) than in context (66.7%) of the sentence

Dentysta zaleca myć zęby dwa razy na dzień.
'The dentist recommends brushing your teeth twice a day.' [1].

This might be explained by two factors. Firstly, the translation of the PL phrase *na dzień* 'per day' is headed by a different preposition in CS – *za den* – or it can be expressed by a single adverb – *denně* 'daily'. Secondly, and in connection with the first factor, the wrong responses include highly similar words that respondents probably thematically associated with the concept of a dentist from the stimulus sentence: *dáseň* 'gum', *díru* 'hole', or *žízeň* 'thirst'. Moreover, in CS, these responses occur often together with the preposition *na* 'on', e.g. *na dáseň* 'for (your) gum', *na žízeň* 'against thirst' and thus might seem perfectly legitimate to the respondents.

Ln Interferences. Effects of another language (Ln) interference occur relatively rarely (with 11 target words) among the responses in context. Out of the 5208 data points for the context condition, 37 responses could be classified as interferences from EN, DE or SK. One of the few obvious interferences was at the target word *drzwi* 'doors' which was translated as EN *drive* by one Czech respondent who indicated to live in Great Britain. Also, *głosu* 'voice [genitive]' was translated as *skla* 'glass [genitive]' by another respondent living in Great Britian. One respondent translated *biurku* 'desk [locative]' as *tužka* 'biro', probably due to the similarity of PL *biurko* and EN *biro*. The target word *ból* – CS *bolest* – 'pain' was translated as *byl* 'he was' by 53.3% of the respondents, probably due to the SK past tense verb form *bol* 'he was'. Another 6.7% translated *ból* as *míč* 'ball', most likely due to the EN *ball*.

One of the responses was most probably a combination of Ln interference and priming: the target word *torcie* 'cake [locative]' in the sentence

Jenny zapaliła świeczki na urodzinowym torcie.

'Jenny lit the candles on the birthday cake.' [1]

was translated as *svícnu* 'candlestick' [genitive/dative/locative] by 16.1% of the respondents. This probably happened though the EN word *torch* and through the successful recognition of *świeczki* 'candles' as the CS *svíčky* 'candles'.

(Perceived) Morphological Mismatches

PL Feminine Accusative Ending –ę in Ns: *swoją rolę* 'her role [accusative]' was translated as *role* [nominative singular or nominative/accusative plural] when the correct equivalent would have been *roli* [accusative] in CS. Nevertheless, *role* was counted as a correct answer as the interpretation of the target word as a plural does not harm the overall understanding of the sentence. 26.7% translated the target word *próbę* 'test, try' in the sentence

Kim chciała iść na sportownię na kurs na próbę.

'Kim wanted to give the workout class a try.' [1]

with words ending with an *-e, -é* or *-ě*: *přírodě* [dative of *příroda* 'nature'], *tance* 'dances', *hřiště* 'sport field, playground', *sondě* [dative of *sonda* 'sond'], *laně* [locative of *lano* 'rope'], *poprvé* 'for the first time', *zkoušce* [dative of *zkouška* 'test'] for which the correct CS translation would have been *zkoušku* [accusative of *zkouška*].

PL Feminine Instrumental Ending of Ns –ą is apparently mistaken for the regular feminine ending in the nominative or accusative case *–a*. A regular PL-CS correspondence of these endings should be *ą:ou*, although other correspondences with PL *–ą* also occur. Typical mistakes were translations of *królową* as *králova* 'the king's', *szczotką* as *šotka* 'Scottish woman', *pocztą* as *pocta* 'honour'.

Verb forms in third person plural, e.g. *kwitną* 'they bloom' in which the ending *-ą* would correspond to the CS verb ending *–ou* were also frequently mistaken for a feminine N ending: 13% translated it with a feminine N, e.g., *teplota* 'temperature', *květina* 'flower' or *kytky* 'flowers' [colloquial] instead of *kvetou*.

Words with Different Grammatical Gender. Target words with different grammatical gender were translated less often correctly when presented without any context than in context. There were 11 target words with divergent grammatical gender between PL stimulus and correct CS translation. In all 11 cases, the greatest percentage of the responses is of the same gender as the stimulus in the condition without context. For instance, for the target word *biurko* 'desk', respondents have entered a number of neuter Ns, such as *pero* 'pen', *pírko* 'little feather', *horko* 'hot weather'. This changed drastically in the condition with context. The percentage of correct responses increased with sentential context in all cases. The difference of correct responses between the two conditions ranges from 3.1% to 73.3% with a mean increase by 28.3% per word pair. The difficulty for the readers here was to consider the possibility that the correct translation might actually be of a different grammatical gender.

Concerning potential misinterpretations of inflectional endings, only the form *napiwku* [genitive] of *napiwek* 'tip' that, if not identified correctly as an inanimate masculine genitive form, might easily be misperceived as a feminine accusative form with the inflectional suffix *–u* in CS. Nevertheless, the percentage of feminine responses for the form *napiwku* in context did not increase when compared to the responses for the base form *napiwek*.

Infinitive Verb Forms Mistaken for Ns. A number of respondents apparently perceived the PL infinitive ending *–ć* for a correspondence to the CS nominal masculine agentive suffix *–č*, while the correct PL-CS correspondence for infinitive verb endings would be *ć:t*. The two suffixes (PL infinitive *–ć* and CS derivational *–č*) are indeed phonetically cloze. One of the prominent examples was the target word *bawić* 'to play' that was translated as *bavič* 'entertainer' by 39.4% when presented without context. Also, other Ns which the respondents most probably associated with the concept of *bavič* were among the responses: *komik* 'comedian' and *zábava* 'amusement'. The verb appeared in two of the sentences, where it was translated as *bavič* significantly less often – 13.3% and 3.2% respectively.

When *padać* was presented without any context, only 62.9% of the respondents translated the target word correctly with its CS cognate *padat*. It was often mistaken for *padák* 'parachute'. When presented in the sentence

Zauważyłam, że nie mam parasola, gdy zaczęło padać.
'I realized I had no umbrella as it began to rain.' [1],

however, 96.7% translated it correctly as *padat* 'to fall' or *pršet* 'to rain'. The share of infinitive forms mistaken for Ns range from 0 in both conditions to 76.7% for target infinitives without context. On the average, 30.4% of all infinitives without context and only 5.9% infinite verb forms in context were mistaken for Ns.

A complete table with a comparative overview of the target verbs together with the frequencies of their misinterpretations as Ns and correct responses with and without context can be found in the data supplement. As a result of the error analysis, a binary variable for difference in grammatical gender was added in the regression model in order to represent the added difficulty of such target words.

4.4 Correlations and Model

With regard to surprisal, only the surprisal values of the target words and of the whole sentences have a low, but significant correlation with the results obtained in the context condition. The correlation of the CS target words' surprisal and target word intelligibility is only slightly higher than the PL surprisal of the target words ($r = .191 > r = .186$). The correlations of the mean and total surprisal values of the whole sentences with the results in context are only significant in the case of the original PL stimuli sentences, not in the case of their closest CS translations. However, when leaving the cognates out of the correlation analysis, the correlation with the total surprisal of the PL sentence increases to $r = .411$ ($p < .01$), even more when correlating only the false friends (all categories) and intelligibility ($r = .443$, $p < .01$).

There is a highly significant covariance between the corresponding surprisal measures (for target, bigram, trigram, and sentence) from the two LMs (the CS and the PL one – see Sect. 3.2), the strongest correlation being that of the total surprisal per sentence in both languages ($r = .732$, $p < .001$).

For the linguistic distance measures, all correlations are highly significant for the target words in context. The correlations are the highest with the linguistic distance of the target. The longer the involved string of words, the lower the correlation between distance and intelligibility of target words gets: target word > bigram > trigram > sentence. The correlation of intelligibility and linguistic distance is higher for the target words without context ($r = .772$, $p < .001$) than in context ($r = .680$, $p < .001$).

All lexical distance and false friends variables proved to be highly significant, the strongest correlation being the total lexical distance of the entire sentence ($r = .508$, $p < .001$). Both lexical distance and false friends correlate stronger with the results ($r = .353$ for the category of false friends, $p < .001$) when they are counted as a total score per sentence than when normalized through the number of words in a sentence. In context, a relatively low, but highly significant correlation was found for the target word having a different gender in the two languages ($r = .272$, $p < .001$). Without context, the correlation of grammatical gender and intelligibility is only slightly higher and highly significant ($r = .281$, $p < .001$). No correlation was found for the number of words in a sentence. A multiple linear regression with the relevant variables distance of target word, PL sum of surprisal for the sentence, and number of non-cognates per sentence results in a highly significant adjusted $R^2 = .496$ ($p < .001$), i.e. this model can account for 49.6% of the variance in the data for all sentences.

5 Discussion and Conclusion

When viewing the whole stimulus set, the results show clearly that context helps to correctly identify highly predictable target words in sentential context as opposed to the same words without context. However, the correlations with surprisal are low, the highest being the sum of surprisal of the PL stimulus sentence (not of the closest translation). Other factors appeared to be at least equally important, most of all linguistic distance of the target word and the target word being of a different gender in the two languages.

The error-analytical observations lead to the conclusion that divergent grammatical gender of words in a related foreign language can be strongly misleading and that readers very often tend to choose a translation with the same grammatical gender, especially when there is no sentential context. As soon as sentential context is available, the role of the different grammatical gender loses its dominance. Czech readers proved to be unlikely to identify the PL ending *-ą* as an instrumental marker similar to the CS *-ou*, but often mistook it for a feminine nominal ending. Accordingly, the PL accusative ending *-ę* was frequently mistaken for a plural marker or an ending similar to the CS *-ě* in feminine dative or locative forms or neuter locative forms. It was shown that predictive context helps to correctly identify infinite verb forms in sentences, since they were significantly more often mistaken for Ns when presented without context.

However, individual cases of wrong associations with a thematically dominant concept in the sentences have shown that even understandable high-constraint sentential context can lead to wrong associations and to a lower number of correct responses than without context, even if the target word is a frequent cognate.

An analysis of results for the different lexical categories of target words reveals different levels of importance of the predictors in for these categories. The differences in correct responses between the context and the context-free condition were significant for all categories of target words except for those identical in base forms, cognates in other contexts, and real false friends. The difference between the two conditions was the greatest for false friends that are cognates and for non-cognates that offer possible associations with the correct translations.

For true cognates, no significant correlation between intelligibility and surprisal was found. However, surprisal as a predictor has a much greater impact if target words are non-cognates or false friends than if they are cognates, which suggests that in disambiguation of these, readers rely more on context than on word similarity. The effect of the predictive context seems to be especially striking with non-clear-cut cases of false friends. Since the correlations with linguistic distance are lower for target words in context than without context, the influence of linguistic distance on intelligibility proved to decrease in predictive sentential context. In the final regression model, the total surprisal of the sentence obtained from the PL model has a low, but significant correlation with the results.

References

1. Block, C.K., Baldwin, C.L.: Cloze probability and completion norms for 498 sentences: behavioral and neural validation using event-related potentials. Behav. Res. Methods **42**(3), 665–670 (2010). https://doi.org/10.3758/BRM.42.3.665
2. Čermák, F., Rosen, A.: The case of InterCorp, a multilingual parallel corpus. Int. J. Corpus Linguist. **13**(3), 411–427 (2012)
3. Crocker, M.W., Demberg, V., Teich, E.: Information Density and Linguistic Encoding (IDeaL). Künstliche Intell. **30**, 77–81 (2016). https://doi.org/10.1007/s13218-015-0391-y
4. Experiment website. http://intercomprehension.coli.uni-saarland.de/en/
5. Golubović, J.: Mutual intelligibility in the Slavic language area. Ph.D. dissertation. University of Groningen (2016)
6. Golubović, J., Gooskens, C.: Mutual intelligibility between West and South Slavic languages. Russ. Linguist. **39**, 351–373 (2015)
7. Gulan, T., Valerjev, P.: Semantic and related types of priming as a context in word recognition. Rev. Psychol. **17**(1), 53–58 (2010)

8. Harley, T.: The Psychology of Language – From Data to Theory, 2nd edn. Psychology Press, Hove (2008). http://www.psypress.com/harley

9. Heeringa, W., Golubovic, J., Gooskens, C., Schüppert, A., Swarte, F., Voigt, S.: Lexical and orthographic distances between Germanic, Romance and Slavic languages and their relationship to geographic distance. In: Gooskens, C., van Bezoijen, R. (eds.) Phonetics in Europe: Perception and Production, pp. 99–137. Peter Lang, Frankfurt a.M. (2013)

10. Heeringa, W., Swarte, F., Schüppert, A., Gooskens, C.: Modeling intelligibility of written Germanic languages: do we need to distinguish between orthographic stem and affix variation? J. German. Linguist. **26**(4), 361–394 (2014). https://doi.org/10.1017/S1470542714000166

11. Heinz, C.: Semantische Disambiguierung von false friends in slavischen L3: die Rolle des Kontexts. ZfSl **54**(2), 147–166 (2009)

12. Jágrová, K.: Reading Polish with Czech Eyes. Ph.D. dissertation. Saarbrücken University, Saarland (to be published)

13. Jágrová, K.: Processing effort of Polish NPs for Czech readers – A+N vs. N+A. In: Guz, W., Szymanek, B. (eds.) Canonical and Non-canonical Structures in Polish. Studies in Linguistics and Methodology, vol. 12, pp. 123–143. Wydawnictwo KUL, Lublin (2018)

14. Jágrová, K., Avgustinova, T., Stenger, I., Fischer, A.: Language models, surprisal and fantasy in Slavic intercomprehension. Comput. Speech Lang. **53**, 242–275 (2019). https://doi.org/10.1016/j.csl.2018.04.005

15. Jágrová, K., Stenger, I., Marti, R., Avgustinova, T.: Lexical and orthographic distances between Czech, Polish, Russian, and Bulgarian – a comparative analysis of the most frequent nouns. In: Olomouc Modern Language Series (5) – Proceedings of the Olomouc Linguistics Colloquium, pp. 401–416. Palacký University, Olomouc (2017)

16. Kneser, R., Ney, H.: Improved backing-off for M-gram language modeling. In: International Conference on Acoustics, Speech, and Signal Processing, Detroit, MI, vol. 1, pp. 181–184 (1995). https://doi.org/10.1109/ICASSP.1995.479394

17. Křen, M., et al.: SYN2015: reprezentativní korpus psané češtiny. Ústav Českého národního korpusu FF UK, Praha (2015). Accessible: http://www.korpus.cz

18. Levenshtein, V.: Binary codes capable of correcting deletions, insertions, and reversals. Cybern. Control Theory **10**, 707–710 (1966)

19. Muikku-Werner, P.: Co-text and receptive multilingualism – Finnish students comprehending Estonian. Nord. J. Linguist. **5**, 99–113 (2014). https://doi.org/10.12697/jeful.2014.5.305

20. Vanhove, J.: Receptive multilingualism across the lifespan. Cognitive and linguistic factors in cognate guessing. Ph.D. dissertation. University of Freiburg (2014). https://doc.rero.ch/record/210293/files/VanhoveJ.pdf. Accessed 17 Jan 2019

Information Extraction

Multi-lingual Event Identification in Disaster Domain

Zishan Ahmad[✉], Deeksha Varshney, Asif Ekbal, and Pushpak Bhattacharyya

Department of Computer Science and Engineering, Indian Institute of Technology Patna,
Bihta, India
{1821cs18,1821cs13,asif,pb}@iitp.ac.in

Abstract. Information extraction in disaster domain is a critical task for effective disaster management. A high quality event detection system is the very first step towards this. Since disaster annotated data-sets are not available in Indian languages, we first create and annotate a dataset in three different languages, namely *Hindi, Bengali* and *English*. The data was crawled from the different news websites and annotated with expert annotators using a proper annotation guidelines. The events in the dataset belong to 35 different disaster classes. We then build a deep ensemble architecture based on Convolution Neural Network (CNN) and Bi-directional Long Short Term Memory (Bi-LSTM) network as the base learning models. This model is used to identify event words and phrases along with its class from the input sentence. Since our data is sparse, the model yields a very low F1-score in all the three languages. To mitigate the data sparsity problem we make use of multi-lingual word embedding so that joint training of all the languages could be done. To accommodate joint training we modify our model to contain language-specific layers so that the syntactic differences between the languages can be taken care of by these layers. By using multi-lingual embedding and training the whole dataset on our proposed model, the performance of event detection in each language improves significantly. We also report further analysis of language-wise and class-wise improvements of each language and event classes.

Keywords: Event identification · Information extraction · Text mining

1 Introduction

Event is an occurrence that happens at a place and a particular time or time interval. Identification of events from textual data is not only an important task in information extraction but has a lot of practical applications as well. Automatically extracting events becomes an interesting task in information extraction because of the complexities involved in the text, and due to the fact that an event description may be spread over several other words and sentences. With the advent of internet, a huge amount of data is created each day in the form of news, blogs, social media posts, etc. and extracting relevant information from these is a huge challenge due to their built-in multi-source nature. The information extraction systems of today have to deal not only with isolated text, but also with large-scale and small-scale repositories occurring in different languages.

© Springer Nature Switzerland AG 2023
A. Gelbukh (Ed.): CICLing 2019, LNCS 13451, pp. 129–142, 2023.
https://doi.org/10.1007/978-3-031-24337-0_10

Event identification entails identifying the word or phrase that represents an event and also deciding the type of event from a set of pre-defined labels. Automatic event extraction system based on machine learning requires a huge amount of data. Collecting and annotating such data is both time consuming and expensive. Creating such a dataset for multiple languages becomes even more challenging, because finding good annotators for diverse languages is also difficult and expensive. In this paper we create and publish a dataset annotated for events in disaster domain in three different languages, namely *Hindi*, *Bengali* and *English*. We then build a deep learning model, based on CNN (Convolutional Neural Network) and Bi-LSTM (Bi-Directional Long Short Term Memory) for the task of event identification.

Since the dataset is small for a particular language, we leverage the information of the other languages while training. In order to achieve this we make use of multi-lingual word embeddings to bring all the language representations to a shared vector space. We show that by training our model in such a way we are able to utilize the dataset of all the three languages and improve the performance of our system for each language. We also build a second model using deep learning with separate language-specific layers. By using this model along with multi-lingual word embedding, we are able to further improve the performance for all the languages. Our empirical results show that, for the task of event identification, sharing of knowledge across the languages helps in improving the overall performance of the system.

1.1 Problem Definition

Event Identification or detection is a sequence labelling task. Given a sentence in *Hindi*, *Bengali* or *English* of the form $w_1, w_2, w_3, ...w_n$ the task is to predict the best label sequence of the form $l_1, l_2, l_3, ...l_n$. In order to properly denote the boundaries, a multi-word event trigger is encoded in terms of IOB [26] notation, where B, I and O denote the beginning, intermediate and outside token of an event. There are 35 different disaster class labels which need to be identified. The example mentioned below depicts the input sentence and output label sequence. Event triggers are boldfaced in the example.

- **Input Hindi Sentence:** गृह मंत्रालय मुंबई के बम विस्फोटों के मद्देनजर इस बात की विशेष तौर पर जांच कर रहा है कि अक्षरधाम मंदिर और १९९३ के मुंबई बम विस्फोटों के फैसलों की प्रतिक्रिया के रूप में तो यह हमले नहीं हुए
- **Transliteration:** grih mantraalay mumbai ke **bam visphoton** ke maddenajar is baat kee vishesh taur par jaanch kar raha hai ki aksharadhaam mandir aur 1993 ke mumbai **bam visphoton** ke phaisalon kee pratikriya ke roop mein to yah **hamale** nahin hue
- **Translation:** In view of the Mumbai **bomb blasts**, the Home Ministry is specially investigating the fact that these **attacks** did not take place as response to the Akshardham Temple and the 1993 Bombay **bomb blasts**
- **Output:** O O O O I_Terrorist_Attack I_Terrorist_Attack O O O O O O O O O O O O O O O O O I_Terrorist_Attack I_Terrorist_Attack O O O O O O O O I_Terrorist_Attack O O.

2   Related Work

Event extraction is a well-researched problem in the area of Information Extraction, particularly for event detection and classification. Some of the well-known works can be found in [4,6,22].

The early approaches were based on pattern matching in which, patterns were created from predicates, event triggers and constraints on its syntactic context [2,3,8,29]. Later, feature based methods were used to learn a better representation for sentences by forming rich feature sets for event detection models ranging from lower-level representations [7,13,30] to higher-level representations such as cross-sentence or cross-event information [10,11,14–16].

The major disadvantages associated with the traditional feature based methods are due to the complexity associated with handcrafted feature engineering. To overcome this, Nguyen and Grishman (2015) [22] used CNN to automatically extract efficient feature representations from the pre-trained embeddings (word, position and entity-type) for event extraction. For multi-event sentences, Chen et al. [4] introduced a dynamic multi-pooling layer according to event triggers and arguments into CNN to capture more crucial information.

In 2016, Nguyen and Grishman [23] proposed a technique that made use of non-consecutive convolution for sentences which skip unnecessary words in word sequences. Feng et al. [6] combined the representations learned from a CNN with a Bidirectional LSTM [9] to learn the continuous representations of a word. Further in the same year, Nguyen and others [21] use a variant of LSTM called the Gated Recurrent Units (GRU) [5] in association with a memory network to jointly predict events and their arguments. Often event detection suffers from data sparseness. To handle this Liu et al. [19] used FrameNet to detect events and mapped the frames to event-types, thus obtaining extra training data. Liu et al. [20] further extended the work by building an extraction model using the information from arguments and FrameNet using supervised attention mechanism.

Recently, neural network models involving dependency trees have also been attempted. Nguyen and Grishman (2018) [24] explored syntactic representations of sentences and examined a convolutional neural network based on dependency trees to perform event detection. Sha et al. [27] came up with dependency bridges over Bi-LSTM for event extraction. Their work was further extended by Orr et al. [25], who used attention to combine syntactic and temporal information. Liu et al. [18] built a Gated Multi-Lingual Attention (GMLATT) framework which used monolingual context attention to gather information in each language and gated cross-lingual attention to combine information of different languages. They produced their results for English and Chinese languages. Feng et al. [6] illustrated the techniques for building a language independent event extraction system. Lin et al. [17] proposed a multi-lingual system for sequence labelling. They used closely related languages and used character CNNs to get the representation of words in languages that share alphabets.

In this paper, instead of using character CNNs to get shared representation of a word, we use multi-lingual embeddings. This gives us the freedom to do joint learning between the languages that do not share the same characters. We also demonstrate that our method can be used for event identification for languages that are diverse and syntactically dissimilar like *English*, *Hindi* and *Bengali*.

3 Methodology

We develop three models based on deep learning for conducting our experiments. First we build a deep learning model based on Bi-Directional Long Short Term Memory (Bi-LSTM) [9] and Convolution Neural Network (CNN) [12]. We train the model separately for each of the three languages using monolingual word embedding. We then leverage the information from the other languages, by following two approaches: i). We use the same deep learning model to train with all the languages simultaneously with multi-lingual word embedding as features; ii). In the second model separate MLP (Multi Layer Perceptron) is used for each language at the final layer. Here too the training is done with all the languages at the same time using multi-lingual word embedding as features.

3.1 Mono-lingual Word Embedding Representation

The pre-trained monolingual word-embeddings used in the experiments are popularly known as fastText [1][1]. This method makes use of, continuous skip-gram model and is used to obtain the word-embedding representation of size 300 for each word. The advantage of using fastText is that even if some words are not available in the training corpus, their representations can still be obtained. Thus a better representation of out-of-vocabulary words are obtained by using fastText. The skip-gram model of fastText is trained on *Wikipedia* data dumps of their respective language.

3.2 Multi-lingual Word Embedding Representation

We use alignment matrices[2] to align the embedding of different languages ($L1,L2$ and $L3$) to a single vector space. The word embedding representations obtained from fastText for language $L1$ are multiplied with the alignment matrices $L1$ to transform the word embedding to a shared vector space. This method of transforming the word embedding was proposed in Smith et al. [28]. They proved that a self-consistent linear mapping between the semantic spaces must be orthogonal. Two parallel dictionaries between languages, D_{L2-L1} and D_{L3-L1} are used for this purpose. By aligning all the three-language embedding to the vector space of $L1$, we obtain multi-lingual word embedding. The following Eq. 1 is used to obtain the alignment matrix.

$$\max_{O} \sum_{i=1}^{n} y_i^T O x_i, \quad \text{Subject to } O^T O = I \tag{1}$$

where x_i and y_i belong to a dictionary $\{x_i, y_i\}_1^n$ of paired words of two languages $L1$ and $L2$ (or $L1$ and $L3$) respectively, and O is the orthogonal transformation matrix.

---

[1] https://github.com/facebookresearch/fastText/blob/master/pretrained-vectors.md.

[2] https://github.com/Babylonpartners/fastText_multilingual.

Single Valued Decomposition (SVD) is then used to accomplish the objective of Eq. 1. Matrices U and V are obtained, such that $O = UV^T$. By applying the transformation V^T to the source language $L1$ and U^T to the target language $L2$, both the languages can be mapped into a single vector space. The same thing is repeated for $L1$ and $L3$. For our experiments we transform *Hindi* and *Bengali* language word embeddings to the vector space of *English* word embeddings.

3.3 Baseline Model for Event Identification

The event identification model used for mono-lingual setting is shown in Fig. 1. The task of event identification is formulated as a sequence labeling problem. For each input token we need to decide if the token belongs to an event and also what event type it belongs to. The input to the model is a sentence, represented by a sequence of word embeddings. Since Bi-LSTM takes fixed input sizes the smaller sentences are padded with zero vectors and brought to equal length. This sequence is passed to Bi-LSTM and CNN of filter sizes 2 and 3. The Bi-LSTM yields output representation of each word and CNN gives convoluted bi-gram and tri-gram features. These features are concatenated and passed through a Multi Layer Perceptron (MLP), followed by a Softmax classifier which gives the probability distribution over the possible tags I_Event_Type or O_Event.

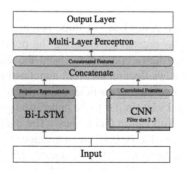

Fig. 1. Architecture of baseline model

3.4 Proposed Multilingual Event Identification Model

We develop a model similar to the baseline model for multilingual event identification. The input sentence to this model is represented by sequence of multi-lingual word-embeddings. We create separate MLP layer for each language, such that for an input data of language $L1$, only the MLP of language $L1$ will be used. The backpropagation will take place only through MLP of language $L1$, and the weights of other language MLPs will not be updated. All the layers before MLPs will be updated for every input, irrespective of the languages. Thus, the Bi-LSTM and CNNs produce a shared representation that is used by all the MLPs. The separate MLPs act as language specific decoders which decodes event representations of each language from the shared representation of languages.

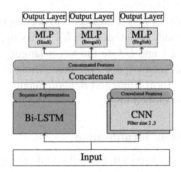

Fig. 2. Architecture of the proposed multilingual event identification model

4   Datasets and Experiments

In this section we provide the details of the datasets used in our experiments along with the experimental setup.

4.1   Datasets

We use a multi-lingual setup, comprising of three languages, namely English, Hindi and Bengali. The dataset was created by crawling the popular news websites in the respective languages. All the news crawled are from the disaster events (both man-made and natural). All the news documents were annotated by three annotators, with good linguistic knowledge and having sufficient knowledge of the related area. The annotation guideline followed was similar to the annotation guideline provided by TAC KBP 2017 Event Sequence Annotation Guidelines[3]. The multi-rater Kappa agreement ratio of the annotators was found to be 0.85 on an average.

The total dataset is comprising of 2,191 documents (*Hindi*: 922, *Bengali*: 999 and *English*: 270). It comprises of 44,615 sentences (*Hindi*: 17,116, *Bengali*: 25,717 and *English*:1,782) and a total of 596,423 words (*Hindi*: 276,155, *Bengali*: 273,115 and *English*: 47,153). Thirty five different disaster types were chosen as class labels for annotation. The classes and the distribution of classes in the three language datasets are detailed in Table 1.

4.2   Experimental Setup

For implementing the deep learning model and conducting the experiments, a Python based library Keras[4] was used. Since both the models are based on Bi-LSTM the input sequence lengths needed to be the same. Padding by zero vector of length 300 was used to make all the sequences equal in length. The sequence length was fixed to 75.

---

[3] https://cairo.lti.cs.cmu.edu/kbp/2017/event/TAC_KBP_2017_Event_Coreference_and_
Sequence_Annotation_Guidelines_v1.1.pdf.

[4] http://keras.io.

Table 1. Event types in the dataset, and number of triggers for each class in *Hindi, Bengali* and *English* datasets

Class	Hindi	Bengali	English	Class	Hindi	Bengali	English
Epidemic	0	256	29	Forest Fire	327	10	34
Hurricane	51	25	23	Terrorist Attack	507	741	31
Drought	5	8	6	Vehicular Collision	482	436	91
Earthquake	145	325	177	Industrial Accident	329	70	35
Shootout	545	942	194	Volcano	238	4	91
Blizzard	140	13	9	Land Slide	225	33	30
Surgical Strikes	2	344	34	Train Collision	325	32	53
Fire	470	600	154	Pandemic	0	513	0
Tsunami	10	56	11	Riots	311	148	14
Storm	601	154	79	Armed Conflicts	44	399	6
Normal Bombing	89	1783	40	Cyclone	148	21	47
Avalanches	130	1	23	Seismic Risk	0	1	0
Suicide Attack	750	613	15	Transport Hazards	379	882	94
Tornado	159	12	39	Aviation Hazard	174	238	41
Floods	231	43	55	Cold Wave	134	19	39
Heat Wave	438	65	11	Hail Storms	184	0	32
Limnic Eruptions	0	0	1	Famine	0	0	1
Rock Fall	0	0	6				

Relu was used for activation and *Dropout* of 0.3 was used for all the intermediate layers in the model. For the baseline event identification model, one MLP with two linear layers were used. The final layer consists of 35 neurons (since there are 35 classes), and *Softmax* is used for classification of the final output. For the multi-lingual event identification model, three MLPs are used in parallel for the three languages. Each MLP contains two linear layers, with final layer containing 35 neurons. *Softmax* is used for the classification of final output from all the three MLPs.

5   Results and Analysis

We present all the experimental results in this section. The first experiment (*Exp-1*) is conducted on our baseline model (c.f Fig. 1), by using monolingual word embedding as the input features. This model is trained separately for each language, thus we have individual models for each of the three languages.

In the second experiment (*Exp-2*) multi-lingual word embeddings are used as the input features. The datasets of all the languages are shuffled and taken together for training. Thus in the second experiment we have only one model that operates for all the three languages. The third experiment (*Exp-3*) is conducted on the multilingual event identification model (c.f Fig. 2). In this experiment we use multi-lingual word embeddings as input feature, and the training is done with all the languages together.

Table 2. Macro-averaged *Precision (P)*, *Recall (R)* and *F1-score (F)* for the three experiments: 5-fold cross-validated

Classes	Exp-1			Exp-2			Exp-3		
	P	R	F	P	R	F	P	R	F
Hindi	0.32	0.25	0.25	0.32	0.25	0.26	0.40	0.37	**0.36**
Bengali	0.22	0.18	0.18	0.33	0.25	0.26	0.35	0.29	**0.30**
English	0.18	0.21	0.18	0.33	0.29	0.28	0.43	0.38	**0.39**

The results of these experiments are shown in Table 3. From the results it can clearly be seen that the performance of all the languages improve in our second setting. The improvement for the *Hindi* is the least while that of *English* is the most. The results clearly show that using multi-lingual embeddings and all the three-language datasets for training helps in training and produces better results, than training individual models using mono-lingual word embeddings. After conducting the third experiment (i.e. Exp-3) the results improve even further for every language. This shows that a separate language layer for each language helps for better learning in the multi-lingual setting. Separate layers act as the special language decoders that take shared representation from the previous layers and learn the best mapping from this to the language output. The best jump in F1-score is seen in *English*. This is because the *English* dataset is the smallest in size, and thus the baseline model for this dataset is under-trained. The F1-score of *English* dataset more than doubles (from 0.18 to 0.39) when using multi-lingual settings and separate language layers. The F1-score for *Hindi* and *Bengali* datasets improve by 11% and 12% respectively, over the baseline.

The class-wise F1-score for each of the three experiments for all the datasets are shown in Table 2. For each class the best F1-score is shown in bold. It can be clearly seen that the third setup outperforms the first and second setups for most of the classes for all the three languages. We observe the most performance improvement for the *English* language. For many classes the initial model could not predict any instance while the final model was able to give a good F1-score. This usually happens when one of the language datasets has a good number of instances for that particular class. For e.g. the class *Blizzard* has 140 instances in *Hindi* language dataset, while only 9 instances are present in the *Bengali* and *English* datasets. The initial prediction of the experiment yields an F1-score of 0 for *English*, while the third experiment boosts the F1-score to 0.88. This clearly shows that sharing of knowledge across the different languages do help each other. For class like *Train Collision* 352 instances are present in *Hindi* dataset while only 32 and 53 instances are present in *Bengali* and *English* dataset, respectively. In this case we see that the system could not predict the class for *Bengali* and *English* dataset for *Exp 1* and *Exp 2*. However by using separate layers the system was able to predict the class *Train collision*. This shows the effectiveness of separate layers even when only one of the datasets has enough instances of a class and others do not. Few class instances like *Limnic Eruptions*, *Rock Fall* and *Famine* are only present in the *English* dataset with a few instances. For such classes we see no improvement in

Table 3. *F1-score* for the three experiments on *Hindi (H)*, *Bengali (B)* and *English (E)* datasets: 5-fold cross-validated ('–' represents the absence of instances of the class for that language)

Language	Exp-1			Exp-2			Exp-3		
	H	B	E	H	B	E	H	B	E
Drought	0	0	0	0.32	0	0	**0.67**	**0.06**	0
Earthquake	0.38	0.80	0.82	**0.53**	0.82	0.93	0.49	**0.83**	**0.96**
Shoot out	0.57	0.65	0.58	**0.58**	**0.65**	0.65	0.55	0.58	**0.67**
Blizzard	**0.57**	0	0	0.17	0	00	0.53	0	**0.88**
Fire	0.12	0.70	0.58	0.2	**0.73**	0.67	**0.29**	0.69	**0.68**
Tsunami	0	0	0	0	0	0	**0.41**	**0.37**	**0.13**
Storm	0.54	**0.66**	0.55	0.48	0.17	0.51	**0.54**	0.4	**0.72**
Normal Bombing	0	0.60	**0.39**	0.17	**0.72**	0.37	**0.22**	0.68	0.32
Avalanches	0.67	0	0.28	0.67	0	0.65	**0.78**	0	**0.87**
Suicide Attack	0.67	0.52	0	0.68	0.63	0.28	**0.68**	**0.68**	**0.54**
Tornado	0.43	0	0.59	0.52	0	0.95	**0.52**	**0.33**	**0.99**
Forest Fire	0.14	0	0	0.04	0	0	**0.38**	0	**0.32**
Terrorist Attack	**0.68**	0.39	0	0.67	**0.46**	0.21	0.66	0.43	**0.29**
Vehicular Collision	0.45	0.33	0.22	0.42	0.44	0.38	0.5	**0.45**	**0.52**
Industrial Accident	0.02	0	0	0	0	**0.30**	**0.20**	0	0
Volcano	0.41	0	0.67	0.41	0	0.63	**0.50**	0	**0.69**
Land Slide	0.64	0	0.13	**0.67**	0.36	0.65	0.65	**0.48**	**0.8**
Train Collision	0.14	0	0	0.06	0	0	**0.38**	**0.12**	**0.17**
Riots	0.14	0	0	0.10	**0.24**	**0.10**	**0.36**	0.23	0.07
Armed Conflict	0	0.37	0	0.03	**0.47**	**0.29**	**0.17**	0.44	0
Cyclone	0.03	0	0.63	0	0	0.37	**0.29**	0	**0.80**
Transport Hazard	0.07	0.53	0.16	0.11	**0.56**	0.25	**0.24**	0.53	**0.40**
Aviation Hazard	0.14	0.08	0	**0.48**	**0.64**	**0.52**	0.42	0.63	0.45
Floods	0.62	0	0.49	**0.62**	**0.65**	0.64	0.60	0.56	**0.69**
Cold Wave	0.24	0	0.24	0.20	0.22	0.72	**0.56**	**0.40**	**0.90**
Heat Wave	0.59	0	0	**0.59**	**0.38**	0.09	0.53	0.28	**0.5**
Hail Storm	0.63	0.64	0.07	0.62	0	0.16	**0.66**	**0.66**	**0.28**
Hurricane	0	0	0	0	0	0.12	**0.15**	**0.09**	**0.79**
Surgical Strike	0	0.02	0	0	0.46	0.07	0	**0.50**	**0.27**
Epidemic	–	0.1	0	–	0.07	0	–	**0.40**	0
Pandemic	–	0.63	**0.63**	–	**0.64**	0	–	0.56	0.56
Seismic Risk	–	0	0	–	0	0	–	0	0
Limnic Eruptions	–	–	0	–	–	0	–	–	0
Rock Fall	–	–	0	–	–	0	–	–	0

performance. Thus improvement is only noticed when enough number of instances are present in at least one of the languages.

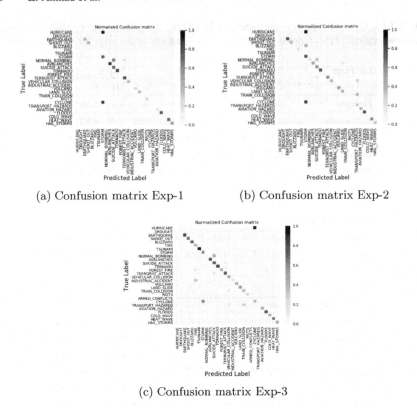

(a) Confusion matrix Exp-1   (b) Confusion matrix Exp-2

(c) Confusion matrix Exp-3

Fig. 3. Confusion matrix of Exp-1, Exp-2 and Exp-3 for *Hindi* language test set

5.1 Error Analysis

We provide analysis of the errors produced by our models in this section. Since the event identification problem classifies the words or phrases into different classes, we use confusion matrices[5] to determine which classes the model is confused with. We observe that miss-classifications reduce for all the languages from *Exp 1* to *Exp 3*. From the confusion matrix of *Hindi*, shown in Fig. 3a, it can be clearly seen that the system was getting confused between the types that were very close to each other. For example *Blizzard* was getting confused with *Storm*. In this situation multilingual event identification model showed significant improvement (c.f Fig. 3c). It was able to identify accurately nearly 60% instances of *Blizzard* in the *Hindi* dataset. Even though the instances of the class *Blizzard* are very few in other languages, the instances of *Storm* are plenty in the other datasets. This helps the model in creating a representation that can better discriminate between the classes *Storm* and *Blizzard*- thus improving the performance for both the classes. In Fig. 3a, we see that the model was almost unable to identify the classes *Fire* and *Tsunami*. However, using the multilingual model it could identify these classes better. This might be attributed because for both of these events, *Bengali* dataset helps

---

[5] The confusion matrices for each language dataset were computed on 80:20 train-test split.

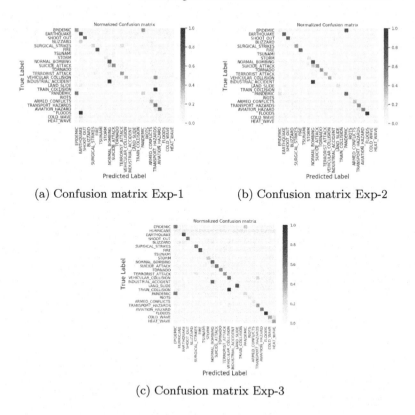

(a) Confusion matrix Exp-1   (b) Confusion matrix Exp-2

(c) Confusion matrix Exp-3

Fig. 4. Confusion matrix of Exp-1, Exp-2 and Exp-3 for *Bengali* language test set

by providing many instances of these classes (c.f Fig. 3c). Similarly, good presence of the events *Landslide* and *Heatwave* in the *Hindi* dataset help in extracting these events in the *Bengali* dataset (c.f Fig. 4c).

The baseline system for the *English* dataset was confused between the classes *Vehicular Collision*, *Train Collision*, *Transport Hazard* and *Aviation Hazard* (c.f Fig. 5a). The source of confusion between these classes was the lack of enough data for these classes in the *English* dataset. However, these classes are very well covered by *Hindi* and *Bengali* datasets as seen in Table 1. This helps in greatly reducing the confusion between these classes (c.f Fig. 5c) and improving the F1-scores of all these classes simultaneously. Furthermore, we observe that for all the languages some missing classes are predicted accurately in our final model. These classes are shown as gray colored rows in Figs. 3c, 4c and 5c.

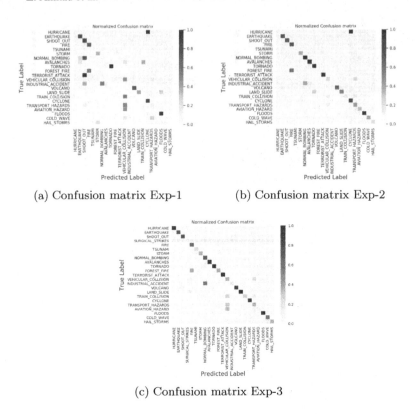

(a) Confusion matrix Exp-1   (b) Confusion matrix Exp-2

(c) Confusion matrix Exp-3

Fig. 5. Confusion matrix of Exp-1, Exp-2 and Exp-3 for *English* language test set

6 Conclusion and Future Works

In this paper we have proposed a deep neural network architecture for multi-lingual event extraction in disaster domain. We have created the event annotated datasets in three different languages, namely *Hindi*, *Bengali* and *English*. We build two different deep learning based models to identify events from a given text. We have used both mono-lingual and multi-lingual word embeddings as input features to our model. We have empirically shown that using multi-lingual word embeddings, we can leverage the information across the different languages using joint training, that eventually improves the performance of each language. We also show that using separate MLPs for each language further improves the performance of the system for all the languages.

We make use of syntactically different languages like *English* and *Hindi* in our experiments, thus we can say that using multi-lingual word embeddings and language specific MLPs, the syntactic differences between languages can be handled, while also improving the performance of the system by joint training. This work can be used as a benchmark setup for future experiments on Event Identification in *Hindi*, *English* and *Bengali*. As future work, it would be interesting to add more languages of diverse nature to the experiments mentioned in this paper and observe the results.

Acknowledgement. The research reported in this paper is an outcome of the project titled "A Platform for Cross-lingual and Multi-lingual Event Monitoring in Indian Languages", supported by IMPRINT-1, MHRD, Govt. of India, and MeiTY, Govt. of India. The *Bengali* and *English* language datasets used in the experiments, were created by the project partners at Indian Institute of Technology Kharagpur, and Anna University - K. B. Chandrashekar Research Centre respectively.

References

1. Bojanowski, P., Grave, E., Joulin, A., Mikolov, T.: Enriching word vectors with subword information. Trans. Assoc. Comput. Linguist. **5**, 135–146 (2017)
2. Cao, K., Li, X., Fan, M., Grishman, R.: Improving event detection with active learning. In: Proceedings of the International Conference Recent Advances in Natural Language Processing, pp. 72–77 (2015)
3. Cao, K., Li, X., Grishman, R.: Improving event detection with dependency regularization. In: Proceedings of the International Conference Recent Advances in Natural Language Processing, pp. 78–83 (2015)
4. Chen, Y., Xu, L., Liu, K., Zeng, D., Zhao, J.: Event extraction via dynamic multi-pooling convolutional neural networks. In: Proceedings of the 53rd Annual Meeting of the Association for Computational Linguistics and the 7th International Joint Conference on Natural Language Processing (Volume 1: Long Papers), vol. 1, pp. 167–176 (2015)
5. Cho, K., et al.: Learning phrase representations using RNN encoder-decoder for statistical machine translation. arXiv preprint arXiv:1406.1078 (2014)
6. Feng, X., Qin, B., Liu, T.: A language-independent neural network for event detection. Sci. China Inf. Sci. **61**(9), 1–12 (2018). https://doi.org/10.1007/s11432-017-9359-x
7. Ferguson, J., Lockard, C., Weld, D., Hajishirzi, H.: Semi-supervised event extraction with paraphrase clusters. In: Proceedings of the 2018 Conference of the North American Chapter of the Association for Computational Linguistics: Human Language Technologies (Volume 2: Short Papers), vol. 2, pp. 359–364 (2018)
8. Grishman, R., Westbrook, D., Meyers, A.: Nyu's English ACE 2005 system description. ACE **5** (2005)
9. Hochreiter, S., Schmidhuber, J.: Long short-term memory. Neural Comput. **9**(8), 1735–1780 (1997)
10. Hong, Y., Zhang, J., Ma, B., Yao, J., Zhou, G., Zhu, Q.: Using cross-entity inference to improve event extraction. In: Proceedings of the 49th Annual Meeting of the Association for Computational Linguistics: Human Language Technologies, vol. 1, pp. 1127–1136. Association for Computational Linguistics (2011)
11. Ji, H., Grishman, R.: Refining event extraction through cross-document inference. In: Proceedings of ACL 2008: HLT, pp. 254–262 (2008)
12. Kim, Y.: Convolutional neural networks for sentence classification. arXiv preprint arXiv:1408.5882 (2014)
13. Li, Q., Ji, H., Huang, L.: Joint event extraction via structured prediction with global features. In: Proceedings of the 51st Annual Meeting of the Association for Computational Linguistics (Volume 1: Long Papers), vol. 1, pp. 73–82 (2013)
14. Li, X., Nguyen, T.H., Cao, K., Grishman, R.: Improving event detection with abstract meaning representation. In: Proceedings of the First Workshop on Computing News Storylines, pp. 11–15 (2015)
15. Liao, S., Grishman, R.: Using document level cross-event inference to improve event extraction. In: Proceedings of the 48th Annual Meeting of the Association for Computational Linguistics, pp. 789–797. Association for Computational Linguistics (2010)

16. Liao, S., Grishman, R.: Acquiring topic features to improve event extraction: in pre-selected and balanced collections. In: Proceedings of the International Conference Recent Advances in Natural Language Processing 2011, pp. 9–16 (2011)

17. Lin, Y., Yang, S., Stoyanov, V., Ji, H.: A multi-lingual multi-task architecture for low-resource sequence labeling. In: Proceedings of the 56th Annual Meeting of the Association for Computational Linguistics (Volume 1: Long Papers), vol. 1, pp. 799–809 (2018)

18. Liu, J., Chen, Y., Liu, K., Zhao, J.: Event detection via gated multilingual attention mechanism. Statistics **1000**, 1250 (2018)

19. Liu, S., Chen, Y., He, S., Liu, K., Zhao, J.: Leveraging framenet to improve automatic event detection. In: Proceedings of the 54th Annual Meeting of the Association for Computational Linguistics (Volume 1: Long Papers), vol. 1, pp. 2134–2143 (2016)

20. Liu, S., Chen, Y., Liu, K., Zhao, J.: Exploiting argument information to improve event detection via supervised attention mechanisms. In: Proceedings of the 55th Annual Meeting of the Association for Computational Linguistics (Volume 1: Long Papers), vol. 1, pp. 1789–1798 (2017)

21. Nguyen, T.H., Cho, K., Grishman, R.: Joint event extraction via recurrent neural networks. In: Proceedings of the 2016 Conference of the North American Chapter of the Association for Computational Linguistics: Human Language Technologies, pp. 300–309 (2016)

22. Nguyen, T.H., Grishman, R.: Event detection and domain adaptation with convolutional neural networks. In: Proceedings of the 53rd Annual Meeting of the Association for Computational Linguistics and the 7th International Joint Conference on Natural Language Processing (Volume 2: Short Papers), vol. 2, pp. 365–371 (2015)

23. Nguyen, T.H., Grishman, R.: Modeling skip-grams for event detection with convolutional neural networks. In: Proceedings of the 2016 Conference on Empirical Methods in Natural Language Processing, pp. 886–891 (2016)

24. Nguyen, T.H., Grishman, R.: Graph convolutional networks with argument-aware pooling for event detection (2018)

25. Orr, J.W., Tadepalli, P., Fern, X.: Event detection with neural networks: a rigorous empirical evaluation. arXiv preprint arXiv:1808.08504 (2018)

26. Ramshaw, L.A., Marcus, M.P.: Text chunking using transformation-based learning. In: Armstrong, S., Church, K., Isabelle, P., Manzi, S., Tzoukermann, E., Yarowsky, D. (eds.) Natural Language Processing Using Very Large Corpora, pp. 157–176. Springer, Dordrecht (1999). https://doi.org/10.1007/978-94-017-2390-9_10

27. Sha, L., Qian, F., Chang, B., Sui, Z.: Jointly extracting event triggers and arguments by dependency-bridge RNN and tensor-based argument interaction. In: Thirty-Second AAAI Conference on Artificial Intelligence (2018)

28. Smith, S.L., Turban, D.H., Hamblin, S., Hammerla, N.Y.: Offline bilingual word vectors, orthogonal transformations and the inverted softmax. arXiv preprint arXiv:1702.03859 (2017)

29. Tanev, H., Piskorski, J., Atkinson, M.: Real-time news event extraction for global crisis monitoring. In: Kapetanios, E., Sugumaran, V., Spiliopoulou, M. (eds.) NLDB 2008. LNCS, vol. 5039, pp. 207–218. Springer, Heidelberg (2008). https://doi.org/10.1007/978-3-540-69858-6_21

30. Yang, B., Mitchell, T.: Joint extraction of events and entities within a document context. arXiv preprint arXiv:1609.03632 (2016)

Detection and Analysis of Drug Non-compliance in Internet Fora Using Information Retrieval Approaches

Sam Bigeard[1(✉)], Frantz Thiessard[2], and Natalia Grabar[1]

[1] CNRS, Univ Lille, UMR 8163 STL - Savoirs Textes Langage, 59000 Lille, France
sam.bigeard@uni-hamburg.de
[2] U Bordeaux, Inserm, Bordeaux Population Health Research Center, team ERIAS,
UMR 1219, 33000 Bordeaux, France

Abstract. In the health-related field, drug non-compliance situations happen when patients do not follow their prescriptions and do actions which lead to potentially harmful situations. Although such situations are dangerous, patients usually do not report them to their physicians. Hence, it is necessary to study other sources of information. We propose to study online health fora with information retrieval methods in order to identify messages that contain drug non-compliance information. Exploitation of information retrieval methods permits to detect non-compliance messages with up to 0.529 F-measure, compared to 0.824 F-measure reached with supervized machine learning methods. For some fine-grained categories and on new data, it shows up to 0.70 Precision.

Keywords: Online health fora · Information retrieval · Text mining

1 Introduction

In the health-related field, drug non-compliance situations happen when patients do not follow indications given by their doctors and by prescriptions. Typically, patients may decide to change the dosage, to refuse to take prescribed drugs, to take drugs without prescriptions, etc.

Misuse of drugs is a specific case of non-compliance, in which the drug is used by patients with a different intent than the one that motivated the prescription. Recreational medication use or suicide attempts are examples of drug misuse, but it has been noticed that the typology of drug misuses includes several other harmful situations [1]. Althought these situations are dangerous for patients and their health, they do not inform their doctors that they are not following the prescription instructions. For this reason, the misuse situations become harder to detect and to prevent. Hence, there is the necessity to study other sources of information. We propose to contribute to this research question and to study social media messages, in which patients willingly and without any particular effort talk about their health and their practices regarding drug use [2].

© Springer Nature Switzerland AG 2023
A. Gelbukh (Ed.): CICLing 2019, LNCS 13451, pp. 143–154, 2023.
https://doi.org/10.1007/978-3-031-24337-0_11

Currently, social media has become an important source of information for various research areas, such as geolocalisation, opinion mining, event extraction, translation, or automatic summarizing [3].

In the medical domain, social media has been efficiently exploited in information retrieval for epidemiological surveillance [4,5], study of patient's quality of life [6], or drug adverse effects [7].

Yet, up to now, few works are focused on drug misuse and non-compliance. We can mention non-supervised analysis of tweets about non-medical use of drugs [8], and creation of a semantic web platform on drug abuse [9]. Both of these works are dedicated to one specific case of non-compliance, which is the drug abuse.

In our work, we propose to address a larger set of non-compliance situations. For this, we propose to exploit information retrieval methods. Our objective is to identify messages related to these situations in health fora in French.

In what follows, we first introduce the methods. We will first describe the method designed for the use of information retrieval system for supervised classification of messages and will compare the results obtained with the state of the art machine learning approaches. Then, we will use the information retrieval system on new non-annotated data and on different topics, and assess whether non-compliance messages are correctly detected. The discussion of the results is proposed. Finally, we conclude with directions for future research.

2   Methods

We propose to adapt an existing information retrieval system, using reference data, to perform the categorization of health fora messages in order to detect messages containing non-compliance information. We will compare these results with the state of the art machine learning methods. Then, we will use information retrieval in a fully unsupervised way to explore new non-annotated data on several topics related to non-compliance. The purpose is to assess if non-compliance messages can be found with information retrieval methods.

We first introduce our reference data and then describe the two ways to use information retrieval methods. We also describe the supervised machine learning approach.

2.1   Reference and Test Data

The reference and test data are built from corpora collected from several health fora written in French:

- Doctissimo[1] is a well known health website to French-speaking people. It is widely used by people with punctual health questions. The main purpose of this website is to provide the platform to patients and to enable them to have discussions with other patients or their relatives. We collected messages from

---

[1] http://forum.doctissimo.fr.

several Doctissimo fora: pregnancy, general drug-related questions, back pain, accidents in sport activities, diabetes. We collected messages written between 2010 and 2015;

- AlloDocteur[2] is a question/answer health service, in which patients can ask questions which are answered by real doctors;
- masante.net[3] is another question/answer service, in which the answers are provided by real doctors;
- Les diabétiques[4] is a specialized forum related to diabetes.

In all these fora, the contributors are mainly sick persons and their relatives, who join the community to ask questions or provide accounts on their disorders, treatments, etc. These people may be affected by chronic disorders or present punctual health problems.

To build the reference data we use two fora from Doctissimo (pregnancy and general drug-related questions). We keep only messages that mention at least one drug. This gives a total of 119,562 messages (15,699,467 words). For the test data, we collect 145,012 messages from other corpora. Messages longer than 2,500 characters are excluded because they provide heterogeneous content difficult to categorize and process, both manually and automatically. The drug names are detected with specific vocabulary containing French commercial drug names built from several sources: *base CNHIM Thériaque*[5], *base publique du médicament*[6], and *base Medic'AM* from *Assurance Maladie*[7]. Each drug name is associated with the corresponding ATC code [10].

The reference data is manually annotated. Three annotators were asked to assign each message to one of the two categories:

- *non-compliance* category contains messages which report on drug misuse or non-compliance. When this category is selected, the annotators are also asked to shortly indicate what type of non-compliance is concerned (overuse, dosage change, brutal quitting...). This indication is written as free text with no defined categories. For instance, the following example shows non-compliance situation due to the forgotten intake of medication: *"bon moi la miss boulette et la tete en l'air je devais commencer mon "utrogestran 200" a j16 bien sur j'ai oublier! donc je l'ai pris ce soir!!!!"* (well me miss blunder and with the head in the clouds I had to start the "utrogestran 200" at d16 and I forgot of course! so I took it this evening!!!!)
- *compliance* category contains messages reporting normal drug use (*"Mais la question que je pose est 'est ce que c'est normal que le loxapac que je prends*

---

[2] http://www.allodocteur.fr.

[3] http://ma-sante.net.

[4] http://www.lesdiabetiques.com.

[5] http://www.theriaque.org.

[6] http://base-donnees-publique.medicaments.gouv.fr.

[7] https://www.ameli.fr/l-assurance-maladie/statistiques-et-publications/donnees-statistiques/medicament/medic-am/medic-am-mensuel-2017.php.

met des heures à agir ???" (Anyway the question I'm asking is whether it is normal that loxapac I'm taking needs hours to do something???)) and messages without use of drugs (*"ouf boo, repose toi surtout, il ne t'a pas prescris d'aspegic nourisson??" (ouch boo, above all take a break, he didn't prescribe aspegic for the baby??))*

When annotators are unable to decide, they can mark up the corresponding messages accordingly. The categorization of these messages, as well as the categorization of annotation disagreements, are discussed later. The three annotators involved in the process are: one medical expert in pharmacology and two computer scientists familiar with medical texts and annotation tasks. Because this kind of annotation is a complicated task, especially concerning the decision on drug non-compliance, all messages annotated as non-compliant are additionally verified by one of the annotators.

The manual annotation process permitted to double-annotate 1,850 messages, among which we count 1,717 messages in the *compliance* category and 133 messages in the *non-compliance* category. These numbers indicate the natural distribution of *non-compliance* messages.

Concerning the annotation into *compliance* and *non-compliance* categories, the inter-annotator agreement [11] is 0.46, which is a moderate agreement [12]. This indicate that is difficult categorisation task.

Within the *non-compliance* category, we count 16 types of non-compliance: they contain between 1 and 29 messages. As example, the *change of weight* type contains 2 messages, *recreational use of drugs* 2 messages, *suicide attempt* 2 messages, and *overuse* 20 messages.

We see that due to the small number of non-adherence messages available and the multiplicity of the types of non-compliance, it may be difficult to obtain sufficient reference data to be used with supervised methods. For these reasons, we propose to use non-supervised methods for the detection of messages with sub-categories of drug non-compliance.

The corpus is pre-processed using Treetagger [13] for tokenization, POS-tagging, and lemmatization. The corpus is used in three versions: (1) in the *forms* corpus, the messages are only tokenized and lowercased; (2) in the *lemmas* corpus, the messages are also lemmatized, the numbers are replaced by a unique placeholder, and diacritics are removed such as in *anxiété/anxiete (anxiety)*; (3) in the *lexical lemmas* corpus, we keep only lemmas of lexical words (verbs, nouns, adjectives, and adverbs). Besides, in each message, the drugs are indexed with the three first characters of the ATC categories [10].

2.2 Categorization with Information Retrieval

We use the Indri information retrieval system [14] to detect non-compliance messages following two ways:

– First, we use the annotated reference data to distinguish between drug compliant and non-compliant messages. The corpus is split in two sets:

1. 44 *non-compliance* messages (one third of the whole *non-compliance* category) are used for the creation of queries, and the query lexicon is weighted proportionally to its frequency in the messages;
2. All 98 *compliance* messages and 89 *non-compliance* messages (two thirds of the whole *non-compliance* category) are used for the evaluation.

The question we want to answer is whether the existing subset of *non-compliance* messages permits to retrieve other *non-compliance* messages. The evaluation is done automatically, using the reference data and computing Precision, Recall, and F-measure with each version of the corpus (forms, lemmas and lexical lemmas). This experiment may give an idea of the performance of this method when searching information in new non-annotated data;

– Then, we look for specific types of non-compliance without using the manual annotations. This exploitation of information retrieval system is fully unsupervised. At the previous steps, we discovered several types of non-compliance that can be found in our corpus. Now, we propose to exploit the existing reference data and our knowledge of the corpus gained at previous steps in order to create the best queries for the detection of similar messages.

These queries are applied to a larger corpus with 20,000 randomly selected messages that contain at least one mention of drugs. The results are evaluated manually computing the Precision.

2.3 Supervised Categorization with Machine Learning

We also perform automatic detection of non-compliant messages with machine learning algorithms. The goal of this step is to provide a state of the art evaluation against which the results provided by the information retrieval methods can be compared.

Supervised machine learning algorithms learn a language model from manually annotated data, which can then be applied to new and unseen data. The categories are drug *compliance* and *non-compliance*. The unit processed is the message. The features are the vectorized text of messages (forms, lemmas and lexical lemmas) and the ATC indexing of drugs. The train set contains 94 non-compliant messages and 93 compliant messages. The test set contains 39 non-compliant messages and 40 compliant messages.

We use the Weka [15] implementation of several supervised algorithms: Naive-Bayes [16], Bayes Multinomial [17], J48 [18], Random Forest [19], and Simple Logistic [20]. These algorithms are used with their default parameters and with the string to wordvector function.

3   Results and Discussion

We present three sets of results obtained when detecting messages with the drug non-compliant information: (1) supervised categorization and evaluation of messages using the information retrieval system (Sect. 3.1); (2) supervised categorization of messages using the machine learning algorithms (Sect. 3.2) (3) unsupervised retrieval of messages using the information retrieval system (Sect. 3.3). We also indicate some limitations of the current work (Sect. 3.4).

3.1   Supervised Categorization and Evaluation with Information Retrieval

Table 1. Information retrieval results for the categorization of messages into the *non-compliance* category

	Precision	Recall	F-measure
Top 10 results			
forms	0.100	0.011	0.020
lemmas	0.400	0.045	0.081
lexical lemmas	0.400	0.045	0.081
Top 100 results			
forms	0.480	0.539	0.508
lemmas	0.480	0.539	0.508
lexical lemmas	**0.500**	**0.561**	**0.529**

The results obtained with the information retrieval system Indri are presented in Table 1. The evaluation values are computed against the reference data for the top 10, 20, 50 and 100 results. With lower cut-off (10, 20, 50) the Recall is limited, since it is impossible to find all 88 non-compliance messages when only 50 messages are processed. With this experiment, the best results are obtained with the *lexical lemmas* corpus. The lemmatization shows an important improvement over the *forms* corpus, which means that lemmatization is important for the information retrieval applications. As expected, the values of Recall and Precision are improved with a larger sample of data: there is more probability that the 89 relevant messages are found among the top 100 messages. Overall, we can see that this information retrieval system can find non-compliant messages although the results are noisy and incomplete.

3.2   Supervised Categorization with Machine Learning

The results of the categorization of messages into the *non-compliance* category obtained with supervised machine learning algorithms are presented in Table 2. We tested several algorithms but show only the results for the best two algorithms, Naive Bayes and Naive Bayes Multinomial. We can observe that the best results (up to 0.824 F-measure) are obtained on the *lexical lemmas* corpus. In all the experiments, Recall is higher than or equal to Precision.

Among the errors observed with NaiveBayes, 12 messages are wrongly categorized as non-compliant and 9 as compliant. Within these 12 messages, four contain terms associated with excess and negative effects (such as *"Je n'imaginais pas que c'était si grave"* (I didn't imagine it was that bad) or *"s'il vous plaît ne faites pas n'importe quoi"* (please don't make a mess)), usually specific to non-compliance messages.

Table 2. Machine learning results obtained for the categorization of messages into the *non-compliance* category

	Precision	Recall	F-measure
NaiveBayes			
forms	0.769	0.769	0.769
lemmas	0.786	0.846	0.815
lexical lemmas	0.761	0.897	**0.824**
NaiveBayesMultinomial			
forms	0.732	0.769	0.750
lemmas	0.795	0.795	0.795
lexical lemmas	0.786	0.846	**0.815**

We conclude that, although information retrieval systems can be adapted to perform categorization of messages thanks to the existing reference data, their results are less competitive comparing to the results obtained with machine learning algorithms. We assume that with larger reference data, information retrieval systems may be more competitive.

3.3 Unsupervised Detection with Information Retrieval

Here, we report a more standard exploitation of the information retrieval system. The experiments are done at a finer-grained level and focus on precise types of non-compliance. Several queries are tested. We will present queries and their results related to important drug non-compliance situations: gain and loose of weight, recreational drug use, suicide attempts or ideation, overdoses, and alcohol consumption. The descriptors used for the creation of queries are issued from the reference data. Their selection is based on their frequency and TFIDF scores [21]. The top 20 results are analyzed for each query.

Gaining/Losing Weight. The keywords used are *poids, kilo, grossir, maigrir (weight, kilo, gain weight, lose weight)*, such as suggested by the manually built reference data. This query is applied to the lemmatized corpus. We expected to find mainly messages related to the use of drugs with the purpose to intentionally lose or gain weight, as well as messages related to weight changes due to side effects of drugs. These expectations are partly verified. Thus, among the top 20 messages, 17 are about weight change as side effects of drugs, one message is about the use of drugs to lose weight intentionally, and 2 messages are about weight loss but with no relation to the consomption of drugs. Overall, among the top 20 messages returned by this query, one new non-compliance message is found. This situation may correspond to the reality (misuse of drugs for weight changes is less frequent than weight change due to drug side effects) or to the corpus used (several messages are concerned with anti-depressant drugs that have as a common side effect weight change). Overall, this gives 0.05 Precision.

Recreational Drug Use. The goal of this set of queries is to retrieve messages where prescription drugs are used for recreational purpose: be "high", reach hallucinations, feel happy, etc. We tried several queries:

- First, the descriptors *drogue, droguer (non-medical drug, to take non-medical drugs)* are used. In French, the word *drogue* usually refers to street drugs, but not to prescription drugs. Yet, in the corpus, people use this word for neuroleptic medication in order to illustrate their feeling that these drugs open the way to addictions and have the same neuroleptic effects as street drugs. Hence, we can find messages such as *J'ai été drogué pendant 3 ans au xanax (I was drugged with xanax for 3 years)* or *Sa soulage mais ses une vrai drogue ce truc !!! (It helps but this stuff is really a drug!!!)* These queries find interesting results (15 out of 20 messages), but may provide different insights than those expected;
- Then, the descriptors *hallu, allu, hallucination (hallucination)* are used. Among the top 20 messages, 2 messages report intentional seeking of the hallucination effects caused by some drugs, 7 messages are about people experiencing hallucinations but as unwanted side effects, 11 messages about people suffering from hallucinations and taking drugs to reduce them. This means that this query provided 2 new messages with non-compliance.
- Finally, the descriptor *planer (to be high from drugs)* is used. Among the top 20 messages, 19 are about the "high" feeling from drugs, be it intentional (9 messages) or non-intentional (10 messages). The 9 intentional messages correspond to non-compliance situations. Like in this message, *"J'ai déjà posté quelques sujets à propos de ce fléau qu'est le stilnox (...) je prends du stilnox, pour m'évader, pour planer"* (I already posted a few topics about this plague that is stilnox (...) I take stilnox, to escape, to get high).

This set of queries illustrate how it is possible to apply an iterative process for the construction of appropriate queries quickly and with little reference data. Overall, the set of queries shows 0.35 Precision.

Suicide. Here, the descriptor used is *suicide (suicide)*. The query is applied to the lemmatized corpus. We expected to find messages in which people report on taking or planning to take drugs (like anti-depressants) with suicidal intentions. Among the top 20 results, 9 messages are about drugs and suicide with no particular relation between them, 5 other messages are about the fact that some drugs may increase the risk of suicide, 5 other messages are critical about the fact that drugs may increase the risk of suicide, and one message reported on a real suicide attempt caused by drug withdrawal. We consider that discussions on relation between drugs and suicide, and of course reporting on suicide attempts, may be important for our research because they represent the importance of these topics in the analyzed fora. This gives 0.7 Precision.

Overuse. The descriptor used is *boites (boxes)* because it often represents the quantity of drugs taken in the reference data. This query is applied to the *forms* corpus because, for this query, it is important to preserve the plural form. Among the top 20 messages, 6 messages are directly related to drug overuse, 3 messages are related to high dosage that may correspond to overuses, 2 messages are related to suicide attempts by ingestion of large amounts of drugs, 2 messages in which people propose to share unused prescription drugs, and, finally, 7 messages are unrelated to drug overuse. This gives 0.65 Precision. Another advantage of this query is that it retrieved various non-compliance situations.

Alcohol. The descriptor used is simply *alcool (alcohol)*. The query is applied to the lemmatized corpus. We expected to find discussions about adverse side effects of alcohol consumption while taking medication. These expectations are partly verified. Hence, among the top 20 results, 12 messages reported about interactions between alcohol and medication, 3 messages discussed about medication prescribed for alcohol withdrawal, while 5 messages were unrelated to the use of medication. Overall, this query gives 0.6 Precision. Additionally, we can notice that 8 out of the 12 messages regarding alcohol-drug interactions are dedicated to the mood disorder medications. This situation may reflect the fact that neuroleptic drugs interact with the effects of alcohol. But this situation may also be the artifact of the corpus containing several messages dedicated to mood disorder drugs.

3.4   Comparison of the Two Categorization Approaches

For the classification task, supervised machine learning shows better results. Yet it requires a too large amount of reference data to be usable to find specific types of non-adherence. For this task we found that information retrieval is able to find non-compliance messages from topics associated with a specific type of non-compliance, such as suicide or overuse. With this method we found 0.45 of average Precision with up to 0.70 for some queries. Besides in the examples of the *boxes* query we were able to discover a new type of non-compliance: people sharing their unused medication with others.

We conclude that these two approaches are complementary: combination of their results may provide an efficient way to enrich the data. The approaches can also be combined: information retrieval queries can quickly provide varied non-compliance messages and thus help supervised machine learning to perform better.

3.5   Limitations of the Current Work

The main limitation regarding our work is the reduced size of the reference data. It contains indeed only 133 messages in the *non-compliance* category. This may limit the performances of the supervised models. Yet, these reference data allow to create quite efficient categorization models, which reach up to 0.824 F-measure in the case of machine learning. We assume that availability of larger reference data will improve the performance of the methods. One of the main motivations to exploit information retrieval methods is the possibility to enrich the reference data with this unsupervised approach.

Another limitation of the work is that messages detected as cases of non-compliance are not currently fully analyzed by medical doctors, pharmacists and pharmacovigilants. Further work will be needed to make the results exploitable by the medical community, who may not be familiar with the methods used in our experiments.

4   Conclusions

This work presents the exploitation of information retrieval methods in two different ways to detect drug non-compliance in Internet fora. We mainly exploit the French forum *Doctissimo*. The messages are first manually assigned into *compliance* and *non-compliance* categories and then used for designing automatic methods and their evaluation.

We adapt Indri, an information retrieval system, for supervised categorisation and evaluation, and obtain up to 0.60 Precision at top 10 results and up to 0.34 Precision at top 50 results. This method can be used to detect non-compliance messages but the noise prevents it from being competitive by comparison with machine learning approaches. Indeed, machine learning algorithms reach up to 0.786 Precision and 0.824 F-measure.

The information retrieval system is also used for a more fine-grained categorization of messages at the level of individual types of non-compliance, where the small number of messages in each category makes supervised learning impossible. This approach is fully unsupervised. Five topics are addressed with different queries suggested by the very few messages available in the reference data. This approach provides on average 0.42 Precision, and up to 0.70 Precision for some queries, such as computed among the top 20 results. This second approach exploiting the information retrieval system can also help in discovering new kinds of non-compliance situations and provide additional insight on topics of concern among patients.

The information gathered thank to these methods can be used by concerned experts (pharmaceutical industry, public health, general practitioners...) to provide prevention and education actions to patients and their relatives. For instance, packaging of drugs can be further adapted to their real use, dedicated brochures and discussions can be done with patients on known and possible drug side effects, on necessary precautions, etc.

The main perspective of the current work is to enrich the reference data and to work more closely with health professionals for the exploitation of the results.

Acknowledgments. This work has been performed as part of the DRUGSSAFE project funded by the ANSM, France and of the MIAM project funded by the ANR, France within the reference ANR-16-CE23-0012. We thank both programs for their funding. We would like also to thank the annotators who helped us with the manual annotation of misuses, Bruno Thiao Layel for extracting the corpus, Vianney Jouhet and Bruno Thiao Layel for building the list with drugs names, and The-Hien Dao for the set of disorders exploited. Finally, we thank the whole ERIAS team for the discussions.

References

1. Bigeard, E., Grabar, N., Thiessard, F.: Typology of drug misuse created from information available in health fora. In: MIE 2018, pp. 1–5 (2018)
2. Gauducheau, N.: La communication des émotions dans les échanges médiatisés par ordinateur: bilan et perspectives. Bulletin de psychologie, pp. 389–404 (2008)
3. Louis, A.: Natural language processing for social media. Comput. Linguist. **42**, 833–836 (2016)
4. Collier, N.: Towards cross-lingual alerting for bursty epidemic events. J. Biomed. Semant. **2** (2011)
5. Lejeune, G., Brixtel, R., Lecluze, C., Doucet, A., Lucas, N.: Added-value of automatic multilingual text analysis for epidemic surveillance. In: Artificial Intelligence in Medicine (AIME) (2013)
6. Tapi Nzali, M.: Analyse des médias sociaux de santé pour évaluer la qualité de vie des patientes atteintes d'un cancer du sein. Thèse de doctorat, Université de Montpellier, Montpellier, France (2017)
7. Morlane-Hondère, F., Grouin, C., Zweigenbaum, P.: Identification of drug-related medical conditions in social media. In: LREC, pp. 1–7 (2016)
8. Kalyanam, J., Katsuki, T., Lanckriet, G.R.G., Mackey, T.K.: Exploring trends of nonmedical use of prescription drugs and polydrug abuse in the twittersphere using unsupervised machine learning. Addict. Behav. **65**, 289–295 (2017)
9. Cameron, D., et al.: PREDOSE: a semantic web platform for drug abuse epidemiology using social media. J. Biomed. Inform. **46**, 985–997 (2013)
10. Skrbo, A., Begović, B., Skrbo, S.: Classification of drugs using the ATC system (anatomic, therapeutic, chemical classification) and the latest changes. Med. Arh. **58**, 138–41 (2004)
11. Cohen, J.: A coefficient of agreement for nominal scales. Educ. Psychol. Measur. **20**, 37–46 (1960)
12. Landis, J., Koch, G.: The measurement of observer agreement for categorical data. Biometrics **33**, 159–174 (1977)
13. Schmid, H.: Probabilistic part-of-speech tagging using decision trees. In: ICNMLP, Manchester, UK, pp. 44–49 (1994)
14. Strohman, T., Metzler, D., Turtle, H., Croft, W.B.: Indri: a language-model based search engine for complex queries. In: Proceedings of the International Conference on Intelligent Analysis (2005)
15. Witten, I., Frank, E.: Data Mining: Practical Machine Learning Tools and Techniques. Morgan Kaufmann, San Francisco (2005)
16. John, G.H., Langley, P.: Estimating continuous distributions in Bayesian classifiers. In: Kaufmann, M. (ed.) Eleventh Conference on Uncertainty in Artificial Intelligence, San Mateo, pp. 338–345 (1995)
17. McCallum, A., Nigam, K.: A comparison of event models for Naive Bayes text classification. In: AAAI Workshop on Learning for Text Categorization, Madison, Wisconsin (1998)

18. Quinlan, J.: C4.5 Programs for Machine Learning. Morgan Kaufmann, San Mateo (1993)
19. Breiman, L.: Random forests. Mach. Learn. **45**, 5–32 (2001)
20. Landwehr, N., Hall, M., Frank, E.: Logistic model trees. Mach. Learn. **95**, 161–205 (2005)
21. Salton, G., McGill, M.J.: Retrieval refinements, pp. 199–206 (1983)

Char-RNN and Active Learning
for Hashtag Segmentation

Taisiya Glushkova and Ekaterina Artemova[(⊠)]

National Research University Higher School of Economics, 20 Myasnitskaya Ulitsa,
Moscow 101000, Russia
toglushkova@edu.hse.ru, echernyak@hse.ru

Abstract. We explore the abilities of character recurrent neural network (char-RNN) for hashtag segmentation. Our approach to the task is the following: we generate synthetic training dataset according to frequent n-grams that satisfy predefined morpho-syntactic patterns to avoid any manual annotation. The active learning strategy limits the training dataset and selects informative training subset. The approach does not require any language-specific settings and is compared for two languages, which differ in inflection degree.

Keywords: Hashtag segmentation · Recurrent neural network · Character level model

1 Introduction

A hashtag is a form of metadata labeling used in various social networks to help the users to navigate through the content. For example, one of the most popular hashtags on Instagram is "#photooftheday" [photo of the day]. Hashtags are written without any delimiters, although some users use an underscore or camel-casing to separate words. Hashtags themselves may be a great source for features for following opinion mining and social network analysis. Basically hashtags serve as keyphrases for a post in social media. By segmenting the hashtags into separate words we may use regular techniques to process them. The problem of hashtag segmentation resembles of another problem, namely word segmentation.

The problem of word segmentation is widely studied in languages like Chinese, since it lacks whitespaces to separate words, or in German to split compound words. In languages like English or Russian, where compounds are not that frequent as in German and where whitespace delimiters are regularly used, the problem of word segmentation arises mainly when working with hashtags.

Formally the problem is stated as follows: given a string of n character $s = s_1 \ldots s_n$ we need to define the boundaries of the substrings $s_{i:j}, i < j$, so that each substring is meaningful (i.e. is a regular word, named entity, abbreviation, number, etc.). The main challenge of this problem is that the segmentation might be ambiguous. For example, a string "somethingsunclear" might be segmented

A. Gelbukh (Ed.): CICLing 2019, LNCS 13451, pp. 155–168, 2023.
https://doi.org/10.1007/978-3-031-24337-0_12

as "something sun clear" or "somethings unclear". To deal with the ambiguity more processing is required, such as POS-tagging, estimation of frequencies of all hashtag constituencies or their co-occurrence frequency. The frequencies can be estimated on a large corpus, such as BNC[1], COCA[2], Wikipedia. However when working with noisy user generated data, such as texts or hashtags from social networks, the problem of unknown words (or out of vocabulary words) arises. In language modeling this problem is solved by using smoothing, such as Laplacian smoothing or Knesser-Ney smoothing. Otherwise additional heuristics can be used to extend the dictionary with word-like sequences of characters. Unlike language modelling, in hashtag segmentation frequency estimation is not only source for defining word boundaries. Otherwise candidate substrings can be evaluated according to length [24].

Several research groups have shown that introducing character level into models help to deal with unknown words in various NLP tasks, such as text classification [7], named entity recognition [14], POS-tagging [8], dependency parsing [6], word tokenization and sentence segmentation [28] or machine translation [4,5]. The character level model is a model which either treats the text as a sequence of characters without any tokenization or incorporates character level information into word level information. Character level models are able to capture morphological patterns, such as prefixes and suffixes, so that the model is able to define the POS tag or NE class of an unknown word.

Following this intuition, we use a character level model for hashtag segmentation. Our main motivation is the following: if the character level model is able to capture word ending patterns, it should also be able to capture the word boundary patterns. We apply a character level model, specifically, a recurrent neural network, referred further as char-RNN, to the task of hashtag segmentation. The char-RNN is trained and tested on the synthetic data, which was generated from texts, collected from social networks in English and Russian, independently. We generate synthetic data for training by extracting frequent N-grams and removing whitespaces. The test data is annotated manually[3]. Since the problem statement is very basic, we use additional techniques, such as active learning, character embeddings and RNN hidden state visualization, to interpret the weights, learned by char-RNN. We address the following research questions and claim our respective contributions:

1. We show that our char-RNN model outperforms the traditional unigram or bigram language models with extensive use of external sources [24,31].
2. What is the impact of high inflection in languages such as Russian on the performance of character-level modelling as opposed to languages with little inflection such as English? We claim that character-level models offer benefits for processing highly inflected languages by capturing the rich variety of word boundary patterns.

---

[1] https://corpus.byu.edu/bnc/.

[2] https://corpus.byu.edu/coca/.

[3] The test data is available at: https://github.com/glushkovato/hashtag_segmentation.

3. As getting sufficient amount of annotated training collection is labor-intensive and error-prone, a natural question would be: can we avoid annotating real-world data altogether and still obtain high quality hashtag segmentations? We approach this problem by using morpho-syntactic patterns to generate synthetic hashtags.
4. A potentially unlimited volume of our synthetic training dataset raises yet another question of whether an informative training subset could be selected. To this extent, we apply an active learning-based strategy to subset selection and identify a small portion of the original synthetic training dataset, necessary to obtain a high performance.

2   Neural Model for Hashtag Segmentation

2.1   Sequence Labeling Approach

We treat hashtag segmentation as a sequence labeling task. Each character is labeled with one of the labels $\mathcal{L} = \{0,1\}$, (1) for the end of a word, and (0) otherwise (Table 1 and 2). Given a string $s = s_1, \ldots, s_n$ of characters, the task is to find the labels $Y^* = y_1^* \ldots, y_n^*$, such that $Y^* = \arg\max_{Y \in \mathcal{L}^n} p(Y|s)$ (Fig. 1).

Table 1. Illustration of sequence labeling for segmentation #ремонтдома ["ремонт дома", house renovation]

р	е	м	о	н	т	д	о	м	а
0	0	0	0	0	1	0	0	0	1

Table 2. Illustration of sequence labeling for segmentation "#photooftheday" [photo of the day]

p	h	o	t	o	o	f	t	h	e	d	a	y
0	0	0	0	1	0	1	0	0	1	0	0	1

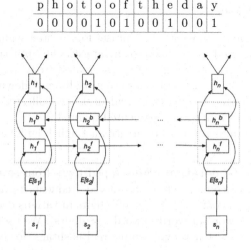

Fig. 1. Neural model for hashtag segmentation

The neural model for hashtag segmentation consists of three layers.

The embedding layer is used to compute the distributed representation of input characters. Each character c_i is represented with an embedding vector $e_i \in \mathbb{R}^{d_e}$, where d_e is the size of the character embedding. E is the look up table of size $|V| \times d_e$, where V is the vocabulary, i.e. the number of unique characters.

The feature layer is used to process the input. We use a bi-directional recurrent layer with LSTM units to process the input in forward and backward directions. The LSTM units we use are default keras LSTM units as introduced by Hochreiter.

The inference layer is used to predict the labels of each character. We use a single dense layer as for inference and *softmax* to predict the probabilities of the labels $\mathcal{L} = \{0, 1\}$.

$$softmax(y_j) = \frac{e^{y_j}}{\sum_{k=1}^{|\mathcal{L}|} e^{y_k}}$$

Each character is assigned with the most probable label.

The parameters of the char-RNN are the following:

1. Embedding layer = 50 input dimensions;
2. Feature layer = 64 bidirectional LSTM units;
3. Inference layer = 2 output neurons with softmax activation function mapped to each of 64 outputs.

3   Dataset

In this section we describe the datasets we used for hashtag segmentation. We experimented with Russian and English datasets to compare the performance of the char-RNN.

3.1   Russian Dataset

To our knowledge there is no available dataset for hashtag segmentation in Russian, so we faced the need to create our own dataset. Our approach to the dataset creation was twofold: the training data was created from social network texts by selecting frequent n-grams and generating hashtags following some hashtag patterns. The test dataset consists of real hashtags collected from vk.com (a Russian social network) and were segmented manually.

We followed the same strategy to create an English language dataset.

Training Dataset Generation. We scraped texts from several pages about civil services from vk.com. Next we extracted frequent n-grams that do not contain stopwords and consist of words and digits in various combinations (such as word + 4 digits + word or word + word + 8 digits). We used several rules to merge these n-grams so that they resemble real hashtags, for example:

- remove all whitespace: wordwordworddigits
  Examples: ЁлкаВЗазеркалье, нескольколетназад
- replace all whitespace with an underscore: word_word_digits
  Examples: увд_юга_ столицы
- remove some whitespace and replace other spaces with an underscore: word_worddigits.
  Examples: ищусвоегогероя_уфпс

A word here might be a word in lower case, upper case or capitalized or an abbreviation. There might be up to four digits.

In general, we introduced 11 types of hashtags, which contain simply constructed hashtags as well as the complex ones. Here are a couple of examples:

- The hashtag consists of two parts: the word/abbreviation in the first part and the number or word in the second. The underscore is a delimiter.
  Examples: word_2017, NASA_2017, word_word
- Two or three words, which are separated by an underscore.
  Examples: Word_Word, word_word_word

Test Dataset Annotation. We segmented manually 2K the most frequent hashtags, extracted from the same collection of the scraped texts.

The resulting size of the Russian dataset is 15k hashtags for training and 2k hashtags for testing.

3.2   English Dataset

We used the dataset, released by [24]. This dataset consists of:

- a collection of tweets, which we used to generate the synthetic training hashtags according to the same rules as for Russian;
- a collection of annotated and separated hashtags, which we used as a testing set. From this test set we excluded ambiguous hashtags, annotated with several possible segmentations.

The resulting size of the English dataset is 15k hashtags for training and 1k hashtags for testing.

4   Active Learning

We followed the strategy for active learning, as in [19]. The training procedure consists of multiple rounds of training and testing of the model. We start by training the model on 1k hashtags, which were randomly selected from the training dataset. Next we test the model on the reminder of the training dataset and select 1k hashtags according to the current model's uncertainty in its prediction of the segmentation. These hashtags are not manually relabelled, since a) they belong to the synthetically generated training dataset and b) the correct

Table 3. Samples from both datasets

	Russian dataset	English dataset
Train	мвдпетровкадети	sunwouldcome
	ПоисковыхРабот09	ThingsGoingWell
	КМССтихи	StartSchool_72
	середины_века	tonightgoodnight
	Будем_Рады	muchloveu
Test	ЛайфхакМЧС	KrispyKreme
	важно_знать	twitteriffic
	ЗавтраБылаВойна	titsuptuesday
	важнаядата	MissUSA
	ФИФА2018	ipv6summit

labeling for these hashtag is already known. In [19] three uncertainty measure are presented, from which we selected the maximum normalized log-probability (MNLP) assigned by the model to the most likely sequence of tags. The model is then retrained on the hashtags it is uncertain about. Note, that here we do not check if the predictions of the model are correct. We are more interested in training the model on hard examples than in evaluating the quality of intermediate results. We refer the reader to [19] for more technical details (Table 3).

5   Experiments

5.1   Baseline

As for baseline algorithm, we consider the [24] system architecture as a state-of-the-art algorithm. Unfortunately, their approach is not straightforwardly applicable to our synthetic Russian dataset, because it requires twofold input: a hashtag and a corresponding tweet or a text from any other social media, which is absent in our task setting due to synthetic nature of the training dataset.

For this reason as a baseline algorithm for English dataset we refer to results from [24], and as for Russian dataset, we used the probabilistic language model, described in [31]. The probability of a sequence of words is the product of the probabilities of each word, given the word's context: the preceding word. As in the following equation:

$$p(w_1, w_2, .., w_n) = \prod_{i=1}^{n} p(w_i | w_{i-1}),$$

where

$$p(w_i | w_{i-1}) = \frac{f(w_{i-1}, w_i)}{f(w_{i-1})}$$

In case there is no such a pair of words (w_{i-1}, w_i) in the set of bigrams, the probability of word w_i is obtained as if it was only an unigram model:

$$p(w_i) = \frac{f(w_i) + \alpha}{\sum_{j=1}^{|W|} f(w_j) + \alpha|V|}$$

where V - vocabulary, $f(w_i)$ - frequency of word w_i, and $\alpha = 1$.

In Table 4 we present three baseline results: LM [31] for Russian and English datasets; context-based LM [24] for English dataset only. We treat a segmentation as correct if prediction and target sequences are the same.

Table 4. Accuracy of the baseline algorithm on the Russian and English datasets

	Accuracy
Russian dataset [31]	0.634
English dataset [31]	0.526
English dataset [24]	0.711

5.2 Neural Model

In our experiments we used 5 epochs to train the char-RNN with LSTM units. For each language we observed three datasets with different number of hashtags. In case of Russian language, the more data we use while training, the higher the accuracy. As for English, the highest accuracy score was achieved on a set of 10k hashtags (Table 5). Due to it's lower morphological diversity and complexity the model starts to overfit on training sets with large sizes. The training showed that mostly the model makes wrong predictions of segmentation on hashtags of complex types, such as "wordword_worddigits".

Table 5. Accuracy of LSTM char-RNN on both Russian and English datasets

	Accuracy, Russian dataset	Accuracy, English dataset
5k	0.9682	0.9088
10k	0.9765	**0.9134**
15k	**0.9811**	0.8733

Our results outperform all choosen baseline both for Russian and English datasets. Note, that we have two baselines for the English dataset: one is purely frequency-based, another is cited from [24], where external resources are heavily used. We show that using significantly less amount of training data, we achieve a boost in quality by switching from statistical word language models to char-RNN. As expected, the results on Russian dataset are higher than for the English dataset due to higher inflection degree in Russian as opposed to English.

5.3 Active Learning

In order to evaluate the efficiency of deep learning with active learning when used in combination, we run the experiments for both languages. As for the datasets, we took the ones on which the highest accuracy was obtained (15k for Russian and 10k for English).

The learning process consists of multiple rounds which are repeated until the test set is finished. At the beginning we train the model on 1k of randomly selected hashtags and predict the probability of segmentation for the remaining hashtags. Then we sort the remaining hashtags in ascending order according to the probability assigned by the model and pick 1k of hashtags which the model is least confident about. Finally, we add these hashtags with the least probable sequence of tags to the training data and continue training the model. This pipeline is repeated till there are no samples left.

In comparison to our initial experiments, application of active learning demonstrates impressive results. The amount of labeled training data can be drastically reduced, to be more specific, in both cases the size of the training set can be reduced by half without any decline in accuracy (see Figs. 2 and 3).

Active learning selects a more informative set of examples in contrast to supervised learning, which is trained on a set of randomly chosen examples. We decided to analyze the updated version of the training data and see if number of morphologically complex types of hashtags is higher than the simple ones. We were able to divide hashatgs into complex and simple as the model is trained on synthetic data and there is a finite number of templates by which each hashtag can be generated.

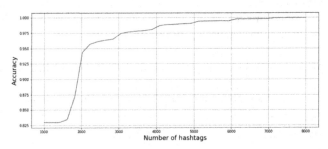

Fig. 2. Accuracy obtained on Russian Dataset

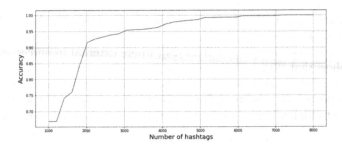

Fig. 3. Accuracy obtained on English Dataset

To better understand the contribution of uncertainty sampling approach, we plot the distribution of different types of hashtags in new training datasets for both languages, Russian and English (see Fig. 4 and 5). According to identified types of hashtags in real data, it can be seen from the plots that in both cases the algorithm added more of morphologically complex hashtags to training data - types 3, 6 and 7. These types mostly consist of hashtags with two or three words in lower case without underscore.

Examples of featured types:

type 3: wordword_2017
type 6: wordword, word2017word
type 7: wordwordword, wordword2017word

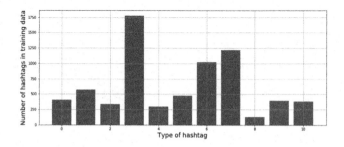

Fig. 4. Distribution of top 7k hashtag types in Russian dataset, chosen by an active learning algorithm

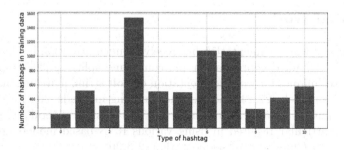

Fig. 5. Distribution of top 7k hashtag types in English dataset, chosen by an active learning algorithm

5.4   Visualization

In order to see if embeddings of similar characters, in terms of string segmentation, appear near each-other in their resulting 50-dimensional embedding space, we have applied one technique for dimensionality reduction: SVD to character embeddings to plot them on 2D space. For both languages meaningful and interpretable clusters can be extracted: capital letters, letters in lower case, digits and underscore, as shown below (Figs. 6 and 7).

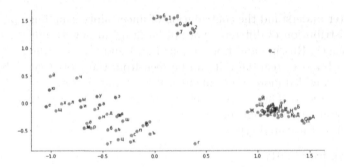

Fig. 6. SVD visualization of character embeddings on Russian dataset

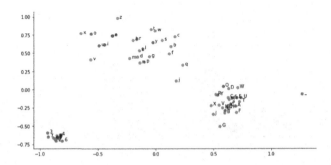

Fig. 7. SVD visualization of character embeddings on English dataset

6   Related Work

The problem of word segmentation has received much attention in Chinese and German NLP for word segmentation and compound splitting [30], respectively. The major techniques for word segmentation exploit string matching algorithms [21], language models [22,24] and sequence labeling methods [30]. Recent trend of deep learning as a major approach for any NLP task in general and sequence labeling in particular resulted in using various RNN-based models and CNN-based model for Chinese word segmentation [2,3,30].

Since [30] Chinese word segmentation is addressed as a character labeling task: each character of the input sequence is labeled with one of the four labels $\mathcal{L} = \{B, M, E, S\}$, which stand for character in Begin, Middle or End of the word or Single character word. [30] uses a maximum entropy tagger to tag each character independently. This approach was extended in [29] to the sequence modeling task, and linear conditional random fields were used to attempt it and receive state of the art results. A neural approach to Chinese segmentation mainly uses various architectures of character level recurrent neural networks [11–13] and very deep constitutional networks [10]. Same architectures are used for dialectal Arabic segmentation [9].

The evolution of German compound splitters is more or less similar to Chinese word segmentation systems. The studies of German compound splitting started with corpus- and frequency-based approaches [2,3] and are now dominated with neural-based distributional semantic models. However, German compound splitting is rarely seen as sequence modeling task.

The problem of hashtag segmentation, analysis and usage in English has been approached by several research groups. As it was shown by [22] hashtag segmentation for TREC microblog track 2011 [23] improves the quality of information retrieval, while [24] shows that hashtag segmentation improves linking of entities extracted from tweets to a knowledge base. Both [22,24] use Viterbi-like algorithm for hashtag segmentation: all possible segmentations of hashtag are scored using a scoring function:

$$\texttt{Score}(S) = \sum_{s_i \in S} \log(P_{Unigram}(s_i)),$$

where $P_{Unigram}$ are probabilities, computed according to the unigram model based on a large enough corpus or any N-gram service.

Following the idea of scoring segmentation candidates, [21] introduces other scoring functions, which include a bigram model (2GM) and a Maximum Unknown Matching (MUM), which is adjustable to unseen words.

[25] attempt to split camel-cased hashtags using rule-based approach and POS-tagging for further semantic classification. WordSegment[4] has been used for sentiment analysis [26,27] and other applications.

To our knowledge there has been little work done for word or hashtag segmentation in Russian.

6.1 Active Learning in NLP

Active learning is machine learning technique which allows efficient use of the available training data. It presumes that, first an initial model is trained on a very little amount of data and next tested on large unlabeled set. Next the model is able to choose a few most difficult examples and ask an external knowledge source about the desired labels. Upon receiving these labels, the model is updated and retrained on the new train set. There might be a few rounds of label querying and model updating. To use active learning strategy, we need a definition of what a difficult example is and how to score its difficulty. One of the most common scoring approaches is entropy-based uncertainty sampling, which selects the examples with the lowest prediction probability.

Active learning is widely used in NLP applications, when there is little annotated data while the amount of unlabeled data is abundant. Being ultimately used for text classification using traditional machine learning classifiers [17,18], active learning is less known to be used with deep learning sequence classifiers. Recent works report on scoring word embeddings that are likely to be updated with the greatest magnitude [20] and on using maximum normalized

---

[4] http://www.grantjenks.com/docs/wordsegment/.

log-probability (MNLP) assigned by the model to the most likely sequence of tags [19]:

$$\frac{1}{n} \max_{y_1 \ldots y_n} \sum_{i=1}^{n} \log P[y_n | y_1, \ldots, y_{n-1}, x_{ij}]$$

6.2   Training on Synthetic Data

The lack of training data is an issue for many NLP applications. There have been attempts to generate and use synthetic data for training question answering systems [15] and SQL2text systems [16]. In [24] synthetic hashtags are generated by removing whitespace characters from frequent n-grams, while in [1] German compounds are synthesized for further machine translation.

7   Conclusions

In this paper we approach the problem of hashtag segmentation by using char-RNNs. We treat the problem of hashtag segmentation as a sequence labeling task, so that each symbol of a given string is labeled with 1 (there should be a whitespace after this symbol) or 0 (otherwise). We use two datasets to test this approach in English and in Russian without any language-specific settings. We compare char-RNN to traditional probabilistic algorithms. To interpret the results we use a few visualization techniques and the strategy of active learning to evaluate the complexity of training data, since we use synthetically generated hashtags for training.

The results show that:

1. When approached on character level, hashtag segmentation problem can be solved using relatively small and simple recurrent neural network model without usage of any external corpora and vocabularies. Such char-RNN not only outperforms significantly traditional frequency-based language models, but also can be trained on synthetic data generated according to morpho-syntactic patterns, without any manual annotation and preprocessing.
2. In languages with high inflection (such as Russian) the char-RNN achieves higher results than in languages with little inflections (such as English) due to the ability of the char-RNN to capture and memorize word boundary patterns, especially word ending patterns (i.e. adjective endings "ый","ая","ое" or verbal endings "ать","еть" in Russian).
3. The amount of generated synthetic training data can be limited by using techniques for active learning which allows to select sufficient training subset without any loss of quality.

Acknowledgements. The paper was prepared within the framework of the HSE University Basic Research Program and funded by the Russian Academic Excellence Project '5-100'.

References

1. Matthews, A., Schlinger, E., Lavie, A., Dyer, C.: Synthesizing compound words for machine translation. In: Proceedings of the 54th Annual Meeting of the Association for Computational Linguistics (Volume 1: Long Papers), pp. 1085–1094 (2016)
2. Riedl, M., Biemann, C.: Unsupervised compound splitting with distributional semantics rivals supervised methods. In: Proceedings of the 2016 Conference of the North American Chapter of the Association for Computational Linguistics: Human Language Technologies, pp. 617–622 (2016)
3. Koehn, P., Knight, K.: Empirical methods for compound splitting. In: Proceedings of the Tenth Conference on European Chapter of the Association for Computational Linguistics, vol. 1, pp. 187–193 (2003)
4. Vaswani, A., et al.: Attention is all you need. In: Advances in Neural Information Processing Systems, pp. 6000–6010 (2017)
5. Chung, J., Cho, K., Bengio, Y.: A character-level decoder without explicit segmentation for neural machine translation (2016)
6. Alberti, C., et al.: SyntaxNet models for the CoNLL 2017 shared task (2017)
7. Joulin, A., Grave, E., Bojanowski, P., Mikolov, T.: Bag of tricks for efficient text classification. In: EACL 2017, p. 427 (2017)
8. Santos, C.D., Zadrozny, B.: Learning character-level representations for part-of-speech tagging. In: Proceedings of the 31st International Conference on Machine Learning, ICML 2014, pp. 1818–1826 (2014)
9. Samih, Y., et al.: A neural architecture for dialectal Arabic segmentation. In: Proceedings of the Third Arabic Natural Language Processing Workshop, pp. 46–54 (2017)
10. Sun, Z., Shen, G., Deng, Z.: A gap-based framework for Chinese word segmentation via very deep convolutional networks (2017)
11. Cai, D., Zhao, H., Zhang, Z., Xin, Y., Wu, Y., Huang, F.: Fast and accurate neural word segmentation for Chinese (2017)
12. Zhang, Q., Liu, X., Fu, J.: Neural networks incorporating dictionaries for Chinese word segmentation (2018)
13. Cai, D., Zhao, H.: Neural word segmentation learning for Chinese. In: Proceedings of the 54th Annual Meeting of the Association for Computational Linguistics (Volume 1: Long Papers), pp. 409–420 (2016)
14. Ma, X., Hovy, E.: End-to-end sequence labeling via bi-directional LSTM-CNNs-CRF (2016)
15. Weston, J., et al.: Towards AI-complete question answering: a set of prerequisite toy tasks (2015)
16. Utama, P., et al.: An end-to-end neural natural language interface for databases (2018)
17. Schohn, G., Cohn, D.: Less is more: active learning with support vector machines. In: ICML, pp. 839–846 (2000)
18. Tong, S., Koller, D.: Support vector machine active learning with applications to text classification. J. Mach. Learn. Res. **2**, 45–66 (2001)
19. Shen, Y., Yun, H., Lipton, Z.C., Kronrod, Y., Anandkumar, A.: Deep active learning for named entity recognition (2017)
20. Zhang, Y., Lease, M., Wallace, B.C.: Active discriminative text representation learning. In: AAAI, pp. 3386–3392 (2017)
21. Reuter, J., Pereira-Martins, J., Kalita, J.: Segmenting Twitter hashtags. Int. J. Nat. Lang. Comput. **5**, 23–36 (2016)

22. Berardi, G., Esuli, A., Marcheggiani, D., Sebastiani, F.: ISTI@ TREC Microblog Track 2011: Exploring the Use of Hashtag Segmentation and Text Quality Ranking. TREC (2011)
23. Ounis, I., Macdonald, C., Lin, J., Soboroff, I.: Overview of the TREC-2011 microblog track. In: Proceedings of the 20th Text REtrieval Conference (TREC 2011) (2011)
24. Bansal, P., Bansal, R., Varma, V.: Towards deep semantic analysis of hashtags. In: Hanbury, A., Kazai, G., Rauber, A., Fuhr, N. (eds.) ECIR 2015. LNCS, vol. 9022, pp. 453–464. Springer, Cham (2015). https://doi.org/10.1007/978-3-319-16354-3_50
25. Declerck, T., Lendvai, P.: Processing and normalizing hashtags. In: Proceedings of the International Conference Recent Advances in Natural Language Processing, pp. 104–109 (2015)
26. Akhtar, Md.S., Sawant, P., Ekbal, A., Pawar, J., Bhattacharyya, P.: IITP at EmoInt-2017: measuring intensity of emotions using sentence embeddings and optimized features. In: Proceedings of the 8th Workshop on Computational Approaches to Subjectivity, Sentiment and Social Media Analysis, pp. 212–218 (2017)
27. Park, J.H., Xu, P., Fung, P.: PlusEmo2Vec at SemEval-2018 Task 1: Exploiting emotion knowledge from emoji and# hashtags (2018)
28. Shao, Y., Hardmeier, C., Nivre, J.: Universal word segmentation: implementation and interpretation. Trans. Assoc. Computat. Linguist. 6, 421–435 (2018)
29. Peng, F., Feng, F., McCallum, A.: Chinese segmentation and new word detection using conditional random field. In: Proceedings of the 20th International Conference on Computational Linguistics, p. 562 (2004)
30. Xue, N.: Chinese word segmentation as character tagging. Int. J. Comput. Linguist. Chin. Lang. Process. 8(1), 29–48 (2003). Special Issue on Word Formation and Chinese Language Processing
31. Norvig, P.: Natural language corpus data. Beautiful Data 219–242 (2009)

Extracting Food-Drug Interactions from Scientific Literature: Relation Clustering to Address Lack of Data

Tsanta Randriatsitohaina[1] and Thierry Hamon[1,2(✉)]

[1] LIMSI, CNRS, Univ. Paris-Sud, Université Paris-Saclay, 91405 Orsay, France
{tsanta,hamon}@limsi.fr
[2] Université Paris 13, Sorbonne Paris Cité, 93430 Villetaneuse, France

Abstract. Food-Drug Interaction (FDI) occurs when food and drug are taken simultaneously and cause unexpected effect. This paper tackles the problem of mining scientific literature in order to extract these interactions. We consider this problem as a relation extraction task which can be solved with classification method. Since Food-Drug Interactions need a fine-grained description with many relation types, we face the data sparseness and the lack of examples per type of relation. To address this issue, we propose an effective approach for grouping relations sharing similar representation into clusters and reducing the lack of examples. Cluster labels are then used as labels of the dataset given to classifiers for the FDI type identification. Our approach, relying on the extraction of relevant features before, between, and after the entities associated by the relation, improves significantly the performance of the FDI classification. Finally, we contrast an intuitive grouping method based on the definition of the relation types and a unsupervised clustering based on the instances of each relation type.

Keywords: Food-Drug interactions · Information extraction · Text mining

1 Introduction

Although knowledge bases or terminologies exist in specialized domains, updating these information often requires to access unstructured data such as scientific literature. The problem deeply occurs when focusing on a new knowledge which has no recording in terminological resources yet. Thus, while drug interactions [1] or drug adverse effects [2] are listed in databases such as DrugBank[1] [14] or Thériaque[2], other information such as interactions between drug and food is barely listed in knowledge and mainly scattered in heterogeneous sources [14]. Besides, information is mainly stored as sentences. Actually, while food-drug interactions can correspond to various types of adverse drug effects and lead to harmful consequences on the patient's health and well-being, they are less known and studied and consequently very sparse in the scientific

---

[1] https://www.drugbank.ca/.

[2] http://www.theriaque.org.

© Springer Nature Switzerland AG 2023
A. Gelbukh (Ed.): CICLing 2019, LNCS 13451, pp. 169–180, 2023.
https://doi.org/10.1007/978-3-031-24337-0_13

literature. Similarly to interactions between drugs, Food-Drug Interaction (FDI) corresponds to the appearance of an unexpected effect. For example, grapefruit is known to inhibit the effect of an enzyme involved in the metabolism of several drugs [7]. Other foods may affect the absorption of a drug or its distribution in the organism [5].

The relation extraction task in biomedical texts generally consists in the identification of the related entities and the recognition of the relation category. In this article, we address the automatic identification of interaction statements between drug and food in abstracts of scientific articles issued from the Medline database.

To extract this information from the abstracts, we face several difficulties: (1) drugs and foods are very variable in the summaries. Drug can be mentioned by its common international name or active drug substances, while foods may be referenced by a particular nutrient, component or food family; (2) the interactions are described in a rather precise way in the texts, which leads to a limited number of examples; (3) the available set of annotations does not include the different types of interaction homogeneously and the learning set is often unbalanced.

Our contributions focus on FDI extraction and improvement of previous classification results by proposing a relation representation which addresses the lack of data, applying clustering method on type of relations, and using cluster labels in a classification step for identification of FDI type.

2   Related Work

Various types of approaches have been explored to extract relations from biomedical texts. Some approaches combine patterns and CRF for recognition of symptoms in biomedical texts [8]. Other approaches generate automatically lexical patterns for processing free text in clinical documents relying on a multiple sequential alignment to identify similar contexts [12]. Sentence's verb is compared to a list of verbs known as indicating relation to determine the relation between entities [15]. Then they construct the syntax dependency tree around the verbs to identify the related entities.

Drug-drug Interaction (DDI) extraction described by [3] is similar to our food-drug interaction extraction problem even if we need to identify much more types of relation (see Sect. 3). Our method joins their two steps approach for DDI detection and classification in which we added a relevant sentences selection step as proposed in [10]. [10] focus on the identification of relevant sentences and abstracts for extraction of pharmacokinetic evidence of DDI. [9] built two classifiers for DDI extraction: a binary classifier to extract interacting drug pair and a DDI type classifier to associate the interacting pairs with predefined relation categories. [4] consider the extraction of protein localization relation as a binary classification. All the protein-location pairs appearing in the same sentence are considered as positive instances if they are related, and negative otherwise. In contrast, we use multi-class classification for relation type recognition. [11] propose a CNN-based method for DDI extraction. In their model, drug mentions in a sentence are normalized in the following way: the two considered drug names are replaced by drug1 and drug2 according to the occurrence order in the sentence, respectively, and all the other drugs are replaced by drug0. Other works use recurrent neural network model with multiple attention layers for DDI classification [17].

3   Dataset

Studies have already been conducted considering Food-Drug Interactions in which the POMELO dataset was developed [6]. This dataset consists of 639 abstracts of scientific articles from the medical field (269,824 words, 5,752 sentences). They were collected from the PubMed portal[3] by the query: (`"FOOD DRUG INTERACTIONS"[MH] OR "FOOD DRUG INTERACTIONS*"`) AND (`"adverse effects*"`). All 639 abstracts were annotated according to 9 types of entities and 21 types of relation in Brat [16] by a pharmacy resident. The annotations focus on information about relation between food, drug and pathologies.

Since we are considering Food-Drug Interactions in this paper, we construct our dataset by taking into account every couple of *drug* and *food* or *food-supplement* from POMELO dataset. The resulting dataset is composed of 831 sentences labelled with 13 types of relations: *decrease absorption, slow absorption, slow elimination, increase absorption, speed up absorption, new side effect, negative effect on drug, worsen drug effect, positive effect on drug, improve drug effect, no effect on drug, without food, non-precised relation*. The statistics of the dataset is given in Table 1. Meaning of the relation type is detailed in Sect. 4.1.

Table 1. Statistics of annotated relations by initial types

Relation type	#	Percentage
Non-precised relation	476	57.3%
Decrease absorption	49	5.9%
Positive effect on drug	19	2.3%
Negative effect on drug	85	10.2%
Increase absorption	38	4.6%
Slow elimination	15	1.8%
Slow absorption	15	1.8%
No effect on drug	109	13.1%
Improve drug effect	6	0.7%
Without food	9	1.1%
Speed up absorption	1	0.1%
Worsen drug effect	5	0.6%
New side effect	4	0.5%
Total	831	100%

4   Grouping Types of Relation

The distribution of our dataset is very unbalanced as shown in the Table 1. For instance, *speed up absorption* relation has only one example, which do not permit efficient generalization of the represented relation. This lack of examples is due to the fine-grained

---

[3] https://www.ncbi.nlm.nih.gov/pubmed/.

description of the relations. To solve this problem, we propose two methods for grouping relations sharing similarities in order to obtain more examples per group of relations. The first method relies on the definition of relation types (intuitive grouping) while the second one is based on unsupervised clustering of the relation instances.

4.1 Intuitive Grouping

In this section, we propose a very intuitive way for grouping Food-Drug relations. FDI identification task presents similarity with Drug-Drug Interaction, where two drugs taken together lead to a modification of their effects. ADME [5] (absorption, distribution, metabolism and excretion) relations are involved, but applying this grouping in POMELO dataset would require supplementary annotation process.

The intuitive grouping is done as below:

1. **Non-precised relation.** Instances labelled with '*non-precised relation*' do not give more precision about the relation involved. While we do not have information that would permit to combine them with another relation, they will be considered as one individual group, especially since they represent more than half of the data.
2. **No effect.** '*No effect on drug*' instances represent food-drug relations in sentences where it is explicitly expressed that the considered food has no effect on the drug, unlike other relations that express actual food-drug interactions. As a result, these instances are represented as one individual group.
3. **Reduction.** Since instances labelled with '*decrease absorption*', '*slow absorption*', '*slow elimination*' express diminution of action of drug under the influence of a food, they are grouped to form the *reduction* relation.
4. **Augmentation.** Similarly to *reduction* relation, instances labelled with '*increase absorption*', '*speed up absorption*' are grouped to form the *augmentation* relation.
5. **Negative.** This group includes instances labelled with '*new side effect*', '*negative effect on drug*', '*worsen drug effect*', '*without food*'. *negative effect on drug* express explicitly a negative effect of food on drug, '*worsen drug effect*' expresses a negative effect of the drug, *side effect* is generally an adverse effect of the drug that join a negative connotation, the same to '*without food*' that prevents from taking food with the considered drug.
6. **Positive.** By analogy with the *negative* relation, '*positive effect on drug*', '*improve drug effect*' are grouped to form the *positive* relation.

For the rest of the paper, we will note this intuitive grouping method ARNP that stands for Augmentation, Reduction, Negative and Positive.

At the end, we get 6 Food-Drug relation types with relatively balanced number of examples. Statistics of this new distribution are given in Table 2.

4.2 Unsupervised Clustering

Clustering is a data mining method that aims at dividing a set of data into different homogeneous groups, in that the data of each subset shares common characteristics, which most often correspond to similarity criteria defined by measures of distance

Table 2. Statistics of annotated relations by grouped types

Relation type	#	Percentage
Non-precised relation	476	57.3%
No effect on drug	109	13.1%
Reduction	79	9.5%
Augmentation	39	4.7%
Negative	103	12.4%
Positive	25	3%
Total	831	100%

between elements. To obtain a good partitioning, it is necessary to minimize intra-class inertia to obtain clusters as homogeneous as possible and maximize inter-class inertia in order to obtain well-differentiated subsets. In this section, we propose to use clustering method to group Food-Drug relations involving food effect on drug.

Relation Representation. In our case, the data to be clustered is Food-Drug relations. For this purpose, each relation should be represented by a set of features such that the resulting data $D = [F_1, F_2, ..., F_n]$, should be a vector of size n, where n is the number of relations to be clustered, F_i is a set of features representing relation R_i. The most natural way to get features F_i is to group every sentences S_i labelled by relation R_i in the initial dataset D_S: F_i = Concatenation(S_i) for S_i in D_S. We assume this representation as the baseline of our task.

To improve the relation representation, we propose a supervised approach to extract the more relevant features for relation R_i by training a n-classes SVM Classifier on the initial dataset D_S. SVM decision is based on an hyperplane that maximizes the margin between the samples and the separator hyperplane represented by $h(x) = w^T x + w_0$, where $x = (\mathbf{x}_1, ..., \mathbf{x}_N)^T x = (\mathbf{x}_1, ..., \mathbf{x}_N)^T$ is the vector of features and $w = (\mathbf{w}_1, ..., \mathbf{w}_N)^T$ the vector of weights. From these weights, we can determine the importance of each feature on SVM decision given by a matrix of feature coefficients C of size $n \times nf$ where n is the number of classes (here relations) and nf the number of features.

We propose to extract the nm most important features for each relation to represent the considered relation such that the relation R_i is represented by a vector of nm features which corresponds to the nm first positive features of the ith vector of C. The resulting dataset is a matrix $D = [F_1, F_2, ..., F_n]$ of size $n \times nm$ where n is the number of classes (here relations), nm is the number of features to extract, and F_i is the feature extracted to represent relation i.

However, relation representation is quite more complicated than word representation since the meaning of the sentence relies entirely on the two related arguments considered. In order to capture more accurately the expression of the relation in a sentence, we propose to use as features, lemmas before the first argument of the relation, lemmas between the two arguments, and lemmas after the second argument for the SVM classification. For the rest of the paper, we will note this method BBA-SVM.

Relation Clustering and FDI-Classification. Following the relation description in Sect. 4.1, we consider *non-precised relation* and *no effect on drug* relation as is, but the 11 others will be grouped into 4 clusters. We apply the approach proposed in Sect. 4.2 on sentences labelled by the 11 effect relations in POMELO dataset. The resulting data is a matrix D of size $11 \times nm$ where nm is the number of features extracted to represent a relation, that is given to an unsupervised clustering algorithm to be grouped into 4 clusters. The results is a vector of cluster labels $Cl = [Cl_1, Cl_2, ..., Cl_{11}]$ that contains 4 unique values where Cl_i is the cluster to which the relation R_i belongs. Once clusters defined, labels of sentences from the initial dataset are replaced by the cluster labels associated to the relation. Finally, we perform a 6-classes classification to identify FDI type.

This pipeline is summarized in Fig. 1. In this paper, we address the issue of lack of number of examples per relation by grouping relations with similar features. We assume an intuitive clustering way on relations involving food effect on drug, leading to 4 relations that are Augmentation, Reduction, Negative and Positive. Then we use clustering method to automatically group such relations according to features selected preliminary by a SVM classifier. Once effect relation clustered, we carry out a 6-classes-classification to identify the type involved in each sentence. So a configuration is composed of:

1. a relation representation step with number of features and features extraction method as parameters
2. a clustering step with clustering algorithm as parameter
3. a classification step with classifier and features as parameters

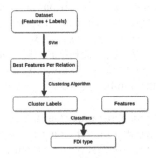

Fig. 1. Architecture of the approach - Relation Representation - Relation Clustering - FDI Classification

5 Experiments

Since our objective is to determine Food-Drug Interaction type, our experiments are focused on the performance of the relation classification from the POMELO dataset.

5.1 Clustering

Relation Representation. We experiment the impact of relation representation on the classification performance by varying the approach used to represent relation: (1) baseline - a relation R is represented by set of words of all sentences labelled by the relation R; (2) a lemma-SVM approach - lemmas are given to a SVM classifier and the more relevant features are extracted; (3) our BBA-SVM approach - lemmas before the first argument of the relation, lemmas between the two arguments, and lemmas after the second argument are given to a SVM classifier and best features are extracted; (4) inflected forms and lemmas, lemmas before the first argument of the relation, lemmas between the two arguments, and lemmas after the second argument are given to a SVM classifier and the more relevant features are extracted (ILBBA).

Clustering Algorithms. To evaluate our approach, we compare the performance of 4 clustering algorithms from Scikit-learn [13] implementation: (1) KMeans - to divide data into k subsets, central points k called centroids of partitions are identified such that the distance between the centroid and the points inside each partition is minimum; (2) Mini Batch K-Means - a variant of the KMeans algorithm which uses mini-batches to reduce the computation time, Mini-batches are subsets of the input data, randomly sampled in each training iteration; (3) Spectral clustering does a low-dimension embedding of the affinity matrix between samples, followed by a KMeans in the low dimensional space; (4) Agglomerative Clustering performs a hierarchical clustering using a bottom up approach: each observation starts in its own cluster, and clusters are successively merged together.

Clustering Evaluation. We use 4 metrics to evaluate the clustering assignement compared with the intuitive ARNP grouping method. Among them, we have (1) Adjusted Rand index that measures the similarity of the two assignments, ignoring permutations and with chance normalization; (2) Homogeneity - each cluster contains only members of a single class; (3) Completeness - all members of a given class are assigned to the same cluster; (4) Calinski-Harabaz Index is used to evaluate the model, where a higher Calinski-Harabaz score relates to a model with better defined clusters.

5.2 FDI Type Classification

Preprocessing. Each sentence of the dataset is preprocessed as following: numbers were replaced by the character '#' as proposed in [10], other special characters are removed, each word is converted to lower case.

Features. To evaluate the efficiency of our proposed approach, features are composed of inflected forms, lemmas, POS-tag of words, lemmas before the first argument of the relation, lemmas between the two arguments, and lemmas after the second argument.

Classification Models. Several classes of classification algorithms exist according to their mode of operation: (1) Linear models make classification decision based on value of a linear combination of the features; (2) Neighbors-based models classify an object according to the vote of common classes of its nearest neighbors; (3) Tree-based models represent features as nodes of a decision tree with leaves as class labels; (4) Ensemble models combine the decision of multiple algorithms to obtain better classification performance; (5) Bayesian models are probabilistic classifiers based on Bayes theorem assuming independence between features. Classification models can be combined with preprocessing methods to improve the quality of the features thus facilitating decision-making.

In this experiment, we evaluate the performance of at least one classifier of each classes from Scikit-learn [13] implementation: (1) a Decision Tree (DTree), (2) a l2-linear SVM classifier (LSVC-l2), (3) a Logistic Regression (LogReg), (4) a Multinomial Naive Bayes (MNB), (5) a Random Forest Classifier (RFC), (6) a K-Nearest-Neighbors (KNN), and (7) a SVM combined with Select From Model feature selection algorithm (SFM-SVM).

Classification Quality. Since our goal is to extract Food-Drug interaction from texts, we evaluate our approach by its ability to identify such relations, which is measured by the score of the classifier in each configuration. 3 types of metrics are used in this case: precision (P), recall (R), F1-score (F_1). Considering that one of the challenge of the task is the imbalance of the numbers of examples per class, we compare the macro-scores, that computes scores per class then globally averages the scores, and the micro-scores, that computes scores over all individual decisions. Scores are obtained from a 10-fold cross-validation process.

6   Results and Discussion

Results presented in this section are the performance of a configuration in FDI type identification task using the POMELO dataset and cluster labels. Here the best result is achieved by 200 BBA-SVM features clustered by a Spectral Clustering algorithm, given as label in a SFM-SVM classifier using as features, lemmas before the first argument of the relation, lemmas between the two arguments, and lemmas after the second argument of the relation.

Table 3. Macro F1-score obtained using different methods for relation representation given to clustering algorithms *KMeans, MiniBatch-KMeans (MBKM), Spectral Clustering, Agglomerative Clustering*

Relation representation	KMeans	MBKM	Spectral	Agglomerative
Baseline	0.362	0.394	**0.522**	0.374
Lemma	0.361	**0.405**	0.373	0.366
BBA-SVM	0.385	0.384	**0.58***	0.361
Inflected+Lemma+BBA	0.473	0.507	0.367	**0.517**

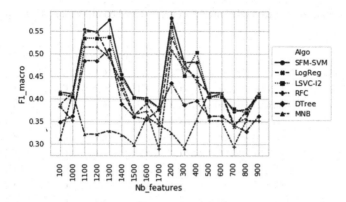

Fig. 2. Macro F1-score obtained on different models while varying the number of features to represent relation for clustering - *Spectral Clustering Model, BBA-SVM Method* - BBA features for classification

Table 4. Macro F1-score obtained while varying features used for relation classification after clustering - *KMeans, MiniBatchKMeans, SpectralClustering, AgglomerativeClustering* - BBA-SVM Representation

Feature	Without	ARNP	KMeans	MBKM	SpecC	AggloC
Inflected	0.379	0.378	0.416	**0.425**	0.573	0.422
Inflected + postag	0.388	0.368	0.397	0.413	0.559	0.406
Lemma	**0.41**	0.34	**0.462**	0.396	0.563	**0.449**
Inflected + lemma	0.38	**0.401**	0.423	**0.425**	0.575	0.439
Lemma + postag	0.403	0.392	0.444	0.386	0.56	0.436
Inflected + lemma + postag	0.387	0.395	0.425	**0.425**	0.567	0.443
Before + between + after	0.364	0.364	0.385	0.384	**0.58***	0.361
Inflected + lemma + before + between + after	0.383	0.384	0.394	0.402	0.576	0.403
Inflected + lemma + before + between + after + postag	0.379	0.396	0.383	0.398	0.566	0.404

Our BBA-SVM relation representation approach achieves the best F1-score 0.58 on FDI Classification (Table 3) with a difference of 0.23 from ARNP grouping and non-clustered data (Table 4). Thus, this score is obtained using only 200 features for relation clustering (Fig. 2) from the 1676 features composed by lemmas before the first argument of the relation, lemmas between the two arguments, and lemmas after the second argument for the SVM classification. This result justifies our assumption that a relation is characterized by specific features found in a particular position according to the 2 arguments of the relation. Joining this idea, the fact that feature selection method applied before SVM (SFM-SVM) (Table 6) produces a better performance suggest that some features are more important than others and focusing on them improve

Table 5. Clusters labels for each relation and scores obtained on different relation representation methods

		ARNP	Baseline	Lemma	BBA-SVM	ILBBA
Cluster labels	Decrease absorption	C1	C1	C1	C1	C1
	Improve drug effect	C2	C1	C3	C1	C3
	Increase absorption	C3	C3	C1	C1	C1
	Negative effect on drug	C4	C1	C1	C1	C1
	New side effect	C4	C1	C1	C4	C4
	Positive effect on drug	C2	C4	C1	C1	C1
	Slow absorption	C1	C1	C1	C1	C1
	Slow elimination	C1	C2	C2	C1	C1
	Speed up absorption	C3	C1	C2	C3	C1
	Without food	C4	C1	C4	C1	C2
	Worsen drug effect	C4	C1	C3	C2	C1
Clustering scores	Adjusted rand score	1.0	0.101	−0.132	−0.043	−0.043
	Calinski-Harabaz score	1.0	0.376	0.999	0.903	1.146
	Completeness score	1.0	0.519	0.312	0.431	0.431
	Homogeneity score	1.0	0.343	0.272	0.284	0.284
Classification scores	Precision macro	0.385	0.547	0.38	**0.589**	0.372
	Recall macro	0.368	0.518	0.381	**0.582**	0.371
	F1 macro	0.364	0.522	0.373	**0.58**	0.367
	F1 micro	0.599	0.656	0.647	**0.67**	0.652

Table 6. Scores obtained using features before + between + after - BBA-SVM representation - Spectral Clustering Algorithm

Model	Precision	Recall	F1-macro	F1-micro
DTree	0.45	0.441	0.435	0.614
LSVC-l2	0.572	0.558	0.56	0.668
LogReg	0.564	0.528	0.536	**0.674**
MNB	0.342	0.344	0.325	0.535
RFC	0.586	0.496	0.508	0.665
SFM-SVM	**0.589**	**0.582**	**0.58**	0.67
kNN	0.393	0.298	0.253	0.589

the decision-making of the classifier. The difference between micro-score and macro-score decrease from 0.13 with ARNP to 0.09, that suggest a reduction of the imbalance of data. Logistic Regression achieves the best result on micro-score but is a little less efficient in macro-score, which means that the model is more sensitive to imbalance of data than SVM models. Besides, the high score of Calinski-Harabaz (Table 5) implying that clusters are dense and well separated support the effectiveness of our approach. Nevertheless, the other clustering scores indicate an independant assignement from the ARNP method. It is explained while analyzing the labels assigned to each relation. In

Table 5, we observe also that 3 relations are represented individually and all others are grouped into one cluster. At first sight, there is no particular reason to explain the grouping. This suggest that the 3 individuals relation are explicitly different from the others but the rest are not sufficiently separable. It is also possible that the POMELO annotated corpus contains mistaken annotations. Actually, the one-annotator annotation can be improved according to our clustering approach, relying on manual validation, and including classification of relations without more precision. Indeed, these data represent more than half of the data, thus creating ambiguities making classification difficult. However, these results show that our approach produces a significant improvement on task of FDI type identification.

7 Conclusion and Future Work

Our paper contributes to the task of extraction of Food-Drug Interaction (FDI) from scientific literature, that we address as a relation extraction task. While applying supervised learning to this purpose, we face the lack of examples because of the high number of relation types. To address this issue, we propose to represent each relation by most important features extracted from SVM classification, then relations are grouped into clusters, and cluster labels are then used as relation labels on the initial dataset. Our approach is based on the assumption that relations are defined by a set of specific feature located in a particular position from the arguments of the relation. Following this idea, we use lemmas before the first argument of the relation, lemmas between the two arguments, and lemmas after the second argument for the SVM classification and extract from them the most important features used by SVM to make a decision of relation assignement. These features are given to clustering algorithms to obtain a cluster label for each relation, that is used as labels of POMELO dataset for FDI identification. Our approach achieves the best performance with 200 features grouped by Spectral Clustering algorithm, and classified by a pipeline of Select From Model feature selection and SVM classification. We get an improvement of 0.23 on F1-score from the ARNP and the non-clustered data. Besides, the decrease in difference between macro and micro average of F1-score suggest a reduction of the imbalance of data. Therefore, experiments results support the effectiveness of our approach. For future work, we will consider FDI type identification as a multilabel classification, using the cluster as first label, or a more domain-based labeling following the ADME classes [5] (Absorption - Distribution - Metabolism - Excretion) of Drug-Drug Interaction by transfer learning.

Acknowledgements. This work was supported by the MIAM project and Agence Nationale de la Recherche through the grant ANR-16-CE23-0012 France.

References

1. Aagaard, L., Hansen, E.: Adverse drug reactions reported by consumers for nervous system medications in Europe 2007 to 2011. BMC Pharmacol. Toxicol. **14**, 30 (2013)
2. Aronson, J., Ferner, R.: Clarification of terminology in drug safety. Drug Saf. **28**(10), 851–70 (2005)

3. Ben Abacha, A., Chowdhury, M.F.M., Karanasiou, A., Mrabet, Y., Lavelli, A., Zweigenbaum, P.: Text mining for pharmacovigilance: using machine learning for drug name recognition and drug-drug interaction extraction and classification. J. Biomed. Inform. **58**, 122–132 (2015)

4. Cejuela, J.M., et al.: LocText: relation extraction of protein localizations to assist database curation. BMC Bioinform. **19**(1), 15 (2018). https://doi.org/10.1186/s12859-018-2021-9

5. Doogue, M., Polasek, T.: The ABCD of clinical pharmacokinetics. Ther. Adv. Drug Saf. **4**, 5–7 (2013). https://doi.org/10.1177/2042098612469335

6. Hamon, T., Tabanou, V., Mougin, F., Grabar, N., Thiessard, F.: POMELO: medline corpus with manually annotated food-drug interactions. In: Proceedings of Biomedical NLP Workshop Associated with RANLP 2017, Varna, Bulgaria, pp. 73–80 (2017)

7. Hanley, M., Cancalon, P., Widmer, W., Greenblatt, D.: The effect of grapefruit juice on drug disposition. Expert Opin. Drug Metab. Toxicol. **7**(3), 267–286 (2011)

8. Holat, P., Tomeh, N., Charnois, T., Battistelli, D., Jaulent, M.-C., Métivier, J.-P.: Weakly-supervised symptom recognition for rare diseases in biomedical text. In: Boström, H., Knobbe, A., Soares, C., Papapetrou, P. (eds.) IDA 2016. LNCS, vol. 9897, pp. 192–203. Springer, Cham (2016). https://doi.org/10.1007/978-3-319-46349-0_17

9. Kim, S., Liu, H., Yeganova, L., Wilbur, W.J.: Extracting drug-drug interactions from literature using a rich feature-based linear kernel approach. J. Biomed. Inform. **55**, 23–30 (2015)

10. Kolchinsky, A., Lourenço, A., Wu, H.Y., Li, L., Rocha, L.M.: Extraction of pharmacokinetic evidence of drug-drug interactions from the literature. PLoS ONE **10**(5), e0122199 (2015)

11. Liu, S., Tang, B., Chen, Q., Wang, X.: Drug-drug interaction extraction via convolutional neural networks. Comput. Math. Methods Med. **2016** (2016). https://doi.org/10.1155/2016/6918381

12. Meng, F., Morioka, C.: Automating the generation of lexical patterns for processing free text in clinical documents. J. Am. Med. Inform. Assoc. **22**(5), 980–986 (2015). https://doi.org/10.1093/jamia/ocv012

13. Pedregosa, F., et al.: Scikit-learn: machine learning in Python. J. Mach. Learn. Res. **12**, 2825–2830 (2011)

14. Wishart, S., et al.: DrugBank 5.0: a major update to the DrugBank database for 2018. Nucleic Acids Res. **46** (2017)

15. Song, M., Chul Kim, W., Lee, D., Eun Heo, G., Young Kang, K.: PKDE4J: entity and relation extraction for public knowledge discovery. J. Biomed. Inform. **57**, 320–332 (2015)

16. Stenetorp, P., Pyysalo, S., Topić, G., Ohta, T., Ananiadou, S., Tsujii, J.: Brat: a web-based tool for NLP-assisted text annotation. In: Proceedings of the Demonstrations at the 13th Conference of the European Chapter of the Association for Computational Linguistics, EACL 2012, Stroudsburg, PA, USA, pp. 102–107. Association for Computational Linguistics (2012). https://dl.acm.org/citation.cfm?id=2380921.2380942

17. Yi, Z., et al.: Drug-drug interaction extraction via recurrent neural network with multiple attention layers. In: Cong, G., Peng, W.-C., Zhang, W.E., Li, C., Sun, A. (eds.) ADMA 2017. LNCS (LNAI), vol. 10604, pp. 554–566. Springer, Cham (2017). https://doi.org/10.1007/978-3-319-69179-4_39

Contrastive Reasons Detection and Clustering from Online Polarized Debates

Amine Trabelsi$^{(\boxtimes)}$ and Osmar R. Zaïane

Department of Computing Science, University of Alberta, Edmonton, Canada
{atrabels,zaiane}@ualberta.ca

Abstract. This work tackles the problem of unsupervised modeling and extraction of the main contrastive sentential reasons conveyed by divergent viewpoints on polarized issues. It proposes a pipeline approach centered around the detection and clustering of phrases, assimilated to argument facets using a novel Phrase Author Interaction Topic-Viewpoint model. The evaluation is based on the informativeness, the relevance and the clustering accuracy of extracted reasons. The pipeline approach shows a significant improvement over state-of-the-art methods in contrastive summarization on online debate datasets.

Keywords: Reason extraction · Information extraction · Text mining

1 Introduction

Online debate forums provide a valuable resource for textual discussions about contentious issues. Contentious issues are controversial topics or divisive entities that usually engender opposing stances or viewpoints. Forum users write posts to defend their standpoint using persuasion, reasons or arguments. Such posts correspond to what we describe as contentious documents [21,22,24]. An automatic tool that provides a contrasting overview of the main viewpoints and reasons given by opposed sides, debating an issue, can be useful for journalists and politicians. It provides them with systematic summaries and drafting elements on argumentation trends. In this work, given online forum posts about a contentious issue, we study the problems of unsupervised modeling and extraction, in the form of a digest table, of the main contrastive reasons conveyed by divergent viewpoints. Table 1 presents an example of a targeted solution in the case of the issue of "Abortion". The digest Table 1 is displayed à la ProCon.org or Debatepedia websites, where the viewpoints or stances engendered by the issue are separated into two columns. Each cell of a column contains an argument facet label followed by a sentential reason example. A sentential reason example is one of the infinite linguistic variations used to express a reason. For instance, the sentence "that cluster of cell is not a person" and the sentential reason "fetus is not a human" are different realizations of the same reason. For

© Springer Nature Switzerland AG 2023
A. Gelbukh (Ed.): CICLing 2019, LNCS 13451, pp. 181–198, 2023.
https://doi.org/10.1007/978-3-031-24337-0_14

Table 1. Contrastive digest table for abortion.

View 1 Oppose		View 2 Support	
Argum. facet label	Reason	Argum. facet label	Reason
1 **Fetus is not human**	What makes a fetus not human?	6 **Fetus is not human**	Fetus is not human
2 **Kill innocent baby**	Abortion is killing innocent baby	7 **Right to her body**	Women have a right to do what they want with their body
3 **Woman's right to control her body**	Does prostitution involves a woman's right to control her body?	8 **Girl gets raped and gets pregnant**	If a girl gets raped and becomes pregnant does she really want to carry that man's child?
4 **Give her child up for adoption**	Giving a child baby to an adoption agency is an option if a woman isn't able to be a good parent	9 **Giving up a child for adoption**	Giving the child for adoption can be just as emotionally damaging as having an abortion
5 **Birth control**	Abortion shouldn't be a form of birth control	10 **Abortion is not a murder**	Abortion is not a murder

convenience, we will also refer to a sentence realizing a reason as a reason. **Reasons** in Table 1 are short sentential excerpts, from forum posts, which explicitly or implicitly express premises or arguments supporting a viewpoint. They correspond to any kind of intended persuasion, even if it does not contain clear argument structures [8]. It should make a reader easily infer the viewpoint of the writer. **An argument facet** is an abstract concept corresponding to a low level issue or a subject that frequently occurs within arguments in support of a stance or in attacking and rebutting arguments of opposing stance [12]. Similar to the concept of reason, many phrases can express the same facet. Phrases in bold in Table 1 correspond to **argument facet labels**, i.e., possible expressions describing argument facets. Reasons can also be defined as realizations of facets according to a particular viewpoint perspective. For instance, argument facet 4 in Table 1 frequently occurs within holders of Viewpoint 1 who oppose abortion. It is realized by its associated reason. The same facet is occurring in Viewpoint 2, in example 9, but it is expressed by a reason rebutting the proposition in example 4. Thus, reasons associated with divergent viewpoints can share a common argument facet. Exclusive facets emphasized by one viewpoint's side, much more than the other, may also exist (see example 5 or 8). Note that in many cases the facet is very similar to the reason or proposition initially put forward by a particular viewpoint side, see examples 2 and 6, 7. It can also be a general aspect like "Birth Control" in example 5.

This paper describes the unsupervised extraction of these argument facets' phrases and their exploitation to generate the associated sentential reasons in a contrastive digest table of the issue. Our first hypothesis is that detecting the main facets in each viewpoint leads to a good extraction of relevant sentences corresponding to reasons. Our second hypothesis is that leveraging the reply-interactions in online debate helps us cluster the viewpoints and adequately organize the reasons.

We distinguish three common characteristics of online debates, identified also by [9] and [4], which make the detection and the clustering of argumentative sentences a challenging task. First, the unstructured and colloquial nature of used language makes it difficult to detect well-formed arguments. It makes it also noisy containing non-argumentative portions and irrelevant dialogs. Second, the use of non-assertive speech acts like rhetorical questions to implicitly express a stance or to challenge opposing argumentation, like examples 1,3 and 8 in Table 1. Third, the similarity in words' usage between facet-related opposed arguments leads clustering to errors. Often a post rephrases the opposing side's premise while attacking it (see example 9). Note that exploiting sentiment analysis solely, like in product reviews, cannot help distinguishing viewpoints. Indeed, Mohammad et al. [14] show that both positive and negative lexicons are used, in contentious text, to express the same stance. Moreover, opinion is not necessarily expressed through polarity sentiment words, like example 6.

In this work, we do not explicitly tackle or specifically model the above-mentioned problems in contentious documents. However, we propose a generic data driven and facet-detection guided approach joined with posts' viewpoint clustering. It leads to extracting meaningful contrastive reasons and avoids running into these problems. Our contributions consist of: (1) the conception and deployment of a novel unsupervised generic pipeline framework producing a contrastive digest table of the main sentential reasons expressed in a contentious issue, given raw unlabeled posts from debate forums; (2) the devising of a novel Phrase Author Interaction Topic Viewpoint model, which jointly processes phrases of different length, instead of just unigrams, and leverages the interaction of authors in online debates. The evaluation of the proposed pipeline is based on three measures: the informativeness of the digest as a summary, the relevance of extracted sentences as reasons and the accuracy of their viewpoint clustering. The results on different datasets show that our methodology improves significantly over two state-of-the-art methods in terms of documents' summarization, reasons' retrieval and unsupervised contrastive reasons clustering.

2   Related Work

The objective of argument mining is to automatically detect the theoretically grounded argumentative structures within the discourse and their relationships [15,18]. In this work, we are not interested in recovering the argumentative structures but, instead, we aim to discover the underpinning reasons behind people's opinion from online debates. In this section, we briefly describe some of the argument mining work dealing with social media text and present a number of important studies on Topic-Viewpoint Modeling. The work on online discussions about controversial issues leverages the interactive nature of these discussions. Habernal and Gurevych [8] consider rebuttal and refutation as possible components of an argument. Boltužić and Šnajder [3] classify the relationship in a comment-argument pair as an attack (comment attacks the argument), a support or none. The best performing model of Hasan and Ng's work [9] on Reason Classification (RC) exploits the reply information associated with the posts. Most of the computational argumentation methods, are supervised. Even the studies focusing on argument identification [13,19], usually, rely on predefined lists of manually extracted arguments. As a first step towards unsupervised identification of prominent arguments from online debates, Boltužić and Šnajder [4] group argumentative statements into clusters assimilated to arguments. However, only selected argumentative sentences are used as input. In this paper, we deal with raw posts containing both argumentative and non-argumentative sentences.

Topic-Viewpoint models are extensions of Latent Dirichlet Allocation (LDA) [2] applied to contentious documents. They hypothesize the existence of underlying topic and viewpoint variables that influence the author's word choice when writing about a controversial issue. The viewpoint variable is also called stance, perspective or argument variable in different studies. Topic-Viewpoint models are mainly data-driven approaches which reduce the documents into topic-viewpoint dimensions. A Topic-Viewpoint pair t-v is a probability distribution over unigram words. The unigrams with top probabilities characterize the used vocabulary when talking about a specific topic t while expressing a particular viewpoint v at the same time. Several Topic-Viewpoint models of controversial issues exist [17,20,23,25]. Little work is done to exploit these models in order to generate sentential digests or summaries of controversial issues instead of just producing distributions over unigram words. Below we introduce the research that is done in this direction.

Paul et al. [16] are the first to introduce the problem of contrastive extractive summarization. They applied their general approach on online surveys and editorials data. They propose the Topic Aspect Model (TAM) and use its output distributions to compute similarity scores between sentences. Comparative LexRank, a modified LexRank [7], is run on scored sentences to generate the summary. Recently, Vilares and He [27] propose a topic-argument or viewpoint model called the Latent Argument Model_LEX (LAM_LEX). Using LAM_LEX, they generate a succinct summary of the main viewpoints from a parliamentary debates dataset. The generation consists of ranking the sentences according to a discriminative score for each topic and argument dimension. It encourages higher ranking of sentences with words exclusively occurring with a particular topic-argument dimension which may not be accurate in extracting the contrastive reasons sharing common words. Both of the studies, cited above, exploit the unigrams output of their topic-viewpoint modeling. In this work, we propose a Topic-Viewpoint modeling of phrases of different length, instead of just unigrams. We believe phrases allow a better representation of the concept of argument facet. They would also lead to extract a more relevant sentence realization of this latter. Moreover, we leverage the interactions of users in online debates for a better contrastive detection of the viewpoints.

3 Methodology

Our methodology presents a pipeline approach to generate the final digest table of reasons conveyed on a controversial issue. The inputs are raw debate text and the information about the replies. Below we describe the different phases of the pipeline.

3.1   Phrase Mining Phase

The inputs of this module are raw posts (documents). We prepare the data by removing identical portions of text in replying posts. We remove stop and rare words. We consider working with the stemmed version of the words.

The objective of the phrase mining module is to partition the documents into high quality bag-of-phrases instead of bag-of-words. Phrases are of different length, single or multi-words. We follow the steps of El-Kishky et al. [6], who propose a phrase extraction procedure for the Phrase-LDA model. Given the contiguous words of each sentence in a document, the phrase mining algorithm employs a bottom-up agglomerative merging approach. At each iteration, it merges the best pair of collocated candidate phrases if their statistical significance score exceeds a threshold which is set empirically (set according to [6] implementation). The significance score depends on the collocation frequency of candidate phrases in the corpus. It measures their number of standard deviation away from the expected occurrence under an independence null hypothesis. The higher the score, the more likely the phrases co-occur more often than by chance.

3.2   Topic-Viewpoint Modeling Phase

In this section, we present the Phrase Author Interaction Topic-Viewpoint model (PhAITV). It takes as input the documents, partitioned in high quality phrases of different length, and the information about author-reply interactions in an online debate forum. The objective is to assign a topic and a viewpoint labels to each occurrence of the phrases. This would help to cluster them into Topic-Viewpoint classes. We assume that A authors participate in a forum debate about a particular issue. Each author a writes D_a posts. Each post d_a is partitioned into G_{da} phrases of different length (≥ 1). Each phrase contain M_{gda} words. Each term w_{mg} in a document belongs to the corpus vocabulary of distinct terms of size W. In addition, we assume that we have the information about whether a post replies to a previous post or not. Let K be the total number of topics and L be the total number of viewpoints. Let θ_{da} denote the probability distribution of K topics under a post d_a; ψ_a be the probability distributions of L viewpoints for an author a; ϕ_{kl} be the multinomial probability distribution over words associated with a topic k and a viewpoint l; and ϕ_B a multinomial distribution of background words. The generative process of a post according to the PhAITV model (see Fig. 1) is the following. An author a chooses a viewpoint v_{da} from the distribution ψ_a. For each phrase g_{da} in the post, the author samples a binary route variable x_{gda} from a Bernoulli distribution σ. It indicates whether the phrase is a topical or a background word. Multi-word phrases cannot belong to the background class. If $x_{gda} = 0$, she samples the word from ϕ_B. Otherwise, the author, first, draws a topic z_{gda} from θ_{da}, then, samples each word w_{mg} in the phrase from the same $\phi_{z_{gda}v_{da}}$.

Note that, in what follows, we refer to a current post with index id and to a current phrase with index ig. When the current post is a reply to a previous post by a different author, it may contain a rebuttal or it may not. If the reply attacks the previous author then the reply is a rebuttal, and Rb_{id} is set to 1 else if it supports, then the rebuttal takes 0. We define the **parent posts** of a current post as all the posts of the author who the current post is replying to. Similarly, the **child posts** of a current post are all the posts replying to the author of the current post. We assume that the probability of a rebuttal $Rb_{id} = 1$ depends on the degree of opposition between the viewpoint v_{id} of the current post and the viewpoints $\mathcal{V}_{id}^{par}$ of its parent posts as the following:

$$p(Rb_{id} = 1|v_{id}, \mathcal{V}_{id}^{par}) = \frac{\sum_{l'}^{\mathcal{V}_{id}^{par}} \mathbf{I}(v_{id} \neq l') + \eta}{|\mathcal{V}_{id}^{par}| + 2\eta}, \qquad (1)$$

where $\mathbf{I}(\text{condition})$ equals 1 if the condition is true and η a smoothing parameter.

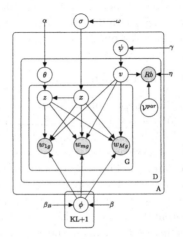

Fig. 1. Plate Notation of The PhAITV model

For the inference of the model's parameters, we use the collapsed Gibbs sampling. For all our parameters, we set fixed symmetric Dirichlet priors. According to Fig. 1, the Rb variable is observed. However, the true value of the rebuttal variable is unknown to us. We fix it to 1 to keep the framework purely unsupervised, instead of guiding it by estimating the reply disagreement using methods based on lexicon polarity [17]. Setting $Rb = 1$ means that all replies of any post are rebuttals attacking all of the parent posts excluding the case when the author replies to his own post. This comes from the observation that the majority of the replies, in the debate forums framework, are intended to attack the previous proposition (see data statistics in Table 2 as an example). This setting will affect the viewpoint sampling of the current post. The intuition is that, if an author is replying to a previous post, the algorithm is encouraged to sample a viewpoint which opposes the majority viewpoint of parent posts (Eq. 1). Similarly, if the current post has some child posts, the algorithm is encouraged to sample a viewpoint opposing the children's prevalent stance. If both parent and child posts exist, the algorithm is encouraged to oppose both, creating some sort of adversarial environment when the prevalent viewpoints of parents and children are opposed. The derived sample equation of current post's viewpoint v_{id} given all the previous sampled assignments in the model $\vec{v}_{\neg id}$ is:

$$p(v_{id} = l|\vec{v}_{\neg id}, \vec{w}, \vec{Rb}, \vec{x}) \propto n^{(l)}_{a,\neg id} + \gamma \times \frac{\prod_t^{W_{id}} \prod_{j=0}^{n^{(t)}_{id}-1} n^{(t)}_{l,\neg id} + j + \beta}{\prod_{j=0}^{n_{id}-1} n^{(.)}_{l,\neg id} + W\beta + j}$$

$$\times p(Rb_{id} = 1|v_{id}, V^{par}_{id}) \times \prod_{c|v_{id} \in V^{par}_c} p(Rb_c = 1|v_c, V^{par}_c). \quad (2)$$

The count $n^{(l)}_{a,\neg id}$ is the number of times viewpoint l is assigned to author a's posts excluding the assignment of current post, indicated by $\neg id$; $n^{(t)}_{l,\neg id}$ is the number of times term t is assigned to viewpoint l in the corpus excluding assignments in current post; $n^{(.)}_{l,\neg id}$ is the total number of words assigned to l; W_{id} is the set of vocabulary of words in post id; $n^{(t)}_{id}$ is the number of time word t occurs in the post. The third term of the multiplication in Eq. 2 corresponds to Eq. 1 and is applicable when the current post is a reply. The fourth term of the multiplication takes effect when the current post has child posts. It is a product over each child c according to Eq. 1. It computes how much would the children's rebuttal be probable if the value of v_{id} is l.

Given the assignment of a viewpoint $v_{id} = l$, we also jointly sample the topic and background values for each phrase ig in post id, according to the following:

$$p(z_{ig} = k, x_{ig} = 1|\vec{z}_{\neg ig}, \vec{x}_{\neg ig}, \vec{w}, \vec{v}) \propto$$

$$\prod_{j=0}^{M_{ig}} n^{(1)}_{\neg ig} + \omega + j \times n^{(k)}_{id,\neg ig} + \alpha + j \times \frac{n^{(w_{jg})}_{kl,\neg ig} + \beta}{n^{(.)}_{kl,\neg ig} + W\beta + j}, \quad (3)$$

$$p(x_{ig} = 0|\vec{x}_{\neg ig}, \vec{w}) \propto \prod_{j=0}^{M_{ig}} n^{(0)}_{\neg ig} + \omega + j \times \frac{n^{(w_{jg})}_{0,\neg ig} + \beta_B}{n^{(.)}_{0,\neg ig} + W\beta_B + j}. \quad (4)$$

Here $n^{(k)}_{id,\neg ig}$ is the number of words assigned to topic k in post id, excluding the words in current phrase ig; $n^{(1)}_{\neg ig}$ and $n^{(0)}_{\neg ig}$ correspond to the number of topical and background words in the corpus, respectively; $n^{(w_{jg})}_{kl,\neg ig}$ and $n^{(w_{jg})}_{0,\neg ig}$ correspond to the number of times the word of index j in the phrase g is assigned to topic-viewpoint kl or is assigned as background; $n^{(.)}$s are summations of last mentioned expressions over all words.

After the convergence of the Gibbs algorithm, each multi-word phrase is assigned a topic k and a viewpoint label l. We exploit these assignments to first create clusters $\mathcal{P}_{kl}$s, where each cluster $\mathcal{P}_{kl}$ corresponds to a topic-viewpoint value kl. It contains all the phrases that are assigned to kl at least one time. Each phrase phr is associated with its total number of assignments. We note it as $phr.nbAssign$. Second, we rank the phrases inside each cluster according to their assignment frequencies.

3.3   Grouping and Facet Labeling

The inputs of this module are Topic-Viewpoint clusters, $\mathcal{P}_{kl}$s, $k = 1..K$, $l = 1..L$, each containing multi-word phrases along with their number of assignments. The outputs are clusters, $\mathcal{A}_l$, of sorted phrases corresponding to argument facet labels for each viewpoint l (see Algorithm 1). This phase is based on two assumptions. (1) Grouping constructs agglomerations of lexically related phrases. which can be assimilated to the notion of argument facets. (2) An argument facet is better expressed with a Verb Expression than a Noun Phrase. A Verbal Expression (VE) is a sequence of correlated chunks centered around a Verb Phrase chunk [10]. Algorithm 1 proposes a second layer of phrase grouping on each of the constructed Topic-Viewpoint cluster $\mathcal{P}_{kl}$ (lines 3–20). It is based on the number of word overlap between stemmed pairs of phrases. The number of groups is not a parameter. First, we compute the number of words overlap between all pairs and sort them in descending order (lines 4–7). Then, while iterating on them, we encourage a pair with overlap to create its own group if both of its phrases are not grouped yet. If it has only one element grouped, the other element joins it. If a pair has no matches, then each non-clustered phrase creates its own group (lines 8–20).

Some of the generated groups may contain small phrases that can be fully contained in longer phrases of the same group. We remove them and add their number of assignments to corresponding phrases. If there is a conflict where two or several phrases can contain the same phrase, then the one that is a Verbal Expression adds up the number of assignments of the contained phrase. If two or more are VE, then the longest phrase, amongst them, adds up the number. Otherwise, we prioritize the most frequently assigned phrase (see lines 21–30 in Algorithm 1). This procedure helps inflate the number of assignments of Verbal Expression phrases in order to promote them to be solid candidates for the argument facet labeling. The final step (lines 32–40) consists of collecting the groups pertaining to each Viewpoint, regardless of the topic, and sorting them based on the cumulative number of assignments of their composing phrases. This will create viewpoint clusters, $\mathcal{C}_l$s, with groups which are assimilated to argument facets. The labeling consists of choosing one of the phrases as the representative of the group. We simply choose the one with the highest number of assignment to obtain Viewpoint clusters, $\mathcal{A}_l$s, of argument facet labels, sorted in the same order of corresponding groups in $\mathcal{C}_l$s.

Algorithm 1. Grouping and Labeling

Require: phrases clusters $\mathcal{P}_{kl}$ for topic $k = 1..K$, view $l = 1..L$

1:   $\mathcal{G}_{kl} \leftarrow \emptyset$ is the set of groups of phrases to create from $\mathcal{P}_{kl}$
2:   **for** each phrase cluster $\mathcal{P}_{kl}$ **do**
3:     $\mathcal{Q} \leftarrow$ set of all phrase-pairs from phrases in $\mathcal{P}_{kl}$
4:     **for** each phrase-pair q in $\mathcal{Q}$ **do**
5:      $q.overlap \leftarrow$ number of word intersections in q
6:     **end for**
7:     Sort pairs in $\mathcal{Q}$ by number of matches in descending order
8:     **for** each phrase-pair q in $\mathcal{Q}$ **do**
9:      **if** $q.overlap \neq 0$ **then**
10:       **if** $\neg(q.phrase1.grouped) \wedge \neg(q.phrase2.grouped)$ **then**
11:        New group $grp \leftarrow \{q.phrase1\} \cup \{q.phrase2\}$
12:        $\mathcal{G}_{kl} \leftarrow \mathcal{G}_{kl} \cup \{grp\}$
13:       **else if** only one phrase of q in existing grp' **then**
14:        $grp' \leftarrow grp' \cup \{\text{non grouped phrase of } q\}$
15:       **end if**
16:      **else if** $\neg q.phrase_j.grouped$, $j = 1, 2$ **then**
17:       New group $grp \leftarrow \{q.phrase_j\}$
18:       $\mathcal{G}_{kl} \leftarrow \mathcal{G}_{kl} \cup \{grp\}$
19:      **end if**
20:     **end for**
21:     **for** each grp in $\mathcal{G}_{kl}$ **do**
22:      Sort phrases in grp by giving higher ranking to phrases corresponding to: (1) Verbal Expression; (2) longer phrases; (3) frequently assigned phrases
23:      **for** each phr in grp **do**
24:       Find phr' of grp s.t. $phr'.wordSet \subset phr.wordSet$
25:       **if** $phr'.nbAssign \neq 0$ **then**
26:        $phr.nbAssign \leftarrow phr.nbAssign + phr'.nbAssign$
27:        $phr'.nbAssign \leftarrow 0$
28:       **end if**
29:      **end for**
30:     **end for**
31: **end for**
32: $\mathcal{C}_l \leftarrow$ set of all groups belonging to any $\mathcal{G}_{*l}$ of view l
33: $\mathcal{A}_l \leftarrow \emptyset$ is the sorted set of all argument facets labels of view l
34: **for** view $l = 1$ to L **do**
35:     Sort groups in $\mathcal{C}_l$ based on $grp.cumulatifNbAssign$
36:     **for** each grp in $\mathcal{C}_l$ **do**
37:      $grp.labelFacet \leftarrow$ phrase with highest $phr.nbAssign$
38:      $\mathcal{A}_l \leftarrow \mathcal{A}_l \cup \{grp.labelFacet\}$
39:     **end for**
40: **end for**
41: **return** all clusters $\mathcal{A}_l$s of sorted facets' labels for $l = 1..L$

3.4   Reasons Table Extraction

The inputs of Extraction of Reasons algorithm are sorted facet labels, $\mathcal{A}_l$, for each Viewpoint l (see Algorithm 2). Each label phrase is associated with its sentences $\mathcal{S}_{label}$ where it occurs, and where it is assigned a viewpoint l. The target output is the digest table of contrastive reasons $\mathcal{T}$. In order to extract a short sentential reason, given a phrase label, we follow the steps described in Algorithm 2: (1) find, $\mathcal{S}_{label}^{fInters}$, the set of sentences with the most common overlapping words among all the sentences of $\mathcal{S}_{label}$, disregarding the set of words composing the facet label (if the overlap set is empty consider the whole set $\mathcal{S}_{label}$), lines 6–12 in Algorithm 2; (2) choose the shortest sentence amongst $\mathcal{S}_{label}^{fInters}$ (line 13). The process is repeated for all sorted facet labels of $\mathcal{A}_l$ to fill viewpoint column $\mathcal{T}_l$ for $l = 1..L$. Note that duplicate sentences within a viewpoint column are removed. If the same sentence occurs in different columns, we only keep the sentence with the label phrase that has the most number of assignments. Also, we restore stop and rare words of the phrases when rendering them as argument facets. We choose the most frequent sequence in $\mathcal{S}_{label}$.

Algorithm 2. Extraction of Reasons Digest Table

Require: all clusters $\mathcal{A}_l$s of sorted argument facets' labels for view $l = 1..L$;
1: $\mathcal{T}$ is the digest table of contrastive reasons with $\mathcal{T}_l$s columns
2: $\mathcal{T}.columns \leftarrow \emptyset$
3: **for** view $l = 1$ to L **do**
4:      $\mathcal{T}_l.cells \leftarrow \emptyset$
5:      **for** each $label$ in $\mathcal{A}_l$ **do**
6:          $\mathcal{S}_{label} \leftarrow$ set of all sentences where $label$ phrase occurs and assigned view l
7:          $fInters \leftarrow$ most frequent set of words overlap among $\mathcal{S}_{label}$ s.t. $fInters \neq label.wordSet$
8:          **if** $fInters \neq \emptyset$ **then**
9:              $\mathcal{S}_{label}^{fInters} \leftarrow$ subset of $\mathcal{S}_{label}$ containing $fInters$
10:         **else**
11:             $\mathcal{S}_{label}^{fInters} \leftarrow \mathcal{S}_{label}$
12:         **end if**
13:         $sententialReason \leftarrow$ shortest sentence in $\mathcal{S}_{label}^{fInters}$
14:         $\mathcal{T}_l.cells \leftarrow \mathcal{T}_l.cells \cup \{cell(label + sententialReason)\}$
15:     **end for**
16:     $\mathcal{T}.columns \leftarrow \mathcal{T}.columns \cup \{\mathcal{T}_l\}$
17: **end for**
18: **return** $\mathcal{T}$

---

4   Experiments and Results

Table 2. Datasets statistics.

Forum	CreateDebate		4Forums		Reddit
Dataset	AB	GR	AB	GM	IP
# posts	1876	1363	7795	6782	2663
# reason labels	13	9	–	–	–
% arg. sent.[a]	20.4	29.8	–	–	–
% rebuttals	67.05	66.61	77.6	72.1	–

[a]Argumentative sentences in the labeled posts.

4.1   Datasets

We exploit the reasons corpus constructed by [9] from the online forum CreateDe-bate.com, and the Internet Argument corpus containing 4Forums.com datasets [1]. We also scraped a Reddit discussion commenting a news article about the March 2018 Gaza clash between Israeli forces and Palestinian protesters[1]. The constructed dataset does not contain any stance labeling. We consider 4 other datasets: Abortion (AB) and Gay Rights (GR) for CreateDebate, and Abortion and Gay Marriage (GM) for 4Forums. Each post in the CreateDebate datasets has a stance label (i.e., support or oppose the issue). The argumentative sentences of the posts have been labeled in [9] with a reason label from a set of predefined reason labels associated with each stance. The reason labels can be assimilated to argument facets or reason types. Only a subset of the posts, for each dataset, has its sentences annotated with reasons. Table 2 presents some statistics about the data. Unlike CreateDebate, 4Forums datasets do not contain any labeling of argumentative sentences or their reasons' types. They contain the ground truth stance labels at the author level. Table 2 reports the percentage of rebuttals as the percentage of replies between authors of opposed stance labels. The PhAITV model exploits only the text, the author identities and the information about whether a post is a reply or not. For evaluation purposes, we leverage the subset of argumentative sentences which is annotated with reasons labels, in CreateDebate, to construct several reference summaries (100) for each dataset. Each reference summary contains a combination of sentences, each from one possible label (13 for Abortion, 9 for Gay Rights). This makes the references exhaustive and reliable resources on which we can build a good recall measure about the informativeness of the digests, produced on CreateDebate datasets.

4.2   Experiments Set up

We compare the results of our pipeline framework based on **PhAITV** to those of two studies aiming to produce contrastive summarization in any type of

---

[1] https://www.reddit.com/r/worldnews/comments/8ah8ys/the_us_was_the_only_un_s ecurity_council_member_to/.

contentious text. These correspond to Paul et al.'s [16] and Vilares and He's [27] works. They are based on Topic-Viewpoint models, **TAM**, for the first, and **LAM_LEX** for the second (see Sect. 2). Below, we refer to the names of the Topic-Viewpoint methods to describe the whole process that is used to produce the final summary or digest. We also compare with a degenerate unigram version of our model, Author Interactive Topic Viewpoint **AITV** [26]. AITV's sentences were generated in a similar way to PhAITV's extraction procedure. The difference is that no grouping is involved and the query of retrieval consists of the top three keywords instead of the phrase. As a weak baseline, we generate **random summaries** from the set of possible sentences. We also create **correct summaries** from the subset of reason labeled sentences. One correct summary contains all possible reason types of argumentative sentences for a particular issue. Moreover, we compare with another degenerate version of our model **PhAITV$_{view}$** which assumes the true values of the posts' viewpoints are given. Note that the objective here is to assess the final output of the framework. Separately evaluating the performance of the Topic Viewpoint model in terms of document clustering has shown satisfiable results. We do not report it here for lack of space.

We try different combinations of the PhAITV's hyperparameters and use the combination which gives a satisfying overall performance. PhAITV hyperparameters are set as follows: $\alpha = 0.1$; $\beta = 1$; $\gamma = 1$; $\beta_B = 0.1$; $\eta = 0.01$; $\omega = 10$; Gibbs Sampling iterations is 1500; number of viewpoints L is 2. We try a different number of topics K for each Topic-Viewpoint model used in the evaluation. The reported results are on the best number of topics found when measuring the Normalized PMI coherence [5] on the Topic-Viewpoint clusters of words. The values of K are 30, 10, 10 and 50 for PhAITV, LAM_LEX, TAM and AITV, respectively. Other parameters of the methods used in the comparison are set to their default values. All the models generate their top 15 sentences for Abortion and their 10 best sentences for Gay Rights and Israel-Palestine datasets.

4.3 Evaluating Argument Facets Detection

The objective is to verify our assumption that the pipeline process, up to the Grouping and Labeling module, produces phrases that can be assimilated to argument facets' labels. We evaluate a total of 60 top distinct phrases produced after 5 runs on Abortion (4Forums) and Gay Rights (CreateDebate). We ask two annotators acquainted with the issues, and familiar with the definition of argument facet (Sect. 1), to give a score of 0 to a phrase that does not correspond to an argument facet, a score of 1 to a somewhat a facet, and a score of 2 to a clear facet label. Annotator are later asked to find consensus on phrases labeled differently. The average scores, of final annotation, on Abortion and Gay Rights are **1.45** and **1.44**, respectively. The percentages of phrases that are not argument facets are **12.9%** (AB) and **17.4%** (GR). The percentages of clear argument facets labels are **58.06%** (AB) and **62.06%** (GR). These numbers validate our assumption that the pipeline succeeds, to a satisfiable degree, in extracting argument facets labels.

Table 3. Averages of ROUGE Measures (in %, stemming and stop words removal applied) on Abortion and Gay Rights of CreateDebate. Bold denotes best values, notwithstanding Correct Summaries.

	Abortion		Gay rights	
	R2-R	R2-FM	R2-R	R2-FM
Rand Summ.	1.0	1.0	0.7	0.8
AITV	3.0	2.8	**2.7**	**2.8**
TAM	1.8	2.1	2.0	2.4
LAM_LEX	1.5	1.0	1.1	0.9
PhAITV	**4.5**	**4.6**	**2.7**	**2.8**
Correct Summ.	5.8	5.4	3.0	2.9

4.4 Evaluating Informativeness

We re-frame the problem of creating a contrastive digest table into a summary problem. The concatenation of all extracted sentential reasons of the digest is considered as a candidate summary. The construction of reference summaries, using annotated reasons of CreateDebate datasets, is explained in Sect. 4.1. The length of the candidate summaries is proportional to that of the references. Reference summaries on 4Forums or Reddit datasets can not be constructed because no annotation, of reasons and their types, exists. We assess all methods, on CreateDebate, using automatic summary evaluation metric ROUGE [11]. We report the results of Rouge-2's Recall (R-2 R) and F-Measure (R-2 FM). Rouge-2 captures the similarities between sequences of bigrams in references and candidates. The higher the measure, the better the summary. All reported ROUGE-2 values are computed after applying stemming and stop words removal on reference and candidate summaries. This procedure may also explain the relatively small values of reported ROUGE-2 measures in Table 3, compared to those usually computed when stop words are not removed. The existence of stop words in candidate and references sentences increases the overlap, and hence the ROUGE measures' values in general. Applying stemming and stop words removal was based on some preliminary tests that we conducted on our dataset. The tests showed that two candidate summaries containing different numbers of valid reasons, would have a statistically significant difference in their ROUGE-2 values when stemming and stop words removal applied.

Table 3 contains the averaged results on 10 generated summaries on Abortion and Gay Rights, respectively. LAM_LEX performs poorly in this task (close to Random summaries) for both datasets. PhAITV performs significantly better than TAM on Abortion, and slightly better on Gay Rights. Moreover, PhAITV shows significant improvement over its degenerate unigram version AITV on Abortion. This shows that phrase modeling and grouping can play a role in extracting more diverse and informative phrases. AITV beats its similar unigram-based summaries on both datasets. This means that the proposed

Table 4. Median values of Relevance Rate (Rel), NPV and Clustering Accuracy Percentages on CreateDebate, FourForums and Reddit Datasets. Bold denotes best results, notwithstanding PhAITV$_{view}$.

| | CreateDebate | | | | | | 4Forums | | | | | | Reddit | | |
| | Abortion | | | Gay Rights | | | Abortion | | | Gay Marriage | | | Isr/Pal | | |
	Rel	NPV	Acc.	Rel	NPV	Acc.	Rel	NPV	Acc.	Rel	NPV	Acc.	Rel	NPV	Acc.
AITV	0.66	58.33	59.09	0.5	75.0	66.66	0.66	66.66	71.42	0.5	50.0	66.66	0.6	55.55	60.00
TAM	0.53	50.00	46.42	0.5	50.0	42.85	0.33	37.50	66.66	0.3	50.0	33.33	0.3	66.66	50.00
LAM_LEX	0.40	50.00	64.44	0.5	50.0	50.00	0.46	37.50	46.60	0.5	50.0	50.00	0.3	25.00	33.33
PhAITV	**0.93**	**75.00**	**73.62**	**0.8**	**75.0**	**75.00**	**0.80**	**69.44**	**71.79**	**0.7**	**80.0**	**71.42**	**0.9**	**75.00**	**77.77**
PhAITV$_{view}$	0.93	87.50	83.33	0.9	100	100	0.80	83.33	81.81	0.9	100	100	–	–	–

pipeline is effective in terms of summarization even without the phrase modeling. In addition, PhAITV's ROUGE measures on Gay Rights are very similar to those of the correct summaries (Table 3). Examples of the final outputs produced by PhAITV framework and the two contenders on Abortion is presented in Table 5. The example digests produce proportional results to the median results reported in Table 4. We notice that PhAITV's digest produces different types of reasons from diverse argument facets, like putting child up for adoption, life begins at conception, and mother's life in danger. However, such informativeness is lacking on both digests of LAM_LEX and TAM. Instead, we remark the recurrence of subjects like killing or taking human life in TAM's digest.

4.5   Evaluating Relevance and Clustering

For the following evaluations, we conduct a human annotation task with three annotators. The annotators are acquainted with the studied issues and the possible reasons conveyed by each side. They are given lists of mixed sentences generated by the models. They are asked to indicate the stance of each sentence when it contains any kind of persuasion, reasoning or argumentation from which they could easily infer the stance. Thus, if they label the sentence, the sentence is considered a relevant reason. The average Kappa agreement between the annotators is 0.66. The final annotations correspond to the majority label. In the case of a conflict, we consider the sentence irrelevant.

We consider measuring the relevance by the ratio of the number of relevant sentences divided by the total number of the digest sentences. Table 4 contains the median relevance rate (Rel) over 5 runs of the models, on all datasets. PhAITV-based pipeline realizes very high relevance rates and outperforms its rivals, TAM and LAM_LEX, by a considerable margin on all datasets. Moreover, it beats its unigram counterpart AITV. These results are also showcased in Table 5's examples. The ratio of sentences judged as reasons given to support a stance is higher for PhAITV-based digest. Interestingly, even the PhAITV's sentences judged as irrelevant are not off-topic, and may denote relevant argument facets like "abortion is murder". Results confirm our hypothesis that phrasal facet argument leads to a better reasons' extraction.

Table 5. Sample Digest Tables Output of sentential reasons produced by the frameworks based on PhAITV, LAM_LEX and TAM when using Abortion dataset from CreateDebate. Sentences are labeled according to their stances as the following: (+) reason for abortion; (-) reason against abortion; and (0) irrelevant.

PhAITV + Grouping + Extraction		
Viewpoint 1		Viewpoint 2
(-)	If a mother or a couple does not want a child there is always the option of putting the child up for adoption.	(+) The fetus before it can survive outside of the mother's womb is not a person.
(-)	I believe life begins at conception and I have based this on biological and scientific knowledge.	(+) Giving up a child for adoption can be just as emotionally damaging as having an abortion.
(-)	God is the creator of life and when you kill unborn babies you are destroying his creations.	(+) you will have to also admit that by definition; abortion is not murder.
(-)	I only support abortion if the mothers life is in danger and if the fetus is young.	(-) No abortion is wrong.
(0)	The issue is whether or not abortion is murder.	(0) I simply gave reasons why a woman might choose to abort and supported that.

LAM_LEX [27]		
Viewpoint 1		Viewpoint 2
(-)	abortion is NOT the only way to escape raising a child that would remind that person of something horrible	(+) if a baby is raised by people not ready, or incapable of raising a baby, then that would ruin two lives.
(+)	I wouldn't want the burden of raising a child I can't raise	(+) The fetus really is the mother's property naturally
(0)	a biological process is just another name for metabolism	(0) Now this is fine as long as one is prepared for that stupid, implausible, far-fetched, unlikely, ludicrous scenario
(0)	The passage of scripture were Jesus deals with judging doesn't condemn judging nor forbid it	(0) you are clearly showing that your level of knowledge in this area is based on merely your opinions and not facts.
(0)	your testes have cells which are animals	(0) we must always remember how life is rarely divided into discreet units that are easily divided

TAM [16]		
Viewpoint 1		Viewpoint 2
(-)	I think that is wrong in the whole to take a life.	(+) Or is the woman's period also murder because it also is killing the potential for a new human being?
(-)	I think so it prevents a child from having a life.	(-) it maybe then could be considered illegal since you are killing a baby, not a fetus, so say the fetus develops into an actuall baby
(+)	Abortion is not murder because it is performed before a fetus has developed into a human person.	(0) In your scheme it would appear to be that there really is no such thing as the good or the wrong.
(0)	He will not obey us.	(0) NO ONE! but God.
(0)	What does it have to do with the fact that it should be banned or not?	(0) What right do you have to presume you know how someone will life and what quality of life the person might have?

All compared models generate sentences for each viewpoint. Given the human annotations, we consider assessing the viewpoint clustering of the relevant extracted sentences by two measures: the Clustering Accuracy and the Negative Predictive Value of pairs of clustered sentences(NPV). NPV consider a pair of sentences as unit. It corresponds to the number of true stance opposed pairs in different clusters divided by the number of pairs formed by sentences in opposed clusters. A high NPV is an indicator of a good inter-clusters opposition i.e., a good contrast of sentences' viewpoints. Table 4 contains the median NPV and Accuracy values over 5 runs. Both AITV and PhAITV-based frameworks achieve very encouraging NPV and accuracy results without any supervision. PhAITV outperforms significantly the competing contrastive summarization methods. This confirms the hypothesis that leveraging the reply-interactions, in online debate, helps detect the viewpoints of posts and, hence, correctly cluster the reasons' viewpoints. Table 5 shows a much better alignment, between the viewpoint clusters and the stance signs of reasons (+) or (-), for PhAITV comparing to competitors. The NPV and accuracy values of the sample digests are close to the median values reported in Table 4. The contrast also manifests when similar facets are discussed but by opposing viewpoints like in "life begins at conception" against "fetus before it can survive outside the mother's womb is not a person". The results of PhAITV are not close yet to the $PhAITV_{view}$ where the true posts' viewpoint are given. This suggests that the framework can achieve very accurate performances by enhancing viewpoint detection of posts.

5   Conclusion

This work proposes an unsupervised framework for the detection, clustering, and displaying of the main sentential reasons conveyed by divergent viewpoints in contentious text from online debate forums. The reasons are extracted in a contrastive digest table. A pipeline approach is suggested based on a Phrase Mining module and a novel Phrase Author Interaction Topic-Viewpoint model. The evaluation of the approach is based on three measures computed on the final digest: the informativeness, the relevance, and the accuracy of viewpoint clustering. The results on contentious issues from online debates show that our PhAITV-based pipeline outperforms state-of-the-art methods for all three criteria. In this research, we dealt with contentious documents in online debate forums, which often enclose a high rate of rebuttal replies. Other social media platforms, like Twitter, may not have rebuttals as common as in online debates. Moreover, a manual inspection of the digests suggests the need for improvement in the detection of semantically similar reasons and their hierarchical clustering.

References

1. Abbott, R., Ecker, B., Anand, P., Walker, M.A.: Internet argument corpus 2.0: an SQL schema for dialogic social media and the corpora to go with it. In: LREC (2016)
2. Blei, D.M., Ng, A.Y., Jordan, M.I.: Latent Dirichlet allocation. J. Mach. Learn. Res. **3**, 993–1022 (2003)
3. Boltužić, F., Šnajder, J.: Back up your stance: recognizing arguments in online discussions. In: Proceedings of the First Workshop on Argumentation Mining, Baltimore, Maryland, pp. 49–58. Association for Computational Linguistics (2014). https://www.aclweb.org/anthology/W14-2107

4. Boltužić, F., Šnajder, J.: Identifying prominent arguments in online debates using semantic textual similarity. In: Proceedings of the 2nd Workshop on Argumentation Mining, Denver, CO, pp. 110–115. Association for Computational Linguistics (2015). https://www.aclweb.org/anthology/W15-0514

5. Bouma, G.: Normalized (pointwise) mutual information in collocation extraction. In: Proceedings of GSCL, pp. 31–40 (2009)

6. El-Kishky, A., Song, Y., Wang, C., Voss, C.R., Han, J.: Scalable topical phrase mining from text corpora. Proc. VLDB Endow. **8**(3), 305–316 (2014). https://doi.org/10.14778/2735508.2735519

7. Erkan, G., Radev, D.R.: LexRank: graph-based lexical centrality as salience in text summarization. J. Artif. Intell. Res. (JAIR) **22**(1), 457–479 (2004)

8. Habernal, I., Gurevych, I.: Argumentation mining in user-generated web discourse. Comput. Linguist. **43**(1), 125–179 (2017)

9. Hasan, K.S., Ng, V.: Why are you taking this stance? Identifying and classifying reasons in ideological debates. In: Proceedings of the 2014 Conference on Empirical Methods in Natural Language Processing (EMNLP), Doha, Qatar, pp. 751–762. Association for Computational Linguistics (2014). https://www.aclweb.org/anthology/D14-1083

10. Li, H., Mukherjee, A., Si, J., Liu, B.: Extracting verb expressions implying negative opinions. In: Proceedings of the AAAI Conference on Artificial Intelligence (2015). https://www.aaai.org/ocs/index.php/AAAI/AAAI15/paper/view/9398

11. Lin, C.Y.: Rouge: a package for automatic evaluation of summaries. In: Marie-Francine Moens, S.S. (ed.) Text Summarization Branches Out: Proceedings of the ACL-04 Workshop, Barcelona, Spain, pp. 74–81. Association for Computational Linguistics (2004)

12. Misra, A., Anand, P., Fox Tree, J.E., Walker, M.: Using summarization to discover argument facets in online idealogical dialog. In: Proceedings of the 2015 Conference of the North American Chapter of the Association for Computational Linguistics: Human Language Technologies, Denver, Colorado, pp. 430–440. Association for Computational Linguistics (2015). https://www.aclweb.org/anthology/N15-1046

13. Misra, A., Oraby, S., Tandon, S., Ts, S., Anand, P., Walker, M.A.: Summarizing dialogic arguments from social media. In: Proceedings of the 21th Workshop on the Semantics and Pragmatics of Dialogue (SemDial 2017), pp. 126–136 (2017)

14. Mohammad, S.M., Sobhani, P., Kiritchenko, S.: Stance and sentiment in tweets. ACM Trans. Internet Technol. **17**(3), 26:1–26:23 (2017). https://doi.org/10.1145/3003433

15. Park, J., Cardie, C.: Identifying appropriate support for propositions in online user comments. In: Proceedings of the First Workshop on Argumentation Mining, Baltimore, Maryland, pp. 29–38. Association for Computational Linguistics (2014). https://www.aclweb.org/anthology/W14-2105

16. Paul, M., Zhai, C., Girju, R.: Summarizing contrastive viewpoints in opinionated text. In: Proceedings of the 2010 Conference on Empirical Methods in Natural Language Processing, Cambridge, MA, pp. 66–76. Association for Computational Linguistics (2010). https://www.aclweb.org/anthology/D10-1007

17. Qiu, M., Jiang, J.: A latent variable model for viewpoint discovery from threaded forum posts. In: Proceedings of the 2013 Conference of the North American Chapter of the Association for Computational Linguistics: Human Language Technologies, Atlanta, Georgia, pp. 1031–1040. Association for Computational Linguistics (2013). https://www.aclweb.org/anthology/N13-1123

18. Stab, C., Gurevych, I.: Identifying argumentative discourse structures in persuasive essays. In: Proceedings of the 2014 Conference on Empirical Methods in Natural Language Processing (EMNLP), Doha, Qatar, pp. 46–56. Association for Computational Linguistics (2014). https://www.aclweb.org/anthology/D14-1006

19. Swanson, R., Ecker, B., Walker, M.: Argument mining: extracting arguments from online dialogue. In: Proceedings of the 16th Annual Meeting of the Special Interest Group on Discourse and Dialogue, Prague, Czech Republic, pp. 217–226. Association for Computational Linguistics (2015). https://aclweb.org/anthology/W15-4631

20. Thonet, T., Cabanac, G., Boughanem, M., Pinel-Sauvagnat, K.: VODUM: a topic model unifying viewpoint, topic and opinion discovery. In: Ferro, N., et al. (eds.) ECIR 2016. LNCS, vol. 9626, pp. 533–545. Springer, Cham (2016). https://doi.org/10.1007/978-3-319-30671-1_39

21. Trabelsi, A., Zaiane, O.R.: Finding arguing expressions of divergent viewpoints in online debates. In: Proceedings of the 5th Workshop on Language Analysis for Social Media (LASM), Gothenburg, Sweden, pp. 35–43. Association for Computational Linguistics (2014). https://www.aclweb.org/anthology/W14-1305

22. Trabelsi, A., Zaïane, O.R.: A joint topic viewpoint model for contention analysis. In: Métais, E., Roche, M., Teisseire, M. (eds.) Natural Language Processing and Information Systems, pp. 114–125. Springer, Cham (2014). https://doi.org/10.1007/978-3-319-07983-7_16

23. Trabelsi, A., Zaiane, O.R.: Mining contentious documents using an unsupervised topic model based approach. In: Proceedings of the 2014 IEEE International Conference on Data Mining, pp. 550–559 (2014)

24. Trabelsi, A., Zaïane, O.R.: Extraction and clustering of arguing expressions in contentious text. Data Knowl. Eng. **100**, 226–239 (2015)

25. Trabelsi, A., Zaïane, O.R.: Mining contentious documents. Knowl. Inf. Syst. **48**(3), 537–560 (2016)

26. Trabelsi, A., Zaïane, O.R.: Unsupervised model for topic viewpoint discovery in online debates leveraging author interactions. In: Proceedings of the AAAI International Conference on Web and Social Media (ICWSM), Stanford, California, pp. 425–433. Association for the Advancement of Artificial Intelligence (2018)

27. Vilares, D., He, Y.: Detecting perspectives in political debates. In: Proceedings of the 2017 Conference on Empirical Methods in Natural Language Processing, Copenhagen, Denmark, pp. 1573–1582. Association for Computational Linguistics (2017). https://www.aclweb.org/anthology/D17-1165

Visualizing and Analyzing Networks of Named Entities in Biographical Dictionaries for Digital Humanities Research

Minna Tamper[1]([✉])[iD], Petri Leskinen[1][iD], and Eero Hyvönen[1,2][iD]

[1] Semantic Computing Research Group (SeCo), Aalto University, Espoo, Finland
{minna.tamper,petri.leskinen,eero.hyvonen}@aalto.fi
[2] HELDIG – Helsinki Centre for Digital Humanities, University of Helsinki, Helsinki, Finland
http://seco.cs.aalto.fi, http://heldig.fi

Abstract. This paper shows how named entity extraction and network analysis can be used to examine biographies individually and in groups to aid historians in biographical and prosopographical research. For this purpose a reference network of 13 100 biographies in the collections of the Biographical Centre of the Finnish Literature Society was created, based on links between the biographies as well as automatically extracted named entities found in the texts. The data was published in a SPARQL endpoint as a Linked Data knowledge graph on top of which network analytic tools were created and analysis were done showing the usefulness of the approach in Digital Humanities. The reference graph has been utilized for network analysis to examine egocentric networks of individual persons as well as networks among groups of people in prosopography. The data and tools presented are in use since autumn 2018 in the semantic portal BiographySampo that has had tens of thousands of users.

Keywords: Named entity extraction · Information extraction · Text mining

1 Introduction

BiographySampo[1] is a semantic portal that is based on a knowledge graph that has been created using natural language processing methods, linked data, and semantic web technologies [15,35]. The graph currently contains ca. 13 100 biographical textual descriptions of notable Finns that can be browsed through using faceted search and a variety of data-analytic tools. 9200 of the entries contain a short, free text biography of the person, created by 977 professional authors. The portal has been built to help historians and scholars in biographical [33] and

---

[1] Online at www.biografiasampo.fi; see project homepage https://seco.cs.aalto.fi/ projects/biografiasampo/en/ for further info and publications.

© Springer Nature Switzerland AG 2023
A. Gelbukh (Ed.): CICLing 2019, LNCS 13451, pp. 199–214, 2023.
https://doi.org/10.1007/978-3-031-24337-0_15

prosopographical research [8,37][2]. A major novelty of BiographySampo is to provide the user with data-analytic and visualization tools for solving research problems in Digital Humanities (DH), based on Linked Data [9,12].

In the biography texts, the authors mention other people they consider significant from an occupational or other relevant perspective. In our case study, the editors of the dictionary of biography at the publisher Finnish Literature Society (SKS) have changed these mentions into internal links to corresponding articles in the dictionary if there is one.[3] A link is added typically only once when a person is mentioned for the first time. These links serve in the original biography collection as a way to browse and move between the biographies.

However, many links are missing from the text. For example, there are mentions of relatives and external people who do not have a biography in the dictionary to be linked to, e.g., William Shakespeare and Richard Wagner. In addition, if a biography A mentions person B, but the biography of B has been added in the collection after editing A, it has not been possible to add the link. The explicit links between people in the biographical texts therefore create a scarcely interlinked reference network of the biographical texts.

This paper argues that making the reference network underlying a biographical dictionary explicit can be useful in biographical and prosopographical research. The idea of using the network analysis of historical people for Digital Humanities research has been suggested before in, e.g., [3,38]. A contribution of our paper is to apply the idea to biography collections, where connections are based on entity mentions. To support the argument, we present a case study using BiographySampo where the reference network underlying its textual biographies was extracted and enriched into a knowledge graph and published as a linked data service, on top of which a set of tools were created for Digital Humanities research. This idea is currently being applied also to a genealogical network extracted from the same texts [20].

In the following, the underlying knowledge graph with its person and place ontologies, and the process of extracting and enriching the reference network is first presented (Sect. 2). After this, application views to study the networks underlying the biographical texts are presented (Sect. 3). Firstly, a network analysis tool is presented for visualizing and studying the *egocentric network* of a protagonist in biographical research. Secondly, this idea is generalized for prosopography where groups of people sharing characteristics (e.g., occupation, gender, or area of living) are studied. Here the user can first separate the target group using faceted search and then visualize the group's *sociocentric network*. Thirdly, when visualizing the networks, it turned out that they often include serendipitous [1] (surprising) connections between people, raising the question: why are these two people interconnected? A tool is clearly needed for explaining

---

[2] Prosopography is a method that is used to study groups of people through their biographical data. The goal of prosopography is to find connections, trends, and patterns from these groups.

[3] Actually, the biographies in our case study come from several separate databases, including the general National Biography of Finland as a core, supplemented with four other thematic dictionaries [16].

the connections, not only showing them. For this purpose, an application view showing the textual contexts in which the connections arise was created. Lastly, the toolset presented also includes an application called contextual reader [24], where the user is able to get information about the extracted linked entities by hovering the mouse on top of the mentions. After presenting the application views, the applications and named entity extraction is evaluated (Sect. 4). In conclusion (Sect. 5), the contributions of the paper are summarized, related works discussed, and the directions of further research suggested.

2   Extracting Named Entities from Biographical Texts

In order to build and integrate network analysis tools, reference analysis tools, and the contextual reader application to BiographySampo, the existing links and named entities need to be extracted from the texts, and the underlying BiographySampo Knowledge Graph (BSKG) be enriched accordingly. In this section the knowledge base, extraction process, and the data transformations that enable the end user applications are discussed.

Knowledge Graph. BSKG includes the biography collections[4] of SKS written by 977 scholars from different fields. The biographies describe the lives and achievements of historical and contemporary figures, containing vast amounts of references to notable Finnish and foreign figures and to historical events, works (e.g., paintings, books, music, and acting), places, organizations, and dates. The graph includes 13 144 people with a biographical description, 51 200 related people mentioned in the biographies, and the 977 authors of the biographies. There are furthermore 225 000 lifetime events of the protagonists including their births, deaths, and other biographical events. The biographical texts also contain manually added 31 500 HTML links between the biographies that were included in the knowledge base [35]. There is also a separate graph of 4970 places, extracted from the Finnish Gazetteer of Historical Places and Maps (Hipla) and data service[5] [14,17]. Foreign place names were linked using the Google Maps APIs[6]. The lifetime events have lots of mentions of other kinds, such as governmental or educational buildings, public places etc. An additional dataset of approximately 2000 resources was extracted for them from Wikidata. The data was also augmented with a list of countries in the world and their capitals [21].

Extraction and Linking Process. The biographical texts [35] were transformed into an RDF dataset and enriched with linguistic information, totaling in 120 million triples. The data can be queried from a SPARQL endpoint. This data contains manually annotated links that have been extracted from the HTML as well as links based on entity linking.

Named entity linking (NEL) tools [25,26,28] typically use a process that can be broken into three tasks [4,7]: 1) named entity recognition (NER), 2) named

---

[4] https://kansallisbiografia.fi/english/national-biography.
[5] http://hipla.fi.
[6] http://developers.google.com/maps/.

entity disambiguation (NED), and 3) NEL. NER identifies the entities from text, NED disambiguates them, and lastly NEL links the mentions to their meanings in ontologies or knowledge bases. Our new linking tool, NELLI, extracts and links entities from texts in a similar manner. However, in addition it combines multiple, in our case three different tools for NER and NEL. The purpose of this approach is to improve disambiguation by utilizing a voting scheme [6,32] where each tool has a vote on the interpretation it makes for the same piece of text. The best candidate is the one with the most votes. For example, to identify a place from the string *Turku Cathedral* the tools return three answers of which one is *Turku* and two are *Turku Cathedral*, the winning interpretation.

NELLI uses the tools FiNER, ARPA, and LINFER. FiNER[7] is a rule-based NER tool for Finnish, ARPA [23] is a NER and NEL tool [10,18] that queries matches from controlled vocabularies. To supplement FiNER and ARPA, a third tool LINFER was implemented utilizing the linguistic RDF data to identify named entities. The parsed linguistic data not only contains part of speech information but also Dependency Grammar relations. With this information a set of rules was created to infer which proper nouns (or nouns) would be most likely place or person names. With this tool entities, e.g., *Åbo Akademin kirjasto* (engl. *Åbo Akademi University Library*) can be identified by analyzing inflected forms and dependencies. These rules were encapsulated in LINFER to utilize the linguistic features of words and their relations.

In addition to each tool having a vote, votes can be earned for entity length, linkage, and by named entity type. Sometimes it may be difficult to correctly identify longer named entities, such as place or organization names, and therefore a vote is given to the longest matching candidates. Also candidates that have found a match in an ontology are favored with a vote. NELLI also has a priority order for named entity types where votes can be added to favor some entity types over others. For example, the address *Konemiehentie 2, Espoo* contains a city name. In order to have the address as the top voted candidate, it will help to give to the address type a higher score than to the more general location.

Once NELLI has all the interpretations and metrics about the candidates, it calculates the votes and writes the results in Turtle format. For this extension of the original data, we used NLP Interchange Format (NIF)[8] [11], Dublin Core Metadata[9], and a custom namespace[10] to supply classes and properties that describe named entity metadata. For recording the results, the application writes *nbf:NamedEntity* class instances that have the basic information about the entity. It has properties to describe the extracted string (*nif:isString*), base form of the string (*nif:lemma*), its named entity type (*nbf:namedEntityType*), where it is linked (*skos:relatedMatch*), the location of the string in text (*nif:beginIndex, nif:endIndex*), and the method that was used to extract the named entity (*nbf:usedNeMethod*). In the source dataset, the texts have been split into doc-

---

[7] https://github.com/Traubert/FiNer-rules/blob/master/finer-readme.md.

[8] http://persistence.uni-leipzig.org/nlp2rdf/specification/core.html.

[9] http://dublincore.org/documents/dcmi-terms/.

[10] Denoted with prefix **nbf**.

uments, paragraphs, sentences, and words. The word-entities are also added a *dct:isPartOf* property referring to the named entity instances they are a part of and similarly the sentences have a *nbf:hasNamedEntity* property. The value of the *nbf:namedEntityType* property is an instance of the *nbf:NamedEntityType* class that is the description of the named entity type. The value of the *nbf:used-NeMethod* property is an instance of the class *nbf:NamedEntityMethod* that has provenance information about the tools used to extract the named entity. In addition to the *nbf:NamedEntity* class, there is also the *nbf:NamedEntityGroup* class that groups the entities in each sentence based on location and possible overlap. Each group has all members indicated with the property *nbf:member* and the top voted entity with *nbf:primary*.

Reference Networks. Network analysis of people [3,38] is a set of methods that can be used to study social networks [30]. In our case, the networks were built from the HTML links and mentions of people in the biographies to create a reference network which is analogous to citation networks [34]. In a reference network, the nodes are people, and when a person A is mentioned in the biography of B, a directed edge is added from B to A. The edges are instances of the class *nbf:Reference* with properties for the source biography *nbf:source*, the mentioned person *nbf:target*, and the type of the reference as *nbf:ManualAnnotation* (for HTML links) or *nbf:AutomaticAnnotation* (for identified named person entities). The number of references to the target person in the source biography is declared as the value of the *nbf:weight* property which for manual HTML links equals one.

The transformed network data can then be used in applications by querying the nodes, e.g., biographical details of people, and the edges, e.g., the links between people. Based on the data, the networks can be generated automatically for an individual or a group.

3 Applications

To test the potential of network analysis in biography and prosopography, a reference network was constructed based on HTML links in ca. 6100 of the 13 144 biographies, enhanced with additional edges from 400 biographies by NELLI. This group was limited to politicians, writers, athletes, lutherans, artists, architects, and musicians because their biographies contain long textual descriptions.

The BSKG included entities for people and places extracted from texts. For place linking, we used the YSO Places ontology[11] of the Finnish Ontology Service Finto that contains contemporary place resources for municipalities, provinces, countries, and continents. The contemporary data was extended with the WarSampo place ontology [13] that includes historical Finnish places. A priority order was set for place and person entities so that more specific place names, for instance, have a higher score. Also, to avoid having people's first and last names mislabeled as places, person named entities were given a higher score.

---

[11] https://finto.fi/yso-paikat/en/.

With this setup, 33 120 entities were extracted and used as a basis for four application views presented next in this section.

1. Egocentric Networks. The egocentric networks are formed from people nodes, i.e., biographical details of people, and edges, i.e., the links between people. The networks are generated to the center of the screen and centered around one person, in this case the protagonist. On the left hand side of the user interface, there are network toggles that can be used to alter the layout of the network in the following ways: Firstly, the user can toggle the amount of nodes to be seen, i.e., limit the size of the network to be visualized. Secondly, the user can select to see the network built using the manual HTML links only, automatically extracted links, or both. In this way, the manual and automatically extracted links can be compared with each other. Thirdly, to emphasize the most significant nodes in the graph, the node size can be determined based on using four distance and centrality measures used in network analysis: distance to the protagonist, degree, in-degree, out-degree, or pagerank [27]. Fourthly, it is possible to color the person nodes based on the gender, occupational area, or distance to the protagonist. The network is generated based on the selected toggle options, and the automatic links option shows the edge weight based on how frequently the person is mentioned in the text.

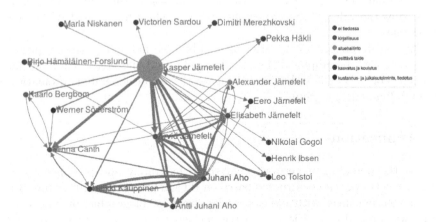

Fig. 1. Kasper Järnefelt's egocentric network where nodes are colored by occupation.

Figure 1 depicts the egocentric network of the Finnish critic, translator, and cultural person Kasper Järnefelt (1859–1941). The network shows, e.g., lots of links to contemporary Finnish cultural persons with a biography in the system (based on the HTML links), as well as connections to external people, such as authors Nikolai Gogol, Henrik Ibsen, and Leo Tolstoi (based on NELLI), who do not have biography in BiographySampo. The linkage is based on the fact that Järnefelt has translated their works into Finnish. The width of the edges indicates the number of references between the biographies and is an indication of potential importance. The legend box in the right upper corner explains the color coding of the occupational areas used for the nodes. The toggles for making the selections for the visualization are not shown in the figure for brevity.

In BiographySampo, the egocentric networks are located under the *Network* tab in the personal home pages of the protagonists.

2. Sociocentric Networks The sociocentric networks are located in their own view in BiographySampo. They can be accessed from the navigation bar under the title *Verkostot* (engl. Networks). In this application view, the user first filters the target group she is interested in studying by using faceted search. For example, people of similar occupation or place of birth can be easily filtered out by selections in corresponding facets.

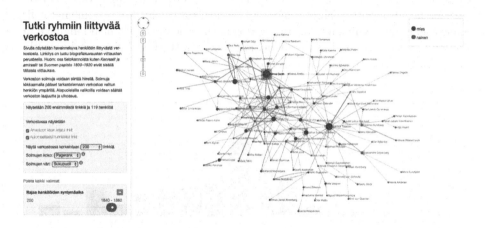

Fig. 2. Minna Canth in a sociocentric network where nodes are colored by gender. (Color figure online)

An example of a group view is presented in Fig. 2. The facets are situated on the left hand side of the screen underneath the general network analysis toggles, but are not visible here. In this case, the user has filtered out Finnish authors of the mid 19th century. Here, the Finnish female playwright and social activist Minna Canth gains the highest pagerank, illustrated by the size of her node. The gender of persons is indicated by red (women) or blue (men), an option selected from the toggles on the left.

3. Explaining References. BiographySampo also contains an application view that explains the edges in the egocentric and sociocentric networks. This reference view can be found for each protagonist in a separate tab on their homepage. The idea is to explain edges by providing the user with the sentences in which

the references to other people are mentioned.[12] The sentences can be retrieved
from the linguistic graph of the underlying SPARQL endpoint [35]. The references have been divided into two groups: 1) Sentences in other bios that make a
reference to the protagonist's biography. 2) Sentences in the protagonist's biography that make reference to other biographies. For example, the references to
Minna Canth include sentences from the biographies of actors, writers, and play-
wrights that were influenced by her, whereas her own biography mentions mainly
contemporary writers and artists.

In addition to listing sentences that include links, BiographySampo also has
a separate statistics application view that depicts *at what time* a person is referenced to. Here time is based on the birth year of the protagonist in the biography
making the reference. The purpose of this temporal view is to be able to see how
a person is referenced through time. For example, as shown in the Fig. 3, Minna
Canth is frequently mentioned over a long time, because, e.g., the actors and
directors of the national biography are using her plays. In comparison, a person
such as the 19th century philosopher and Finnish statesman Johan Vilhelm Snellman, who had a significant role in improving the role of the Finnish language in
the 19th century Finland, is mentioned, as shown in the Fig. 4, frequently mostly
in the biographies of his contemporaries.

Fig. 3. The references made to Minna Canth.

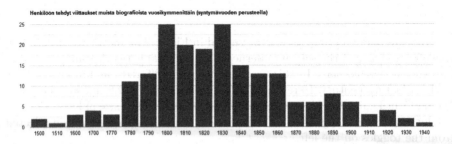

Fig. 4. The references made to Johan Snellman.

---

[12] The view currently lists only sentences that contain manually added HTML links.

This view can not only be used to identify the influences of these notable Finns in history, but also to study the edges that exist in the networks. This helps the user to see why an edge exists between two people and what kind of semantic meaning each edge holds in the network. For example, in the references page of Minna Canth, the user can see that in most cases she is mentioned because of her literary work, and by people who have acted, directed, or visited her salon to discuss and exchange thoughts on literature and ideologies, such as Darwinism.

4. Contextual Reader. Contextual reader is yet another application of NELLI data in BiographySampo. The idea here is to show the text annotated with links to the named entities, such as people, places, and organizations. It enhances the reading experience of the user by providing contextual linked information about the named entities in the biographies when the mouse is hovered over the text. The application is in work in Fig. 5, where the mouse is over *Nikolai Gogol*.

To achieve this, NELLI was configured to link named entities to contextual background information in the BSKG and other datasets available in SPARQL endpoints, such as biographies, map services, and ontology services. This was done to interlink biographical texts to each other and to help the user to understand and learn better from the texts based on their context [22, 26].

The system is based on the CORE tool [24], where entity mentions in texts can be linked to linked data resources in real time. Here string-based semantic disambiguation is used and only one interpretation is always selected. In contrast, in BiographySampo annotations are created in a pre-processing phase facilitating deeper analysis and disambiguation of entities, where challenging multiple interpretations can be given to the end-user for final human disambiguation.

This application was integrated into the biography tab of a person's homepage. The user can read the biography and gain more understanding through the links to people (indicated in blue color), places (green), and organizations (gray) (cf. Fig. 5). The links are also indicated by a symbol showing the type of the link. By hovering on top of an internal link (to BSKG) or an external link (to, e.g., Wikidata), as in Fig. 5 to the Russian playwright Nikolai Gogol, the user gets more information about that person (here from the Wikidata SPARQL endpoint). The place links lead to a map view to provide information related to that place and a map marking the location.

4   Assessment and Evaluation

In this section, lessons learned in developing the applications are first discussed and their usefulness assessed from an end user perspective. After this, an evaluation of the NEL tool NELLI follows.

Assessing Applications. The network analysis views have been built for individuals and for groups of people. The egocentric networks for individuals are often smaller in size and therefore facets are not included in the view for filtering out related people. However, in some cases it may be interesting to scale

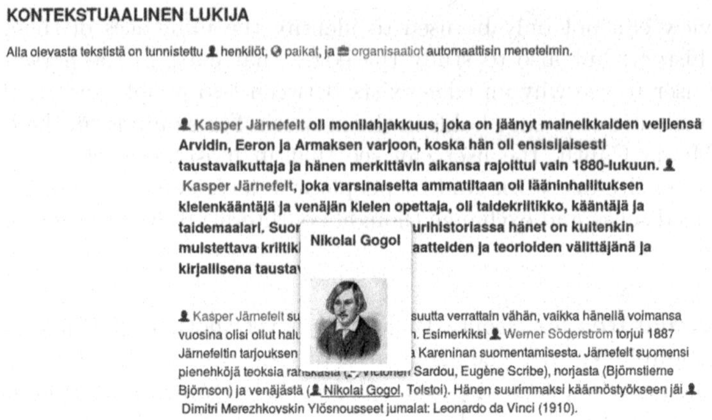

Fig. 5. Contextual Reader application used on Kasper Järnefelt's biography. (Color figure online)

egocentric networks to include only occupational references or people who have lived at the same time. The basic network toggling tools are provided for both views and can be used to color the nodes by occupation or gender. Also, it is possible to compare networks based on manual and automatic links.

The reference explanation view adds textual context to the links and in most cases is a helpful tool for understanding relations between nodes. However, the view currently only shows the sentences with manual HTML links. In addition, in some cases one sentence does not have enough context for an explanation. For example, in the biography of *Aale Tynni* (a poet) there is a highly serendipitous surprising link to *Tapio Rautavaara* (a singer, actor, and athlete). It turns out that both of them got a gold medal in the Olympic Games in London 1948, but in different categories: Tynni in lyrics and Rautavaara in javelin throw. However, the sentence with the link does not explain this. The information is in the previous sentence, and it would be useful in this case to show more than one sentence to explain the relation.

The contextual reader application visualizes the extracted named entities in the text and adds more contexts through linking to BSKG and external datasets and ontologies. There are currently only three types of named entities visible in the contextual reader to provide context but more could be added, e.g., named works of art. Also, it would be useful to add images or maps for places as has been done in [13]. The extracted named entities from the texts are often people that the authors of the biographies consider significant occupationally. The networks and the reference analysis reflects these choices creating biases similarly to [38]. Reference networks in our case are not actual social networks. For example, in the biography of *Jutta Urpilainen*, the former Prime Minister *Jyrki Katainen* is mentioned because *Urpilainen* worked as the Minister of Finance in *Katainen's* Cabinet of Finland, but in the biography of *Katainen*, *Urpilainen* is not mentioned. It is important to keep in mind that these networks only give insight to who are considered by authors and their sources to be significant to the protagonist [15].

Evaluation of Named Entity Extraction. In order to measure the quality of NELLI in the task of identifying named entities, we inspected place and person links for 50 biographical texts. Self-references to the protagonist were ignored in calculations because the idea was to identify information that helps the reader to understand better about this person and the references to self do not add value in this task. In addition, we calculated organization names containing a linked place name (e.g., The National Museum of Finland) as false positive. The linking of places and people was evaluated using precision, recall, and F1-score as shown in Table 1; the identification of organizations was ignored in the test as this is still ongoing work.

Table 1. Results for recognition and linking places and people.

	Entities	TP	FP	FN_{all}	FN_{out}	Precision	$Recall_{out}$	$Recall_{all}$	$F1_{out}$	$F1_{all}$
Places	823	655	168	77	43	80%	94%	89%	86%	84%
People	348	339	9	227	119	97%	74%	60%	84%	74%

The results in the Table 1 for places and people have been counted in two ways: 1) to exclude false negatives that cannot be found from the ontology (FN_{out}) and 2) to include all false negatives (FN_{all}). By comparing these two counts, it can be seen how entities missing from the used ontologies impacts the results for places and especially people. In most cases, the tool is dependent on the chosen ontology due to having only a few people with the same names. However, the overall $F1_{all}$-scores are good for people 74% and places 84%. The precision ($Precision$) for places is lower than the recall, causing a drop in the F1-score. In comparison, the precision for people is nearly perfect. However, the recall ($Recall_{out}$, $Recall_{all}$) for people is lower than the recall for places. This is because some people cannot be found from the ontology due to tool errors, incorrect data (missing maiden or married names, badly formed data labels), or problems with baseforming foreign names.

The precision for places suffers also due to mixing last names with place names when the names are not identified from the text. In order to reduce mixing of place and person names, the last names could be identified using the extracted full person names. Often people are referenced in the text first with a full name and later with only the last name. By using the last names from the full names, most references could be extracted and mix-ups with place names avoided. The place recognition often mixes place names and regular words, such as adjectives as places. For example, when the initial word of a sentence is the infected form of the word *oma* (engl. own), it is understood as the place Oman. By adding a rule that only considers entities that are written with a capital letter can help to reduce these issues. However, it alone is not enough and utilization of linguistic information can help to filter initial words that are not proper nouns.

5   Conclusions

In this case study, a total of 31 500 manually created links between biographies were utilized to visualize and study a reference network underlying a dictionary of biographies. In addition, the application of NELLI to the data added a total of 33 120 named entity links in the network of which some 12 800 were for places and some 20 800 for people. This data was utilized to enrich the networks with additional references to people cataloged in the dictionary of biography and with new external nodes in the network. NELLI succeeded in identifying people with 74% accuracy and places with 84% accuracy. Four application views were added in BiographySampo to support analysis of the networks for the end user.

The selection of ontologies has a role in the success of the work. The place names in biographical texts were distinctive and easy to link to comprehensive ontologies with low granularity. It was helpful, too, that the BSKG contained only a handful of namesakes. By adding fixes to prevent the linking of adjectives and postpositions to places it is possible to increase the success rates. The disambiguation scheme enabled successful linking of person names, which prevented most of the mix-ups between people and places.

The applications presented were based on reference networks. Unlike in [3,31, 38], the user can study the networks of different groups through facet selections and visualize the networks in a variety of ways, such as re-sizing the nodes based on their topological properties or by coloring the nodes based on occupational area or gender. The networks of individuals can be studied in the egocentric network to see, for instance, the spreading of influences. The foreign influences of notable writers, politicians, and philosophers are prominent in the automatically enriched networks, and a full view of their reach can be seen through the egocentric networks. The networks, complemented with the reference view to study the explanations for the edges, gives more insight into the impact of individuals in groups. The contextual reader application enhances the reading experience by providing information about the linked entities. These applications facilitate novel, more diverse usage of BiographySampo in biographical and prosopographical research.

However, the applications also raise new questions and problems of source criticism regarding the quality of the automatically extracted content and semantic interpretation of the networks [15]. It is clear, for example, that the people selected in the dictionary do not necessarily constitute a homogeneous prosopographical group but were selected by the editors, and people mentioned in the texts reflect the decisions made by the authors and the sources they have used.

Related Work. Representing and analyzing biographical data is a new research and application field [15,36]. The network analysis based on biographical data has been studied in [3,19,38] where networks were created using a variety methods to extract named entities and their relations from text. In BiographySampo, the networks were created using the NEL approach [25,26,28].

The network analysis views were constructed to study individuals and groups of people. Several related works [3,5,19,31,38] and network analysis and visualization methods [27] have influenced the tools presented in this paper. The tools in BiographySampo extend traditional systems by adding user controls that can be used to scale and toggle the layout of the networks. In addition, the sociocentric network analysis allows the user to use facets (such as gender, vocation, birth and death places) to form groups of people and study their networks. To extend the network analysis tools, BiographySampo also includes a reference analysis view explaining the links, which is similar to KORP's[13] [2] keywords in context view but provides context for the edges in the network similarly to LinkedJazz's[14] [31] relationship view. Unlike in LinkedJazz, the view shows all relations and how a person is referenced throughout time to show how a person's work influences his or her contemporaries and other generations of notable people. The view is constructed using text that has been transformed into RDF [35] and by querying the sentences with manually crafted links from the SPARQL service.

In order to visualize the named entities, a contextual reader application [24] was created. Similar visualizations of named entity data have been used in, e.g., DBpedia Spotlight[15] [26] and Gate Cloud[16] [25]. The WarSampo [13] portal and the Semantic Finlex portal [29] include contextual reader applications that have been configured to link text into ontologies in real-time. These applications have influenced the creation of the BiographySampo's contextual reader. However, in our case the entities are not extracted in real time but in a preprocessing phase for more robust semantic disambiguation.

Acknowledgments. Our research was part of the Severi project (http://seco.cs.aalto. fi/projects/severi), funded mainly by Business Finland. Thanks to Mikko Kivelä for inspirational discussions and CSC - IT Center for Science for computational resources.

References

1. Aylett, R.S., Bental, D.S., Stewart, R., Forth, J., Wiggins, G.: Supporting serendipitous discovery. In: Digital Futures (Third Annual Digital Economy Conference), Aberdeen, UK, 23–25 October 2012 (2012)
2. Borin, L., Forsberg, M., Roxendal, J.: Korp – the corpus infrastructure of Språkbanken. In: Proceedings of LREC 2012, Istanbul: ELRA, pp. 474–478 (2012)
3. Brouwer, J., Nijboer, H.: Golden agents. A web of linked biographical data for the Dutch Golden Age. In: BD2017 Biographical Data in a Digital World 2017, Proceedings, vol. 2119, pp. 33–38. CEUR Workshop Proceedings (2018). https://ceur-ws.org/Vol-2119/paper6.pdf
4. Bunescu, R.C., Pasca, M.: Using encyclopedic knowledge for named entity disambiguation. In: EACL 2006, 11st Conference of the European Chapter of the Association for Computational Linguistics, vol. 6, pp. 9–16 (2006)

---

[13] https://korp.csc.fi/.
[14] https://linkedjazz.org/.
[15] https://www.dbpedia-spotlight.org/demo/.
[16] https://cloud.gate.ac.uk/.

5. Elson, D.K., Dames, N., McKeown, K.R.: Extracting social networks from literary fiction. In: Proceedings of the 48th Annual Meeting of the Association for Computational Linguistics, pp. 138–147. Association for Computational Linguistics (2010)
6. Ferragina, P., Scaiella, U.: TAGME: on-the-fly annotation of short text fragments (by Wikipedia entities). In: Proceedings of the 19th ACM International Conference on Information and Knowledge Management, pp. 1625–1628. ACM (2010)
7. Hachey, B., Radford, W., Nothman, J., Honnibal, M., Curran, J.R.: Evaluating entity linking with Wikipedia. Artif. Intell. **194**, 130–150 (2013)
8. Hakosalo, H., Jalagin, S., Junila, M., Kurvinen, H.: Historiallinen elämä - Biografia ja historiantutkimus. Suomalaisen Kirjallisuuden Seura (SKS), Helsinki (2014)
9. Heath, T., Bizer, C.: Linked Data: Evolving the Web into a Global Data Space. Synthesis Lectures on the Semantic Web: Theory and Technology. Morgan & Claypool (2011)
10. Heino, E., et al.: Named entity linking in a complex domain: case second world war history. In: Gracia, J., Bond, F., McCrae, J.P., Buitelaar, P., Chiarcos, C., Hellmann, S. (eds.) LDK 2017. LNCS (LNAI), vol. 10318, pp. 120–133. Springer, Cham (2017). https://doi.org/10.1007/978-3-319-59888-8_10
11. Hellmann, S., Lehmann, J., Auer, S., Brümmer, M.: Integrating NLP using linked data. In: Alani, H., et al. (eds.) ISWC 2013. LNCS, vol. 8219, pp. 98–113. Springer, Heidelberg (2013). https://doi.org/10.1007/978-3-642-41338-4_7
12. Hyvönen, E.: Publishing and Using Cultural Heritage Linked Data on the Semantic Web. Morgan & Claypool, Palo Alto (2012)
13. Hyvönen, E., et al.: WarSampo data service and semantic portal for publishing linked open data about the second world war history. In: Sack, H., Blomqvist, E., d'Aquin, M., Ghidini, C., Ponzetto, S.P., Lange, C. (eds.) ESWC 2016. LNCS, vol. 9678, pp. 758–773. Springer, Cham (2016). https://doi.org/10.1007/978-3-319-34129-3_46
14. Hyvönen, E., Ikkala, E., Tuominen, J.: Linked data brokering service for historical places and maps. In: Proceedings of the 1st Workshop on Humanities in the Semantic Web (WHiSe), vol. 1608, pp. 39–52. CEUR Workshop Proceedings (2016). https://ceur-ws.org/Vol-1608/paper-06.pdf
15. Hyvönen, E., et al.: BiographySampo – publishing and enriching biographies on the semantic web for digital humanities research. In: Hitzler, P., et al. (eds.) ESWC 2019. LNCS, vol. 11503, pp. 574–589. Springer, Cham (2019). https://doi.org/10.1007/978-3-030-21348-0_37
16. Hyvönen, E., Leskinen, P., Tamper, M., Tuominen, J., Keravuori, K.: Semantic national biography of Finland. In: Proceedings of the Digital Humanities in the Nordic Countries 3rd Conference (DHN 2018), vol. 2084, pp. 372–385. CEUR Workshop Proceedings (2018). https://ceur-ws.org/Vol-2084/short12.pdf
17. Ikkala, E., Tuominen, J., Hyvönen, E.: Contextualizing historical places in a gazetteer by using historical maps and linked data. In: Proceedings of Digital Humanities 2016 (DH 2016), Krakow, Poland, pp. 573–577 (2016). https://dh2016.adho.org/abstracts/39
18. Kettunen, K., Mäkelä, E., Ruokolainen, T., Kuokkala, J., Löfberg, L.: Old content and modern tools-searching named entities in a Finnish OCRed historical newspaper collection 1771–1910. arXiv preprint arXiv:1611.02839 (2016)
19. Langmead, A., Otis, J., Warren, C., Weingart, S., Zilinski, L.: Towards interoperable network ontologies for the digital humanities. Int. J. Hum. Arts Comput. **10**, 22–35 (2016)

20. Leskinen, P., Hyvönen, E.: Extracting genealogical networks of linked data from biographical texts. In: Hitzler, P., et al. (eds.) ESWC 2019. LNCS, vol. 11762, pp. 121–125. Springer, Cham (2019). https://doi.org/10.1007/978-3-030-32327-1_24

21. Leskinen, P., Hyvönen, E., Tuominen, J.: Analyzing and visualizing prosopographical linked data based on biographies. In: BD2017 Proceedings of the Second Conference on Biographical Data in a Digital World 2017, vol. 2119, pp. 39–44. CEUR Workshop Proceedings (2018). https://ceur-ws.org/Vol-2119/paper7.pdf

22. Lindquist, T., Long, H.: How can educational technology facilitate student engagement with online primary sources? A user needs assessment. Libr. Hi Tech 29(2), 224–241 (2011)

23. Mäkelä, E.: Combining a REST lexical analysis web service with SPARQL for mashup semantic annotation from text. In: Presutti, V., Blomqvist, E., Troncy, R., Sack, H., Papadakis, I., Tordai, A. (eds.) ESWC 2014. LNCS, vol. 8798, pp. 424–428. Springer, Cham (2014). https://doi.org/10.1007/978-3-319-11955-7_60

24. Mäkelä, E., Lindquist, T., Hyvönen, E.: CORE - a contextual reader based on linked data. In: Proceedings of Digital Humanities 2016, Krakow, Poland, pp. 267–269 (2016). https://dh2016.adho.org/abstracts/4

25. Maynard, D., Roberts, I., Greenwood, M.A., Rout, D., Bontcheva, K.: A framework for real-time semantic social media analysis. J. Web Semant. 44, 75–88 (2017)

26. Mendes, P.N., Jakob, M., García-Silva, A., Bizer, C.: DBpedia spotlight: shedding light on the web of documents. In: Proceedings of the 7th International Conference on Semantic Systems, pp. 1–8. ACM (2011)

27. Newman, M.: Networks. Oxford University Press, Oxford (2018)

28. Nguyen, D.B., Hoffart, J., Theobald, M., Weikum, G.: AIDA-light: high-throughput named-entity disambiguation. In: Proceedings of LDOW, Linked Data on the Web, vol. 1184. CEUR Workshop Proceedings (2014). https://ceur-ws.org/Vol-1184/ldow2014_paper_03.pdf

29. Oksanen, A., Tuominen, J., Mäkelä, E., Tamper, M., Hietanen, A., Hyvönen, E.: Semantic Finlex: transforming, publishing, and using Finnish legislation and case law as linked open data on the web. In: Knowledge of the Law in the Big Data Age. Frontiers in Artificial Intelligence and Applications, vol. 317, pp. 212–228. IOS Press (2019)

30. Otte, E., Rousseau, R.: Social network analysis: a powerful strategy, also for the information sciences. J. Inf. Sci. 28(6), 441–453 (2002)

31. Pattuelli, M.C., Miller, M., Lange, L., Thorsen, H.K.: Linked Jazz 52nd street: a LOD crowdsourcing tool to reveal connections among Jazz artists. In: Proceedings of Digital Humanities 2013, pp. 337–339 (2013)

32. Piccinno, F., Ferragina, P.: From TagME to WAT: a new entity annotator. In: Proceedings of the First International Workshop on Entity Recognition & Disambiguation, pp. 55–62. ACM (2014)

33. Roberts, B.: Biographical Research. Understanding Social Research. Open University Press (2002)

34. Small, H.: Co-citation in the scientific literature: a new measure of the relationship between two documents. J. Am. Soc. Inf. Sci. 24(4), 265–269 (1973)

35. Tamper, M., Leskinen, P., Apajalahti, K., Hyvönen, E.: Using biographical texts as linked data for prosopographical research and applications. In: Ioannides, M., et al. (eds.) EuroMed 2018. LNCS, vol. 11196, pp. 125–137. Springer, Cham (2018). https://doi.org/10.1007/978-3-030-01762-0_11

36. Tuominen, J., Hyvönen, E., Leskinen, P.: Bio CRM: a data model for representing biographical data for prosopographical research. In: Biographical Data in a Digital World 2017, Proceedings, vol. 2119. CEUR Workshop Proceedings (2018). https://ceur-ws.org/Vol-2119/paper7.pdf
37. Verboven, K., Carlier, M., Dumolyn, J.: A short manual to the art of prosopography. In: Prosopography Approaches and Applications. A Handbook, pp. 35–70. Unit for Prosopographical Research (Linacre College) (2007)
38. Warren, C.N., Shore, D., Otis, J., Wang, L., Finegold, M., Shalizi, C.: Six degrees of francis bacon: a statistical method for reconstructing large historical social networks. DHQ: Digit. Hum. Q. **10**(3) (2016)

Unsupervised Keyphrase Extraction
from Scientific Publications

Eirini Papagiannopoulou$^{(\boxtimes)}$ and Grigorios Tsoumakas

Aristotle University of Thessaloniki, University Campus, 54124 Thessaloniki, Greece
{epapagia,greg}@csd.auth.gr

Abstract. We propose a novel unsupervised keyphrase extraction approach that filters candidate keywords using outlier detection. It starts by training word embeddings on the target document to capture semantic regularities among the words. It then uses the minimum covariance determinant estimator to model the distribution of non-keyphrase word vectors, under the assumption that these vectors come from the same distribution, indicative of their irrelevance to the semantics expressed by the dimensions of the learned vector representation. Candidate keyphrases only consist of words that are detected as outliers of this dominant distribution. Empirical results show that our approach outperforms state-of-the-art and recent unsupervised keyphrase extraction methods.

Keywords: Unsupervised keyphrase extraction · Outlier detection · MCD estimator

1 Introduction

Keyphrase extraction aims at finding a small number of phrases that express the main topics of a document. Automated keyphrase extraction is an important task for managing digital corpora, as keyphrases are useful for summarizing and indexing documents, in support of downstream tasks, such as search, categorization and clustering [11].

We propose a novel approach for unsupervised keyphrase extraction from scientific publications based on outlier detection. Our approach starts by learning vector representations of the words in a document via GloVe [28] trained solely on this document [26]. The obtained vector representations encode semantic relationships among words and their dimensions correspond typically to topics discussed in the document. The key novel intuition in this work is that we expect non-keyphrase word vectors to come from the same multivariate distribution indicative of their irrelevance to these topics. As the bulk of the words in a document are non-keyphrase we propose using the Minimum Covariance Determinant (MCD) estimator [30] to model their dominant distribution and consider its outliers as candidate keyphrases.

© Springer Nature Switzerland AG 2023
A. Gelbukh (Ed.): CICLing 2019, LNCS 13451, pp. 215–229, 2023.
https://doi.org/10.1007/978-3-031-24337-0_16

Figure 1 shows the distribution of the Euclidean distances among vectors of non-keyphrase words, between vectors of non-keyphrase and keyphrase words, and among vectors of keyphrase words for a subset of 50 scientific publications from the Nguyen collection [25]. We notice that non-keyphrase vectors are closer together (1st boxplot) as well as the keyphrase vectors between each other (3rd boxplot). However, the interesting part of the figure is the 2nd boxplot where the non-keyphrase vectors appear to be more distant from the keyphrase vectors, which is in line with our intuition.

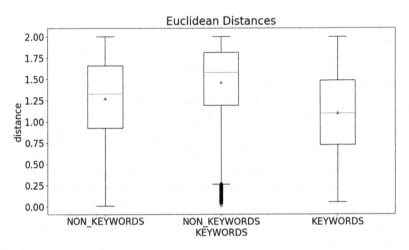

Fig. 1. Distribution of Euclidean distances among non-keywords (1st boxplot), between non-keywords and keywords (2nd boxplot), and among keywords (3rd boxplot).

Figure 2 plots 5d GloVe representations of the words in a computer science article from the Krapivin collection [18] on the first two principal components. The article is entitled *"Parallelizing algorithms for symbolic computation using MAPLE"* and is accompanied by the following two golden keyphrases: logic programming, computer algebra systems. We notice that keyphrase words are on the far left of the horizontal dimension, while the bulk of the words are on the far right. Similar plots, supportive of our key intuition, are obtained from other documents.

The rest of the paper is organized as follows. Section 2 gives a review of the related work in the field of keyphrase extraction as well as a brief overview of multivariate outlier detection methods. Section 3 presents the proposed keyphrase extraction approach. Section 4 describes experimental results highlighting different aspects of our method. We also compare our approach with other state-of-the-art unsupervised keyphrase extraction methods. Finally, Sect. 5 presents the conclusions and future directions of this work.

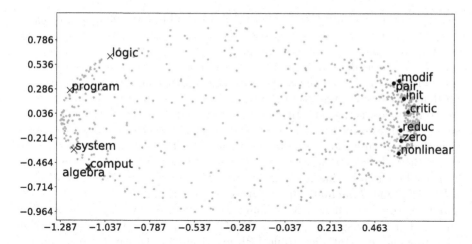

Fig. 2. PCA 2d projection of the 5d GloVe vectors in a document. Keyphrase words are the "x" in black color, while the rest of the words are the gray circle points. Indicatively, we depict a few non-keywords with black circle points.

2   Related Work

In this section, we present the basic unsupervised methodologies (Sect. 2.1). Then, we briefly review basic multivariate outlier detection methods (Sect. 2.2).

2.1   Keyphrase Extraction

Most keyphrase extraction methods have two basic stages: a) the selection of candidate words or phrases, and b) the ranking of these candidates. As far as the first one is concerned, most techniques detect the candidate lexical units or phrases based on grammar rules and syntax patterns [11]. For the second stage, supervised and unsupervised learning algorithms are employed to rank the candidates. Supervised methods can perform better than unsupervised ones, but demand significant annotation effort. For this reason, unsupervised methods have received more attention from the community. In the rest of this sub-section, we briefly review the literature on unsupervised keyphrase extraction methods.

TextRank [23] builds an undirected and unweighted graph of the nouns or adjectives in a document and connects those that co-occur within a window of W words. Then, the PageRank algorithm [4] runs until it converges and sorts the nodes by decreasing order. Finally, the top-ranked nodes form the final keyphrases. Extensions to TextRank are *SingleRank* [35] which adds a weight to every edge equal to the number of co-occurrences of the corresponding words, and *ExpandRank* [35] which adds as nodes to the graph the words of the k-nearest neighboring documents of the target document. Additional variations of TextRank are *PositionRank* [8] that uses a biased PageRank that considers word's positions in the text, and *CiteTextRank* [10] that builds a weighted graph considering information from citation contexts. Moreover, in [36,37] two similar graph-based ranking models are proposed that take into account information from *pre-trained* word embeddings. In addition, local word embeddings/semantics, i.e., embeddings trained from the single document under consideration are used by the Reference Vector Algorithm (RVA) [26]. RVA computes the mean vector of the words in the document's title and abstract and, then, candidate keyphrases are extracted from the title and abstract, ranked in terms of their cosine similarity with the mean vector, assuming that the closer to the mean vector is a word vector, the more representative is the corresponding word for the publication.

Topic-based clustering methods such as *KeyCluster* [21], *Topical PageRank (TPR)* [20], and *TopicRank* [3] aim at extracting keyphrases that cover all the main topics of a document utilizing only nouns and adjectives and forming noun phrases that follow specific patterns. KeyCluster groups candidate words using Wikipedia and text statistics, while TPR utilizes Latent Dirichlet Allocation and runs PageRank for every topic changing the PageRank function so as to take into account the word topic distributions. Finally, TopicRank creates clusters of candidates using hierarchical agglomerative clustering. It then builds a graph of topics with weighted edges that consider phrases' offset positions in the text and runs PageRank. A quite similar approach to TopicRank, called MultipartiteRank, has been recently proposed in [2]. Specifically, the incoming edge weights of the nodes are adjusted promoting candidates that appear at the beginning of the document.

Finally, we should mention the strong baseline approach of *TfIdf* [16] that scores the candidate n-grams of a document with respect to their frequency inside the document, multiplied by the inverse of their frequency in a corpus.

2.2   Multivariate Outlier Detection Methods

Outlier detection methods are categorized into three different groups based on the availability of labels in the dataset [9]: *Supervised methods* assume that the dataset is labeled and train a classifier, such as a support vector machine (SVM) [38] or a neural network [5]. However, having a labeled dataset of outliers is rare in practice and such datasets are extremely imbalanced causing difficulties to machine learning algorithms. *One-class classification* [24] assumes that training data consist only of data coming from one class without any outliers. In this case, a model is trained on these data that infers the properties of normal examples. This model can predict which examples are abnormal based on these properties. State-of-the-art algorithms of this category are One-class SVMs [33] and autoencoders [12]. The one-class SVM model calculates the support of a distribution by finding areas in the input space where most of the cases lie. In particular, the data are nonlinearly projected into a feature space and are then separated from the origin by the largest possible margin [6,34]. The main objective is to find a function that is positive (negative) for regions with high (low) density of points. Finally, *unsupervised methods*, which are the most popular ones, score the data based only on their innate properties. Densities and/or distances are utilized to characterize normal or abnormal cases.

A popular unsupervised method is Elliptical Envelope [13,14,27], which attempts to find an ellipse that contains most of the data. Data outside of the ellipse are considered as outliers. The Elliptical Envelope method uses the Fast Minimum Covariance Determinant (MCD) estimator [31] to calculate the ellipse's size and shape. The MCD estimator is a highly robust estimator of multivariate location and scatter that can capture correlations between features. Particularly, given a data set D, MCD estimates the center, $\bar{x}_J^*$, and the covariance, S_J^*, of a subsample $J \subset D$ of size h that minimizes the determinant of the covariance matrix associated to the subsample:

$$(\bar{x}_J^*, S_J^*) : \det S_J^* \leq \det S_K, \forall K \subset D, |K| = h$$

Another popular unsupervised technique is Isolation Forest (IF) [19], which builds a set of decision trees and calculates the length of the path needed to isolate an instance in the tree. The key idea is that isolated instances (outliers) will have shorter paths than *normal* instances. Finally, the scores of the decision trees are averaged and the method returns which instances are inliers/outliers.

In this work, we are interested in detecting the outliers that do not fit the model well (built by the majority of the non-keyphrase words in a text document) or do not belong to the dominant distribution of those words. We expect that the keyphrases and a minority of words that are related to keyphrase words would be the outliers with respect to the dominant non-keyphrase words' distribution or the corresponding model built based on them.

3   Our Approach

Our approach, called *Outlying Vectors Rank* (OVR), comprises four steps that are detailed in the following subsections.

3.1   Learning Vector Representations

Inspired by the graph-based approaches where the vertices added to the graph are restricted with syntactic filters (e.g., selection only nouns and adjectives in order to focus on relations between words of such part-of-speech tags), we remove from the given document all punctuation marks, stopwords and tokens consisting only of digits. In this way, GloVe does not take common/unimportant words into account that are unlikely to be keywords during the model training. Then, we apply stemming to reduce the inflected word forms into root forms. We use stemming instead of lemmatization as there are stemmers for various languages. However, we should investigate the possibility of using lemmas, in the future.

Subsequently, we train the GloVe algorithm solely on the resulting document. As training takes place on a single document, we recommend learning a small number of dimensions to avoid overfitting. It has been shown in [26] that such local vectors perform better in keyphrase extraction tasks than global vectors from larger collections. The GloVe model learns vector representations of words such that the dot product of two vectors equals the logarithm of the probability of co-occurrence of the corresponding words [28]. At the same time, the statistics of word-word co-occurrence in a text is also the primary source of information for graph-based unsupervised keyphrase extraction methods. In this sense, the employed local training of GloVe on a single document and the graph-based family of methods can be considered as two alternative views of the same information source. Particularly, in previous unsupervised graph-based keyphrase extraction approaches, the limited number of keyphrases (minority) is often assumed to be densely connected to other words in the document and often lies in the center of the word graph. In this vein, the local word vectors capture in a more expressive and alternative way (vector representation) this type of special behavior that the most important words of a document (including the keywords) often present.

3.2 Filtering Non-keyphrase Words

The obtained vector representation encodes semantic regularities among the document's words. Its dimensions are expected to correspond loosely to the main topics discussed in the document. We hypothesize that the vectors of non-keyphrase words can be modeled with a multivariate distribution indicative of their irrelevance to the document's main topics.

We employ the fast algorithm of [31] for the MCD estimator [30] in order to model the dominant distribution of non-keyphrase words. In addition, this step of our approach is used for filtering non-keyphrase words and we are therefore interested in achieving high, if not total, recall of keyphrase words. For the above reasons, we recommend using a quite high (loose) value for the proportion of outliers. Then, we apply a second filtering mechanism to the words whose vectors are outliers of the distribution of non-keyphrase words that was modeled with the MCD estimator. Specifically, we remove any words with length less than 3. We then rank them by increasing position of the first occurrence in the document and consider the top 100 as candidate unigrams, in line with the recent research finding that keyphrases tend to appear closer to the beginning of a document [7].

Notice that OVR does not have to consider further term frequency thresholds or syntactic information, e.g. part-of-speech filters/patterns, for the candidate keyphrases identification. The properties of the resulting local word vectors capture the essential information based on the flow of speech and presentation of the key-concepts in the article.

3.3 Generating Candidate Keyphrases

We adopt the paradigm of other keyphrase extraction approaches that extract phrases up to 3 words [15,22] from the original text, as these are indeed the most frequent lengths of keyphrases that characterize documents. As valid punctuation mark for a candidate phrase we consider the hyphen ("-"). Candidate bigrams and trigrams are constructed by considering the candidate unigrams (i.e. the top 100 outliers mentioned earlier) that appear consecutively in the document.

3.4 Scoring Candidate Keyphrases

As a scoring function for candidate unigrams, bigrams, and trigrams we use the TfIdf score of the corresponding n-gram. However, we prioritize to bigrams and trigrams by doubling their TfIdf score, since such phrases are more descriptive and accompany documents as keyphrases more frequently than unigrams [29].

4 Empirical Study

We first present the setup of our empirical study, including details on the corpora, algorithm implementations, and evaluation frameworks that were used (Sect. 4.1). Then, we study the performance of our approach based on the proportion of the outlier vectors that is considered (Sect. 4.2), and we compare the performance of the MCD estimator with other outlier detection methods (Sect. 4.3). In Sect. 4.4, we compare OVR with other keyphrase extraction methods and we discuss the results. Finally, we give a qualitative (Sect. 4.5) evaluation of the proposed approach.

4.1 Experimental Setup

Our empirical study uses 3 popular collections of scientific publications: a) Krapivin [18], b) Semeval [17] and c) Nguyen [25], containing 2304, 244 and 211 articles respectively, along with author- and/or reader-assigned keyphrases.

We used the implementation of GloVe from Stanford's NLP group[1], initialized with default parameters ($x_{max} = 100$, $\alpha = \frac{3}{4}$, $window\ size = 10$), as set in the experiments of [28]. We produce 5-dimensional vectors with 100 iterations. Vectors of higher dimensionality led to worse results. We used the NLTK suite[2] for preprocessing. Moreover, we used the EllipticEnvelope, OneClassSVM, and IsolationForest classes from the scikit-learn library[3] [27] for the MCD estimator, One-Class SVM (OC-SVM), and Isolation Forest (IF), respectively, with their default parameters. We utilize the PKE toolkit [1] for the implementations of the other unsupervised keyphrase extraction methods as well as our method. The code for the OVR method will be uploaded to our GitHub repository, in case the paper gets accepted.

We adopt two different evaluation approaches. The first one is the strict *exact match* approach, which computes the F_1-score between golden keyphrases and candidate keyphrases, after stemming and removal of punctuation marks,

---

[1] https://github.com/stanfordnlp/GloVe.
[2] https://www.nltk.org/.
[3] https://scikit-learn.org.

such as dashes and hyphens. However, we also adopt the more loose *word match* approach [29], which calculates the F_1-score between the set of words found in all golden keyphrases and the set of words found in all extracted keyphrases after stemming and removal of punctuation marks. We compute $F_1@10$ and $F_1@20$, as the top of the ranking is more important in most applications.

4.2   Evaluation Based on the Proportion of Outlier Vectors

In Tables 1 and 2, we give the $F_1@10$ and $F_1@20$ of the OVR method using different proportion of outlier vectors, from 10% up to 49%, on the three data sets according to the exact match as well as the word match evaluation framework, respectively. Generally, we notice that the higher the outliers' percentage the better is the performance of OVR method. Particularly, in almost all cases (except for the $F_1@10$ of MR based on the word match evaluation), our approach with outlier percentages equal or higher than 30% outperforms the other competitive keyphrase extraction approaches that their performance is presented in Sect. 4.4 (Tables 5 and 6). We set the proportion of outliers for the rest of our experimental study to 0.49 for the Elliptical Envelope method as well as the other outlier detection methods, whose results are given below (Sect. 4.3), as with this proportion we achieve the highest F_1-scores.

Table 1. $F_1@10$ and $F_1@20$ of the OVR method using different proportion of outlier vectors on the three datasets according to exact match evaluation framework.

% Outliers	Semeval		Nguyen		Krapivin	
	$F_1@10$	$F_1@20$	$F_1@10$	$F_1@20$	$F_1@10$	$F_1@20$
10	0.130	0.124	0.164	0.139	0.113	0.086
20	0.172	0.179	0.204	0.192	0.149	0.122
30	0.184	0.194	0.225	0.209	0.164	0.137
40	0.190	**0.200**	0.230	0.212	0.169	0.143
49	**0.194**	**0.200**	**0.237**	**0.214**	**0.174**	**0.145**

The loose value for the proportion of the outliers helps us in order to apply an effective filtering approach on the candidate keywords that form the keyphrases. We consider that the weak majority of the vocabulary (51% of inliers) represent a common vocabulary that is used by the author during writing the article, while the strong minority (49% of outliers) represents the keywords and an accompanying vocabulary that goes hand in hand with the discussion and the description of the keywords (the core concepts of the document). Such information is captured through the co-occurrence statistics, which are utilized by GloVe.

4.3   Evaluation Based on the Type of Outlier Detection Method

We have designed 2 additional different versions of the proposed OVR approach using 2 alternative outlier detection techniques, which are described previously

Table 2. F_1@10 and F_1@20 of the OVR method using different proportion of outlier vectors on the three datasets according to word match evaluation framework.

% Outliers	Semeval		Nguyen		Krapivin	
	F_1@10	F_1@20	F_1@10	F_1@20	F_1@10	F_1@20
10	0.262	0.283	0.315	0.308	0.286	0.256
20	0.330	0.379	0.383	0.393	0.342	0.326
30	0.349	0.408	0.414	0.426	0.369	0.353
40	0.358	0.417	0.425	0.435	0.384	0.346
49	**0.364**	**0.424**	**0.433**	**0.438**	**0.390**	**0.375**

in Sect. 2.2, One-class SVM (OC-SVM) and Isolation Forest (IF). In Tables 3 and 4, we provide the F_1@10 and F_1@20 of the different variants of OVR method according to the exact match as well as the word match evaluation framework. Once more, the results confirm that the MCD estimator successfully captures correlations between the vectors' dimensions.

Table 3. F_1@10 and F_1@20 of the OVR method using various outlier detection techniques on the three data sets according to exact match evaluation framework.

Method	Semeval		Nguyen		Krapivin	
	F_1@10	F_1@20	F_1@10	F_1@20	F_1@10	F_1@20
OC-SVM	0.127	0.127	0.141	0.117	0.109	0.086
IF	0.167	0.171	0.192	0.167	0.153	0.126
MCD	**0.194**	**0.200**	**0.237**	**0.214**	**0.174**	**0.145**

Table 4. F_1@10 and F_1@20 of the OVR method using various outlier detection techniques on the three data sets according to word match evaluation framework.

Method	Semeval		Nguyen		Krapivin	
	F_1@10	F_1@20	F_1@10	F_1@20	F_1@10	F_1@20
OC-SVM	0.279	0.309	0.286	0.275	0.268	0.247
IF	0.337	0.380	0.369	0.366	0.351	0.335
MCD	**0.364**	**0.424**	**0.433**	**0.438**	**0.390**	**0.375**

This happens as the classical methods such as OC-SVM can be affected by outliers so strongly that the resulting model cannot finally detect the outlying observations (masking effect) [32]. Moreover, some normal data points may appear as outlying observations. On the other hand, robust statistics, such as the ones that the MCD estimator uses, aim at finding the outliers searching for the model fitted by the majority of the word vectors. Then, the identification of the outliers is defined with respect to their deviation from that robust fit.

4.4   Comparison with Other Approaches

We compare OVR to the baseline TfIdf method, four state-of-the-art graph-based approaches SingleRank (SR), TopicRank (TR), PositionRank (PR), and MultipartiteRank (MR) with their default parameters, as finally set in the corresponding papers. We also compare our approach to the RVA method which also uses local word embeddings. All methods extract keyphrases from the full-text articles except for RVA which uses the full-text to create the vector representation of the words, but returns keyphrases only from the abstract.

Table 5 shows that OVR outperforms the other methods in all datasets by a large margin, followed by TfIdf (2nd) and MR (3rd), based on the exact match evaluation framework. TR and PR follow in positions 4 and 5, alternately for the two smaller datasets, but without large differences between them in Krapivin. RVA achieves generally lower scores according to the exact match evaluation framework, as it extracts keyprases only from the titles/abstracts, which approximately contain half of the gold keyphrases on average [26]. SR is the worst-performing method in all datasets.

Table 5. F_1@10 and F_1@20 of all competing methods on the three data sets according to exact match evaluation framework.

Method	Semeval		Nguyen		Krapivin	
	F_1@10	F_1@20	F_1@10	F_1@20	F_1@10	F_1@20
SR	0.036	0.053	0.043	0.063	0.026	0.036
TR	0.135	0.143	0.126	0.118	0.099	0.086
PR	0.132	0.127	0.146	0.128	0.102	0.085
MR	0.147	0.161	0.147	0.149	0.116	0.100
TfIdf	0.153	0.175	0.199	0.204	0.126	0.113
RVA	0.094	0.124	0.097	0.114	0.093	0.099
OVR	**0.194**	**0.200**	**0.237**	**0.214**	**0.174**	**0.145**

Moreover, Table 6 confirms the superiority of the proposed method based on the word match evaluation framework. Once more, OVR outperforms the other methods in all datasets by a large margin except for Semeval where MR slightly outperforms OVR.

Based on statistical tests, OVR is significantly better than the rest of the methods in all datasets (besides the MR in Semeval with respect to word match evaluation approach) at the 0.05 significance level. As far as the statistical significance tests concerned, we performed two-sided paired t-test or two-sided Wilcoxon test based on the results of the normality test on the differences of the F_1-scores across the three datasets' articles.

Table 6. F_1@10 and F_1@20 of all competing methods on the three data sets according to word match evaluation framework.

Method	Semeval		Nguyen		Krapivin	
	F_1@10	F_1@20	F_1@10	F_1@20	F_1@10	F_1@20
SR	0.285	0.299	0.322	0.309	0.290	0.256
TR	0.347	0.380	0.376	0.351	0.312	0.277
PR	0.296	0.318	0.371	0.350	0.342	0.302
MR	**0.365**	0.403	0.407	0.383	0.342	0.303
TfIdf	0.308	0.368	0.370	0.394	0.309	0.305
RVA	0.333	0.366	0.374	0.379	0.348	0.337
OVR	0.364	**0.424**	**0.433**	**0.438**	**0.390**	**0.375**

4.5 Qualitative Results

In this section, we use OVR to extract the keyphrases of a publication. This scientific article belongs to the Nguyen data collection. We quote the publication's title and abstract below in order to get a sense of its content:

> **Title: Interestingness of Frequent Itemsets Using Bayesian Networks as Background Knowledge**
> **Abstract:** The paper presents a method for pruning frequent itemsets based on background knowledge represented by a Bayesian network. The interestingness of an itemset is defined as the absolute difference between its support estimated from data and from the Bayesian network. Efficient algorithms are presented for finding interestingness of a collection of frequent itemsets, and for finding all attribute sets with a given minimum interestingness. Practical usefulness of the algorithms and their efficiency have been verified experimentally.
> **Gold Keyphrases:** *association rule, frequent itemset, background knowledge, interestingness, Bayesian network, association rules, emerging pattern*

In Fig. 3, we give the PCA 2d projection of the 5d GloVe vectors of the document as well as the Euclidean distances distribution among non-keywords, between non-keywords and keywords, and among keywords. Moreover, for evaluation purposes, we transform the set of "gold" keyphrases into the following one (after stemming and removal of punctuation marks, such as dashes and hyphens):

> {(associ, rule), (frequent, itemset), (background, knowledg), (interesting), (bayesian, network) (emerg, pattern)}

The OVR's result set is given in the first box below, by decreasing ranking score, followed by its stemmed version in the second box. The words that are both in the golden set and in the set of our candidates are highlighted with bold typeface:

> {attribute sets, **bayesian networks**, **interestingness**, **itemsets**, **background knowledge**, **bayesian**, attribute, **frequent itemsets**, interesting attribute, interesting attribute sets, **interestingness** measure, interesting **patterns**, **association rules**, data mining, probability distributions, given minimum, minimum **interestingness**, given minimum **interestingness**, minimum support, **knowledge** represented}

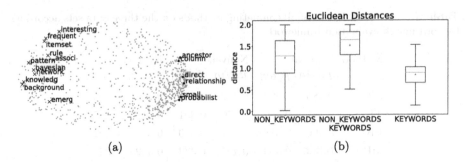

Fig. 3. (a) gives the PCA 2d projection of the 5d GloVe vectors of the document, while (b) shows the Euclidean distances distribution among non-keywords (1st boxplot), between non-keywords and keywords (2nd boxplot), and among keywords (3rd boxplot)

{(attribut, set), (**bayesian, network**), (**interesting**), (**itemset**), (**background, knowledg**), (**bayesian**), (attribut), (**frequent, itemset**), (interest, attribut), (interest, attribut, set), (**interesting**, measur), (interest, **pattern**), (**associ, rule**), (data, mine), (probabl, distribut), (given, minimum), (minimum, **interesting**), (given, minimum, **interesting**), (minimum, support), (**knowledg**, repres)}

According to the exact match evaluation technique, the top-20 returned candidate keyphrases by OVR include 5 True Positives (TPs), 15 False Positives (FPs) and 1 False Negative (FNs). Hence, precision = 0.25, recall = 0.83 and $F_1 = 0.38$. However, according to the word match evaluation technique, the top-20 returned candidate keyphrases by OVR include 10 TPs, 12 FPs and 1 FNs. Hence, precision = 0.45, recall = 0.91, and $F_1 = 0.60$.

5 Conclusions and Future Work

We proposed a novel unsupervised method for keyphrase extraction from scientific publications, called Outlying Vectors Rank (OVR). Our method learns vector representations of the words in a target document by locally training GloVe on this document and then filters non-keyphrase words using the MCD estimator to model their distribution. The final candidate keyphrases consist of those lexical units whose vectors are outliers of the non-keyphrase distribution and appear closer to the beginning of the text. Finally, we use TfIdf to rank the candidate keyphrases.

In the next steps of this work, we aim to delve deeper into the local vector representations obtained by our approach and their relationship with keyphrase and non-keyphrase words. We plan to study issues such as the effect of the vector size and the number of iterations for the convergence of the GloVe model, as well as look into alternative vector representations. In addition, we aim to investigate the effectiveness of the Mahalanobis distance in the scoring/ranking process.

References

1. Boudin, F.: PKE: an open source python-based keyphrase extraction toolkit. In: Proceedings of the 26th International Conference on Computational Linguistics, COLING 2016, Proceedings of the Conference System Demonstrations, Osaka, Japan, pp. 69–73 (2016). https://aclweb.org/anthology/C/C16/C16-2015.pdf
2. Boudin, F.: Unsupervised keyphrase extraction with multipartite graphs. In: Proceedings of the 16th Annual Conference of the North American Chapter of the Association for Computational Linguistics Proceedings of NAACL, NAACL 2018, New Orleans (2018)
3. Bougouin, A., Boudin, F., Daille, B.: TopicRank: graph-based topic ranking for keyphrase extraction. In: Proceedings of the 6th International Joint Conference on Natural Language Processing, IJCNLP 2013, Nagoya, Japan, pp. 543–551 (2013). https://aclweb.org/anthology/I/I13/I13-1062.pdf
4. Brin, S., Page, L.: The anatomy of a large-scale hypertextual web search engine. Comput. Netw. 30(1–7), 107–117 (1998)
5. Das, S.: Elements of artificial neural networks [book reviews]. IEEE Trans. Neural Netw. 9(1), 234–235 (1998)
6. Dreiseitl, S., Osl, M., Scheibböck, C., Binder, M.: Outlier detection with one-class SVMs: an application to melanoma prognosis. In: AMIA Annual Symposium Proceedings. AMIA Symposium 2010, pp. 172–176 (2010). https://www.ncbi.nlm.nih.gov/pmc/articles/PMC3041295/
7. Florescu, C., Caragea, C.: A position-biased pagerank algorithm for keyphrase extraction. In: Proceedings of the Thirty-First AAAI Conference on Artificial Intelligence, San Francisco, California, USA, pp. 4923–4924 (2017). https://aaai.org/ocs/index.php/AAAI/AAAI17/paper/view/14377
8. Florescu, C., Caragea, C.: PositionRank: an unsupervised approach to keyphrase extraction from scholarly documents. In: Proceedings of the 55th Annual Meeting of the Association for Computational Linguistics, ACL 2017, Vancouver, Canada, pp. 1105–1115 (2017). https://doi.org/10.18653/v1/P17-1102
9. Goldstein, M., Uchida, S.: A comparative evaluation of unsupervised anomaly detection algorithms for multivariate data. PLoS ONE 11(4), e0152173 (2016)
10. Gollapalli, S.D., Caragea, C.: Extracting keyphrases from research papers using citation networks. In: Proceedings of the 28th AAAI Conference on Artificial Intelligence, Québec, Canada, pp. 1629–1635 (2014). https://www.aaai.org/ocs/index.php/AAAI/AAAI14/paper/view/8662
11. Hasan, K.S., Ng, V.: Automatic keyphrase extraction: a survey of the state of the art. In: Proceedings of the 52nd Annual Meeting of the Association for Computational Linguistics, ACL 2014, (Volume 1: Long Papers), Baltimore, MD, USA, pp. 1262–1273 (2014). https://aclweb.org/anthology/P/P14/P14-1119.pdf
12. Hawkins, S., He, H., Williams, G.J., Baxter, R.A.: Outlier detection using replicator neural networks. In: Kambayashi, Y., Winiwarter, W., Arikawa, M. (eds.) DaWaK 2002. LNCS, vol. 2454, pp. 170–180. Springer, Heidelberg (2002). https://doi.org/10.1007/3-540-46145-0_17
13. Hubert, M., Debruyne, M.: Minimum covariance determinant. Wiley Interdisc. Rev.: Comput. Stat. 2(1), 36–43 (2010)
14. Hubert, M., Debruyne, M., Rousseeuw, P.J.: Minimum covariance determinant and extensions. Wiley Interdisc. Rev.: Comput. Stat. 10(3), e1421 (2018)
15. Hulth, A.: Improved automatic keyword extraction given more linguistic knowledge. In: Proceedings of the 2003 Conference on Empirical Methods in Natural

Language Processing, EMNLP 2003, Stroudsburg, PA, USA, pp. 216–223 (2003). https://doi.org/10.3115/1119355.1119383

16. Jones, K.S.: A statistical interpretation of term specificity and its application in retrieval. J. Documentation **28**(1), 11–21 (1972)

17. Kim, S.N., Medelyan, O., Kan, M., Baldwin, T.: SemEval-2010 task 5: automatic keyphrase extraction from scientific articles. In: Proceedings of the 5th International Workshop on Semantic Evaluation, SemEval@ACL 2010, Uppsala, Sweden, pp. 21–26 (2010). https://aclweb.org/anthology/S/S10/S10-1004.pdf

18. Krapivin, M., Autayeu, A., Marchese, M.: Large dataset for keyphrases extraction. In: Technical Report DISI-09-055, Trento, Italy (2008)

19. Liu, F.T., Ting, K.M., Zhou, Z.: Isolation forest. In: Proceedings of the 8th IEEE International Conference on Data Mining (ICDM 2008), Pisa, Italy, 15–19 December 2008, pp. 413–422 (2008). https://doi.org/10.1109/ICDM.2008.17

20. Liu, Z., Huang, W., Zheng, Y., Sun, M.: Automatic keyphrase extraction via topic decomposition. In: Proceedings of the 2010 Conference on Empirical Methods in Natural Language Processing, EMNLP 2010, Massachussets, USA, pp. 366–376 (2010). https://www.aclweb.org/anthology/D10-1036

21. Liu, Z., Li, P., Zheng, Y., Sun, M.: Clustering to find exemplar terms for keyphrase extraction. In: Proceedings of the 2009 Conference on Empirical Methods in Natural Language Processing, EMNLP 2009, Singapore, pp. 257–266 (2009). https://www.aclweb.org/anthology/D09-1027

22. Medelyan, O., Frank, E., Witten, I.H.: Human-competitive tagging using automatic keyphrase extraction. In: Proceedings of the 2009 Conference on Empirical Methods in Natural Language Processing, EMNLP 2009, Singapore, pp. 1318–1327 (2009). https://www.aclweb.org/anthology/D09-1137

23. Mihalcea, R., Tarau, P.: TextRank: bringing order into text. In: Proceedings of the 2004 Conference on Empirical Methods in Natural Language Processing, EMNLP 2004, Barcelona, Spain, pp. 404–411 (2004). https://www.aclweb.org/anthology/W04-3252

24. Moya, M.M., Hush, D.R.: Network constraints and multi-objective optimization for one-class classification. Neural Netw. **9**(3), 463–474 (1996)

25. Nguyen, T.D., Kan, M.-Y.: Keyphrase extraction in scientific publications. In: Goh, D.H.-L., Cao, T.H., Sølvberg, I.T., Rasmussen, E. (eds.) ICADL 2007. LNCS, vol. 4822, pp. 317–326. Springer, Heidelberg (2007). https://doi.org/10.1007/978-3-540-77094-7_41

26. Papagiannopoulou, E., Tsoumakas, G.: Local word vectors guiding keyphrase extraction. Inf. Process. Manag. **54**(6), 888–902 (2018). https://doi.org/10.1016/j.ipm.2018.06.004

27. Pedregosa, F., et al.: Scikit-learn: machine learning in Python. J. Mach. Learn. Res. **12**, 2825–2830 (2011). https://dl.acm.org/citation.cfm?id=2078195

28. Pennington, J., Socher, R., Manning, C.D.: Glove: global vectors for word representation. In: Proceedings of the 2014 Conference on Empirical Methods in Natural Language Processing, EMNLP 2014, Doha, Qatar, pp. 1532–1543 (2014). https://aclweb.org/anthology/D/D14/D14-1162.pdf

29. Rousseau, F., Vazirgiannis, M.: Main core retention on graph-of-words for single-document keyword extraction. In: Hanbury, A., Kazai, G., Rauber, A., Fuhr, N. (eds.) ECIR 2015. LNCS, vol. 9022, pp. 382–393. Springer, Cham (2015). https://doi.org/10.1007/978-3-319-16354-3_42

30. Rousseeuw, P.J.: Least median of squares regression. J. Am. Stat. Assoc. **79**(388), 871–880 (1984). https://doi.org/10.1080/01621459.1984.10477105

31. Rousseeuw, P.J., van Driessen, K.: A fast algorithm for the minimum covariance determinant estimator. Technometrics **41**(3), 212–223 (1999)
32. Rousseeuw, P.J., Hubert, M.: Robust statistics for outlier detection. Wiley Interdisc. Rev.: Data Min. Knowl. Discov. **1**(1), 73–79 (2011). https://doi.org/10.1002/widm.2
33. Schölkopf, B., Platt, J.C., Shawe-Taylor, J., Smola, A.J., Williamson, R.C.: Estimating the support of a high-dimensional distribution. Neural Comput. **13**(7), 1443–1471 (2001)
34. Schölkopf, B., Williamson, R.C., Smola, A.J., Shawe-Taylor, J., Platt, J.C.: Support vector method for novelty detection. In: Advances in Neural Information Processing Systems 12, NIPS Conference, Denver, Colorado, USA, 29 November–4 December 1999, pp. 582–588 (1999). https://papers.nips.cc/paper/1723-support-vector-method-for-novelty-detection
35. Wan, X., Xiao, J.: Single document keyphrase extraction using neighborhood knowledge. In: Proceedings of the 23rd AAAI Conference on Artificial Intelligence, AAAI 2008, Chicago, Illinois, USA, pp. 855–860 (2008). https://www.aaai.org/Library/AAAI/2008/aaai08-136.php
36. Wang, R., Liu, W., McDonald, C.: Corpus-independent generic keyphrase extraction using word embedding vectors. In: Software Engineering Research Conference (2014)
37. Wang, R., Liu, W., McDonald, C.: Using word embeddings to enhance keyword identification for scientific publications. In: Sharaf, M.A., Cheema, M.A., Qi, J. (eds.) ADC 2015. LNCS, vol. 9093, pp. 257–268. Springer, Cham (2015). https://doi.org/10.1007/978-3-319-19548-3_21
38. Wille, L.T.: Review of "Learning Kernel Classifiers: Theory and Algorithms by Ralf Herbrich". MIT Press, Cambridge (2002). 13–17, ISBN 026208306x, p. 384; and review of "learning with kernels: support vector machines, regularization optimization and beyond by Bernhard Scholkopf and Alexander J. Smola". IT Press, Cambridge (2002). ISBN 0262194759, p. 644. SIGACT News **35**(3) (2004). https://doi.org/10.1145/1027914.1027921

Information Retrieval

Retrieving the Evidence of a Free Text Annotation in a Scientific Article: A Data Free Approach

Julien Gobeill[1,2]([✉]), Emilie Pasche[1], and Patrick Ruch[1,2]

[1] SIB Text Mining Group, Swiss Institute of Bioinformatics, Geneva, Switzerland
`julien.gobeill@hesge.ch`
[2] HES-SO/HEG Geneva, Information Sciences, Geneva, Switzerland

Abstract. The exponential growth of research publications provides challenges for curators and researchers in finding and assimilating scientific facts described in the literature. Therefore, services that sup-port the browsing of articles and the identification of key concepts with minimal effort would be beneficial for the scientific community. Reference databases store such high value scientific facts and key concepts, in the form of annotations. Annotations are statements assigned by curators from an evidence in a publication. Yet, if annotated statements are linked with the publication's references (e.g. PubMed identifiers), the evidences are rarely stored during the curation process. In this paper, we investigate the automatic relocalization of biological evidences, the Gene References Into Function (GeneRIFs), in scientific articles. GeneRIFs are free text statements extracted from an article, and potentially reformulated by a curator. De facto, only 33% of geneRIFs are copy-paste that can be retrieved by the reader with the search tool of his reader. For automatically retrieving the other evidences, we use an approximate string matching algorithm, based on a finite state automaton and a derivative Levenshtein distance. For evaluation, two hundred candidate sentences were evaluated by human experts. We present and compare results for the relocalization in both abstracts and fulltexts. With the optimal setting, 76% of the evidences are retrieved with a precision of 97%. This data free approach does not require any training data nor a priori lexical knowledge. Yet it remarkable how it handles with complex language modifications such as reformulations, acronyms expansion, or anaphora. In the whole MEDLINE, 350,000 geneRIFs were retrieved in abstracts, and 15,000 in fulltexts; they are currently available for highlighting in the Europe PMC literature browser.

Keywords: Natural language processing · Information retrieval · Approximate string matching

1 Introduction

Structured databases have become important resources for integrating and accessing scientific facts [1]. Yet, the normalized and integrated content still lags behind the current knowledge contained in the literature [2, 3]. Entities' properties of, such as gene

A. Gelbukh (Ed.): CICLing 2019, LNCS 13451, pp. 233–246, 2023.
https://doi.org/10.1007/978-3-031-24337-0_17

functions, are usually characterized in experiments conducted by re-search teams, then reported in natural language published in scientific articles. These properties need to be located, extracted and then integrated in normalized annotations by curators of reference gene databases, in order to be exploited by other researchers or databases. In biology, reference databases are the Gene Database maintained by the National Center for Biotechnology Information (NCBI) [4], or the UniProt database maintained by the Swiss Institute of Bioinformatics [5]. Manual curation of these scientific articles is labor intensive, but produces consistent and high-quality annotations for populating reference biological databases. In 2007, it was estimated that, despite the fact that the mouse genome is now fully sequenced, its functional annotation will not be complete in databases before 2047 [6].

Statements about gene functions, known as Gene References into Function (GeneRIFs), are collected by the NCBI. They are short statements extracted by life science experts from scientific articles. GeneRIFs are intended to facilitate access to publications documenting experiments on gene functions. Yet, geneRIFs are simply linked to an article, but not localized in the fulltext. This loss of information during the curation process is harmful, as curators want to learn about new functions as quickly as they can, preferably without having to scan all the paper for retrieving the evidence. GeneRIFs relocalization can thus be performed by automated approaches in order to help curators to keep up with the growing flow of publications.

Nowadays, machine learning and expert systems are popular for automatic tasks in biomedical articles [7, 8]. Yet, both approaches need a critical amount of a priori knowledge, respectively learning data and hand-written language rules, before being able to manage edit modifications or reformulations. Such a priori knowledge can be costly to gather, when it is sometimes simply not available. In contrast, approximate string matching provides a data-free approach for estimating the similarity between an evidence and a passage.

In this paper, we investigate the abilities of an approximate string matching algorithm for retrieving evidences (geneRIFs) in scientific publications. This paper is organized as follows. Section 2 reports related works on approximate string matching in biomedical literature, and on geneRIFs retrieval. Section 3 describes the data, the methods, and the benchmarks used for evaluation. Lastly, Sect. 4 reports evidence retrieval results on abstracts, then on fulltexts.

2 Related Work

In biology, approximate string matching is a popular approach for gene name recognition [9]. For this specific task, survey studies compare and optimize different algorithms in terms of performance and computation time [10]. Approximate string matching was also investigated for more general named entity recognition [11], alignment of DNA sequences [12], optical character recognition [13], or approximate text search in literature [14].

Dealing with geneRIFs, the Genomics Track in the Text Retrieval Conferences (TREC) addressed in 2003 the issue of extracting geneRIFs statements from a scientific publication [15]. Participating teams were asked to maximize the lexical overlap between the geneRIF and a passage, measured by the Dice coefficient. The track overview reported two best approaches. The first team [7] investigated sentences normalization thanks to stemming, gene names dictionaries and thesaurus, then trained a Bayesian classifier. The second team [16] also trained a Bayesian classifier, but input sentences were only abstracts title and last sentences, and features were only normalized gene names and verbs. No groups obtained much improvement compared to the baseline, which consisted in choosing titles. Yet, the lexical overlap was pointed out as a weak measure of equivalence, as geneRIFs can be para-phrases of articles sentences.

In [17], generated evidences were compared to authentic geneRIFs. For text representation, authors produced several features based on sentence position, sentence discourse, gene normalization or ontology terms mapping. Furthermore, 3,000 sentences were annotated by experts for building training and test sets. Despite these numerous efforts, authors concluded that their machine learning approach did not produce results comparable to sentence position.

3   Methods

In this section, we first focus on data: the geneRIFs dataset, and the corresponding articles (abstracts accessed via MEDLINE, and fulltexts accessed via PubMed Central). We then deal with the methods: we introduce the Levenshtein distance, and the derivative distance used on this work. Finally, we detail the evaluation: benchmarks, judgements and metrics.

3.1   GeneRIFs

GeneRIFs are freely available in the NCBI server (ftp://ftp.ncbi.nih.gov/gene/GeneRIF/). The whole dataset was acquired on August 2018, and contained 1.2 million of geneRIFs.

According the NCBI website, geneRIFs allow any scientist to add the functional annotation of a gene contained in the Gene database. The geneRIFs must be linked to an existing gene entry. Three information are mandatory for completing a submission:

1. A concise phrase describing a function (less than 425 characters in length).
2. A published paper describing that function, implemented by supplying the PubMed identifier (PMID) of a citation in MEDLINE.
3. The curator's e-mail address.

The concise phrase is free text: the geneRIF curator is free to copy-paste, edit or reformulate the evidence described in the paper. Moreover, a geneRIF is linked to a publication, not to the specific passage that describes the function. Yet, it is stated that the title is not accepted.

Once submitted, the text of the GeneRIF is reviewed for inappropriate content and typographical errors, but not otherwise edited. Finally, it is stated that most of the GeneRIFs are provided by the staff of the National Library of Medicine's Index Section, who have advanced degrees in the life sciences.

Here is one example of geneRIF in json format: {"geneID": 2827861, "PMID": 15664975, "text: "Strains with mutations in either the prcA-prtP or the msp region showed altered expression of the other locus. (msp = major outer sheath protein)"}. In this example, the geneRIF annotator has copy pasted a sentence from the abstract, and has developed the acronym "msp" (which is not developed in the abstract).

3.2 Publications

All geneRIFs are provided with a PubMed identifier (PMID), which is linked with a specific publication contained in the MEDLINE database. Thus, for all papers referred in geneRIFs, titles and abstracts are freely accessible via MEDLINE. The open access fulltexts can be freely accessed via the PubMed Central database.

Abstracts from MEDLINE. MEDLINE is a bibliographic database of life sciences. In January 2018, it contained 27 million records of scientific papers from more than 5,500 selected journals in the domains of biology and health. MEDLINE is maintained by the United States National Library of Medicine (NLM), and is accessible via the NLM search engine (PubMed), or freely downloadable via services and FTP.

A MEDLINE record contains the title and the abstract, and also metadata such as authors' information, journal's information, the publication date, or keywords (Medical Subject Headings) added by the NLM's indexers. All MEDLINE records are uniquely identified by a PMID.

Fulltexts from PubMed Central. PubMed Central is a free fulltext database of publicly accessible literature published in life sciences. In January 2018, it contained 4.6 million records. Yet, only a subset of 1.4 million of publications is Open Access and linked with a PMID. 444,000 geneRIFs are linked with a paper contained in PubMed Central. Out of these 444,000, only 153,000 (35%) are linked with a paper contained in the Open Access subset.

The BioMed Platform. For accessing papers, we used BioMed, a local resource for literature access and enrichment. BioMed provides access to Open Access abstracts and fulltexts, thanks to a synchronized mirror of MEDLINE and PubMed Central. It also provides some text services such as information extraction [18], question answering [19, 20], or sentence splitting. BioMed is currently used in the workflow of protein curators at the Swiss Institute of Bioinformatics [21].

3.3 Algorithm

The Levenshtein Distance. The Levenshtein distance [22] quantifies the similarity between two strings; for this purpose, it computes the number of single-characters operations required to transform one string into the other. The available operations are:

- d: deletion of a character
- s: substitution of a character with another
- i: insertion of a character.

For instance, three operations are requested for transforming the string "kitten" into the string "sitting": (1) substitution of "k" with "s", (2) substitution of "e" with "i", (3) insertion of "g". The number of operations is seen as the distance: the less operations are requested, the most similar strings are. An exact match results in a distance of zero. Finally, each operation can be weighted in order to tune the algorithm for a specific task.

The Evidence Retrieval Distance. Our approach for retrieving a geneRIF in a given paper is sentence-centric. First, the abstract (or fulltext when it is available) is split into sentences, thanks to local services provided by the BioMed platform. Then, all sentences are compared with the geneRIF using a derivative Levenshtein distance. There are different available implementations of the Levenshtein algorithm, including recursive, or iterative with matrix [23]. For this study, we have used a library from the Comprehensive Perl Archive Network (Text::Fuzzy).

Preliminary analysis revealed some limitations in the default Levenshtein distance for evidence retrieval. First, the deletion of characters is actually not an issue: a lot of geneRIFs curators choose to cut several words from a sentence when producing the statement (such as "We show that" or "These results indicate that"). We have thus decided to ignore deletion operations. Second, the Levenshtein distance has no normalization according to the string length: a number of ten operations is obviously better for a thirty words evidence than for a five words one. We have thus introduced a normalization with the total number of characters in the geneRIF. We finally obtained a derivative Levenshtein distance, called Evidence Retrieval (ER) distance, as given below:

$$ER\ Distance(str_1, str_2) = \frac{S + I}{length(str_2)}$$

In this formula, str1 is a candidate sentence, str2 is the evidence to retrieve, S and I are the total number of substitutions and insertions required to transform str1 into str2 (deletions are not counted), and length(str2) is the number of characters in str2. The ER distance between a geneRIF and a candidate sentence can be seen as the percentage of characters in the geneRIF that required to be introduced or substituted in the sentence. An ER distance of zero means that the geneRIF is contained in the sentence, but that the sentence may contain some supplementary characters that were not chosen by the curator.

3.4 Evaluation

Test Sets. For evaluation purposes, we designed two test datasets of 100 geneRIFs. The first test set was sampled in the set of the 444,000 geneRIFs linked to publications available in MEDLINE. The task evaluated with this first test set was evidence retrieval in abstracts. The second test set was sampled in the set of 153,000 geneRIFs linked to publications contained in the PMC Open-Access subset. The task evaluated with this second test set was evidence retrieval in fulltexts.

For all geneRIFs in the benchmarks, ER distances were computed between the geneRIF and all publication's sentences. Then, only the sentence with the smallest ER distance was selected: we call it the candidate sentence. Results were analyzed according intervals and thresholds of ER distance values.

Equivalence Judgements. Two experts evaluated the equivalence between the geneR-IFs and the candidate sentences. Experts were bioinformaticians, one having an advanced degree in information sciences, the other one in biology. They were asked to judge the equivalence between a geneRIF and a candidate sentence, according to three possible values:

- 100%: the candidate sentence contains all the annotatable in-formation contained in the geneRIF. The expert thinks that a curator could make an annotation only with this sentence.
- 50%: the candidate sentence contains some annotatable in-formation contained in the geneRIF. The expert thinks that this sentence could help a curator, but also that other useful information are contained in other sentences.
- 0%: the candidate sentence contains no annotatable information contained in the geneRIF. The expert thinks that this sentence would be of no help for a curator.

The pairs (geneRIF, candidate sentence) were presented to the experts in a random order. ER distances were hidden from experts. For each candidate sentence, the final equivalence is the average judgement of both experts. For instance, if one expert gave a 100% equivalence and the other one 50%, the final retained equivalence for this candidate sentence is 75%. Moreover, for each candidate sentence, an Inter-Annotator (IA) Agreement was computed. The IA Agreement is 100% minus the absolute difference between both judgements. For instance, if one expert gave a 100% equivalence and the other one 50%, the final IA Agreement for this candidate sentence is 50%.

4 Results and Discussion

First of all, we present some preliminary results and statistics on geneRIFs and papers' sentences. Next, we focus on evidence retrieval in abstracts, then on evidence retrieval in fulltexts. For each benchmark, we present some interesting examples of candidate sentences and edit modifications handled by our approach.

4.1 Preliminary Results

GeneRIFs' Length. We used regular expressions in order to compute and compare the lengths of geneRIFs, and the lengths of sentences in an abstract or a fulltext. Lengths were computed with samples of 20,000 geneRIFs, 4,000 abstracts and 400 fulltexts. Mean and quartiles are showed in Table 1.

Table 1. Length of geneRIFs, and sentences in abstracts and fulltexts.

	Mean	1st q	Median	3rd q
GeneRIFs	21	14	20	27
Abstracts sentences	24	16	22	29
Fulltexts sentences	20	9	18	33

In terms of length, GeneRIFs are quite similar to sentences from abstracts. For fulltexts, mean and median are similar to geneRIFs, but the distribution is sparser: there seems to be more short and long sentences in fulltext. Finally, we observe a mean of ten sentences per abstract, versus five hundred per fulltext. This strengthens our sentence-centric approach. The sparser distribution in fulltexts could be explained by the fact that abstracts sentences aim at summarizing the fulltext content. Fulltext sentences are thus more likely to be short or long (such as long explanations). In this perspective, geneRIFs seem to be more factual evidences than detailed explanations.

GeneRIFs' Words Presence in Abstracts. We then studied what proportions of geneRIFs words can be found in abstracts. In Fig. 1, 440 sampled geneRIFs are plotted according to their proportion of words present in the corresponding abstract.

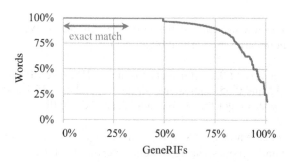

Fig. 1. GeneRIFs words present in abstracts.

50% of geneRIFs have all their words contained in the abstract, including 33% that are found in exact match. The next 25% have between 90% and 100% of their words in the abstract, while the proportion quickly drops for the last 25%. This leads to several assumptions. First, a huge proportion of geneRIFs seems to be extracted from the abstracts. Second, simple exact match already retrieves one third of the evidences, which seems to be a fair baseline. The second third has above 95% of words present in the abstract; these evidences could be retrieved by approximate string matching. Finally, the last third could be less reachable by our approach (at least for abstracts), as fewer words are present.

4.2 Evidence Retrieval in Abstracts

We now present the results of evidence retrieval in abstracts. For one hundred geneRIFs, two experts were asked to judge the equivalence of the best candidate sentence extracted from the corresponding abstract.

Equivalence According to Distance Intervals. Equivalences and IA Agreements are given in Fig. 2, according to ER distance intervals.

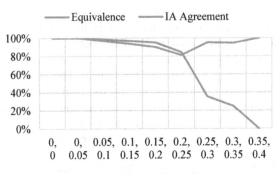

Fig. 2. Equivalence according to distance intervals. (Color figure online)

It is remarkable how the equivalence curve quickly drops. We can split the curve into three parts (boundaries are vertical blue lines):

- for values between 0 and 0.2, equivalence is above 95% (even 100% for [0, 0.05])
- then, between 0.2 and 0.35, equivalence drops from 84% to 35% and 25%
- finally, for distances bigger than 0.35, equivalence falls to 0%.

It is also remarkable how the IA Agreement curve behaves according to these three parts. For the first and last ones, the agreement is very high (above 90%, and even 100% for 100% or 0% equivalence), while the central part seems to be more uncertain for judges.

Thus, the ER distance shows high abilities to produce two distinct sets of true positives (100% equivalence) and true negatives (0% equivalence), with 100% agreement in both cases. Between both is a set of more uncertainty for middle ER distances, with smaller equivalence values and less agreement between judges.

Equivalence According to Distance Thresholds. We know consider equivalence according to proportion of retrieved geneRIFs. We did not focus anymore on ER distance intervals but on ER distance thresholds, in order to know what proportion of geneRIFs are retrieved in abstracts with a given threshold. Results are given in Fig. 3.

Considering the previous first part, 68% of geneRIFs are retrieved with an equivalence mean of 99% with an ER distance threshold of 0.2. With a threshold of 0.25, 76% of geneRIFs are retrieved with an equivalence mean of 97%. Considering the drop in Fig. 2, these thresholds are to be considered for delivering an optimal output.

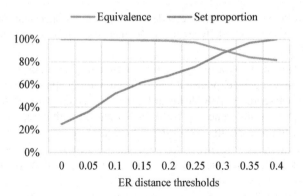

Fig. 3. Equivalence and set proportion for abstracts.

Interesting Examples. We now present some remarkable retrieved evidences.

Words Addition or Deletion. Beyond copy-pastes, the geneRIFs curators are likely to delete some unnecessary words, or to add words in order to bring some precision that is not explicit in the sentence.

GeneRIF: "PfSir2a fine-tunes ribosomal RNA gene transcription." Abstract sentence (ER distance 0): "Here we investigate the nucleolar function of PfSir2a and demonstrate that PfSir2a fine-tunes ribosomal RNA gene transcription." In this case, the geneRIF curator chose to cut the evidence introduction.

GeneRIF: "Transgenic flies, expressing the human ERRa-G allele, constitutively over-express Cyp12d1, Cyp6g2 and Cyp9c1." Abstract sentence (ER distance 0.05): "Transgenic flies, expressing the ERRa-G allele, constitutively over-expressed Cyp12d1, Cyp6g2 and Cyp9c1." In this case, the geneRIF curator simply adds "human" in order to add context and to specify the related organism.

Acronyms Expansion. Authors often use acronyms, which are developed only the first time they mention them in the article (e.g. "EC" for "Endothelial cells"). Thus, geneRIFs curators are likely to develop the acronyms used in the extracted evidence. Moreover, Greek letters are often used in gene names, while geneRIF curator can prefer full letter names (e.g. "alpha" instead of "α"). These kinds of modifications can be handled by pattern substitutions and numerous rules, especially for acronym expansions [24]. In contrast, approximate string matching can handle with these modifications without linguistic knowledge.

GeneRIF: "interaction of alpha6beta1 in embryonic stem cells (ECCs) with laminin-1 activates alpha6beta1/CD151 signaling which programs ESCs toward the endothelial cells lineage fate" Abstract sentence (ER distance 0.28): "Thus, interaction of α6β1 in ESCs with LN1 activates α6β1/CD151 signaling which programs ESCs toward the EC lineage fate." In this case, the geneRIF curator developed two Greek letters, and three acronyms. Even if the ER distance is quite high, it is remarkable how approximate string matching dealt with so many substitutions and selected the good sentence.

Anaphora Resolution. As evidences can deal with just mentioned patterns in the argumentative flow, authors are likely to use pronouns (e.g. "it") or anaphora instead of writing repetitions.

GeneRIF: "Under hypoxia reoxygenation or ischemia and reperfusion, StAR and CYP11A1 protein and gene expression was reduced without apparent relation to TSPO changes". Abstract sentence (ER distance 0.20): "Under the same conditions, StAR and CYP11A1 protein and gene expression was reduced without apparent relation to TSPO changes". In this case, the author did not repeat the experiment conditions in his sentence; highlighting the evidence in the paper allows a curator to quickly check in the neighborhood of the sentence the sentence if the reported conditions are the same.

4.3   Evidence Retrieval in Fulltexts

We now present the results of evidence retrieval in fulltexts.

Equivalence According to Distance Intervals. Equivalences and IA Agreements are given in Fig. 4, according to ER distance intervals.

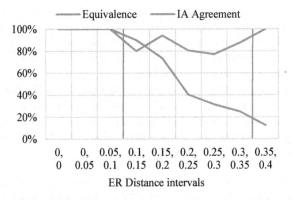

Fig. 4. Equivalence according to distance intervals.

Equivalence curves share the same shape for fulltexts than previously for abstracts:

– for values between 0 and 0.1, equivalence is 100%
– then, between 0.1 and 0.35, equivalence slowly drops from 90% to 25%
– finally, for distances bigger than 0.35, equivalence falls to 13%.

As for abstracts, the IA Agreement is 100% when equivalence is maximum or minimum. Yet, for the central part, we observe more uncertainty for judges in order to determine if the sentence are equivalent or not.

Equivalence According to Distance Thresholds. Results linking ER distance thresholds and proportions of retrieved geneRIFs in fulltexts are given in Fig. 5.

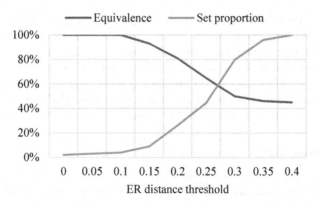

Fig. 5. Equivalence and set proportion for abstracts.

Considering the previous first part, only 4% of geneRIFs are retrieved in fulltext with an equivalence of 100% with an ER distance threshold of 0.1. Going up to a threshold of 0.15, 9% of geneRIFs are retrieved with an equivalence mean of 90%. Considering the drop observed in Fig. 4, these thresholds seem to be considered for delivering an optimal output. Bigger values for thresholds could be risky in a non-reviewed production.

It is difficult to evaluate the complementarity of evidence retrieval in fulltexts and abstracts. Further analyses reveal that for the five true positives in fulltexts with an ER distance inferior to 0.1, four were also true positives in abstracts (with an average ER distance of 0.03), while only one was a true negative in the abstract (with an ER distance of 0.32). GeneRIFs curators tend to preferably choose their evidence in abstracts. For geneRIFs that were missed in abstracts, a small portion could be retrieved in fulltext.

Another aspect to consider is that the abstract is a summary of fulltext. Thus, when the geneRIFs are created from an abstract sentence, there is probably a passage in the fulltext where the evidence comes from, probably differently formulated, and more detailed. This could explain why ER distances are larger in fulltexts than in abstracts. In other words, the approximate string matching would retrieve fulltext sentences that were summarized in the abstract and used as geneRIFs.

Yet, for approximately one quarter of geneRIFs, our approach is not able to retrieve the evidence neither in the abstract nor in the fulltext. In the below section, we present some of these missed geneRIFs.

Interesting Examples. We now present some remarkable retrieved evidences.

Reformulated Evidences. GeneRIF: "Three FNR proteins (ANR, PP_3233, and PP_3287) respond to O2 differently." Fulltext sentence (ER distance 0.15): "Thus, the response of ANR was similar to that reported previously for E. coli FNR, further confirming the similarities between these two proteins, but PP_3233 and PP_3287 were less responsive with both O2 and nitric oxide compared with ANR (25)." In this case, the sentence contains the same information than the geneRIF, and was selected by our algorithm even if the interaction was interpreted and reformulated into "respond to O2 differently".

GeneRIF: "These data support the role of proconvertase PCSK5 in the process-ing of ovarian inhibin subunits during folliculogenesis." Fulltext sentence (ER distance 0.18): "We demonstrate that the spatial and temporal expression of the proconvertase PCSK5 overlaps with the expression and processing of mature inhibin subunits in the ovary during follicle expansion." In this case, a good sentence was retrieved, while "we demonstrate that the [...] expression of the proconvertase PCSK5 overlaps..." was reformulated into "These data support the role of proconvertase PCDK5 in...".

4.4 Missed GeneRIFs

Here are some examples of missed geneRIFs.

GeneRIFs: "Observational study and genome-wide association study of gene-disease association." "Clinical trial of gene-disease association and gene-environment interac-tion" "Observational study of gene-disease association, gene-environment interaction, and pharmacogenomic/toxicogenomic."

These three examples taken in the abstracts benchmark are more descriptions of the studies than facts REWRITE. Thus, it is not surprising that no good sentence was retrieved in the articles.

In other geneRIFs that were not successfully retrieved, we observed several long geneRIFs with multiple sentences, probably gathering facts from different parts of the article. At last, some geneRIFs are statements inferred by the curator, thus are beyond the scope of statistical or linguistic systems.

5 Conclusion

We investigated approximate string matching for the task of retrieving evidences expressed in geneRIFs in their corresponding papers. We defined an Evidence Retrieval distance derived from the Levenshtein distance. When selecting the abstract sentence having the smallest distance value, this approach can retrieve 76% of evidences with a very high precision (97%). While 33% of geneRIFs are just copy-pastes, it is remarkable how this approach can handle with complex editing modifications, including reformula-tions, acronym expansions, or anaphora resolutions. Beyond evidence retrieving, approx-imate string matching could be used in fulltext for retrieving sentences that contain the information summarized in an abstract.

It is worth reminding that, in contrast with machine learning or rule-based approaches, which need a substantial amount of training data or a priori knowledge, such classic approaches are on the shelf, and are still very effective for concrete text mining applications.

In 2017, within the Elixir Excelerate project, which aims at ensuring the delivery of world-leading life-science data services, the whole geneRIFs dataset was treated. 337,000 were retrieved in abstracts, and 13,770 in fulltexts. These retrieved geneRIFs were then used in order to fill the Europe PMC database [1, 25], maintained by the Euro-pean Bioinformatics Institute (EBI). These geneRIFs are now available for highlighting when a user reads an article. We hope that this will help the curators to navigate in the literature in their daily workflow, thus facilitating to bridge the gap between information contained in literature and in databases.

Acknowledgments. This research was supported by the Elixir Excelerate project, funded by the European Commission within the Research Infrastructures programme of Horizon 2020, grant agreement number 676559. The authors thank their colleagues from the SIB Swiss Institute of Bioinformatics (Core-IT), in particular Daniel Texeira and Heinz Stockinger, who provided insight and expertise that greatly assisted the research. The authors also thank the European Bioinformatics Institute, in particular Johanna McEntyre and Aravind Venkatesan, for the integration of retrieved geneRIFs into EuropePMC.

References

1. Venkatesan, A., et al.: SciLite: a platform for displaying text-mined annotations as a means to link research articles with biological data. Wellcome Open Res. **1**, 25 (2016). https://doi.org/10.12688/wellcomeopenres.10210.1

2. Howe, D., et al.: Big data: the future of biocuration. Nature **455**(7209), 47–50 (2008). https://doi.org/10.1038/455047a

3. Gobeill, J., Pasche, E., Vishnyakova, D., Ruch, P.: Managing the data deluge: data-driven GO category assignment improves while complexity of functional annotation increases. Database (Oxford) (2013). https://doi.org/10.1093/database/bat041

4. Brown, G.R., et al.: Gene: a gene-centered information resource at NCBI. Nucl. Acids Res. **43**(D1), D36–D42 (2015). https://doi.org/10.1093/nar/gku1055

5. Bultet, L.A., Aguilar-Rodriguez, J., Ahrens, C.H., Ahrne, E.L., Ai, N., et al.: The SIB Swiss Institute of Bioinformatics' resources: focus on curated databases. Nucl. Acids Res. **44**, D27–D37 (2016). https://doi.org/10.1093/nar/gkv1310

6. Baumgartner, W.A., Cohen, K.B., Fox, L.M., Acquaah-Mensah, G., Hunter, L.: Manual curation is not sufficient for annotation of genomic databases. Bioinformatics **23**(13), i41–i48 (2007). https://doi.org/10.1093/bioinformatics/btm229

7. Jelier, R., et al.: Searching for geneRIFs: concept-based query expansion and Bayes classification. In: TREC Proceedings, pp. 225–233 (2003)

8. Obermeyer, Z., Emanuel, E.J.: Predicting the future - big data, machine learning, and clinical medicine. New Engl. J. Med. **375**(13), 1216 (2016). https://doi.org/10.1056/NEJMp1606181

9. Tsuruoka, Y., Tsujii, J.I.: Improving the performance of dictionary-based approaches in protein name recognition. J. Biomed. Inform. **37**(6), 461–470 (2004)

10. Papamichail, D., Papamichail, G.: Improved algorithms for approximate string matching. BMC Bioinform. **10**(1), S10 (2009)

11. Wang, W., Xiao, C., Lin, X., Zhang, C.: Efficient approximate entity extraction with edit distance constraints. In: Proceedings of the 2009 ACM SIGMOD International Conference on Management of Data, pp. 759–770 (2009)

12. Buschmann, T., Bystrykh, L.V.: Levenshtein error-correcting barcodes for multiplexed DNA sequencing. BMC Bioinform. **14**(1), 272 (2013)

13. Lasko, T.A., Hauser, S.E.: Approximate string matching algorithms for limited-vocabulary OCR output correction. In: Photonics West 2001-Electronic Imaging, pp. 232–240 (2000)

14. Wang, J., et al.: Interactive and fuzzy search: a dynamic way to explore MEDLINE. Bioinformatics **26**(18), 2321–2327 (2010)

15. Hersh, W.R., Bhupatiraju, R.T.: TREC genomics track overview. In: TREC Proceedings, pp. 14–23 (2003)

16. Bhalotia, G., Nakov, P., Schwartz, A.S., Hearst, M.A.: BioText Team report for the TREC 2003 Genomics Track. In: TREC Proceedings, pp. 612–621 (2003)

17. Jimeno-Yepes, A.J., Sticco, J.C., Mork, J.G., Aronson, A.R.: GeneRIF indexing: sentence selection based on machine learning. BMC Bioinform. **14**(1), 171 (2013)

18. Gobeill, J., Ruch, P., Zhou, X.: Query and document expansion with medical subject headings terms at medical Imageclef 2008. In: Peters, C., et al. (eds.) CLEF 2008. LNCS, vol. 5706, pp. 736–743. Springer, Heidelberg (2009). https://doi.org/10.1007/978-3-642-04447-2_95
19. Gobeill, J., et al.: Deep Question Answering for protein annotation. Database (Oxford) (2015). https://doi.org/10.1093/database/bav081
20. Pasche, E., Teodoro, D., Gobeill, J., Ruch, P., Lovis, C.: QA-driven guidelines generation for bacteriotherapy. In: AMIA Annual Symposium Proceedings, pp. 509–513 (2009)
21. Mottin, L., et al.: neXtA5: accelerating annotation of articles via automated approaches in neXtProt. Database **2016**, baw098 (2016)
22. Levenshtein, V.I.: Binary codes capable of correcting deletions, insertions, and reversals. In: Soviet Physics Doklady, vol. 10, no. 8, pp. 707–710 (1966)
23. Wagner, R.A., Fischer, M.J.: The string-to-string correction problem. J. ACM (JACM) **21**(1), 168–173 (1974)
24. Pustejovsky, J., Castano, J., Cochran, B., Kotecki, M., Morrell, M.: Automatic extraction of acronym-meaning pairs from MEDLINE databases. Stud. Health Technol. Inform. **1**, 371–375 (2001)
25. Europe PMC Consortium: Europe PMC: a full-text literature database for the life sciences and platform for innovation. Nucl. Acids Res. (2014). https://doi.org/10.1093/nar/gku1061

Salience-Induced Term-Driven Serendipitous Web Exploration

Yannis Haralambous[1]([✉])[ID] and Ehoussou Emmanuel N'zi[1,2]

[1] IMT Atlantique & UMR CNRS 6285 Lab-STICC, CS 83818, 29238 Brest Cedex 3,
France
yannis.haralambous@imt-atlantique.fr, manuenzi@gmail.com
[2] La mètis, 22 rue d'Aumale, 75009 Paris, France

Abstract. The Web is a vast place and search engines are our sensory organs to perceive it. To be efficient, Web exploration ideally should have a high serendipity potential. We present a formalization of Web search as a linguistic transformation, evaluate its stability, and apply it to produce *serendipity lattices*, containing suggestions of term chains to be used as exploration paths. We show experimentally that these lattices conform to two of the three serendipity criteria: relatedness and unexpectedness.

Keywords: Web search · Information objects · Information retrieval

According to www.statista.com, around 4.2 billion people are active Internet users. The Web is a vast place (it contains at least 5.4 billion pages according to www.worldwidewebsize.com) and therefore Web search engines are pivotal tools in daily human activities, all around the world. In the same way as our physical sensory organs process a huge amount of information but transmit only a relevant part of it to the brain, Web search engines provide us with a very limited number of pages, the criterion of choice being (or supposedly being) *relevance*.

Considering Web search as a sort of perception, we redefine *salience* as the appearance of some information object (information objects include URLs, text segmented and analyzed as terms, synsets, concepts, etc.[1]) in a given amount of top-ranked results of a Web search. When an information object I "appears" (we will define exactly what we mean by this) in a Web search based on a query q, we will say that I is q-*salient*.

One can explore the Web by following links, or by using terms (*term-driven* Web exploration) provided as queries to Web search engines. When these queries are built automatically out of terms included in the documents obtained by previous queries, we call this *Automatic Query Expansion* [5] based on *Pseudo-Relevance Feedback* [2,11]. In the literature, this kind of exploration is usually oriented by semantic relatedness.

---

[1] Information objects obtained by the search engine or contained in the documents pointed to by the URLs can also include images, videos and other multimodal objects, in this paper we consider only textual objects.

© Springer Nature Switzerland AG 2023
A. Gelbukh (Ed.): CICLing 2019, LNCS 13451, pp. 247–262, 2023.
https://doi.org/10.1007/978-3-031-24337-0_18

On the other hand, in the last years more and more importance is given to *serendipity* as scientific progress vector. Information systems are studied with respect to their serendipitous potential (called *serendipity factor* by [9]), and serendipity plays also an important role in recommendation systems [8,10].

When it comes to Web exploration, total randomness is not a viable strategy since it would distance the user more and more from eir centers of interest; on the other hand, keeping a semantic relatedness between terms used as queries would keep the user in a narrow domain but would not allow for any potential for surprise. This is where *serendipity-oriented exploration* proves useful [3]. According to [7] there are three aspects of serendipity:

1. *relatedness*: the new terms must be related somehow with the previous ones, for the exploration to make sense;
2. *unexpectedness*: they should be unexpected (where "expected" terms can be either those who are semantically related to the previous ones, or those that users commonly combine in queries);
3. *interestingness*: they should represent entities that are important to the user.

We present an algorithm that will, out of two terms, produce a *lattice* of terms (synsets, concepts, etc.), the paths of which can be used for serendipitous Web exploration: if t and t' are the initial two terms, we obtain terms t_i which are *related* to t and t' since they are $(t \wedge t')$-salient, and *unexpected* with respect to t since they are not t-salient. Every term that is included in the lattice in this way is automatically attached to one or more preexisting paths, and all paths end at t': we call them *serendipitous exploration paths*.

To choose candidate terms t_i inside information objects, our approach uses linguistic methods (POS tagging, dependency relations, . . .), therefore we consider the mapping from a term pair to a term lattice as a linguistic transformation. But for that to make sense, we first need to prove that the process is stable enough to produce consistent results in a reasonable time window.

This paper is structured as follows: we first build a formal framework to study Web search and describe the information objects involved; we then experimentally prove stability of the Web search operation in our framework; and finally we present the algorithm of lattice construction and experimentally show that it meets the criteria of relatedness and unexpectedness.

1   The Framework

On an abstract level, Web search is an act of communication between a human and a machine (the server). The motivation of this act for the human is to find information or knowledge on the Web, the message sent by the human is encoded in a formal language (the language of *queries*), it is transmitted through HTTP protocol to the server of the search engine, and the latter replies by a Web page containing a ranked collection of information objects. A Web search session is a sequence of query/results pairs, where the human user refines queries until the results fulfill eir needs. A Web exploration session is a search session where the user has a more global need: to explore a domain, where surprise can work as a trigger for serendipitous ideas, inventions or discoveries.

We use the term "information objects" to keep a certain level of generality: they can be URLs, page titles and snippets (containing selected excerpts from the pages pointed to by the URLs), but also contents of documents pointed to by URLs, various metadata provided by the search engine, etc. This amounts to a collection of heterogeneous data that can be analyzed in various ways. We will restrict ourselves to textual contents, in order to be able to use natural language processing methods.

Formalizing Web search as a linguistic transformation carries some inherent difficulties:

1. Web search results, which are ranked collections of information objects, contain *heterogeneous* and *engine-dependent* information; URLs are often accompanied by text snippets extracted from the URL target, as well as other information;
2. Web search results are partly *random*; indeed, even repeating the same query in a short lapse of time will *not* systematically produce the same result (depending of course on the measures used to compare information objects);
3. information object collections are *ranked*, and this rank—even though the algorithm the search engine has used to obtain it may not be totally impartial—has to be taken into account in the evaluation of results;
4. disambiguation of queries is a difficult process because the more terms one sends (to disambiguate by collocation) the less pages contain all terms of the query, and these pages are not necessarily the most interesting ones;
5. depending on the requirements of a specific task, the nature of the information objects needed by the user can vary;
6. information objects need to be compared, and for this we need specific measures which should ensure convergence of the searching process: despite their randomness, information objects need some stability properties so that the pseudo-relevance used to obtain a given exploration strategy remains valid in a reasonable time window.

In the following we will consider separately the different facets of the problem.

1.1   Queries

The Web search communication process is asymmetric: while the information objects received from the search engine have a rich multilevel structure, the *query* required to obtain them is a mostly short utterance in a very restricted formal language, the vocabulary of which belongs to a given natural language. We define the *query language* as:

Definition 1 (Query language). *Let $\mathfrak{E}$ be a sufficiently large set of English language terms (in the sense of noun phrases having a high termhood value in a given knowledge domain, cf. [6]). We define the **query language** $\mathfrak{Q}$ as being the* *regular formal language based on the alphabet $\mathfrak{E} \cup \{$ "(", ")", "∧", "∨", "¬"$\}$, and the following rules:*

$$S \rightarrow W,$$
$$W \rightarrow W \wedge W \mid W \vee W \mid \neg W \mid (W),$$
$$W \rightarrow m,$$

where $m \in \mathfrak{E}$ and S is the start symbol.

Note that we do not include the (obvious) rule of term concatenation $W \rightarrow W\ W$ because it is ambiguous and search engine-dependent.

Words of the formal language $\mathfrak{Q}$ have to be encoded before transmission: this involves surrounding elements of $\mathfrak{E}$ by ASCII double quotes "..."[2] and replacing symbols $\wedge, \vee, \neg$ by character strings AND, OR, NOT. Furthermore there are constraints on the character lengths of queries: in the search engine we will consider for experimentation, the maximal length of an encoded query is of 1,500 characters.

There are many possible enhancements to the definition of $\mathfrak{Q}$ which we will not consider in this paper: use of proximity operators, (upper or lower) casing, use of search engine user preferences, use of HTTP headers, memorization of previous queries by the search engine, etc. In our experiments we requested English text results and deactivated memorization of query history and adult content filtering.

2   Information Objects

The definition of information object should be flexible enough to be adapted to various search engines and user needs. We proceed as follows. Let N (*repetition rate*) and M (*dimension*) be numbers and q a query in $\mathfrak{Q}$:

1. we systematically repeat the query q N times is a short time window and merge the results as a ranked weighted collection of information objects;
2. we restrict this collection to the M top-ranked results;
3. we define *salience* as the occurrence of an information object in this set of results;
4. information objects are annotated by their *weight vectors*: these M-dimensional vectors store a weight computed out of the their presence at the various ranks of the collection, and other characteristics (e.g., as dependency distance from some other term).

In this paper we will use $N = 10$ and $M = 50$.

There are two kinds of information objects, the distinction depending on the web search engine features:

---

[2] Double quotes are necessary for some search engines to avoid replacement of the English term due to spelling correction on the search engine side.

1. **local** information objects, which are returned directly by the search engine (URLs, page titles and snippets, metadata) or extracted from them: terms, synsets, concepts, and more;
2. **remote** information objects which are extracted from documents pointed to by the URLs.

The difference between the two is of practical nature: documents to download can be of very variable size, and analyzing them can be complex and error-prone. It should be noted that (local) text snippets are of relatively uniform size and have been chosen by the search engine as being representative of the document—normally they contain the individual query terms.

As an example of local information objects, here are the URL, Web page title and text snippet obtained with rank 21 when sending the query `misapplied` to a major search engine:

URL: https://eu.usatoday.com/story/opinion/policing/spotlight/2018/
05/08/justice-system-false-imprisonment-policing-usa/587723002/

Title: 'Misapplied justice' causes innocent men and...

Snippet: 08/05/2018 · '**Misapplied** justice' causes innocent to suffer. It's estimated that about 20,000 people in prison have been falsely convicted, with blacks wrongfully incarcerated at higher rate than whites

As we can see in the example, neither the title nor the snippet need to be grammatical sentences, and the snippet contains the query term.

In this paper we consider only local information objects, but it is always possible to switch to remote ones when search strategy requires it (formalization and algorithms are the same). More specifically we will consider only URLs, terms, synsets and concepts, even though the framework is valid for many more types.

Definition 2 (Information Object). *We call **information object** of dimension M a pair (t, v) where t is an URL (including the empty URL ε), or a term, or a synset, or a concept, and $v \in \mathbb{R}^M$.*

*We call t the **content** and v the **weight vector** of the object. The order of elements of v represents rank in the collection of information objects returned by the search engine.*

Let us denote by $\mathbb{O}_U$, $\mathbb{O}_T$, $\mathbb{O}_S$, $\mathbb{O}_C$ the sets of information objects (where contents are respectively URLs, terms, synsets, concepts).

Definition 3 (Web Search). *Let $q \in \mathfrak{Q}$ be a word of the query language, $*$ an information object type in $\{U, T, S, C\}$, and $t_{1...N} \in \mathbb{R}^N_+$ an N-tuple of time values. Let $\phi_{*, t_{1...N}}(q)$ be the result of a (N-repeated) Web search of q at times $t_{1...N}$, where the i-th elements of weight vectors are defined as $v_i := \frac{1}{N} \sum_{j=1}^N \lambda_{i,j}(I)$, where $\lambda_{i,j}(I)$ can be either the presence (or the result of some linguistic computation) of some $I \in \mathbb{O}_*$ at the i-th rank of the collection of information objects returned by the search engine at the j-th iteration of the query.*

The value of $\phi_{*,t_1...N}(q)$ can be considered as an element of $2^{\mathbb{O}*}$ (that is a set of arbitrary size of information objects of type $*$). In other words, we have, for each $*$:

$$\phi : \mathbb{R}_+ \times \mathfrak{Q} \to 2^{\mathbb{O}*}$$
$$(t_{1...N}, q) \mapsto \varphi_{*,t_1...N}(q).$$

Notice whenever we will consider $\{(t_1, v_1), \dots, (t_n, v_n)\}$ as being the image of $\phi_{*,t_1...N}(q)$, we will consider that (a) $t_i \neq t_j$ for all $i \neq j$, and (b) there is at least one nonzero v_i.

We will show in Sect. 3 that φ is statistically independent of $t_{1...N}$, in a reasonable time window, and hence can be considered as an application of queries into information object sets.

2.1   URLs

URLs [4, §3] are the most fundamental information object contents. The URL of a Web page is divided in five parts: (a) the *scheme* (`http` or `https`); (b) the *authority* (machine name, domain, domain extension); (c) the *path* to the remote file (in Unix notation); (d) following a question mark, the *query*: one or more attributes in name-value form; (e) the *fragment*: whatever follows the #. For the sake of simplicity we will keep only parts (b) and (c). Let us call $\mathbb{U}$ the set of parts (b) and (c) of all valid URLs according to [4], to which we add the empty URL ε.

The number of URLs returned by search engines varies depending on the frequency of the terms queried in the Web corpus and on the search engine's resources. As a reasonable compromise between an insufficiently small number of URLs and an overwhelming amount of increasingly irrelevant URLs, we have chosen to keep only the 50 top-ranked URLs in each result provided by the search engine, and this has lead to the definition of information content with URL content of dimension $M = 50$.

2.2   Terms

If we restrict ourselves to local information objects, terms are obtained in the following way: we parse the Web page titles and text snippets provided by the search engine in order to obtain lexemes (standard forms), POS tags and syntactic dependencies. Then,

- in the simplest case we keep only nouns and named entities;
- if required, we keep also noun phrases with specific POS tag combinations (adjective + noun, etc.) that can be considered as terms;
- in specific knowledge domains we can also filter these noun phrases by calculating their *termhood* coefficient (cf. [6]).

In other words we have a spectrum of methods ranging from simple noun filtering to full terminological extraction.

The value of each coordinate of the weight vector of a term information object can be binary (present or not at a given rank) or a weight calculated as the termhood or some other characteristic of the occurrence of the given term in the given title+snippet (for example, if there is a dependency chain between the term and one of the terms of the query, a weight depending on the length of the chain and the nature of its dependencies). In the case of remote information objects, one can also consider tfidf and similar relevance metrics.

2.3 Synsets

A synset is a sense given (a) as the common sense of a group of words and (b) by a short glose. WordNet, for example, is a graph of synsets. Using word disambiguation methods we can attach synset IDs to terms (if there is ambiguity, we keep either the most probable synset or all probable synsets). The weight can then be the probability that the term belongs to the current synset.

2.4 Concepts

By semantically annotating terms, we can attach them to specific concepts in standard or ad hoc ontologies. As semantic annotation will be done on-the-fly, there can be many matchings for a given term, and hence we can either keep the most probable matching or keep all of them as different information objects. The weight will then be the relevance of the matching.

The difference between "synsets" and "concepts" is of practical nature in this framework: in the former case we consider WordNet or a similar resource as a single knowledge resource and weight information objects by the probability of belonging to a given synset (resulting from word disambiguation methods); the latter case corresponds to search strategies in specific knowledge domains where (standard or ad hoc) ontologies exist or are in the process of being built. The weight can correspond to the relation between the concept and the concept(s) represented by the term(s) of the query.

3 Stability of Web Search

To study the stability of $\varphi_{*,t_1...N}$ we start by the most fundamental information objects, namely URL information objects: $\varphi_{U,t_1...N}$. Let $q \in \mathfrak{Q}$. The result of $\varphi_{U,t_1...N}(q)$ is an element I of $2^{\mathfrak{O}_U}$, that is a set $\{(t_1, v_1), \ldots, (t_k, v_k)\}$, where t_i are URLs and v_i weight vectors.

If $I = \{(t_1, v_1), \ldots, (t_k, v_k)\}$, we will denote by $\mathcal{C}(I) = \{t_1, \ldots, t_k\}$ the set of content elements of I.

Definition 4 (Information Object Distances). *Let* $I = \{(t_1, v_1), \ldots, (t_k, v_k)\}$ *and* $I' = \{(t'_1, v'_1), \ldots, (t'_{k'}, v'_{k'})\}$ *be URL information objects.*

Let $K = \#(\mathcal{C}(I)) + \#(\mathcal{C}(I') \setminus \mathcal{C}(I))$. Let us renumber I' by a function ℓ such that $\ell(i) = j$ if there is a t_j in I such that $t_j = t'_i$, and otherwise ℓ is an injective function into $k + 1, \ldots, K$.

We can consider I as a vector of $\mathbb{R}^K$ (dimensions $1, \ldots, k$) and I' as a vector of the same $\mathbb{R}^K$ (some dimensions between 1 and k, and all dimensions between $k + 1$ and K). The coefficients of these vectors cannot be the v_i (resp. $v'_{\ell(i)}$) because these are already vectors in $\mathbb{R}^M$—therefore we "flatten" them using a function f.

Let $f : \mathbb{R}^M \to \mathbb{R}_+$ a function such that if $f(v) = 0 \Rightarrow v = \mathbf{0}$, then we define

$$d_f(I, I') := \frac{\sum_{i=1}^{K} |f(v_i) - f(v'_{\ell(i)})|}{\sum_{i=1}^{K} \max(f(v_i), f(v'_{\ell(i)}))},$$

which is the Soergel distance [12, p. 987].

We will define two distances:

1. d_{ranked} when $f(v_i) = \sum_{j=1}^{M} \frac{v_{i,j}}{j}$, where $v_{i,j}$ is the j-th weight of the weight vector v_i;
2. d_{unranked} when $f(v_i) = 1$ if there is at least one j such that $v_{i,j} > 0$ and $f(v_i) = 0$ if the weight vector v_i is a zero vector.

Notice that $d_{\text{unranked}}(I, I')$ is the Jaccard distance of sets $\mathcal{C}(I)$ and $\mathcal{C}(I')$.

Definition 5 (d-Stability). *Let d a 1-bounded distance on $\mathbb{O}_U$, that is an application $d : \mathbb{O}_U \times \mathbb{O}_U \to [0,1]$ such that $d(o,o) = 0$, $d(o,o') = d(o',o)$ and $d(o,o'') \leq d(o,o') + d(o',o'')$ for all $o, o', o'' \in \mathbb{O}_U$.*
*We call a querying application $\varphi_{U,t_1,\ldots,t_N}$ d-**stable** if*

$$\max_{\substack{t_1 < \cdots < t_N \\ t'_1 < \cdots < t'_N}} d(\varphi_{U,t_1,\ldots,t_N}(q), \varphi_{U,t'_1,\ldots,t'_N}(q)) \lesssim 0.05,$$

for any $q \in \mathfrak{Q}$. In that case we will write φ_U instead of $\varphi_{U,t_1,\ldots,t_N}$.

Hypothesis 1. *The Web search application φ_U is both d_{ranked}-stable and d_{unranked}-stable.*

Experimental Validation of Hypothesis 1. Our corpus will consist of the fifty first nouns of *Moby Dick*: year, money, purse, nothing, shore, part, world, way, spleen, circulation, mouth, damp, soul, coffin, warehouse, rear, funeral, hypos, hand, principle, street, people, hat, time, sea, substitute, pistol, ball, sword, ship, man, degree, feeling, ocean, city, wharf, isle, reef, commerce, surf, downtown, battery, mole, wave, breeze, hour, sight, land, crowd, water.

Let $N = 10$ and $M = 50$. For each w in this set we have calculated a first series of ten $\phi_U(w)$ and then a second series, that makes a total of 200 queries for each word. For each w we calculated the distances d_{unranked} and d_{ranked} of the i-th elements of the two series, and then measured the average and standard deviation of the distance measurements.

The results can be seen in the following two diagrams where the x axis represents average $d_{(un)ranked}$ and the y axis standard deviation of $d_{(un)ranked}$:

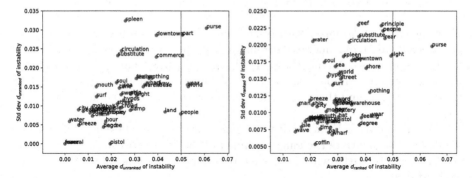

As we see, in the unranked case most distances are under the 0.05 barrier, with a few exceptions: *purse, year, world, part*; in the ranked case only the word *purse* has an instability higher than 0.05. The difference between the two distances is due to the fact that changes in results occur are rather located in lower ranks (near the bottom of the Web page). This is not the case for *purse* where the results seem to oscillate between two versions differing also in higher ranks. Note that the two distances are significantly different: their correlation coefficient is not higher than 0.476.

One could wonder whether instability is related to the frequency of words. Indeed, there is a small positive correlation ($\rho_{unranked} = 0.184$) between the frequency of words in English language[3] and unranked instability, but this correlation almost disappears when ranked distance is measured ($\rho_{ranked} = 0.037$):

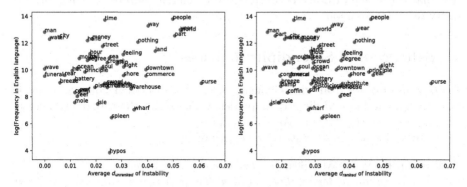

Differences between the two diagrams depend on the word (e.g., they are important for *rear* and unimportant for *people*) and show whether instability lies in lower ranks or can be found at all ranks.

---

[3] Taken from the Corpus of Contemporary American English https://corpus.byu.edu/coca/.

The next step is to prove the stability of φ_T, φ_S, φ_C. As these rely entirely on the titles and text snippets accompanying URLs, it suffices to show that title and snippet for a given URL do not vary when queries are repeated.

Experimental Validation of the Invariance of Title and Snippet
On a corpus of 1,784 queries (108 unique queries) we obtained 86,493 URLs (7,211 unique ones). Among the 86,493 URLs, only 618 (that is 0.7%) had more than one different snippet (out of which 577 had two different snippets, 29 had three, 9 had four, and there were single URLs with 6, 7 and 10 different snippets respectively).

We calculated a 1-bounded distance (more precisely, the Levenshtein distance divided by the length of the longest snippet of the two) between the snippets for the 618 cases of URLs with more than one snippet. Here is the distribution of this distance:

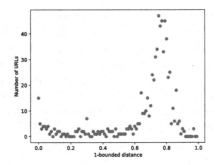

There is a peak at approx. 0.8 and the most plausible hypothesis explaining it is that this is the distance between entirely different snippets having only the query terms in common.

4   Salience and Serendipity Lattices

In the following we deal with elements of $\mathbb{O}_T$ because their content elements (i.e., terms) can be used directly in queries, which is not the case of elements of $\mathbb{O}_S$ and $\mathbb{O}_C$ (see the "Future Works" section).

We first define a more restrictive version of φ_T:

Definition 6 (Valid Web Search). *Let a query q be equal to $t \wedge t' \wedge \cdots \wedge t^{(n)}$, i.e., a conjunction of terms. Let $\varphi_T(q)$ be the Web search with query q as an application $\mathfrak{Q} \to 2^{\mathbb{O}_T}$.*

*We call the search $\varphi_T(q)$ **valid** if (a) there are elements $(t, v), \ldots, (t^{(n)}, v^{(n)})$ (corresponding to the terms $t, t', \ldots, t^{(n)}$ in the query) in $\varphi_T(q)$, and (b) these elements have weight vectors v such that if for some j_0, $v_{j_0} = 0$ then all information objects of $\varphi_T(q)$ have weight vectors with zero j_0-th dimension.*

In other words, among the results returned by the search engine, we consider only those the snippet of which contains *all* query terms. On a practical level, to obtain a valid Web search one has to consider as many URLs as necessary to obtain M snippets containing the query terms. In theory, one has to examine *all* URL snippets until obtaining the M first ones fulfilling our need. But in reality search engines tend to display those URLs first, and it is highly probable that when a snippet missing some query term is displayed then all following snippets also miss at least one query term.

Every Web search can be transformed into a valid Web search if we remove the URLs the snippets of which do not contain *all* query terms. Therefore we can always assume that a given Web search is valid.

Let us now define the notion of salience:

Definition 7 (Salience). *Let q be a query and $\varphi_T(q)$ a valid Web search. Let t be a term. We say that t is q-**salient** if there is a (nonzero) occurrence vector v such that $(t, v) \in \varphi_T(q)$.*

Obviously every t belonging to q is q-salient, since the search is valid. This constraint is very important because, once exhausted the results containing all query terms, search engines will provide "partial snippets", i.e., results in which *not all* terms of the query appear. It is this constraint that guarantees our algorithm stopping after a finite number of steps: by adding more and more terms to the query, the number of pages containing *all of them* will necessarily decrease until reaching zero.

Definition 8 (Serendipity lattice of t, t'). *Let t, t' be two terms. We will call **serendipity lattice** of t, t' a lattice (the partial order of which we denote by $\succ$) of terms with t as supremum and t' as infimum, with the following properties:*

1. *all paths $t = t_1 \succ \cdots \succ t_m \succ t'$ are of equal length m (the height of the lattice);*
2. *for any element t_i in such a path, t_i is not $(t_1 \wedge \cdots \wedge t_{i-1})$-salient;*
3. *t' is $(t_1 \wedge \cdots \wedge t_m)$-salient.*

Intuitively, every path $t_2, \ldots, t_m$ is a sequence of terms to add to t so that t' becomes salient in the resulting query, but not in any smaller query. The fact that no term is salient with respect to the terms preceding it in the path guarantees the minimality of the path: every intermediate term is necessary. All paths are of the same length because in our algorithm we stop when a path makes t' salient, after having finished examining all paths of the same length.

Algorithm 1 (p. 12) provides the serendipity lattice S of a pair of terms t, t'. Some explanations:

- we consider S as a graph whose edges (s, s') represent orderings $s \succ s'$. In the first part of the algorithm, edges are added to S, but S has not yet a lattice structure because some paths may not lead to t'. In the second part of the algorithm we remove the unnecessary edges by a bottom-up approach: we start from t' and keep in S only the edges we encounter while going upwards;

Algorithm 1: Calculation of the serendipity lattice S of the pair of terms t, t'.

Data: Terms t, t', (valid) Web search application φ_T

Result: A serendipity lattice of t, t', that is a lattice S with t as supremum and t' as infimum

```
/* Calculating a superset of S                                      */
if t' is t-salient then
 │  S = {(t, t')}
else
 │  Q ← {t};
 │  Q' ← {t ∧ t'};
 │  MaxLength ← ∞;
 │  for q' ∈ Q' do
 │   │  if |q'| ≤ MaxLength then
 │   │   │  for x ∈ Filter(φ_T(q')) do
 │   │   │   │  for q ∈ Q do
 │   │   │   │   │  if |q| ≤ MaxLength then
 │   │   │   │   │   │  if x is not q-salient then
 │   │   │   │   │   │   │  Q ← Q ∪ {q ∧ x};
 │   │   │   │   │   │   │  /* last(q) means last conjunctive term in q */
 │   │   │   │   │   │   │  S ← S ∪ {(last(q), x)};
 │   │   │   │   │   │   │  if t' is not (q ∧ x)-salient then
 │   │   │   │   │   │   │   │  Q' ← Q' ∪ {q' ∧ x}
 │   │   │   │   │   │   │  else
 │   │   │   │   │   │   │   │  S ← S ∪ {(x, t')};
 │   │   │   │   │   │   │   │  if MaxLength = ∞ then
 │   │   │   │   │   │   │   │   │  MaxLength ← |q|
 │   │   │   │   │   │   │   │  end
 │   │   │   │   │   │   │  end
 │   │   │   │   │   │  end
 │   │   │   │   │  end
 │   │   │   │  end
 │   │   │  end
 │   │  end
 │  end
end
/* Removing unnecessary edges from S                                */
A ← {t'};
S' ← ∅;
while S ≠ ∅ do
 │  for (s, s') ∈ S do
 │   │  if s' ∈ A then
 │   │   │  S' ← S' ∪ {(s, s')};
 │   │   │  A ← A ∪ {s};
 │   │   │  S ← S \ {(s, s')};
 │   │  end
 │  end
end
S ← S';
```

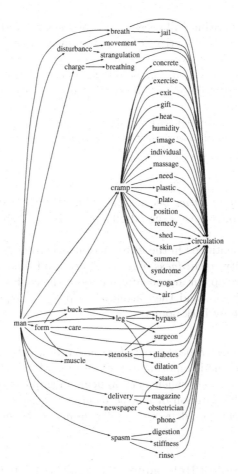

Fig. 1. Serendipity lattice for the pair *man – circulation*.

- queries q and q' are built by progressively adding conjunctive terms and the function "last" returns the rightmost conjunctive term. Note that conjunction is not commutative, in the sense that $\varphi_U(q \wedge q')$ is not necessarily equal to $\varphi_U(q' \wedge q)$;
- $|q|$ denotes the number of conjunctive terms in q, in particular $|t| = 1$;
- by ∞ we represent a sufficiently large number;
- the "Filter" function will limit the amount of terms used to create further queries and avoid combinatorial explosion. We weight terms in $\varphi_T(q')$ as follows: $w(t_0) = \sum_{i=1}^{M} \frac{r_i}{i+\varepsilon}$ where r_i is the binary value of the i-th coordinate of the weight vector of t_0 and $0 < \varepsilon < 0.1$ is a random number, whose goal is to keep the weight values distinct, so that w is an injective function. After weighting the terms we filter their set by keeping only a limited number (1,000) of top-weight terms.

Note that the algorithm will necessarily stop because of the constraint of the snippet containing all conjunctive terms of the query: when the query becomes complex enough, either we will attain some topmost result of the query $t \wedge t'$ and hence t' will become salient, or the search engine will not be able to produce any result with snippet containing all query terms and hence the algorithm will stop.

See Fig. 1 for an example of serendipity lattice with 50 vertices and 100 edges, for the pair *man – circulation* from the "Moby Dick" corpus.

4.1   Discussion

We call S a *serendipity lattice* because we claim that the paths between supremum and infimum of S have a serendipity potential as Web exploration paths, let us see why.

As we already mentioned in the introduction, according to [7] there are three aspects of serendipity: *relatedness*, *unexpectedness* and *interestingness*.

We will first develop our argumentation about the conformity of our method to the two first criteria, and then attempt to prove it experimentally on the corpus of 50 first Moby Dick nouns.

Argumentation. Let t, t' be the pair of terms out of which we create a serendipity lattice S. Our method conforms to the criterion of *relatedness* since every x in the lattice is $(t \wedge t' \wedge \cdots)$-salient, i.e., appears in snippets together with t and t'. It conforms to the criterion of *unexpectedness* because every x in the lattice is *not* $(t \wedge \cdots)$-salient, i.e., does *not* appear in any of the snippets obtained by the previous query.

Experimental Proof of Relatedness. To evaluate relatedness of serendipity lattices we will use the FrameNet v1.7 corpus (https://framenet.icsi.berkeley.edu/). In this context, a *frame* is "a schematic representation of a situation involving various participants, props, and other conceptual roles" (Wikipedia). Version 1.7 of FrameNet comes with 1,073 frames and 4,235 distinct nouns. We consider that *being in the same frame is a strong relatedness indicator between two words*.

Our "Moby Dick" corpus (the lattices obtained out of the 50 first nouns of *Moby Dick*) contains 2,331 lattices, which, in turn, contain 9,630 distinct words and 26,779 distinct lattice edges.

2,607 words of our corpus appear in FrameNet, and the extremities of 406 lattice edges belong to the same FrameNet frame, which means that they are strongly related. As a comparison basis, we took the same number of FrameNet words, but this time arbitrarily chosen, created random edges between them and counted how many of them happened to be in the same frame. We repeated the operation 100 times. Here are the results:

Method	# FN words	# pairs of FN words	# pairs in the same frame
In serendipity lattices	2,607	11,097	406
Randomly chosen	2,607	11,097	6.35 ± 2.35

In other words, if the extremities of our lattice edges were unrelated then we would expect to find an average of 6.35 of them (with standard deviation 2.35) in the same frames. But we find 406, that is more than **sixty times as many** lattice edge extremities belonging to the same FrameNet frame. Hence, these extremities are *related*.

As for *unexpectedness*, we will use the *Word Associations Network* (https://wordassociations.net) by Yuriy A. Rotmistrov (see also [1]). We consider that this resource provides all "expected" terms that are mentally associated with a given term. We consider that, *the fact that a lattice edge does not appear in WAN is a strong indicator of unexpectedness.*

For the 9,630 words of the lattices of our "Moby Dick" corpus we retrieved 1,938,534 associations. From the 26,779 lattices edges of our corpus, only 1,782 appear among WAN associations, therefore we can conclude that **93,4%** of our lattice edges can be considered as "unexpected".

5   Conclusion and Future Works

We have presented a formalization of the Web search operation considered as a linguistic transformation. We have proven experimentally the stability of this operation. Then we have introduced two new notions: *q-salience* (whether an information object appears in the results of a Web search of query q) and *serendipity lattice*, which is the result of an algorithm we introduce, and aims at providing paths for serendipitous Web exploration. Finally, we have shown experimentally that our lattices conform to two of the three serendipity criteria: relatedness and unexpectedness.

Future work will consist into extending the approach to be concept-based instead of term-based and to explore evaluation of serendipity potential in more depth. Our goal is to build an interface for assisting the user in exploring the Web efficiently—serendipity lattices will be part of the building blocks of this interface.

Acknowledgments. The authors would like to thank the staff of La Mètis for their precious help in implementing the algorithms and performing massive online searches.

References

1. Acar, S., Runco, M.A.: Assessing associative distance among ideas elicited by tests of divergent thinking. Creat. Res. J. **26**, 229–238 (2014)

2. ALMasri, M., Berrut, C., Chevallet, J.-P.: A comparison of deep learning based query expansion with pseudo-relevance feedback and mutual information. In: Ferro, N., et al. (eds.) ECIR 2016. LNCS, vol. 9626, pp. 709–715. Springer, Cham (2016). https://doi.org/10.1007/978-3-319-30671-1_57

3. Beale, R.: Supporting serendipity: using ambient intelligence to augment user exploration for data mining and web browsing. Int. J. Hum.-Comput. Stud. **65**, 421–433 (2007)

4. Berners-Lee, T.: RFC 3986. Uniform Resource Identifier (URI): Generic Syntax (2005)

5. Carpineto, C., Romano, G.: A survey of automatic query expansion in information retrieval. ACM Comput. Surv. **44**, 1–49 (2012)

6. Frantzi, K.T., Ananiadou, S., Tsujii, J.: The *C-value/NC-value* method of automatic recognition for multi-word terms. In: Nikolaou, C., Stephanidis, C. (eds.) ECDL 1998. LNCS, vol. 1513, pp. 585–604. Springer, Heidelberg (1998). https://doi.org/10.1007/3-540-49653-X_35

7. Huang, J., et al.: Learning to recommend related entities with serendipity for web search users. ACM TALLIP **17**, 25:1–25:22 (2018)

8. Kotkov, D., Wang, S., Veijalainen, J.: A survey of serendipity in recommender systems. Knowl.-Based Syst. **111**, 180–192 (2016)

9. McCay-Peet, L., Toms, E.G.: The serendipity quotient. Proc. Am. Soc. Inf. Sci. Technol. **48**, 1–4 (2011)

10. Meng, Q., Hatano, K.: Visualizing basic words chosen by latent Dirichlet allocation for serendipitous recommendation. In: Proceedings of the 3rd International Conference on Advanced Applied Informatics, pp. 819–824 (2014)

11. Vaidyanathan, R., Das, S., Srivastava, N.: Query expansion strategy based on pseudo relevance feedback and term weight scheme for monolingual retrieval. Int. J. Comput. Appl. **105**, 1–6 (2015)

12. Willett, P., Barnard, J.M., Downs, G.M.: Chemical similarity searching. J. Chem. Inf. Comput. Sci. **38**, 983–996 (1998)

Language Modeling

Two-Phased Dynamic Language Model: Improved LM for Automated Language Translation

Debajyoty Banik[1]([⊠]), Asif Ekbal[2], and Pushpak Bhattacharyya[3]

[1] Kalinga Institute of Industrial Technology, Bhubaneswar, India
`debajyoty.banik@gmail.com`
[2] Indian Institute of Technology Patna, Bihta, India
`asif@iitp.ac.in`
[3] Indian Institute of Technology Bombay, Mumbai, India
`pb@cse.iitb.ac.in`

Abstract. We discuss the importance of domain specific language model in statistical machine translation system. Both the structures and phrase selection are not the same for different domains. So, the language model trained with the general domain data or other domain data can not provide better accuracy. Moreover, there may have some specific focus in different texts of the same domain. Hence, the language model trained with data from the default domain may not yield significant output. In this paper, we learn our system dynamically based on the better matches with the input text. Instead of directly selecting pre-trained language model we prepare the prioritized language model according to the situation. The proposed model is evaluated for Hindi-English translation. It shows a significant improvement on the translated output in terms of the BLEU score. Our evaluation shows that automated domain adoption to predict better language model improves the translation quality.

Keywords: Natural language processing · Statistical method · Language model · Web scraping

1 Introduction

Domain adaption in the statistical machine translation (SMT) system has resulted in improved translations once the domain is properly classified [1]. In order to improve the translation quality, we adapt context wise language models for better translation. There is a significant improvement in translation instead of just using a general monolingual corpus.

We want to calculate the probability of a string $W = w_1, w_2, ..., w_n$. Intuitively, $p(W)$ is the probability of a sequence of English (target language) tokens which turns out to be W. To find out $p(W)$, the typical statistical approach estimates the counting how often W occurs in the training data. Most long sequences of words aren't in the text at all, however. Therefore, we need to split down the

© Springer Nature Switzerland AG 2023
A. Gelbukh (Ed.): CICLing 2019, LNCS 13451, pp. 265–279, 2023.
https://doi.org/10.1007/978-3-031-24337-0_19

$p(W)$ calculation into smaller stages, for which we can obtain adequate data and approximate distributions of probabilities. In n-gram language modeling, it is decomPoSed the probability using the chain rule:

$$p(w_1, w_2, ..., w_n) = p(w_1)p(w_2|w_1)...p(w_n|w_1, w_2, ..., w_{n-1}) \qquad (1)$$

The language model probability $p(w_1, w_2, ..., w_n)$ is a product of word probabilities where given a history of preceding words as discussed in Eq. 3. To be able to estimate these word probability distributions, we limit the history to m words:

$$p(w_n|w_1, w_2, ..., w_{n-1}) \simeq p(w_n|w_{n-m}, ..., w_{n-2}, w_{n-1}) \qquad (2)$$

In Statistical machine translation, language models are used to increase the fluency of translated texts. It is a common concept that the language model using large data-set may be more trustable since we would get more histories (the basis of language modeling). But it may leads to over fitting if that is not domain specific. To make it domain specific, data sparsity problem can be noticeable. In this paper, we are trying to consider both the problems. Depends on a belief that a language model that embraces a larger context provides better prediction ability, researchers in [2] presented two different language models: A backward language model improving the traditional forward language model and a shared knowledge control model catching long distance relations outside the reach of regular n-gram language models. The traditional language model cannot provide best fluency for different type of the texts from various domains. We reconstruct the language models in different way by scrapping the web for different domains/context. Finally, suitable language model is used for the translation based on the input text.

1.1 Motivation

Developing improved language models also outcomes in models that further serve their expected function of interpreting the natural language. It is the encouragement to create ever more reliable models of expression. Our motivation is drawn from the fact that usually a document is based on a topic and has a particular domain that it is discussing. The topics may be related to technology, politics, sports, health etc. Thus if the language models are built specifically for a particular domain and used for translation, the results are expected to be better. This is useful in speech recognition [3] where the input speech is mostly based on a particular topic. For a particular domain, some words or phrases are more commonly used in texts, which help us to analyze what sort of topic is being spoken about. When we mostly use a particular set of words or phrases, we become sure of the topic expressed. These phrases (cue words) can also be used to check the domain of the input text and extract data from the web to create the domain-wise language model. This can be used to adapt the language model and use it for better translation.

The workflow of the paper is like: the overall methodology of creation of the domain-wise language model is described in Sect. 2. Section 3 deals with the experimental setup used in this process and analyzes the results obtained and finally Sect. 4 gives the conclusions for domain-wise language modeling for Hi-En translations.

1.2 Related Work

From the literature survey, we observe that there has not been much detailed discussion as to how statistical machine translations (SMT) system can be improved by improving Language Model, more specifically on domain adaption by using n-gram modelling.

In [4] the article says about how to build a large-scale language model that allows a huge amount of training data to be scaled, and how much translation performance improves as the language model increases in size.

In [5], they have tried a variety of sizes and domains of training data; but this has been tried on the parallel corpora and not on the language models to investigate the effect on translation. They investigated the behavior of an exPoSed SMT system to train data of varying sizes and types. Their research findings revealed that only a parallel corpora of small sizes could be used to train purPoSes without reducing too far the evaluation scores for both the experiments considering two language pairs: English-Romanian and German-Romanian in both communication directions.

Recently there has been work in using Neural Language Models (NLMs). In [6] they have demonstrated that deep NLMs having three or four layers which perform well than the fewer layers in terms of both the perplexity and the translation quality. They have combined various techniques to successfully train deep NLMs that jointly condition on both the source and target contexts. They also use this concept for domain adaption on the sms-chat domain. But the adaption is done to tune the decoder weight by using a tune set of 260K words in the newswire, web, and the sms-chat domains to tune the decoder weights and a separate small, 8K words set to tune re-ranking weights. To train adapted NLMs, they have used trained models on general in-domain data and further fine-tune with out-domain data for about four hours.

A significant work to compare language model has been carried out in [7]. They discussed the variations among the language models compiled from the initial target-language texts and those compiled from manually interpreted texts into the target language. Substantiating the results of the Translation Studies, they found that the above are slightly stronger predictors of the translated sentences than the former, and therefore suit the better reference. Furthermore, translated texts yield better language models for statistical machine translation than the original text.

Like we have stated earlier, the significant work in developing topic wise language models has been done only to benefit automated speech recognition. In [3] they have used a similar process, generate queries and obtain outputs from the World Wide Web, to develop language models. They have described an iterative web crawling approach which uses a competitive set of adaptive models comprised of a generic topic independent background language model, a noise model representing spurious text encountered in web-based data (Web data), and a topic specific model to generate query strings using a relative entropy based approach for WWW search engines and to weight the downloaded Web data appropriately for building topic specific language models. They also seen how this method can be used to easily construct language models for a specific domain, with only an initial collection of sample statements and how it can tackle the different issues associated with Web data. They managed to obtain a 20% decrease in perplexity for the therapeutic goal area. The perplexity gains converted into an increase of 4% in the ASR word error rate (absolute) in the ASR word error rate (absolute) leading to a relative gain of 14% percent.

In [8] adaptive language modelling has been done for the domain of blogging and micro-blogging. They have used information retrieval (IR) language modeling based on the large volumes of constantly changing data. This data fulfill the needs to frequently integrating and removing data from the model. They have identified a set of matches from a corpus given a query sentence, then the likelihood estimation are calculated for that vary query. The IR-LM can be beneficial when the language model needs to be updated with adding and removing the data which works for the social data where new content is constantly generated.

Recently researchers use simple dynamic language model to design the transcription system in healthcare [9].

Neural based dynamic language model is also introduced in [10].

Finally, exploring knowledge and its analysis in language model is discussed in [11].

2   Domain Adoption in the Language Model

To prepare a language model we usually use n-gram modeling. n-gram language models are based on assumptions on the likelihood that terms would imitate each other. If we review a significant volume of language, we can find that the term 'home' more frequently accompanies the term 'moving' than does the word 'room' [12]. This information is learned by the language model to assure fluency.

Language modeling (LM) is the production of probability distributions capable of predicting the next term throughout the sequence, provided the preceding ones. There are two challenges to prepare best language model which fit for specific text: 1. Domain specific language model cannot provide better accuracy for other specific domain. We may not familiar with the domain of the upcoming input text. 2. The language model trained with General domain data may lead over-fitting problem. Considering both issues, we prepare a prioritized language model based on the specific topic of input text addition with general language model.

The Primary concept of this prioritized language model is as follows:

Let, $w_1^L = w_1, w_2, ..., w_L$ denotes a string of L tokens over a fixed vocabulary. An n-gram language model assigns a probability to W_1^L.

$$(w_1^L) = \prod_{L}^{i=1} P(w_i|w_1^{i-1}) \approx \bar{P}(w_i|w_{i-n+1}^{i-1}) \tag{3}$$

The Markov assumption is the foundation for the modeling of the language. Let, f (w_i^j) describes the occurrence frequency of the $w^j i$ sub-string within a wide target language string $w^L 1$, called the training results. The maximum likelihood (ML) estimate approximates for the n-grams by using their relative frequencies.

$$r(w_i|w_i^{i-1} - n + 1) = \frac{f(w_{i-n+1}^i)}{f(w_{i-n+1}^{i-1})} \tag{4}$$

The Markov assumption states that only a limited number of previous words affect the probability of the next word. It is technically wrong, and it is not too hard to come up with counter examples which demonstrate that a long history is required. However, limited data restricts the collection of reliable statistics to short histories.

Typically, we choose a number of words in the history based on the training data we have. More training data allows for longer histories. Most commonly, trigram language models are used. They consider a two-word history to predict the third word. We have also used tri-gram language model in all the cases. The estimation of trigram word prediction probabilities $p(w_3|w_1, w_2)$ is straightforward. We count how often in our training corpus the sequence w_1, w_2 is followed by the word w_3, as opPoSed to other words. According to maximum likelihood estimation, we compute:

$$p(w_3|w_1, w_2) = count(w_1, w_2, w_3)|sum(w)count(w_1, w_2, w_3) \tag{5}$$

Finally, individual models (domain specific language model and general language model) combine using joint probability model based on their priority. It is quite obvious that the domain specific language model gets more priority than the general language model. The primary reason to keep the general language model to cover all situations which miss at domain specific model.

The first step is devoted to construct the general language model. The second step is devoted to retrieve cue words from the input text (test data). Based on those cue words we identify the online documents using the matching ratio. We don not use Term frequency-inverse document frequency (TF-IDF) [13] calculation to mine the documents because the input text may be a new copy and it will take much to cover all variety documents searching. These documents are used to prepare the domain specific language model and merge with the general language model to get final language model which is best fit for the machine translation and other application. The detailed flowchart is shown in Fig. 1. As discussed earlier, cue words are extracted based on the TF-IDF computing. The term frequency $tf(t, d)$ is the easiest way in a system to use the actual estimate of a phrase, i.e., the amount of occasions the word t appears in document d. If we denote the raw count by $f_{t,d}$, then the simplest tf scheme is $tf(t, d) = f t, d$. The term frequency adjusted for document length: $\frac{f_{t,d}}{(number of words in d)}$. The augmented frequency, to prevent a bias towards longer documents. $tf(t, d)$ is shown in Eq. 6:

$$tf(t, d) = 0.5 + 0.5(\frac{f_{t,d}}{max(f_{t'd}) : t' \in d}) \tag{6}$$

The inverse document frequency is a calculation of how much knowledge the term contains, i.e. whether it is normal or uncommon in all documents. It is indicated in Eq. 7.

$$idf(t, D) = log\frac{N}{|D \in d : t \in d|} \tag{7}$$

where, N represents the total number of documents in the $N = |D|N = |D||\{d \in D : t \in d\}||\{d \in D : t \in d\}|$: amount of documents in which the token t occurs (i.e., $tf(t, d) \neq 0 \; tf(t, d) \neq 0$)). If the word is not in the corpus, otherwise division-by-zero will follow. Thus, it is common to adjust the denominator to $1 + |\{d \in D : t \in d\}|1 + |\{d \in D : t \in d\}|$.

We prepare the language models for translations from Hindi to English. A language model is created in the target language. Hence, here we will be making language models in English.

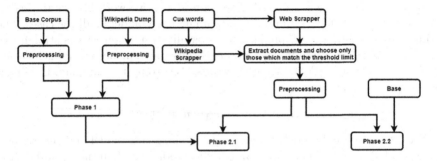

Fig. 1. Flowchart for the adaptive language model

2.1 Preprocessing

Documents should be preprocessed so that relevant and clear information can be obtained. Preprocessing helps to improve the quality of texts and makes all the documents to be at par with each other. The steps of preprocessing followed for all documents are:

Tokenization: Tokenization is the process of splitting something into bits, called tokens, even tossing out other characters, including punctuation, at the same time. We cut and chuck away punctuation characters on white room. True casing: True casing means converting all the characters to lowercase.

2.2 Base Corpus

For any language model creation, we need a huge amount of texts. By analyzing patterns in that text, n-gram probabilities are calculated.

2.3 Phase-One

Here, we calculate the n-gram probability for base corpus plus Wikipedia dump [14]. Wikipedia has a variety of domains and topics so it can be a valuable addition to the language model. It can cover various types of sentences with more confident and able to solve the data sparsity problem. The language model, which is prepared using the general data-sets, is waiting to take participation at the phase-two after preparing the domain specific language model.

2.4 Phase-Two

In this section, we define the process used to make the adaptive language model; i.e. one which is made strictly for the topic which is being spoken of. No need to have any prior knowledge about the topic or domain. This real life problem is solved in real time using the proposed approach. The primary processes are 1. cue-word extraction, 2. web-scrapping and wikipedia scrapping based on the cue-word, then 3. combining the general language model with the prepared language model based on the extracted documents (with assigned priority) to create the language model.

Cue Words: Cue words are the significant or important words present in the text. Cue words express the ideas of topics that provide some information about the content. Instead of the individual word we consider phrases as cue-word for our work which makes the search more realistic and contextual. We have parsed each sentence into trees based on phrases and to obtain the required cue words we have extracted the NOUN PHRASE (NP) and VERB PHRASE (VP). We have not extracted ADJECTIVE PHRASES (ADJP) separately because they are present inside the VP already. Considering phrases are better than words because if we use just some specific PoS tags we might miss out some contextual information. If we use PoS tags for cue word extraction and fix the tags for noun forms to just NNS (Noun Plural) and NN (Noun Singular), they may not be adequate enough for improvisation. NP for all forms of nouns are taken into consideration in this case. Other tags such as adverbs, prepositions do not contribute to spot the cue words and add only to the noise in retrieving relevant documents. All of the abbreviations of the PoS tag follow standard form (i.e. penn tree bank [15]). Some examples of phrases is shown in Table 1.

Table 1. Examples of tags

Tags	Text
NP	Vegan children
VP	Are not suitable in for infants
ADJP	Suitable for infants
NNS	Infants
JJ	Suitable

We realized that all the retrieved phrases would not be helpful in retrieving more information from the Web and Wikipedia. Based on the observation and imperial analysis, we have set the threshold for size of the cue-word phrases as six words because too big query will not produce any suitable result from the web at all and will be just a loss of time. We have used the Stanford Parser for parsing [16].

Web Scraper and Wikipedia Scraper: The cue words obtained in Sect. 2.4 are sent as a query to the Web. The World Wide Web consists of a lot more textual information. The amount of information related to a topic increases steeply in the web. So if cue words are sent as a query, it would retrieve documents relevant to topic and domain. We have used Webhose-python [17] as our web Scraper. It gives us the news domain pages related to the query.

We have further used a python program to return just those wikipedia pages for which cue words/phrases are sent in as queries. This also helps us to find the related pages as required.

Setting Up Threshold: Not all the pages which are received from the web or the wikipedia may be relevant to one domain. For example, while retrieving pages from Wikipedia, we observed that for the same cue word "Air", we received a no a no pages ranging from "Air guitar" to "air". As we are dealing with health corpus, we do not need "air guitar" to be included. So, we set up a threshold which would help in the inclusion of files. It (R) is computed in Eq. 8.

$$R = \frac{|K|}{|T|} \tag{8}$$

where K refers to the keywords, present in the document, which match the list of keywords except the stop words and T denotes a total number of words in the document except stop words. This equation is also used to fine-tune the threshold value δ. Then we compute R for different files which are obtained from the web or the input text to compare with δ.

Table 2 shows the file type and obtained R for the different texts. These texts can be treated as a specialized version of the different domains. So, computed R for different files crawled from the web are shown here altogether.

In Table 3, the "air" expelled in one section is a misleading title. It is about a video game and not related to health as the title assumes us to think; thus the threshold 0.2 is good. Hence, we set this ratio to 0.22 for inclusion in case of webhose files and 0.2 in the case of wikipedia files.

2.5 Final Language Model

All these files are then combined to form language models. We have done this in two ways.

1. Base+Wikipedia dump+webhose files+specific wikipedia pages

Table 2. Matching ratio (R) of different crawled files

Filename	Ratio (R)
social2.txt	0.189247311828
new3.txt	0.1483198146
new_danielpipes.en.txt	0.149747834533
new_emille.en.txt	0.237342128443
tourism1.txt	0.175356615839
social1.txt	0.194812069878
tech1.txt	0.185867895545
health1.txt	0.242798353909
tech2.txt	0.183632734531
health2.txt	0.235188509874
new1.txt	0.155754651964
health3.txt	0.227241615332

Table 3. Table for checking ratios of wikipedia pages32

Filename	Ratio (R)
are filled with airen_wiki_article.txt	0.2000664673
cognitive impairment_en_wiki_article.txt	0.2082551595
clean air_en_wiki_article.txt	0.2476382416
the air expelled in one second_en_wiki_article.txt	0.1226463104
Orcas Island Airport_en_wiki_article.txt	0.1226415094
spreads in the lungs and airways_en_wiki_article.txt	0.226668751
Mayor Buenaventura Vivas Airport_en_wiki_article.txt	0.0379746835

2. Base+webhose files+specific wikipedia pages

We call them phase-two part 1 and phase-two part 2, respectively. We will then test which of the language models,the base,phase-one or phase-two give us the best results.

The main application of this experiment lies in identifying the domain of incoming test file and using that language model which matches the domain. For this, we use concept the match ratio (R). We keep another set of cue words, in the source language and match the incoming test file to each of these lists. The maximum match ratio will provide us the idea about the domain the input test file. The preferred domain specific language model get more priority to built the final language model. To compute the domain of the test file, we calculate the ratio of different domain wise cuewords list with its content. The hi-level architecture is shown in Fig. 2.

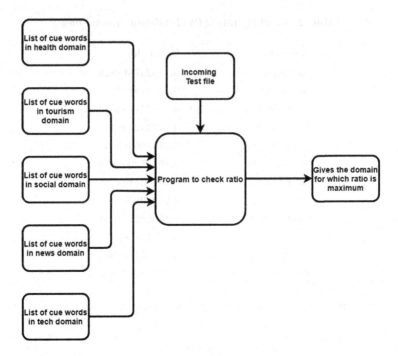

Fig. 2. Flowchart demonstrating checking of domain of test file

3   Experimental Evaluation

We detail the experiments in this Section to check which out of the four models would perform best for a domain. We train 3-gram LMs of each of the models for the following domains: health, news, social, tech and tourism. We have performed the experiment on Hindi to English Translations.

The base corpora has been obtained from monolingual sets of News Crawl articles and Europarlv7 [12,18]. The combination of these gives us 1492827 sentences. The Wikipedia Dump has been downloaded from the web [19]. We have taken the pages and articles of 2017-05-20. The processing of wikipedia files from xml format to plain text has been done with the help of BeautifulSoup and PHP for the conversion.

Preprocessing of the texts also needed to be done. For English, the Moses tokenizer was used and For Hindi, the Indic NLP tokenizer [20] was used. Some other independent python programs were also written to get rid of unwanted patterns and to clean the corpus.

For cue word extraction, we have used Stanford Parser [21]. It converts the text into tree format from which we can extract the required phrases using python programs. The following corpora are used for the cue-word extraction:

The cue words for health are extracted from HindEnCorp corpus [22] which contains the parallel corpus consisting of 200,000 words of English text (8.9k sentences) and IIT-Bombay annotated corpus [23] which consists of 15,589 sentences. The cue words for news, social, tech are extracted from HindEnCorp corpus.

The news domain cue words are extracted from the Tides and DanielPipes Commentary section which consist of 50k and 6.6k sentences, respectively. The text used for social domain consists of 39.8k sentences and for tech domain contains 66.7k sentences. The cue words for tourism are extracted from IIT-Bombay annotated corpus [23] which contains 16222 sentences and The TDIL Program and the Indian Language Corpora Initiative [24] which contains 25000 sentences.

After the filtering of cue words (threshold filtering of 6 words maximum), for the health domain we are left with around 42,000 phrases which are then sent as queries to Webhose and a web crawler which returns the pages as found in the news in text format and to the wikipedia Scraper.

Each of the pages are then checked for keyword match ratio and the ones not fulfilling in the criteria are discarded.

Then the pages are combined in two formats as mentioned in Sect. 2.5.

Then Language Models are made for each of the four formats described in Sect. 2 using the SRILM toolkit.

For the data on health domain, the prepared parallel corpus was from the corpus of the TDIL Program and the Indian Language Corpora Initiative. It was trained using Moses. It contains 25,000 sentences in the parallel corpora which are divided accordingly:

We then translate and evaluate the quality using the BLEU score and RIBES score to see which of the four models worked better for the test file which was of health domain. We also compared the scores for test file to a widely available existing Language Model, Gigaword LM [25]. It is trained using the newswire text provided in the English Gigaword corpus (1200M words of NYT, APW, AFE, XIE).

RIBES is an automatic evaluation metric for machine translation, developed in NTT Communication Science Labs. Its implementation is commonly distributed in Python.

Although, BLEU (biligual evaluation understudy) is the score, mostly used for MT evaluation in the last couple of years. Its measures the accuracy based on the n-grams match.

Table 4. Table showing BLEU and RIBES score for health-based test data

Language model type	BLEU	RIBES
BASE [12]	17.82	0.703778
PHASE 1	17.83	0.706083
BASE+WIKI_DUMP+WEB+WIKI	18.37	0.703172
BASE+WEB+WIKI	18.47	0.706040
GIGA LM	13.83	0.673181

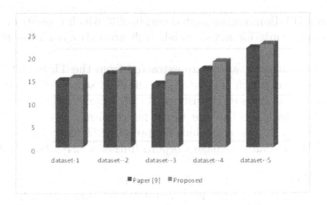

Fig. 3. Comparison analysis 1 [12]

Table 5. Snapshot of the IIT Bombay English-Hindi corpus.

	Language	Train	Dev	Test
#Types	hin	343,601	2,625	8,489
	eng	250,782	2,569	8,957
#Sentences	hin/eng	1,492,827	520	2,507
#Tokens	hin	22,171,543	10,174	63,853
	eng	20,667,259	10,656	57,803

In Table 4, we see that for a test file which is in health domain, the maximum BLEU score is obtained for the language model which is prepared using the base+selective webhose pages+selective wikipedia pages. It is higher than the one with base+selective webhose pages+selective wikipedia pages+ Entire Wikipedia Dump, even though the latter has more data. We find 3.65% increase in BLEU point if we use the language model which is prepared by the part-two of the phase-two architecture rather than the traditional language model. After considering the widely available Gigaword language model [27] we find 33.5% improvement in BLEU with the proposed approach.

The second part of the experiment comprises of choosing of language model; i.e. the understanding of which domain the test file belongs to. The cue words

Table 6. Statistics of HindEnCorp data-set

Set	#Sentences	#Tokens	
		En	Hi
Training	64724 × 2	458831	532985
Test	1002	7229	8519
Development	1001 × 2	7241	8271

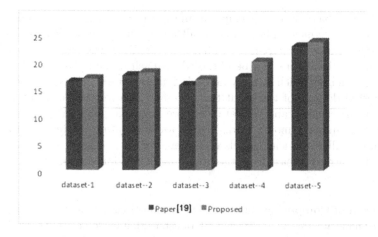

Fig. 4. Comparison analysis 2 [26]

come in handy again, though this time we use a translated version of the previously obtained cue words. We have directly translated the list file to source and kept them ready for use. Using the similar ratio formulation, we print the name of the list which shows the domain for which the ratio is maximum. We then decode our test file using that particular LM only.

Also, we have used the IIT Bombay English-Hindi corpus [28] and the HindEnCorp [22] for more experiment which is based on miscellaneous domain. The statistics of the IIT Bombay English-Hindi and HindEnCorp data-sets are shown in Table 5 and Table 6, respectively. We call these data-sets as data-set-1 and data-set-2, respectively. Moreover, we have collected some random sentences and prepare data-set-3, data-set-4, and data-set-5. The comparison analysis of the translated outputs between the proposed method and the phrase-based SMT [12] for these data-sets are shown Fig. 3. Similar comparison with hierarchical phrase-based SMT [26] is shown in Fig. 4. In every case, we incorporate the proposed methodology with that specific approach to prepare the language model. It shows a significant improvement in every cases using our method.

4   Conclusions and Future Works

In this paper, we have experimented the usage of domain-wise language models for Hindi to English translations. We identify the keywords and use them to incorporate the relative keywords. We have proposed a two-phased architecture to adapt a language model in such a way that the language model has a close proximity to different domains. By using this two-phased architecture, we showed a wide improvement in the health-domain related data. It is also proved that just because an LM contains a larger number of histories, does not mean that it will be performing a better translation as seen with the inclusion of wikipedia dump. The extraction of phrases as keywords not only give a lot of related data to add to the base, but also helps us finding the domain of test file if used in the source language. There was an although out increase in BLEU for the new adapted model of part-two of phase-two architecture. On the other hand, if enough parallel corpus can be obtained for each domain, we can also keep a language model set which is ready for each domain and use it for translation with the preferred domain of the input text. We can also test this concept on more languages and confirm and validate for more language pairs that the two-phase approach of language model works better for translation. In recent future, we will combine all statistical models to achieve better accuracy in the real time. This approach may release from the bounding of the domain specific translation problem.

Declaration of Competing Interest. The authors declare that they have no known competing nancial interests or personal relationships that could have appeared to influence the work reported in this paper.

References

1. Banerjee, P., Du, J., Li, B., Kumar Naskar, S., Way, A., van Genabith, J.: Combining multi-domain statistical machine translation models using automatic classifiers. In: AMTA 9th Conference of the Association for Machine Translation in the Americas, USA (2010)
2. Xiong, D., Zhang, M., Li, H.: Enhancing language models in statistical machine translation with backward n-grams and mutual information triggers. In: Proceedings of the 49th Annual Meeting of the Association for Computational Linguistics, pp. 1288–1297 (2011)
3. Sethy, A., Georgiou, P.G., Narayanan, S.S.: Building topic specific language models from webdata using competitive models (2005)
4. Brants, T., Popat, A.C., Xu, P., Och, F.J., Dean, J.: Large language models in machine translation. In: Proceedings of the Joint Conference on Empirical Methods in Natural Language Processing and Computational Language Learning (2007)
5. Gavrila, M., Vertan, C.: Training data in statistical machine translation - the more, the better? In: Proceedings of Recent Advances in Natural Language Processing Hissar, Bulgaria, 12–14 September 2011, pp. 551–556 (2011)
6. Luong, T., Kayser, M., Manning, C.D.: Deep neural language models for machine translation. In: Proceedings of the 19th Conference on Computational Natural Language Learning, CoNLL 2015, Beijing, China, 30–31 July 2015, pp. 305–309 (2015)
7. Lembersky, G., Ordan, N., Wintner, S.: Language models for machine translation: original vs. translated texts. In: Proceedings of the 2011 Conference on Empirical Methods in Natural Language Processing (2011)
8. Huerta, J.M.: An information-retrieval approach to language modeling: applications to social data. In: Proceedings of the NAACL HLT 2010 Workshop on Computational Linguistics in a World of Social Media, pp. 7–8 (2010)
9. Sorkey, A.J., Conrad, S.A.: Medical transcription with dynamic language models. US Patent 10,658,074, 19 May 2020
10. Delasalles, E., Lamprier, S., Denoyer, L.: Dynamic neural language models. In: Gedeon, T., Wong, K.W., Lee, M. (eds.) ICONIP 2019. LNCS, vol. 11955, pp. 282–294. Springer, Cham (2019). https://doi.org/10.1007/978-3-030-36718-3_24

11. Rosset, C., Xiong, C., Phan, M., Song, X., Bennett, P., Tiwary, S.: Knowledge-aware language model pretraining. arXiv preprint arXiv:2007.00655 (2020)
12. Koehn, P., Och, F.J., Marcu, D.: Statistical phrase-based translation. In: Proceedings of the 2003 Conference of the North American Chapter of the Association for Computational Linguistics on Human Language Technology, vol. 1. Association for Computational Linguistics, pp. 48–54 (2003)
13. Hiemstra, D.: A probabilistic justification for using tf × idf term weighting in information retrieval. Int. J. Digit. Libr. **3**(2), 131–139 (2000)
14. https://www.wikidata.org/wiki/wikidata:database_download (2020)
15. https://www.ling.upenn.edu/courses/fall_2003/ling001/penn_treebank_pos.html (2020)
16. De Marneffe, M.C., MacCartney, B., Manning, C.D., et al.: Generating typed dependency parses from phrase structure parses. In: Proceedings of LREC, Genoa Italy, vol. 6, pp. 449–454 (2006)
17. https://github.com/webhose/webhoseio-python (2020)
18. Koehn, P.: Europarl: a parallel corpus for statistical machine translation. In: Proceedings of MT Summit X, Phuket, Thailand, pp. 79–86 (2005)
19. https://dumps.wikimedia.org/ (2020)
20. https://anoopkunchukuttan.github.io/indic_nlp_library/ (2020)
21. Klein, D., Manning, C.D.: Accurate unlexicalized parsing (2003)
22. Bojar, O., et al.: Hindencorp-Hindi-English and Hindi-only corpus for machine translation. In: LREC, pp. 3550–3555 (2014)
23. Khapra, M.M., Kulkarni, A., Sohoney, S., Bhattacharyya, P.: All words domain adapted WSD: finding a middle ground between supervision and unsupervision. In: Conference of Association of Computational Linguistics (ACL 2010) (2010)
24. Jha, G.N.: The TDIL program and the Indian Language Corpora Initiative (ILCI). In: LREC (2010)
25. https://www.keithv.com/software/giga (2020)
26. Chiang, D.: Hierarchical phrase-based translation. Comput. Linguist. **33**(2), 201–228 (2007)
27. https://catalog.ldc.upenn.edu/ldc2003t05 (2020)
28. Kunchukuttan, A., Mehta, P., Bhattacharyya, P.: The IIT Bombay English-Hindi parallel corpus. arXiv preprint arXiv:1710.02855 (2017)

Composing Word Vectors for Japanese Compound Words Using Dependency Relations

Kanako Komiya[✉], Takumi Seitou, Minoru Sasaki, and Hiroyuki Shinnou

Ibaraki University, 4-12-1 Nakanarusawa, Hitachi-shi, Ibaraki 316-8511, Japan
{kanako.komiya.nlp,14t4037r,minoru.sasaki.01,
hiroyuki.shinnou.0828}@vc.ibaraki.ac.jp

Abstract. The use of distributed representations, e.g., via word2vec, has become popular in recent years. However, Japanese has many compound words and we often face the situation where meanings of a word and a compound word should be compared. Therefore, in the current study, we composed compound word vectors from those of constituent word vectors. We took into consideration the dependency relations of compound words to compose word vectors of them. The experiments revealed that, when we consider dependency relations, (1) we could obtain better representations for compound words when we separately learn models for each dependency relation, (2) each model could obtain good representations with fewer epochs, and (3) the learned weights for a model of compound words with one dependency relation could be used for fine-tuning for models for compound words of other dependency relations.

Keywords: Word vectors · Compound words · Word embeddings

1 Introduction

The use of distributed representations, e.g., via word2vec [6–8], has become popular in recent years. Distributed representations are vector representations of meanings, which are calculated on the basis of their contexts, and are used to investigate the similarity in meaning of two individual language units. There is much research on distributed representations of various language units, such as on a word, a phrase or a document level (see Sect. 2).

However, Japanese has many compound words and sometimes it makes it difficult to compare two words. For example, UniDic[1] [4], a dictionary developed by National Institute for Japanese Language and Linguistics, defined "いちご狩り, ichigo-gari, strawberry picking" as one word (short-term unit) and "ぶどう狩り, budou-gari, grape picking" as a compound word (long-term unit)[2]. In this

---

[1] http://unidic.ninjal.ac.jp/(In Japanese).
[2] いちご means strawberries, ぶどう means grapes, and 狩り means picking or hunting in Japanese.

© Springer Nature Switzerland AG 2023
A. Gelbukh (Ed.): CICLing 2019, LNCS 13451, pp. 280–292, 2023.
https://doi.org/10.1007/978-3-031-24337-0_20

case, a morphological analyzer using UniDic treats" いちご狩り , ichigo-gari, strawberry picking" as one word and " ぶどう狩り, budou-gari, grape picking" as two words, which makes it impossible to directly compare the word meanings of these two words via word vectors: the distributed representation of words.

In addition, word boundaries in Japanese are unspecific because Japanese does not have word delimiters between words. Therefore, Japanese dictionary individually defines words; Japanese has different definitions of words according to each dictionary or each tagger. That is why a word for a dictionary or a tagger could be a compound word for another dictionary or another tagger. Therefore, we believe that a method to compare meanings of a word and a compound word is necessary.

In the current study, we composed a compound word vector from those of constituent word vectors. We focused on UniDic that defined two language units, the short-term unit and the long-term unit, and composed word vectors of long-term units from word vectors of two short-term units. We took into consideration the dependency relations of compound words to compose word vectors of them. Specifically, we classified the compound words by the dependency relations and separately trained models for each dependency relation (see Sect. 3). We utilized 13 dependency relations (described in Sect. 4).

The experiments revealed that we could obtain better representations for compound words when we separately learn models for each dependency relation than when we learn a model for all the relations together. In addition, the experiments showed that each model could obtain good representations with fewer epochs when we took into consideration the dependency relations (see Sects. 5 and 6).

We examined other classifications of dependency relations and investigated the effectiveness of fine-tuning (described in Sect. 7). Finally, we conclude our work in Sect. 8.

2 Related Work

There has been much research on composing phrase representations from multiple word representations in recent years [1–3,9]. [1,9] used dependency relations. [9] showed that the model using different composition matrices for different dependency relations was the best. [2] proposed an implicit tensor factorization method for learning the word vectors of transitive verb phrases. [3] proposed a method for jointly learning compositional and noncompositional word vectors for phrases by adaptively weighting both types of word vectors using a compositionality scoring function.

At the same time, as described in Sect. 1, it is necessary to compose distributed representations of compound words. We believe that we can compose word vectors of compound words from those of constituent word vectors using the method to compose word vectors of phrases. Therefore, in the current study, we compose word vectors of Japanese compound words following a method of composing phrase vectors.

3   Composing Word Vectors of Japanese Compound Words Using Dependency Relation

We followed the method of [9], the method for composition of phrase vectors, to compose word vectors of compound words. We employed UniDic that defined two language units, the short-term unit and the long-term unit and composed word vectors for long-term units from two word vectors of short-term units. We only utilized the long-term units that consists of two short-term units although some long-term units consist of more than two short-term units, because we believe that these long-term units can be recursively composed using the same way to compose long-term units that consist of two short-term units.

[9] composed word vectors for phrases taking word vectors as inputs. They utilized the method proposed by [10,11], *Relfunc*, the method that can compose a vector depending on the relation r between two inputs. We used this model to compose word vectors for long-term units from two vectors of short-term units. A word vector of a long-term unit, l, can be composed from two word vectors of short-term units, s_1 and s_2, using the following model.

$$l = f(s_1, s_2, r) = \sigma \left(W_r \begin{bmatrix} s_{1r} \\ s_{2r} \\ b_r \end{bmatrix} \right) \tag{1}$$

where, $\sigma(.)$ denotes an element-wise sigmoid function, and $W_r \in \mathbb{R}^{d \times (2d+1)}$ and $b_r \in \mathbb{R}$ are parameters trained for each relation r. Therefore, the weights and biases of neural networks varies according to the dependency relations.

We employed the same loss function as [9]. It is the square errors between composed vectors and gold word vectors:

$$J(\theta) = \frac{1}{T} \sum_{i=1}^{N} \frac{1}{2} ||l_i - t_i||^2 + \frac{\lambda}{2} ||\theta||^2 \tag{2}$$

where, vector l_t denotes a long-term-unit vector composed by Eq. 1 and vector t_i denotes a gold word vector for a long-term unit. θ is all the other parameters.

We classified the compound words by the dependency relations and separately trained models for each dependency relation, following [9].

4   Compound Words and Dependency Relations

We defined the classes of dependency relations based on the structure of Japanese compound words. First, we extracted the long-term units that consists of two short-term units from 23,000 compound words in Balanced Corpus of Contemporary Japanese (BCCWJ) [5]. They were 3,500 examples and we manually classified them. A class of dependency relation has at least 30 examples in 3,500 compound words[3]. The definitions of the classes are as follows.

---

[3] The relations that had less than 30 examples were stuck in'others' class.

Table 1. Dependency relations and number of examples.

Dependency	1	2	3	4	5	6	7	8	9	10	11	12	13
Number of Data	1010	107	34	38	650	259	49	422	307	428	71	37	88

1. Combinations where the former short-term unit explains the latter short-term unit, c.f. " 講習会, kousyu- kai, lecture-class" when " 講習" means lectre and " 会" means class or meating.
2. Combinations of an object and a predicate, c.f. " 債務放棄, saimu-houki, debt-waiver" when " 債務" means debt and " 放棄" means waive, waiver or waiving,
3. Combinations of a complement and a predicate, c.f. " 法的整理, houteki-seiri, legal-liquidation (liquidating)" when " 法的" means legal and "整理 " means liquidating,
4. Combinations of a subject and a predicate, c.f. "画面割れ , gamen-ware, screen-cracking" when " 画面" means screen and " 割れ" means cracking,
5. Combinations includes a unit, c.f. "1 ドル , 1 doru, 1 dollar" when " ドル" means dollar,
6. Combinations of a main word and a suffix, c.f. "具体的 ", gutai-teki, concrete (combination of a word concrete (具体) and a suffix that makes an adjective verb (的)),"
7. Combinations of a prefix and a main word, c.f. " 副代表, fuku-daihyou, sub-delegate" when "副 " means sub and "代表 " means delegate,
8. Combinations that includes a particle, c.f. " ための , tame-no, for (combination of for(ため) and a particle "no(の)",)"
9. Combinations of a proper noun and a general noun, c.f. "茨城県 , Ibaraki-ken, Ibaraki-prefecture" when "茨城 " means Ibaraki (place name) and "県 " means prefecture.
10. Combinations of a noun and a verb. The combination makes a verb, c.f. "応募する , oubo-suru, apply (combination of do(する) and application(応募),)"
11. Numbers, c.f. "三二 , sanjyu-ni, 32 (combination of 3(三) and 2(二),)"
12. Combinations where each short-term unit has no meaning, c.f. "だが , daga, however," and
13. Others, c.f. a four-character idiom like "意気揚々 , iki-youyou, high-spirits." Here, "意気 " means spirits, "揚 " means upraise and " 々" means the same charcter as the last character, or repetition[4].

Table 1 shows the number of examples of long-term units by each dependency relation in manually-classified 3,500 examples.

We trained a model with manually-tagged examples using support vector machine (liblinear[5]). After that, SVM model classified the dependency relations of the other compound words in BCCWJ.

---

[4] " 揚々" is the same as "揚揚 .".
[5] https://www.csie.ntu.edu.tw/~cjlin/liblinear/.

5   Experiment

We generated word vectors of long-term units and short-term units as follows.

1. Separate long-term-unit words in Japanese in a corpus with spaces.
2. Separate short-term-unit words in Japanese in a corpus with spaces.
3. Add a file of long-term-unit words to a file of short-term-unit words.
4. Generate word vectors from a file that contains long-term-unit words and short-term-unit words.

We employed BCCWJ as a corpus. We used word2vec[6] to generate word vectors. We employed a skip-gram algorithm and the vector size was set to 100 to generate the word vectors. We used default settings for other parameters. We utilized only long-term units that consists of two short-term units. Word vectors of 169,736 long-term units and 339,472 short-term units, those of constituent word vectors, were obtained in this way.

We employed a feed-forward neural network to compose word vectors for long-term units. Here, inputs are two word vectors for short-term units and an output is a word vector for long-term unit. We set maximum epoch number to 2,000 and tried three types of unit number of a hidden layer: 100, 300, and 500. The initial weights of the network were randomly generated.

We carried out two-fold cross validation, specifically, we used half of word vectors we generated as the training data and used the rest of them as the test data. We evaluated the cosine similarities between a composed word vector and a word vector for a long-term unit directly generated by the word2vec program. Here, the composed vector is better when the cosine similarity is higher.

6   Results

6.1   Unit Number of a Hidden Layer

We tried three types of unit number of a hidden layer, 100, 300, and 500, to find the best settings for the neural network to compose the word vectors for compound words. We examined the best unit number using the network without considering dependency relations, namely, the model trained by all the word vectors. Table 2 shows the result of the experiment.

Table 2 shows that the best unit number for a hidden layer is 500. However, it takes time to train a model with 500 and the training data decrease when the models were separately learned. Therefore, we used 300 units for a hidden layer when we separately learn the models.

6.2   Effect of Dependency Relations

We compared the composed word vectors for compound words with and without taking into consideration dependency relations. Micro-averaged cosine

---

[6] https://radimrehurek.com/gensim/.

Table 2. Comparison of performances by unit numbers for a hidden layer.

Epochs	100	300	500
200	0.263	0.284	0.337
400	0.300	0.328	0.393
600	0.333	0.387	0.435
800	0.412	0.425	0.475
1,000	0.433	0.468	0.501
1,200	0.460	0.491	0.522
1,400	0.476	0.515	0.538
1,600	0.498	0.528	0.552
1,800	0.510	0.545	0.562
2,000	0.523	0.555	0.569

Table 3. Comparison of cosine similarities of models with and without taking into consideration dependency relations.

Epochs	− Dependency	+ Dependency
200	0.337	0.415
400	0.393	0.500
600	0.435	0.548
800	0.475	0.575
1,000	0.501	0.596
1,200	0.522	0.607
1,400	0.538	0.615
1,600	0.552	0.621
1,800	0.562	0.626
2,000	0.569	0.629

similarities over the whole dataset of word vectors were evaluated. Table 3 summarizes the cosine similarities between the composed word vectors and the gold word vectors for the compound words. In this table, we compared word vectors of compound words when the dependency relations were taken into consideration and those when the dependency relations were not taken into consideration. The cosine similarities were calculated from 200 epochs to 2000 epochs. Table 3 shows that cosine similarity is 0.060 higher when the dependency relations were taken into consideration at 2,000 epochs. In addition, the difference between cosine similarities tend to become higher when the numbers of epochs are smaller. These results indicate that we can obtain the better composed word vectors with fewer epochs when we separately learn models for each dependency relation than when we learn a model for all the relations together.

Table 4. Cosine similarities of each dependency relation.

Dependency	Cos-Sim	Epochs
1	0.608	2,000
2	0.656	1,600
3	0.610	600
4	0.668	800
5	0.617	2,000
6	0.630	2,000
7	0.619	2,000
8	0.620	400
9	0.619	2,000
10	0.722	2,000
11	0.666	600
12	0.470	200
13	0.632	2,000

6.3 Cosine Similarities of Each Dependency Relation

Table 4 shows the cosine similarities of each dependency relation. The table also shows the number of epochs when the best performances were obtained. This table shows that the dependency relation with the highest performance was *Dependency Relation 10*, the combination of a noun and a verb, and the dependency relation with the lowest performance was *Dependency Relation 12*, the combination where each short-term unit has no meaning. we compared the difference between the cosine similarity averaged over all the dependency relation at the best epoch and the cosine similarity averaged over all the dependency relation at 2,000 epochs. They were rarely different from each other (difference was only 0.001) because word vectors were the best at 2,000 epochs for most of dependency relations.

7 Discussion

7.1 Performances of the Models and Classification Accuracy of SVM

In the current study, the classification accuracy of SVM greatly affects the performances of the models to compose word vectors for compound words. Therefore, we evaluated the classification accuracy of SVM. We extracted 100 compound words of each dependency relation and manually checked if they were right or wrong. Table 5 shows the accuracy and most frequent errors of each dependency relation. **MFE** in the table stands for most frequent errors.

Table 5. Accuracy and most frequent errors of each dependency relation.

Dependency	MFE	Accuracy
1	6	84
2	1	42
3	1	8
4	1	16
5	6	99
6	13	91
7	1,6,13	94
8	1,6	85
9	1	91
10	1	99
11	1	86
12	1	86
13	1	84

Table 5 shows the classification accuracies of *Dependency Relation 3* and *Dependency Relation 4* are much lower than other classes of dependency relations. *Dependency Relation 3* is combinations of a complement and a predicate and *Dependency Relation 4* is combinations of a subject and a predicate. The most frequent error of these classification was misclassification to *Dependency Relation 1*, combinations where the former short unit explains the latter short unit, and they are similar each other. Therefore, we combined *Dependency Relation 3* and *Dependency Relation 4* to *Dependency Relation 1*, refer to it as *Dependency Relation 1'*, and learned the models again. Table 6 summarizes the results of the experiment. **Whole** and **Dependency 1'** in the table indicate the cosine similarities of the whole dataset and the class of *Dependency Relation 1'*, respectively.

When **+Dependency** in Table 3 and **Whole** in Table 6 are compared, the results of original setting, **+Dependency** in Table 3, was slightly better. We think this is because the cosine similarities for *Dependency Relation 3* and *Dependency Relation 4* were higher than *Dependency Relation 1*. The examples of *Dependency Relation 3* and *Dependency Relation 4* did not contribute the cosine similarities of *Dependency Relation 1* because their examples were much fewer than *Dependency Relation 1*. In addition, the classification accuracies of *Dependency Relation 3* and *Dependency Relation 4* were low because their examples were few and they misclassified to *Dependency Relation 1* because it was the class with the most examples.

Table 6. Experiment with Dependency 1'

Epochs	Whole	Dependency 1'
200	0.402	0.327
400	0.489	0.415
600	0.541	0.480
800	0.571	0.521
1,000	0.590	0.547
1,200	0.603	0.567
1,400	0.613	0.583
1,600	0.620	0.594
1,800	0.624	0.602
2,000	0.629	0.609

Table 7. Experiment with Dependency 13'

Epochs	Whole	Dependency 13'
200	0.415	0.497
400	0.500	0.570
600	0.548	0.599
800	0.575	0.616
1,000	0.596	0.622
1,200	0.608	0.627
1,400	0.616	0.630
1,600	0.622	0.622
1,800	0.627	0.632
2,000	0.630	0.630

7.2   Error Analysis

The class of dependency relation with the lowest performance was *Dependency Relation 12*, the combination where each short-term unit has no meaning. The properties of *Dependency Relation 12* are as follows.

1. It has fewer examples.
2. It contains particles and verbal auxiliaries, which are not usually contained in other classes, as short-term units that constitutes long-term units.

To address a problem caused from the first property, we combined *Dependency Relation 12* to *Dependency Relation 13*, refer to it as *Dependency Relation 13'*, and learned the models again. Table 7 summarizes the results of the experiment. **Whole** and **Dependency 13'** in the table indicate the cosine similarities of the whole dataset and the class of *Dependency Relation 13'*, respectively.

Table 8. Fine-tuning using weights of the model of *Dependency Relation 1*.

Epochs	Cos-Sim	Difference from Random Initial Values
200	0.491	+0.080
400	0.541	+0.041
600	0.573	+0.025
800	0.591	+0.017
1,000	0.606	+0.009
1,200	0.612	+0.005
1,400	0.620	+0.005
1,600	0.624	+0.002
1,800	0.628	+0.001
2,000	0.630	+0.001

When **+Dependency** in Table 3 and **Whole** in Table 7 are compared, the results of new setting, **Whole** in Table 7, was slightly better. However, the difference was very small because the number of examples of *Dependency Relation 12* was quite few.

We believe that the second property is the main reason of the low performance of the class of *Dependency Relation 12*. We conduct the composition assuming that the meanings of compound words could be composed from those of constituent words. Therefore, we believe that the composition mechanism does not work if the constituent words had no meanings.

7.3   Fine-Tuning of Models

Fine-tuning is a method to improve the performance of one task by re-learning using the weights of a model learned for another task as initial values. It is useful to speed up the learning of the target task and is effective when the data size of the target task is small. In the current study, we investigated two fine-tuning models, fine-tuning using the weights of the model of *Dependency Relation 1* and that using the weights of the model of *Dependency Relation 13*. We selected them because *Dependency Relation 1* had the most examples in all the relations and we believe that *Dependency Relation 13* is independent from the other classes because it contains "other" dependency relations.

Tables 8 and 9 shows that cosine similarities improved at all epochs when the models were fine-tuned. In addition, the cosine similarities are relatively high with fewer epochs. The performances of the models with fine-tuning converged at approximately 2,000 epochs, which is almost identical to the model without fine-tuning.

When Tables 8 and 9 are compared, fine-tuning with weights of the model of *Dependency Relation 13* is better than *Dependency Relation 1*. When Tables 10 and 11 are compared, fine-tuning with weights of the model of *Dependency Rela-*

Table 9. Fine-tuning using weights of the model of *Dependency Relation 13*.

Epochs	Cos-Sim	Difference from Random Initial Values
200	0.609	+0.193
400	0.623	+0.123
600	0.630	+0.082
800	0.635	+0.060
1,000	0.637	+0.041
1,200	0.639	+0.032
1,400	0.641	+0.025
1,600	0.642	+0.020
1,800	0.643	+0.016
2,000	0.643	+0.014

Table 10. Cosine similarities of each dependency relation with fine-tuning using weights of the model of *Dependency Relation 1*.

Dependency	Cos-Sim	Epochs	Difference from Table 4
1	0.608	2000	±0
2	0.659	600	+0.003
3	0.626	200	+0.061
4	0.674	400	+0.014
5	0.619	1,400	+0.002
6	0.637	2,000	+0.007
7	0.621	1,000	+0.002
8	0.646	200	+0.026
9	0.624	2,000	+0.005
10	0.727	2,000	+0.005
11	0.680	200	+0.014
12	0.506	200	+0.034
13	0.634	1,200	+0.002

tion 13 tends to obtain better results for the class of word vectors for compound words with dependency relations with more examples. We think that fine-tuning with weights of the model of *Dependency Relation 13* is better because it could improve the result for the class of word vectors for compound words with *Dependency Relation 1*, which is the class with the most examples.

Table 11. Cosine similarities of each dependency relation with fine-tuning using weights of the model of *Dependency Relation 13*.

Dependency	Cos-Sim	Epochs	Difference from Table 4
1	0.635	2,000	+0.026
2	0.661	600	+0.005
3	0.630	200	+0.019
4	0.678	400	+0.009
5	0.620	1,400	+0.003
6	0.638	2,000	+0.008
7	0.622	800	+0.003
8	0.650	200	+0.034
9	0.728	2,000	+0.007
10	0.682	2,000	+0.005
11	0.524	200	+0.017
12	0.632	200	+0.051

8   Conclusions

In this paper, we composed compound word vectors from those of constituent word vectors. We took into consideration the dependency relations of compound words and separately learn models for each dependency relation to compose word vectors of compound words. The experiments revealed that, when we took into consideration dependency relations, we could obtain better representations for compound words with fewer epochs, and the learned weights for a model of compound words with one dependency relation could be used for fine-tuning for models for compound words of other dependency relations. Error analysis of the class of the dependency relation with the lowest performance indicated that the composition mechanism does not work if the constituent words had no meanings because we conducted the composition assuming that the meanings of compound words could be composed from those of constituent words.

Acknowledgments. This work was partially supported by JSPS KAKENHI Grant Number 18K11421 and research grant of Woman Empowerment Support System of Ibaraki University.

References

1. Baroni, M., Zamparelli, R.: Nouns are vectors, adjectives are matrices: representing adjective-noun constructions in semantic space. In: Proceedings of EMNLP 2010, pp. 1183–1193 (2010)
2. Hashimoto, K., Tsuruoka, Y.: Learning embeddings for transitive verb disambiguation by implicit tensor factorization. In: Proceedings of the 3rd Workshop on Continuous Vector Space Models and their Compositionality, pp. 1–11 (2015)

3. Hashimoto, K., Tsuruoka, Y.: Adaptive joint learning of compositional and non-compositional phrase embeddings. In: Proceedings of the 54th Annual Meeting of the Association for Computational Linguistics (Volume 1: Long Papers), pp. 205–215. Association for Computational Linguistics (2016). https://arxiv.org/abs/1603.06067

4. Maekawa, K., et al.: Design, compilation, and preliminary analyses of balanced corpus of contemporary written Japanese. In: Proceedings of the Seventh International Conference on Language Resources and Evaluation (LREC 2010), pp. 1483–1486 (2010)

5. Maekawa, K., et al.: Balanced corpus of contemporary written Japanese. In: Language Resources and Evaluation, vol. 48, pp. 345–371 (2014)

6. Mikolov, T., Chen, K., Corrado, G., Dean, J.: Efficient estimation of word representations in vector space. In: Proceedings of ICLR Workshop 2013, pp. 1–12 (2013)

7. Mikolov, T., Sutskever, I., Chen, K., Corrado, G., Dean, J.: Distributed representations of words and phrases and their compositionality. In: Proceedings of NIPS 2013, pp. 1–9 (2013)

8. Mikolov, T., Yih, W., Zweig, G.: Linguistic regularities in continuous space word representations. In: Proceedings of NAACL 2013, pp. 746–751 (2013)

9. Muraoka, M., Shimaoka, S., Yamamoto, K., Watanabe, Y., Okazaki, N., Inui, K.: Finding the best model among representative compositional models. In: Proceedings of the 28th Pacific Asia Conference on Language, Information and Computation (PACLIC 2014), pp. 65–74 (2014)

10. Socher, R., Bauer, J., Manning, C.D., Ng, A.Y.: Parsing with compositional vector grammars. In: Proceedings of ACL 2013, pp. 455–465 (2013)

11. Socher, R., Karpathy, A., Le, Q.V., Manning, C.D., Ng, A.Y.: Grounded compositional semantics for finding and describing images with sentences. Trans. Assoc. Comput. Linguistics 2, 207–218 (2014)

Microtext Normalization for Chatbots

Ranjan Satapathy[1,2], Erik Cambria[2(✉)], and Nadia Magnenat Thalmann[1]

[1] Institute for Media Innovation, Nanyang Technological University, Singapore, Singapore
{ranjan.satapathy,nadia}@ntu.edu.sg
[2] School of Computer Science and Engineering, Nanyang Technological University,
Singapore, Singapore
cambria@ntu.edu.sg

Abstract. With the current upsurge in the usage of social media platforms, the trend of using short text, or microtext, in place of standard English has witnessed a significant rise. This work incorporates microtext normalization into a robot's chatbot. The work leverages the fact that humans tend to write in different unconstrained ways. This work also involves a binary classifier to detect microtext, which helps in reducing the execution time of the microtext normalization module. The results show an improvement in the chatbot's understanding and performance increase to most forms of unconstrained languages available on social media. The BLEU score is used to evaluate the efficiency before and after the normalization of sentences. Results show that the microtext normalization technique promises to increase unconstrained text understanding in a pre-trained chatbot.

Keywords: Microtext normalization · Chatbot · Dialogue systems

1 Introduction

Building a dialogue system which understands human language is not an easy task as humans interact socially in enormous different ways. Communicating using unconstrained natural language is an intuitive and flexible way for humans to interact. Understanding this kind of linguistic input is challenging for machines because of the diversity found in words and phrases used over different social media platforms. There are many abbreviations and non-standard words used in SMSs and tweets [17]. These type of communications are usually performed in real time and over platforms which impose limits on the length of the messages, as in the case of Twitter and the traditional SMS system. Due to these constraints, the writing format of these messages clearly differs from normal standards. Features such as word shortenings, contractions and abbreviations are commonly used both to gain writing speed and circumvent the length limitations.

In recent years, the rise and expansion of social media has enabled users to share their views and interests in an impromptu manner. For example, they write terms or sentences such as "c u 2morrow" (see you tomorrow), "tgif" (thank God it's Friday) and "abt" (about) which may not be found in standard English but are widely seen in

© Springer Nature Switzerland AG 2023
A. Gelbukh (Ed.): CICLing 2019, LNCS 13451, pp. 293–303, 2023.
https://doi.org/10.1007/978-3-031-24337-0_21

SMS, tweets, Facebook posts, blogs, discussion forums and chat logs. These unconstrained ways of writing text is called microtext. Microtext became one of the most widespread communication forms among users due to its casual writing style and colloquial tone [20].

The rise of social media usage has also led to the unconstrained generation of sentences in speech such as "wassup" (what is up), "howz" (how is) and interjections like ahem, aw, etc. which has emotions attached to them. Given that most data today is mined from the web, microtext analysis is key for many natural language processing (NLP) and data mining tasks, as most text classifiers are trained in plain English. In the context of sentiment analysis [4,6,11], microtext normalization is a necessary step for pre-processing text before polarity detection is performed.

The challenge arises when systems try to automatically rectify and replace them with the standard words [21,26]. Microtext normalization could be thought of as a simple find-and-replace pre-processing [15] step. For instance, a sampling of Twitter studied in [21] found over 4 million OOV words where new spellings were created constantly, both voluntarily and accidentally.

- **Input Text** : Wassup Nadine

  - **chatbot's actual answer** : I could not find an answer to that.

  - **Expected chatbot's answer** : I'm doing good. How about you

- **Input Text** : Howz you doing

  - **chatbot's actual answer** : I could not find an answer to that.

  - **Expected chatbot's answer** : I am doing good. How about you?

- **Input Text** : Talk to you later

  - **chatbot's actual answer** : Talk to you later

  - **Expected chatbot's answer** : Talk to you later

We integrate a microtext normalization module in robot's chatbot. The microtext normalization module consists of a binary classifier and a microtext lexicon. Binary classifier classifies a sentence as microtext or non-microtext. A microtext classified sentence is sent to lexicon to transform OOV to IV sentence. The Sentence BLEU[1] is used to score the similarity between normalized sentences output from the proposed framework and human annotated sentences. The proposed work is a step towards curbing the gap between the humans and chatbot by leveraging on a microtext lexicon to transform

---

[1] https://www.nltk.org/_modules/nltk/translate/bleu_score.html.

out-of-vocabulary (OOV) words to their in-vocabulary (IV) or human readable counterparts. The rest of the paper is structured as follows: Sect. 2 explains the related work, Sect. 3 explains the proposed framework, Sect. 3.1 explains the Datasets used, Sect. 4 explains the results and discussions and finally the Sect. 5 explains the conclusion and future work.

2 Related Work

Opinions and its associated concepts such as sentiments, emotions, attitudes, and evaluations are the center of study of sentiment analysis [32,35]. This section discusses through the related work in microtext normalization and dialogue systems.

2.1 Microtext Analysis

Microtext has become ubiquitous in today's communication. This is partly a consequence of Zipf's law, or principle of least effort (for which people tend to minimize energy cost at both individual and collective levels when communicating with one another), and it poses new challenges for NLP tools which are usually designed for well-written text [13]. Normalization is the task of transforming unconventional words to their respective standard counterpart.

In [19], authors present a novel unsupervised method to translate Chinese abbreviations. It automatically extracts the relation between a full-form phrase and its abbreviation from monolingual corpora, and induces translation entries for the abbreviation by using its full-form as a bridge. [12] uses a classifier to detect OOV words, and generates correction candidates based on morphophonemic similarity. The types and features of microtext are reliant on the nature of the technological support that makes them possible. This means that microtext will vary as new communication technologies emerge. In our related work, we categorized normalization into three well-known NLP tasks, namely: spelling correction, SMT, and automatic speech recognition (ASR).

Spelling Correction. Correction is executed on a word-per-word basis seen as a spelling checking task. This model gained extensive attention in the past and a diversity of correction practices have been endorsed by [3,8,18,24,33]. Instead, [31] and [9] proposed a categorization of abbreviation, stylistic variation, prefix-clipping, which was then used to estimate their probability of occurrence. Thus far, the spelling corrector became widely popular in the context of SMS, where [7] advanced the hidden Markov model whose topology takes into account both "graphemic" variants (e.g., typos, omissions of repeated letters, etc.) and "phonemic" variants (e.g., spellings that resemble the word's pronunciation).

Statistical Machine Translation. Statistical Machine Translation (SMT) outlooks microtext as a foreigner language that has to be translated to plain English, meaning that normalization is done through a SMT task. When compared to the previous task, this method appears to be rather straightforward and better since it has the possibility to model (context-dependent) one-to-many relationships which were out-of-reach previously [16]. Some examples of works include [1, 14, 25]. However, the SMT still overlooks some features of the task, particularly the fact that lexical creativity verified in social media messages is barely captured in a stationary sentence board.

Automatic Speech Recognition. ASR considers that microtext tends to be a closer approximation of the word's phonemic representation rather than its standard spelling. As follows, the key of microtext normalization becomes very similar to speech recognition which consists of decoding a word sequence in a (weighted) phonetic framework. [16] proposed to handle normalization based on the observation that text messages present a lot of phonetic spellings, while more recently [15] proposed an algorithm to determine the probable pronunciation of English words based on their spelling. Although the computation of a phonemic representation of the message is extremely valuable, it does not solve entirely all the microtext normalization challenges (e.g., acronyms and misspellings do not resemble their respective IV words' phonemic representation). Authors in [2] have merged the advantages of SMT and the spelling corrector model.

Table 1. Sample Lexicon incorporated in Nadine's system

OOV word	Class	Polarity	IV word
a3	OTHER	NEUTRAL	Anytime, any place, anywhere
ru/18	OTHER	NEUTRAL	Are you over 18?
AAF	ACR	POSITIVE	As A Friend
bestie	OTHER	POSITIVE	Best friend
ne1	PHON	NEUTRAL	Anyone
urz	PHONETIC	NEUTRAL	Yours
b3	OTHER	NEGATIVE	Blah, blah, blah
aight	CLP	NEUTRAL	all right

2.2 Dialogue System

Authors in [37] built an open-domain end-to-end human-computer conversational agent to integrate a large commonsense knowledge base into end-to-end conversational models. [30] investigated the limitations of building a Generative Hierarchical Neural Network Models based dialogue system and show how it outperforms state-of-the-art neural language models. Emotion detection in conversations [22] is a necessary step for a number of applications, including opinion mining over chat history, social media threads, debates, argumentation mining, understanding consumer feedback in live conversations, etc. Currently systems do not treat the parties in the conversation individually by adapting to the speaker of each utterance. There are social media based chatbot [10, 36] which do not take microtexts into account. So, our main motive is to include the microtexts to train the system, so that the chatbot learns the intrinsic linguistic patterns and generate a response accordingly.

3   Proposed Framework for Chatbot

Composite nature of the NLP problem is addressed by the CLSA model [5]. In this regard, microtext module is the first step. The syntactics layer aims at preprocessing text so that informal text is reduced to human readable format (any language), inflected forms of verbs and nouns are normalized, and basic sentence structure is made explicit. In order to prepare and build the lexicon, we crawled popular acronyms from NetLingo[2], MakeUseOf[3], Slangs[4], and Internet Slang[5].

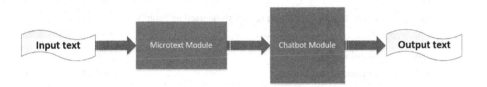

Fig. 1. Proposed framework for Chatbot

The proposed model incorporates microtext understanding in the chatbot. It helps the chatbot to understand the unconstrained languages as shown in Table 2. The framework shown in Fig. 2 has a binary classifier which classifies a text into OOV or IV, based on the learned features. The classifier employs a n-gram model with several machine learning techniques as shown in Table 3a and Table 3b.

Table 1 shows the sample lexicon which helps social robot's NLP module understand the social media language. The proposed framework is shown in Fig. 1. The text is passed through microtext module for normalization and then passed on to Nadine's NLP module.

3.1   Datasets

This section discusses the datasets used in the evaluation of the binary classifier. If a sentence contains at least one microtext token, the sentence will be classified as

---

[2] Reproduced by Permission©1995–2018 NetLingo®The Internet Dictionary at http://www.netlingo.com.

[3] http://makeuseof.com/tag/30-trendy-internet-acronyms.

[4] http://acronymsandslang.com/.

[5] http://internetslang.com/.

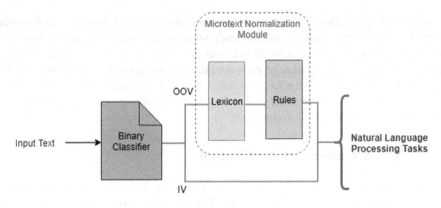

Fig. 2. Proposed framework

Table 2. Examples of unconstrained language with emotions associated with it

Microtext	Meaning	Polarity
aah	Fright	NEGATIVE
aha	Understanding, triumph (can also be used as "ahh")	POSITIVE
duh	Expresses annoyance over something stupid or obvious	NEGATIVE
haha	Regular laughter	POSITIVE
wow	Impressed, astonished	POSITIVE

microtext. The two classes in both the datasets are equally distributed. Table 3b and Table 3a shows the accuracy of machine learning algorithms on different datasets.

Table 3. Evaluation results on different datasets

(a) 10-fold Accuracy on NUS SMS dataset

Classifier	10-fold (%)
NuSVC	85.14
Linear SVC	*92.95*
Original Naive Bayes	89.62
Multinomial Naive Bayes	89.92
Bernoulli Naive Bayes	89.24
Logistic Regression	91.05
SGDC	91.42

(b) 10-fold Accuracy on Normalized tweets dataset

Classifier	10-fold (%)
NuSVC	84.2
Linear SVC	*87.4*
Original Naive Bayes	83.5
Multinomial Naive Bayes	81.8
Bernoulli Naive Bayes	82.7
Logistic Regression	84.9
SGDC	83.4

NUS SMS Corpus. This corpus (Table 4) has been created from the NUS English SMS corpus[6], the authors [34] randomly selected 2,000 messages. The messages were first normalized into standard English and then translated into standard Chinese. For our evaluation purposes, we only used the actual messages and their normalized English version (leaving out their Chinese counterparts).

Table 4. Sample real time tweets/SMS

Social media texts	Expanded forms
I'll meet u b4 lec then	I will meet you before the lecture then.
Where r u	Where are you
Hey are we going out tmr	Hey are we going out tomorrow
So u stayin in d hostel ?	So you are staying in the hostel ?
R u going to b done anytime soon ?	Are you going to be done anytime soon ?

Normalized Tweet Dataset. Authors in [27], built a lexicon which consists of real time tweets and their IV counterparts. The dataset is available on request.

4 Results and Discussion

This section discusses the results. The evaluation of the framework is done at time complexity and sentence similarity (BLEU Score).

4.1 Dataset Collection and Annotation

The dataset in [27] was available on request. The dataset consists of tweets crawled from Twitter streaming API[7]. The data was preprocessed using following rules:

1. removal of usernames (starting with),
2. urls (eg., https://www.Twitter.com),
3. Removal of punctuation marks,

4.2 Time Complexity

The results in Table 3a and Table 3b shows different algorithms applied on both the datasets. The models are trained on the unigram features as microtexts work at the word-level [27]. The result shows **Linear SVC** to be the best binary classifier for both the datasets. The binary classifier reduces the time complexity and makes the overall framework run faster. The framework ran on Python 3.5 on Ubuntu operating system with 64 GB RAM and 30 GB 1080 T_i Nvidia Graphics. It took 11.4 s to run without the binary classifier and only 8.8 s with the binary classifier. The binary classifier works as a filter, which reduces the overall execution time of the framework.

---

[6] http://github.com/kite1988/nus-sms-corpus.

[7] https://developer.twitter.com/en/docs.

4.3 BLEU Score

BLEU score [23] is employed as an evaluation task. It is used to evaluate the quality of text which has been machine-translated from one natural language to another. It's strength is that it correlates highly with human judgements by averaging out individual sentence judgment errors. Figure 3a and Fig. 3b shows the BLEU score for the normalized Tweet and NUS SMS data respectively. The results show **Mean BLEU score** of more than **0.8** is achieved for both the dataset. The model's output is compared against the human annotated text as provided in the datasets.

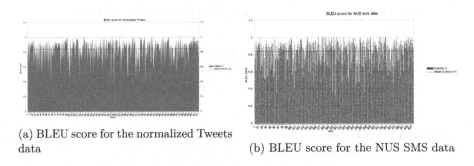

(a) BLEU score for the normalized Tweets data

(b) BLEU score for the NUS SMS data

Fig. 3. Evaluation of datasets based on BLEU score

5 Conclusion and Future Work

The proposed framework consists of a binary classifier that classifies a given sentence into either microtext or non-microtext. If a sentence contains at least one microtext token, the sentence will be classified as microtext. The binary classifier takes syntactic features to determine a class label. In the experiments performed, Linear SVC gives an accuracy of 87.4% on Normalized Tweet Dataset and 92.95% on NUS SMS data. The addition of the binary classifier also improves the overall execution time of the task. The detected microtexts are then passed through the lexicon. Lexicon transforms the out-of-vocabulary texts to their in-vocabulary counterparts. BLEU score was taken as an evaluation metric and showed a mean of more than 0.8 in both datasets.

Future work will focus on experimenting whether lexicons could be replaced by a more cognitive approach which is a phonetic system (e.g., International Phonetic Alphabet) [29] and integrating more sophisticated models [28] with the robot without hampering the robot's response time. It will improve the generalization of the proposed rules based on more cognitive qualities of speech such as phones, phonemes, intonation and separation of words and syllables. Integrating the microtext with dialogue systems and chatbots could help understand the unconstrained style.

Acknowledgment. This research is supported by the BeingTogether Centre, a collaboration between Nanyang Technological University (NTU) Singapore and University of North Carolina (UNC) at Chapel Hill. The BeingTogether Centre is supported by the National Research Foundation, Prime Minister's Office, Singapore under its International Research Centres in Singapore Funding Initiative.

References

1. Aw, A., Zhang, M., Xiao, J., Su, J.: A phrase-based statistical model for SMS text normalization. In: 21st International Conference on Computational Linguistics and 44th Annual Meeting of the Association for Computational Linguistics, pp. 33–40 (2006)
2. Beaufort, R., Roekhaut, S., Cougnon, L.A.l., Fairon, C.D.: A hybrid rule/model-based finite-state framework for normalizing SMS messages. In: ACL, pp. 770–779. Association for Computational Linguistics (2010)
3. Brill, E., Moore, R.C.: An improved error model for noisy channel spelling correction. In: Proceedings of the 38th Annual Meeting on Association for Computational Linguistics, pp. 286–293 (2000)
4. Cambria, E., Hussain, A.: Sentic album: content-, concept-, and context-based online personal photo management system. Cogn. Comput. **4**(4), 477–496 (2012)
5. Cambria, E., Poria, S., Bisio, F., Bajpai, R., Chaturvedi, I.: The CLSA model: a novel framework for concept-level sentiment analysis. In: Gelbukh, A. (ed.) CICLing 2015. LNCS, vol. 9042, pp. 3–22. Springer, Cham (2015). https://doi.org/10.1007/978-3-319-18117-2_1
6. Cambria, E., Song, Y., Wang, H., Howard, N.: Semantic multi-dimensional scaling for open-domain sentiment analysis. IEEE Intell. Syst. **29**(2), 44–51 (2014)
7. Choudhury, M., Saraf, R., Jain, V., Sarkar, S., Basu, A.: Investigation and modeling of the structure of texting language. Int. J. Doc. Anal. Recogn. **10**(3–4), 157–174 (2007)
8. Church, K.W., Gale, W.A.: Probability scoring for spelling correction. Stat. Comput. **1**(2), 93–103 (1991)
9. Cook, P., Stevenson, S.: An unsupervised model for text message normalization. In: Proceedings of the Workshop on Computational Approaches to Linguistic Creativity, pp. 71–78 (2009)
10. Cui, L., Huang, S., Wei, F., Tan, C., Duan, C., Zhou, M.: SuperAgent: a customer service chatbot for e-commerce websites. Proceedings of ACL 2017, System Demonstrations, pp. 97–102 (2017)
11. Grassi, M., Cambria, E., Hussain, A., Piazza, F.: Sentic web: a new paradigm for managing social media affective information. Cogn. Comput. **3**(3), 480–489 (2011)
12. Han, B., Baldwin, T.: Lexical normalisation of short text messages: Makn sens a# twitter. In: Proceedings of the 49th Annual Meeting of the Association for Computational Linguistics: Human Language Technologies, vol. 1, pp. 368–378 (2011)
13. Hutto, C.J., Gilbert, E.: VADER: a parsimonious rule-based model for sentiment analysis of social media text. In: Eighth international AAAI Conference on Weblogs and Social Media, pp. 216–225 (2014)
14. Kaufmann, M., Kalita, J.: Syntactic normalization of Twitter messages. In: International conference on natural language processing, Kharagpur, India (2010)
15. Khoury, R.: Microtext normalization using probably-phonetically-similar word discovery. In: 2015 IEEE 11th International Conference on Wireless and Mobile Computing, Networking and Communications (WiMob), pp. 392–399 (2015)
16. Kobus, C., Yvon, F., Damnati, G.: Normalizing SMS: are two metaphors better than one? In: Proceedings of the 22nd International Conference on Computational Linguistics, vol. 1, pp. 441–448. Association for Computational Linguistics (2008)
17. Li, C., Liu, Y.: Normalization of text messages using character-and phone-based machine translation approaches. In: Thirteenth Annual Conference of the International Speech Communication Association, pp. 2330–2333 (2012)

18. Li, M., Zhang, Y., Zhu, M., Zhou, M.: Exploring distributional similarity based models for query spelling correction. In: Proceedings of the 21st International Conference on Computational Linguistics and the 44th Annual Meeting of the Association for Computational Linguistics, pp. 1025–1032. ACL-44, Association for Computational Linguistics (2006)
19. Li, Z., Yarowsky, D.: Unsupervised translation induction for Chinese abbreviations using monolingual corpora. In: Proceedings of ACL-08: HLT, pp. 425–433. Association for Computational Linguistics (2008)
20. Liu, F., Weng, F., Jiang, X.: A broad-coverage normalization system for social media language. In: Proceedings of the 50th Annual Meeting of the Association for Computational Linguistics: Long Papers, vol. 1, pp. 1035–1044. Association for Computational Linguistics (2012)
21. Liu, F., Weng, F., Wang, B., Liu, Y.: Insertion, deletion, or substitution? Normalizing text messages without pre-categorization nor supervision. ACL-HLT 2011 - Proceedings of the 49th Annual Meeting of the Association for Computational Linguistics: Human Language Technologies, vol. 2, pp. 71–76 (2011)
22. Ma, Y., Nguyen, K.L., Xing, F., Cambria, E.: A survey on empathetic dialogue systems. Inf. Fusion **64**, 50–70 (2020)
23. Papineni, K., Roukos, S., Ward, T., Zhu, W.J.: BLEU: a method for automatic evaluation of machine translation. In: Proceedings of the 40th annual meeting on association for computational linguistics, pp. 311–318. Association for Computational Linguistics (2002)
24. Pennell, D., Liu, Y.: A character-level machine translation approach for normalization of SMS abbreviations. In: Proceedings of 5th International Joint Conference on Natural Language Processing, pp. 974–982 (2011)
25. Pennell, D.L., Liu, Y.: Normalization of informal text. Comput. Speech Lang. **28**(1), 256–277 (2014)
26. Petrović, S., Osborne, M., Lavrenko, V.: The Edinburgh Twitter corpus. In: Proceedings of the NAACL HLT Workshop on Computational Linguistics in a World of Social Media, pp. 25–26 (2010)
27. Satapathy, R., Guerreiro, C., Chaturvedi, I., Cambria, E.: Phonetic-based microtext normalization for Twitter sentiment analysis. In: 2017 IEEE International Conference on Data Mining Workshops (ICDMW), pp. 407–413. IEEE (2017)
28. Satapathy, R., Li, Y., Cavallari, S., Cambria, E.: Seq2seq deep learning models for microtext normalization. In: 2019 inTernational Joint Conference on Neural Networks (IJCNN), pp. 1–8. IEEE (2019)
29. Satapathy, Ranjan, Singh, Aalind, Cambria, Erik: PhonSenticNet: a cognitive approach to microtext normalization for concept-level sentiment analysis. In: Tagarelli, Andrea, Tong, Hanghang (eds.) CSoNet 2019. LNCS, vol. 11917, pp. 177–188. Springer, Cham (2019). https://doi.org/10.1007/978-3-030-34980-6_20
30. Serban, I.V., Sordoni, A., Bengio, Y., Courville, A., Pineau, J.: Building end-to-end dialogue systems using generative hierarchical neural network models. In: Proceedings of the Thirtieth AAAI Conference on Artificial Intelligence, pp. 3776–3783. AAAI Press (2016)
31. Sproat, R., Black, A.W., Chen, S., Kumar, S., Ostendorf, M., Richards, C.: Normalization of non-standard words. Comput. Speech Lang. **15**(3), 287–333 (2001)
32. Susanto, Y., Livingstone, A., Ng, B.C., Cambria, E.: The hourglass model revisited. IEEE Intell. Syst. **35**(5), 96–102 (2020)
33. Toutanova, K., Moore, R.C.: Pronunciation modeling for improved spelling correction. In: Proceedings of the 40th Annual Meeting on Association for Computational Linguistics, pp. 144–151. Association for Computational Linguistics (2002)
34. Wang, P., Ng, H.T.: A beam-search decoder for normalization of social media text with application to machine translation. In: NAACL, pp. 471–481 (2013)

35. Wang, Z., Ho, S., Cambria, E.: A review of emotion sensing: categorization models and algorithms. Multimed. Tools Appl. **79**, 35553–35582 (2020)
36. Xu, A., Liu, Z., Guo, Y., Sinha, V., Akkiraju, R.: A new chatbot for customer service on social media. In: Proceedings of the 2017 CHI Conference on Human Factors in Computing Systems, pp. 3506–3510. ACM (2017)
37. Young, T., Xing, F., Pandelea, V., Ni, J., Cambria, E.: Fusing task-oriented and open-domain dialogues in conversational agents. In: Proceedings of AAAI, pp. 11622–11629 (2022)

Building Personalized Language Models Through Language Model Interpolation

Milton King[(✉)] and Paul Cook[(✉)]

University of New Brunswick, Fredericton, Canada
{milton.king,paul.cook}@unb.ca

Abstract. Social media users differ in how they write such as writing style and topics. This suggests that personalized language models—language models tailored to a specific person—could outperform a single generic language model. One challenge, however, is that language models typically require a large volume of text to train on, but for many people such a volume of text is not available. In this paper, we train n-gram and neural language models on relatively large in-domain background corpora, and on relatively small amounts of text from individual social media users, specifically authors of blogs. In experiments with interpolated language models, we find that, although user-specific language models trained on a small amount of text from a user perform relatively poorly, they can be interpolated with language models trained on a large background corpus to give improvements over either approach on its own. We further find that n-gram and neural language models are complementary, and can be interpolated to give improvements over either approach used individually. Our evaluation considers perplexity, and two evaluation measures motivated by next word suggestion on smart-phones. We find that although perplexity is widely used for intrinsic evaluation of language models, it is a poor indicator of performance in terms of these other measures.

Keywords: Language models · User-specific language model · Personalization

1 Introduction

The difference in the writing styles of people on social media platforms, such as blogs, suggests that language models tailored towards individuals can outperform generic language models. However, language models often require a large amount of text to train on, which is often not available for many people. Therefore, we interpolate language models that we trained on a large in-domain corpus with language models that are trained on far less text from a single person by averaging the probabilities from each model. Our evaluation considers perplexity, and two other evaluation measures motivated by next word suggestion on smart-phones. We find that although perplexity is widely used for intrinsic evaluation

© Springer Nature Switzerland AG 2023
A. Gelbukh (Ed.): CICLing 2019, LNCS 13451, pp. 304–315, 2023.
https://doi.org/10.1007/978-3-031-24337-0_22

of language models, it is a poor indicator of performance in terms of these other measures. We show that even though the language model that was trained on a single user performs fairly poorly on its own, it can increase the performance of the generic language model that was trained on a large background corpus. We look at two types of language models, a long short-term memory neural network (LSTM) and an n-gram language model which uses Kneser-Ney smoothing [3] and found that using an LSTM as a background model outperformed all other models by themselves. However, we achieve better performance when interpolating an LSTM-based background model with a user-level model that was trained on only 1000 tokens for two of our three evaluation metrics. Furthermore, we show that the type of model used as the background model largely affects the performance of the interpolated models. Nevertheless, we find that both types of background language models achieve better results when interpolated with a user-level language model. We also look at the effects that the volume of text from a user has on the language models' performance.

2   Related Work

Personalizing language models can be viewed as a domain adaptation problem, which often involves models that train on a large out-of-domain corpus and then adapted to a specific domain. Such approaches for language modeling often involve a recurrent neural network at its base with a method to perform the adaptation such as [11], who used a topic distribution vector fed into the hidden and output layers of their network. Another common way to achieve domain adaptation is to initially train on a background corpus and then continue training on text from the target domain [8]. A model can also leverage metadata about the author such as age, gender, and personality, to assist in personalized NLP tasks [9]. [7] used metadata about the author as well, along with two LSTM models—with one model being used to model an author and another model being used to model how the author's text changes given a specific addressee. Although metadata can be helpful, it is often not available and therefore we only consider text from authors in our following experiments.

[5] achieved domain adaptation, with the domains being categories of YouTube videos, by combining multiple LSTMs, where each LSTM represents a specific domain and are then combined with a "mixer" LSTM that determines how much weight to give to each domain-specific LSTM. Similarly, we combine the output of multiple language models with the difference that we average our outputs across all language models instead of using an LSTM to combine them and are using text from individual users. Language models are also commonly used in an extrinsic evaluation such as text classification. This includes domain-adapted language models in sentiment analysis, question classification, and topic classification [4], or determining if a tweet is relevant to a natural disaster [2].

There have been many different domains that have been approached with domain adaptation techniques such as Youtube speech recognition [5] and newspaper sections [6]. Our work differs from many of the previous works because our background corpus and test set are within the same domain and therefore, we are adapting to the user's writing style and not the domain itself. This is a more subtle change, where the only thing that differs is the authors themselves.

3   Data and Evaluation

3.1   Dataset

The dataset that we use contains blogs and was used in [13], which contains $19,320$ users. We use the sentence splitter from [10] to generate one sentence per line and prepended each sentence with a start-of-sentence token and append each sentence with an end-of-sentence token. We hold out the 10 users with the most tokens and divide their text into *USER*—a set of 10 user-specific corpora, which will be used to train our user-level models—and *TEST*, which will be our test set. The partition is approximately a 25/75 split. *TEST* consists of a $38,219$ sentences, containing a total of $765,614$ tokens with the number of tokens from a single author ranging from $52,633$ to $112,049$. We hold out the next 20 users with the most tokens for future analysis. We take a maximum of $30,000$ tokens (not including start and end-of-sentence tokens) from each of the other $19,290$ user to form our background corpus *BACKGROUND*. The $30,000$ threshold is implemented to avoid biases and promotes a more evenly distributed corpus across multiple users and is based on the token count for each user. We further modify *BACKGROUND* by replacing all words that occur less that 10 times with the unknown token *unk*. This gives us a background corpus of $6,668,281$ sentences containing $129,549,606$ tokens (including start and end-of-sentence tokens) with a vocabulary of $92,578$ types. The number of sentences from any single user ranges from 1 to 4024 and number of tokens from a user ranging from 112 to 38053, including start and end-of-sentence tokens. All corpora have numerals replaced with a *num* token and were casefolded and tokenized by Stanford Core NLP [10].

3.2   Evaluation

One of the most common language models used in language modelling is an LSTM due to its often superior performance [14] and therefore the two types of language models that we look at include an LSTM implemented in Pytorch[1] and an n-gram model that uses Kneser-Ney smoothing known as Kenlm [3]. We trained each type of language model exclusively on either a single user's corpus from *USER* or on the collection of blog posts that we call *BACKGROUND* and will be denoted as *language_model-corpus* from hereon. For example, an LSTM trained on *USER* would be *LSTM-USER*. The language models are trained across sentence boundaries. The language models are tested on *TEST*, with the models trained on *USER* only testing on their corresponding test sentences from the same user.

---

[1] Implementation is based off the code from https://github.com/yunjey/pytorch-tut orial/blob/master/tutorials/02-intermediate/language_model/main.py#L30-L50.

3.3   Evaluation Metrics

There are a few common metrics that are used to evaluate the performance of a language model and some may be preferred over others depending on the target application for the language model. We use three different evaluation metrics in order to achieve a better understanding of how our models compare. In this section, we discuss the evaluation metrics that we use to evaluate our models.

Adjusted Perplexity. Perplexity is defined below in Eq. 1

$$\text{perp} = -\frac{1}{N} \sum_{i=1}^{N} \log(p(w_i)) \tag{1}$$

with N being the size of the vocabulary. It is one of the most common intrinsic evaluation metrics used to evaluate language models that are trained on the same corpus, but does not give a fair comparison of language models that have different vocabulary. This is due to the fact that language models can artificially inflate their score by decreasing their vocabulary size and predict most words as *unk*. Adjusted perplexity was proposed by [15], which penalizes a language model that predicts *unk* and therefore allows us to fairly compare models that were trained on different corpora. The overall perplexity calculation does not change, but the probability of *unk* is calculated using Eq. 2, defined below as

$$p(\text{unk}) = \frac{p(\text{unk})}{|\text{UNK-TYPES}|} \tag{2}$$

with $UNK - TYPES$ being the set of types that are converted to unk in the test file. During our evaluation, the start-of-sentence type is never predicted as the target word but the LSTMs generate probabilities for all words in its vocabulary, including the start-of-sentence type and therefore we remove the probability given to it before applying our softmax function making the size of our output layer $v - 1$, where v is the size of the vocabulary. The *n-gram* models do not give a probability for the start-of-the-sentence type as a default. All models are still given the start-of-sentence token as the starting token. The models are expected to predict the end-of-sentence token for this metric.

Accuracy at k. Accuracy at k is a metric that is closely associated with the down stream application of next word prediction on smart-phones where many soft keyboards provide three suggestions for the next word. Accuracy at k captures the number of times that a model predicts the target word in the top k words of its vocabulary. We look at k in a range from 1 to 5. Unlike adjusted perplexity, we do not evaluate a model on their ability to predict the end-of-sentence token. This is to reflect the use of word suggestion on smart-phones where the end-of-sentence is not useful to the user. The probabilities for *unk* and end-of-sentence are set to 0 because these will never be the target word.

Accuracy at k Given c Keystrokes. We take the accuracy at k one step closer to how many smart-phones perform next word prediction by allowing the language model to look at the first c characters of the target word. This is similar to the task of query completion on the character level, which was approached by [6] and [12]. This evaluation metric simulates that the user types the first c characters of the target word. We set k to 3 because this is common for most smart-phones and look at c on a range from 0 to 3. If c is equal to or larger than the target word then we say the model predicted the word with a probability of 1 for that c. Similar to accuracy at k, we do not evaluate the model's ability to predict the end-of-sentence token. Again, the probabilities for *unk* and end-of-sentence are set to 0 because we never expect them to be the target word in a real-world environment, such as on a smart-phone.

4   Experimental Results

In this section we first tune our LSTM models that are trained on *user* text. We then evaluate our models using our three evaluation metrics. We look at the models' performances by themselves—allowing *user* models to train on different volumes of text from a single user—followed by interpolations involving two models. The *user* models that we use for interpolations are trained on 1000 tokens. We then discuss the performance of interpolations involving three and four models.

4.1   Tuning User-Trained Language Models

We use text from five random users from the *BACKGROUND* corpus to tune our parameters for our LSTM that is trained on a user via grid search. Each model is only given 1000 tokens from a single user. For example, we will have five models with the same parameters, but each one will be trained on text from a different user. The parameters for the LSTMs that we tune are number_of_layers $(1, 2)$; number_of_hidden_units $(128, 256, 512, 1024, 2048)$; embeddings_size $(64, 128, 512, 1024)$; number_training_epochs $(1, 2, 3)$; batch_size $(1, 2, 5, 15, 30, 454)$. Our final user-level LSTM models are single layer with 256 hidden units, an embedding size of 256, trained using a batch size of 2 for 1 epoch.

Our LSTM that was trained on *BACKGROUND* uses the default parameters of an embeddings size of 128, 1024 hidden units, 1 layer, a batch size of 45, and 1 training epoch. They were trained using a cross entropy loss function. We applied the same final parameters that we use for *LSTM-user*, but preliminary experiments showed that the default values achieved better results. Our n-gram model uses the trigram implementation of the KenLM model.

4.2   Impact of Volume of Data on User-Level Language Models

In this section, we observe the effects of allowing the language models to have access to a larger amount of training data from a user. We randomly select sets containing approximately 1, 000; 10, 000; 100, 000; and 200, 000 tokens from each user from *USER*. We train each type of model on each of these sets—with each model only training on text from one user—and evaluate them on *TEST*.

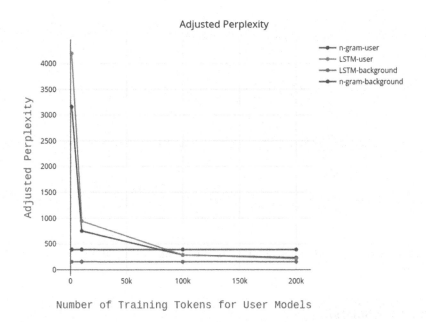

Fig. 1. Adjusted perplexity for different amounts of training text (lower is better).

We show the adjusted perplexity for each language model and the effects that the amount of training text has on our *user* models in Fig. 1. The amount of training text for *LSTM-background* and *n-gram-background* does not change. We see that both *n-gram-user* and *LSTM-user* perform relatively poorly when given less than 100*k* tokens to train on but quickly achieve a better score than *n-gram-background* around 100*k* tokens and approaches *LSTM-background* at 200*k* tokens. The fact that the models achieve similar score with far less training text shows the importance of training on text from the user and not a generic corpus even when generic text is within domain.

Next, we compare the accuracy at k for the four different models. Again, *LSTM-background* achieves the highest score. We also see, that the *user* models outperform *n-gram-background* with only 10*k* tokens instead of the roughly 100*k* tokens needed for adjusted perplexity. The findings for accuracy at k show that *user* models do not compete with large models trained on in-domain background corpora, which is not in line with the findings when evaluating with adjusted perplexity.

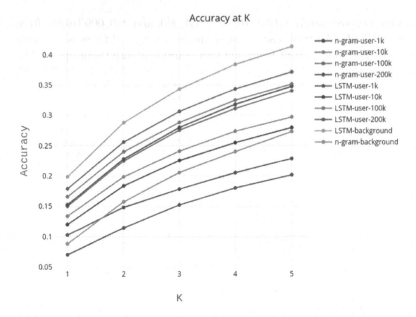

Fig. 2. Accuracy at k for different amounts of training text with k as the x-axis (higher is better).

4.3   Interpolation

In this section, we show the performance of our interpolated models using the different evaluation metrics. Since we are interested in users with a small amount of text, our *user* models that are involved with interpolations are trained on only 1000 tokens from a single user. As discussed in Sect. 3.2, we interpolate our models by averaging their probabilities.

We first look at the effects that two-way interpolation has on adjusted perplexity shown in Fig. 3. We see that any interpolation involving a *background* model outperforms *n-gram-background* without interpolation. This result is interesting since both *user* models performed relatively poorly but are still able to assist the *n-gram-background* model. Furthermore, each *background* model improves its performance when interpolated with either a *user* model or another *background* model of a different type, with *LSTM-background* and *n-gram-background* achieving the best score. This supports the findings of [1], who found that neural language models and n-gram language models are complimentary. Although, *LSTM-user* interpolated *n-gram-user* does not outperform *n-gram-user* by itself.

Next, we look at interpolating models with respect to accuracy at k shown in Fig. 4. For this metric, language models generate probabilities for each word in its vocabulary, but we only look at the top 50 probabilities—to reduce the computational cost. If a word that is present in the top 50 of one model but not in another, then the probability is set to 0 for the model that does not

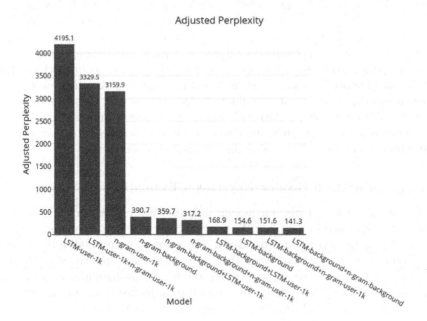

Fig. 3. Evaluating interpolated models using adjusted perplexity (lower is better).

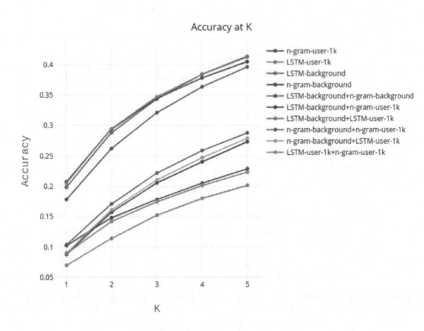

Fig. 4. Evaluating accuracy at k for interpolated models (higher is better).

contain the word and the interpolation is performed the same as before. Here we see that any interpolation that involves a *background* model outperforms either model by itself, which contradicts the findings from adjusted perplexity in Fig. 3. Furthermore, *LSTM-background* interpolated with *n-gram-background* achieves the best performance in terms of adjusted perplexity, but is outperformed by *LSTM-background* by itself or when it is interpolated with either *user* model when using the accuracy at k evaluation. We achieve our highest scores with *LSTM-background* interpolated with either *user* model. Interestingly, the two *user* models are not complimentary when evaluating with accuracy at k, which supports the findings when evaluating with adjusted perplexity.

4.4   Accuracy at 3 Given c Keystrokes Evaluation

In this section, we evaluate our models using accuracy at 3 given c keystrokes, which was explained in Sect. 3.3. We selected 3 because it is common for smartphones to suggest up to 3 words. Given c keystrokes, we normalize the model's probabilities across all word types that begin with the c keystrokes. Similar to accuracy at k, during testing we average the top 50 probabilities for any given c keystrokes with c being on a range from 0 to 3, inclusive. Again, we will first look at the models by themselves and then the interpolations.

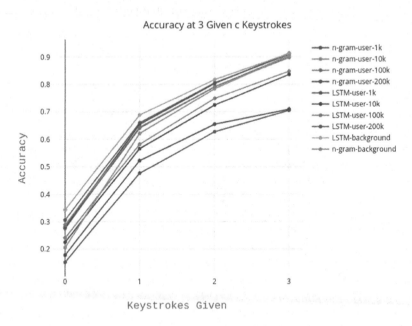

Fig. 5. Evaluating models using accuracy at 3 given c keystrokes (higher is better).

Figure 5 shows accuracy at 3 given c for models without interpolation while allowing *user* models to train on different amounts of text from a single user.

We see similar findings to accuracy at k in Fig. 2, where neither model by itself beats *LSTM-background* and either *user* model trained on at least 10k tokens outperforms *n-gram-background*.

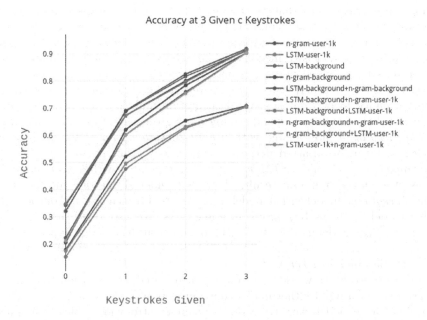

Fig. 6. Evaluating interpolated models using accuracy at 3 given c keystrokes (higher is better).

Next, we evaluate our interpolated models with respect to accuracy at 3 given c keystrokes shown in Fig. 6. It shows that the only model to outperform *LSTM-background* by itself is *LSTM-background* interpolated with *n-gram-background* for $c \geq 1$. This finding is different than the findings shown when evaluating with accuracy at k where *LSTM-background* interpolated with either *user* model outperforms *LSTM-background* interpolated with *n-gram-background*.

Three-Way and Four-Way Interpolations. In an attempt to better our results using our models that were trained on 1000 tokens from a user, we perform interpolations involving three and four models. We consider all possible three-way interpolations. However, none of the three-way or four-way interpolations outperform our previously best models for all our metrics, which were *LSTM-background* interpolated with *n-gram-background* for both adjusted perplexity and accuracy at 3 given c keystrokes, and *LSTM-background* interpolated with either *user* model for accuracy at k.

5   Conclusions

In this work, we looked at interpolating language models trained on an in-domain background corpus with language models trained on a small amount of text from a single user. We showed that we can outperform a large background model by interpolating it with a language model that was trained on as little as 1000 tokens from a user in terms of adjusted perplexity and accuracy at k. However, interpolating two models trained on two large background corpora achieved our best results for adjusted perplexity and accuracy at 3 given c keystrokes. We showed that adjusted perplexity does not reflect the performance of a language model's ability to perform next-word suggestion and does not give the same findings as both accuracy at k and accuracy at k given c keystrokes. For example, *LSTM-background* interpolated with *n-gram-background* outperforms all other models in terms of adjusted perplexity, but is outperformed by both *LSTM-background* by itself and *LSTM-background* interpolated with a *user* model in terms of accuracy at k. Our results show that under all metrics, interpolating either *background* model with a model trained on text from a user will outperform either model by themselves except in a few cases. We also found that the type of the language model used for the background corpus has a large impact on the final results, with models including *LSTM-background* outperforming models that include *n-gram-background*.

In future work, we would like to apply a weighted interpolation that determines how much input either model contributes to a final prediction. Also, our neural language models were trained using a cross entropy loss which is designed for a model to favour a lower perplexity, but we would like to train the neural language models with a loss function that takes into account the ranking—making it more suitable for accuracy at k and accuracy at k given c keystrokes evaluation metrics.

References

1. Adams, O., Makarucha, A., Neubig, G., Bird, S., Cohn, T.: Cross-lingual word embeddings for low-resource language modeling. In: Proceedings of the 15th Conference of the European Chapter of the Association for Computational Linguistics: Volume 1, Long Papers, pp. 937–947. Association for Computational Linguistics (2017). https://aclweb.org/anthology/E17-1088
2. Alam, F., Joty, S., Imran, M.: Domain adaptation with adversarial training and graph embeddings. In: Proceedings of the 56th Annual Meeting of the Association for Computational Linguistics (Volume 1: Long Papers), Melbourne, Australia, pp. 1077–1087 (2018). https://aclweb.org/anthology/P18-1099
3. Heafield, K., Pouzyrevsky, I., Clark, J.H., Koehn, P.: Scalable modified Kneser-Ney language model estimation. In: Proceedings of the 51st Annual Meeting of the Association for Computational Linguistics, Sofia, Bulgaria, pp. 690–696 (2013). https://kheafield.com/papers/edinburgh/estimate_paper.pdf
4. Howard, J., Ruder, S.: Universal language model fine-tuning for text classification. In: Proceedings of the 56th Annual Meeting of the Association for Computational

Linguistics (Volume 1: Long Papers), Melbourne, Australia, pp. 328–339 (2018). https://aclweb.org/anthology/P18-1031

5. Irie, K., Kumar, S., Nirschl, M., Liao, H.: RADMM: recurrent adaptive mixture model with applications to domain robust language modeling. 2018 IEEE International Conference on Acoustics, Speech and Signal Processing (ICASSP), pp. 6079–6083 (2018)

6. Jaech, A., Ostendorf, M.: Personalized language model for query auto-completion. In: Proceedings of the 56th Annual Meeting of the Association for Computational Linguistics (Volume 2: Short Papers), Melbourne, Australia, pp. 700–705 (2018). https://aclweb.org/anthology/P18-2111

7. Li, J., Galley, M., Brockett, C., Spithourakis, G., Gao, J., Dolan, B.: A persona-based neural conversation model. In: Proceedings of the 54th Annual Meeting of the Association for Computational Linguistics (Volume 1: Long Papers), pp. 994–1003. Association for Computational Linguistics (2016). https://doi.org/10.18653/v1/P16-1094, https://aclweb.org/anthology/P16-1094

8. Lin, Z., Sung, T., Lee, H., Lee, L.: Personalized word representations carrying personalized semantics learned from social network posts. In: 2017 IEEE Automatic Speech Recognition and Understanding Workshop (ASRU), pp. 533–540 (2017)

9. Lynn, V., Son, Y., Kulkarni, V., Balasubramanian, N., Schwartz, H.A.: Human centered NLP with user-factor adaptation. In: Proceedings of the 2017 Conference on Empirical Methods in Natural Language Processing, Copenhagen, Denmark, pp. 1157–1166 (2017). https://www.aclweb.org/anthology/D17-1120

10. Manning, C., Surdeanu, M., Bauer, J., Finkel, J., Bethard, S., McClosky, D.: The stanford CoreNLP natural language processing toolkit. In: Proceedings of the 52nd Annual Meeting of the Association for Computational Linguistics: System Demonstrations, Baltimore, USA, pp. 55–60 (2014)

11. Mikolov, T., Zweig, G.: Context dependent recurrent neural network language model. In: 2012 IEEE Spoken Language Technology Workshop (SLT), pp. 234–239. IEEE (2012)

12. Park, D.H., Chiba, R.: A neural language model for query auto-completion. In: Proceedings of the 40th International ACM SIGIR Conference on Research and Development in Information Retrieval, SIGIR 2017, pp. 1189–1192. ACM, New York (2017). https://doi.org/10.1145/3077136.3080758, https://doi.acm.org/10.1145/3077136.3080758

13. Schler, J., Koppel, M., Argamon, S., Pennebaker, J.W.: Effects of age and gender on blogging. In: AAAI Spring Symposium: Computational Approaches to Analyzing Weblogs, vol. 6, pp. 199–205 (2006)

14. Sundermeyer, M., Schlüter, R., Ney, H.: LSTM neural networks for language modeling. In: INTERSPEECH 2012, pp. 194–197 (2012)

15. Ueberla, J.P.: Analyzing and improving statistical language models for speech recognition. CoRR abs/cmp-lg/9406027 (1994)

dpUGC: Learn Differentially Private Representation for User Generated Contents (Best Paper Award, Third Place, Shared)

Xuan-Son Vu[1]($\boxtimes$)(iD), Son N. Tran[2](iD), and Lili Jiang[1]($\boxtimes$)(iD)

[1] Department of Computing Science, Umeå University, Umeå, Sweden
{sonvx,lili.jiang}@cs.umu.se
[2] ICT Discipline, University of Tasmania, Hobart, Australia
sn.tran@utas.edu.au

Abstract. This paper firstly proposes a simple yet efficient generalized approach to apply differential privacy to text representation (i.e., word embedding). Based on it, we propose a user-level approach to learn personalized differentially private word embedding model on user generated contents (UGC). To our best knowledge, this is the first work of learning user-level differentially private word embedding model from text for sharing. The proposed approaches protect the privacy of the individual from re-identification, especially provide better trade-off of privacy and data utility on UGC data for sharing. The experimental results show that the trained embedding models are applicable for the classic text analysis tasks (e.g., regression). Moreover, the proposed approaches of learning differentially private embedding models are both framework- and data-independent, which facilitates the deployment and sharing. The source code will be available when the paper is published.

Keywords: Private word embedding · Differential privacy · UGC

1 Introduction

Word embedding, also known as word representation, represents a word as a vector capturing both syntactic and semantic information, so that the words with similar meanings should have similar vectors [15]. This representation has two important advantages: efficient representation due to dimensionality reduction, and semantic contextual similarity due to a more expressive representation. Thanks for these advantages, word embedding is widely used to learn text representation for text analysis tasks. Some commonly used word embedding models include Word2Vec [18], GloVe [22], and FastText [5] and successfully applied in a variety of tasks like parsing [2], topic modeling [3], and document classification [25]. Training word embedding model on big data requires high performance computing resources. For example, Word2Vec model was learned

© Springer Nature Switzerland AG 2023
A. Gelbukh (Ed.): CICLing 2019, LNCS 13451, pp. 316–331, 2023.
https://doi.org/10.1007/978-3-031-24337-0_23

on 100 billion words from Google News corpus, and the FastText model of Facebook was learned from 840 billion words. Thus, once an efficient word embedding model was trained, it is most likely to be widely shared among researchers and communities. However, since word embedding models preserve pretty much semantic relations between words, the shared pre-trained models may lead to privacy breaches especially when they were trained from UGC data such as tweets and Facebook posts. For instance, user *first name* (e.g., "John"), *last name* ("Smith") and *disease* (e.g., "prostatitis") may be represented as similar vectors in word embedding model. Even user real name is absent from the pre-trained models, other available information such as *username, address, city name, occupation*, could be represented with similar vectors, with/without auxiliary data, leading to re-identification risk to discover the individual to which the data belongs to, by using some approaches like author identification [19], age and gender prediction [9]. Even further, the latent privacy breaches may cause a follow-up security issue. Figure 1 shows a prank on Facebook to get other users' passwords. In case this type of information is learned and embedded in the embedding model, there exists a possible risk that one can exploit *user* as a query to the shared embedding model and get their *password*. One might argue that the sensitive information likes *user, password* should not be leaked out and should have been removed from the embedding model. However, the purpose of learning from sensitive data is to learn the model without privacy leakage for facilitating research on sensitive data. To protect privacy, we statistically guarantee the chance to re-identify individuals by using output from the pre-trained

Fig. 1. A prank causes user credentials leak in FB data

models. Thanks to that, further research on the sensitive data **at large scale** can be possible such as "what is the common patterns between users when they configure their passwords?" (to analyze security risks) or "what diseases are normally unspeakable but get shared online?" (to analyze user behaviours on social networks). Similarly, this approach can be applied to user-level medical text data, which is very sensitive, to make research on medical data possible. Figure 2 shows our approach to learn data distribution from private UGC data to facilitate studies on down-stream tasks.

As discussed above, it is critical to protecting privacy when learning embedding model for UGC data sharing. To address the challenge of revealing information about an individual in the training data, Dwork et al. [6,7] proposed differential privacy technique which provides a strong guarantee of privacy, and soon became a well known standard in privacy preservation. However, differential privacy is a general mechanism and how to apply to different data type is a non-trivial problem. Some previous work applying differential privacy on text data [16,24,30] was either dependent on pre-defined sensitive features or applicable on federated framework instead of centralized data. The main difference is that they applied differential privacy to prevent the text data from personal data breaches, while this paper applies differential privacy to learn a shareable word embedding model from text data. Therefore, more challenges on privacy-budget control and data utility preservation have to be addressed.

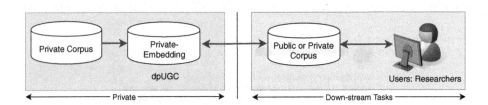

Fig. 2. Overview of our safe-to-share embedding model that can be used to facilitate research on sensitive data with privacy-guarantee.

1.1   Goal of the Paper

The goal of this paper is to develop effective and efficient approaches to apply differential privacy on text data to learn differentially private word embedding models. The ultimate purpose is to share the trained word embedding model, which prevents the highly latent risk of privacy breaches in word embedding models learning from UGC data and meanwhile maintains reasonable data utility. The main contributions of this paper are:

- We propose a simple yet efficient generalized approach of applying differential privacy on text data to learn embedding model for UGC data sharing.
- We apply user-level privacy-guarantee on above differentially private word embedding model to maintain better data utility.

– We conduct extensive experiments to evaluate the effectiveness of our proposed approach to preserve data utility, especially we test the approaches on text analysis task (i.e., regression).

The rest of this paper is organized as follows. Subsection 1.2 presents related work. Section 2 shows some preliminaries of differential privacy and word embedding. Section 3 presents the proposed approaches to learn differentially private word embedding. Experiments settings and evaluation results are discussed in Sect. 4 and 5. Section 6 concludes the paper followed by future work.

1.2 Previous Work

Anonymization [4] and sanitization [28] have been widely used in privacy protection. **Differential Privacy** later emerged as the key privacy guarantee by providing rigorous, statistical guarantees against any inference from an adversary [6]. Differential privacy has been applied in many research on different types of data including images [1,10,21,29], network [20], text [16,24,30], and general neural network architectures [23]. There is a family of algorithms called Private Aggregation of Teacher Ensembles (PATE), which becomes popular and contributes to research on differential privacy for machine learning. The advantage of PATE is to achieve private learning by coordinating the activity and sharing weights between different learning models [21]. Two limitations of PATE include: 1) it lacks flexibility and modularity when integrating to other frameworks, and 2) PATE was only trained on image data, which is not applicable to text data. Due to the different representation formats, differential privacy on text data reserves more difficulties. To apply differential privacy on text data, [16,24,30] transformed the problem of differential privacy on text to language modeling problem, which aims to protect privacy for next word suggestion task on user devices. They are different from what we are addressing in this paper since their federated models require both clients and servers. Whereas our differentially private word embedding model is learned from centralized data, and further used for sensitive data sharing.

2 Preliminaries

2.1 Differential Privacy

To address the challenge of revealing information about an individual in the training data, **differential privacy** [6,7,13,14] essentially hides any individual by ensuring that the resulting model is nearly indistinguishable from the one without that individual. Differential privacy provides a strong guarantee of privacy even when the adversary has arbitrary external knowledge. The basic idea is to add enough noise to the outcome (e.g., the model resulting from training) to hide the contribution of any single individual to that outcome. Let D be a collection of data records, and one record corresponds to an individual. A mechanism $\mathcal{M} : D \rightarrow \mathbb{R}^d$ is a randomized function mapping database D to a probability distribution over some range. $\mathcal{M}$ is said to be differentially private if adding or removing a single data record in D only affects the probability of any outcome within a small multiplicative factor. The formal definition of (ϵ, δ) differential privacy is:

Definition 1 *[(ϵ-δ)-differential privacy].* A randomized mechanism $\mathcal{M}$ is (ϵ, δ)-differential privacy where $\epsilon \geq 0, \delta \geq 0$, if for all data records in D and D' differing on at most one record, and $\forall S \subseteq \text{Range}(\mathcal{M})$:

$$\Pr[\mathcal{M}(D) \in S] \leq e^{\epsilon} \times \Pr[\mathcal{M}(D') \in S] + \delta$$

The values of (ϵ, δ) here are called **privacy-budget**. They control the level of the privacy, i.e., smaller values of (ϵ, δ) guarantee better privacy but lower data utility.

Privacy-Budget: There are typically two types of privacy-budget: (1) global privacy-budget [17], and (2) personalized privacy-budget [8]. The main difference is that, the global budget is counted to all users while the personalized budget is counted based on different users. Therefore, personalized privacy-budget is a better way to control data utilities and privacy due to the fine-grained privacy budget.

As introduced above, D and D' are adjacent datasets differing on at most one record, McMahan et al. [16] introduced a *user-level differential privacy*, where D and D' are adjacent datasets differing on at most one user's all records. This definition will be used to form our *Personalized DP-Embedding* algorithm in Sect. 3.

2.2 Word Embedding

Word embedding is one of the most popular representations of document vocabulary. Simply speaking, they are vector representations of particular words. It is capable of capturing the context of a word in a document, semantic similarity, and relation with other words. Word2Vec [18] developed by Google, is one of the most popular technique to learn word embeddings using a shallow neural network. Specifically, they propose a neural network architecture (i.e., the skip-gram model) that consists of an input layer, a projection layer, and an output layer to predict nearby words. Given a sequence of words $w_1, \ldots, w_T$ in a corpus, each word vector is trained to maximize the following log probability of neighboring words:

$$\frac{1}{T} \Sigma_{t=1}^{T} \Sigma_{j \in nb(t)} \log p(w_j | w_t)$$

where $nb(t)$ is the set of neighbouring words of word w_t and $p(w_j | w_t)$ is the normalized exponential probability (i.e., hierarchical softmax) of the associated word vectors $\vec{w_j}$ and $\vec{w_t}$.

3 Methodologies: Differentially Private Word Embedding

This section describes our proposed approaches toward differentially private word embedding on text data. We start by introducing a generalized approach to learn differentially private word embedding (Subsect. 3.1). After that, we reformulate the word embedding learning problem to user-level for personalized differentially private embedding (Subsect. 3.2).

3.1   Differentially Private (DP-) Embedding

Differently from most of the previous studies, which apply differential privacy to image data, we implement algorithms for learning differentially private word embedding on text data. Compared with image data (represented by pixel positions, sizes of geometric forms, shapes etc.), text data captures more semantic and ambiguity, which makes it harder to preserve both privacy-guarantee and data utilities. Before introducing our approach, we formulate the problem of learning word embedding on text data as follows.

Word Embedding Learning: Given a document corpus $D = \{d_1, \ldots, d_n\}$, each document $d \in D$ contains a sequence of words $\{w_1, \ldots, w_{|m|}\}$ from a fixed dictionary V. We use distributed representation and map every word in V to a k-dimensional vector. The goal is to learn an embedding function f that outputs a fixed length k embedding for every $w \in V$. k is called embedding size and typically from 50 to 300 dimensions.

As explained in Sect. 2.1, the basic idea of differential privacy is to inject noise to a model in order to make more difficulty of predictableness, thus more difficult for hackers to predict the actual inputs. However, direct injection of noise to a trained embedding model will deteriorate the model's quality. Technically, it would be better to insert noise during the learning process as we can optimize both the performance of the model and its privacy by treating the noise as a constraint. In the case of word2vec model, we can learn a differentially private embedding matrix W by using a differential privacy optimizer such as DP-SGD (differentially private stochastic gradient descent) [1]. In particular, we apply noise to "gradient" during the training. As shown in Algorithm 1-a, at each training step, a single training lot L is used (line 3). Each training lot might have several minibatch B. But to make it simple, we consider $L = B$ in this case. A lot L is a random set of training samples in D with a predefined lot size $|L|$. Afterwards, we compute gradients (line 4–5) which then will be added with noise (line 6–7) and applied to the standard gradient descent method (line 8–9). At line 10, a privacy accountant is used to accumulate privacy spending during the training process to ensure privacy guarantee. In the following, we will give more details regarding loss function and privacy accountant.

Loss Function: *Cross entropy loss* is used as our loss function, which measures the probabilistic distance between the predicted probabilities p and the true binary labels y. In our case, using one-hot encoding, the true label y_i is 1 only when w_i is the output word; y_i is 0 otherwise. The loss function $\mathcal{L}_\theta$ of the model with parameter config θ aims to minimize the cross entropy between the prediction and the ground truth, as lower cross entropy indicates high similarity between two distributions.

$$\mathcal{L}_\theta = -\sum_{i=1}^{V} y_i \log p(w_i|w_I) = -\log p(w_O|w_I) \tag{1}$$

In the skip-gram model, the embedding matrix W and output matrix W' are a collection of input vectors and context vectors, respectively. Given one word w_i, its embedding vector $\vec{w}_i$ is one row of W. Correspondingly, its context (output) vector $\vec{w}_i'$ is a column of the output matrix W'. The final output layer applies softmax to compute the probability of predicting the output word w_O given w_I, and therefore:

$$p(w_O|w_I) = \frac{\exp(\vec{w_O}^\top \vec{w_I})}{\sum_{i=1}^{V} \exp(\vec{w_O}^\top \vec{w_I})}$$

Apply above to Eq. (1), we have new loss function:

$$\mathcal{L}_\theta = -\log \frac{\exp(\vec{w_O}^\top \vec{w_I})}{\sum_{i=1}^{V} \exp(\vec{w_i'}^\top \vec{w_I})} = -\vec{w_O}^\top \vec{w_I} + \log \sum_{i=1}^{V} \exp(\vec{w_i'}^\top \vec{w_I})$$

In above loss function, the complexity of computing $\nabla \log p(w_O|w_I)$ is proportional to V, which is often large (10^5 to 10^7 terms). In what follows, we will reduce the training cost by using *Negative Sampling (**NEG**)* [18], which was employed to train Google word2vec model. *NEG* focuses on learning high-quality word embedding rather than modeling the word distribution in natural language. *NEG* loss approximates the binary classifier's output with sigmoid functions. Given an input word w_I, the correct output word is known as w. In the meantime, we sample M other words from the noise sample distribution Q, denoted as $\tilde{w}_1, \tilde{w}_2, \ldots, \tilde{w}_M \sim Q$. We label the decision of the binary classifier as d, which can only take a binary value ($d = 1$ for positive samples, $d = 0$ for negative samples). Thus, the final *NEG* loss function looks like:

$$\mathcal{L}_\theta = -[\log p(d = 1|w, w_I) + \sum_{i=1}^{M} \mathbb{E}_{\tilde{w}_i \sim Q} \log p(d = 0|\tilde{w}_i, w_I)] \tag{2}$$

Differentially Private Word Embedding Model Training: We employ DP-SGD [1] to train the model using back-propagation. At each step of the differentially private SGD (see Algorithm 1-a), we compute the gradient $\nabla_\theta f(\theta, x_i)$ for a random minibatch B (we consider $L = B$ in this case). Then we clip the 2-norm of each gradient belonging to the minibatch and compute their average. In the final step, noise is added in order to protect privacy before taking a gradient descent step using this "noisy" gradient. For dataset D, mechanism $\mathcal{M}$ (explained in Sect. 2.1) is then given by:

$$\mathcal{M}(D) = \Sigma_{i \in B} \tilde{\nabla}(f(x_i)) + \mathcal{N}(0, C^2\sigma^2 \mathbf{I})$$

where $\tilde{\nabla}(f(x_i))$ denotes the gradients clipped with a constant $C > 0$. The clipping is important since it helps to control gradient exploding and vanishing problem [11].

Privacy-Accountant: Privacy-accountant keeps track of the privacy spendings through the whole training procedure. It is an important part of differentially private SGD. We applied an "accountant" procedure that computes the privacy spendings at each access to the training data and the accountant accumulates the cost at each step.

Thoughts: Our experimental studies proved that the proposed approach above is an efficient way to learn differentially private word embedding to guarantee privacy. In this approach, the privacy-budget (ϵ, δ) is either a hyperparameter (i.e., must set before training) or must be accumulated after each training epoch. The privacy budget, therefore, is dependent on the number of training epochs, as it introduces noise into "gradients" of parameters in every training step [23]. It is observed that when there is a small privacy budget, only a small number of epochs can be used to train the model [1]. While when the number of training epochs needs to be large to guarantee the model accuracy, the above approach may potentially sacrifice a portion of model utility. This observation motivates us to further improve the proposed approach above.

One intuitive solution is to inject noise differently to each part of the training data (e.g., add more noise into features which are less relevant to the model output, and vice-versa [23]). However, in reality, it is not always easy to reason what words in UGC are more significant/sensitive than others. For instance, political opinion in Asia countries is a sensitive topic and even forbidden by law in some countries (e.g., China). Conversely, they are less sensitive in the USA. Based on the above observations and thoughts, we propose **Personalised DP-Embedding** to control privacy-budget based on user privacy concerns.

3.2 Personalised DP-Embedding

To achieve differential privacy in learning word embedding, we need user-level privacy [16]. Thus, we reformulate the problem of word embedding to accommodate user-level privacy, given the fact that in UGC, the mapping from user to data is known.

User-Level Word Embedding Learning: Given a collection of user-level data $\{D_1, \ldots, D_u, \ldots, D_k\}$ where each user-level data D_u contains a number of documents about user u. Without loss of generality, we define that a data collection $\mathcal{D}$ contains a set of n document $\{d_1, d_2, \ldots, d_n\}$ and $D_u \subseteq D$ ($1 \leq u \leq k$). Each document d contains a sequence of m words $\{w_1, \ldots, w_m\}$ from a fixed dictionary V. The different part from the original formalization (in Sect. 3.1) is that, each user-level data D_u has its own privacy-budget $(\epsilon, \delta)_u$. We do not have to set a predefined budget before learning or redistribute noise based on features as shown in [23]. Alternatively, we learn and protect people privacy based on their needs [26]. During the training process, if privacy-budget $(\epsilon, \delta)_u$ of user u is used up, the user-level data D_u will no-longer be used (see Algorithm 1-b). In the algorithm (line 3), we firstly have to get a list of valid user-level data D (by checking list of valid user $\mathcal{U}$). Then L samples are drawn with probability L/K (line 4). At line 5, we get list of users $\mathcal{U}_{L_t}$ where the sampled examples were taken. After this, we compute gradient (line 6–7), add noise & apply gradient clipping (line 8–9), and go descent (line 10–11). At line 12, we compute the current privacy spending using privacy-accountant API of [1]. From line 13 to line 17, we update privacy spending for all users $\mathcal{U}_{L_t}$, who get involved in the training step t, by the mean of privacy spending at that training step t. Then we exclude any user that got out of privacy-budget from $\mathcal{U}$. One of the challenges in this approach is to obtain user privacy concerns. Based on a previous work [26,27], we found that the privacy-budget can be predicted using a strong correlation between user personality and their privacy concerns. Thus, we employed the model [26] to decide privacy-budget of user-level data. Though the model was tested on a derived data, it is sufficient to predict privacy concern degree for cold-start users (i.e., having no user-defined privacy concern degree).

Personalized Privacy-Accountant: Based on the privacy accountant of the DP-Embedding algorithm, we implement a personalized privacy-accountant for the Personalized DP-Embedding algorithm to control privacy-budget of user-level data. Privacy-accountant can be used to 1) predict privacy-budget of all users if the information is not available (mentioned above), and 2) keep track of privacy-budget of each user to decide whether or not to use their data.

Remarks: The advantage of this Personalized DP-Embedding algorithm is that the user privacy-concerns will not be violated since they are defined by either user or algorithm (in case of cold-start users). Traditionally, the differential privacy based algorithm was learned based on a predefined (ϵ, δ)-budget to protect user data. In this way, the user level of privacy concerns was not considered and satisfied. Therefore, the proposed personalized DP-Embedding approach addresses this problem to fulfill the user needs of privacy.

4 Experimental Settings

4.1 Evaluation Criteria

We test our learned word embedding models on two criteria: 1) **word similarity**: it is a standard measurement for evaluating word embedding models [15]. The purpose is to detect the changes in semantic space. More similarity means less change in semantic space, which proves better word embedding model. One simple example is calculating similarity between the query word ($wo\vec{m}an$) and the predicted word $qu\vec{e}en$ given an embedding model (trained on $ki\vec{n}g$ and $m\vec{a}n$); 2) **data utilities**: the main purpose of developing DP-Embedding is to preserve privacy when sharing the model for other scholars, and especially preserve data utility to facilitate their research. Therefore, we evaluate the data utility of our learned embedding models by applying them to a downstream task - regression (more in Sect. 5.2).

4.2 Datasets

Two datasets are used for experimental evaluation: Text8[1] dataset and myPersonality.org (myPer) dataset. Text8 dataset is commonly used to evaluate the

---

[1] http://mattmahoney.net/dc/textdata.html.

Require: Examples $\{x_1, \ldots, x_N\}$, loss function $\mathcal{L}(\theta)$, embed dimension k
Ensure: return optimized θ to calculate $W^{(k)}$ - a learned DP-Embedding.
    // **Algorithm 1-a: DP-Embedding**
1: Initialize θ_0 randomly
2: **for all** round $t = 0, 1, 2, \ldots, T$ **do**
3:    Take a random sample L_t with sampling probability L_t/N
4:    **Compute gradient**
5:    For each $i \in L_t$, compute $g_t(x_i) \leftarrow \nabla_{\theta_0} \mathcal{L}(\theta_t, x_i)$ // $\mathcal{L}$ is from (2)
6:    **Add noise**
7:    $\tilde{g}_t \leftarrow \frac{1}{L}(\Sigma_i \tilde{g}_t(x_i) + \mathcal{N}(0, \sigma^2 C^2 \mathbf{I})$
8:    **Descent**
9:    $\theta_{t+1} \leftarrow \theta_t - \eta_t \tilde{g}_t$
10:    $\mathcal{M}.\text{accum\_priv\_spending}(z)$
11: **end for**
12: _____

Require: Examples $\{x_1, \ldots, x_N\}$, loss function $\mathcal{L}(\theta)$, embed dimension k
Ensure: return optimized θ to calculate $W^{(k)}$ - a learned DP-Embedding.
    // **Algorithm 1-b: Personalized DP-Embedding**
1: Initialize θ_0 randomly
2: **for all** round $t = 0, 1, 2, \ldots, T$ **do**
3:    $K \leftarrow$ (get list of samples from valid users $\mathcal{U}$)
4:    Take a random sample $L_t \in K$ with sampling probability L_t/K.
5:    $\mathcal{U}_{L_t} \leftarrow$ the set of users where the sample L_t come from.
6:    **Compute gradient**
7:    For each $i \in L_t$, compute $g_t(x_i) \leftarrow \nabla_{\theta_0} \mathcal{L}(\theta_t, x_i)$ // $\mathcal{L}$ is from (2)
8:    **Add noise**
9:    $\tilde{g}_t \leftarrow \frac{1}{L}(\Sigma_i \tilde{g}_t(x_i) + \mathcal{N}(0, \sigma^2 C^2 \mathbf{I})$
10:    **Descent**
11:    $\theta_{t+1} \leftarrow \theta_t - \eta_t \tilde{g}_t$
12:    $(\epsilon_t, \delta_t) = \mathcal{M}.\text{get\_priv\_spending}(z)$
13:    **Update privacy spending for each user**
14:    **for all** user $u \in \mathcal{U}_{L_t}$ **do**
15:        $(\epsilon, \delta)_u \leftarrow (\epsilon, \delta)_u + \frac{(\epsilon_t, \delta_t)}{L}$
16:        If user u gets out of privacy-budget: $\mathcal{U} \leftarrow \mathcal{U} \setminus \{u\}$
17:    **end for**
18: **end for**

Algorithm 1. Algorithms of DP-Embedding and Personalized DP-Embedding. $\mathcal{M}$ is the privacy account API of Abadi et al. [1].

quality of embedding models trained in different manners (e.g., normal embedding versus differentially private embedding). myPer dataset was used for data utility evaluation because of two reasons. Firstly, myPer dataset contains both a public set with 250 users and a private set with more than 153K users. Since early 2018, the private part of the myPersonality data is no longer available for scholars to apply, therefore, it increases the need for sharing information from the private data with privacy-guarantee than ever. Secondly, it fulfills the scenario this paper addresses, where the sensitive data has to be shared somehow for research benefits, and the urgency to guarantee privacy for data sharing. Lastly, myPer dataset was widely used[2] [12] and we can conduct an evaluation by comparing the performance with previous works. Table 1 summaries some statistics of the two datasets.

---

[2] https://goo.gl/M8iQ6m.

Table 1. A simple statistics of the myPersonality dataset and Text8 corpus.

Dataset	#users	#documents	#words
myPer (private)	153,727	22,043,394	416,862,367
myPer (public)	250	9,917	144,616
Tex8 corpus	–	–	17,005,207

4.3 Experiment Design

Two sets of experiments are designed to prove the effectiveness of the proposed DP-Embedding models regarding semantic space and regression task. **Changes in Semantic Space** is detected to prove the effectiveness of preserving semantic relations of our *DP Embedding* in comparison with a standard embedding. We firstly train a standard implementation of Word2Vec embedding (we refer it as *Gold model*) on Text8 corpus. Secondly, we used our proposed approach to train the following two different word embedding models (1) DP-Embedding using the DP-Embedding and (2) None-DP Embedding model (without privacy guarantee) on Text8 corpus. Regarding evaluation, we issue the same set of queries to the DP-Embedding and None-DP Embedding and compare them with the returned top words from *Gold model*. The None-DP Embedding model is needed since we need to have a comparable learning pattern to compare to the DP Embedding, i.e., having a clipping gradient function. Regarding evaluation metric, we used MAP (mean-average-precision) to calculate word similarity (more details in Sect. 5.1). MAP is widely used in information retrieval to evaluate results based on the top K returned results. Given a list of queries Q and their correct answers, MAP metric calculates the mean of the average precision scores for each query, $MAP = \frac{\Sigma_{q=1}^{Q} AvgP(q)}{Q}$. Here we apply two different types of MAP called MAP-Word and MAP-Char. The MAP-Word evaluates the top similar words at word level, and the MAP-Char evaluates at character level. The difference between them is that, at word-level, MAP-Word will only capture exact words in the top results. However, during the training process, some similar words are at the top too but the MAP-Word cannot capture this information (e.g., "there" and "that"). Inversely, the MAP-Char can capture very nicely this information at character level (see Table 2 for example of MAP-Word and MAP-Char).

Regression Task: This experiment is used to prove that our differentially private models can preserve good data utility when they are shared for other scholars to use in a downstream application (e.g., regression). Here the regression task is to predict the extrovert personality score of people from the myPer(public) dataset. Given 250 users and the ground truth of extrovert scores, we divided the data to 80% for training and the other 20% for testing. As we mentioned in Sect. 4.2, there are two parts in myPer dataset (i.e., public and private). We first set up a experiment - E(public), where trained the public available word embedding model (Word2Vec) from Google and a character embedding model[3] for feature representation. Meanwhile, we set up another experiment - E(private), where we trained our *DP-Embedding* and *None-DP Embedding* on myPer(private). Based on the hypothesis that a regression task R on features extracted from both public and private dataset will perform better than that only on public dataset (i.e., $R_{E(Private)+E(Public)} \geq R_{E(public)}$.), we compare the regression performance based on model from E(public) with that from both E(public) and E(private). To prove above hypothesis by evaluation, we implemented the following regression methods:

- Baseline-SVR: it is a regression baseline using Support Vector Machine-Regression (SVR) method, where only E(public) is used for feature extraction.
- Baseline-LR: it is a regression baseline using linear regression (LR), where only E(public) is used for feature extraction.
- DP-(SVM and LR): it is similar to the above methods except both E(private) and E(public) are used for feature extraction. E(private) here is trained using DP-Embedding.
- NoneDP-(SVM and LR): it is similar to the above methods except we used E(private) and E(public) for feature extraction. E(private) here is trained without differential privacy.

Table 2. Top similar words of DP-Embedding (a), and Non-DP Embedding (b) models given three queries "three", "eight", and "they" at 100K learning step. The second column shows the best results from the Gold model. MAP(W, C) denotes (MAP-Word, MAP-Char).

Query	Gold model	DP-Embedding (top 4)	MAP (W,C)	Topic
three	four:two:five:seven	zero:one:feeder:nine	(0, 3.814)	Numbers
eight	seven:nine:six:four	cornerback:four:stockholders:zero	(0.5, 0.1347)	Numbers
they	we:there:you:he	morgan:century:contentious:ferroelectric	(0, 0.4237)	Pronouns

(a) Top 4 on DP-Embedding model

Query	Gold model	Non-DP Embedding (top 4)	MAP (W, C)	Topic
three	four:two:five:seven	one:in:UNK:zero	(0, 0.1288)	Numbers
eight	seven:nine:six:four	integrator:transfection:four:one	(0.33, 0.3561)	Numbers
they	we:there:you:he	that:monorail:it:lesbian	(0, 0.2341)	Pronouns

(b)Top 4 on Non-DP Embedding model

5 Evaluation Results

5.1 Evaluation #1: Changes in Semantic Space

Figure 3 and Table 2 show the evaluation on DP-Embedding and Non-DP Embedding regarding semantic space change. Given 11 word samples as input

---

[3] https://github.com/minimaxir/char-embeddings/.

Table 3. Regression performance on public embedding with and without privacy guarantee in comparison with not using public embedding. Evaluation score is RMSE. † marks good checkpoints to publish the DP-Embedding model, and LS stands for Learning-Step.

LS	SVR			LR			Privacy-Budget (0.125, δ)
	Baseline-SVR	DP-SVR	NoneDP-SVR	Baseline-LR	DP-LR	NoneDP-LR	
20	2.6563	**1.7881**	3.5942	1.2903	**1.2616**	1.2642	0.0184 †
200	2.6563	2.4983	**2.0198**	1.2903	**1.2589**	1.2717	0.0189
500	2.6563	**2.7795**	3.6231	1.2903	**1.2514**	1.2909	0.0197 †
1K	2.6563	3.2146	**2.0206**	1.2903	**1.2611**	1.262	0.0211
5K	2.6563	6.1596	**2.7472**	1.2903	**1.2577**	1.2642	0.0372
10K	2.6563	**1.6396**	3.9155	1.2903	1.2768	**1.2574**	0.0755
50K	2.6563	2.9438	**2.5769**	1.2903	1.2574	**1.2556**	0.5929
90K	2.6563	**2.4033**	2.5175	1.2903	1.2585	**1.258**	0.7681
100K	2.6563	2.6043	**2.0215**	1.2903	**1.2548**	1.262	0.7926

queries, we obtain the top 100 returned words for them from both models, and compare them with the results from *gold model* by calculating MAP score ($K = 100$, $Q = 11$).

As shown in Fig. 3, there is significant difference between DP-Embedding and Non-DP Embedding when comparing to the *Gold model*. In Fig. 3-(a), it clearly shows that DP-Embedding performs slightly lower performance compared to None-DP Embedding. This is understandable due to the injected noise into the model for differential privacy. However, one interesting fact we can observe in Fig. 3-(a) is that even the performance at word-level (using the MAP-Word metric) is lower, at character-level (using the MAP-Char metric), DP-Embedding performs better than None-DP Embedding. This observation gives a hint that the reasonable noise we inject into the model actually helps the model to improve at character level. Intuitively, injecting noise is similar to modifying characters of words. It is worth to notice that, up to date, this observation is very new and has not reported in any work before. Thus, further verification is worthwhile in future.

Table 2 presents an example with top 4 results from both models given three queries ("three", "eight", and "they"). Two types of MAP scores (MAP-Word and MAP-Char) are calculated. As shown in the third column, some irrelevant concepts (e.g., "feeder", or "stockholders") are being mixed up at the top in DP-Embedding, while for the None-DP Embedding (the fifth column), the relevant concepts (i.e., "four", "one", "it") are climbing up to the top. Though relevant concepts are always expected to get closer over each training step for word embedding model, the added noise to the DP-Embedding model over a learning step can create the distance between sensitive concepts further. In this way, privacy is guaranteed for word embedding model. However, too much noise might destroy the model quality, thus we will evaluate data utility.

5.2 Evaluation #2: Regression Task

As explained in Sect. 4.3, we address the regression problem in this experiment by using DP-Embedding model and None-DP Embedding model respectively. RMSE (root mean square error) is used to evaluate the regression task. Lower RMSE proves better regression performance. Table 3 shows that the usage of DP-Embedding gets better or slightly different results than the None-DP Embedding. This clearly shows that, with (ϵ, δ) privacy guarantee at some settings, such as (0.125, 0.0184)-DP and (0.125, 0.0197)-DP, we achieve the optimized trade-off of privacy-guarantee and data utilities.

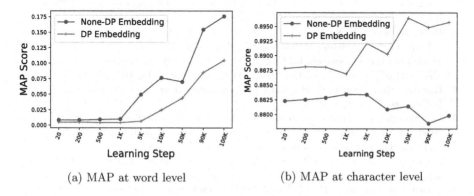

(a) MAP at word level   (b) MAP at character level

Fig. 3. Semantic space changes when learning embedding model with and without differential privacy compared to the *Gold model*. Learning step is number of minibatch steps

6 Conclusions

In this work, we proposed algorithms for learning differentially private text representation (i.e., word embeddings) for user generated contents (UGC) sharing. We empirically evaluated the algorithms on a realistic UGC dataset and demonstrated that the proposed embedding model benefits from sensitive data while maintaining user-privacy. The differentially private word embedding allows information from sensitive data to be shared independently. Differently from the previous works on differential privacy, which simply preserve privacy against adversaries on sensitive data, we trained word embedding model on potentially sensitive user generated contents, and our trained model is applied for data sharing with privacy-guarantee. As the very first work on publicly shared embedding models, this work highlights the new direction of publicly shared embedding models on sensitive text data. Much future work remains. For example, one promising direction would be exploring strategies to detect what are exactly sensitive contents to certain users (e.g., building a knowledge base) to apply our proposed personalized-privacy guarantee.

Acknowledgement. This work is supported by the Federated Database project funded by Umeå University, Sweden. The computations were performed on resources provided by the Swedish National Infrastructure for Computing (SNIC) at HPC2N center. The authors also thank the myPersonality project for data contribution.

References

1. Abadi, M., et al.: Deep learning with differential privacy. ArXiv e-prints (2016)
2. Bansal, M., Gimpel, K., Livescu, K.: Tailoring continuous word representations for dependency parsing. In: Proceedings of the 52nd Annual Meeting of the Association for Computational Linguistics (Volume 2: Short Papers), pp. 809–815. Association for Computational Linguistics (2014). https://doi.org/10.3115/v1/P14-2131, http://aclweb.org/anthology/P14-2131
3. Batmanghelich, K.N., Saeedi, A., Narasimhan, K., Gershman, S.: Nonparametric spherical topic modeling with word embeddings. In: Proceedings of the 54th Annual Meeting of the Association for Computational Linguistics (Volume 2: Short Papers) abs/1604.00126, pp. 537–542 (2016). http://arxiv.org/abs/1604.00126
4. Bayardo, R.J., Agrawal, R.: Data privacy through optimal k-anonymization. In: ICDE, pp. 217–228 (2005)
5. Bojanowski, P., Grave, E., Joulin, A., Mikolov, T.: Enriching word vectors with subword information. Trans. Assoc. Comput. Linguist. **5**, 135–146 (2017)
6. Cynthia, D.: Differential privacy, pp. 1–12. ICALP (2006)
7. Dwork, C., Smithy, A.: Differential privacy for statistics: what we know and what we want to learn (2009)
8. Ebadi, H., Sands, D., Schneider, G.: Differential privacy: now it's getting personal. In: Proceedings of the 42Nd Annual ACM SIGPLAN-SIGACT Symposium on Principles of Programming Languages, POPL 2015, pp. 69–81. ACM, New York (2015). https://doi.org/10.1145/2676726.2677005, http://doi.acm.org/10.1145/2676726.2677005
9. Flekova, L., Gurevych, I.: Can we hide in the web? Large scale simultaneous age and gender author profiling in social media notebook for PAN at CLEF 2013. In: Working Notes for CLEF 2013 Conference, Valencia, Spain, 23–26 September 2013 (2013)
10. Fredrikson, M., Jha, S., Ristenpart, T.: Model inversion attacks that exploit confidence information and basic countermeasures. In: Proceedings of the 22nd ACM SIGSAC Conference on Computer and Communications Security, CCS 2015, pp. 1322–1333 (2015)
11. Goodfellow, I., Bengio, Y., Courville, A.: Deep Learning. MIT Press (2016). http://www.deeplearningbook.org
12. Kosinski, M., Matz, S., Gosling, S., Popov, V., Stillwell, D.: Facebook as a social science research tool. Am. Psychol. (2015)
13. Lee, J., Clifton, C.: How much is enough? Choosing ϵ for differential privacy. In: Lai, X., Zhou, J., Li, H. (eds.) ISC 2011. LNCS, vol. 7001, pp. 325–340. Springer, Heidelberg (2011). https://doi.org/10.1007/978-3-642-24861-0_22
14. Lee, J., Clifton, C.: Differential identifiability. In: Proceedings of KDD (2012)
15. Levy, O., Goldberg, Y.: Linguistic regularities in sparse and explicit word representations. In: Proceedings of the Eighteenth Conference on Computational Natural Language Learning, pp. 171–180. Association for Computational Linguistics (2014). https://doi.org/10.3115/v1/W14-1618, http://aclweb.org/anthology/W14-1618

16. McMahan, H.B., Ramage, D., Talwar, K., Zhang, L.: Learning differentially private language models without losing accuracy. CoRR abs/1710.06963 (2017). http://arxiv.org/abs/1710.06963
17. McSherry, F.D.: Privacy integrated queries: an extensible platform for privacy-preserving data analysis. In: SIGMOD (2009)
18. Mikolov, T., Chen, K., Corrado, G., Dean, J.: Efficient estimation of word representations in vector space. CoRR abs/1301.3781 (2013). http://arxiv.org/abs/1301.3781
19. Mohsen, A.M., El-Makky, N.M., Ghanem, N.: Author identification using deep learning. In: 2016 15th IEEE International Conference on Machine Learning and Applications (ICMLA), pp. 898–903 (2016). https://doi.org/10.1109/ICMLA.2016.0161
20. Nguyen, H.H., Imine, A., Rusinowitch, M.: Detecting communities under differential privacy. In: Proceedings of the 2016 ACM on Workshop on Privacy in the Electronic Society, WPES 2016, pp. 83–93. ACM, New York (2016). https://doi.org/10.1145/2994620.2994624, http://doi.acm.org/10.1145/2994620.2994624
21. Papernot, N., Song, S., Mironov, I., Raghunathan, A., Talwar, K., Erlingsson, Ú.: Scalable private learning with PATE. In: Sixth International Conference on Learning Representation (ICLR 2018) (2018)
22. Pennington, J., Socher, R., Manning, C.D.: GloVe: global vectors for word representation. In: Empirical Methods in Natural Language Processing (EMNLP), pp. 1532–1543 (2014), http://www.aclweb.org/anthology/D14-1162
23. Phan, N., Wu, X., Hu, H., Dou, D.: Adaptive laplace mechanism: differential privacy preservation in deep learning. CoRR abs/1709.05750 (2017). http://arxiv.org/abs/1709.05750
24. Popov, V., Kudinov, M., Piontkovskaya, I., Vytovtov, P., Nevidomsky, A.: Distributed fine-tuning of language models on private data. In: International Conference on Learning Representations (2018). https://openreview.net/forum?id=HkgNdt26Z
25. Taddy, M.: Document classification by inversion of distributed language representations. In: Proceedings of the 53rd Annual Meeting of the ACL and the 7th International Joint Conference on Natural Language Processing (Volume 2: Short Papers), pp. 45–49. Association for Computational Linguistics (2015). https://doi.org/10.3115/v1/P15-2008, http://aclweb.org/anthology/P15-2008
26. Vu, X.S., Jiang, L.: Self-adaptive privacy concern detection for user-generated content. In: Proceedings of the 19th International Conference on Computational Linguistics and Intelligent Text Processing (CICLing), Volume 1: Long papers (2018)
27. Vu, X.S., Jiang, L., Brändström, A., Elmroth, E.: Personality-based knowledge extraction for privacy-preserving data analysis. In: Proceedings of the Knowledge Capture Conference, K-CAP 2017, pp. 45:1–45:4. ACM, New York (2017). https://doi.org/10.1145/3148011.3154479, http://doi.acm.org/10.1145/3148011.3154479
28. Wang, R., Wang, X., Li, Z., Tang, H., Reiter, M.K., Dong, Z.: Privacy-preserving genomic computation through program specialization, pp. 338–347. CCS (2009)
29. Wu, Z., Wang, Z., Wang, Z., Jin, H.: Towards privacy-preserving visual recognition via adversarial training: a pilot study. In: Ferrari, V., Hebert, M., Sminchisescu, C., Weiss, Y. (eds.) ECCV 2018. LNCS, vol. 11220, pp. 627–645. Springer, Cham (2018). https://doi.org/10.1007/978-3-030-01270-0_37
30. Zhang, Y., Ding, N., Soricut, R.: SHAPED: shared-private encoder-decoder for text style adaptation. In: The 16th Annual Conference of the North American Chapter of the Association for Computational Linguistics: Human Language Technologies (NAACL) (2018)

Multiplicative Models for Recurrent Language Modeling

Diego Maupomé and Marie-Jean Meurs[(✉)]

Université du Québec à Montréal, Montréal, QC, Canada
maupome.diego@courrier.uqam.ca, meurs.marie-jean@uqam.ca

Abstract. Recently, there has been interest in multiplicative recurrent neural networks for language modeling. Indeed, simple Recurrent Neural Networks (RNNs) encounter difficulties recovering from past mistakes when generating sequences due to high correlation between hidden states. These challenges can be mitigated by integrating second-order terms in the hidden-state update. One such model, multiplicative Long Short-Term Memory (mLSTM) is particularly interesting in its original formulation because of the sharing of its second-order term, referred to as the intermediate state. We explore these architectural improvements by introducing new models and testing them on character-level language modeling tasks. This allows us to establish the relevance of shared parametrization in recurrent language modeling.

Keywords: Language modeling · Recurrent neural networks · Multiplicative recurrent neural networks

1 Introduction

One of the principal challenges in computational linguistics is to account for the word order of the document or utterance being processed [6]. Of course, the numbers of possible phrases grows exponentially with respect to a given phrase length, requiring an approximate approach to summarizing its content. Recurrent Neural Networks (RNNs) are such an approach, and they are used in various tasks in Natural Language Processing (NLP), such as machine translation [16], abstractive summarization [20] and question answering [11]. However, RNNs, as approximations, suffer from numerical troubles that have been identified, such as that of recovering from past errors when generating phrases. We take interest in a model that mitigates this problem, multiplicative RNNs (mRNNs), and how it has been and can be combined for new models. To evaluate these models, we use the task of *recurrent language modeling*, which consists in predicting the next token (character or word) in a document. This paper is organized as follows: RNNs and mRNNs are introduced respectively in Sects. 2 and 3. Section 4 presents new and existing multiplicative models. Section 5 describes the datasets and experiments performed, as well as results obtained. Sections 6 discusses and concludes our findings.

© Springer Nature Switzerland AG 2023
A. Gelbukh (Ed.): CICLing 2019, LNCS 13451, pp. 332–341, 2023.
https://doi.org/10.1007/978-3-031-24337-0_24

2   Recurrent Neural Networks

RNNs are powerful tools of sequence modeling that can preserve the order of words or characters in a document. A document is therefore a sequence of words, $x_1, \ldots, x_T$. Given the exponential growth of possible histories with respect to the sequence length, the probability of observing a given sequence needs to be approximated. RNNs will make this approximation using the product rule,

$$P(x_1, \ldots, x_T) = P(x_1)P(x_2|x_1) \ldots P(x_T|x_1, \ldots, x_{T-1}),$$

and updating a *hidden state* at every time step. This state is first null,

$$h_0 = 0.$$

Thereafter, it is computed as a function of the past hidden state as well as the input at the current time step,

$$h_t = f(h_{t-1}, x_t),$$

known as the *transition function*. f is a learned function, often taking the form[1]

$$h_t = \tanh(Ux_t + Wh_{t-1}).$$

This allows, in theory, for straightforward modeling of sequences of arbitrary length.

In practice, RNNs encounter some difficulties that need some clever engineering to be mitigated. For example, learning long-term dependencies such as those found in language is not without its share of woes arising from numerical considerations, such as the well-known vanishing gradient problem [2]. This can be addressed with gating mechanisms, such as Long Short-Term Memory network (LSTM) [9] and Gated Recurrent Unit (GRU) [3].

A problem that is more specific to generative RNNs is their difficulty recovering from past errors [7], which [15] argue arises from having hidden-state transitions that are highly correlated across possible inputs. One approach to adapting RNNs to have more input-dependent transition functions is to use the multiplicative "trick" [23]. This approximates the idea of having the input at each time synthesize a dedicated kernel of parameters dictating the transition from the previous hidden state to the next. These two approaches can be combined, as in the multiplicative LSTM (mLSTM) [15].

We begin by contending that, in making RNNs multiplicative, sharing what is known as the *intermediate state* does not significantly hinder performance when parameter counts are equal. We verify this with existing as well as new gated models on several well-known language modeling tasks.

---

[1] Additive biases are omitted throughout the paper for concision.

3   Multiplicative RNNs

Most recurrent neural network architectures, including LSTM and GRU share the following building block:

$$\tilde{h}_t = Ux_t + Wh_{t-1}. \tag{1}$$

$\tilde{h}_t$ is the *candidate* hidden state, computed from the previous hidden state, h_{t-1}, and the current input, x_t, weighted by the parameter matrices W and U, respectively. This candidate hidden state may then be passed through gating mechanisms and non-linearities depending on the specific recurrent model.

Let us assume for simplicity that the input is a one-hot vector (one component is 1, the rest are 0 [22] [see p. 45]), as it is often the case in NLP. Then, the term Ux_t is reduced to a single column of U and can therefore be thought of as an input-dependent bias in the hidden state transition. As the dependencies we wish to establish between the elements of the sequences under consideration become more distant, the term Wh_t will have to be significantly larger than this input-dependent bias, Ux_t, in order to remain unchanged across time-steps. This will mean that from one time-step to the next, the hidden-to-hidden transition will be highly correlated across possible inputs. This can be addressed by having more input-dependent hidden state transitions, making RNNs more expressive.

In order to remedy the aforementioned problem, each possible input i can be given its own matrix $W^{(i)}$ parameterizing the contribution of h_t to $\tilde{h}_t$.

$$\tilde{h}_t = Ux_t + \underbrace{\left(\sum_i W^{(i)}x_t^{(i)}\right)}_{\mathbf{W}^{(x_t)}} h_{t-1}. \tag{2}$$

This is known as a tensor RNN (tRNN) [23], because all the matrices can be stacked to form a rank 3 tensor, $\mathbf{W}$. The input x_t selects the relevant slice of the tensor in the one-hot case and a weighted sum over all slices in the dense case. The resulting matrix then acts as the appropriate W.

However, such an approach is impractical because of the high parameter count such a tensor would entail. The tensor can nonetheless be approximated by factorizing it [24] as follows:

$$\mathbf{W}^{(x_t)} = V\mathrm{diag}(W_x x_t)W_h, \tag{3}$$

where W_x and W_h are weight matrices, and diag is the operator turning a vector v into a diagonal matrix where the elements of v form the main diagonal of said matrix. Replacing $\mathbf{W}^{(x_t)}$ in Eq. (2) by this tensor factorization, we obtain

$$\tilde{h}_t = Ux_t + Vm_t, \tag{4}$$

where m_t is known as the *intermediate state*, given by

$$m_t = (W_x x_t) * (W_h h_{t-1}). \tag{5}$$

Here, $*$ refers to the Hadamard or element-wise product of vectors. The intermediate state is the result of having the input apply a learned filter via the new parameter kernel W to the factors of the hidden state. It should be noted that the dimensionality of m_t is free and, should it become sufficiently large, the factorization becomes as expressive as the tensor. The ensuing model is known as a mRNN [23].

4 Sharing Intermediate States

While mRNN outperform simple RNNs in character-level language modeling, they have been found wanting with respect to the popular LSTM [9]. This prompted [15] to apply the multiplicative "trick" to LSTM resulting in the mLSTM, which achieved promising results in several language modeling tasks [15].

4.1 mLSTM

Gated RNNs, such as LSTM and GRU, use *gates* to help signals move through the network. The value of these gates is computed in much the same way as the candidate hidden state, albeit with different parameters. For example, LSTM uses two different gates, i and f in updating its memory cell, c_t,

$$c_t = f_t * c_{t-1} + i_t * \tanh(\tilde{h}_t). \tag{6}$$

It uses another gate, o, in mapping c_t to the new hidden state, h_t,

$$h_t = o_t * \sigma(c_t), \tag{7}$$

where σ is the sigmoid function, squashing its input between 0 and 1. f and i are known as forget and input gates, respectively. The forget gates allows the network to ignore components of the value of the memory cell at the past state. The input gate filters out certain components of the new hidden state. Finally, the output gates separates the memory cell from the actual hidden state. The values of these gates are computed at each time step as follows:

$$i_t = \sigma(U_i x_t + W_i h_{t-1}) \tag{8}$$

$$f_t = \sigma(U_f x_t + W_f h_{t-1}) \tag{9}$$

$$o_t = \sigma(U_o x_t + W_o h_{t-1}). \tag{10}$$

Each gate has its own set of parameters to infer. If we were to replace each W_* by a tensor factorization as in mRNN, we would obtain a mLSTM model. However, in the original formulation of mLSTM, there is no factorization of each would-be $\mathbf{W}_*$ *individually*. There is no separate intermediate state for each gate, as one would expect. Instead, a single intermediate state, m_t, is computed to replace

h_{t-1} in *all* equations in the system, by Eq. 5. Furthermore, each gate has its own $V_\star$ weighting m_t. Their values are computed as follows:

$$i_t = \sigma(W_i h_{t-1} + V_i m_t) \tag{11}$$

$$f_t = \sigma(W_f h_{t-1} + V_f m_t) \tag{12}$$

$$o_t = \sigma(W_o h_{t-1} + V_o m_t). \tag{13}$$

The model can therefore no longer be understood as an approximation of the tRNN. Nonetheless, it has achieved empirical success in NLP. We therefore try to explore the empirical merits of this shared parametrization and apply them to other RNN architectures.

4.2   True mLSTM

We have presented the original mLSTM model with its shared intermediate state. If we wish to remain true to the original multiplicative model, however, we have to factorize every would-be $W_\star$ tensor separately. We have:

$$i_t = \sigma(U_i x_t + V_i m_{i,t}) \tag{14}$$

$$f_t = \sigma(U_f x_t + V_f m_{f,t}) \tag{15}$$

$$o_t = \sigma(U_o x_t + V_o m_{o,t}), \tag{16}$$

with each $m_{\star,t}$ being given by a separate set of parameters:

$$m_{\star,t} = (W_{\star,x} x_t) * (W_{\star,h} h_{t-1}). \tag{17}$$

We henceforth refer to this model as true mLSTM (tmLSTM). We sought to apply the same modifications to the GRU model, as LSTM and GRU are known to perform similarly [4,8,12]. That is, we build a true multiplicative GRU (tmGRU) model, as well as a multiplicative GRU (mGRU) with a shared intermediate state.

4.3   GRU

The GRU was first proposed by [3] as a lighter, simpler variant of LSTM. GRU relies on two gates, called, respectively, the *update* and *reset* gates, and no additional memory cell. These gates intervene in the computation of the hidden state as follows:

$$h_t = (1 - z_t)h_{t-1} + z_t \tanh(\tilde{h}_t), \tag{18}$$

where the candidate hidden state, $\tilde{h}_t$, is given by:

$$\tilde{h}_t = U_h x_t + W_h(r_t * h_{t-1}). \tag{19}$$

The update gate deletes specific components of the hidden state and replaces them with those of the candidate hidden state, thus updating its content. On the other hand, the reset gate allows the unit to start anew, as if it were reading the first symbol of the input sequence. They are computed much in the same way as the gates of LSTM:

$$z_t = \sigma(U_z x_t + W_z h_{t-1}), \tag{20}$$

$$r_t = \sigma(U_r x_t + W_r h_{t-1}). \tag{21}$$

4.4 True mGRU

We can now make GRU multiplicative by using the tensor factorization for z and r:

$$z_t = \sigma(U_z x_t + V_z m_{z,t}), \tag{22}$$

$$r_t = \sigma(U_r x_t) + V_r m_{r,t}, \tag{23}$$

with each $m_{*,t}$ given by Eq. 17. There is a subtlety to computing $\tilde{h}_t$, as we need to apply the reset gate to h_{t-1}. While h_t itself is given by Eq. 4, $m_{h,t}$ is not computed the same way as in mLSTM and mRNN. Instead, it is given by:

$$m_{h,t} = (W_x x_t) * (W_h(r_t * h_{t-1})). \tag{24}$$

4.5 mGRU with Shared Intermediate State

Sharing an intermediate state is not as immediate for GRU. This is due to the application of r_t, which we need in computing the intermediate state that we want to share. That is, r_t and m_t would both depend on each other. We modify the role of r_t to act as a filter on m_t, rather than a reset on individual components of h_{t-1}. Note that, when all components of r_t go to zero, it amounts to having all components of h_{t-1} at zero. We have

$$z_t = \sigma(U_z x_t + V_z m_t) \tag{25}$$

and

$$r_t = \sigma(U_r x_t + V_r m_t). \tag{26}$$

$\tilde{h}_t$ is given by

$$\tilde{h}_t = U_h x_t + V_h(r_t * m_t), \tag{27}$$

with m_t the same as in mRNN and mLSTM this time, i.e. Eq. 5. The final hidden state is computed the same way as in the original GRU (Eq. 18).

5 Experiments in Character-Level Language Modeling

Character-level language modeling (or character prediction) consists in predicting the next character while reading a document one character at a time. It is a common benchmark for RNNs because of the heightened need for shared parametrization when compared to word-level models. We test mGRU on two well-known datasets, the Penn Treebank and Text8.

5.1 Penn Treebank

The Penn Treebank dataset [17] comes from a series of Wall Street Journal articles written in English. Following [18], sections 0–20 were used for training, 21–22 for validation and 23–24 for testing, respectively, which amounts to 5.1M, 400K and 450K characters, respectively.

The vocabulary consists of 10K lowercase words. All punctuation is removed and numbers were substituted for a single capital N. All words out of vocabulary are replaced by the token <unk>.

The training sequences were passed to the model in batches of 32 sequences. Following [15], we built an initial mLSTM model of 700 units. However, we set the dimensionality of the intermediate state to that of the input in order to keep the model small. We do the same for our mGRU, tmLSTM and tmGRU, changing only the size of the hidden state so that all four models have roughly the same parameter count. We trained it using the Adam optimizer [13], selecting the best model on validation over 10 epochs. We apply no regularization other than a checkpoint, keeping the best model over all epochs. The performance of the model is evaluated using cross entropy in bits per character (BPC), which is log_2 of perplexity.

All models outperform previously reported results for mLSTM [15] despite lower parameter counts. This is likely due to our relatively small batch size. However, they perform fairly similarly. Encouraged by these results, we built an mGRU with both hidden and intermediate state sizes set to that of the original mLSTM (700). This version highly surpasses the previous state of the art while still having fewer parameters than previous work.

For the sake of comparison, results as well as parameter counts (where available) of our models (bold) and related approaches are presented in Table 1. mGRU and larger mGRU, our best models, achieved respectively an error of 1.07 and 0.98 BPC on the test data, setting a new state of the art for this task.

5.2 Text8

The Text8 corpus [10] comprises the first 100M plain text characters in English from Wikipedia in 2006. As such, the alphabet consists of the 26 letters of the English alphabet as well as the space character. No vocabulary restrictions were put in place. As per [18], the first 90M and 5M characters were used for training and validation, respectively, with the last 5M used for testing.

Encouraged by our results on the Penn Treebank dataset, we opted to use similar configurations. However, as the data is one long sequence of characters, we divide it into sequences of 200 characters. We pass these sequences to the model in slightly larger batches of 50 to speed up computation. Again, the dimensionality of the hidden state for mLSTM is set at 450 after the original model, and that of the intermediate state is set to the size of the alphabet. The size of the hidden state is adjusted for the other three models as it was for the PTB experiments. The model is also trained using the Adam optimizer over 10 epochs.

Table 1. Test set error on Penn Treebank and parameter counts in character-level language modeling

Model	Parameter count	Error(BPC)
GRU [1]	3M	1.53
mRNN [18]	–	1.41
LSTM [5]	–	1.38
batch-normalized LSTM [5]	–	1.32
mLSTM [15]	–	1.27
fast-slow LSTM [19]	7.2M	1.19
mLSTM	**292K**	**1.11**
tmLSTM	**292K**	**1.09**
tmGRU	**292K**	**1.08**
mGRU	**292K**	**1.07**
larger mGRU	**2.1M**	**0.98**

Table 2. Test set error on Text8 and parameter counts in character-level language modeling

Model	Parameter count	Error (BPC)
GRU [1]	5M	1.53
mRNN [18]	–	1.54
LSTM [5]	–	1.43
mLSTM [15]	20M	1.42
mLSTM	**133K**	**1.37**
batch-normalized LSTM [5]	–	1.36
tmGRU	**133K**	**1.35**
tmLSTM	**133K**	**1.35**
mGRU	**133K**	**1.35**
large mLSTM [15]	46M	1.27
larger mGRU	**877K**	**1.21**
LSTM [14]*	45M	1.19

The best model as per validation data over 10 epochs achieves 1.40 BPC on the test data, slightly surpassing an mLSTM of smaller hidden-state dimensionality (450) but larger parameter count. Our results are more modest, as are those of the original mLSTM. Once again, results do not vary greatly between models.

As with the Penn Treebank, we proceed with building an mGRU with both hidden and intermediate state sizes set to 450. This improves performance to 1.21 BPC, setting a new state of the art for this task and surpassing a large mLSTM of 1900 units from [15] despite having far fewer parameters (45M to 5M).

For the sake of comparison, results as well as parameter counts of our models and related approaches are presented in Table 2. It should be noted that some of these models employ *dynamic evaluation* [7], which fits the model further during evaluation. We refer the reader to [14]. These models are indicated by a star.

6 Conclusion

We have found that competitive results can be achieved with mRNNs using small models. We have not found significant differences in the approaches presented, despite added non-intuitive parameter-sharing constraints when controlling for model size. Our results are restricted to character-level language modeling. Along this line of thought, previous work on mRNNs demonstrated their increased potential when compared to their regular variants [15,21,23]. We therefore offer other variants as well as a first investigation into their differences. We hope to have evinced the impact of increased flexibility in hidden-state transitions on RNNs sequence-modeling capabilities. Further work in this area is required to transpose these findings into applied tasks in NLP.

Acknowledgements. This research was enabled by support provided by Calcul Québec and Compute Canada. MJM acknowledges the support of the Natural Sciences and Engineering Research Council of Canada [NSERC Grant number 06487-2017] and the Government of Canada's New Frontiers in Research Fund (NFRF), [NFRFE-2018-00484].

References

1. Bai, S., Kolter, J.Z., Koltun, V.: An empirical evaluation of generic convolutional and recurrent networks for sequence modeling. CoRR abs/1803.01271 (2018). http://arxiv.org/abs/1803.01271

2. Bengio, Y., Simard, P., Frasconi, P.: Learning long-term dependencies with gradient descent is difficult. IEEE Trans. Neural Netw. 5(2), 157–166 (1994)

3. Cho, K., et al.: Learning phrase representations using RNN encoder-decoder for statistical machine translation. arXiv preprint arXiv:1406.1078 (2014)

4. Chung, J., Gulcehre, C., Cho, K., Bengio, Y.: Empirical evaluation of gated recurrent neural networks on sequence modeling. arXiv preprint arXiv:1412.3555 (2014)

5. Cooijmans, T., Ballas, N., Laurent, C., Courville, A.C.: Recurrent batch normalization. CoRR abs/1603.09025 (2016). http://arxiv.org/abs/1603.09025

6. Ghodsi, A., DeNero, J.: An analysis of the ability of statistical language models to capture the structural properties of language. In: Proceedings of the 9th International Natural Language Generation Conference, pp. 227–231 (2016)

7. Graves, A.: Generating sequences with recurrent neural networks. CoRR abs/1308.0850 (2013). http://arxiv.org/abs/1308.0850

8. Greff, K., Srivastava, R.K., Koutník, J., Steunebrink, B.R., Schmidhuber, J.: LSTM: a search space odyssey. IEEE Trans. Neural Netw. Learn. Syst. (2016)

9. Hochreiter, S., Schmidhuber, J.: Long short-term memory. Neural Comput. **9**(8), 1735–1780 (1997)
10. Hutter, M.: Human knowledge compression contest (2006). http://prize.hutter1.net/
11. Iyyer, M., Boyd-Graber, J., Claudino, L., Socher, R., Daumé III, H.: A neural network for factoid question answering over paragraphs. In: Proceedings of the 2014 Conference on Empirical Methods in Natural Language Processing (EMNLP), pp. 633–644 (2014)
12. Jozefowicz, R., Zaremba, W., Sutskever, I.: An empirical exploration of recurrent network architectures. In: Proceedings of the 32nd International Conference on Machine Learning (ICML-15), pp. 2342–2350 (2015)
13. Kingma, D.P., Ba, J.: Adam: a method for stochastic optimization. CoRR abs/1412.6980 (2014). http://arxiv.org/abs/1412.6980
14. Krause, B., Kahembwe, E., Murray, I., Renals, S.: Dynamic evaluation of neural sequence models. CoRR abs/1709.07432 (2017). http://arxiv.org/abs/1709.07432
15. Krause, B., Lu, L., Murray, I., Renals, S.: Multiplicative LSTM for sequence modelling. arXiv preprint arXiv:1609.07959 (2016)
16. Luong, T., Pham, H., Manning, C.D.: Effective approaches to attention-based neural machine translation. In: Proceedings of the 2015 Conference on Empirical Methods in Natural Language Processing, pp. 1412–1421 (2015)
17. Marcus, M.P., Marcinkiewicz, M.A., Santorini, B.: Building a large annotated corpus of English: the Penn treebank. Comput. Linguist. **19**(2), 313–330 (1993)
18. Mikolov, T., Sutskever, I., Deoras, A., Le, H.S., Kombrink, S., Cernocky, J.: Subword language modeling with neural networks. Preprint (2012). http://www.fit.vutbr.cz/imikolov/rnnlm/char.pdf
19. Mujika, A., Meier, F., Steger, A.: Fast-slow recurrent neural networks. In: Advances in Neural Information Processing Systems, pp. 5917–5926 (2017)
20. Paulus, R., Xiong, C., Socher, R.: A deep reinforced model for abstractive summarization. CoRR abs/1705.04304 (2017). http://arxiv.org/abs/1705.04304
21. Radford, A., Józefowicz, R., Sutskever, I.: Learning to generate reviews and discovering sentiment. CoRR abs/1704.01444 (2017)
22. Socher, R., Bengio, Y., Manning, C.: Deep learning for NLP. Tutorial at Association of Computational Logistics (ACL), 2012, and North American Chapter of the Association of Computational Linguistics (NAACL) (2013)
23. Sutskever, I., Martens, J., Hinton, G.E.: Generating text with recurrent neural networks. In: Proceedings of the 28th International Conference on Machine Learning (ICML-11), pp. 1017–1024 (2011)
24. Taylor, G.W., Hinton, G.E.: Factored conditional restricted Boltzmann machines for modeling motion style. In: Proceedings of the 26th Annual International Conference on Machine Learning, pp. 1025–1032. ACM (2009)

Impact of Gender Debiased Word Embeddings in Language Modeling

Christine Basta[1,2(✉)] and Marta R. Costa-jussà[1]

[1] TALP Research Center, Universitat Politècnica de Catalunya, Barcelona, Spain
{marta.ruiz,christine.raouf.saad.basta}@upc.edu
[2] Institute of Graduate Studies and Research, Alexandria University,
Alexandria, Egypt

Abstract. Gender, race and social biases have recently been detected as evident examples of unfairness in applications of Natural Language Processing. A key path towards fairness is to understand, analyse and interpret our data and algorithms. Recent studies have shown that the human-generated data used in training is an apparent factor of getting biases. In addition, current algorithms have also been proven to amplify biases from data.

To further address these concerns, in this paper, we study how an state-of-the-art recurrent neural language model behaves when trained on data, which under-represents females, using pre-trained standard and debiased word embeddings. Results show that language models inherit higher bias when trained on unbalanced data when using pre-trained embeddings, in comparison with using embeddings trained within the task. Moreover, results show that, on the same data, language models inherit lower bias when using debiased pre-trained emdeddings, compared to using standard pre-trained embeddings.

Keywords: Recurrent neural language model · Word embeddings · Debiased word embeddings

1 Introduction

Natural Language Processing techniques and applications have been acquiring increasing attention in the last decade. Artificial Intelligence approaches have recently been highly utilized in such techniques. Unfortunately, these approaches have proven to have some fairness problems [3]. Bias towards a particular gender, sex orientations or race is a common problem arising in most of machine learning applications. As a consequence, there is an open debate on the topic [7].

Scientists define bias to happen when a model outcomes differently given pairs of individuals that only differ in a targeted concept, like gender [8]. Therefore, more research is carried out to address this issue. There are approaches, e.g. [2,22], that focus on neutralizing word embeddings (which is the task of learning numerical representation of words) to solve the amplification of bias from data.

© Springer Nature Switzerland AG 2023
A. Gelbukh (Ed.): CICLing 2019, LNCS 13451, pp. 342–350, 2023.
https://doi.org/10.1007/978-3-031-24337-0_25

Other techniques, e.g. [10,18], work on debiasing the data at the source before it is consumed in training. Moreover, bias (and gender bias in particular) has been addressed in coreference resolution (which is the task of relating expressions that refer to the same thing), e.g. [24] and [20], showing the effectiveness of measuring and correcting these biases in such tasks.

In this paper, we approach understanding gender bias effect on language modeling. The goal of language modeling is to model the distribution of word sequences. Experiments were carried to understand the influence of data and previously debiased word embeddings on the language model.

The rest of the paper is organized as follows. Section 2 describes the techniques used to make this paper self-contained, which includes word embeddings, their debiased version and recurrent neural language models. Section 3 discusses the research questions that we are formulating in this study. Afterwards, Sect. 4 illustrates the experimental framework including data and system parameters. Results and conclusions are finally reported in Sect. 5 and 6, respectively.

2 Background

In this section, we are providing a brief overview of the Natural Language Processing techniques used along the paper which are word embeddings and their debiased version and recurrent neural language models.

2.1 Words Embeddings

Word embeddings are numerical representations of words and they can be computed by different approaches [16,17]. One of the most recent popular approaches is using word2vec [16]. Word2vec has two learning models: Continuous Bag of Words (CBOW) and Skip-gram. In CBOW, the word representation is predicted given its context, whereas in Skip-gram, the context is predicted given a word. Word2vec first builds a vocabulary from training corpus that is fed into the learning model and thus, learns the vector representations of each word. Word2Vec has an advantage of clustering similar sentences due to the ability to calculate the cosine distance among each word [9].

2.2 Debiased Word Embeddings

While word embedding models have become essential in Natural Language Processing applications, they have shown some shortcomings. Given that word embeddings are learned from human-generated corpora, they have been shown to learn and, even worse, amplify social biases [2,6].

Authors in [2] were the first to pay attention to the fact that the embeddings themselves have a kind of bias that should be resolved. To reduce the bias in an embedding, the embeddings of gender neutral words were changed, by removing their gender associations. A post processing method is applied that projects gender-neutral words to a subspace which is perpendicular to the gender dimension, defined by a set of gender definition words [24].

2.3 Recurrent Language Model

Language model is the task of computing the probability of sequences of words. One popular strategy of computing language modeling has been by using ngram models [21], which is mainly based on computing statistics of the appearances of chunks of words. An alternative methodology has been using feed-forward [1] or recurrent neural networks [15]. The main advantage of the neural networks approaches is employing word embeddings, which means their ability to use classes of words instead of using the words themselves. While the significant advantage of the recurrent networks is taking into account an unlimited number of words, whereas classical ngram models or feed forward networks have to use Markov assumptions limiting the context of a word to a size of a pre-defined window. In practice, vanilla recurrent neural networks are not enough to keep the information of long sequences, that is why gated methods such as long-short term memories (LSTMs) [5] or gated recurrent units (GRUs) [4] have been recently preferred.

3 Research Questions

Language modeling is trained on a large monolingual corpora. State-of-the-art neural language models either learn word embeddings inside the model or use pre-trained word embeddings [11]. Figure 1 outlines LSTM language modeling when using pre-trained embeddings as input to the model.

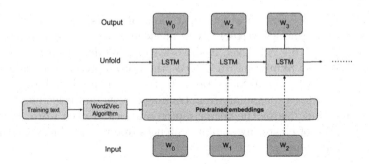

Fig. 1. LSTM language model using pre-trained embeddings

In this work, we approach answering the following questions:

1. Do standard language modeling training sets contain bias?
2. Does language modeling show gender bias if it is trained on a biased corpora?
3. Do current debiasing word embedding techniques reduce gender bias in language modeling?

To investigate these questions, we propose to train a recurrent language model and contrast its performance on different test sets with and without stereotypes. Next section describes our experiments.

4   Experimental Framework

This section reports details on the experimental framework of our study. We report details on the data and implementation used.

4.1   Datasets

One standard training set for language modeling is Wikitext-103, in which text is extracted from Wikipedia, consisting of 103,227,021 tokens and 267,735 vocabulary size [14].

To answer the first research question of this study (*Do standard language modeling training sets contain bias?*), pronouns counting is computed and shown in Fig. 2. The counting shows a clear under-representation of female pronouns. Female pronouns include *she, her, herself*; and male pronouns include *he, his, himself*.

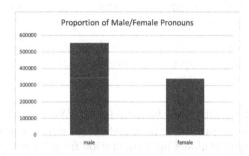

Fig. 2. Proportion of male/female pronouns in the training data

For testing, we used the stereotypes present in occupations. We composed 6 different test sets taking information from [2] and which are defined in Table 1.

For further evaluation, the models were experimented on a recent balanced dataset: the GAP-Coreference test-set [23]. This set is available from Google on github[1]. As reported, GAP is a gender-balanced dataset containing 8,908 coreference-labeled pairs. We extracted the text from the uploaded raw text to be suitable for our evaluations. Statistics of this set are shown in Table 2, given number of sentences, words, vocabulary (Vocab), and out-of-vocabulary (OOV).

4.2   Parameters

Regarding the words embeddings and the corresponding debiasing, we used the Gensim library [19] for training word2vec CBOW algorithm [16] on the training

---

[1] https://github.com/google-research-datasets/gap-coreference.

Table 1. Test sets composition

Test	Structure	Examples
1	Definitional male	He is a father, He is a male
2	Definitional female	She is a mother, She is a female
3	Stereotypical nouns for male	He is a surgeon, he is a ballplayer
4	Swapping the male's stereotypes with female pronouns	She is a surgeon, she is a ball player
5	Stereotypical nouns for female	she is a hairdresser, She is a ballerina
6	Swapping the female's stereotypes with male pronouns	He is a hairdresser, he is a ballerina

Table 2. Test sets statistics

Test set	Sentences	Words	Vocab	OOV
1, 2	37	150	42	4
3, 4	207	828	212	21
5, 6	60	240	65	5
GAP	2000	143155	33221	15917

corpus to create the related words embeddings. We used the implementation from Bolukbasi et al. [2] available from github[2] to debias these embeddings.

We used the language model implementation from Socher et al. [12,13] available from github[3]. Parameters are reported in Table 3.

Table 3. Parameters in language model.

Parameters	Value
Batch size	100
Embedding size	267734
Embeddings' dimension	400
Dropping-out rate	0.2
Sequence length	70

5   Results

Since we are using the same vocabulary for all our experiments, we relied on the perplexity (PP) which is an standard measure to evaluate the quality of the

---

[2] https://github.com/tolga-b/debiaswe.
[3] https://github.com/salesforce/awd-lstm-lm.

language model. Table 4 shows the results over the different test sets presented in previous section. This table allows us to answer the two remaining research questions, previously mentioned. Regarding the question: *Does language modeling show gender bias if it is trained on a biased corpora?*, the answer is affirmative as shown by the increase in perplexity from definitional *she* (test 2) compared to definitional *he* (test 1). Regarding the question: *Do current debiasing word embeddings reduce gender bias, compared to using pre-trained biased word embeddings?*, the answer seems to be affirmative again. This is shown by the lower relative increment in perplexity from definitional *he* to definitional *she* and the lower difference in perplexity when using the counterpart of either feminine, or masculine stereotypes -i.e. differences between test 1 and 2, test 3 and 4, and test 5 and 6, respectively. It is worth mentioning that the lowest relative increment occurs when using non-pretrained embeddings at all, which corresponds also to the best performing model by far.

Table 4. Perplexity on different test sets

System	Test 1	Test 2	Test 3	Test 4	Test 5	Test 6
Non Pre-train Emb	204.7	238.8 (+0,17)	247.3	307.6 (+0,24)	311.8	282.6 (−0,10)
Biased Pre-train Emb	345.7	524.6 (+0,52)	402.3	515.1 (+0,28)	598	469.8 (−0,22)
DeBiased Pre-train Emb	331.9	499.1 (+0,50)	377.8	481.3 (+0,27)	566.5	447.2 (−0,21)

For further analysis of test 4 (swapping the male's stereotypes with female pronouns) and test 6 (swapping the female's stereotypes with male pronouns), we show examples in Table 5. Table 5 (top) shows sentences from both test 3 and 4 with the corresponding perplexities before and after debiasing embeddings, while Table 5 (bottom) shows sentences from test 5 and 6. The perplexity for these sentences gets lower when using the pre-trained biased word embeddings and this reduction in perplexity can be interpreted as a neutralization of stereotypes. In both tables, the reduction of perplexity, in test 4 and 6, is higher than the reduction in test 3 and 5, respectively. However, the reduction in perplexity in test 4 is higher than that of test 6, affecting more stereotypes, emphasizing the conclusion that debiasing embeddings has a higher effect on females than on males. This correlates to the fact that females are underrepresented in the training corpus as shown in Fig. 2, making the debias in word embeddings more relevant in this case.

Finally, evaluation of the models on the GAP test-set is shown in Table 6. Perplexity results show the biased and debiased word2vec embeddings perform similarly on a balanced data.

Table 5. Perplexity variation in male and female stereotyped sentences from biased to debiased embedding

Sentence (M)	Bias	DeBias	Sentence (F)	Bias	DeBias
He is an archaeologist	740.0	687.7	She is an archaeologist	937.9	878.6
He is a ballplayer	1094.7	973.3	She is a ballplayer	1328.3	1253.6
He is a broadcaster	467.8	428.2	She is a broadcaster	593.7	542.2
He is a cardiologist	949.6	883.6	She is a cardiologist	1222.5	1120.6
He is a custodian	578.5	543.3	She is a custodian	753.8	700.1
He is an economist	716.8	662.5	She is an economist	926.5	865.9
He is a lawmaker	818	747.3	She is a lawmaker	1057.4	962.1
He is a parishioner	930.9	855.2	She is a parishioner	1174.5	1068.4
He is a photojournalist	914.5	829.2	She is a photojournalist	1224.5	1127.2
He is a protege	534	493.4	She is a protege	684.3	622.5
He is a provost	603.2	562.9	She is a provost	787.4	731
Sentence (F)	Bias	DeBias	Sentence (M)	Bias	DeBias
She is a dermatologist	1161.8	1094.7	He is a dermatologist	921.9	887.7
She is an organist	907.6	861.2	He is an organist	743.8	693.6
She is a paralegal	1181.2	1096.4	He is a paralegal	912.6	839.7
She is an observer	786.5	745.1	He is an observer	624.9	581.6

Table 6. Perplexity on a balanced test set

System	Perplexity
Non Pre-train Emb	340.51
Biased Pre-train Emb	1343.7
DeBiased Pre-train Emb	1343.6

6   Conclusions and Further Work

Generally, human-generated data under-represents females. Debiased pre-trained word embeddings have a decreasing effect on gender bias, when compared to the corresponding biased pre-trained embeddings. However, using self-trained word embeddings in language modeling results in the less biased and best-performing system.

Further directions in research include studying effect of debiasing embeddings on other Natural Language Processing applications. Moreover, studying ways to balance data to equally represent males and females. Another approach to be considered, is to investigate how to debias Natural Language Processing techniques. First, these techniques should discover if there is any type of bias, then should start balancing this bias within the system itself.

Acknowledgments. This work is supported in part by the AGAUR through the FI PhD Scholarship; the Spanish Ministerio de Economía y Competitividad, the European Regional Development Fund and the Agencia Estatal de Investigación, through the postdoctoral senior grant Ramón y Cajal, the contract TEC2015-69266-P (MINECO/FEDER,EU) and the contract PCIN-2017-079 (AEI/MINECO).

References

1. Bengio, Y., Ducharme, R., Vincent, P., Janvin, C.: A neural probabilistic language model. J. Mach. Learn. Res. **3**, 1137–1155 (2003). http://dl.acm.org/citation.cfm?id=944919.944966

2. Bolukbasi, T., Chang, K.W., Zou, J.Y., Saligrama, V., Kalai, A.T.: Man is to computer programmer as woman is to homemaker? debiasing word embeddings. In: Lee, D.D., Sugiyama, M., Luxburg, U.V., Guyon, I., Garnett, R. (eds.) Advances in Neural Information Processing Systems, vol. 29, pp. 4349–4357. Curran Associates, Inc. (2016). http://papers.nips.cc/paper/6228-man-is-to-computer-programmer-as-woman-is-to-homemaker-debiasing-word-embeddings.pdf

3. Chiappa, S., Gilliam, T.P.: Path-specific counterfactual fairness. arXiv:1802.08139 (2018)

4. Cho, K., et al.: Learning phrase representations using RNN encoder-decoder for statistical machine translation. In: Proceedings of the Conference on EMNLP, pp. 1724–1734 (2014). http://aclweb.org/anthology/D/D14/D14-1179.pdf

5. Hochreiter, S., Schmidhuber, J.: Long short-term memory. Neural Comput. **9**(8), 1735–1780 (1997). https://doi.org/10.1162/neco.1997.9.8.1735, http://dx.doi.org/10.1162/neco.1997.9.8.1735

6. Islam, A.C., Bryson, J.J., Narayanan, A.: Semantics derived automatically from language corpora necessarily contain human biases. Science **356**, 183–186 (2017)

7. Leavy, S.: Gender bias in artificial intelligence: the need for diversity and gender theory in machine learning. In: Proceedings of the 1st International Workshop on Gender Equality in Software Engineering, pp. 14–16. ACM (2018)

8. Lu, K., Mardziel, P., Wu, F., Amancharla, P., Datta, A.: Gender bias in neural natural language processing. CoRR abs/1807.11714 (2018), http://arxiv.org/abs/1807.11714

9. Ma, L., Zhang, Y.: Using word2vec to process big text data. In: 2015 IEEE International Conference on Big Data (Big Data), pp. 2895–2897. IEEE (2015)

10. Madaan, N., Singh, G., Mehta, S., Chetan, A., Joshi, B.: Generating clues for gender based occupation de-biasing in text. arXiv preprint arXiv:1804.03839 (2018)

11. Makarenkov, V., Shapira, B., Rokach, L.: Language models with glove word embeddings. CoRR abs/1610.03759 (2016)

12. Merity, S., Keskar, N.S., Socher, R.: Regularizing and optimizing LSTM language models. arXiv preprint arXiv:1708.02182 (2017)

13. Merity, S., Keskar, N.S., Socher, R.: An analysis of neural language modeling at multiple scales. arXiv preprint arXiv:1803.08240 (2018)

14. Merity, S., Xiong, C., Bradbury, J., Socher, R.: Pointer sentinel mixture models. arXiv preprint arXiv:1609.07843 (2016)

15. Mikolov, T., Karafiát, M., Burget, L., Cernocký, J., Khudanpur, S.: Recurrent neural network based language model. In: Kobayashi, T., Hirose, K., Nakamura, S. (eds.) INTERSPEECH, pp. 1045–1048. ISCA (2010). http://dblp.uni-trier.de/db/conf/interspeech/interspeech2010.html#MikolovKBCK10

16. Mikolov, T., Sutskever, I., Chen, K., Corrado, G.S., Dean, J.: Distributed representations of words and phrases and their compositionality. In: Burges, C.J.C., Bottou, L., Welling, M., Ghahramani, Z., Weinberger, K.Q. (eds.) Advances in Neural Information Processing Systems, vol. 26, pp. 3111–3119. Curran Associates, Inc. (2013). http://papers.nips.cc/paper/5021-distributed-representations-of-words-and-phrases-and-their-compositionality.pdf

17. Pennington, J., Socher, R., Manning, C.: Glove: global vectors for word representation. In: Proceedings of the 2014 Conference on Empirical Methods in Natural Language Processing (EMNLP), pp. 1532–1543. Association for Computational Linguistics, Doha, Qatar (2014). http://www.aclweb.org/anthology/D14-1162

18. Rao, S., Tetreault, J.: Dear sir or madam, may I introduce the YAFC corpus: corpus, benchmarks and metrics for formality style transfer. arXiv preprint arXiv:1803.06535 (2018)

19. Řehůřek, R., Sojka, P.: Software framework for topic modelling with large corpora. In: Proceedings of the LREC 2010 Workshop on New Challenges for NLP Frameworks, pp. 45–50. ELRA, Valletta, Malta (2010). http://is.muni.cz/publication/884893/en

20. Rudinger, R., Naradowsky, J., Leonard, B., Van Durme, B.: Gender bias in coreference resolution. arXiv preprint arXiv:1804.09301 (2018)

21. Stolcke, A.: SRILM - an extensible language modeling toolkit. In: INTERSPEECH. ISCA (2002)

22. Vera, M.F.: Exploring and mitigating gender bias in glove word embeddings (2018)

23. Webster, K., Recasens, M., Axelrod, V., Baldridge, J.: Mind the GAP: a balanced corpus of gendered ambiguous pronouns. CoRR abs/1810.05201 (2018)

24. Zhao, J., Wang, T., Yatskar, M., Ordonez, V., Chang, K.W.: Gender bias in coreference resolution: Evaluation and debiasing methods. arXiv preprint arXiv:1804.06876 (2018)

Initial Explorations on Chaotic Behaviors of Recurrent Neural Networks

Bagdat Myrzakhmetov[1,2]([✉]), Rustem Takhanov[1], and Zhenisbek Assylbekov[1]

[1] School of Science and Technology, Nazarbayev University, Astana, Kazakhstan
{bagdat.myrzakhmetov,rustem.takhanov,zhassylbekov}@nu.edu.kz
[2] National Laboratory Astana, Nazarbayev University, Astana, Kazakhstan

Abstract. In this paper we analyzed the dynamics of Recurrent Neural Network architectures. We explored the chaotic nature of state-of-the-art Recurrent Neural Networks: Vanilla Recurrent Network and Recurrent Highway Networks. Our experiments showed that they exhibit chaotic behavior in the absence of input data. We also proposed a way of removing chaos from Recurrent Neural Networks. Our findings show that initialization of the weight matrices during the training plays an important role, as initialization with the matrices whose norm is smaller than one will lead to the non-chaotic behavior of the Recurrent Neural Networks. The advantage of the non-chaotic cells is stable dynamics. At the end, we tested our chaos-free version of the Recurrent Highway Networks (RHN) in a real-world application. In the language modeling task, chaos-free versions of RHN perform on par with the original version.

Keywords: Chaos theory · Recurrent neural networks · Recurrent highway networks · Language modeling

1 Introduction

The dynamics of the Neural Networks has been studied in recent papers [2,4]. Laurent and Brecht [7] proposed to design architecture of a Recurrent Neural Network (RNN) cell in such a way that it is not chaotic. The concept of chaos [6, 8] comes from the theory of nonlinear dynamical systems and essentially means that wide divergence in outcomes of a system is due to small differences in initial conditions (such as those due to rounding errors in numerical computation). So, Laurent and Brecht show that the widely-used RNN cells, LSTM [5] and GRU [3], are chaotic. Depending on the initialization of the weights, LSTM and GRU might show a chaotic behavior. The proposed Chaos Free Network (CFN) architecture is devoid of chaos and is not inferior to LSTM. Recently, there were two main advancements over ubiquitous LSTM architecture: 1) Zoph and Le [20] used LSTM to generate a new RNN cell, which they refer to as a 'Neural Architecture Search' (NAS) cell; 2) Zilly et al. [1] extended the success of Highway networks [12] to recurrent networks and suggested a new RNN cell, which they refer to as a 'Recurrent Highway Network' (RHN) cell. Both, NAS

© Springer Nature Switzerland AG 2023
A. Gelbukh (Ed.): CICLing 2019, LNCS 13451, pp. 351–363, 2023.
https://doi.org/10.1007/978-3-031-24337-0_26

and RHN cells significantly outperform the LSTM cell in language modeling tasks when evaluated on a traditional PTB dataset [11]. Therefore the following questions arise: Are these new state of the art architectures chaotic? If so, then according to Laurent and Brecht [7] there should be non-chaotic alternatives that do not underperform significantly. And if there are no such analogs, can chaos be necessary after all? We will try to answer these questions in this paper.

We explored the state of the art Recurrent Highway Networks (RHN, [1]) and vanilla RNN [16] for chaotic behavior. Our experiments showed that both RHN and vanilla RNN, depending on the initialization of the weights, might show a chaotic behavior. This chaotic behavior may lead to a high degree of sensitivity to the initial state in the long-term behavior of forward orbits. We found out that the initialization of the weight matrices heavily affects chaoticity of Neural Networks.

2   Chaotic Nature of Simple Vanilla RNN

2.1   Vanilla RNN in $1D$ Case

In this section we consider the dynamics of the Vanilla Recurrent Neural Network. Before analyzing the complex neural network architectures of Recurrent Highway Network (RHN) and Neural Architecture Search (NAS), we started by analyzing the simple Recurrent Neural Network (RNN), proposed by Elman in 1990 [16], for chaoticity. So, for the Simple RNN architecture we want to discuss the nonlinear map $h_{t+1} = \tanh(Wh_t + Ux_t + b)$, where W and U are the weight matrices. We assume that there is no input data is provided, and the bias term is zero, so our system will become: $h_{t+1} = \tanh(Wh_t)$.

In this subsection, we consider the simple RNN architecture in 1D case, we assume that our values $h, W \in \mathbb{R}$, i.e. our state and weight are scalars.

Claim 1. *A dynamical system induced by Simple RNN:*

$$h_{t+1} = \tanh(Wh_t), \quad h_t, W \in \mathbb{R} \tag{1}$$

is non-chaotic when $W \in (-1, 1)$.

Fixed Point and Bifurcation Analysis. One of the main goals of bifurcation theory [13] is to find the fixed points and the periodic points of maps and then look for the region of their stability. The fixed points of the mappings are calculated by solving the equation $f(x) = x$. For our case, $h = \tanh(Wh)$, for $W \leq 1$ there is only one solution: $h = 0$, for other values, there are 3 solutions.

The implicit plot of $h = \tanh(Wh)$ is given in Fig. 1. Plots of h and $\tanh(Wh)$ for $W = 1, 2$ are provided in Fig. 2a and 2b.

To discuss the stability of the above fixed points, we can use the stability criterion which say that if $|f'(x)|_{x=x^*} < 1$ then the fixed point $x = x^*$ is stable, otherwise it is unstable [13].

In our case, we have fixed points $h = 0$, $h_1(W)$ and $h_2(W)$. For the first fixed point $h = 0$: $|f'(h)|_{h=0} = W(1 - \tanh^2(Wh)) = W(1 - \tanh^2(0)) = W(1 - 0) =$

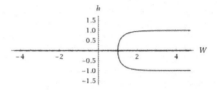

Fig. 1. Implicit plot of $h = \tanh(Wh)$.

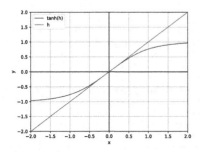

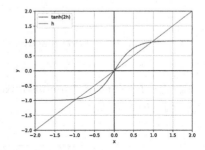

(a) One solution of $h = \tanh(h)$.   (b) Three solutions of $h = \tanh(2h)$.

Fig. 2. Solutions of $h = \tanh(Wh)$ for $W = 1$ and $W = 2$

W. So, by using the above notation, fixed point $h = 0$ is stable when $|W| < 1$, otherwise it is unstable. Hence $h = 0$ remains as a stable fixed point when $-1 < W < 1$.

As W crosses the value 1, the stable fixed point $h = 0$ becomes an unstable one. Thus, a qualitative change in the behavior of the fixed point occurs at $W = 1$ of the parameter value. So we consider $W = 1$ as the first bifurcation point. Also at $W = -1$, there also occurs a bifurcation.

To analyze the fixed points $h_1(W)$ and $h_2(W)$, we have to solve the inequality: $|W(1 - \tanh^2(Wh))| < 1$, where $h_1 \in (0, 1)$ and $h_2 \in (-1, 0)$. If this inequality holds, then the fixed points are stable. The solutions of this inequality will be $|W|\text{sech}^2(Wh) < 1$. Most of the values of h_1 and h_2 are close to 1 and -1. If we put these values of h into the inequality $|W|\text{sech}^2(Wh) < 1$, then this inequality holds for all W, therefore we do not consider these fixed points.

The bifurcation diagram of the $1D$ RNN is given in Fig. 3.

Since our main aim is to study the long term behavior of the map, so, after understanding the behavior of the fixed points of $f = \tanh(Wh)$, we now consider the periodic points of period 2 and higher and look into their stability property. The period 2 points are fixed points of the second order iteration of the map. So, let us consider the iterated map $f^2(h)$.

If we draw the graph of $f^2(h) = \tanh(\tanh(h))$ for $W = 1$ with the line $x = y$, there will be only one point of intersection, which is $h = 0$, which is already our first order fixed point of f. This graph is shown in Fig. 4.

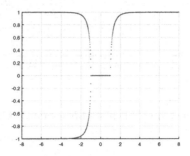

Fig. 3. Bifurcation diagram of the $1D$ RNN $h_{n+1} = \tanh(Wh_n)$.

The fixed points of f are also fixed points of f^2 as $f(h) = h \Rightarrow f(f(h)) = f(h) \Rightarrow f^2(h) = h$.

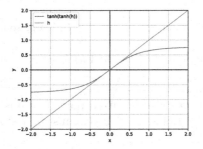

Fig. 4. Solutions of $h = \tanh(\tanh(h))$.

The period 2 points of the map are given by the solution of the equation: $f^2(h) = f(f(h)) = \tanh(W(\tanh(Wh))) = h$. We find the solutions $h \approx -1$, $h = 0$ and $h \approx 1$ for other values of W (except $W = 1$).

Let us see how the derivatives of the second iterate function change at the bifurcation value.

$$\frac{\partial}{\partial h}(\tanh(W\tanh(Wh))) = W^2\text{sech}^2(Wh)\text{sech}^2(W\tanh(Wh)).$$

Now let's analyze the bifurcation points:

$$|f'(h)| = W^2\text{sech}^2(Wh)\text{sech}^2(W\tanh(Wh)) \Rightarrow$$
$$|f'(h)|_{h=1} = W^2\text{sech}^2(W)\text{sech}^2(W\tanh(W)).$$

The value of the above equation is always less than the absolute values of $|1|$ for any values of W. So all points of W are stable on the second order periods.

So for all values of W, $W^2\text{sech}^2(W)\text{sech}^2(W\tanh(W))$ will be between -1 and 1.

Also for the second fixed point, we have

$$|f'(h)|_{h=-1} = W^2 \text{sech}^2(-W)\text{sech}^2(W\tanh(-W)).$$

Here also $-1 < W^2\text{sech}^2(-W)\text{sech}^2(W\tanh(-W)) < 1$ is for any values of W. This means that these two fixed points of f^2 are stable fixed points for all values of W and they will not become unstable. Periodic points of period 2 will not occur.

Lyapunov Exponent Analysis. As said before, a chaotic system is sensitive to initial conditions. Lyapunov exponent is the rate at which nearby trajectories diverge from each other with time [15] and a measure for identifying the chaoticity of the system [14]. Now let's consider two iterations of our map starting from two values of x which are very close to each other, i.e. with the very small difference δ: x_0 and $x_0 + \delta x_0$. Under the rule of the map let these points be shifted to $x_1, ..., x_n$ and $x_1 + \delta x_1, ..., x_n + \delta x_n$. If we expand $f(x)$ about x_n we have $\delta x_n = f'(x_{n-1})\delta x_{n-1}$ assuming that δx_n is sufficiently small. Hence the divergence of the two trajectories after n steps, δx_n, is related to their initial separation, δx_0, by $\left|\frac{\delta x_n}{\delta x_0}\right| = \prod_{i=0}^{n-1}|f'(x_i)|$.

We expect that this will vary exponentially at large n, $\left|\frac{\delta x_n}{\delta x_0}\right| = e^{\lambda n}$. So the Lyapunov exponent is defined by $\lambda = \lim_{n\to\infty} \frac{1}{n}\sum_{i=0}^{n-1}\ln|f'(x_i)|$. Obviously, if $\lambda > 0$, neighboring trajectories separate from each other at large n, which corresponds to chaos. However, if trajectories converge to a fixed point or a limit cycle they will get closer together, which corresponds to $\lambda < 0$. Hence, we can determine whether or not the system is chaotic by the sign ($+$ or $-$) of the Lyapunov exponent. Below, we have given the calculated values of the Lyapunov exponent for some values of the parameter W in case of the Simple RNN map. We have considered iteration size of 100000 to get the values.

In our case, $f(h) = \tanh(Wh)$ and $f'(h) = W(1 - \tanh^2(Wh))$.

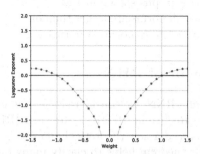

Fig. 5. Lyapunov coefficient versus W value.

In Fig. 5 we have shown the values of Lyapunov Exponent versus the weight value W. From this Figure we can see that the values of -1.1 and 1.1 of W the

Lyapunov Exponents become positive, showing the beginning of a chaotic region. Also this Fig. 5 further supports the first two bifurcation points as -1.0 and 1.0 where the Lyapunov Exponent is almost zero. Interestingly, after attaining the chaotic region at $W = 1.0$ and $W = -1.0$, we see the negative Lyapunov exponent values. They signify that within the chaotic region also, at certain values of the parameter, there are regular behaviors. This is supported also by the bifurcation diagram which we have drawn in Fig. 3 in the previous section. (After some values of W, it will stabilize).

2.2   Multidimensional Case for Vanilla RNN

Now, consider the high dimension cases. For vanilla RNN:

$$h_{(t+1)} = \tanh(Wh_t), \ h_t \in R^d \tag{2}$$

Claim 2. *If $||W|| < 1$, then for (2) we have the following statement: for any h_0 we have $\lim_{(n \to \infty)} h_t = 0$.*

Proof. $||h_{t+1}|| = ||\tanh(Wh_t)|| \leq ||Wh_t|| \leq ||W||||h_t||$. Therefore, $||h_t|| \leq ||W||^t||h_0|| \to 0$, $t \to \infty$.

Claim 3. *There exists W with $||W|| > 1$, such that induced dynamical system (2) is chaotic (means that there should be at least 1 nontrivial attractor, i.e. attractor which is not a point).*

When do we have these Claims 2 and 3? Let $||h||$ be a norm on $\mathbb{R}^d$ such that $||\tanh(h)|| \leq ||h||$. Examples of such norm are:

a) $||h||_p = (\sum_{i=1}^{d} h_i^p)^{1/p}$ is a l_p norm. For any such norm let us define corresponding matrix norm as $||W||_p = \max_{h:||h||_p=1} ||Wh||_p$.

Now, we tested the weight matrix, the norm of which is greater than 1. Lets consider the weight matrix $W = \begin{bmatrix} -1 & -4 \\ -3 & -2 \end{bmatrix}$. If we plot the graph of $h^{(1)}$ vs. t and $h^{(2)}$ vs. t, then we get the graphs shown in Figs. 6a and 6b.

In the above example, all norms are larger than one (Frobenius norm, nuclear norm (trace norm), max norm, l_1 norm, l_2 norm).

3   Chaotic Behavior of RHN

After exploring the vanilla RNN, we considered the dynamics of the state of the art RNN architectures. Recurrent Highway Network (RHN) was proposed by Zilly et al. [1] and introduced a new theoretical analysis based on the Geršgorins circle theorem [17]. This theorem helps to clarify many optimization issues and modeling. Their approach allows transition depths to be larger than one.

The main idea behind increasing the depth of the step-to-step recurrent state transition is to allow the RNN tick for several time steps per step of the sequence [18,19]. By using this technique we can adapt the recurrence depth to the problem.

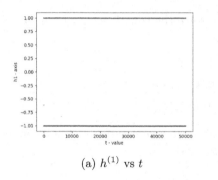

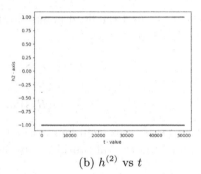

(a) $h^{(1)}$ vs t (b) $h^{(2)}$ vs t

Fig. 6. State vs. time graphs for $2D$ case when the norm is larger than one

RHN architecture is given in the following form: the Highway layer computation is defined as:

$$s_{t+1} = p \odot (h - s_t) + s_t \qquad (3)$$

where

$$p := \sigma(W_p x + R_p s + b_p); \qquad (4)$$

$$h := \tanh(W_h x + R_h s + b_h); \qquad (5)$$

$\odot$ denotes Hadamard product.

3.1 RHN Chaoticity in 1D

In this subsection we analyze the dynamics of the Recurrent Highway Network (RHN) in 1D case. First, we can start with the analysis of fixed points and check the region of their stability.

If we assume that no input is provided, then the induced form of the RHN will become:

$$p := \sigma(R_p s); \qquad (6)$$

$$h := \tanh(R_h s); \qquad (7)$$

$$s_{t+1} = p \odot (h - s_t) + s_t. \qquad (8)$$

If we put everything together, we will get the following equation:

$$s_{t+1} = \sigma(R_p s_t) \odot (\tanh(R_h s_t) - s_t) + s_t. \qquad (9)$$

For the Recurrent Highway Networks we have the following claim.

Claim 4. *A dynamical system induced by RHN in Eq. 9 shows non-chaotic behavior when $R \in (-1, 1)$, as thus follows from the properties of Vanilla RNN.*

Proof. To find the fixed points, we have to solve the equation: $x = f(x)$, so for the Eq. 9 we will have:

$$s = \sigma(R_p s) \odot (\tanh(R_h s) - s) + s \Rightarrow$$
$$0 = \sigma(R_p s) \odot (\tanh(R_h s) - s) \Rightarrow$$
$$0 = \sigma(R_p s) \text{ and } 0 = (\tanh(R_h s) - s)$$

$0 = \sigma(R_p s) \Rightarrow$ no solutions exists, as the values of sigmoid function lies between 0 and 1. So we will consider only the second part.

$$0 = \tanh(R_h s) - s \Rightarrow$$
$$s = \tanh(R_h s)$$

This equation $s = \tanh(R_h s)$ is the case for the simple RNN. We already considered the fixed point analysis of the simple RNN in Sect. 2.1. So the fixed points of the Recurrent Highway Networks are the same as the vanilla RNN and the fixed point analysis of the Vanilla RNN can be applied for the RHN. This proves our Claim 4.

3.2   RHN Chaoticity in 2D

Claim 5. *There exists R: $\|R\| > 1$ such that a dynamical system induced by RHN in Eq. 9 is chaotic.*

We performed experiments to check the chaotic behavior of the RHN in 2D. We show that in the absence of the input data RHN can lead to dynamical systems $s_{t+1} = \Phi(s_t)$ that are chaotic [6]. Again, we assume that there is no input data is provided. Then the dynamical system induced by a two-dimensional RHN with weight matrices: $R_p = \begin{bmatrix} 0 & 1 \\ 1 & 0 \end{bmatrix}$ and $R_h = \begin{bmatrix} -5 & -8 \\ 8 & 5 \end{bmatrix}$ and zero bias for the model. s can be initialized with any values. If we assume that no input data is provided and all bias terms are zero, then the induced RHN architecture will become as in Eqs. 6, 7 and 8.

Now we plot the RHN state values $s_t^{(1)}$ vs. $s_t^{(2)}$ for $t = 100000$ iterations. The resulting plot is shown in Fig. 7. Most trajectories converge toward the depicted attractor. We can get above pictures for any initial values of s (we can initialize with zeros or any values) and for any number of highway layers (we tried 1, 5, 10 highway layers). This picture shows the strange attractor as in LSTM and GRU given in Laurent and Brecht [7].

Now we studied time series analysis of this system. If we plot s^1 vs. t we can notice that the values of s^1 will jump from one place to another in the chaotic manner. There is no convergence. This is given in Fig. 8a. This is also true for s^2 vs. t (given in Fig. 8b). Then, if we plot the graph s^1 vs. s^2, we can get the strange attractor as shown in Fig. 7.

Next we tested chaoticity of the RHN by using the Lyapunov instability of Bernoulli shift [8] as in Sect. 2.1. We consider the two points which are initially

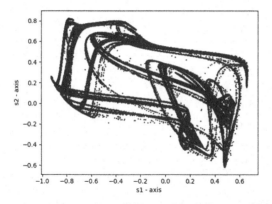

Fig. 7. Strange attractor of chaotic behavior of RHN for the weight matrices: R_p=[[0, 1],[1, 0]] and R_h=[[−5, −8],[8, 5]]

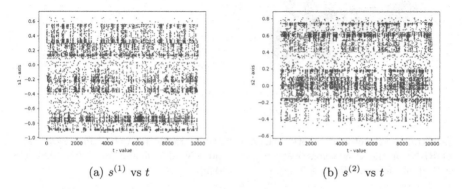

(a) $s^{(1)}$ vs t (b) $s^{(2)}$ vs t

Fig. 8. State vs. time graphs for 2D case RHN

very close to each other, with δs_0 "infinitesimally small" differences: $\delta s_0 :=$ $|\hat{s}_0 - s_0|$. Then we iterate these two points through our induced RHN map $s_{t+1} = \Phi(s_t)$, in Eq. 8, 100 times and calculated the Euclidean distance between $|\hat{s}_t - s_t|$ these points. The graph is given in Fig. 9. From this graph we can see that after some iteration, two trajectories diverge exponentially despite the fact that initially these two points are highly localized, with the distance no more than 10^{-7}.

Also we tested the weight matrices

$R_p = \begin{bmatrix} -2 & 6 \\ 0 & -6 \end{bmatrix}$ and $R_h = \begin{bmatrix} -5 & -8 \\ 8 & 5 \end{bmatrix}$. The norm of these matrices are larger

than one. If we plot the graph of s^1 vs. s^2 with $t = 100000$, then we again explore the strange attractor as shown in Fig. 10a. Here again, for any initial value of s and for any number of highway layers we will get this picture.

After exploring the chaotic behavior in RHN, we now tried to build chaos-free Neural Networks. For RHN, we again use our Claim 2 which was applied

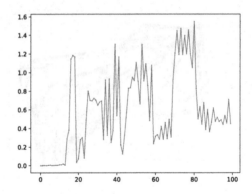

Fig. 9. $|\hat{s}_t - s_t|$ for 2 trajectories: s_0 and $s_0 + \delta s_0$

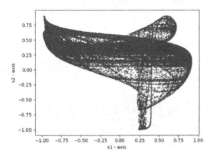

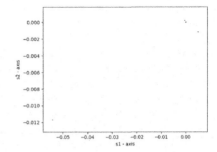

(a) Strange attractor of chaotic behavior of RHN for the weight matrices: $R_p =$ [[-2, 6], [0, -6]] and R_h = [[-5, -8], [8, 5]]

(b) Attractor for weight matrices: $R_p =$ [[0, 0.5], [0.5, 0]] and R_h=[[-0.5, -0.8], [0.8, 0.5]]

Fig. 10. Strange and regular attractor of chaotic and non-chaotic RHN for 2D case

in vanilla RNN. Here, if we initialize the weights with matrices whose norm is smaller than one, then again we can have the non-chaotic behavior in RHN.

In the above cases, the norm of the weight matrices are larger than one and we explored the chaotic behavior. Now, let's analyze the case when the norm of the matrices are smaller than 1. For example, we can test these weight matrices:

$$R_p = \begin{bmatrix} 0 & 0.5 \\ 0.5 & 0 \end{bmatrix} \text{ and } R_h = \begin{bmatrix} -0.5 & -0.8 \\ 0.8 & 0.5 \end{bmatrix}.$$

The norm of these two matrices are smaller than one. If we explore the values of s^1 and s^2 for $t \to \infty$, then, both values of s will go to zero. We also plotted the graph of s^1 vs. s^2. The plot is given in Fig. 10b. From this, we can see that we can get non-chaotic RHN when we initialize the weight matrices with the values whose norm is smaller than one.

4   Experiments

In this section, we tested our non-chaotic neural cells in real-world applications. Our aim is to identify, how non-chaotic version will affect on the performance. Is a non-chaotic behaviour good in a real-world application? Do we need a chaoticity? Or is it good to have a chaotic systems? To answer these questions, we performed experiments.

We examined Recurrent Highway Networks on the language modeling task. We use Penn Tree Bank (PTB) [11] corpus, which was pre-processed by Mikolov et al. [10]. First we reproduced the initial results from Zilly et al. [1] without weighting (WT) of input and output mappings and got the 68.355 perplexity on the validation set and 65.506 perplexity on the test set. These results are similar to the results in the paper (In the paper it was 67.9 and 65.4).

Then we tested our chaos-free version. We initialized the weight matrix in a way, such that their Frobenius norm do not exceed 1. We use TensorFlow [9] to perform our experiments. We first created a matrix whose norm is smaller than one and feed it during the initialization. We used the same hyper-parameters as in Zilly et al. [1] during the training. On PTB dataset, our non-chaotic neural cells showed 68.715 perplexity on the validation set and 66.290 perplexity on the test set. Full results and results of Chaos Free Network (CFN) [7] are given in Table 1. From this, we can see that the chaos free version of RHN showed similar results as the chaotic version and that chaos-free initialization will not lead to a decrease in performance.

Table 1. Perplexity on the PTB set.

Model	Validation perplexity	Test perplexity
Variational RHN + WT [1]	68.355	65.506
Non-chaotically initialized RHN	**68.715**	**66.290**
CFN (2 layers)+dropout [7]	79.7	74.9

5   Conclusion and Future Work

In this paper we analyzed the dynamics of the Recurrent Neural Networks. Our analyses showed that the vanilla RNN and the most recent RHN architecture exhibit a chaotic behavior in the absence of input data. We found out that, depending on the initialization of the weight matrices, we can have non-chaotic systems. Our experiments showed that the initialization of the weights with the matrices whose norm is less than one can lead to non-chaotic behavior. The advantage of non-chaotic cells is stable dynamics. We also performed experiments with non-chaotic RHN cells. Our experiments on language modeling with the PTB dataset showed similar results as an RHN cell with chaos by using the same hyper-parameters. In the future, we are going to test non-chaotic RHN cells for other tasks: speech processing, image processing. Also for NAS architecture, at this moment, generating the architecture is an expensive process for us, as there are not enough resources. We will test our chaos-free initialization for NAS architectures again.

Acknowledgement. This work has been funded by the Committee of Science of the Ministry of Education and Science of the Republic of Kazakhstan, IRN AP05133700. The work of Bagdat Myrzakhmetov partially has been funded by the Committee of Science of the Ministry of Education and Science of the Republic of Kazakhstan under the research grant AP05134272. The authors would like to thank Professor Anastasios Bountis for his valuable feedback.

References

1. Zilly, J.G., Srivastava, R.K., Koutník, J., Schmidhuber, J.: Recurrent highway networks. In: International Conference on Machine Learning, pp. 4189–4198 (2017)
2. Sussillo, D., Barak, O.: Opening the black box: low-dimensional dynamics in high-dimensional recurrent neural networks. Neural Comput. **25**(3), 626–649 (2013)
3. Cho, K., et al.: Learning phrase representations using RNN encoder-decoder for statistical machine translation. arXiv preprint arXiv:1406.1078 (2014)
4. Pascanu, R., Mikolov, T., Bengio, Y.: On the difficulty of training recurrent neural networks. In: International Conference on Machine Learning, pp. 1310–1318 (2013)
5. Hochreiter, S., Schmidhuber, J.: Long short-term memory. Neural Comput. **9**(8), 1735–1780 (1997)
6. Strogatz, S.H.: Nonlinear Dynamics and Chaos: With Applications to Physics, Biology, Chemistry, and Engineering. Westview press, Boulder (2014)
7. Laurent, T., von Brecht, J.: A recurrent neural network without chaos. arXiv preprint arXiv:1612.06212 (2017)
8. Ott, E.: Chaos in Dynamical Systems. Cambridge University Press, Cambridge (2002)
9. Abadi, M., et al.: Tensorflow: a system for large-scale machine learning. In: OSDI, vol. 16, pp. 265–283 (2016)
10. Mikolov, T., Karafiát, M., Burget, L., Černocký, J., Khudanpur, S.: Recurrent neural network based language model. In: Eleventh Annual Conference of the International Speech Communication Association (2010)
11. Marcus, M.P., Marcinkiewicz, M.A., Santorini, B.: Building a large annotated corpus of English: the Penn treebank. Comput. Linguist. **19**(2), 313–330 (1993)
12. Srivastava, R.K., Greff, K., Schmidhuber, J.: Training very deep networks. In: Advances in Neural Information Processing Systems (NIPS), pp. 2377–2385 (2015)
13. Kuznetsov, Y. A.: Elements of Applied Bifurcation Theory, vol. 112. Springer, Cham (2013)
14. Wolf, A., Swift, J.B., Swinney, H.L., Vastano, J.A.: Determining Lyapunov exponents from a time series. Physica D **16**(3), 285–317 (1985)
15. Lyapunov, A.M.: The general problem of the stability of motion. Int. J. Control **55**(3), 531–534 (1992)
16. Elman, J.L.: Finding structure in time. Cogn. Sci. **14**(2), 179–211 (1990)
17. Geršgorin, S.: Über die Abgrenzung der Eigenwerte einer Matrix Bulletin de l'Académie des Sciences de l'URSS. Classe des sciences mathématiques et na, no. 6, 749–754 (1932)

18. Srivastava, R.K., Steunebrink, B.R., Schmidhuber, J.: First experiments with POWERPLAY. Neural Netw. Official J. Int. Neural Netw. Soc. **41**, 130–136 (2013)
19. Graves, A.: Adaptive computation time for recurrent neural networks. arXiv preprint arXiv:1603.08983 (2016)
20. Zoph , B., Le, Q.V.: Neural architecture search with reinforcement learning. arXiv preprint arXiv:1611.01578 (2016)

Lexical Resources

LingFN: A Framenet for the Linguistic Domain

Shafqat Mumtaz Virk[1]([⊠])⬤, Per Klang[2]⬤, Lars Borin[1]⬤, and Anju Saxena[3]⬤

[1] Språkbanken, University of Gothenburg, Gothenburg, Sweden
virk.shafqat@gmail.com, lars.borin@svenska.gu.se
[2] Department of Scandinavian Languages, Uppsala University, Uppsala, Sweden
per.klang@nordiska.uu.se
[3] Department of Linguistics and Philology, Uppsala University, Uppsala, Sweden
anju.saxena@lingfil.uu.se

Abstract. Frame semantics is a theory of meaning in natural language, which defines the structure of the lexical semantic resources known as framenets. Both framenets and frame semantics have proved useful for a number of natural language processing (NLP) tasks. However, in this connection framenets have often been criticized for their limited coverage. A proposed reasonable-effort solution to this problem is to develop domain-specific (sublanguage) framenets to complement the corresponding general-language framenets for particular NLP tasks, and in the literature we find such initiatives covering domains such as medicine, soccer, and tourism. In this paper, we report on building a framenet to cover the terms and concepts encountered in descriptive linguistic grammars (written in English) i.e. a framenet for the linguistic domain (LingFN) to complement the general-language BFN.

Keywords: Frame semantics · Framenets · Semantics

1 General Background and Introduction

Frame semantics is a theory of meaning introduced by Charles J. Fillmore and colleagues [8–10]. The backbone of the theory is a conceptual structure called **a semantic frame**, which is a script-like description of a prototypical situation that may occur in the real world, and can refer to a concept, an event, an object, or a relation. The major idea is that word meanings are best understood when studied from the point of view of the situations to which they belong, rather than looking at them separately and analyzing them individually. For example, consider a situation in which someone (the perpetrator) carries and holds someone (the victim) against his/her will by force – a kidnapping situation. With reference to this situation, words like *kidnap*, *abduct*, *nab*, *snatch*, etc. are easier to understand, which also involves identification and linking of the participants of the situation (perpetrator, victim, purpose, place, etc. in the case of the *kidnapping* situation). In the world of frame semantics, the words which evoke semantic frames are called *lexical units* (LU) or *triggers*, and the participants and props in those frames are known as frame elements (FE). There are two types of FEs: core and non-core. Frame elements which are obligatory for the situation to make sense are *core*, while

© Springer Nature Switzerland AG 2023
A. Gelbukh (Ed.): CICLing 2019, LNCS 13451, pp. 367–379, 2023.
https://doi.org/10.1007/978-3-031-24337-0_27

others are optional, i.e. *non-core*. For example, in the case of a KIDNAPPING frame, the PERPETRATOR and VICTIM are core, while peripheral elements like PURPOSE, PLACE, MANNER, etc., are non-core FEs.[1]

FrameNet [1] – also commonly known as Berkeley FrameNet (BFN) – is a lexical semantic resource for English which is based on the theory of frame semantics. BFN contains descriptions of real world situations (i.e. semantic frames) along with the participants of those situations. Each of the semantic frames has a set of associated words (i.e. LUs) which evoke a particular semantic frame. The participants of the situations (i.e. FEs) are also identified for each frame. In addition, each semantic frame is coupled with example sentences taken from spoken and written discourse. FrameNet 1.5 has 1,230 semantic frames, 11,829 lexical units, and 173,018 example sentences. Further, a set of documents annotated with semantic frames and their participant information is also provided. In naturally occurring language, words mostly do not stand on their own, rather are interlinked through various syntactic, semantic, and discourse relations. To cope with this property of natural language, FrameNet has defined a set of frame-to-frame relations, and has proposed connections of certain frames to certain other frames thus providing a network of frames (hence the name FrameNet).

BFN and its annotated data have proved to be very useful for automatic shallow semantic parsing [11], which itself has applications in a number of natural language processing (NLP) tasks including but not limited to information extraction [18], question answering [17], coreference resolution [16], paraphrase extraction [13], and machine translation [21]. Because of their usefulness, framenets have also been developed for a number of other languages (Chinese, French, German, Hebrew, Korean, Italian, Japanese, Portuguese, Spanish, and Swedish), using the BFN model. This long standing effort has contributed extensively to the investigation of various semantic characteristics of many languages at individual levels, even though most crosslinguistic and universal aspects of the BFN model and its theoretical basis still remain to be explored.[2] Though framenets and the frame-annotated data have proved to be very useful both for the linguistic and the NLP communities, they have often been criticized for their lack of cross-linguistic applicability and limited coverage. In an ongoing project,[3] attempts are now being made to align framenets for many languages in order to investigate and test the cross-linguistic aspects of the FrameNet and its underlying theoretical basis. A proposed reasonable-effort solution to the coverage issue is to develop domain-specific (sublanguage) framenets to augment and extend the general-language framenet. In the literature, we can find such initiatives where domain-specific framenets are being developed, e.g.: (1) medical terminology [4]; (2) *Kicktionary*,[4] a soccer language framenet;

---

[1] Labels of FEs and frames are conventionally set in small caps.

[2] Most of the framenets – including BFN – have been developed in the context of linguistic lexicology, even if several of them have been used in NLP applications (again including BFN). The Swedish FrameNet (SweFN) forms a notable exception in this regard, having been built from the outset as a lexical resource for NLP use and only secondarily serving purposes of linguistic research [2,6].

[3] http://www.ufjf.br/ifnw/.

[4] http://www.kicktionary.de/.

and (3) the *Copa 2014* project, covering the domains of soccer, tourism and the World Cup in Brazilian Portuguese, English and Spanish [19].

Like many others, the area of linguistics has developed a rich set of domain specific terms and concepts (e.g. *inflection, agreement, affixation*, etc.). Such terms have been established by linguists over a period of centuries in the course of studying, investigating, describing and recording various linguistic characteristics of a large number of different languages at phonological, morphological, syntactic, and semantic levels. For various computational purposes, attempts have been made to create inventories of such terms and keep a record of them, e.g.: (1)he *GOLD*[5] ontology of linguistic terms; (2) the *SIL glossary of linguistic terms;*[6] (3) the *CLARIN concept registry;*[7] and (4) *OLiA* [5].

A minority of the terms in the collections above are used only in linguistics (e.g., *tense* in its noun sense), and in many cases, non-linguistic usages are either rare (e.g., *affixation*) or specific to some other domain(s) (e.g., *morphology*). Others are polysemous, having both domain-specific and general-language senses. For example, in their usage in linguistics the verb *agree* and the noun *agreement* refer to a particular linguistic (morphosyntactic) phenomenon, where a syntactic constituent by necessity must reflect some grammatical feature(s) of another constituent in the same phrase or clause, as when adjectival modifiers agree in gender, number and case with their head noun. This is different from the general-language meaning of these words, implying that their existing FN description (if available) cannot be expected to cover their usage in linguistics, which we will see below is indeed the case. Naturally, we need to build new frames, identify their LUs and FEs, and find examples in order to cover them and make them part of the general framenet if we are to extend its coverage. This is one of the major objectives of the work we report in this paper. The other objective is to investigate the relational aspects of the resulting linguistic frames. In the GOLD ontology, attempts were made to divide and organize the linguistic concepts into various groups. This organization is not without problems. GOLD seems to lack a theoretical foundation, and the validity of the organization remains untested. Also there is only one type of default relation – IS-A – between the terms/concepts. We aim to extend the relational structure of linguistic frames by exploring new relation types between the linguistic terms/concepts and building a network (i.e. Linguistic FrameNet – LingFN) of them.

The rest of the paper is organized as follows: Sect. 2 briefly describes the data sets that we have used, Sect. 3 outlines the architecture of the framenet, and the methodology is given in Sect. 4. This is followed by the application of the linguistic framenet (Sect. 5) and the conclusion and future work (Sect. 6).

2 The Data

The *Linguistic Survey of India* (LSI) [12] presents a comprehensive survey of the languages spoken in South Asia. It was conducted in the late nineteenth and the early twentieth century by the British government, under the supervision of George A. Grierson.

---

[5] http://linguistics-ontology.org/.

[6] http://glossary.sil.org.

[7] https://www.clarin.eu/ccr.

The survey resulted in a detailed report comprising 19 volumes of around 9,500 pages in total. The survey covered 723 linguistic varieties representing the major language families and some unclassified languages, of almost the whole of nineteenth-century British-controlled India (modern Pakistan, India, Bangladesh, and parts of Burma). Importantly for our purposes, for each major variety it provides a grammatical sketch (including a description of the sound system).

The LSI grammar sketches provide basic grammatical information about the languages in a fairly standardized format. The focus is on the sound system and the morphology (nominal number and case inflection, verbal tense, aspect, and argument indexing inflection, etc.), but there is also syntactic information to be found in them. Despite its age, it is the most comprehensive description available of South Asian languages, and since it serves as the main data source in a large linguistic project in which we are involved, it is natural for us to use it as a starting point for the development of LingFN, but in the future we plan to extend our range and use other publicly available digital descriptive grammars.

3   The General Architecture of LingFN

Figure 1 shows the general architecture of LingFN.

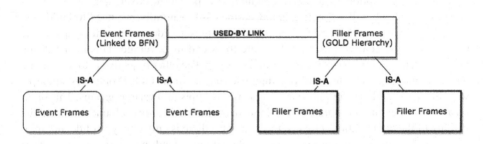

Fig. 1. The basic structure of LingFN

3.1   Frame Types

As shown in Fig. 1, there are two main types of frames: *filler frames* and *event frames*. The frames which are comparatively a bit more complex in their structure and represent eventful types of scenes (or concepts) are called event frames. In contrast filler frames are simple in their structure and may act as slot fillers to the event frames. To explain these two types of frames further, let us take an example frame from the BFN. The BORROW frame holds information about a situation involving a BORROWER that takes possession of a THEME belonging to a LENDER under the tacit agreement that the THEME be returned after a DURATION of time. Observe the annotated sentence below:

(1)   Does [BORROWER my Mum] [LU borrow] [THEME money] [LENDER off you]? (BFN)

The LU in Example 1 evokes an entire scene containing various roles, or FEs. These type of complex frames are referred to as *event frames* in LingFN. A corresponding example from the linguistic domain could be the AGREEMENT frame. Consider the following annotated example:

(2) [The PARTICIPANT-1 participle] [LU agrees] in [GRAMMATICAL_CATEGORY gender and number] with [PARTICIPANT_2 the object] [CONDITION if the latter is in the form of the nominative].

These stand in contrast to frames with a less eventful structure (i.e. *filler frames*, which usually fill in the roles of the event frames). An annotated example of a filler frame (the MONEY frame) is given below in bold, followed by the PARTICIPLE frame from the linguistic domain.

(3) Does [BORROWER my Mum] [LU borrow] [THEME [MONEY **money**]] [LENDER off you]?   (BFN)

(4) [The PARTICIPANT-1 [PARTICIPLE **participle**]] [LU agrees] in [GRAMMATICAL_CATEGORY gender and number] with [PARTICIPANT_2 the object] [CONDITION if the latter is in the form of the nominative].

3.2 Frame-to-Frame Relations

The above described frame types are ordered hierarchically by the (frame-to-frame) IS-A relation. This relation is intended to preserve the GOLD ontology structure by linking a top-level frame to a frame at the next lower level. For example to preserve the fact that a clitic is a morpheme, the corresponding CLITIC frame is linked to the MORPHEME frame by an IS-A relation. Traditionally, the IS-A relation is used for a much richer inheritance type of linkage, where the lower level entity inherits certain attributes from the upper level entity. At this stage we are using IS-A in a simpler sense to preserve only the structural level information but in future, we intend to enhance this type of linking.

As mentioned above, filler frames may appear as FEs of the event frames. Any observation of a filler frame occurring in an event frame is documented in LingFN by means of connecting the frames with a used-by link. For example, consider the annotated example sentence from the LINGUISTIC PLACING frame below.

(5) [FORM Adverbs] [COP are] [LU placed] [MORPHOSYNTACTIC_POSITION before adjectives and after verbs] .   (LINGUISTIC_PLACING)

The FEs in Example 5 contain some words that are relevant for the linguistic domain, such as, *adverb*, *adjective* and *verb*. These words evoke certain filler frames, e.g. GOLD_VERBAL, GOLD_ADJECTIVAL, and GOLD_PART_OF_SPEECH_PROPERTY. When an event frame FE contains an LU that is found in the filler frames, this connection is recorded in LingFN as used-by link.

4 Methodology

The development of a domain specific framenet involves (1) selection of a framenet development strategy; (2) identification and construction of new domain specific frames; and (3) identification and development of FEs of the newly identified frames. In this section, we describe the methodology for each of these stages.

4.1 Framenet Development

At the framenet level, there are at least four different types of methodologies which have been previously discussed in the literature, namely *Lexicographic Frame-by-Frame*; *Corpus-Driven Lemma-by-Lemma*; *Full-Text*; and *Domain-by-Domain*.

In our case, we have used a hybrid of the lexicographic and the corpus-driven approach. The choice is largely driven by the available resources and the objectives of the project for which LingFN is being developed. As mentioned above, there are available inventories of linguistic terms and concepts. This means one could start with entries in one of those inventories (e.g. GOLD) and build semantic frames for them (i.e., the frame-by-frame strategy). This is actually how we started. As can be expected, GOLD's coverage is limited. Also, the objectives of the project require that we cover the LSI corpus. So, we develop new frames when encountering corpus data that is unattested in GOLD (i.e., the corpus-driven approach).

4.2 Frame Identification and Construction

Frame Identification. Before we can construct a frame, we need to decide when and what domain-specific frames we need to design. In the first round, we started with the GOLD list of linguistic terms and developed corresponding frames. In the second round, we scanned through the LSI corpus, and began developing frames for the linguistic terms/concepts found in the corpus but not in GOLD. Here, we were faced with an additional issue of deciding which terms/concepts are specific to the linguistic domain and resolving the ambiguous cases. Since we are using a domain specific corpus, an assumption in this regard could be that the terms within a domain-specific corpus are mostly related to that particular domain. Since this can not be guaranteed, we have to deal with the polysemous occurrence of the terms. For this purpose, we have used *semi-automatic uniqueness differentiation* (SUDi) [14], a corpus-driven statistical method to judge the polysemous nature of the lemmas.

Frame Construction. Once we have chosen a frame (in the case of the GOLD list), or identified a frame (in the case of scanning through the LSI corpus), to be developed, we construct the corresponding frame and put it into a frame repository. Constructing a frame requires three things: (1) identification of frame LUs; (2) identification of FEs; and (3) creation and storage of the frame as a lexical entry in a dedicated lexical database.

For lexical unit identification, we have largely relied on linguistic intuition and general linguistic knowledge. For example, for the semantic frame AFFIXATION, the identified list of lexical units is *{affix.v, affixed.a, infix.v, infixed.a, postfixed.a, prefix.v, suffix.v,* *suffixed.a}*. The FE identification involves the recognition of various semantic roles of the frame. We have used a corpus-driven approach for the identification purpose, which is described in detail in Sect. 4.3. For the third part, we have used a web-based frame editor provided by Språkbanken's Karp infrastructure [3]. The tool was built as part of the Swedish FrameNet++ project [2,7], and was used for creation of Swedish frames. Figure 2 shows a snapshot of the editor, and as can be seen, the creation of a new frame means filling in various fields making part of the structure of the frame. Most of the fields shown in the figure should be self-explanatory, but the following may require explanation:

Fig. 2. A frame in the Karp editor

Used-By Links: This entry holds the names of the event frames that uses the filler frames. For instance, take the LINGUISTIC PLACING frame. This frame contains various verbal and adjectival words denoting acts of placing things. An annotated example follows:

(6)   It will be seen that [THEME the personal pronoun which we translate as a possessive] is [FREQUENCY often] [LU put] [GRAMMATICAL_CATEGORY in the nominative] [GOAL before such prefixes].

The following result could be seen, if the example above were annotated with the filler frames (fillers in boldface):

(7)   It will be seen that [THEME the **[PRONOUN pronoun]** which we translate as a possessive] is [FREQUENCY often] [LU put] [GRAMMATICAL_CATEGORY in the **[CASE nominative]]** [GOAL before such **[AFFIX prefixes]]**.

So, from the example above, it is seen that the LUs from the following GOLD frames are used by the event frame:

- Frame: GOLD_PRONOUN = [SUBCLASS possessive] [SUBCLASS personal] [LU pronoun]
- Frame: GOLD_CASE = [LU nominative]
- Frame: GOLD_AFFIXATION = [LU prefixes]

The Core Elements and Peripheral Elements: These two fields are supposed to contain a list of core and non-core FEs belonging to the frame. At this stage we do not distinguish between these FE types, but in the future the plan is to make use of this distinction. The process of identifying the FEs for a given frame is described in Sect. 4.3. Below we list various FEs together with a brief description of the type of information they contain.

- Language_Variety: This FE is intended to record the name of a language variety.
- Reference_Language: Often, while describing a particular aspect of a language a reference is made to another language (e.g. in the sentence 'The verb agrees with its subject in gender, number and person as in Hindostani' Reference_Language is 'Hindostani'.)
- Data: Often, examples are given to describe a particular aspect, this FE is intended to record those examples.
- Data_Translation: The English glossing of examples recorded by the Data FE.
- Subclass: The subclass of the entity/phenomenon being described by the frame e.g. 'interrogative' in the 'interrogative adjective'.
- Position: The position of the entity being described e.g. 'initial' in 'initial soft consonants....'
- Frequency: The frequency of the phenomenon being described e.g. 'often' in 'Adjectives are often followed by a suffix...'
- Degree: The degree of the phenomenon being described e.g. 'strong' in 'Initial soft consonants are pronounced with a very strong aspiration.'
- Condition: The condition in which the phenomenon being described occurs e.g. 'if the object is of the second person' in 'subject of the first person is not separately marked if the object is of the second person.'
- Means: The means by which the described phenomenon is materialized e.g. 'by means of postpositions' in 'The pronouns are inflected like nouns by means of postpositions '
- Manner: The manner in which the described phenomenon takes place e.g. 'without any suffix' in 'The genitive is apparently formed by prefixing the governed to the governing word without any suffix'.
- Certainty: The certainty of the phenomenon being described e.g. 'may' in 'Both the hard and the soft sounds aspirant may be either aspirated or unaspirated'.

Examples: A list of example sentences from the LSI corpus annotated with LUs and FEs is provided in this field. We have provided at least 20 annotated examples per frame. Some annotated sentences are provided below. The parentheses to the right contain the name of the frames where these peripheral FEs appear.

(8) a. [LANGUAGE_VARIETY Burmese] [LU orthography]   (ORTHOGRAPHIC_SYSTEM)

   b. [LU sentence] [DATA m'tā teku bri no], [DATA_TRANSLATION he rice to-buy wishes], [DATA_TRANSLATION he wants to buy rice]   (ORTHOGRAPHIC_PHRASE)

   c. [NATIONALITY Danish] [LU philologist] [NAME Rask]   (SPECIALIZED_LINGUISTS)

   d. [PLACE final] [LU vowels]   (SEGMENT)

   e. [AUTHOR Mr. Godwin Austen' s] [LU vocabulary]   (LINGUISTIC_DATA_STRUCTURE)

   f. [BRANCH comparative] [LU philology]   (LINGUISTIC_SUBFIELD)

Table 1. Frame elements of the AGREEMENT frame

Head	Dependent	Sentence
Relation: advcl (adverbial clause)		
'agrees'	'if the latter is in the form of the nominative'	'The participle agrees in gender and number with the object if the latter is in the form of the nominative'
'agrees'	'when mutable'	'in the case of Transitive verbs, the subject is put in the agent case, and, when mutable, the verb agrees in gender and number with the object'
Relation: advmod (adverbial modifier)		
'agree'	'apparently'	'The personal pronouns apparently also agree'
'agree'	'often'	'The numerals often agree very closely with those in use in the Kuki-Chin group'
'agrees'	'closely'	'The Gondi of Mandla closely agrees with the preceding sketch'
Relation: nmod (nominal modifier)		
'agrees'	'in the singular'	'That latter language agrees with Gondi in the singular, but uses the masculine and not the neuter form to denote the plural of nouns which denote women and goddesses'
'agrees'	'in gender and number'	'In the former case, the participle which forms the tense agrees in gender and number with the object'
'agrees'	'with Gujarati'	'The oblique form agrees with Gujarati'
'agrees'	'as in Gondi'	'In Bhili we find forms such as innen bala, thy son, where the possessive pronoun agrees with the qualified noun in the same way as in Gondi'
Relation: nsubj (nominal subject)		
'agree'	'Adjectives'	'Adjectives agree with their nouns in gender and number, but do not alter with the case of the noun'
'agree'	'Other forms'	'Other forms mainly agree with Konkani'
'agree'	'The numerals'	'The numerals often agree very closely with those in use in the Kuki-Chin group'

4.3 Frame Element Identification and Development

After we have realized the need for a new domain specific frame (i.e. after dealing with polysemy), we have to design the structure of the frame which involves identification of different FEs of the frame and its relation to other frames. For the purpose of identification of FEs, we have relied on a usage based data driven approach. The procedure we have opted for is as follows:

As a first step, a set of lexical units were identified manually for each of the frame candidates (e.g. agree.v is one potential LU for the AGREEMENT frame). Next, all example sentences from the LSI data containing any of the potential lexical units for a given frame were gathered in a text file. These examples were then parsed using the Stanford Dependency Parser [15], and the constituents within parses were then grouped by the relation type e.g. all (head, dependent) constituent pairs for the relation type 'advmod (i.e. adverbial modifier)' were grouped together. From these groupings, the entries where the head of the relation constituent contained any of the lexical units of a particular frame were separated and saved in a text file. Such a grouping basically gives us all string segments which were used at a particular relational position for a given relation in the whole corpus for each semantic frame. Next, the string constituents were manually observed to determine if an FE is required to capture the information contained within a particular string segment.

Lets take an example to better explain the above given procedure. For the AGREE-MENT frame, all sentences containing the lemma 'agree' (considering that agree.v, agreement.n are two lexical units) were collected and parsed as explained above. Table 1 below lists a few selected entries of (Head, Dependent, Sentence) for various relation types.

In addition to general understanding of the frame, the usage based examples given in Table were used to design the following FEs for the AGREEMENT frame.

- Participant_1: Based on the 'nsub' relation and intended to record the first participant of the agreement.
- 'Participant_2': Based on the relation nmod (the argument followed by the keyword 'with') and intended to record the second participant of the agreement.
- 'Grammatical_Category': Based on the relation nmod (the argument usually followed by the keyword 'in') and intended to record the grammatical category on which the two participants agree.
- 'Degree': Based on the relation 'advmod' (e.g. the modifiers 'strongly', 'loosely', etc.) and intended to record the degree of agreement.
- 'Frequency': Based on the relation 'advmod' (e.g. the modifiers 'often', 'sometimes', etc.) and intended to record the frequency of agreement.
- 'Language_Variety': General understanding, intended to record the language/variety for which the agreement is being talked about.
- 'Reference_Language': Based on the relation nmod (the argument usually followed by the keyword 'as'), and intended to record the reference language if any.
- 'Condition': Based on the relation 'advcl', and intended to record the condition of agreement if any.

The same procedure was used to identify FEs for all the developed frames.

4.4   Current Status of LingFN

Table 2 shows the current status of LingFN in figures. As can be seen we have developed around 100 frames in total with 32 FEs, around 360 LUs, and more than 2,800 annotated example sentences.

Table 2. LingFN statistics

Frame type	# frames	Used-by links	Peripheral FEs	LUs	Annotated examples
Event frames	5	171	16	25	1 858
Filler frames	94	154	16	335	948
Total	99	325	32	360	2806

5   Applications of LingFN

On the application side, among others, we intend to use LingFN in two ongoing linguistic research projects for automatic extraction of the information encoded in descriptive grammars in order to build extensive typological databases to be used in large-scale comparative linguistic investigations.

The area of automatic linguistic information extraction is very young, and very little work has been previously reported in this direction. Virk et al. [20] report on experiments with pattern and syntactic parsing based methods for automatic linguistic information extraction. Such methods seem quite restricted and cannot be extended beyond certain limits. We believe a methodology based on the well-established theory of frame semantics is a better option as it offers more flexibility and has proved useful in the area of information extraction in general. The plan is to develop a set of frames for the linguistic domain, annotate a set of descriptive grammars with BFN frames extended by the newly built frame set, train a parser using the annotated data as a training set, and then use the parser to annotate and extract information from the other, unannotated descriptive grammars.

However, in the present paper we limit ourselves to a description of the first part (i.e., development of new frames), and we leave the other tasks (annotations of grammars, training of a parser, and information extraction) as future work.

6   Conclusions and Future Work

Using a combination of the frame-by-frame and lemma-by-lemma framenet development approaches, we have reported on the development of LingFN, an English framenet for the linguistic domain. Taking it as a three stage process, we have described the methodologies for the frame identification, FE identification, and frame development processes. The frames were developed using a frame editor, and the resulting frames have been stored as lexical semantic entries accessible through a general-purpose computational lexical infrastructure.

This is ongoing work, and in the future we plan initially to build additional frames to cover all of the LSI corpus, and subsequently to extend the LingFN beyond this data in order to build a wider-scale resource covering basically the whole linguistic domain.

Acknowledgments. The work presented here was funded partially by the Swedish Research Council as part of the project *South Asia as a linguistic area? Exploring big-data methods in areal and genetic linguistics* (2015–2019, contract no. 421-2014-969), and partially by the *Dictionary/Grammar Reading Machine: Computational Tools for Accessing the World's Linguistic Heritage* (DReaM) Project awarded 2018-2010 by the Joint Programming Initiative in Cultural Heritage and Global Change, Digital Heritage and Riksantikvarieämbetet, Sweden.

References

1. Baker, C.F., Fillmore, C.J., Lowe, J.B.: The Berkeley FrameNet project. In: Proceedings of ACL/COLING 1998, pp. 86–90. ACL, Montreal (1998). https://doi.org/10.3115/980845.980860
2. Borin, L., Dannélls, D., Forsberg, M., Kokkinakis, D., Toporowska Gronostaj, M.: The past meets the present in Swedish FrameNet++. In: 14th EURALEX International Congress, pp. 269–281. EURALEX, Leeuwarden (2010)
3. Borin, L., Forsberg, M., Olsson, L.J., Uppström, J.: The open lexical infrastructure of Språkbanken. In: Proceedings of LREC 2012, pp. 3598–3602. ELRA, Istanbul (2012)
4. Borin, L., Toporowska Gronostaj, M., Kokkinakis, D.: Medical frames as target and tool. In: FRAME 2007: Building Frame Semantics Resources for Scandinavian and Baltic Languages. (Nodalida 2007 Workshop Proceedings), pp. 11–18. NEALT, Tartu (2007)
5. Chiarcos, C.: Ontologies of linguistic annotation: survey and perspectives. In: Proceedings of LREC 2012, pp. 303–310. ELRA, Istanbul (2012)
6. Dannélls, D., Borin, L., Forsberg, M., Friberg Heppin, K., Gronostaj, M.T.: Swedish FrameNet. In: Dannélls, D., Borin, L., Friberg Heppin, K. (eds.) The Swedish FrameNet++: Harmonization, Integration, Method Development and Practical Language Technology Applications, pp. 37–65. John Benjamins, Amsterdam (forthcoming)
7. Dannélls, D., Borin, L., Friberg Heppin, K. (eds.): The Swedish FrameNet++: Harmonization, Integration, Method Development and Practical Language Technology Applications. John Benjamins, Amsterdam (forthcoming)
8. Fillmore, C.J.: Frame semantics and the nature of language. In: Annals of the New York Academy of Sciences, vol. 280, no. 1, pp. 20–32 (1976). https://doi.org/10.1111/j.1749-6632.1976.tb25467.x
9. Fillmore, C.J.: Scenes-and-frames semantics. In: Zampolli, A. (ed.) Linguistic Structures Processing, pp. 55–81. North Holland, Amsterdam (1977)
10. Fillmore, C.J.: Frame semantics. In: Linguistic Society of Korea. Linguistics in the Morning Calm, pp. 111–137. Hanshin Publishing Co., Seoul (1982)
11. Gildea, D., Jurafsky, D.: Automatic labeling of semantic roles. Comput. Linguist. **28**(3), 245–288 (2002). https://doi.org/10.1162/089120102760275983
12. Grierson, G.A.: A Linguistic Survey of India, vol. I-XI. Government of India, Central Publication Branch, Calcutta (1903–1927)
13. Hasegawa, Y., Lee-Goldman, R., Kong, A., Akita, K.: FrameNet as a resource for paraphrase research. Constructions Frames **3**(1), 104–127 (2011)
14. Malm, P., Ahlberg, M., Rosén, D.: Uneek: a web tool for comparative analysis of annotated texts. In: Proceedings of the IFNW 2018 Workshop on Multilingual FrameNets and Constructicons at LREC 2018, pp. 33–36. ELRA, Miyazaki (2018)
15. Manning, C.D., Surdeanu, M., Bauer, J., Finkel, J., Bethard, S.J., McClosky, D.: The Stanford CoreNLP natural language processing toolkit. In: Proceedings of ACL 2014, pp. 55–60. ACL, Baltimore (2014). http://www.aclweb.org/anthology/P/P14/P14-5010
16. Ponzetto, S.P., Strube, M.: Exploiting semantic role labeling, WordNet and Wikipedia for coreference resolution. In: Proceedings of HLT 2006, pp. 192–199. ACL, New York (2006). http://www.aclweb.org/anthology/N/N06/N06-1025
17. Shen, D., Lapata, M.: Using semantic roles to improve question answering. In: Proceedings of EMNLP-CoNLL 2007, pp. 12–21. ACL, Prague (2007). http://www.aclweb.org/anthology/D/D07/D07-1002
18. Surdeanu, M., Harabagiu, S., Williams, J., Aarseth, P.: Using predicate-argument structures for information extraction. In: Proceedings of ACL 2003, pp. 8–15. ACL, Sapporo (2003). http://www.aclweb.org/anthology/P03-1002

19. Torrent, T.T., et al.: Multilingual lexicographic annotation for domain-specific electronic dictionaries: the Copa 2014 FrameNet Brasil project. Constructions Frames **6**(1), 73–91 (2014)
20. Virk, S.M., Borin, L., Saxena, A., Hammarström, H.: Automatic extraction of typological linguistic features from descriptive grammars. In: Ekštein, K., Matoušek, V. (eds.) TSD 2017. LNCS (LNAI), vol. 10415, pp. 111–119. Springer, Cham (2017). https://doi.org/10.1007/978-3-319-64206-2_13
21. Wu, D., Fung, P.: Semantic roles for SMT: a hybrid two-pass model. In: Proceedings of HLT-NAACL 2009, pp. 13–16. ACL, Boulder (2009). http://dl.acm.org/citation.cfm?id=1620853.1620858

SART - Similarity, Analogies, and Relatedness for Tatar Language: New Benchmark Datasets for Word Embeddings Evaluation

Albina Khusainova[1]($\boxtimes$), Adil Khan[1], and Adín Ramírez Rivera[2]

[1] Innopolis University, Universitetskaya, 1, 420500 Innopolis, Russia
{a.khusainova,a.khan}@innopolis.ru
[2] University of Campinas (UNICAMP), Campinas 13083-970, Brazil
adin@ic.unicamp.br

Abstract. There is a huge imbalance between languages currently spoken and corresponding resources to study them. Most of the attention naturally goes to the "big" languages—those which have the largest presence in terms of media and number of speakers. Other less represented languages sometimes do not even have a good quality corpus to study them. In this paper, we tackle this imbalance by presenting a new set of evaluation resources for Tatar, a language of the Turkic language family which is mainly spoken in Tatarstan Republic, Russia.

We present three datasets: Similarity and Relatedness datasets that consist of human scored word pairs and can be used to evaluate semantic models; and Analogies dataset that comprises analogy questions and allows to explore semantic, syntactic, and morphological aspects of language modeling. All three datasets build upon existing datasets for the English language and follow the same structure. However, they are not mere translations. They take into account specifics of the Tatar language and expand beyond the original datasets. We evaluate state-of-the-art word embedding models for two languages using our proposed datasets for Tatar and the original datasets for English and report our findings on performance comparison.

The datasets are available at https://github.com/tat-nlp/SART.

Keywords: Word embeddings · Evaluation · Analogies · Similarity · Relatedness · Low-resourced languages · Turkic languages · Tatar language

1 Introduction

Word embeddings have become almost an intrinsic component of NLP systems based on deep learning. Therefore, there is a need for their evaluation and comparison tools. It is not always computationally feasible to evaluate embeddings

© Springer Nature Switzerland AG 2023
A. Gelbukh (Ed.): CICLing 2019, LNCS 13451, pp. 380–390, 2023.
https://doi.org/10.1007/978-3-031-24337-0_28

directly on the task they were built for; that is why there is a need for inexpensive preliminary evaluations. For example, there are similarity/relatedness tests, where human judgements are obtained for pairs of words, such that each pair is rated based on the degree of similarity/relatedness between the words, and then model scores are compared to humans judgements. Another type is analogies test—questions of the form A:B::C:D, meaning A to B is as C to D, and D is to be predicted. While such tests exist for widespread languages, e.g., SimLex-999 [8] was translated to four major languages [9], less represented languages suffer from the absence of such resources. In this work, we attempt to close this gap for Tatar, an agglutinative language with rich morphology, by proposing three evaluation datasets. In general, their use is not limited to embeddings evaluation; they can benefit any system which models semantic/morphosyntactic relationships. For example, they can be used for automated thesauri, dictionary building, machine translation [7], or semantic parsing [2].

2   Related Works

The most well-known similarity/relatedness datasets for English are RG [13], WordSim-353 [6], MEN [5], and SimLex-999 [8]. The problem with WordSim-353 and MEN is that there's no distinction between similarity and relatedness concepts, and we tried to address it in our work. The analogies task was first introduced by Mikolov et al. [10] and then adapted for a number of languages. The adaptation process is not trivial since analogies should examine specifics of a given language while too much customization would make datasets incomparable. We aimed at finding a compromise between these two extremes.

3   Proposed Datasets

3.1   Similarity Dataset

For constructing the Similarity dataset we used the WordSim-353 dataset; namely, it's version by Agirre et al. [1], in which they split the original dataset into two subsets, one for similarity, and the other for relatedness evaluation. We took the first subset (for similarity), consisting of a total of 202 words, and manually translated it to Tatar. Then we removed or replaced with analogies when possible pairs containing rare words, and those which needed to be adapted to account for cultural differences, e.g., the Harvard-Yale pair was replaced with *KFU-KAI*, which are acronyms of two largest Tatarstan university names. We also filtered out most of the pairs with loanwords from Russian.

We defined the distribution of synonymy classes we want to be present in the dataset as follows:

- Strong synonyms, 22%;
- Weak synonyms, 30%;
- Co-hyponyms, 17%;

- Hypernym-hyponyms, 15%;
- Antonyms, 5%; and
- Unrelated, 11%.

These categories and percentages were chosen to represent the diversity in synonymy relations and to focus on not-so-obvious pairs, which constitute the majority (67%) of the dataset—everything apart from strong synonyms and unrelated words. After the described preprocessing, we split the remaining pairs between these categories and added our own pairs such that the total is still 202 and all categories are full. We used SimLex-999 to find some of the new word pairs. From the part-of-speech point of view, the dataset is mostly build up from nouns, 87%, a small fraction of adjectives, 12%, and 1% of mixed pairs.

3.2   Relatedness Dataset

The Relatedness dataset was constructed using a similar procedure - we took the second relatedness subset of WordSim-353, and translated it adapting/replacing pairs with rare or irrelevant words. For this dataset we kept more loanwords from Russian to keep it close to the original dataset for comparability. The dataset contains 252 words, 98% of which are nouns, and 2% are mixed ones.

Annotation. Here we explain the process of obtaining human scores for the datasets described above. We constructed a survey for each dataset and provided a set of instructions to respondents. For Similarity dataset it was motivated by SimLex-999: we showed the examples of synonyms (life-existence), nearly synonyms (hair-fur), and clearly explained the difference between similarity and relatedness concepts using such examples as car-road. For Relatedness dataset instructions explained different association types (by contrast, by causation, etc.). Then annotators were asked to rate each pair by assigning it to one of the four categories. Depending on the survey, the options were different, below are the versions for *similarity* and (relatedness):

1. Words are absolutely *dissimilar* (unrelated)
2. Weak *similarity* (relatedness)
3. Moderate *similarity* (relatedness)
4. Words are *very similar or identical* (strongly related)

Later this scale was converted to 0–10 to match the existing datasets ($1 \rightarrow 0$; $2 \rightarrow \frac{10}{3}$; $3 \rightarrow \frac{20}{3}$; $4 \rightarrow 10$). A total of 13 respondents rated each dataset, all native Tatar speakers. Inter-annotator agreement measured as average Spearman's ρ between pairwise ratings equals 0.68 for Similarity and 0.61 for Relatedness dataset.

3.3   Analogies Dataset

We used several existing analogies datasets to identify common categories to include to our new dataset, namely, the original one [10], and ones for Czech [14] and Italian [3]. We identified 8 such categories (marked below as †). We also included new categories, most of which explore the morphological richness of the Tatar language, and some account for cultural/geographic characteristics. We applied a frequency threshold: we did not include pairs where any of the words was not in the top 100 000 of most frequent words.

The final list of the categories is as follows:

Semantic Categories

Capital-country†: capitals of countries worldwide, mostly taken from the original dataset, plus four additional countries, e.g., "Prague"-"Czech Republic", the latter expressible as a single word in Tatar. 51 pairs.
Country-currency†: national currencies worldwide, e.g., "Turkey"-"lira". 11 pairs.
Capital-republic inside Russia: capitals and names of republics, which are federal subjects of Russia, e.g., "Kazan"-"Tatarstan". 14 pairs.
Man-woman†: family relations, like brother-sister, but also masculine/feminine forms of professions and honorifics. e.g., *afande-khanym*, which can be translated as "mister"-"missis". 27 pairs.
Antonyms (adjectives): e.g. "clean"-"dirty". Differs from 'Opposite' category in the original dataset: words here do not share roots. 50 pairs.
Antonyms (nouns): e.g., "birth"-"death", roots also differ. Both antonym categories were built using the dictionary of Tatar antonyms[1]. 50 pairs.
Name-occupation: e.g., "Tolstoi"-"writer", for people famous in/associated with Tatarstan Republic. 40 pairs.

Syntactic Categories

Comparative†: positive and comparative forms of adjectives, e.g. "big"-"bigger". 50 pairs.
Superlative†: positive and superlative forms of adjectives. Superlatives in Tatar are usually formed by adding separate *in'* "most" word before the adjective, but sometimes the first part of the word is repeated twice, e.g., *yashel-yam-yashel*, "green"-"greenest". We include 30 such pairs.
Opposite†: basically antonyms sharing the root, e.g., "tasty"-"tasteless". 45 pairs.
Plural†: singular and plural word forms, e.g., "school"-"schools". Contains two subcategories: for nouns, 50 pairs, and pronouns, 10 pairs.
Cases: specific for Tatar, which has 6 grammatical cases: nominative, possessive, dative, accusative, ablative, and locative. We pair 30 words in nominative

---

[1] Safiullina, ISBN 5-94113-178-X.

case with their forms in each of other cases, resulting in 5 subcategories. The example of nominative-possessive subcategory can be "he"-"his". Words are a mix of nouns and pronouns.

Derivation (profession): nouns and derived profession names, e.g., "history"-"historian". 25 pairs.

Derivation (adjectives): nouns and derived adjectives, e.g., "salt"-"salty". 30 pairs.

The following 12 categories explore different verb forms. Tatar, an agglutinative language, is extremely rich w.r.t. word forms. When it comes to verbs, the form depends on negation, mood, person, number, and tense, among others. We explore only some combinations of these aspects, which we think are most important. To make it easier to comprehend, we only provide examples (for the verb "go"), and the above-mentioned details can be implied from them, while main aspect is described by category name. We picked 21 verbs and put them in different forms as required by the following categories:

Negation: "he goes"-"he doesn't go".

Mood (imperative): "to go"-"go!".

Mood (conditional): "go!"-"if he goes".

Person1-2: "I go"-"you go".

Person1-3: "I go"-"he goes".

Plural, 1 person: "I go"-"we go".

Plural, 2 person: "you go (alone)"-"you go (as a group)".

Plural, 3 person: "he goes"-"they go".

Tense, past (definite)†: "he goes"-"he went".

Tense, past (indefinite): "he goes"-"he probably went".

Tense, future (definite): "he goes"-"he will go".

Tense, future (indefinite): "he goes"-"he will probably go".

Verbal adverbs, type 1: imperative and adverb, e.g. "go!"-"while going". For the same 21 verbs.

Verbal adverbs, type 2: imperative and adverb, e.g. "go!"-"on arrival". For the same 21 verbs.

Passive voice: two verbs, second being a passive voice derivation from the first, e.g., "he writes"-"it is being written". 25 pairs.

So, in total we constructed 34 categories: 7 semantic and 27 syntactic ones. For each category/subcategory we generated all possible combinations of pairs belonging to it, e.g., the first category (capital-country) contains 51 unique pairs, hence, $50 \cdot 51 = 2550$ combinations. So, we have 10004 semantic and 20140 syntactic questions, in total 30144.

4   Experiments and Evaluation

In this section we evaluate three word embedding models, Skip-gram with negative sampling SG [11], FastText [4], and GloVe [12], with the proposed datasets. These models were chosen for evaluation for their popularity and ease of training. As for Fasttext, it was chosen also because it works with n-grams, hence, was expected to handle complex morphology of Tatar better. We trained SG and FastText using gensim[2] library and GloVe using the original code[3]. Private 126 M tokens corpus, kindly provided by Corpus of Written Tatar[4], was used to train these models. The corpus was obtained primarily from web-resources and is made up of texts in different genres, such as news, literature, official.

All models were trained with 300 dimensions and window size 5. We used different versions of SG and FastText in our experiments, all trained for 10 epochs using batch size 128. For other parameters, we will refer to minimum word count as mc, subsampling threshold as sub, negative samples number as neg, minimum/maximum n-gram length as $gram\_l$. When tuning these parameters we were focused on Analogies task, and for Similarity and Relatedness tasks we chose best results among all trained models.

As for GloVe, there is only one version for all experiments—trained for 100 epochs with x_max parameter set to 100, and other parameters set to default.

4.1 Similarity and Relatedness Results

For evaluation we calculate Spearman's ρ correlation between average human score and cosine similarity between embeddings of words in pairs. We report Spearman's ρ for SG, FastText, and GloVe for both tasks and parameter configuration which led to the best performance in Table 1.

Table 1. Spearman's ρ for Similarity and Relatedness and model parameters.

Model	Similarity	Relatedness
SG	0.52	0.60
	$mc = 5$, $sub = 0$, $neg = 64$	$mc = 5$, $sub = 0$, $neg = 20$
FastText	**0.54**	**0.62**
	$mc = 2$, $sub = 1e^{-4}$, $neg = 64$, $gram\_l = 3/6$	$mc = 2$, $sub = 1e^{-4}$, $neg = 64$, $gram\_l = 3/6$
GloVe	0.48	0.53

As we see, FastText does better on both tasks and GloVe performs substantially worse, while for all models the similarity task appears to be trickier, which does not positively correlate with the inter-annotator agreement.

4.2 Analogies Results

We follow the same evaluation procedure as in the original work [10], so, we report accuracy, where true prediction means exact match (1st nearest neighbor).

---

[2] https://radimrehurek.com/gensim.
[3] https://github.com/stanfordnlp/GloVe.
[4] http://www.corpus.tatar/en.

We used same parameters for both SG and FastText: $mc = 5$, $sub = 0$, $neg = 20$, initial/minimal learning rates were set to 0.05 and $1e^{-3}$ respectively. For FastText $gram\_l$ was set to 3/8. The results are presented in Table 2, where we compare performance of SG, FastText, and GloVe for each category of analogies. For summaries (bold lines), we calculate average over categories instead of overall accuracy, to cope with imbalance in categories' sizes: this way all categories have the same impact on summary score.

We observe that syntactic questions are substantially easier to answer for all models, which indicates they better capture morphological relationships than semantic ones. If we examine syntactic categories, we see that analogy questions with nouns and adjectives (from the top to *noun-adj*) are on average more challenging for models to answer than questions with verbs (from *negation* till the end). Let's take, for example, plural categories for nouns and verbs—the difference is drastic, accuracy for plural verb categories is 2–4 times higher than that for nouns. This suggests that more emphasis should be made on modeling grammatical aspects of nouns and adjectives in the future.

Overall, categories' complexity varies a lot, e.g., all models show more than 80% accuracy in *present-past-indefinite* category, whereas *country-currency* and *noun-adj* are especially complex. Among other categories which challenge our models we see some types of noun cases, as well as antonyms: whether sharing the root (*opposite*) or not (*adj-antonym*).

As expected, SG and FastText exhibit more similar behaviour, when compared to GloVe, due to their architectural commonalities. GloVe performs much worse in syntactic categories; for most of them the results are incomparably lower. We see that SG beats FastText in semantic questions, but is inferior to it in syntactic ones, which is as expected because FastText by construction should learn more about morphology, and Tatar is morphologically rich language.

4.3   Comparison with English

To make a comparison between languages, we took a 126 M tokens snippet of English News Crawl 2017 corpus[5], trained models with same parameters as described in Sect. 4.2, and evaluated them against the original English datasets.

Similarity and Relatedness. First, we measured Spearman's ρ for similarity and relatedness splits of WordSim-353 by Agirre et al. [1] for trained English models, and compared with corresponding results for Tatar in Table 3.

For the English language, performance of SG and FastText on similarity and relatedness tasks is the same, and for GloVe the numbers are lower. If we compare with Tatar, we see that the values lie in the same range which is good because this indicates that the datasets are probably comparable, as it was designed. Interestingly, though, in contrast to Tatar, relatedness task appears to be more difficult for English models.

---

[5] http://www.statmt.org/wmt18/translation-task.html.

Table 2. Accuracy (%) per analogy category.

Category	SG	FastText	GloVe
Semantic categories			
capital-country	40.51	32.31	15.53
country-currency	4.55	5.45	5.45
capital-republic-rf	33.52	23.08	30.22
man-woman	40.46	38.32	41.03
adj-antonym	8.61	7.43	6.73
noun-antonym	8.78	7.39	7.67
name-occupation	27.95	12.76	15.71
Semantic average	**23.48**	**18.11**	**17.48**
Syntactic categories			
comparative	76.04	76.78	44.65
superlative	26.90	33.33	5.29
opposite	15.15	13.08	5.30
plural-nouns	38.04	41.59	16.94
plural-pronouns	23.33	27.78	16.67
cases-possessive	32.07	39.66	5.40
cases-dative	14.83	12.53	3.33
cases-accusative	30.80	36.09	7.70
cases-ablative	5.98	12.64	1.15
cases-locative	12.99	15.17	4.94
profession	20.33	17.33	8.50
noun-adj	7.01	4.14	6.09
negation	52.14	57.62	19.76
imperative	40.24	44.52	8.33
conditional	40.00	52.14	7.14
person1-2	65.71	81.90	28.57
person1-3	74.29	71.67	64.52
plural-1person	91.43	93.57	65.71
plural-2person	63.57	95.24	35.24
plural-3person	91.90	86.43	75.24
present-past-def	82.62	88.33	65.48
present-past-indef	87.86	86.43	84.29
present-future-def	45.24	70.24	26.90
present-future-indef	55.95	55.71	28.81
verbal-adv-1	23.81	29.05	9.76
verbal-adv-2	40.24	50.71	5.48
passive-voice	34.50	35.33	14.67
Syntactic average	**44.18**	**49.22**	**24.66**
All average	**39.92**	**42.82**	**23.18**

Table 3. Spearman's ρ for Similarity and Relatedness for Tatar and English.

Model	Task	Tatar	English
SG	Similarity	0.52	**0.68**
	Relatedness	0.60	**0.55**
FastText	Similarity	**0.54**	0.68
	Relatedness	**0.62**	0.55
GloVe	Similarity	0.48	0.46
	Relatedness	0.53	0.39

Analogies. Second, we evaluated trained English models on the original analogies dataset. We compare average accuracies in semantic and syntactic categories for two languages in Table 4.

Table 4. Comparison of average accuracies (%) for Tatar and English.

Model		Tatar	English
SG	Semantic	**23.48**	**53.11**
	Syntactic	44.18	56.39
FastText	Semantic	18.11	41.27
	Syntactic	**49.22**	**65.48**
GloVe	Semantic	17.48	47.47
	Syntactic	24.66	36.41

English FastText falls behind English SG in semantic questions (41% vs. 53%) but outperforms in syntactic questions (65% vs. 56%), the same tendency as it was observed for the Tatar models and in the original work [4]. Notice that regardless of language GloVe shows poor results in syntactic categories.

What's most interesting, however, is that English models perform much better: they are from 12% to 30% more accurate than their Tatar counterparts. We think that the overall better performance of English models can be partially explained by easier questions: our dataset includes such nontrivial categories as antonyms, cases, and derivations.

To analyze it further, we can compare results for Tatar and English category-wise. We mentioned in Sect. 3.3 that 8 common categories were selected to be included in Analogies dataset to make it comparable with existing ones. Now in Table 5 we report accuracy values for these common categories for Tatar and English languages tested on FastText model, as it showed better overall results for both languages.

We see from the results that there's some correlation between Tatar and English models. Both models perform well in comparatives, worse in opposites,

Table 5. Accuracy (%) of FastText for common categories.

Category	Tatar	English
Semantic categories		
country-capital	32.31	62.04
country-currency	5.45	4.02
man-woman	38.32	52.37
Syntactic categories		
comparative	76.78	86.94
superlative	33.33	75.58
opposite	13.08	35.71
plural-nouns	41.59	76.43
present-past-def	88.33	56.60

and very poorly in *country-currency* category. Surprisingly, Tatar model performs better in past tense. For other categories gaps are too huge to compare. For *country-capital* the better performance may be explained by the fact that the news dataset used for training English models is probably an especially good resource for such data compared to Tatar corpus. For superlatives the probable reason is the earlier explained characteristic—superlatives in Tatar are usually formed by adding a separate word, and adjectives included in this category represent a very small subset of adjectives, which have one-word superlative form.

Overall, the difference may also be due to dissimilarities of training corpora. There are no large parallel corpora for English and Tatar, unfortunately; if this was the case, the comparison would be more accurate.

5 Conclusion

In this paper, we introduced three new datasets for evaluating and exploring word embeddings for the Tatar language. The datasets were constructed accounting for cultural specificities and allow in-depth analysis of embedding performance w.r.t. various language characteristics. We examined the performance of SG, FastText and GloVe word embedding models on introduced datasets, showing that for all three tests there is much room for improvement. Cross-language comparison demonstrated that models' performance varies greatly with language; nonetheless, similar trends were observed across languages.

Acknowledgments. We thank Mansur Saykhunov, the main author and maintainer of Corpus of Written Tatar (http://www.corpus.tatar/en), for providing us data; and all respondents of the surveys for constructing Similarity/Relatedness datasets. This research was partially supported by the Brazilian National Council for Scientific and Technological Development (CNPq grant # 307425/2017-7).

References

1. Agirre, E., Alfonseca, E., Hall, K., Kravalova, J., Pasca, M., Soroa, A.: A study on similarity and relatedness using distributional and wordnet-based approaches. In: Proceedings of Human Language Technologies: The 2009 Annual Conference of the North American Chapter of the Association for Computational Linguistics, pp. 19–27. Association for Computational Linguistics (2009). http://aclweb.org/anthology/N09-1003
2. Beltagy, I., Erk, K., Mooney, R.: Semantic parsing using distributional semantics and probabilistic logic. In: Proceedings of the ACL 2014 Workshop on Semantic Parsing, pp. 7–11. Association for Computational Linguistics (2014). https://doi.org/10.3115/v1/W14-2402, http://aclweb.org/anthology/W14-2402
3. Berardi, G., Esuli, A., Marcheggiani, D.: Word embeddings go to italy: a comparison of models and training datasets. In: IIR (2015)
4. Bojanowski, P., Grave, E., Joulin, A., Mikolov, T.: Enriching word vectors with subword information. arXiv preprint arXiv:1607.04606 (2016)
5. Bruni, E., Boleda, G., Baroni, M., Tran, N.K.: Distributional semantics in technicolor. In: Proceedings of the 50th Annual Meeting of the Association for Computational Linguistics (Volume 1: Long Papers), pp. 136–145. Association for Computational Linguistics (2012). http://aclweb.org/anthology/P12-1015
6. Finkelstein, L., et al.: Placing search in context: the concept revisited. In: Proceedings of the 10th International Conference on World Wide Web, pp. 406–414. WWW 2001, ACM, New York, NY, USA (2001). https://doi.org/10.1145/371920.372094
7. He, X., Yang, M., Gao, J., Nguyen, P., Moore, R.: Indirect-hmm-based hypothesis alignment for combining outputs from machine translation systems. In: Proceedings of the 2008 Conference on Empirical Methods in Natural Language Processing, pp. 98–107. Association for Computational Linguistics (2008). http://aclweb.org/anthology/D08-1011
8. Hill, F., Reichart, R., Korhonen, A.: Simlex-999: evaluating semantic models with (genuine) similarity estimation. Comput. Linguist. 41(4), 665–695 (2015). https://doi.org/10.1162/COLI_a_00237, http://aclweb.org/anthology/J15-4004
9. Leviant, I., Reichart, R.: Judgment language matters: multilingual vector space models for judgment language aware lexical semantics. CoRR, abs/1508.00106 (2015)
10. Mikolov, T., Chen, K., Corrado, G., Dean, J.: Efficient estimation of word representations in vector space. arXiv preprint arXiv:1301.3781 (2013)
11. Mikolov, T., Sutskever, I., Chen, K., Corrado, G., Dean, J.: Distributed representations of words and phrases and their compositionality. In: Proceedings of the 26th International Conference on Neural Information Processing Systems, vol. 2, pp. 3111–3119. NIPS 2013, Curran Associates Inc. USA (2013). http://dl.acm.org/citation.cfm?id=2999792.2999959
12. Pennington, J., Socher, R., Manning, C.D.: Glove: global vectors for word representation. In: Empirical Methods in Natural Language Processing (EMNLP), pp. 1532–1543 (2014). http://www.aclweb.org/anthology/D14-1162
13. Rubenstein, H., Goodenough, J.B.: Contextual correlates of synonymy. Commun. ACM 8(10), 627–633 (1965)
14. Svoboda, L., Brychcín, T.: New word analogy corpus for exploring embeddings of Czech words. In: Gelbukh, A. (ed.) CICLing 2016. LNCS, vol. 9623, pp. 103–114. Springer, Cham (2018). https://doi.org/10.1007/978-3-319-75477-2_6

Cross-Lingual Transfer for Distantly Supervised and Low-Resources Indonesian NER

Fariz Ikhwantri[1,2(✉)]

[1] Kata Research Team, Kata.ai, Jakarta, Indonesia
research@kata.ai, ikhwantri.f.aa@m.titech.ac.jp
[2] Tokyo Institute of Technology, Tokyo, Japan

Abstract. Manually annotated corpora for low-resource languages are usually small in quantity (gold), or large but distantly supervised (silver). Inspired by recent progress of injecting pre-trained language model (LM) on many Natural Language Processing (NLP) task, we proposed to fine-tune pre-trained language model from high-resources languages to low-resources languages to improve the performance of both scenarios. Our empirical experiment demonstrates significant improvement when fine-tuning pre-trained language model in cross-lingual transfer scenarios for small gold corpus and competitive results in large silver compare to supervised cross-lingual transfer, which will be useful when there is no parallel annotation in the same task to begin. We compare our proposed method of cross-lingual transfer using pre-trained LM to different sources of transfer such as mono-lingual LM and Part-of-Speech tagging (POS) in the downstream task of both large silver and small gold NER dataset by exploiting character-level input of bi-directional language model task.

Keywords: Cross-lingual · Low resource languages · Named entity recognition

1 Introduction

Building large named entity gold corpus for low-resource languages is challenging because time consuming, limited availability of technical and local expertise. Thus, manually annotated corpora for low-resource languages are usually small, or large but automatically annotated. In most cases, the former are used as a test set to evaluate models trained on the latter one.

To reduce the annotation efforts, previous works [19] utilized parallel corpus to project annotation from high-resource languages to low-resources languages using word-alignment. Another promising approach is to use knowledge base e.g DBPedia [1,2] or semi-structured on multi-lingual documents e.g Wikipedia [20] to generate named entity seed.

Work was done when working at Kata.ai.

A. Gelbukh (Ed.): CICLing 2019, LNCS 13451, pp. 391–405, 2023.
https://doi.org/10.1007/978-3-031-24337-0_29

Previous works on multi-lingual Wikipedia with motivation to acquire general corpus [20] and knowledge alignment between high–resource and low–resource languages encounter low recall problem because of incomplete and inconsistent alignments [22]. Some work on monolingual data with intensive rule labelling [1] and label validation [2] to create automatic annotation also face the same problem.

Our contribution in this paper consists of two parts. First, we propose to improve NER performance of a low-resource language, namely Indonesian, trained on noisily annotated Wikipedia data by (1) fine-tuning English NER model, and (2) using contextual word representations derived from either English (EN), Indonesian (ID), or Cross-lingual (EN to ID) fine-tuning of pre-trained language models which exploit character-level input. Second, we analyze why using pre-trained English language model from [26] yields improvement compare to monolingual Indonesian language model by looking at the dataset size, shared characteristic such as orthography, and its different like grammatical and morphological different to source language (English). We show that fine-tuning ELMo in unsupervised cross-lingual transfer can improve the performance significantly from baseline Stanford-NER [8], CNN-LSTM-CRF [18] and previous works using state-of-the-art multi-task NER with language modeling as an auxiliary task [16,29] trained on conversational texts, and its monolingual counterpart that is trained on different dataset size in the target language, which in our case is Indonesian unlabeled corpora retrieved from Wikipedia and news dataset [33].

2   Related Works

Recently, Peters et al., [26] proposed to use pre-trained embedding from language model (ELMo) of large corpora for many NLP tasks such as NER [34], semantic role labeling [21], textual entailment [5], question answering [27] and sentiment analysis [31]. Motivated by deep character embedding for word representation that is useful in many linguistic probing and downstream tasks [24] and trained on large corpora using language model objective, we chose to investigate ELMo embedding as weight-initialization for NER task in a low-resource languages.

2.1   Deep Character Embedding

Character embedding is important to handle out-of-vocabulary problem such as in out-of-domain data [16] or another language with shared orthography [7]. The input words to Bidirectional LM, are computed by using concatenation of multiple convolution filters over sum of characters sequences of length [11,12], 2 depth highway layers [32] and a linear projection.

The input to highway layers y_k is the concatenation of $y_{k,1}, ..., y_{k,h}$ from $H_1, ..., H_h$ as $y_k = [y_{k,1}, ..., y_{k,h}]$. The output x_h of highway layers of depth h are computed as in Equation (1), where $T = \sigma(W_T x_{h-1} + b_T)$ and, $x_0 = y_k$ as an input to the first highway layer.

$$x_h = T \odot (W_H x_{h-1} + b_H) + (1 - T) \odot x_{h-1} \qquad (1)$$

2.2   Bidirectional Language Models (BiLM)

Language modeling (LM) computes the probability of token t_k in sequence of tokens length N given the preceding tokens $(t_1, t_2, ..., t_{k-1})$ as $\log p(t_1, t_2, ..., t_N) = \sum_{k=1}^{N} \log p(t_k|t_1, t_2, ..., t_{k-1})$. Reversed order LM, computes the probability of token t_k in a sequence of tokens of length N given the succeeding tokens in $\log p(t_{k+1}, t_{k+2}, ..., t_N)$ as $p(t_1, t_2, ..., t_N) = \sum_{k=1}^{N} \log p(t_k|t_{k+1}, t_{k+2}, ..., t_N)$.

$$\sum_{k=1}^{N} (\log p(t_k|t_1, t_2, ..., t_{k-1}|\theta_x, \overrightarrow{\theta}_{LSTM}, \theta_s) +$$

$$\log p(t_k|t_{k+1}, t_{k+2}, ..., t_N|\theta_x, \overleftarrow{\theta}_{LSTM}, \theta_s)) \tag{2}$$

In downstream task such as NER sequence labeling, the output of ELMo [26] used for contextual word representation is the concatenation of projected highway layer [32] of Deep Character Embedding output [11,12], forward and backward output of LM-LSTM output of hidden layer. There are several ways to use ELMo layer for sequence labeling task, one of them is to use only last layers output of BiLM-LSTM. In this research, we only explore using last hidden layer of BiLM-LSTM [25].

2.3   Cross-Lingual Transfer via Multi-task Learning

Cross-lingual transfer learning aims to leverage high–resources languages for low-resource languages. Yang et al., (2016) [36] proposed to transfer character embedding from English to Spanish because they shared same alphabet, while Cotterell et al., (2017) [7] study several languages transfer within the same family and orthographic representation using character embedding as shared input representation. In their proposed model, they shared character convolutions for composing words but not the LSTM layer. In the previous works above, the training process minimizes the joint loss of low-resource and high-resource languages as supervised multi-task learning (MTL) objective. However we found that due to grammatical and morphological different, it is more significant to do pre-training scenario (INIT) instead of joint-training objective.

3   Proposed Method

In this section we explain briefly our two proposed method. Our first proposed method extend supervised cross-lingual transfer using ELMo (Fig. 1, left image). Our second proposed method fine-tune ELMo from English to Indonesian News dataset to use on distantly supervised and small gold Indonesian NER dataset.

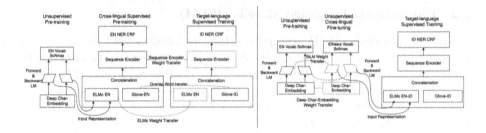

Fig. 1. Cross-lingual Transfer Learning by using Character-level pre-training. Left image, our proposed Unsupervised-Supervised Cross-lingual Transfer where we fine-tune ELMo on target task NER but on source language. Right image, our proposed Cross-lingual Language Model fine-tuning where we fine-tune ELMo on target language Indonesian

3.1 Supervised Cross-Lingual Transfer with ELMo

Alfina et al. [2] observed that automatically annotated corpora fail to tag many orthographically similar entity of "America" to "Amerika" in Indonesian. We also confirmed that, there are many cases of false negative in orthographically similar LOCATION alias such as "Pacific" to "Pasifik" in Indonesian Wikipedia. Intuitively, we proposed to increase the recall performance due to many false-negative error by supervised cross-lingual transfer [36] using pre-trained weights from state-of-the-arts NER model that uses Bidirectional language model. In the experiment result Table 4, the model corresponds to [English NER Sources] ELMo EN-1B Tokens from "Supervised CL Transfer with ELMo" scenario.

3.2 Unsupervised Cross-Lingual Transfer via ELMo Fine-Tuning

We proposed to use a pre-trained language model of high-resource languages such as English in order to initialize better weights for low-resource languages. The cross-lingual transfer in our research is simple and almost the same as [10] with language modeling objectives but we replace English target vocab with Indonesian by random initialization (Fig. 1, right image).

Our motivation to propose this method is because we observed that there are only marginal improvement using monolingual Indonesia LM of 82M Tokens from Wikipedia compared to using English LM trained on 1B Tokens on applying ELMo to Distantly Supervised NER dataset. This might be attributed due to large difference of publicly available unlabeled corpus size, such as 82M in Indonesia Wikipedia[1] vs 1B Tokens of language model benchmark or 2.9B English Wikipedia available to train. In the experiment result Table 4, the model corresponds to ELMo EN-ID Transfer from one of the "CL via ELMo EN" group scenario.

---

[1] as of 20-08-2018 Wikipedia Database dump.

4   Dataset

In this research, we used gold and silver annotation named entity corpus in English as sources in transfer learning. For target language, we used large silver annotation Indonesian as training dataset. We use two set of small clean < 40k tokens and ≤ 1.2k sentences as testing data in model comparison scenarios and another one as training data in ablation scenario for analysis, in addition of unlabeled data from Wikipedia and newswire.

4.1   Gold Named Entity Corpus

CoNLL 2003. Dataset is well known shared task benchmark dataset in many NLP experiment. We follow the standard training, validation (testa), and test (testb) split scenario. The label consist of PERSON, LOCATION, ORG, and MISC. We experiment additional scenarios for cross-lingual transfer which ignore MISC labels.

Clean 1.2K DBPedia. Human annotations for a subset of the silver annotation corpus are important to measure the quality of that automatic annotation. Thus, we asked an Indonesian linguist to re-label the subset of data and compute the metrics for DEE, MDEE and +Gazz silver annotation dataset. The precision, recall and F1 score of the subset w.r.t our clean annotation can be found in Table 2. The clean annotation can be found at Github Link[2]. We used this in-house annotation to do ablation analysis after training distantly supervised NER. We will made this subset of cleaned DBPedia Entity from noisy annotation publicly available in order to allow others to replicate our results in low-resources (gold) scenario.

4.2   Noisy Named Entity Corpus

Wikipedia Named Entity. WP2 and WP3 are two version of dataset [20]. The corpus obtained from this github repository[3], because the initial link mentioned in the [20] is down. In this research we use these 2 version that corresponding to WP2 and WP3 of this silver standard named entity recognition dataset. We evaluate this dataset on CoNLL test [34] and WikiGold [3].

DBPedia Entity Expansion. Our research used publicly available DBPedia Entity Expansion (DEE, Gold) [1] and Modified Rule (MDEE, +Gazetteers) [2] dataset for Indonesian. Interested readers should check the original references for further details. The dataset label statistics can be found in Table 1. We used the same test (Gold) in silver annotation Indonesian NER dataset. However, due to entity expansion technique, previous works [1, 2] only considers Entity without

---

[2] https://github.com/kata-ai/wikiner.

[3] https://github.com/dice-group/FOX/tree/master/input/Wikiner.

Table 1. Dataset statistics used in our experiments. #Tok: numbers of tokens. #Sent: numbers of sentences. Alfina et. al. [1,2] use **Gold as their test set**. Clean 1.2K are used to measure noisy percentage of DEE, MDEE, and +Gazz and low-resources scenario

Dataset	PER	LOC	ORG	#Tok	#Sent
DEE	13641	16014	2117	599600	20240
MDEE	13336	17571	2270	599600	20240
+Gazz	13269	22211	2815	599600	20240
Gold (Test)	569	510	353	14427	737
Clean 1.2K	1068	1773	720	38423	1220

Table 2. 1.2K instances of silver annotation performance with respect to the Clean 1.2k annotation. Clean 1.2k annotation is **subset** of DEE, MDEE and +Gazz

Annotation	Prec	Recall	F1
DEE (1.2K)	60.85	33.08	42.86
MDEE (1.2K)	61.77	35.07	44.74
+Gazz (1.2K)	63.83	40.44	49.51

their span (BIO) labels. In order to alleviate this difference, we transform the contiguous Entity with same label into BIO span. This rule based conversion does not seem affecting exact match span-based F1-metrics in distantly supervised scenarios when we reproduce the model in the same configuration.

4.3   ID-POS Corpus

The ID-POS corpus [28] contains 10K sentences of 250K tokens from news domain. There are 23 labels in the dataset. For POS tagging model, we train 5 model of 5-fold cross-validation following split dataset by [15]. For each fold of the models, we transfer the pre-trained weights into all NER train dataset in both large distantly supervised and low-resources gold NER scenarios.

4.4   Unlabeled Corpus for Language Model

Total number of vocabulary in Wikipedia Indonesia are 100k unique tokens from 2 millions total sentences with 82 millions total tokens. While total number of vocabulary in Kompas & Tempo dataset [33] are 130k tokens from 85k total sentences with 11 millions total tokens.

5   Experiments

Our main experiment for cross-lingual settings is Austronesian language, Indonesian. We choose Indonesian due to its language characteristics such as morphological distance from Indo-European family but same Latin alphabet orthography to English. It contains many loanwords for verb and named entity words from several languages. Most of the named entity are kept in the same form as the original language lexicon. It also categorized as low-resources as there is no large scale standardized and publicly available gold annotated dataset for NER task.

We use AllenNLP [9] implementation for Baseline BiLSTM-CRF and extend our own implementation based on Supervised Cross-lingual Transfer, Cross-lingual using ELMo from EN, Monolingual ELMo and Unsupervised-Supervised Cross-lingual Transfer. We make our extension and pre-trained bi-LM of monolingual and cross-lingual available on Github Links (Anonymous). We do not tune the model hyper-parameter such as dropout or learning rate, as there is no gold validation on comparable scenario with [2]. In addition, we found that tuning hyper-parameter to noisy validation do not improve and can even lead to worse result such as over-fitting to false negative.

General Model Configuration. We initialize all NER neural models on both monolingual and cross-lingual of Indonesian as target by using pre-trained word embedding with Glove [23] on our Wikipedia dumps. The Glove-ID vectors are freeze during training on DEE, MDEE and +Gazz data. All the Indonesian NER models on distantly supervised data are trained for 10 epochs using Adam [13] with learning rate 0.001 for Optimization of batch size 32. For model using ELMo module, we use dropout rate 0.5 after the last layer output and before concatenation with word embedding and l2 regularization [14] on ELMo weights to prevent model over-fitting and retain pre-trained knowledge. We use 2 layer Bi-LSTM-CRF layer with hidden size 200 and the word embedding dimension 50.

Unsupervised Cross-Lingual NER Transfer via ELMo. In cross-lingual bi-directional LM using CL via ELMo EN scenario, we use pre-trained weights from English 1B tokens[4] to Indonesian News dataset (IDNews) [33]. We use implementation of bidirectional language model by Peters et al., (2018) [25,26][5] and modified it for cross-lingual transfer scenario. We fine-tune the model for 3 epochs by replacing the Softmax vocab layer with randomly initialized weight. We only fine-tune language model in cross-lingual scenarios on 3 epochs instead of 10 is to prevent catastrophic forgetting [10,30].We called this model ELMo EN-ID Transfer. As a baseline, we use ELMo EN-1B Tokens model directly in the CL via ELMo EN scenario.

Supervised Cross-Lingual NER Transfer. For the cross-lingual transfer learning baseline scenario, we use WP2, WP3 [20] and CoNLL 2003 dataset [34] of English language to train standard BiLSTM-CRF without ELMo initializer on 1B Language Model benchmarks. The models are trained on English

---

[4] model-checkpoint.

[5] https://github.com/allenai/bilm-tf.

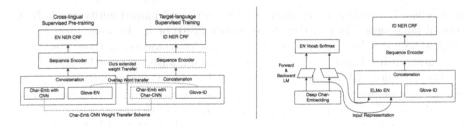

Fig. 2. Left image, Baseline scenario for supervised cross-lingual transfer learning. Right image, Baseline scenario for directly using ELMo 1B Tokens EN initializer

languages and then the pre-trained weights are used as initalizer for both supervised and unsupervised transfer learning on DEE, MDEE, and +Gazz dataset. For the pre-trained English model, we report our reproduced baseline, recent state-of-the-arts NER and ELMo LSTM-CRF on WikiNER dataset [20] to show the improvement on noisy mono-lingual data and use as pre-trained model. We train the English NER models for 75 epochs with patience 25 epochs for early stopping during training. In the experiment result Table 4, the model corresponds to [Sources] `BiLSTM-CRF` in "`Supervised CL NER Transfer`" scenarios.

Mono-lingual ELMo. In this scenarios, we use directly Pre-trained bi-LM on a mono-lingual corpus such as 1 billions word English [6], 82 millions Indonesian Wikipedia or 11 millions Indonesian News [33] dataset which illustrated on Fig. 2 on the right. In the experiment result Table 4, the model corresponds to `ELMo` ([Unlabeled corpus]) in "`Mono-lingual ELMo`"

POS Tagging Transfer. In this scenarios, we train a standard Bi-LSTM model using Softmax with Cross-entropy loss function to Indonesian POS tagging dataset. The transfer procedure almost the same as Supervised Cross-lingual NER Transfer as illustrated in Fig. 2 on the right, while there are 2 differences i) the top-most layer is Linear with Softmax Activation instead of CRF, and ii) the sources task is POS tagging instead of English NER. We train 5 models based on 5-fold cross-validation split provided by Kurniawan et al., (2018) [15], we report the averaged F1 of each k-th-fold model as pre-trained weights in both large silver and small clean annotation. In the experiment result Table 4, the model corresponds to `ID-POS BiLSTM-CRF` in "`POS Tagging Transfer`" scenario.

This experiment scenario serve as comparison of transfer learning from different but related task in Yang et al., (2017) [36]. In addition, previous work by Blevins et al. (2018) [4] show that LM contains syntactic information thus serve as comparison to pre-trained monolingual bidirectional LM.

Multi-task NER with BiLM. We also train and evaluate using recent state-of-the-arts model in Indonesian conversational dataset such as Multi-Task NER with BiLM auxiliary task (BiLM-NER) [17]. In the experiment Table 4, the model corresponds to `BiLM-NER` in "`Baseline`" scenarios.

6   Results and Analysis

In this research, we reports our English dataset results which mainly used to show improvement of pre-trained BiLM and as source weights in transfer learning. We reports our main experiments in several version of large silver for model comparison and a small clean annotation in ablation scenarios. Finally, we analyzed our proposed method of supervised cross-lingual transfer with BiLM and Cross-lingual Transfer via Language Model.

6.1   English Dataset Results

From Table 3, model trained using pre-trained ELMo and random Word Embedding initialization (WE+ELMo LSTM-CRF) are better with an average of 4.925 % F1 score in four WikiNER scenarios compare to Word embedding initialized with Glove 6B words and character-CNN (WE+CharEmb) on CoNLL dataset. However, it is tie on WikiGold test where Glove+CharEmb without MISC labels perform are better than WE+ELMo, whereas the latter are better with MISC labels than the former. Overall, combining both Glove and ELMo yields best results except when using WP2 as training data when tested in CoNLL test.

6.2   Indonesian Dataset Results

We reproduce around the same results of [2] using Stanford NER. Our experiment using a recent state-of-the-arts model in Indonesian conversational dataset namely Multi-Task NER with BiLM auxiliary task (BiLM-NER) [17] (`BiLM-NER`) obtain comparable performance with log-linear model but lower than BiLSTM-CRF [18].

The mono-lingual pre-trained BiLM on 1B English words (ELMO EN-1B Tokens) performs comparable with pre-trained BiLM on 82 millions tokens in (`ELMo (ID-Wiki)`) and 11 millions news tokens (`ELMo (ID-News)`). All of the mono-lingual Embedding from Pre-trained BiLM on silver standard annotation perform worse than baseline supervised cross-lingual with & without BiLM scenarios.

6.3   Cross-Lingual Transfer Analysis

We hypotheses that the performance of using ELMo on cross-lingual settings despite a little counter-intuitive are not entirely surprising can be addressed to i) Most named entities which available on multi-lingual documents are orthographically similar. For instance "America" is "*Amerika*" in Indonesian, while "Obama" is still "*Obama*", "President Barack Obama" is still "*Presiden* Barack Obama"; ii) Due to the orthographic similarities of many entity names, the fact

Table 3. F1 score performance results on WikiGold and CoNLL test set. English NER model w/o (without) MISC and pre-trained weight Glove 6B & ELMo 1B used as pre-train model for cross-lingual transfer scenarios

Train data	WikiGold	CoNLL	Pre-Init
Glove+CharEmb LSTM-CRF			
WP2	71.75	61.78	Glove 6B
WP3	71.40	62.51	Glove 6B
CoNLL	58.00	90.47	Glove 6B
WP2-w/o MISC	75.12	65.35	Glove 6B
WP3-w/o MISC	75.02	63.69	Glove 6B
CoNLL-w/o MISC	58.30	91.37	Glove 6B
WE (Random Init) +ELMo LSTM-CRF			
WP2	76.96	**71.48**	ELMo 1B
WP3	74.95	68.54	ELMo 1B
CoNLL	74.07	90.18	ELMo 1B
WP2-w/o MISC	73.47	66.50	ELMo 1B
WP3-w/o MISC	72.91	66.51	ELMo 1B
CoNLL-w/o MISC	74.52	91.59	ELMo 1B
Glove +ELMo LSTM-CRF			
WP2	**77.14**	69.91	Glove 6B & ELMo 1B
WP3	**76.92**	**70.31**	Glove 6B & ELMo 1B
CoNLL	**75.12**	**91.98**	Glove 6B & ELMo 1B
WP2-w/o MISC	**80.55**	**73.05**	Glove 6B & ELMo 1B
WP3-w/o MISC	**81.09**	**75.60**	Glove 6B & ELMo 1B
CoNLL-w/o MISC	**79.49**	**93.53**	Glove 6B & ELMo 1B

that English and Indonesian languages are typologically different (e.g. in terms of S-V-O word order and Determiner-Noun word order) is not relevant on noisy data, as long as the character sequences of named entities are similar in both languages [7,35].

We confirm our first hypothesis by looking up the percentage of unique word (vocabulary) overlap rate between the Gold ID-NER [1] and three English dataset, namely WP2, WP3 [20] and CoNLL training [34]. The overall vocabulary overlap rate between Gold ID-NER and the three dataset are 26.77%, 25.70%, 15.24% respectively. Furthermore, we checked WP2 per word-tag join overlap rate are PER 51.09%, LOC 60.9%, ORG 60.54%, and O 16.56% percentage. While CoNLL word-tag joins overlap rate are PER 37.53%, LOC 27.54%, ORG 39.46%, and O 9.23%. More details of unique word overlap rate between Indonesian DBPedia Entity, WP2, WP3 and CoNLL can be seen on Table 4. in Supervised Cross-lingual Transfer which only utilized character-embedding and pre-trained monolingual word-embedding trained from CoNLL dataset perform worse on both MDEE and +Gazz dataset than trained on WP2 and WP3 dataset.

We support our second hypothesis by doing ablation on clean annotation (Table 5). Our clean annotation show that, ELMo (ID-Wiki) outperformed ELMo (EN-1B Tokens) on small clean annotation data, but ELMo EN nonetheless still outperformed BiLSTM-CRF especially when combined with Supervised pre-training on CoNLL 2003 English NER [18] (Fig. 3).

Table 4. Experiment on silver standard annotation of Indonesian NER **evaluated on Gold test set** [1] in **large distantly supervised NER scenario.** Bold F1 scores are best result per scenarios (Baseline, Supervised Cross-lingual Transfer, Cross-lingual using ELMo from EN, Mono-lingual ELMo and Unsupervised-Supervised Cross-lingual Transfer). * is the best model on a dataset (DEE, MDEE, or +Gazz) on all model scenarios

Model	DEE	MDEE	+Gazz
Previous Works			
Alfina et al., [2]	41.33	41.87	51.61
BiLM-NER	40.36	41.03	51.77
Baseline			
Stanford-NER-BIO [2]	40.68	41.17	51.01
BiLSTM-CRF	**46.09**	**45.59**	**52.04**
POS Tagging Transfer			
ID-POS BiLSTM-CRF	52.58	51.07	60.57
Supervised CL NER Transfer			
WP2 BiLSTM-CRF	49.88	**52.35**	62.57
WP3 BiLSTM-CRF	51.21	50.95	**62.90**
CoNLL BiLSTM-CRF	**52.56**	50.75	60.81
CL via ELMo EN			
ELMo EN-1B Tokens	51.08	53.19	60.66
ELMo EN-ID Transfer	**52.63**	**54.74**	**63.02**
Mono-lingual ELMo			
ELMo (ID-Wiki)	**50.68**	**52.38**	60.51
ELMo (ID-News)	49.49	51.91	**60.73**
Supervised CL Transfer with ELMo			
WP2 ELMo (EN)	52.99	**55.39***	63.99
WP3 ELMo (EN)	**54.15***	55.28	63.84
CoNLL ELMo (EN)	53.52	53.48	**64.35***

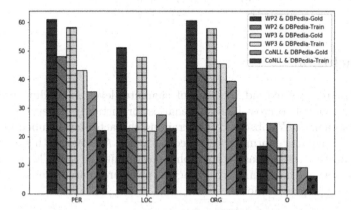

Fig. 3. Word-tag overlap rate breakdown between mono-lingual and cross-lingual corpora. (-) horizontal line: WP2 & DBPedia Gold, right slope: WP2 & DBPedia Train, (+) cross: is overlap between WP3 & DBPedia Gold, (—) vertical: overlap between WP3 & DBPedia Train, (/) left slope: CoNLL Train and DBPedia Gold, (o) dot: CoNLL Train and DBPeida Train

Table 5. Ablation experiment results using Clean 1.2K as training data in **small clean (human annotated) scenario** also **evaluated on Gold test set**. W: Word embedding (Random Init), C: Char-CNN (+EN if INIT from CoNLL 2003) embedding, E: ELMo (EN), G: Glove-ID(+EN if in cross-lingual transfer from English) [23], I: ELMo (ID-Wiki), J: ELMo (EN-ID-News) Transfer

Model	Prec	Rec	F1
Stanford-NER	71.42	53.84	61.39
BiLM-NER	63.65	63.29	63.47
BiLSTM-CRF			
W+C+E	76.42	56.32	64.85
W+C	56.23	56.39	56.31
W+E	73.53	53.32	61.81
C+E	69.13	68.60	68.86
G	63.65	48.50	55.05
G+C	69.17	62.31	65.56
G+E	75.30	65.32	69.96
G+C+E	72.05	68.73	70.35
E	76.27	55.41	64.19
G+C+I	74.53	78.43	76.43
G+I	75.57	77.94	76.74
I	78.55	73.62	76.00
G+C+J	83.26	82.62	82.94
G+J	83.77	83.60	**83.68**
J	82.36	83.74	83.04
INIT from ID-POS			
W+C	72.97	78.97	75.68
INIT from CoNLL 2003			
W+C	66.23	56.25	60.83
G+C	70.18	65.87	67.96
C+E	71.84	64.27	67.85
W+C+E	73.63	65.46	69.30
G+E	73.38	69.08	71.17
G+C+E	72.63	72.99	72.85

7   Conclusion

In this research, we extend the idea of character-level embedding pre-trained on language model to cross-lingual scenarios for distantly supervised and low-resources scenarios. We observed that training character-level embedding of language model requires enormous size of corpora [26]. Addressing this problem, we demonstrate that as long as orthographic constraint and some lexical words in target language such as loanwords to act as pivot are shared, we can utilize the high-resource languages model.

Acknowledgments. We also would like to thank Samuel Louvan, Kemal Kurniawan, Adhiguna Kuncoro, and Rezka Aufar L. for reviewing the early version of this work. We are also grateful to Suci Brooks and Pria Purnama for their relentless support.

References

1. Alfina, I., Manurung, R., Fanany, M.I.: DBpedia entities expansion in automatically building dataset for Indonesian NER. In: 2016 International Conference on Advanced Computer Science and Information Systems (ICACSIS), pp. 335–340 (2016)
2. Alfina, I., Savitri, S., Fanany, M.I.: Modified DBpedia entities expansion for tagging automatically NER dataset. In: 2017 International Conference on Advanced Computer Science and Information Systems (ICACSIS), pp. 216–221 (2017)
3. Balasuriya, D., Ringland, N., Nothman, J., Murphy, T., Curran, J.R.: Named entity recognition in Wikipedia. In: Proceedings of the 2009 Workshop on The People's Web Meets NLP: Collaboratively Constructed Semantic Resources, pp. 10–18. People's Web 2009, Association for Computational Linguistics, Stroudsburg, PA, USA (2009)
4. Blevins, T., Levy, O., Zettlemoyer, L.: Deep RNNs encode soft hierarchical syntax. In: Proceedings of the 56th Annual Meeting of the Association for Computational Linguistics (Volume 2: Short Papers), pp. 14–19. Association for Computational Linguistics (2018). http://aclweb.org/anthology/P18-2003
5. Bowman, S.R., Angeli, G., Potts, C., Manning, C.D.: A large annotated corpus for learning natural language inference. In: EMNLP (2015)
6. Chelba, C., et al.: One billion word benchmark for measuring progress in statistical language modeling (2013)
7. Cotterell, R., Duh, K.: Low-resource named entity recognition with cross-lingual, character-level neural conditional random fields. In: Proceedings of the Eighth International Joint Conference on Natural Language Processing (Volume 2: Short Papers), pp. 91–96. Asian Federation of Natural Language Processing (2017). http://aclweb.org/anthology/I17-2016
8. Finkel, J.R., Grenager, T., Manning, C.: Incorporating non-local information into information extraction systems by Gibbs sampling. In: Proceedings of the 43rd Annual Meeting of the Association for Computational Linguistics (ACL 2005), pp. 363–370. Association for Computational Linguistics (2005). http://www.aclweb.org/anthology/P05-1045
9. Gardner, M., et al.: AllenNLP: a deep semantic natural language processing platform (2017). arXiv:1803.07640
10. Howard, J., Ruder, S.: Universal language model fine-tuning for text classification. In: Proceedings of the 56th Annual Meeting of the Association for Computational Linguistics (Volume 1: Long Papers), pp. 328–339. Association for Computational Linguistics (2018). http://aclweb.org/anthology/P18-1031
11. Jozefowicz, R., Vinyals, O., Schuster, M., Shazeer, N., Wu, Y.: Exploring the limits of language modeling (2016). https://arxiv.org/pdf/1602.02410.pdf
12. Kim, Y., Jernite, Y., Sontag, D., Rush, A.M.: Character-aware neural language models. In: Proceedings of the Thirtieth AAAI Conference on Artificial Intelligence (AAAI 2016), pp. 2741–2749. AAAI Press (2016)
13. Kingma, D.P., Ba, J.: Adam: a method for stochastic optimization. CoRR abs/1412.6980 (2014)
14. Krogh, A., Hertz, J.A.: A simple weight decay can improve generalization. In: Proceedings of the 4th International Conference on Neural Information Processing Systems (NIPS 1991), pp. 950–957. Morgan Kaufmann Publishers Inc., San Francisco, CA, USA (1991). http://dl.acm.org/citation.cfm?id=2986916.2987033

15. Kurniawan, K., Aji, A.F.: Toward a standardized and more accurate Indonesian part-of-speech tagging (2018)

16. Kurniawan, K., Louvan, S.: Empirical evaluation of character-based model on neural named-entity recognition in Indonesian conversational texts. In: Proceedings of the 2018 EMNLP Workshop W-NUT: The 4th Workshop on Noisy User-Generated Text, pp. 85–92. Association for Computational Linguistics (2018). http://aclweb.org/anthology/W18-6112

17. Kurniawan, K., Louvan, S.: Empirical evaluation of character-based model on neural named-entity recognition in Indonesian conversational texts. CoRR abs/1805.12291 (2018). http://arxiv.org/abs/1805.12291

18. Ma, X., Hovy, E.: End-to-end sequence labeling via bi-directional LSTM-CNNS-CRF. In: Proceedings of the 54th Annual Meeting of the Association for Computational Linguistics (Volume 1: Long Papers), pp. 1064–1074. Association for Computational Linguistics (2016). https://doi.org/10.18653/v1/P16-1101, http://aclweb.org/anthology/P16-1101

19. Ni, J., Dinu, G., Florian, R.: Weakly supervised cross-lingual named entity recognition via effective annotation and representation projection. In: Proceedings of the 55th Annual Meeting of the Association for Computational Linguistics (Volume 1: Long Papers), pp. 1470–1480. Association for Computational Linguistics (2017). https://doi.org/10.18653/v1/P17-1135, http://aclweb.org/anthology/P17-1135

20. Nothman, J., Curran, J.R., Murphy, T.: Transforming Wikipedia into named entity training data. In: 2008 Proceedings of the Australasian Language Technology Association Workshop, pp. 124–132 (2008). http://www.aclweb.org/anthology/U08-1016

21. Palmer, M., Kingsbury, P., Gildea, D.: The proposition bank: an annotated corpus of semantic roles. Comput. Linguist. **31**, 71–106 (2005)

22. Pan, X., Zhang, B., May, J., Nothman, J., Knight, K., Ji, H.: Cross-lingual name tagging and linking for 282 languages. In: Proceedings of the 55th Annual Meeting of the Association for Computational Linguistics (Volume 1: Long Papers), pp. 1946–1958. Association for Computational Linguistics (2017). https://doi.org/10.18653/v1/P17-1178, http://aclweb.org/anthology/P17-1178

23. Pennington, J., Socher, R., Manning, C.: Glove: global vectors for word representation. In: Proceedings of the 2014 Conference on Empirical Methods in Natural Language Processing (EMNLP), pp. 1532–1543. Association for Computational Linguistics (2014). https://doi.org/10.3115/v1/D14-1162, http://www.aclweb.org/anthology/D14-1162

24. Perone, C.S., Silveira, R., Paula, T.S.: Evaluation of sentence embeddings in downstream and linguistic probing tasks. CoRR abs/1806.06259 (2018)

25. Peters, M., Ammar, W., Bhagavatula, C., Power, R.: Semi-supervised sequence tagging with bidirectional language models. In: Proceedings of the 55th Annual Meeting of the Association for Computational Linguistics (Volume 1: Long Papers), pp. 1756–1765. Association for Computational Linguistics (2017). https://doi.org/10.18653/v1/P17-1161, http://aclweb.org/anthology/P17-1161

26. Peters, M.E., et al.: Deep contextualized word representations. In: Proceedings of the NAACL (2018)

27. Rajpurkar, P., Zhang, J., Lopyrev, K., Liang, P.: Squad: 100,000+ questions for machine comprehension of text. In: Proceedings of the 2016 Conference on Empirical Methods in Natural Language Processing, pp. 2383–2392. Association for Computational Linguistics (2016). https://doi.org/10.18653/v1/D16-1264, http://www.aclweb.org/anthology/D16-1264

28. Rashel, F., Luthfi, A., Dinakaramani, A., Manurung, R.: Building an Indonesian rule-based part-of-speech tagger. In: 2014 International Conference on Asian Language Processing (IALP), pp. 70–73 (2014)
29. Rei, M.: Semi-supervised multitask learning for sequence labeling. In: Proceedings of the 55th Annual Meeting of the Association for Computational Linguistics (Volume 1: Long Papers), pp. 2121–2130. Association for Computational Linguistics (2017). https://doi.org/10.18653/v1/P17-1194, http://www.aclweb.org/anthology/P17-1194
30. Robins, A.V.: Catastrophic forgetting, rehearsal and pseudorehearsal. Connect. Sci. **7**, 123–146 (1995)
31. Socher, R., et al.: Recursive deep models for semantic compositionality over a sentiment treebank. In: Proceedings of the 2013 Conference on Empirical Methods in Natural Language Processing, pp. 1631–1642. Association for Computational Linguistics (2013). http://www.aclweb.org/anthology/D13-1170
32. Srivastava, R.K., Greff, K., Schmidhuber, J.: Highway networks (2015)
33. Tala, F.Z.: A study of stemming effects on information retrieval in Bahasa Indonesia. Language and Computation, Universiteit van Amsterdam, The Netherlands, Institute for Logic (2003)
34. Tjong Kim Sang, E.F., De Meulder, F.: Introduction to the CoNLL-2003 shared task: language-independent named entity recognition. In: 2003 Proceedings of the Seventh Conference on Natural Language Learning at HLT-NAACL (CONLL 2003), vol. 4, pp. 142–147. Association for Computational Linguistics, Stroudsburg, PA, USA (2003)
35. Xie, J., Yang, Z., Neubig, G., Smith, N.A., Carbonell, J.: Neural cross-lingual named entity recognition with minimal resources. In: Proceedings of the 2018 Conference on Empirical Methods in Natural Language Processing, pp. 369–379. Association for Computational Linguistics (2018). http://aclweb.org/anthology/D18-1034
36. Yang, Z., Salakhutdinov, R., Cohen, W.W.: Transfer learning for sequence tagging with hierarchical recurrent networks. CoRR abs/1703.06345 (2016)

Phrase-Level Simplification
for Non-native Speakers

Gustavo H. Paetzold[1(✉)] and Lucia Specia[2,3(✉)]

[1] Universidade Tecnológica Federal do Paraná, Curitiba, Brazil
ghpaetzold@utfpr.edu.br
[2] The University of Sheffield, Sheffield, UK
l.specia@imperial.ac.uk
[3] Imperial College London, London, UK

Abstract. Typical Lexical Simplification systems replace single words with simpler alternatives. We introduce the task of Phrase-Level Simplification, a variant of Lexical Simplification where sequences of words are replaced as a whole, allowing for the substitution of compositional expressions. We tackle this task with a novel pipeline approach by generating candidate replacements with lexicon-retrofitted POS-aware phrase embedding models, selecting them through an unsupervised comparison-based method, then ranking them with rankers trained with features that capture phrase simplicity more effectively than other popularly used feature sets. We train and evaluate this approach using BenchPS, a new dataset we created for the task that focuses on annotations on the needs of non-native English speakers. Our methods and resources result in a state-of-the-art phrase simplifier that correctly simplifies complex phrases 61% of the time.

Keywords: Phrase simplification · Lexical simplification · Text simplification

1 Introduction

Text simplification strategies can take various forms: lexical simplifiers (LS) replace complex words – referred to as target words – with simpler alternatives, syntactic simplifiers (SS) apply sentence-level transformations such as sentence splitting, and data-driven simplifiers (DDS) learn lexico-syntactic simplification operations from parallel data.

While LS is usually addressed using word embeddings and machine learning models that rank candidate replacements, sentence simplification is more often tackled using translation-based DDS approaches such as sequence-to-sequence neural models, which learn transformations from complex-simple parallel corpora. To date, using DDS methods is the only general alternative to simplifying sequences longer than individual words, i.e. phrases [31,35,37,40], except for some domain-specific expressions, such as medical terms, which have been addressed in early LS strategies [8,9].

© Springer Nature Switzerland AG 2023
A. Gelbukh (Ed.): CICLing 2019, LNCS 13451, pp. 406–431, 2023.
https://doi.org/10.1007/978-3-031-24337-0_30

Using DDS methods to simplify phrases is, however, "risky", as they can also perform other, spurious transformations in the sentence. In addition, although in theory DDS simplifiers can learn how to replace or remove complex phrases, complex-to-simple modifications are bounded by the often limited coverage of phrase simplifications present in the training corpus available, which tend to be much smaller than traditional bilingual parallel corpora used areas such as machine translation [14,36]. Moreover, DDS models offer little flexibility during simplification: it is not easy to control which phrases should be simplified. In user-driven systems [1,7], where the user chooses which portions of text to simplify, this approach would not be suitable.

LS approaches [23] can be adapted to address phrase simplification. By "phrase" we mean any sequence with more than one word. LS approaches usually simplify a complex word through a pipeline of four steps:

- **Complex Word Identification (CWI):** Consists in finding the words in the text that would challenge the target audience being addressed. There have been some efforts in creating effective supervised complex word identifiers [25,30], but most LS approaches do not explicitly address this step, and choose instead to perform it implicitly, such as by considering the target complex word as a candidate substitution for itself [13], or by checking whether or not the replacement produced by the lexical simplifier at the end of the pipeline is actually simpler than the target complex word [12,20].
- **Substitution Generation (SG):** Consists in finding a set of candidate substitutions for a target complex word. Different ways to generate these candidates have been devised, such as through lexicons and thesauri [5,6], complex-to-simple parallel corpora [2,13,20], and word embedding models [12,26]. Generators often produce candidate substitutions without taking into account the context of the target complex word, which means that they are rarely able to differentiate between the multiple senses of ambiguous complex words.
- **Substitution Selection (SS):** Consists in choosing among the candidate substitutions generated in the previous step best fit the context of the target complex word. Much like CWI, this step is rarely addressed explicitly, the exception being some word sense disambiguation [18] and unsupervised approaches [2,26]. Many lexical simplifiers usually skip this step and instead train context-aware Substitution Ranking approaches.
- **Substitution Ranking (SR):** Consists in ranking candidate substitutions, either all candidates or those remaining after the SS step, according to their simplicity. Ranking candidates by their frequencies in large corpora is a very popular way of doing so [6,26], but most recent lexical simplifiers use more elaborate supervised strategies [13,20,22].

Pipelined approaches are inherently flexible as they allow to target specific words. They have led to substantial performance improvements over early approaches [12,13,20]. However, since they only target single words, these strategies cannot be directly applied to simplify phrases, particularly in the case of phrases with compositional meaning.

In this paper we address the task of **Phrase-Level Simplification (PS)**, which has a clear motivation: many phrases cannot be simplified word for word. Take the phrase "*real estate*" in the sentence "*John had to hire a company to manage his real estate.*", as example. Using a single-word LS to replace *real* and *estate* individually for, say *factual* and *land*, the phrase's original meaning would be lost. A phrase simplifier should be able to replace "*real estate*" with a simpler alternative, say *property*, or "*land*".

In an effort to address this problem, we propose an approach to PS that builds on the inherently flexible pipelined LS approaches. Similar to previous LS work, we do not address the challenge of identifying complex phrases, under the assumption this will be done by the user. The main **contributions** of this paper are:

- A novel phrase embeddings model for Substitution Generation that incorporates part-of-speech tags and lexicon retrofitting and outperforms well-known resources (Sect. 2);
- A cost-effective comparison-based technique for configurable Substitution Selection (Sect. 3);
- A study on phrase-level features that can improve the performance of state-of-the-art Substitution Ranking models (Sect. 4); and
- BenchPS: an annotated dataset for PS targeting the needs of non-native English speakers, which maximizes simplification coverage and efficiently handles the problem of ranking large candidate sets (Sect. 5). BenchPS contains 400 instances composed by a complex phrase in a sentence, and gold replacements ranked by simplicity. Between 1,000 and 24,000 annotations were collected in each annotation step, totaling 36,170.

These contributions as well as our experiments (Sect. 6) are described in what follows.

2   Phrase-Level Substitution Generation

The most recent LS approaches employ word embeddings for SG. [12] use a typical GloVe [29] model to create an unsupervised approach that performs comparably to other strategies that use complex-to-simple parallel corpora [13]. [26] introduce another unsupervised approach that uses embeddings trained over a corpus annotated with universal part-of-speech (POS) tags. [20] improve on the latter by incorporating lexicon retrofitting the embeddings. To generate candidates using these models, they simply extract the words with the highest cosine similarity with a given target complex word, then filter any morphological variants amongst them.

These models cannot be used for our purposes since they only contain embeddings for single words. Obtaining phrase-level embeddings has also been addressed in previous work. [34] surveys supervised phrase composition models. These models are usually tested on phrase similarity tasks, and in order to perform well at those, they are trained over large resources, such as PPDB-2.0 [28], which are used as phrase similarity databases. However, preliminary experiments with such models did not yield promising results since, as pointed out by [21], models that perform well in similarity tasks do not necessarily perform well in LS.

In contrast, simpler unsupervised approaches are a promising fit for PS. For example, [39] and [38] simply join words that form phrases extracted from lexicons in the corpus with a special symbol (e.g. underscore) and use off-the-shelf word embedding models. We go a step further to obtain better informed phrase embeddings by using POS tag annotation and retrofitting, which have been shown to improve the performance of word embeddings in LS [20].

2.1   Retrofitted POS-Aware Phrase Embeddings

To create our phrase embeddings, we take the following approach: annotate a corpus with universal POS tags and phrases, train a typical word embeddings model over the corpus, and then retrofit the embeddings over a lexicon of synonyms, such as WordNet [11]. As we will show in Sect. 6, this improves over existing phrase embedding models.

The corpus we annotated has 7 billion words taken from the SubIMDB corpus [24], UMBC webbase[1], News Crawl[2], SUBTLEX [3], Wikipedia and Simple Wikipedia [14]. We POS-tagged the corpus using the Stanford Tagger [33]. To identify phrases, we resorted to SimplePPDB: a corpus containing complex-to-simple English phrase pairs [27]. From SimplePPDB we extracted all bigrams and trigrams as our phrase set, totalling 409,064 phrases. We chose SimpleP-PDB because it contains a wide variety of phrase types, ranging from non-compositional phrasal constructs to compound nouns and multi-word expressions.

Many of our phrases share n-grams. The phrases *"administrative council body"* and *"council body representative"*, for example, have distinct meanings, and should hence compose individual units in our corpus. The challenge is how to annotate both of them in a sentence such as *"The administrative council body representative resigned"* without turning them into a single unit. To do so, we introduce Algorithm 1, which takes a sentence S and a set of phrases P, and returns as output a set R containing various copies of S, each annotated with a different subset of phrases in P that can be found in S. Function copy (S) returns a copy of a sentence, and join (p, S) replaces the spaces separating the tokens of phrase p in sentence S with underscores ("_").

According to Algorithm 1, if the phrase *"administrative council body"* is annotated onto a copy of the sentence *"The administrative council body representative resigned"*, it will result in the sentence *"The administrative_council_body representative resigned"*, which no longer allows for the annotation of *"council body representative"*, given that it does not exactly match with *"council_body representative"*. Consequently, another copy of the sentence will be made for this annotation.

---

[1] http://ebiquity.umbc.edu/resource/html/id/351.

[2] http://www.statmt.org/wmt11/translation-task.html.

Algorithm 1: Phrase Annotation

input: S, P;
output: R;

$R \leftarrow \{S\}$;
$S_c \leftarrow \text{copy}(S)$;

while $\|P\| > 0$ **do**
   foreach $p \in P \cap \text{n-grams}(S_c)$ **do**
      $\text{join}(p, S_c)$;
      $P.\text{remove}(p)$;
   end
   $R.\text{add}(S_c)$;
   $S_c \leftarrow \text{copy}(S)$;
end
return R;

After phrase annotation, we tagged each content word that was not part of a phrase with its universal POS tag (V for verbs, N for nouns, J for adjectives and A for adverbs). Phrases were not annotated with sequences of POS tags to reduce sparsity. The previously mentioned example would be POS annotated as *"The administrative_council_body representative-N resigned-V"*. We then trained a continuous bag-of-words model (CBOW) with 1,300 dimensions using WORD2VEC [16]. We chose these settings based on findings in previous work for LS [21]. The final step was the retrofitting the model. Through retrofitting, it is possible to approximate vectors of words that share linguistic relationships such as synonymy, which are useful in simplification. We retrofitted the model over the synonym relations between all words and phrases in WordNet using the algorithm of [10]. We refer to this retrofitted POS-aware model as **Embeddings-PR**.

For comparison purposes, we built three other models:

- **Embeddings-B**: A base model, trained without POS-annotation nor retrofitting.
- **Embeddings-R**: A model enhanced with retrofitting only.
- **Embeddings-P**: A model enhanced with POS annotations only.

We explain how we apply these models for SG in the following section.

2.2 Generating Candidates with Phrase Embeddings

To find candidate substitutions for a given target complex phrase with our embeddings, we use a three-step process: vocabulary pruning, vocabulary ranking, and ranking pruning.

Vocabulary Pruning. The first step of SG strategy is to discard any words and phrases from the embedding model's vocabulary that are too unlikely to yield useful candidate substitutions. We employ a simple heuristic pruning approach for that. Given a target complex phrase in a sentence, we discard from the model's vocabulary any words/phrases that:

- have numbers or spurious characters, e.g. @#$%,
- are a substring of the target complex phrase, or vice-versa,
- lead to a grammatical error with determiners, e.g. *a* vs *an* preceding candidate,
- lead to a comparative/superlative accordance error, e.g. *more* vs *most* preceding candidate.

Vocabulary Ranking. Once the model's vocabulary is pruned, we then rank the remaining words and phrases according to their cosine similarity with the target complex phrase. This is a very intuitive way of determining which candidates in a CBOW model are most suitable for the simplification of a target phrase, since this type of model tends to group synonyms together.

Ranking Pruning. The final step is to choose the value of α, which determines how many of the best ranking candidates will be passed onto SS. This step is crucial in ensuring the quality of the phrase simplifications produced by our PS approach. If too few candidates are pruned, then the tasks of SS and SR will be more complex to filter and rank the remaining unsuitable candidates, which can lead to frequent ungrammatical and/or meaningless replacements. If too many candidates are pruned, this could compromise the simplicity of the output produced, since the ranker would not have enough options to choose from.

[12,26] and [20] achieve good results by using $\alpha = 10$, but provide no explanation with respect to how they arrived at this number. If there is no training/tuning data available, keeping only the 10 best candidates is a good choice, since it at least allows for more meaningful comparisons with previous work. If there is training/tuning data available, however, one can simply perform an exhaustive search with a large range of pruning settings to determine the best value. In the following section, we discuss how we optimise pruning jointly with the parameters of the SS approach.

3   Comparison-Based Substitution Selection

With few exceptions [2,18,26], most lexical simplifiers do not perform explicit SS.

[18] use typical word sense disambiguation methods trained over thesauri. These methods are inherently unsuitable for PS, since these thesauri rarely contain phrases. [2] introduces an unsupervised approach that does not rely on thesauri. They instead use a word co-occurrence model to calculate the similarity between the target word and the candidates, then discard any candidates that

are either too similar or not similar enough to the target word. Although this approach could be adapted to our purposes, it has been shown not to suitable for embedding-based candidate generators [23]. To address this problem, [26] present a method called Unsupervised Boundary Ranking (UBR). They calculate features of the target word and generated candidates, create a binary classification dataset by assigning label 1 to the target word and 0 to all candidates, train a linear model over the data, rank the generated candidates themselves (which were used for training) based on the boundary between 1's and 0's, and select a proportion of the highest ranking ones. [26] show that this approach yields noticeable improvements for embedding-based models.

The SS approach of [26] has two main strong points: it does not require manually annotated data for training, and it is configurable, allowing one to decide how conservative the selector should be depending on the type of generator used. We devised an unsupervised comparison-based SS approach to further improve this selector.

The intuition behind our approach is as follows: the more evidence there is that a candidate fits the context of a target complex phrase even better than the target phrase itself, the more likely it is to be a valid candidate substitution. Given a set of generated candidate substitutions for a target complex phrase in a sentence, our approach:

1. Calculates the following 40 features for the target phrase and each candidate:
   - The n-gram frequency of all n-grams composed by the target/candidate and the i preceding words and j succeeding words in the sentence for all possible combination of values of i and j between 0 and 2 (9 in total). We calculate these features using frequencies from 4 distinct corpora: SubIMDB [24], SUBTLEX [3], Wikipedia [14], and OpenSubtitles 2016 [15] ($9 * 4 = 36$ total features of this kind).
   - The language model probability of the sentence in its original form (for the target), and with the target replaced by each candidate. We train 3-gram language models for all four of the aforementioned corpora using SRILM (totalling 4 features of this kind).

   The features were chosen based on the findings of [19], where they proved effective in capturing grammaticality and meaning preservation in simplification.
2. Compares the feature values of the candidates and the target phrases, and, for each candidate phrase, calculates the proportion of features that yield a higher value for the candidate than for the target phrase (which we call "winning features").
3. Discards any candidates for which the proportion of winning features is smaller than a selected β value.

If there is no training/tuning data available, one can choose β empirically depending on the type of simplifier and/or the type of candidate generator that is used. If it is known that the generator tends to produce a large number of spurious candidates, a large β can be used, for example. If there is training/tuning data available, exhaustive search over all possible β values is not costly. The experiments in Sect. 6.1 show how β can be optimised in conjunction with the number of pruned candidates during SG.

Once the candidate substitutions are filtered by our comparison-based selector, they are passed onto our SR approach, the last step of the phrase simplification process.

4   Ranking with Phrase-Level Features

The SR approaches of [20] and [12] have been shown the most effective supervised and unsupervised LS ranking strategies, respectively.

The rank averaging approach by [12] ranks candidates according to various features, then averages their ranks in order to produce a final ranking. [20], on the other hand, employ a multi-layer perceptron that quantifies the simplicity difference between two candidate substitutions. To train their model, they calculate features for the gold candidates ranked by simplicity in a manually produced training set, pass them in pairs to the multi-layer percetron, and use as output the rank difference between the two. Given an unseen set of candidate substitutions for a complex word in a sentence, they calculate the simplicity difference between every pair of candidates, then average the differences for each candidate to create a full candidate ranking list.

However, these approaches cannot be directly used in our work, since the features used, which are in their majority n-gram frequencies, are not enough to capture phrase simplicity. We propose a set of 16 additional features:

- the phrase's number of characters,
- the phrase's number of tokens,
- the phrase's raw frequency in the corpus of 7 billion words described in Sect. 2.1, as well as SubIMDB [24], Wikipedia and Simple Wikipedia [14],
- the sentence probability after replacing the target phrase with a candidate phrase according to 3-gram language models trained on the same corpora,
- the maximum, minimum and average raw frequency of the phrase's tokens in the 7 billion word corpus,
- the maximum, minimum and average cosine similarity between the target phrase and each of the candidate's tokens according to our retrofitted POS-aware phrase embeddings model (Embeddings-PR).

In order to evaluate this pipeline for phrase simplification, we created a new dataset, as we describe in what follows.

5   BenchPS: A New Dataset

Since PS has not yet been formally addressed, there are no dedicated datasets for the training and/or evaluation of PS systems. We created a dataset that follows the same format of typical LS datasets [4,13,23,26], where each instance is composed of:

- a sentence,
- a target complex word in that sentence, and
- a set of gold candidates ranked by simplicity.

An important step in creating a dataset of this kind is deciding on which target audience to focus. We chose non-native English speakers, which has been a popular target audience in previous work, including the SemEval 2016 shared task on LS. Four challenges were involved in creating such a resource: finding complex phrases that can be potentially replaced, finding sentences containing such complex phrases, collecting simpler alternatives for the complex phrases, and ranking them according to their simplicity.

5.1 Collecting Complex Replaceable Phrases

This step involved finding a set of phrases that can pose a challenge to non-native English speakers, but also are replaced by alternatives. If the phrases selected do not have any semantically equivalent alternatives, it would be impossible to gather candidate simplifications.

In order to find replaceable complex phrases, we first collected an initial set of complex phrases from SimplePPDB. From the complex side of SimplePPDB, we gathered all two and three-word phrases that did not contain any number. After the initial filtering, we ranked the remaining phrases according to how many distinct simpler phrases they were aligned to in the SimplePPDB database with a probability ≥ 0.7, and then selected the 1,000 highest ranking phrases as an initial set of target complex phrases. We chose 0.7 based on a manual inspection of phrase pairs in SimplePPDB. This heuristic also serves as an initial clue as to how replaceable the phrases are.

Manual inspection of a sample selected phrases and their given simplifications in the database showed that the alignments alone were not enough to determine whether the phrases were indeed replaceable, given that a lot of the simplification pairs are very similar to each other. The word *establish*, for example, is aligned to "*as to determine*", "*in order to determine*", "*determine whether*" and "*determine that*", among many others. To remedy this, we conducted a user study on the replaceability of phrases. We presented annotators with a set of complex phrases, each accompanied by ten candidate replacements: the five alignments in SimplePPDB with the highest probability, and the five words/phrases with the highest cosine similarity with the complex phrase in our Embeddings-B phrase embeddings model (trained without POS annotation or retrofitting). We did so to minimise any biases in our subsequent experiments with the retrofitted POS-aware embedding models, which is one of our main contributions. Annotators where then tasked to judge each complex phrase as being either:

- **Highly replaceable:** You can think of at least three replacements for this phrase.
- **Mildly replaceable:** You can think of up to two replacements for this phrase.
- **Non-replaceable:** You can think of no replacements for this phrase.

The candidates presented serve as suggestions on how the complex phrase could be replaced, but annotators could think of different replacements. It is important to highlight the fact that this annotation methodology was used for the purpose of fulfilling our ultimate goal of finding phrases that could be replaced by others, rather than to quantify phrase replaceability in general.

We hired four volunteers who are computational linguists to annotate the phrases. Each annotator judged 250 phrases. Figure 1 illustrates the interface used for annotation through an example. This user study, as well as all others described henceforth, were conducted using Google Forms[3].

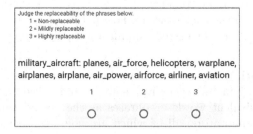

Fig. 1. Annotation interface used for our user study on phrase replaceability

Table 1 shows counts and examples of each annotations on each category. While the phrases judged either highly or mildly replaceble are quite apparently so, not all of the candidates judged non-replaceable by the annotators are, indeed, non-replaceable. For example, the phrase *"shall be able"* could be replaced with *"will be able"* in the sentence *"One of our attendants shall be able to help you with any outstanding matters"*. However, this is not a problem since we are strictly looking for replaceable target phrases to compose our dataset, so high precision is more important than high recall.

Table 1. Replaceability degrees and respective example phrases

Type	Frequency	Examples
Highly replaceable	285	Hazardous substance, walk away, formidable challenge, very beautiful, proceed along, critical phase
Mildly replaceable	297	Principal aim, more apparent, alternative form, main tool, fully understand, agricultural product
Non-replaceable	418	Establish clear, shall be able, more widespread, question remain, provides opportunity, exist means

---

[3] https://www.google.co.uk/forms/about.

To compose a final set of target complex phrases, we selected all 285 very replaceable phrases along with 115 randomly selected mildly replaceable phrases, totalling 400. 351 (87.7%) of them are two-word phrases, while 49 (12.3%) are three-word phrases. The next step in our dataset creation process was to find sentences which contain these complex phrases. To do so, we used the 7 billion word corpus described in Sect. 2.1, whose sentences have 32 words on average, to search for one distinct sentence for each of the 400 complex phrases selected.

5.2   Collecting Phrase Simplifications

With sentences and target complex phrases at hand, we began the process of collecting gold simplifications for them. For that, we employed a two-step annotation process: open and suggested simplification judgements.

Open Simplification Judgements. In this step we presented annotators with the complex phrase in its respective sentence and asked them to suggest as many simpler equivalent words or phrases as they could think of. 307 fluent English speakers annotators, all of which are volunteer students or academic staff of various universities, participated in the process. Each annotator received 10 complex phrases to simplify, and each complex phrase was annotated by at least 5 annotators. Figure 2 illustrates the annotation interface used for this step.

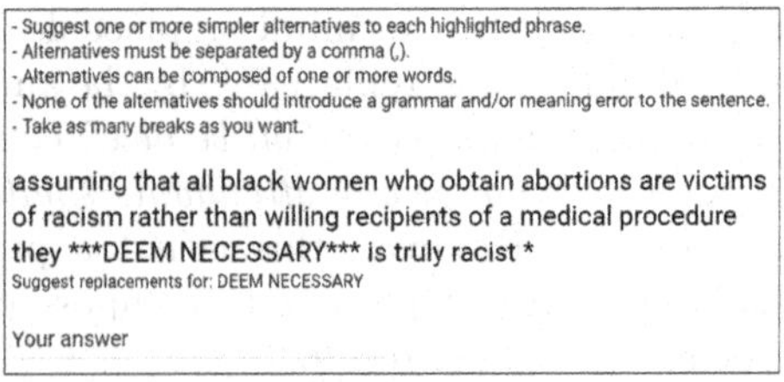

Fig. 2. Annotation interface used for the open simplification step

In total, 3,070 annotations were produced ($307 * 10 = 3,070$) and 3,367 gold simplifications suggested. The average number of gold simplifications suggested per complex phrase is 8.4 (± 3.1), ranging from 1 to 23. Table 2 shows some examples of annotations produced.

While inspecting the simplifications suggested by annotators along with those automatically produced for our replaceability user study, we noticed that there were many simpler, grammatical and meaningful alternatives among the candidates that were not suggested by any annotators. To address this problem and hence complement the simplifications in our dataset, we conducted a second annotation step.

Table 2. Example annotations from the open simplification step

Sentence with target phrase	Suggestions
This is not an obscure philosophical argument but a practical issue of {considerable importance}	Significant importance, serious importance, considerable significance, great relevance
If elected, he would also be eligible for immunity from {criminal prosecution}	Criminal judgment, facing trial, legal action, being punished
He said a special {educational curriculum} is being prepared for the youngsters	Learning timetable, educational timetable, educational programme
The IRS shall file a {fairly significant} claim against R. Allen Stanford he said	Big, crucial, quite significant, important, significant
Mr Hutchinson suggested that issues, such as dealing with the past, could be dealt with {more speedily} in devolved setting	More effectively, quicker, more rapidly, faster

Suggested Simplification Judgements. In this step we presented annotators with the complex phrase in its context along with five automatically produced candidate replacements, and asked them to select those, if any, that could simplify the complex phrase without compromising the meaning or grammaticality of the sentence. For each complex phrase, we randomly selected up to five out of the ten candidate replacements used in our replaceability user study described in Sect. 5.1 which had not yet been suggested by any annotators in the previous annotation step.

400 fluent English speakers with the same profile as in the previous step (Sect. 5.2) participated in the annotation process. Each annotator received 12 complex phrases with five candidates each to judge ($12 * 5 = 60$ candidates in total), and each candidate was judged by at least 8 annotators. Figure 3 shows the annotation interface used for this step.

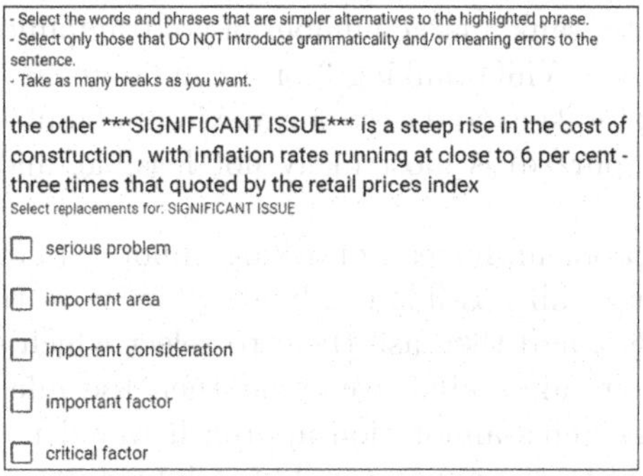

> - Select the words and phrases that are simpler alternatives to the highlighted phrase.
> - Select only those that DO NOT introduce grammaticality and/or meaning errors to the sentence.
> - Take as many breaks as you want.
>
> the other ***SIGNIFICANT ISSUE*** is a steep rise in the cost of construction , with inflation rates running at close to 6 per cent - three times that quoted by the retail prices index
> Select replacements for: SIGNIFICANT ISSUE
>
> ☐ serious problem
>
> ☐ important area
>
> ☐ important consideration
>
> ☐ important factor
>
> ☐ critical factor

Fig. 3. Annotation interface used for the given simplification step

In total, 24,000 judgements were produced $(400 * 60 = 24,000)$. To complement the pool of simplifications of a given complex phrase in our dataset, we selected only the candidates which had been judged positively by at least 4 annotators. We found through manual inspection that this number of judgements best separated good from spurious candidates. A total of 697 new simplifications were produced through this step. After adding these new simplifications to our dataset, the average number of gold simplifications per complex phrase moved from 8.4 ($\pm$3.1) to 9.7 ($\pm$3.4), ranging from 1 to 24. Amongst our gold candidates, 1,161 (29.8%) are single words, 2,213 (56.8%) phrases with two words, 392 (10%) phrases with three words, and 94 (2.4%), 25 (0.6%) and 14 (0.4%) are phrases with four, five and six words, respectively. Table 3 shows some examples of annotations produced in this step.

Table 3. Example annotations from the given simplification step

Sentence with target phrase	Selected suggestions
This is not an obscure philosophical argument but a practical issue of **{considerable importance}**	Great interest, great significance, particular interest, significance
If elected, he would also be eligible for immunity from **{criminal prosecution}**	Disciplinary action, prosecutions
He said a special **{educational curriculum}** is being prepared for the youngsters	Course of study, education program, syllabus
The IRS shall file a **{fairly significant}** claim against R. Allen Stanford he said	Sizable, substantial, very big
Mr Hutchinson suggested that issues, such as dealing with the past, could be dealt with **{more speedily}** in devolved setting	Right away, very quickly, without delay

5.3 Simplification Ranking

The last step in our dataset creation process was to rank our gold simplifications. This is arguably the most challenging step in this process. It is very difficult for a human to confidently rank a large number of gold simplifications such as what we have in our dataset. While ranking 3 or 4 candidates is not too challenging, providing a full rank for 24 candidates (our maximum number of gold candidates for a given complex phrase) is most likely not feasible, and hence would yield unreliable data.

To address this problem, instead of asking annotators to produce full rankings, we could generate all possible candidate pairs from the set, present each pair to the annotators, and then ask them to select which of the candidates is simpler. With all pairs annotated, we could then use inference algorithms to produce full rankings. This annotation approach would reduce the amount of information presented to the annotator, but it introduces a new problem: The number of instances that have to be annotated increases in combinatorial fashion with respect to the number of candidates available. For a phrase with 24 candidate substitutes, for example, 276 pairs would have to be annotated. Because of this limitation, we conceived a new ranking-by-comparison annotation approach. We reduce the number of comparisons in two ways: by presenting the annotator with more than two candidates at a time, and by sampling the space of comparisons based on the minimum number of comparisons in which each candidate must be featured.

Suppose we have a set of five candidates $\langle a, b, c, d, e \rangle$ and are willing to compare three candidates at a time such that all of them appear in at least three comparisons. To do so, we could present annotators with six comparison tasks featuring the following sets of candidates:

$$\langle a, b, c \rangle \ \langle c, d, e \rangle \ \langle a, c, d \rangle$$
$$\langle a, b, e \rangle \ \langle b, d, e \rangle \ \langle a, b, d \rangle$$

By doing so, we would reduce the number of annotations required from 10, which is the number of distinct pairs that can be extracted from $\langle a, b, c, d, e \rangle$, to only six, which is 40% less. Notice also that this approach is flexible, given that one can adjust the number of candidates presented at a time and the minimum number of comparisons per candidate based on the number of candidates to be ranked, degree of informativeness desired and annotation budget.

For this annotation step, we presented annotators with three candidates at a time and required each candidate to appear in at least five comparisons. Each instance presented to annotators was composed of a sentence with a gap in place of the complex phrase and three candidate simplifications. Annotators were asked to select which candidate, when placed in the gap, would make the sentence easiest to understand, and which candidate would make the sentence most difficult to understand. By asking for both the easiest and most difficult of the candidates, we get a full local ranking between the three candidates presented, which allows us to more accurately quantify the simplicity differences between them. In order to produce a set of comparisons to present to the annotators, we randomly selected comparisons from the set of possible 3-candidate combinations that could be extracted from the candidate set until every candidate was featured in at least five comparisons. The annotation interface we used is illustrated in Fig. 4.

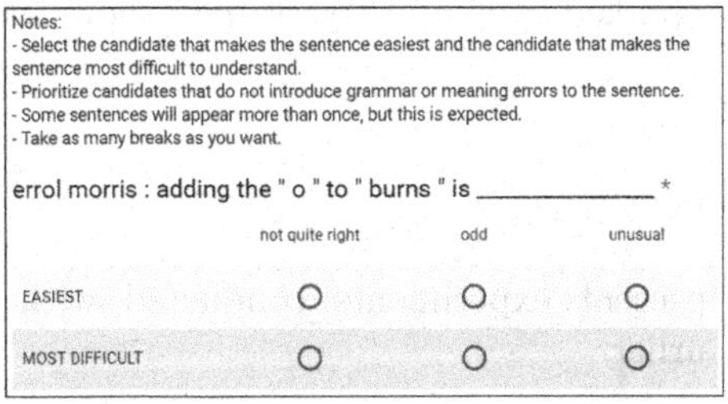

Fig. 4. Annotation interface used for the candidate ranking step

300 non-native English speakers took part in this annotation step (same profile as in Sect. 5.2). Each annotator received 27 comparisons with three candidates each to judge. Due to the limited annotation budget available, each comparison was judged by only one annotator. In total, 8,100 judgements were produced $(300 * 27 = 8,100)$. To generate a final ranking of candidates, we first assigned score 3 to the candidates judged easiest, 1 to the candidates judged most complex, and 2 to the remaining candidates. Then we averaged the scores of each candidate across all comparisons in which they appeared, and used the average to rank them in decreasing order (ties were allowed). Table 4 shows some examples of annotations produced in this step.

Table 4. Example annotations from the candidate ranking step. Candidates are ranked from simplest (1) to most complex (3).

Sentence with target phrase	Ranked candidates
This is not an obscure philosophical argument but a practical issue of {____}	1 - significant importance 2 - great relevance 3 - concern
If elected, he would also be eligible for immunity from {____}.	1 - being punished 2 - criminal judgment 3 - criminal prosecution
He said a special {____} is being prepared for the youngsters.	1 - teaching method 2 - educational curriculum 3 - learning timetable
The IRS shall file a {____} claim against R. Allen Stanford.	1 - significant 2 - very big 3 - quite significant
Mr Hutchinson suggested that issues, such as dealing with the past, could be dealt with {____} in devolved setting	1 - faster 2 - more effectively 3 - more speedily

The full process resulted in a new dataset – **BenchPS**. It contains 400 instances composed of a sentence, a target complex phrase, and an average of 9.7 gold replacements ranked by simplicity. Table 5 shows examples of instances in BenchPS.

6   Experiments

In this section we present experiments conducted with the approach and resources presented earlier.

Table 5. Example instances from the completed BenchPS. Candidate substitutions are ranked from simplest to most complex.

Sentence with target phrase	Gold replacements
A {**medical professional**} in the government recommended testing on entry, but was told it was too expensive to test all	1: physician, 2: doctor, 2: health care provider, 3: medical advisor, 4: medical doctor
The solution for these symptoms is {**very straightforward**} – just go back on the antidepressant and they go away	1: simple, 2: relatively simple, 3: easy, 4: very easy, 4: fairly simple, 5: not at all difficult, 6: pretty simple, 7: fortright
In response, Coelho has digitalized all 16 of his titles in Farsi and posted them today on the internet for anyone to download, {**free of charge**}	1: free-of-charge, 2: for free, 3: freely, 3: free, 4: at no cost, 5: without payment, 5: for no pay, 6: gratis
And the painful irony is that the source of their money woes is exactly what makes them {**most happy**}: cars	1: happiest, 2: pleased, 3: glad
My father died when I was 13 and my mother later remarried, to a man much more like her in temperament and who was the {**most wonderful**} stepfather imaginable	1: best, 2: nicest, 3: greatest, 4: finest

6.1 Candidate Phrase Generation

In our first experiment we conducted an exhaustive search for the best values of the hyperparameters of our SG and SS approaches jointly:

- α: The number of candidates to generate with our retrofitted POS-aware embeddings (Embeddings-PR) for SG.
- β: The proportion of winning comparisons required by our comparison-based SS approach to keep a candidate substitution.

For α, we checked all integer values between 1 and 50, and for β, all values between 0.0 and 1.0 in intervals of 0.01 (100 values), totalling 5,000 value pairs. We also performed a joint exhaustive search between α and a third parameter γ, which is the proportion of candidates discarded by an unsupervised boundary ranking selector (UBR). We do this so we can compare our performance against the current state-of-the-art approach. For γ, we also search for all values between 0.0 and 1.0 in intervals of 0.01. Both SS approaches use the same set of 40 features described in Sect. 3.

We split BenchPS into two equally sized portions (200 instances): One for training, and one for testing. The unsupervised boundary ranking selector is trained using the same procedure described by [26] over the training set. For each possible combination of $\alpha \times \beta$ and $\alpha \times \gamma$, we calculate the following three metrics over the training set:

- **Precision:** Proportion of selected candidates that are in the gold simplifications.
- **Recall:** Proportion of gold simplifications that are in the set of selected candidates.
- **F-measure:** Harmonic mean between Precision and Recall.

The values obtained are illustrated in Fig. 6 in form of heatmaps. Brighter colors indicate better metric scores. While the behaviours of our comparison-based approach and UBR are rather similar for Recall, the same cannot be said for Precision and F-measure. The top left corner of the Precision map of our approach is much brighter-colored than that of UBR, which suggests that our approach is more suitable for the creation of more conservative phrase simplifiers. And as it can be noticed, the F-measure heatmap of our approach is overall much brighter colored than that of UBR, specially in the α range between 5 and 20 (Fig. 5).

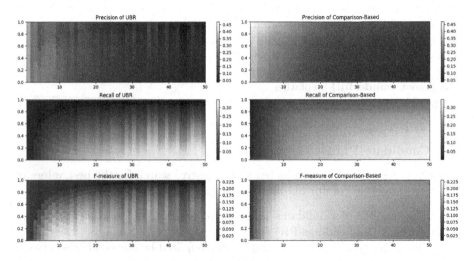

Fig. 5. Results of exhaustive hyperparameter search for SS approaches with respect to our SG approach. The horizontal axes represent α values. Vertical axes on the left and right column represent γ and β values, respectively.

Interestingly, the α value that yielded the highest F-measure scores for both approaches was 11. The β and γ values, on the other hand, were 0.32 and 0.0, respectively. In other words, our SG approach benefits from being paired with our comparison-based SS strategy, but performs better on its own than when paired with an unsupervised boundary ranking selector.

With these hyperparameter values selected, we compared our approach to others. For SG, we include nine strategies in the comparison:

- **WordNet:** Extracts synonyms, hypernyms, and hyponyms of phrases as candidates from WordNet.

- **SimplePPDB:** Extracts aligned paraphrases as candidates from SimpleP-PDB. We include four variants of SimplePPDB generators: one that extracts all aligned paraphrases as candidates (SimplePPDB), and three others that consider only paraphrases aligned with a probability >0.5 (SimplePPDB-0.5), >0.7 (SimplePPDB-0.7), and >0.9 (SimplePPDB-0.9).
- **Embeddings:** Extracts candidates using our SG approach. We create one system for each of the embedding models described in Sect. 2.1: the base model (Embeddings-B), the base model retrofitted over WordNet (Embeddings-R), the POS-aware model (Embeddings-P) and the POS-aware retrofitted model (Embeddings-PR).

We pair each of these nine generators with three SS methods:

- no selection,
- unsupervised boundary ranking (UBR), and
- our comparison-based approach (CB).

To find the best values of β and γ, we perform exhaustive search over the training set with respect to the F-measure. To evaluate all 27 combinations of SG and SS approaches we use the same previously introduced Precision (P), Recall (R), and F-measure (F) metrics over the test set.

The results in Table 6 reveal that, regardless of the selector used, our POS-tagged retrofitted phrase-annotated embeddings outperformed all other generators. Pairing them with our comparison-based selector yielded the highest scores overall. The WordNet generator could not produce a single valid candidate substitution due to its low coverage for phrases. Finally, due to the large number of spurious candidates produced, none of the selectors managed to consistently improve the Precision scores of SimplePPDB generators without detrimental compromises in Recall.

Table 6. Candidate generation and selection results: Precision (P), Recall (R) and F-measure (F)

	No selection			UBR			CB		
	P	R	F	P	R	F	P	R	F
WordNet	0.000	0.000	0.000	0.000	0.000	0.000	0.000	0.000	0.000
SimplePPDB	0.055	0.173	0.084	0.087	0.135	0.106	0.061	0.163	0.089
SimplePPDB-0.5	0.083	0.149	0.107	0.095	0.123	0.107	0.089	0.140	0.109
SimplePPDB-0.7	0.120	0.098	0.108	0.119	0.069	0.088	0.120	0.098	0.108
SimplePPDB-0.9	0.175	0.018	0.032	0.177	0.013	0.024	0.178	0.018	0.032
Embeddings-B	0.140	0.155	0.147	0.163	0.132	0.146	0.163	0.143	0.153
Embeddings-R	0.154	0.171	0.162	0.181	0.146	0.162	0.177	0.157	0.166
Embeddings-P	0.190	0.210	0.200	0.177	0.143	0.158	0.205	0.203	0.204
Embeddings-PR	**0.216**	**0.240**	**0.227**	**0.197**	**0.158**	**0.175**	**0.232**	**0.231**	**0.232**

6.2 Phrase Simplicity Ranking

In our second experiment, we assessed the performance of 20 rankers in the task of capturing phrase simplicity:

- One for each of the 16 features described in Sect. 4.
- **Glavas:** The rank averaging approach of [12], as described in Sect. 4, while using only their original features.
- **Glavas+P:** The same ranker, but with the original feature set complemented with our 16 phrase-level features described in Sect. 4.
- **Paetzold:** The supervised neural ranker of [20], as described in Sect. 4, trained over the training portion of BenchPS using only with the features introduced by [20].
- **Paetzold+P:** The same ranker, but with the original feature set complemented with our 16 phrase-level features described in Sect. 4.

We evaluated the rankers on the test set using three evaluation metrics: Spearman (s) and Pearson (r) correlation between gold and produced rankings, as well as TRank [32], which measures the proportion of times in which the simplest gold candidate was ranked first.

The results in Table 7 confirm that the proposed features help rankers capture phrase simplicity: adding them to the Glavas and Paetzold rankers resulted in substantial performance gains. The supervised neural ranker complemented with our features (Paetzold+P) obtained the highest correlation scores, but was outperformed in TRank by the neural ranker without our features (Paetzold). We believe this is due to the fact that the Paetzold ranker is better at placing simpler candidates at the top rank, but worse than the Paetzold+P ranker in ordering all candidates properly. Additionally, the length and token count of a phrase do not seem to play a significant role in its simplicity. And contrary to what was observed in previous benchmarks for single-word LS [23], language model probabilities tend to capture simplicity at phrase-level more effectively than raw frequencies. This suggests that the context has a stronger influence in determining the simplicity of phrases than the simplicity of single words.

6.3 Full Pipeline Evaluation

In this experiment, we evaluated various combinations of candidate generators and rankers in the creation of full phrase simplifiers. We include two top performing generators from the ones tested in the previous experiments: SimplePPDB-0.5 and Embeddings-PR. We pair both generators with the three selectors described in the previous experiment, and hence consider six different ways of producing candidates.

We paired each of the aforementioned generator/selector pairs with the 20 rankers described in Sect. 6.2. The metric used here is **Accuracy**: the proportion of instances in which the simplifier replaced the complex phrase with one of the gold simplifications from the dataset. In this context, Accuracy encompasses all

Table 7. Simplicity correlation results

	TRank	r	s
Length	0.070	0.003	0.071
Token Count	0.090	0.066	0.123
SimpleWiki-PF	0.230	0.288	0.305
SubIMDB-PF	0.235	0.257	0.270
Wikipedia-PF	0.220	0.254	0.280
7billion-PF	0.225	0.252	0.275
SimpleWiki-LM	0.250	0.319	0.342
SubIMDB-LM	0.250	0.305	0.326
Wikipedia-LM	0.245	0.325	0.366
7billion-LM	0.275	0.338	0.377
Min. Frequency	0.195	0.293	0.319
Max. Frequency	0.140	0.179	0.213
Avg. Frequency	0.185	0.255	0.279
Min. Similarity	0.155	0.162	0.198
Max. Similarity	0.130	0.175	0.219
Avg. Similarity	0.155	0.174	0.221
Glavas	0.255	0.281	0.291
Glavas+P	0.275	0.348	0.372
Paetzold	**0.365**	0.375	0.345
Paetzold+P	0.355	**0.420**	**0.392**

aspects of simplification quality by assuming that the gold candidates in our dataset ensure grammaticality, meaning preservation and simplicity.

The results are illustrated in Table 8. While the addition of our phrase-level features did not have a significant impact on the performance of the rankers for the SimplePPDB-0.5 generator, it led to noticeably higher Accuracy scores for our Embeddings-PR model. The differences observed are statistically significant (10-fold bootstrapping significance tests with $p < 0.01$).

Interestingly, although Table 8 confirms the finding of Sect. 6.1 that our comparison-based approach is indeed more reliable than unsupervised boundary ranking, neither of these SS approaches proved more effective than not performing selection at all. Inspecting the results we found that, although both selectors do manage to improve the precision of the SG approaches evaluated, they discard too many useful candidates, sometimes leaving none behind.

Overall, the most accurate simplifier combines our phrase and POS-aware retrofitted embeddings (Embeddings-PR), no selection (No Sel.), and the supervised neural ranker with our phrase-level features (Paetzold+P).

To further illustrate the effectiveness of our phrase-level features for SR, we compared the performance of the Glavas/Paetzold and Glavas+P/Paetzold+P rankers in thousands of different settings. To do so, we paired each of these four rankers with all the 5,000 settings described in Sect. 6.1 of Embeddings-PR and our comparison-based selector. For the α value (number of candidates generated by Embeddings-PR), we considered all integer values between 1 and 50, and for β (number of winning comparisons necessary for a candidate to be kept), all values between 0.0 and 1.0 in intervals of 0.01. In total, the performance of 20,000 phrase simplifiers were tested (4 rankers * 5,000 generator/selector pairs).

Figure 6 shows heatmaps for the Accuracy scores obtained in all these settings. The brighter (whiter) the color of a given spot, the higher the Accuracy score obtained for that setting. The heatmaps reveal that, although the Accuracy of the rankers is very similar for α values no larger than 12, the rankers trained with our phrase-level features (Glavas+P/Paetzold+P) achieve much higher Accuracy scores when more candidates are generated. This suggests that adding these features to the ranker is crucial when creating phrase simplifiers that aim to maximise the recall of candidates generated, and hence increase the simplicity of the output.

Table 8. Accuracy of full pipeline phrase simplification approaches

	SimplePPDB-0.5			Embeddings-PR		
	No sel.	UBR	CB	No sel.	UBR	CB
Length	0.000	0.115	0.020	0.100	0.100	0.165
Token Count	0.020	0.135	0.035	0.115	0.115	0.155
SimpleWiki-PF	0.130	0.290	0.130	0.350	0.290	0.355
SubIMDB-PF	0.105	0.280	0.105	0.325	0.270	0.325
Wikipedia-PF	0.135	0.285	0.140	0.325	0.275	0.335
7billion-PF	0.115	0.295	0.120	0.320	0.260	0.325
SimpleWiki-LM	0.130	0.315	0.130	0.350	0.310	0.350
SubIMDB-LM	0.145	0.320	0.150	0.365	0.290	0.365
Wikipedia-LM	0.120	0.300	0.125	0.375	0.325	0.375
7billion-LM	**0.160**	0.325	**0.165**	0.385	0.325	0.375
Min. Frequency	0.130	0.205	0.130	0.270	0.215	0.270
Max. Frequency	0.130	0.125	0.130	0.040	0.065	0.080
Avg. Frequency	0.130	0.155	0.130	0.110	0.100	0.130
Min. Similarity	0.05	0.290	0.060	0.405	0.320	**0.430**
Max. Similarity	0.05	0.195	0.060	0.270	0.210	0.320
Avg. Similarity	0.05	0.310	0.060	0.390	0.310	0.410
Glavas	0.120	0.300	0.125	0.380	0.295	0.380
Glavas+P	0.110	**0.355**	0.115	0.390	0.325	0.395
Paetzold	0.135	0.305	0.140	0.365	0.305	0.345
Paetzold+P	0.135	0.350	0.140	**0.460**	**0.370**	0.425

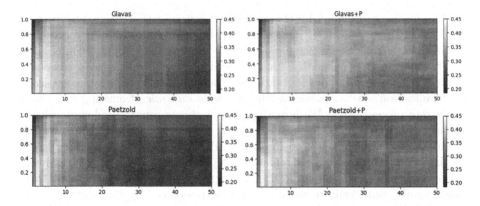

Fig. 6. Accuracy scores for of exhaustive performance comparison. The horizontal axes represent α values. Vertical axes on the left and right column represent γ values.

6.4 Error Analysis

To complement our quantitative analyses, we conducted a manual inspection of the mistakes made by two of the simplifiers featured in the experiments of Sect. 6.3:

- **Embeddings-PS:** the best performing simplifier that uses our Embeddings-PR generator, which employs no SS approach and the Paetzold+P ranker.
- **SimplePPDB-PS:** the best performing simplifier that uses the SimplePPDB-0.5 generator, which employs unsupervised boundary ranking and the Glavas+P ranker.

The authors of this paper, who are fluent non-native English speakers, conducted the analyses. They were presented with the sentence, target complex phrase and simplification produced for each instance of the BenchPS dataset in which these simplifiers failed to produce a simplification that was amongst the gold simplifications available. The annotator was tasked with deciding which of the simplifications produced were grammatical, meaning preserving, and simpler than the target phrase. This type of analysis allows us to truly quantify the proficiency of these simplifiers, since one can never be sure that every conceivable simpler alternative to the target complex phrases of a dataset were captured during its creation. An individual analysis was conducted for each of the aforementioned simplifiers.

We found that, out of the 111 simplification "errors" made by the Embeddings-PS simplifier, 33 (29.72% of 111) were grammatical, meaning preserving, and simpler than the target. And for the SimplePPDB-PS simplifier, the proportion was 47 out of 129 (36.43% of 129). Adding these numbers to the scores in Sect. 6.3 results in that Embeddings-PS has an actual Accuracy of 61% (89+33 out of 200), and SimplePPDB-PS 59% (71+47 out of 200).

428 G. H. Paetzold and L. Specia

The feedback provided by the annotator indicated that Embeddings-PS is more effective than SimplePPDB-PS at finding simplifications for more challenging complex phrases in the dataset. Some examples are "lodge a complaint", which was replaced by "notify" in "... the club says it will *lodge a complaint* against Dyfed-Powys police...", and "purchase and sale", which was replaced by "transaction" in "... as well as the *purchase and sale* of existing residential real estate properties...". We hypothesise that this is caused predominantly by the fact that our Embeddings-PR model is trained over a vocabulary that is much larger than that featured in SimplePPDB. However, we also found that the Embeddings-PS simplifier replaces complex phrases with antonyms much more frequently than SimplePPDB-PS. Some examples are the phrases "provides details" and "quite rapidly", which were replaced by "summarizes" and "gradually", respectively. This is caused by the fact that this type of embeddings model tends to group antonyms close together [17].

7 Final Remarks

We introduced resources and approaches that can be used to address the task of Phrase-Level Simplification. We presented a way of training and employing POS-aware retrofitted phrase embedding models for SG, introduced a new unsupervised comparison-based approach for SS, and proposed a set of phrase-level features that can complement consolidated supervised and unsupervised SR strategies.

To train and evaluate phrase simplifiers, we created BenchPS: a new dataset produced using a flexible annotation methodology that aims to maximize the recall of gold simplifications available and minimize the costs of producing simplicity rankings. Through experimentation, we found that our enhanced phrase embeddings provide a more reliable source of simplifications for complex phrases than stand-alone SimplePPDB. We also found that our comparison-based SS approach can be more effective in discarding inappropriate candidates than the former state-of-the-art strategy. Our experiments with candidate ranking show that adding our phrase-level features to state-of-the-art rankers can increase their performance significantly, specially when a large amount of candidate substitutes are produced during SG and SS. We also found that context plays a more important role in the simplification of phrases than of single words. Through a manual inspection of the mistakes made by our most reliable phrase simplifier, we found that it correctly simplifies complex phrases 61% of the time.

In future work, we intend to address the challenges of automatically identifying complex phrases, as well as simplifying out of vocabulary phrases. The BenchPS dataset[4] is already available online. The code for the phrase simplification approaches introduced will be made available once this paper is published.

---

[4] http://ghpaetzold.github.io/data/BenchPS.zip.

References

1. Azab, M., Hokamp, C., Mihalcea, R.: Using word semantics to assist English as a second language learners. In: Proceedings of the 2015 NAACL (2015)
2. Biran, O., Brody, S., Elhadad, N.: Putting it simply: a context-aware approach to lexical simplification. In: Proceedings of the 49th ACL, Portland, Oregon, USA, June 2011, pp. 496–501. Association for Computational Linguistics (2011)
3. Brysbaert, M., New, B.: Moving beyond Kučera and Francis: a critical evaluation of current word frequency norms and the introduction of a new and improved word frequency measure for American English. Behav. Res. Methods **41**, 977–990 (2009)
4. De Belder, J., Moens, M.-F.: A dataset for the evaluation of lexical simplification. In: Gelbukh, A. (ed.) CICLing 2012. LNCS, vol. 7182, pp. 426–437. Springer, Heidelberg (2012). https://doi.org/10.1007/978-3-642-28601-8_36
5. Devlin, S.: Simplifying natural language for aphasic readers. Ph.D. thesis, University of Sunderland (1999)
6. Devlin, S., Tait, J.: The use of a psycholinguistic database in the simplification of text for aphasic readers. Linguist. Databases 161–173 (1998)
7. Devlin, S., Unthank, G.: Helping aphasic people process online information. In: Proceedings of the 8th SIGACCESS, pp. 225–226 (2006)
8. Elhadad, N.: Comprehending technical texts: predicting and defining unfamiliar terms. In: Proceedings of the 2006 AMIA (2006)
9. Elhadad, N., Sutaria, K.: Mining a lexicon of technical terms and lay equivalents. In: Proceedings of the 2007 BioNLP, pp. 49–56 (2007)
10. Faruqui, M., Dodge, J., Jauhar, S.K., Dyer, C., Hovy, E., Smith, N.: Retrofitting word vectors to semantic lexicons. In: Proceedings of the 2015 NAACL, Denver, Colorado, May–June 2015, pp. 1606–1615. Association for Computational Linguistics (2015)
11. Fellbaum, C.: WordNet: An Electronic Lexical Database. Bradford Books (1998)
12. Glavaš, G., Štajner, S.: Simplifying lexical simplification: do we need simplified corpora? In: Proceedings of the 53rd ACL, Beijing, China, July 2015, pp. 63–68. Association for Computational Linguistics (2015)
13. Horn, C., Manduca, C., Kauchak, D.: Learning a lexical simplifier using Wikipedia. In: Proceedings of the 52nd ACL, Baltimore, Maryland, June 2014, pp. 458–463. Association for Computational Linguistics (2014)
14. Kauchak, D.: Improving text simplification language modeling using unsimplified text data. In: Proceedings of the 51st ACL, Sofia, Bulgaria, August 2013, pp. 1537–1546. Association for Computational Linguistics (2013)
15. Lison, P., Tiedemann, J.: Opensubtitles 2016: extracting large parallel corpora from movie and TV subtitles. In: Proceedings of the 10th LREC (2016)
16. Mikolov, T., Chen, K., Corrado, G., Dean, J.: Efficient estimation of word representations in vector space. arXiv preprint arXiv:1301.3781 (2013)
17. Mikolov, T., Yih, W.-T., Zweig, G.: Linguistic regularities in continuous space word representations. In: Proceedings of 2013 NAACL, pp. 746–751 (2013)
18. Nunes, B.P., Kawase, R., Siehndel, P., Casanova, M., Dietze, S.: As simple as it gets - a sentence simplifier for different learning levels and contexts. In: Proceedings of the 13th ICALT, pp. 128–132 (2013)
19. Paetzold, G., Specia, L.: Understanding the lexical simplification needs of non-native speakers of English. In: Proceedings of the 26th COLING, Osaka, Japan, pp. 717–727. The COLING 2016 Organizing Committee (2016)

20. Paetzold, G., Specia, L.: Lexical simplification with neural ranking. In: Proceedings of the 15th EACL, pp. 34–40. Association for Computational Linguistics (2017)
21. Paetzold, G.H.: Lexical simplification for non-native English speakers. Ph.D. thesis, University of Sheffield (2016)
22. Paetzold, G.H., Specia, L.: LEXenstein: a framework for lexical simplification. In: Proceedings of ACL-IJCNLP 2015 System Demonstrations, Beijing, China, July 2015, pp. 85–90. Association for Computational Linguistics and The Asian Federation of Natural Language Processing (2015)
23. Paetzold, G.H., Specia, L.: Benchmarking lexical simplification systems. In: Proceedings of the 10th LREC, Portoroz, Slovenia. European Language Resources Association (ELRA) (2016)
24. Paetzold, G.H., Specia, L.: Collecting and exploring everyday language for predicting psycholinguistic properties of words. In: Proceedings of the 26th COLING, Osaka, Japan, December 2016, pp. 1669–1679 (2016)
25. Paetzold, G.H., Specia, L.: SemEval 2016 task 11: complex word identification. In: Proceedings of the 10th International Workshop on Semantic Evaluation (SemEval-2016), San Diego, California, June 2016, pp. 560–569. Association for Computational Linguistics (2016)
26. Paetzold, G.H., Specia, L.: Unsupervised lexical simplification for non-native speakers. In: Proceedings of the 13th AAAI, pp. 3761–3767. AAAI Press (2016)
27. Pavlick, E., Callison-Burch, C.: Simple PPDB: a paraphrase database for simplification. In: Proceedings of the 54th ACL, pp. 143–148 (2016)
28. Pavlick, E., Rastogi, P., Ganitkevitch, J., Van Durme, B., Callison-Burch, C.: PPDB 2.0: better paraphrase ranking, fine-grained entailment relations, word embeddings, and style classification. In: Proceedings of the 53rd ACL, pp. 425–430. Association for Computational Linguistics (2015)
29. Pennington, J., Socher, R., Manning, C.D.: Glove: global vectors for word representation. In: Proceedings of the 2014 EMNLP, pp. 1532–1543 (2014)
30. Shardlow, M.: A comparison of techniques to automatically identify complex words. In: Proceedings of the 51st ACL Student Research Workshop, pp. 103–109 (2013)
31. Specia, L.: Translating from complex to simplified sentences. In: Computational Processing of the Portuguese Language, pp. 30–39 (2010)
32. Specia, L., Jauhar, S.K., Mihalcea, R.: SemEval-2012 task 1: English lexical simplification. In: Proceedings of the 1st SemEval, Montréal, Canada, pp. 347–355. Association for Computational Linguistics (2012)
33. Toutanvoa, K., Manning, C.: Enriching the knowledge sources used in a maximum entropy part-of-speech tagger. In: Proceedings of the 2000 SIGDAT, Hong Kong, China, October 2000, pp. 63–70. Association for Computational Linguistics (2000)
34. Wang, S., Zong, C.: Comparison study on critical components in composition model for phrase representation. ACM Trans. Asian Low-Resour. Lang. Inf. Process. 16(3), 16 (2017)
35. Wubben, S., van den Bosch, A., Krahmer, E.: Sentence simplification by monolingual machine translation. In: Proceedings of the 50th ACL, pp. 1015–1024 (2012)
36. Xu, W., Callison-Burch, C., Napoles, C.: Problems in current text simplification research: new data can help. Trans. Assoc. Comput. Linguist. 3, 283–297 (2015)
37. Xu, W., Napoles, C., Pavlick, E., Chen, Q., Callison-Burch, C.: Optimizing statistical machine translation for text simplification. Trans. Assoc. Comput. Linguist. 4, 401–415 (2016)

38. Yin, W., Schütze, H.: An exploration of embeddings for generalized phrases. In: Proceedings of the ACL 2014 Student Research Workshop, Baltimore, Maryland, USA, June 2014, pp. 41–47. Association for Computational Linguistics (2014)
39. Zhao, Y., Liu, Z., Sun, M.: Phrase type sensitive tensor indexing model for semantic composition. In: Proceedings of the 2015 AAAI, pp. 2195–2202 (2015)
40. Zhu, Z., Bernhard, D., Gurevych, I.: A monolingual tree-based translation model for sentence simplification. In: Computational Linguistics, pp. 1353–1361 (2010)

Automatic Creation of a Pharmaceutical Corpus Based on Open-Data

Cristian Bravo⬤, Sebastian Otálora^(✉)⬤, and Sonia Ordoñez-Salinas⬤

Universidad Distrital Francisco José de Caldas, Gesdatos, Bogotá, Colombia
gesdatos@udistrital.edu.co
https://comunidad.udistrital.edu.co/gesdatos/

Abstract. A large amount of information related to the pharmaceutical industry is published through the web, but there are few tools that allow for automatic analysis. The pharmaceutical area at international level, especially for English language, has a large number of virtual lexical tools, that support not only the research but the software development, while for Spanish language, are few lexical tools and less for the particularities of a country as Colombia. This paper presents the pharmaceutical corpus generation based on open data of Colombian medicines published monthly by the National Institute of Medicines and Food Surveillance (INVIMA). A model has been developed that combines the concepts of corpus and ontology, and the model is structured through a multi-related graph. This model is implemented in a graph-oriented database, because it has been shown to manage this type of structures and since they are based on a mathematical theory, graph-oriented database allows to find patterns and relationships that would otherwise not be possible. For the creation of the corpus, a Crawler was developed to download and control the documents, and through text processing and the proper algorithm are stored in the graph-oriented database (Neo4j).

Keywords: Crawler · Neo4j · Natural language processing · Open data Colombia · Pharmaceutical corpus

1 Introduction

Lexical aids are widely used and they are essential elements for natural language analysis, software development, and general research. An analogy could be made with lexical resources between the human conscience and the software applications, although an individual in the construction of ideas requires consulting, ordering and filtering the accumulated information during his life, the software systems need resources lexicons to search for concepts, meanings, synonyms, behaviours rules, among others. The process of building such lexical aids is very costly in terms of the time and knowledge of expert, so today it is used computational algorithms to develop some of the phases.

Universidad Distrital Francisco José de Caldas.

© Springer Nature Switzerland AG 2023
A. Gelbukh (Ed.): CICLing 2019, LNCS 13451, pp. 432–450, 2023.
https://doi.org/10.1007/978-3-031-24337-0_31

On the other hand, governments in different countries and organizations are publishing open-data for the community use. However, such data can be published raw, with some inherent or clustered knowledge, which can be very helpful for the citizen who seeks to do a particular analysis, but most times use data requires computational techniques.

Taking advantage of the open data published with inherent knowledge about the license granted by INVIMA to medicines in Colombia. In the paper presented, a structured corpus is constructed under a multi-related graph that allows finding relationships and solving queries that are latent in the information.

Consequently, this article presents the following structure: General concepts of the elements used within the framework of the research are contextualized; Next presents a review of similar works developed both practical and proposed; Next the corpus development; Next the results obtained and an analysis thereof and finally the conclusions.

2   General Concepts

A description of some concepts related to the work such as Ontology, Corpus and the relationship between these, graph-oriented data bases, theory of graphs and crawler is included below.

Web Crawler. A web crawler is defined as a software that is able to collect information published on the Internet automatically [33]. In its basic form, its function is to start from a set or list of URLs, to visit the web pages, to identify, to capture and to store certain type of information required for its visualization and later use. Generally, a Crawler Web also identifies hyperlinks in web pages in order to access these and broaden the search and collection range of information [16].

On the other hand, a web crawler can be periodic or incremental. A Crawler is periodic when it performs its task of visiting pages in a predefined time interval and it is incremental when it defines variables that decide how many pages to visit and how often it should do it [33]. The construction of a web crawler type will depend on the type of pages you want to visit and how often they are updated. Web crawlers are used in data mining and indexing web pages.

Graph Theory. A graph is a mathematical model that is useful for the representation and solution of real life situations or problems. Graphs are typically represented graphically and are composed of a non-empty set of vertices and a set of edges or incidences. This representation allows to find relations (edges) between different elements (vertex) and to do analyses or calculations on them [7].

Graph theory, as a topic of mathematics, has been used in a large number of areas, such as mathematics, physics, computation, linguistics, among others, to study different situations under the graphs approach [3]. In this way,

Graph theory brings together a large number of elements such as probability, arithmetic, algebra, topology and whose application covers fields such as process optimization, flow realization, search algorithms, shortest path identification, Construction of circuits, analysis of networks, among others.

The graph-oriented databases are one of the applications of the graphs in the computer science field, where the information is structured through graphs, and through different algorithms the information is analysed in search of relations between elements that under Relational database models are not convenient. The latter is due to the property of graphs known as transversely, which in synthesis is the ability of these to study and identify relationships between vertices that are not related to edges directly.

NoSQL Databases. NoSQL databases are characterized by not fully guaranteeing the ACID properties (Atomicity, Consistency, Isolation and Durability) and to differ in their structures from the classic relational model. In general, a NoSQL model allows solving some type of specific problem by trying to optimize elements such as search performance, storage space, reduction in complexity, optimization of processing and computational cost [5, 19]

The NoSQL databases are mainly divided into [29]: a) Documentaries, store the information under an abstract notion of "document", some examples are MongoDB [1] and CouchDB [2]; b) Key-value, implemented under the simplistic model of an identifier and its corresponding value, stand out BigTable [6] and DynamoDB [32]; c) Oriented to columns, its structure is represented by columns and families of columns, Cassandra [28] and HiperTable [34] are two of its main exponents and d) Graph-oriented databases (GODB) are based on graph theory. Later this type of databases is extended.

NoSQL databases have generally been developed under unique contexts, this means that their goal is not to give a solution to the general problems but instead offer a better alternative to each individual problem [43]. Therefore, the development and use of a NoSQL database efficiently will depend on the requirements and the type of solution that is needed.

Graph-Oriented Databases. Graph-oriented databases (GODB) are based on graph theory, and are supported by strong theoretical foundations. In them, the vertices usually represent an object and the edges relations between them, in addition, both elements can have properties [5]. The set of nodes and attributes form a graph similar to a network.

Graph theory provides a mathematical model for a GODB to do basic operations such as inserting, deleting and updating a node or a relationship. It also defines the transversal of graphs, with which you can execute search operations to nodes and their adjacent elements, in other words, a network scan is performed to find elements with non-direct relationships [19].

The GODB allow complex or wide queries in a simpler way, due to their nature they present dynamic models that can vary the structure of the elements of the same set thus avoiding waste of space, the data processing is optimized by using different algorithms of Graph theory and are an alternative when it is more important to do analysis of relations between objects than objects as such [5].

Neo4j is an example of GODB in the market, being of free type, written in the Java language, is a highly scalable database with real-time performance ability and designed to store and process large amounts of information especially for tasks related to Big Data.[1]

Corpus. A corpus is a collection of texts stored electronically [23], these texts may be from one or more areas of knowledge, generally selected and ordered according to some criteria and conform a model that facilitates linguistic analysis Some main features of a corpus are: being in electronic format, containing authentic data, having selection criteria, being representative and a fixed size [45]. Specifically, a corpus is a tool that allows the storage, search, retrieval, classification and analysis of texts associated with a language and a dialect.

Ontologies. An ontology commonly refers to a set of primitive representations with which a domain of knowledge or discourse is modelled and in particular for the area of computation, an ontology forms a semantic and abstract model of data in a specific area of knowledge [15]. The basic structure of an ontology is composed of classes, attributes and relations [13], classes are sets representing objects, attributes are characteristics or properties inherent to objects and relations interactions between different objects [14]. There are some other elements in an ontology, such as axioms, rules, constraints, among others, but in essence, all ontologies share the basic structure.

Medicines Data Standard in Colombia. In Colombia, the government institution known as the Ministry of Health provides a standard for the structuring, coding, updating, and use of medicines related data. The standard has three levels of description [22]: Level 1. Medication in its common description, it has 4 key attributes: active principle, concentration, pharmaceutical form and route of administration; Level 2. Commercial medicines are the medicines with common description and additionally have a trade mark and Level 3. Medication with the commercial presentation is the commercial medicine and additionally owns a unit of content and a commercial presentation.

The medicine data used in performing this work have a level 3 of description.

Colombian Legislation. In Colombia, the Ministry of Health regulates the definition and implementation of the data standard for medicinal products for human use in Colombia by Resolution 3166 of 2015. This country standard and other administrative documents of INVIMA set up fundamental concepts for pharmacological language, Defined below [22]:

---

[1] Neo Technology, Inc. 2017. "Neo4j". https://neo4j.com/.

- Active substance: compound or mixture of compounds having a pharmacological action.
- Concentration: Amount of active substance present in the pharmaceutical form of the medicine that was approved and registered by INVIMA.
- Pharmaceutical form: Medicine delivery system.
- Route of administration: Description of the route of administration of the medicinal product. Medicaments may have several descriptions of the routes of administration.
- Single Medicine Quantity: Identifier that indicates medicine status: Active, inactive, or eliminated.
- Unit of content: It is the entity in which the active substance is contained. It does not necessarily coincide with the pharmaceutical form.
- ATC code: Code of anatomical, therapeutic and chemical classification of the medication. Code corresponding to the highest level assigned by OMS (up to level 5). The code must correspond to the active principle. In case of associations of active principles that have an ATC code, the association code must be entered. Each active principle may have several ATC codes associated with it.
- Common medicine Description: Common description of each medicine has the following structure: Principle(s) active(s), concentration(s), dosage form and administration route.
- Commercial Description: Correspond to the common description of the medicine added by the brand name that appears on the product packaging, which will go to the end and in parentheses. In the case of medicines marketed in their generic description, in parentheses the expression, word or words that in the package differentiates it from the other products marketed in the same way.

3 Related Work

The construction of corpus in medicine is a task that seeks primarily to facilitate the extraction of knowledge in medical texts. This task consists of elements such as natural language processing, data mining and the use of algorithms for data collection.

As for medicines, several papers that used and developed a corpus have been created under different approaches: extracting information associated with medicines in social media and medical literature [8,11,24,35], identification of adverse drugs effects in medical reports [17,18,39] and identification of interactions between different drugs [10,20,21,31]. In the Table 1 there is a summary of relevant corpus in the area. It is important to note that most of the work is in English and very few studies in Spanish [10,44].

In terms of ontologies in medicine, there is extensive work and the largest example of this is the Unified Medical Language System (UMLS), that is an integration of key terminology, classification and standards of biomedical information. Formally, it is a set of files and software that combines biomedical and

Table 1. Medicine related corpus

Corpus	Size	Use
Corpus with information of clinical reports [35]	About 565000 texts and records. But they work with approximately 5% of the documents	It is a corpus focused on research and can be used under a framework that performs capture, integration and presentation of clinical information
Corpus of pharmacological substances and drug interactions [20, 21]	273 summaries and 792 texts	It provides elements to conduct research in pharmacology through natural language processing techniques. This paper also show the development of a medical ontology in conjunction with the corpus
Corpus with adverse effects of drugs in clinical records [18]	2972 summaries	It focused on the automation of the search and identification of adverse drug effects in reports written in natural language
Corpus of drugs, diseases and their relations [31]	300 summaries	The corpus was created under an automatic system of identification of drugs and their adverse effects using data mining
Corpus that contains adverse effects of drugs [24]	1253 records	Like ADE, but its main characteristic is that the information comes from social media and is written in colloquial language. Used on studies of automatic extraction and data mining
Semantic corpus that contains clinical records and patient records [36]	150 records	400 summaries. Used for the development and evaluation of systems that extract clinical information in an automated way of patient records
Contains biomedical and health terminology and vocabulary [30]	More than 1 million of biomedical concepts and 5 million of other concepts	It is directed to the academic and research field. It has several tools to access and treat medical information
Corpus that brings together different tweets with medical information [11]	Approximately 10.822 tweets	Implemented in data mining with the objective of finding information about drugs and adverse effects. It uses machine learning algorithms
Corpus that contains information on adverse drug effects [17]	2972 medical reports	Built from extracting medical reports in MEDLINE (U. S. N. L. of Medicine 2016) using a methodology based in ontologies

health vocabulary and standards to allow interoperability between computer systems [42]. It was created by the National Library of Medicine in the USA and updated by them since the year 1986. UMLS has several databases in 20 languages, but mainly in English and has three basic tools to access information: Methatesaurus: terminology and codes for vocabulary in different languages; Semantic networks: categories and relations of Methatesaurus; Specialist Lexicon: Tool for processing natural language.

Access to these tools can be through web browsers, web APIs or downloading files. With this, researchers can process terms, perform lexical analyzes, categorize, conceptualize, identify attributes and relationships, and in general, any research process in the field. Therefore, it provides a frame of reference of significant impact in the moment of developing corpus and ontologies related to the medicine being a major source of knowledge on the area.

For Pharmaceutical Ontologies, there are examples of work under the following perspectives: construction and modeling of medicine ontologies [37,40,42], support to the medicine prescription process [25–27] and development of systems for the search and storage of medical related information [9,12,41]. A description of the related work is included below:

- A Drug-Therapeutic Ontology called OntoFIS for Spain. Use PLN techniques to analyze clinical records [37].
- Lexical-semantic approach to OntoFIS. Mapped with natural language processing tools [9].
- Semantic model for drug prescription, contains metadata related to their content and their relationships with other drugs [25].
- Ontology implemented along with a framework to make suggestions in drug prescription according to the profile of the patient, a knowledge domain and the inference logic [27].
- Ontology with knowledge related to drugs and a repository of the main generic and safe drug prescriptions. It includes characteristics of drugs as contraindications and the interaction between drugs [12].
- Ontology and tool that facilitates medical prescriptions based on aspects such as interactions and side effects. Through PLN it joins heterogeneous information of drugs and organizes it into a unified graph model [26].
- Methodology for automatic recognition of drug-related information in text descriptions [38].
- Model to improve clinical decision support. The ontology is a scheme of attributes of drugs [40].
- Development of MedXN, a system for extraction and standardization of drug related information in clinical records [41].
- Ontology for pharmacological interactions. Integrates information from various sources into a single model [20].

All these ontological models seek to optimize the health care processes in one way or another using different tools and methods. Besides, they can focus and explore medicine information to create formal models in which computer systems and databases can develop, evolve and improve research processes in medicine.

4   ABA Corpus Construction

The creation of the Corpus, named ABA, consisted of 4 phases: the first was the development of a Crawler for data collection, the second compilation and pre-processing of data, the third design of the models for handling and storage of data and finally the assembly of the cor-pus. Each of these phases is detailed below.

Crawler. To systematize the collection of data, a periodical Crawler was created. It enters monthly to the page of the INVIMA and download the files of approved medicines (in format of spreadsheet) and stores them in a database under a CSV format, for further processing. The Fig. 1 describes a flowchart of the Crawler in which it is observed that the Crawler takes the URL, enters the page, downloads the document, updates the format of data and stores it.

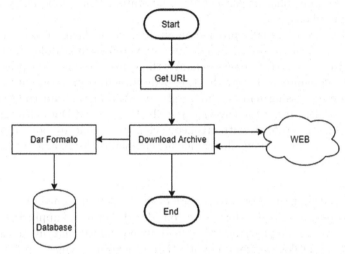

Fig. 1. Data flow crawler

The Crawler performs its task monthly because the data is updated in that same period. A pseudocode of the Crawler is shown in Fig. 2.

When the Crawler performs its task, it also stores a history of changes in which updates of data are controlled with a corresponding date.

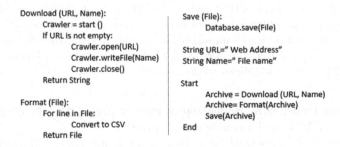

Fig. 2. Pseudocode web crawler

Data Collection and Pre-processing. The corpus is based on open data published by the National Institute of Surveillance of Medicines and Foods of Colombia, INVIMA, regulator of pharmaceutical products.

The Unique Code of Active Medicine list is a data file that the INVIMA updates monthly according to the rules and regulations of products of this nature. The general structure of the list consists of expedient, product, title, health record, expedition date, expiration date, quantity, commercial description, state, current date, inactive date, medical sample, unit, ATC (Anatomical Therapeutic Chemical Classification System), ATC description, administration route, concentration, active ingredient, medical unit, amount of active ingredient, reference unit, pharmaceutical form, laboratory name, laboratory type and laboratory modality.

INVIMA records usually provide about 180,000 medicine records. It is valid to clarify that each record does not necessary represent a different medicine, because within its attributes can change thing like the forms of presentation, amount of sale, amount of ingredient, expedition and expiration dates.

Data pre-processing had two main phases: Automatic data collection, performed by implementing the Crawler described above and Data cleaning, which included correcting records with erroneous characters that are not supported by the database engine. Among the erroneous characters, for example @,$,â,ì, ™, ® among others.

Model Construction. For the creation of the model, the concepts of the ontology and corpus described before were considered. With the applied research of these two terms in the pharmaceutical field, we concluded that due to the nature of corpus and ontologies, from a synthetic and lexical re-sources point of view, both offer tools whose purpose can be similar and complementary. An example of this is that the study of ontologies inevitably entails the analysis of a terminology, to later formalize and conceptualize [4], task that a corpus under its philosophy and reason to be can provide a basis and a domain. Then, there is a dividing line between the concepts of corpus and ontology that turns out to be not so obvious. Under this premise, in this work it is proposed to follow both corpus and ontology guidelines to create a model as an ontology represented by a graph that allows to structure and control the data in the corpus for its subsequent storage, the obtained model is shown in Fig. 3 and whose specification is given by the definition of components and axioms. Even though the model is very simple to represented a medicine, all medicines together conforms a complex network with great possibilities to do research.

Component Definition. Nodes, represented by circles, describe a medicine, a laboratory, an ACT, and a commercial description.

Relationships, represented by arrows, describe interactions between nodes. In this way, a medicine is manufactured by a laboratory, a medicine is specified by its commercial description and is also composed of an ACT.

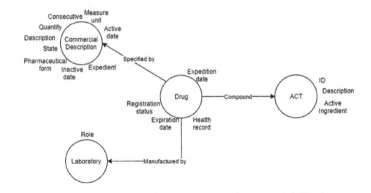

Fig. 3. Model proposed

The attributes describe the characteristics and properties of each node. Thus, the "Business Description" node has attributes consecutive, quantity, pharmaceutical form, active date, measure unit, inactive date and expedient. The "Laboratory" node only has the role attribute. The "Med" node expedition date, expiration date, health record, registration status and expedient. Finally, the "ACT" node has identification, active ingredient and a description.

Axioms Definition. Based on the model, a set of rules are defined in order to control the data: a) if any entry contains information already specified, it is not entered again; b) if an entry contains a medicine that already exists previously under another commercial description, the commercial description node will be created and will generate a new relation of the old node (medicine) to the new commercial description node; c) a laboratory may manufacture one or more medicines; d) a medicine may have several commercial descriptions; e) all nodes should always be related to another node; And f) all the attributes associated to a node must exist, because for the model we select the data that could be relevant in some form or another.

Corpus Creation. From the model a set of Cypher statements for Neo4J was created to automate the entry of information to the database. This automation process is achieved with the help of the ontology, previously developed cleaning and pre-processing algorithms and the file downloaded by the Crawler. In addition, a comparison is made between the data to avoid redundancy in the database. As a result, each record attempts to construct a set of Cypher statements according to the proposed model. Each statement is stored in a file that will have all the scripts generated by the data. These activities are shown in the pseudocode in Fig. 4.

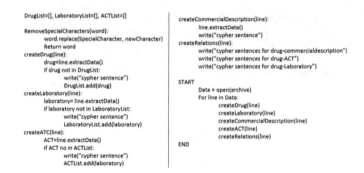

```
DrugList=[], LaboratoryList=[], ACTList=[]

RemoveSpecialCharacters(word):
        word.replace(SpecialCharacter, newCharacter)
        Return word
createDrug(line):
        drug=line.extractData()
        if drug not in DrugList:
                write("cypher sentence")
                DrugList.add(drug)
createLaboratory(line):
        laboratory= line.extractData()
        if laboratory not in LaboratoryList:
                write("cypher sentence")
                LaboratoryList.add(laboratory)
createATC(line):
        ACT=line.extractData()
        If ACT no in ACTList:
                write("cypher sentence")
                ACTList.add(laboratory)

createCommercialDescription(line):
        line.extractData()
        write("cypher sentence")
createRelations(line):
        write("cypher sentences for drug-commercialdescription")
        write("cypher sentences for drug-ACT")
        write("cypher sentences for drug-Laboratory")

START
        Data = open(archive)
        For line in Data:
                createDrug(line)
                createLaboratory(line)
                createCommercialDescription(line)
                createACT(line)
                createRelations(line)
END
```

Fig. 4. Pseudocode - data cleaning and processing

5   Analysis and Results

As a result of the above procedures, a) 197,252 nodes were created, discriminated in 12,340 medicines; 1,208 laboratories: 1,208; 1411 ATC and 182,293 Commercial Descriptions; b) 369,444 relationships and c) a total of 578,154 Cypher sentences. The Fig. 5 shows a fragment of 2,000 nodes of the resulting graph in Neo4j.

Fig. 5. Fragment of corpus graph

Some queries were made that allowed not only to conclude generalities of the dataset, but also to know the query potential of the corpus according to the characteristics and tools with which it was constructed. The following are some of the conclusions:

Although 180,000 commercial descriptions are accumulated in the data log, some of them are repeated because INVIMA uses several records to specify medicine components for some commercial descriptions. Thus, only 68,867 different commercial descriptions have associated one or more medicines.

The number of different trade descriptions that relate to each medicine varies from 1 to 256. However, the average number of medicine descriptions is about 5. As described above, the commercial description corresponds to the common description of the medicine added by the brand that appears on the product packaging, such as some descriptions have brand, flavor and target (for example, kids). The Fig. 6 shows the 5 medicines with the most different commercial descriptions by January 2017

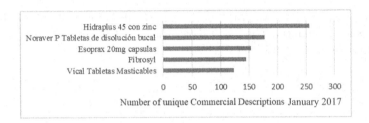

Fig. 6. Medicines with more related commercial descriptions (January 2017)

In order to compare how the medicines are being updated monthly by government, the Fig. 7 shows the 5 medicines with the most different commercial descriptions by August 2017. Some medicines reduced the number of related commercial descriptions and others increased that amount of relations. For example, Hidraplus by August is in 9th position.

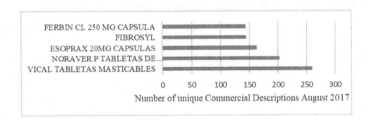

Fig. 7. Medicines with more related commercial descriptions (August 2017)

There are 1208 laboratories that have registered medicines in Colombia, however, 0.4% of the laboratories have registered 12.7% of the medicines in the country. The Fig. 8 shows the 5 laboratories with the most different medicines related.

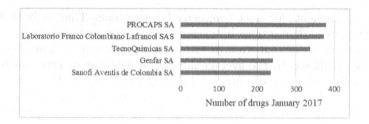

Fig. 8. Laboratories with most different medicines related (January 2017)

In the other hand, by August 2017 the same 5 laboratories distribute the most medicines in the country. The Fig. 9 shows that just Genfar SA Laboratories has reduced its amount of medicines.

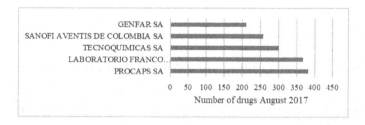

Fig. 9. Laboratories with most different medicines related (January 2017)

Remembering, an active substance is a mixture of compounds having a pharmacological action. Each medicine has one major active substance.

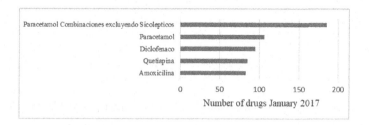

Fig. 10. ATC more frequent in medicines (January 2017)

The 5 active substances that are most often present in the medicines are visualized in the Fig. 10 and which are usually part of the anti-influenza medicines.

Paracetamol Combinaciones is present in medicines such as PAX Caliente, Desenfriolito, Antalgin, Noxpirin and Excedrin. Paracetamol is present in medicines such as Acetaminofen, Dolex and Dixicol. Diclofenaco is present in medicines such as DLFAM, Diozaflex and Quimae.

By August 2017, some new active substances increased their presence in medicines. Figure 11 show the two new active principles in August 2017 list are Montelukast and Xiclomelan.

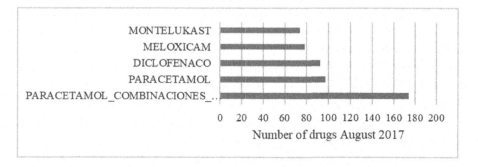

Fig. 11. ATC more frequent in medicines (January 2017)

On the other hand, the web portal Vademecum in its dictionary of active substances, contains moren 5 active substances that can cause tachycardia to those who consume it. This adverse effect is described in Fig. 12 which includes the quantity of medicines that possess those active principles. It is necessary to clarify that the mentioned active principles are not the only ones that can generate tachycardia and that for any affirmation a deeper study would be required.

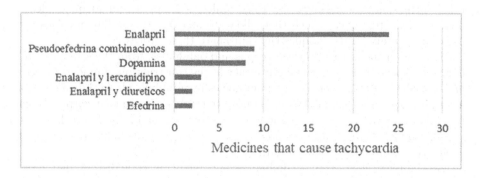

Fig. 12. Some active substances that cause tachycardia

Athletes have restricted the use of some specific substances because they could affect their sports performance. The World Anti-Doping Agency annually publishes the list of banned active sub-stances for athletes. The Fig. 13 shows 15 active substances with restriction in the sport, with the amount of medicines that contain it.

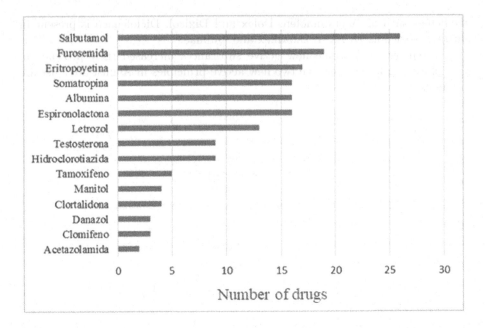

Fig. 13. Some active principles prohibited for athletes

6   Conclusions

Open data has condensed information in thousands or millions of records. Although it has the implicit element of the organization according to the expert, this information is not always easy to get unless it is applied an adequate cleaning and data treatment, which then allows to analyses them. This involves the use of computational techniques that automate these processes.

The complementarity between corpus and ontologies, from the semantic and lexical perspective, generates an advantage in the construction of models like the one described, because the study of ontologies inevitably leads to the analysis of a terminology, later formalized and conceptualized [4], hence that corpus under its philosophy and rationale, can give a basis and a domain for this task.

The model, being graph-oriented presents an important advantage since it avoids the redundancy between data.

By January 2017, the Colombian single medicine code listing had more than 180,000 records, with the application of the described model, a total of 12,340 med nodes, 1,208 laboratory nodes, 1,411 ATC nodes, 182,293 commercial description nodes were represented in the corpus, of which 68,867 are unique with respect to their description, but varying values as the active substance.

There are more than 160 medicines registered in Colombia and that athletes cannot consume, according to the regulations of the World Anti-Doping Agency, these medicines represent about 1.5% of available medicines.

The model is flexible and extensible since, such as, it is possible to add the modules of adverse effects and substances prohibited by doping to extend the scope of the consultations.

7 Future Work

As future work, the development of the corpus can be expanded in search of adverse effects on medicines and active principles. Such a task would entail, in the first instance, the creation of a second Crawler web that would enter online medicine encyclopedias; Another possible expansion is to the area of posology or medical prescriptions, where consumption analyses could be performed to make predictive models for adverse effects. All the above described applied to Colombia. On the other hand, the development of an API to make queries to the ABA corpus is considered. API that would be publicly available on the web for the use of anyone.

References

1. Alexander, L., Allen, S., Bindoff, N.L.: MongoDB Applied Design Patterns. O'Reilly (2013). https://doi.org/10.1017/CBO9781107415324.004
2. Anderson, J.C., Lehnardt, J., Slater, N.: CouchDB: The Definitive Guide (2010)
3. Balakrishnan, R., Ranganathan, K.: A Textbook of Graph Theory. Universitext (1979). Springer (2000). https://books.google.com.co/books?id=ERgLpPQgGQx4cC
4. Bautista-Zambrana, M.R.: Creating corpus-based ontologies: a proposal for preparatory work. Proc. Soc. Behav. Sci. (2015). https://doi.org/10.1016/j.sbspro.2015.11.314
5. Celko, J.: Complete Guide To NoSQL, 1st edn. Elsevier Inc., Amsterdam (2014)
6. Chang, F., et al.: Bigtable: a distributed storage system for structured data. Trans. Comput. Syst. **26**, 1–26 (2008). https://doi.org/10.1145/1365815.1365816
7. Chartrand, G., Zhang, P.: A First Course in Graph Theory. Dover Books on Mathematics. Dover Publications (2012). https://books.google.com.co/books?id=ocIr0RHyI8oC
8. Coden, A., Gruhl, D., Lewis, N., Tanenblatt, M., Terdiman, J.: SPOT the drug! An unsupervised pattern matching method to extract drug names from very large clinical corpora. In: Proceedings - 2012 IEEE 2nd Conference on Healthcare Informatics, Imaging and Systems Biology, HISB 2012, pp. 33–39 (2012). https://doi.org/10.1109/HISB.2012.16
9. Cruanes Vilas, J.: Una aproximación léxico-semántica para el mapeado automático de medicamentos y su aplicación al enriquecimiento de ontologías farmacoterapéuticas. Doctoral, Universidad de Alicante (2014). http://hdl.handle.net/10045/42146
10. Duque, A., Martínez-Romo, J., Araujo, L.: Extracción no supervisada de relaciones entre medicamentos y efectos. Procesamiento Lenguaje Nat. **55**(83–90), 1135–5948 (2015). ISSN 1135-5948
11. Ginn, R., et al.: Mining Twitter for adverse drug reaction mentions: a corpus and classification benchmark. In: proceedings of the Fourth Workshop on Building and Evaluating Resources for Health and Biomedical Text Processing (BioTxtM), no. 1 (2014)

12. Grando, A., Farrish, S., Boyd, C., Boxwala, A.: Ontological approach for safe and effective polypharmacy prescription. In: AMIA ... Annual Symposium proceedings/AMIA Symposium. AMIA Symposium 2012, pp. 291–300 (2012). http://www.pubmedcentral.nih.gov/articlerender

13. Gruber, T.R.: A translation approach to portable ontology specifications. Knowl. Acquisit. **5**(2), 199–220 (1993). https://doi.org/10.1006/knac.1993.1008, http://www.sciencedirect.com/science/article/pii/S1042814383710083

14. Gruber, T.R.: Toward principles for the design of ontologies used for knowledge sharing. Int. J. Hum.-Comput. Stud. **43**(5–6), 907–928 (1995). https://doi.org/10.1006/ijhc.1995.1081, http://www.sciencedirect.com/science/article/pii/S1071581985710816

15. Guarino, N.: Understanding, building and using ontologies. Int. J. Hum.-Comput. Stud. **46**(2), 293–310 (1997). https://doi.org/10.1006/ijhc.1996.0091, http://www.sciencedirect.com/science/article/pii/S1071581996900919

16. Guichard, D.: An Introduction to Combinatorics and Graph Theory. Creative Commons (2016). http://www.freetechbooks.com/an-introduction-to-combinatorics-and-graph-theory-t1079.html

17. Gurulingappa, H., Mateen-Rajpu, A., Toldo, L.: Extraction of potential adverse drug events from medical case reports. J. Biomed. Semant. **3**(1), 1–10 (2012). https://doi.org/10.1186/2041-1480-3-15, http://link.springer.com/article/10.1186/2041-1480-3-15

18. Gurulingappa, H., Rajput, A.M., Roberts, A., Fluck, J., Hofmann-Apitius, M., Toldo, L.: Development of a benchmark corpus to support the automatic extraction of drug-related adverse effects from medical case reports. J. Biomed. Inform. **45**(5), 885–892 (2012). https://doi.org/10.1016/j.jbi.2012.04.008, https://dx.doi.org/10.1016/j.jbi.2012.04.008

19. Harrison, G.: Next Generation Databases: NoSQL, NewSQL, and Big Data. Springer, New York (2015). https://doi.org/10.1007/978-1-4842-1329-2, http://link.springer.com/10.1007/978-1-4842-1329-2

20. Herrero-Zazo, M.: Semantic resources in pharmacovigilance: a corpus and an ontology for drug-drug interactions. Ph.D. thesis, Universidad Carlos II (2015). http://sphynx.uc3m.es/lmoreno/tesisMariaHerrero.pdf

21. Herrero-Zazo, M., Segura-Bedmar, I., Martínez, P., Declerck, T.: The DDI corpus: an annotated corpus with pharmacological substances and drug-drug interactions. J. Biomed. Inform. **46**(5), 914–920 (2013). https://doi.org/10.1016/j.jbi.2013.07.011

22. Invima, Ministerio de Salud: Manual de Normas Técnicas de Calidad. Normas de Calidad Y Guia de Analisis (2015). https://www.invima.gov.co/images/normas_tecnicas.pdf

23. Jones, C., Waller, D.: Corpus Linguistics for Grammar. Routledge, London and New York (2015). https://doi.org/10.1017/CBO9781107415324.004

24. Karimi, S., Metke-Jimenez, A., Kemp, M., Wang, C.: CADEC: a corpus of adverse drug event annotations. J. Biomed. Inform. **55**, 73–81 (2015). https://doi.org/10.1016/j.jbi.2015.03.010, https://dx.doi.org/10.1016/j.jbi.2015.03.010

25. Khalili, A., Sedaghati, B.: Semantic medical prescriptions - towards intelligent and interoperable medical prescriptions. In: Proceedings - 2013 IEEE 7th International Conference on Semantic Computing, ICSC 2013, pp. 347–354 (2013). https://doi.org/10.1109/ICSC.2013.66

26. Khemmarat, S., Gao, L.: Supporting drug prescription via predictive and personalized query system. In: 2015 9th International Conference On Pervasive Computing Technologies For Healthcare (PervasiveHealth), pp. 9–16 (2015). https://doi.org/10.4108/icst.pervasivehealth.2015.259130

27. Kostopoulos, K., Chouvarda, I., Koutkias, V., Kokonozi, A., Van Gils, M., Maglaveras, N.: An ontology-based framework aiming to support personalized exercise prescription: application in cardiac rehabilitation. In: Proceedings of the Annual International Conference of the IEEE Engineering in Medicine and Biology Society, EMBS pp. 1567–1570 (2011). https://doi.org/10.1109/IEMBS.2011.6090456

28. Lakshman, A., Malik, P.: Cassandra - a decentralized structured storage system. In: SIGOPS (2010). https://doi.org/10.1145/1773912.1773922

29. Matías, I., Antiñanco, J., Bazzocco, M.J.: Bases de Datos NoSQL: escalabilidad y alta disponibilidad a través de patrones de diseño. Ph.D. thesis, Universidad Nacional de La Plata (2013)

30. National Library of Medicine: UMLS (2016). https://www.nlm.nih.gov/research/umls/

31. van Mulligen, E.M., et al.: The EU-ADR corpus: annotated drugs, diseases, targets, and their relationships. J. Biomed. Inform. **45**(5), 879–884 (2012). https://doi.org/10.1016/j.jbi.2012.04.004, https://dx.doi.org/10.1016/j.jbi.2012.04.004

32. Niranjanamurthy, M., Archana, U.L., Niveditha, K.T., Abdul Jafar, S., Shravan, N.S.: The research study on DynamoDB-NoSQL database service. Int. J. Comput. Sci. Mob. Comput. **3**, 268–279 (2014)

33. Pant, G., Srinivasan, P.: Learning to crawl: comparing classification schemes. ACM Trans. Inf. Syst. **23**(4), 430–462 (2005). https://doi.org/10.1145/1095872.1095875, http://doi.acm.org/10.1145/1095872.1095875

34. Pokorny, J.: New database architectures: steps towards big data processing. In: IADIS European Conference Data Mining 2013 (2013)

35. Roberts, A., et al.: The CLEF corpus: semantic annotation of clinical text. In: AMIA ... Annual Symposium proceedings/AMIA Symposium. AMIA Symposium, pp. 625–629 (2007)

36. Roberts, A., et al.: Building a semantically annotated corpus of clinical texts. J. Biomed. Inform. **42**(5), 950–966 (2009). https://doi.org/10.1016/j.jbi.2008.12.013, https://dx.doi.org/10.1016/j.jbi.2008.12.013

37. Romá-Ferri, M.: OntoFIS: tecnología ontológica en el dominio farmacoterapéutico. Doctoral, Universidad de Alicante (2009). http://rua.ua.es/dspace/handle/10045/14216

38. Rubrichi, S., Quaglini, S., Spengler, A., Russo, P., Gallinari, P.: A system for the extraction and representation of summary of product characteristics content. Artif. Intell. Med. **57**(2), 145–154 (2013). https://doi.org/10.1016/j.artmed.2012.08.004, https://dx.doi.org/10.1016/j.artmed.2012.08.004

39. Sánchez-Cisneros, D., Lana, S., Moreno, A., Martínez, P., Campillos, L., Segura-Bedmar, I.: Prototipo buscador de información médica en corpus multilingües y extractor de información sobre fármacos. Procesamiento Lenguaje Nat. **49**, 209–212 (2012)

40. Senger, C., Seidling, H.M., Quinzler, R., Leser, U., Haefeli, W.E.: Design and evaluation of an ontology-based drug application database. Methods Inf. Med. **50**(3), 273–284 (2011). https://doi.org/10.3414/ME10-01-0013

41. Sohn, S., Clark, C., Halgrim, S.R., Murphy, S.P., Chute, C.G., Liu, H.: MedXN: an open source medication extraction and normalization tool for clinical text. J. Am. Med. Inform. Assoc. JAMIA 1–8 (2014). https://doi.org/10.1136/amiajnl-2013-002190, http://www.ncbi.nlm.nih.gov/pubmed/24637954

42. UMLS: Unified Medical Language System (2016). https://www.nlm.nih.gov/research/umls/
43. Valbuena, S.J., Londoño, J.M.: Sistemas Para Almacenar Grandes Volúmenes De Datos. Rev. Gti **13**(37), 17–28 (2015)
44. Vázquez, E.: Prospectos medicamentosos: macroestructura comparada aplicada a la traducción (inglés<>español). Skopos (2014)
45. Villayandre Llamazares, M.U.D.L.: Internet como corpus: el caso de bibidí. Contextos **XXI-XXII**(41–44), 205–231 (2003)

Fool's Errand: Looking at April Fools Hoaxes as Disinformation Through the Lens of Deception and Humour

Edward Dearden and Alistair Baron[✉]

Lancaster University, Lancaster, UK
{e.dearden,a.baron}@lancaster.ac.uk

Abstract. Every year on April 1st, people play practical jokes on one another and news websites fabricate false stories with the goal of making fools of their audience. In an age of disinformation, with Facebook under fire for allowing "Fake News" to spread on their platform, every day can feel like April Fools' day. We create a dataset of April Fools' hoax news articles and build a set of features based on past research examining deception, humour, and satire. Analysis of our dataset and features suggests that looking at the structural complexity and levels of detail in a text are the most important types of feature in characterising April Fools'. We propose that these features are also very useful for understanding Fake News, and disinformation more widely.

Keywords: Disinformation · Deception · April Fools · Fake news

1 Introduction

People celebrate April Fools' day each year on April 1st by playing pranks on each other for hilarity's sake. This tradition has transferred over to the traditional media, the most famous example of which is the BBC's 1957 'Swiss Spaghetti Harvest' film[1], which tricked many UK television viewers into believing that a farm in Switzerland grew spaghetti as a crop. With the rise of the web, news sites and companies started releasing annual hoaxes.

In today's world of disinformation and 'Fake News', understanding different forms of deception in news and online media is an important venture. In the 2016 US Presidential Election, dissemination of 'Fake News' was pointed to as one of the crucial factors leading up to Donald Trump's victory and subsequent ongoing tenure as 45th President of the United States of America. If it is true that deceptive news articles swayed a major democratic election, it is certainly important for research towards better understanding and solving of the problem.

One of the main differences between April Fools' articles and typical deceptive texts is the author's intent. The author of an April Fool is not trying to deceive

---

[1] http://news.bbc.co.uk/onthisday/hi/dates/stories/april/1/newsid_2819000/2819261.stm.

© Springer Nature Switzerland AG 2023
A. Gelbukh (Ed.): CICLing 2019, LNCS 13451, pp. 451–467, 2023.
https://doi.org/10.1007/978-3-031-24337-0_32

so much as amuse. In this way April Fools' hoaxes are similar to Irony and Satire, which expect the reader to understand based on context that what is literally being said is not true. By looking at April Fools' news hoaxes, we investigate whether the change of intent affects the linguistic features of deception in April Fools' compared to in 'Fake News'.

By using April Fools' news hoaxes, we can look at a dataset of verifiable false bodies of text spanning back 14 years. Similar work with satirical news articles has yielded interesting results [26]. While it is true April Fools' hoaxes are not completely similar to 'Fake News', mainly in terms of motivation, our hypothesis is that they will provide insight into the linguistic features put on display when an author is writing something fictitious as if it is factual.

The main contributions of this work are:

- Introducing a new dataset of hoax April Fools' articles.
- Investigating the linguistic features of April Fools' hoaxes, particularly how they relate to features of deception and humour.
- Discussing how these features may be useful in the detection of Fake News.

2   Background

As April Fools' reside in a space somewhere between deception and humour, we will provide a brief background in the areas of deception detection and humour recognition. We will also discuss current NLP approaches to Satire and 'Fake News' detection.

2.1   Deception Detection

Deception research often focusses on 'non-verbal' cues to deception, e.g. eye movement. However, we are interested in the verbal cues to deception, i.e. the features hidden within the text. Without non-verbal cues, humans identify deception with very low degrees of success [12]. Much of the research on verbal cues of deception has been completed in the context of Computer Mediated Communications (CMC). This type of communication can be either spontaneous (synchronous) or preplanned structured prose (asynchronous), such as news, which is of interest for the present research.

Works in synchronous deception detection have involved looking at text from spoken and written answers to questions [17], email [10], and chat-based communication [6,8]. Carlson et al. [4] provide a good overview of how different factors can affect the deception model, such as medium, the liar's social ability, and the author's motivation. There are certain groups of features that these works suggest are present in deception. One of these groups is 'Cognition Features'. Lying requires a higher level of cognition than telling the truth so often lies seem to be less complicated and more vague. There is also a tendency towards negative emotional language because liars feel guilty about lying. Certain features suggest that liars have more distance from the story they are telling, e.g. reduced pronouns and details.

These works are useful for looking at the linguistic behaviour of liars, but they do not carry over too well to asynchronous communication, where a deceiver can edit and revise what they have written. Toma and Hancock [27], looking at fake online dating profiles, found that certain features of synchronous deception were not present in asynchronous deception. Liars also more frequently exhibited exaggeration of their characteristics. Other works have looked at fake hotel reviews [2,18]. Features relating to understandability, level of details, writing style, and cognition indicators provided useful clues for identifying fake reviews, though some features may have been genre-dependent. Markowitz and Hancock [11] looked at fraudulent academic writing, an area similar to the news domain where a formalised writing style may mask certain stylistic features of deception. Fake works exhibited overuse of scientific genre words, as well as less certainty and more exaggeration. Stylometric approaches to looking at deception have included Afroz et al. [1] who found that certain style features seemed to leak out even when an author was trying to hide them or imitate someone else. In some ways April Fools' articles are an example of imitation, in which an author is writing a fictional article, mimicking the style of real news.

2.2 Fake News

Conroy et al. [5] provide an overview of computational methods that can be used to tackle the problem of Fake News, including linguistic approaches. Current linguistic research into detecting fake news includes Pérez-Rosas et al. [20] who used features from LIWC [19] for the detection fake news. They found that fake news contained more function words and negations as well as more words associated with insight, differentiation and relativity. Fake News also expressed more certainty and positive language. These results are interesting, but it must be considered that the dataset used was crowdsourced using Amazon Mechanical Turk, meaning the authors of this news were unlikely to be accustomed to writing news articles. Horne and Adali [9] found fake news to be a lot more similar to satire than normal news and also that the title structure and use of proper nouns were very useful for detecting it. Rashkin et al. [21] found that features relating to uncertainty and vagueness are also useful for determining a text's veracity.

2.3 Humour Recognition

Unlike most deceptive texts, April Fools' articles have a motivation of humour. Bringing ideas in from the area of humour recognition therefore may help us characterise hoax articles. Much of the work in humour recognition has focused on detecting humour in shorter texts such as one-liner jokes.

Mihalcea and Strapparava [14] showed that classification techniques can be used to distinguish between humourous and non-humourous texts. They used features such as alliteration, antonymy, and adult slang in conjunction with content features (bag-of-words). Mihalcea and Pulman [13] discussed the significance of 'human-centeredness' and negative polarity in humourous texts. Reyes et al.

[24] looked at a corpus of one-liners and discussed their features. Reyes et al. [25] investigated the features of humour and contrasted to those of irony.

2.4 Irony

Irony is a particular type of figurative language in which the meaning is often the opposite of what is literally said and is not always evident without context or existing knowledge. Wallace [30] suggest that to create a good system for irony detection, one cannot rely on lexical features such as Bag of Words, and one must consider also semantic features of the text. Reyes et al. [25] created a dataset generated by searching for user-created tags and attempted to identify humour and irony. The features used to detect irony were polarity, unexpectedness, and emotional scenarios. More recently, Van Hee et al. [28] investigated annotated ironic tweet corpora and suggested that looking at contrasting evaluations within tweets could be useful for detecting irony. Van Hee et al. [29] also created a system to detect ironic tweets, looking beyond text-based features, using a feature set made up of lexical, syntactic, sentiment, and semantic features.

2.5 Satire

Satire is a form of humour which pokes fun at society and current affairs, often trying to bring something to account or criticise it. This is often achieved using irony and non-sequitur. Satire is similar to April Fools' in that the articles are both deceptive and humourous. The only difference is that satire does tend to be political, whereas April Fools' are often more whimsical.

Burfoot and Baldwin [3] created a system to identify newswire articles as true or satirical. They looked at bag-of-words features combined with lexical features and 'semantic validity'. Rubin et al. [26] used linguistic features of satire to build an automatic classifier for satirical news stories. Their model performed well with an F1-Score of 87% using a feature set combining absurdity, grammar, and punctuation.

3 Hoax Feature Set

The purpose of this work is to identify the features of April Fools' articles, and to see if what we learn is also true of fake news, and possibly disinformation more generally. We want to avoid highly data-driven methods such as bag-of-words because these will learn content and topic-based features of our specific dataset meaning we would not necessarily learn anything about April Fools' or deception more generally. We specifically look at the use of features from the areas of deception detection and humour recognition.

Some previous works have used LIWC [19] to capture Neurolinguistic features of deceptive texts. While we did not use LIWC directly, we did consider important LIWC features from previous work when devising our own features.

For many of our features, we utilise tokenisation and annotations from the CLAWS Part-of-Speech (PoS) tagger [7] and the UCREL Semantic Annotation System (USAS) [22]. The code we used for extracting features, including the output from CLAWS and USAS, are available for reproducibility purposes with the rest of our code[2].

The features we used have been split into seven categories so as to logically group them together to aid analysis and understanding of the results. These categories are: Vagueness, Detail, Imaginative Writing, Deception, Humour, Complexity, and Formality. All features were normalised between 0 and 1.

Vagueness features aim to capture the idea that hoax articles may be less detailed and more ambiguous because the stories are fabricated. Ambiguity was captured by calculating the proportion of words in a text for which there were multiple candidates for annotation. Three types of ambiguity were used: Part-of-Speech Ambiguity, Semantic Ambiguity, and WordNet Synset Ambiguity. Vague descriptions might use more comparative and superlative words as opposed to hard, factual statements [18]. Groups of PoS and Semantic tags were gathered to represent exaggeration, degree, comparative, and superlative words.

Detail features are almost the opposite of vagueness. Genuine news article should contain more details because the events described actually happened. Increased cognition is needed to invent names and places in a text. For this reason we look at the number of proper nouns in a text. Similarly, a fake article may avoid establishing minute details such as dates. We therefore look at Dates, numbers, and Time-related words. Motion words, spatial words, and sense words also establish details that may be less present in deceptive texts.

Imagination features have been used in deception research by Ott et al. [18], based on the work of Rayson et al. [23], which involved comparing informative to imaginative texts. It is worth noting that we are comparing informative texts to pseudo-informative texts, rather than informative to openly imaginative texts. However, they were previously useful in detecting deceptive opinion spam [18], so we evaluate their use here. Rayson et al. [23] identify different PoS tags that are more present in imaginative and informative writing. We used tags that were highlighted from the following PoS groups: conjunctions, verbs, prepositions, articles, determiners, and adjectives.

Deception features are the features of synchronous verbal deception. We include them to investigate if any of the features of spontaneous deception are preserved in spite of a change in medium. Features of asynchronous deception are more relevant to this task and have been distributed between more specific categories, such as Complexity and Details. These synchronous deception features are: First-person pronouns, Negative Emotional Language, and Negations.

Humour features are those from the area of humour recognition. As with deception, some humour features (notably ambiguity) fit better into other categories. The humour features used were: Positive emotion, Relationships, Contextual Imbalance, Alliteration, and Profanity. Contextual Imbalance is characterised as being the average similarity of all adjacent content words in the text.

---

[2] https://doi.org/10.17635/lancaster/researchdata/512.

Similarity was calculated by comparing the vectors of words using the in-built similarity function of spaCy[3]. Positive Emotions and Relationships were both gathered using USAS semantic categories. Profanity was gathered from a list of profanities banned by Google[4]. Alliteration was measured by calculating the proportion of bigrams in the text that began with the same letter.

Formality features aim to capture elements of style in news documents that may show how formal they are. April Fools' may be generally less formal or have less editorial oversight. We used three features based on aspects of the Associated Press (AP) style book: AP Number, AP Date, and AP Title Features. These features checked if the text obeyed AP guidelines in their writing of numbers, dates, and titles. An example of an AP guideline is that all numbers under 10 must be spelled out (e.g. 'four' as opposed to '4'). Spelling mistakes were also counted and used as a feature, using the enchant spell checker[5].

Complexity features represent the structure and complexity of an article. They comprise: punctuation, reading difficulty, lexical diversity, lexical density, average sentence length, and proportion of function words. Punctuation was the number of punctuation marks in the text, found using a regular expression. To calculate the reading difficulty feature, we used the Flesch Reading Ease index. We used a list of function words from Narayanan et al. [16].

4   Data Collection

4.1   April Fools Corpus

When building our dataset, the first challenge we faced was finding news articles that were definitely April Fools' hoaxes. One cannot simply collect all news articles from April 1st as the majority of news from this date is still genuine. It is also infeasible to manually go through all news published on this day every year. So instead we utilised a website that archives April Fools' each year[6]. The collection of links published on this site is crowd-sourced so there are some issues arising from the fact that only the popular/amusing hoaxes are uploaded. However, this problem is fairly minor; in fact crowd sourcing may serve to diversify the kinds of website from which hoaxes are sampled. The site archives April Fools' articles from 2004 onwards, providing 14 years of hoaxes.

We used Beautiful Soup [15] to scrape all of the hoax links. We performed some preprocessing to remove hoaxes that one could tell did not constitute a news story from the URL. Next we processed all of the linked webpages, extracting the headline and body of each hoax separately. The wide range of sites in the corpus made automatic scraping too error-prone, so the final approach was largely manual. Efforts were made to ensure no boilerplate or artefacts from the website were included as these could have caused the classifier to pick up features

---

[3] https://spacy.io/.

[4] https://github.com/RobertJGabriel/Google-profanity-words/blob/master/list.txt.

[5] https://github.com/rfk/pyenchant.

[6] aprilfoolsdayontheweb.com.

such as the date as being features of April Fools. For the same reason, we also removed any edits to the article disclosing its April Fools' nature.

There were various categories of April Fools' articles found, the most common of which were news stories and press releases. News stories are distinct from press releases which we classed as texts that are self referential; usually taking the form of announcements or product reveals. For example, a press release might be a website announcing that they have been bought out by Google, whereas a news story might be an article by the BBC saying that Google has bought out said company. Press releases were manually filtered out for the present study in order to keep the focus on news, and to avoid the features of press releases obscuring those of April Fools' articles. This resulted in a final April Fools' (AF) corpus comprising of 519 unique texts, spread across 371 websites.

4.2   News Corpus

To create a comparable corpus of genuine news articles, Google News was utilised to automatically scrape news articles from the 4th–5th April of the same years (2004–2018) This time range was chosen so the kinds of topics in the news would be of a similar nature. We will refer to these articles as "NAF" articles. The stories were found using 6 search terms that aimed to catch similar topics to those represented in the AF articles. We did this to avoid learning about the differences in topics of articles rather than whether or not an article is a hoax. These search terms were: "news", "US", "sport", "technology", "entertainment", and "politics". Despite our efforts, it was difficult to match the topic distribution exactly: not all the websites in the AF corpus have archived articles going back to 2004. We acknowledge this is a problem but do not consider it too critical as the features we are looking for are not data-driven and so should not be influenced by topic. We then took all of these URLs and automatically scraped the text using the newspaper python package[7]. Using this method we scraped 2,715 news articles.

For each year (2004–2018), we selected the same number of articles as there were in the AF corpus. The 519 AF articles were spread over 371 websites, the most common of which occurred 19 times. To try and match this distribution, we capped the number of articles that could be taken from any given site at 20. Once we had selected our genuine (NAF) articles, we manually checked the text of each article to ensure that the full text was scraped correctly and that the text only contained the news article itself, without boilerplate noise. We went through the same process as for the AF articles of removing any texts that did not fit in the category of News, such as personal blogs. When an article was removed, we replaced it by choosing a news article from a later page of the Google search that found it. Once this process was finished, we had an NAF corpus of 519 articles spread over 240 websites. Table 1 shows a summary of the corpus, which is made available for further research. April Fools' articles contain

---

[7] http://newspaper.readthedocs.io/en/latest/.

Table 1. Summary of April Fools (AF) and Non-April Fools (NAF) corpora.

	Articles	Websites	Avg words	Std words
AF	519	371	411.9	326.9
NAF	519	240	664.6	633.2

fewer words on average. Both AF and NAF articles vary significantly from the mean in their lengths.

4.3   Limitations

This is a small dataset and has various notable limitations. The genuine articles tend to be from a smaller pool of more established websites as it is these websites that are more prominent when searching for news online. Only news articles are contained in the dataset. Further work may extend to blogs and press releases. Sometimes the distinction between blogs and news is arbitrary but we tried to be consistent. Multiple genuine news articles occasionally cover the same story, but this was rare and no one story was ever repeated more than twice. A minimum length of 100 characters was enforced to remove anomalous texts such as video descriptions, however this may have removed some genuine articles. While we bear them in mind, we do not see these limitations as major barriers to the research. We will analyse the data using both quantitative and qualitative techniques that allow us to take a deep dive into the data and understand the language being in April Fools' articles for the first time. We do not believe a significantly larger corpus could be built in a reasonable time period.

5   Analysis

5.1   Classifying April Fools'

To evaluate the comparative strength of our feature groups for predicting hoaxes, we used a Logistic Regression classifier with 10 fold cross-validation. We used default parameters of Logistic Regression (from scikit-learn), with standardization to zero-mean and scaling to unit variance ($x' = x - \bar{x}/\sigma$). A basic Logistic Regression classifier serves our needs as we are primarily concerned with investigating the behaviour of features with an interpretable model, and not maximising classification accuracy through tuning or more elaborate classifiers. The results of these classifications can be seen in Fig. 1.

From the classification results, we can see that our features provide some information to differentiate between April Fools' hoaxes and genuine news articles. The results are not as high as the F_1-Score of 0.87 found by Rubin et al. [26] for the related task of satire detection, though they are similar to results from fake news detection, such as those of Horne and Adali [9] who achieved an accuracy of 71% using Bodies of text and 78% using headlines.

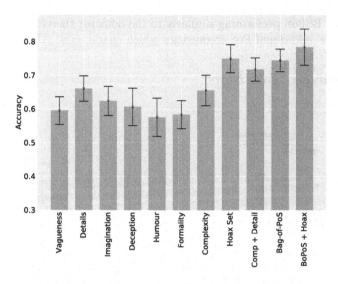

Fig. 1. Mean accuracies of Logistic Regression classifiers across 10 Fold Cross-Validation. Error bars show standard deviation of accuracies across the 10 folds.

Looking at the individual feature groups, Complexity and Detail Features perform best, though not as well as the full Hoax Set. Deception literature suggests that deceptive accounts contain fewer specific details and are generally less complex [4]. Humour performing badly is not surprising as understanding the joke of an AF hoax requires a lot of context and pre-existing knowledge. The features of humour we used in the Humour feature-set were relatively simplistic, more complex, context-aware features may be needed to identify the humour in April Fools' hoaxes. The poor performance of Formality features could suggest that AF Hoaxes are still written to the same journalistic guidelines and standards as their genuine counterparts.

Given the success of the Complexity and Detail features, we classified articles using only these features, achieving an accuracy of 0.718, not far from that of the entire Hoax Set (0.750). This further suggests that looking at details and complexities within a text are crucial when trying to determine if an article is a hoax.

We looked at a non-tailored Bag-of-Part-of-Speech (BoPoS) approach, to compare our curated features to a more data-driven approach. Each PoS tag in CLAWS is used as an individual feature, the occurrences of which are counted for each document. BoPoS was chosen over the more standard Bag-of-Words (BoW) approach because BoW is prone to identifying differences in content and topic, rather than style. BoPoS achieved an accuracy of 0.745, similar to the hoax set. This is not overly surprising as many of the hoax features were part-of-speech counts. These sets do not completely overlap, however. When the hoax set was added to the BoPoS features, the classifier improved its accuracy. This suggests that the non-part-of-speech features in the hoax set provide useful additional information. BoPoS performing similarly to the hoax set therefore suggests that there must be additional PoS frequencies which characterise AF hoaxes.

5.2 Classifying "Fake News"

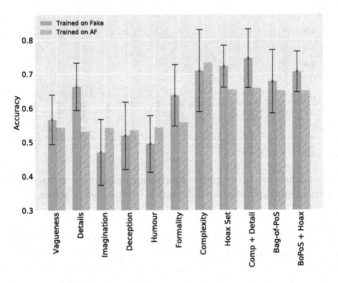

Fig. 2. Accuracies of Logistic Regression classifiers for detecting fake news, trained on Fake News using 10 fold cross-validation and April Fools. Error bars show standard deviation of accuracies across the 10 folds.

Next, we aim to see if we can use the same feature set to effectively identify Fake News. For this we used the fake news dataset introduced by Horne and Adali [9]. This dataset consists of a mixture of articles gathered from well-known fake news outlets and legitimate sites as well as articles gathered by Buzzfeed for an article about fake news in the 2017 election[8]. This is a small dataset (250 articles) split evenly between real and fake. The classification results, again using logistic regression and cross-validation, for fake news can be seen in Fig. 2. For each feature set, one classifier was trained on fake news and evaluated using 10 fold cross-validation, and another trained on April Fools' and tested on fake news.

The classifier trained on fake news using the hoax features achieved an accuracy of 0.722, similar to that achieved by the classifier trained on the hoax features for April Fools' (0.750). This suggests that at least some of the features useful for detecting April Fools' hoaxes are also useful in the identification of deceptive news. Complexity features performed well on the fake news dataset

---

[8] https://tinyurl.com/jlnd3yb.

(0.709), performing almost as well as the full Hoax Set. Details were useful as before but vagueness features performed significantly less well.

When trained on April Fools and predicting fake news with the Hoax Set, an accuracy of 0.653 was achieved. It is possible that some of the same features are useful but their behaviour is different for fake news. Still, the accuracy is not far off the Hoax Set, so there may be some features that manifest themselves similarly for both AF hoaxes and fake news. Finding these features could provide insight into deception and disinformation more generally.

BoPoS performed less well on Fake News, with an accuracy of 0.677, suggesting that PoS tags are not as important when looking at fake news. This, combined with the fact that the hoax features maintained a similar accuracy and complexity features did almost as well as the entire feature set, suggests that the structural features are more important when identifying fake news. BoPoS also did worse when trained on AF and tested on fake, and its drop in accuracy was similar to that of the hoax set. This suggests that there are some PoS tags that are distributed similarly for April Fools and fake news.

5.3   Individual Feature Performances

To see how important individual features were to the classifier, we looked at Logistic Regression weights as shown in Fig. 3. For some features it is interesting to see how they are distributed. To this end, frequency density plots for some of our features are provided in Fig. 4.

There are differences in structural complexity between AF and genuine articles. Lexical Diversity is the most highly weighted feature. As we can see in Fig. 4a, the feature separates hoaxes from genuine articles quite significantly. This could mean hoax texts use more unique words, but it could also be down to the difference in length. High values of lexical diversity correlate to shorter texts and, as we can see in Table 1, the AF articles are shorter, on average. This does still show, however, a difference in complexity. Average sentence length and readability being important features also suggests a difference in complexity. Genuine articles slightly tend towards a shorter average sentence length. NAF articles also tend towards being slightly more difficult to read, though again the difference is not huge.

The story is similar with Fake News – with structural complexity providing key features. Lexical Diversity behaves the same as in April Fools (Fig. 4a). This could again be something to do with average document length, but also could suggest a higher proportion of unique words. Reading difficulty also remains important, though the difference in distribution between fake and real is far more prominent, with genuine articles generally more difficult to read. This means that they generally contain longer sentences and words with more syllables. This difference could suggest that fake news articles are more simplistic than genuine texts. Body punctuation, a feature not weighted as highly for AF, appears to be very important for identifying fake news. More punctuation implies complex structures such as clauses and quotes.

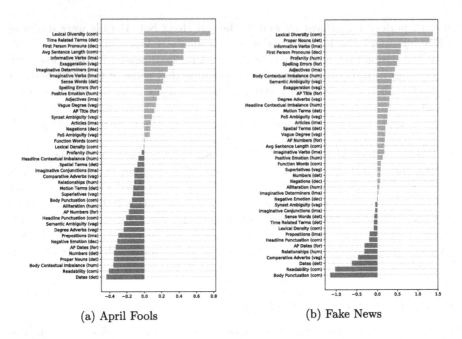

(a) April Fools   (b) Fake News

Fig. 3. Logistic Regression weights for the Hoax Set. A large positive weight suggests an important feature of April Fools/Fake News and a large negative weight suggests an important feature of genuine news.

There are also differences in the level of detail between AF and NAF. Genuine articles tend to contain fewer time-related terms. This seems to go against the idea that genuine articles contain more detail. However, if you look at the occurrences of this feature in the text, the most frequent time-based term is 'will'. This combined with the fact that April Fools tended towards fewer dates (Fig. 4b) and numbers suggests that AF hoaxes refer to events that will happen, but do so in vague terms. This backs up the idea that April Fools are less detailed and more vague. There are also more references to the present. AF hoaxes seem to be more interested in the present and future than the past. AF hoaxes containing fewer dates is interesting, as one might expect that an AF article would mention the date more than a regular article. This is true as far as references to April are concerned, April Fools had more of those. However, the number of references to the month was roughly the same (April Fools actually had slightly fewer overall), though for genuine news it was spread across more months. This may be because real news stories are the culmination of multiple past events that need to be referenced in the story. More significant than references to months, were references to days of the week. Genuine articles contained many more, which backs up the idea of real texts building more detailed stories.

The distribution of proper nouns between AF and NAF is fairly similar (Fig. 4d), possibly skewing towards fewer in April Fools. This could suggest fewer details, i.e. names and places, being established in the fake documents.

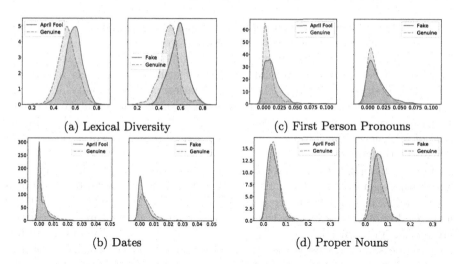

(a) Lexical Diversity

(c) First Person Pronouns

(b) Dates

(d) Proper Nouns

Fig. 4. Density plots of notable features.

Similarly to complexity, the differences in details are not huge, but do seem to be present.

The detail features do not quite behave the same in Fake News articles as in AF. Proper nouns are one of the most important features for characterising fake articles (Fig. 4d). However, unlike AF hoaxes, fake articles tend towards containing more proper nouns than genuine articles. This does not suggest less detail. When looking at the corpus, Fake News articles seem to use a lot of names, often the same ones, such as "Trump" and "Hillary". Interestingly they massively over use the name "Hillary", both suggesting that they are less formal (using the lone forename of a politician), and also that they may have an obsession. Dates are the only other detail feature to be weighted highly for fake news. Fig. 4b shows that this feature behaves similarly as it did for AF hoaxes, though not as dramatically so. Fake articles are more likely to contain very few, or no, dates. These findings suggest that there are not the same types of difference in detail between AF and fake news, though detail does still hold some significance: it was the second best performing feature group.

Not all the important features link to detail and complexity. First person pronouns were an important feature for both AF hoaxes and fake news. The word 'we' was overused in particular by April Fools and to a lesser extent by fake news. This goes against the ideas from traditional deception detection [4] that suggest liars use fewer first person pronouns. In our data, the fake texts use more self-references. This could point towards false articles being more personal and less formal, rather than a feature of deception.

Some of the highly weighted features of fake news are not in common with April Fools. For example, profanity and spelling errors. Both could point towards a reduced level of formality. This would make sense as not being a feature of April Fools. AF writers are usually writing for outlets that publish genuine news, and so likely conform to many of the same standards as genuine news. Fake news, however, comes from less journalistically formal websites.

One of the most obvious differences between April Fools and Fake News in Fig. 3 is that Fake News has a smaller group of features that are very important. Lexical diversity, proper nouns, body punctuation, and readability are significantly higher weighted than anything else. Three of these four features relate to structural complexity and the other to detail. This could suggest that, in the case of fake news, the 'fakeness' lies in the structure of the words rather than the words themselves.

Our results suggest that April Fools and Fake News articles share some similar features, mostly involving structural complexity. The level of detail of a document is also important for both AF hoaxes and fake news, though these features do not behave exactly the same way. Some of the features of deception are present in April Fools, notably those relating to complexity and detail but also first person pronouns, though their behaviour is reversed. The basic features of humour we gathered seem to be less important. A more advanced study of the humour would be required to try and identify it within the AF hoaxes. A successful approach would likely require substantial context and world knowledge.

To compare them to the findings from our feature set, and demonstrate how we can gain new insight by looking at features prominent in the data, as well as those from past literature, we looked at some of the PoS tags that were highly weighted by the BoPoS classifier. Some familiar features show up. Certain time-related tags such as 'quasi nominal adverbs of time' (e.g. "now", "tomorrow") and singular weekday nouns (e.g. "Monday") are highly weighted. Proper Nouns are also highly weighted for fake news in particular. Coordinating conjunctions (e.g. "and", "or") are a prominent feature of NAF articles. More coordinating conjunctions implies more detail and complexity. It is good to see that some of the most highly weighted parts of speech back up our finding that detail and complexity are important in defining April Fools' articles and Fake News.

6   Conclusion

In this paper we have introduced a new corpus of April Fools' hoax news articles. We also created a feature set based on past work in deception detection, humour recognition, and satire detection. Using this feature set, we built a system to classify news articles as either April-Fools' hoaxes or genuine articles. The resulting accuracy of 0.750 suggests that the features we identified are useful in identifying April Fools' hoaxes, though not without room for improvement. We then tested our system on a small dataset of fake news to see if April Fools' hoaxes are similar enough to fake news that similar features can be used to detect both. An accuracy of 0.722 was achieved on the Fake News dataset, suggesting that these features are useful for both tasks.

We analysed our features using a combination of qualitative and quantitative techniques to observe the differences between April Fools' hoaxes and genuine articles. This analysis suggests that the structural complexity and level of detail of a text are important in characterising April Fools. This was also the case for Fake News, though structural complexity seemed more important and the changes in details differed slightly from those in April Fools. Our findings suggest that there are certain features in common between different forms of disinformation and that by looking at multiple varieties, we can hope to learn more about the language of disinformation in general. We also showed that by using a mixture of analysis techniques, we can gain far more insight than we can purely from classification. The corpus we have introduced will also be useful in wider fake news research by providing a dataset of news articles which are completely untrue, similar to how satirical news articles are already being used.

Despite similar features being effective at classifying both April Fools' hoaxes and Fake News, we showed that not all these features behave the same way between the two text types. It is possible that some of these differences in feature behaviour come down to the deceptive intent of the texts. April Fools' are an interesting form of disinformation because the author does not believe what they are writing and is not trying to deceive anybody. By looking at a wider variety of false texts, we can further understand the way that the author's motivation and belief affect the way false information is written.

This early work has provided a new dataset for use in the area of Fake News detection and has highlighted directions for future work, describing features useful for detecting April Fools' articles and showing that they may also be present in fake news. Our findings may provide important insight into deceptive news going into the future.

References

1. Afroz, S., Brennan, M., Greenstadt, R.: Detecting hoaxes, frauds, and deception in writing style online. In: Proceedings of the 2012 IEEE Symposium on Security and Privacy, SP 2012 (2012)
2. Banerjee, S., Chua, A.Y.K., Kim, J.J.: Using supervised learning to classify authentic and fake online reviews. In: Proceedings of the 9th International Conference on Ubiquitous Information Management and Communication, IMCOM 2015 (2015)
3. Burfoot, C., Baldwin, T.: Automatic satire detection: are you having a laugh? In: Proceedings of the ACL-IJCNLP 2009 Conference Short Papers, ACLShort 2009 (2009)
4. Carlson, J.R., George, J.F., Burgoon, J.K., Adkins, M., White, C.H.: Deception in computer-mediated communication. Group Decis. Negotiat. 13, 5–28 (2004)
5. Conroy, N.J., Rubin, V.L., Chen, Y.: Automatic deception detection: methods for finding fake news. In: Proceedings of the 78th ASIS&T Annual Meeting: Information Science with Impact: Research in and for the Community (2015)
6. Derrick, D.C., Meservy, T.O., Jenkins, J.L., Burgoon, J.K., Nunamaker, Jr., J.F.: Detecting deceptive chat-based communication using typing behavior and message cues. ACM Trans. Manage. Inf. Syst. 4(2), 1–21 (2013)
7. Garside, R.: The claws word-tagging system (1987)
8. Hancock, J.E., Curry, L., Goorha, S., Woodworth, M.: On lying and being lied to: a linguistic analysis of deception in computer-mediated communication. Discour. Process. 45, 1–23 (2008)

9. Horne, B.D., Adali, S.: This just. in: Fake news packs a lot in title, uses simpler, repetitive content in text body, more similar to satire than real news. CoRR (2017)
10. Keila, P.S., Skillicorn, D.B.: Detecting unusual email communication. In: Proceedings of the 2005 Conference of the Centre for Advanced Studies on Collaborative Research, CASCON 2005, pp. 117–125 (2005)
11. Markowitz, D.M., Hancock, J.T.: Linguistic traces of a scientific fraud: the case of Diederik Stapel. PLOS One **9**, e105937 (2014)
12. Masip, J., Bethencourt, M., Lucas, G., SEGUNDO, M.S.S., Herrero, C.: Deception detection from written accounts. Scand. J. Psychol. **53**(2), 103–111 (2012)
13. Mihalcea, R., Pulman, S.: Characterizing humour: an exploration of features in humorous texts. In: Gelbukh, A. (ed.) CICLing 2007. LNCS, vol. 4394, pp. 337–347. Springer, Heidelberg (2007). https://doi.org/10.1007/978-3-540-70939-8_30
14. Mihalcea, R., Strapparava, C.: Making computers laugh: investigations in automatic humor recognition. In: Proceedings of the Conference on Human Language Technology and Empirical Methods in Natural Language Processing, HLT 2005 (2005)
15. Nair, V.G.: Getting Started with Beautiful Soup (2014)
16. Narayanan, A., et al.: On the feasibility of internet-scale author identification. In: 2012 IEEE Symposium on Security and Privacy (2012)
17. Newman, M.L., Pennebaker, J.W., Berry, D.S., Richards, J.M.: Lying words: predicting deception from linguistic styles. Pers. Soc. Psychol. Bull. **29**(5), 665–675 (2003)
18. Ott, M., Choi, Y., Cardie, C., Hancock, J.T.: Finding deceptive opinion spam by any stretch of the imagination. In: Proceedings of the 49th Annual Meeting of the Association for Computational Linguistics: Human Language Technologies, HLT 2011, vol. 1 (2011)
19. Pennebaker, J.W., Francis, M.E., Booth, R.J.: Linguistic inquiry and word count: Liwc 2001. Mahway Lawrence Erlbaum Assoc. **71**, 2001 (2001)
20. Pérez-Rosas, V., Kleinberg, B., Lefevre, A., Mihalcea, R.: Automatic detection of fake news. CoRR (2017)
21. Rashkin, H., Choi, E., Jang, J.Y., Volkova, S., Choi, Y.: Truth of varying shades: analyzing language in fake news and political fact-checking. In: Proceedings of the 2017 Conference on Empirical Methods in Natural Language Processing (2017)
22. Rayson, P., Archer, D., Piao, S., McEnery, A.M.: The UCREL semantic analysis system (2004)
23. Rayson, P., Wilson, A., Leech, G.: Grammatical word class variation within the British national corpus sampler. Lang. Comput. **36**, 295–306 (2002)
24. Reyes, A., Rosso, P., Buscaldi, D.: Humor in the blogosphere: first clues for a verbal humor taxonomy. J. Intell. Syst. **18**(4), 311–332 (2009)
25. Reyes, A., Rosso, P., Buscaldi, D.: From humor recognition to irony detection: the figurative language of social media. Data Knowl. Eng. **74**, 1–12 (2012)
26. Rubin, V., Conroy, N., Chen, Y., Cornwell, S.: Fake news or truth? Using satirical cues to detect potentially misleading news. In: Proceedings of the Second Workshop on Computational Approaches to Deception Detection (2016)
27. Toma, C.L., Hancock, J.T.: Reading between the lines: linguistic cues to deception in online dating profiles. In: Proceedings of the 2010 ACM Conference on Computer Supported Cooperative Work, CSCW 2010 (2010)
28. Van Hee, C., Lefever, E., Hoste, V.: Exploring the realization of irony in twitter data. In: LREC (2016)

29. Van Hee, C., Lefever, E., Hoste, V.: Monday mornings are my fave : #not exploring the automatic recognition of irony in English tweets. In: Proceedings of COLING 2016, 26th International Conference on Computational Linguistics, pp. 2730–2739. ACL (2016)

30. Wallace, B.C.: Computational irony: a survey and new perspectives. Artif. Intell. Rev. **43**(4), 467–483 (2015)

Russian Language Datasets in the Digital Humanities Domain and Their Evaluation with Word Embeddings

Gerhard Wohlgenannt[✉], Artemii Babushkin, Denis Romashov,
Igor Ukrainets, Anton Maskaykin, and Ilya Shutov

Faculty of Software Engineering and Computer Systems, ITMO University,
St. Petersburg, Russia
gwohlg@itmo.ru

Abstract. In this paper, we present Russian language datasets in the digital humanities domain for the evaluation of word embedding techniques or similar language modeling and feature learning algorithms. The datasets are split into two task types, word intrusion and word analogy, and contain 31362 task units in total. The characteristics of the tasks and datasets are that they build upon small, domain-specific corpora, and that the datasets contain a high number of named entities. The datasets were created manually for two fantasy novel book series ("A Song of Ice and Fire" and "Harry Potter"). We provide baseline evaluations with popular word embedding models trained on the book corpora for the given tasks, both for the Russian and English language versions of the datasets. Finally, we compare and analyze the results and discuss specifics of Russian language with regards to the problem setting.

Keywords: Word embedding datasets · Language model evaluation · Russian language · Digital humanities · Word analogy

1 Introduction

Distributional semantics base on the idea, that the meaning of a word can be estimated from its linguistic context [1]. Recently, with the work on word2vec [2], where prediction-based neural embedding models are trained on large corpora, word embedding models became very popular as input to solve many natural language processing (NLP) tasks. In word embedding models, terms are represented by low-dimensional, dense vectors of floating-point numbers. While distributional language models are well studied in the general domain when trained on large corpora, the situation is different regarding specialized domains, and term types such as *proper nouns*, which exhibit specific characteristics [3,4]. Datasets in the digital humanities domain, which include some of these aspects, were presented by Wohlgenannt [5]. In this work, those datasets are translated to Russian language, and we provide baseline evaluations with popular word embedding models, and analyze differences between English and Russian experimental results.

© Springer Nature Switzerland AG 2023
A. Gelbukh (Ed.): CICLing 2019, LNCS 13451, pp. 468–479, 2023.
https://doi.org/10.1007/978-3-031-24337-0_33

The manually created datasets contain *analogies* and *word intrusion* tasks for two popular fantasy novel book series: "A Song of Ice and Fire" (ASOIF, by GRR Martin) and "Harry Potter" (HP, by JK Rowling). The *analogy* task is a well-known method for the intrinsic evaluation of embedding models, the *word intruder* task is related to word similarity and used to solve the "odd one out" task [6].

The basic question is how well Russian language word embedding models are suited for solving such tasks, and what are the differences to English language datasets and corpora. More specifically, what is the performance on the two task types, which word embedding algorithms are more suitable for the tasks, and which factors are responsible for any differences between English and Russian language results?

In this work, we manually translated the datasets into Russian. In total, we present 8 datasets, for both book series, the two task types, and the distinction between unigrams and n-gram datasets. Word2vec [2] and FastText [7] with different settings were trained on the Russian (and English) book corpora, and then evaluated with the given datasets. The evaluation scores are sufficiently lower for Russian, but the application of lemmatization on the Russian corpora helps to partly close the gap. Other issues such as ambiguities in the translation of the datasets, and inconsistencies in the transliteration of English named entities are analyzed and discussed. As an example, the best accuracy scores for the ASOIF unigram dataset for Russian are 32.7% for the *analogy* task, and 73.3% for word intrusion, while for English the best results are 37.1%, and 86.5%, resp.

The main contributions include the eight Russian language datasets with 31362 task units in total, translated by two independent teams, baseline evaluations with various word2vec and FastText models, comparisons between English and Russian, and the analyses of the results, specifically with regards to corpus word frequency and typical issues in translation and transliteration.

The paper is structured as follows: After an overview of related work in Sect. 2, the two task types and the translated datasets are introduced in Sect. 3. Section 4 first elaborates the evaluation setup, for example the details of the corpora, and the model settings used in the evaluations. Subsequently, evaluation results, both aggregated and fire-grained, are presented. We discuss the findings in Sect. 5, and then provide conclusions in Sect. 6.

2   Related Work

Word embedding vectors are used in a many modern NLP applications to represent words, often by applying pre-trained models trained on large general-purpose text corpora. Ghannay et al. [8] compare the performance of model types such as word2vec CBOW and skip-gram [2], and GloVe [9]. For example, FastText [7] provides pre-trained models for many language for download. But there are also language-specific efforts, eg. RusVectōrēs [10] include a number of models trained on various Russian corpora. Workshops on Russian language semantic similarity [11] emphasized the importance of the research topic.

For the intrinsic evaluation of word embedding models, researchers often use existing word similarity datasets like WordSim-353 [12] or MEN [13], or analogy datasets like Google [2] or BATS [14].

In specialized domains, large text corpora for training are often not available. Sahlgren and Lenci [15] evaluate the impact of corpus size and term frequency on accuracy in word similarity tasks, and as expected, corpus size has a strong impact. The datasets that we translated contain a high percentage of named entities such as book characters and locations. Herbelot [3] discusses various aspects of instances (like named entities) versus kinds, such as the detection of instantiation relations in distributional models. Distributional models are shown to be better suited for categorizing than for distinguishing individuals and their properties [16]. This is also reflected by our results, esp. the analysis of task difficulty in word intrusion (see Sect. 4.2).

The work on using distributional methods in the digital humanities domain is limited. More efforts have been directed at dialog structure and social network extraction [17,18] or character detection [19].

3   Tasks and Datasets

This section introduces the task types (analogy and word intrusion), discusses the dataset translation process, and briefly describes the word embedding algorithms used, as well as the basics of implementation.

3.1   Task Types

The datasets and evaluations focus on two task types: word intrusion and word analogy. Term analogy is a popular method for the evaluation of models of distributional semantics, applied for example in the original word2vec paper [2]. Word intrusion is a task similar to *word similarity*, which is a popular intrinsic evaluation method for word embedding models (see Sect. 2).

The word analogy task captures semantic or syntactic relations between words, a well-known example is "*man* is to *king*, like *woman* is to *queen*". The model is given the first three terms as input, and then has to come up with the solution (*queen*). Word embeddings models can, for example, apply simple linear vector arithmetic to solve the task, with $vector(man) - vector(woman) + vector(king)$. Then, the term closest (eg. measured by cosine similarity) to the resulting vector is the candidate term.

In the second task type, word intrusion, the goal is to find an intruding word in a list of words, which have a given characteristic. For example, find the intruder in: *Austria Spain Tokyo Russia*. In our task setup, the list always includes four terms, where one is the intruder to be detected.

3.2   Dataset Translation

The given datasets are based on extended versions of English language datasets presented in [5]. Those datasets were manually created inspired by categories and relations of online Wikis about "A Song of Ice and Fire" and "Harry Potter". The goal was to provide high quality datasets by filtering ambiguous and very-low frequency terms. The three dimensions (2 book series, 2 task types, unigram and n-gram datasets) led to eight published datasets.

In this work, the datasets were translated to Russian language. Two separate teams of native speakers translated the datasets to Russian. We found, that multiple book translations exist for the Harry Potter book series, and decided to work on two different book translations in this case. As many terms, esp. named entities like book characters and location names, have a slightly different translation or transliteration from the English original to Russian, we ended up with two independent datasets.

For ASOIF, both translation teams based their translations on the same Russian book version, and there where only slight differences – mostly regarding terms which are unigrams in English, but n-grams in Russian, and a few words which can have multiple translations into Russian.

3.3   Word Embedding Models

For the baseline evaluation of the datasets, we apply two popular word embedding models. Firstly, word2vec [2] uses a simple two-layer feed-forward network to create embeddings in an unsupervised way. The simple architecture facilitates training on large corpora. Basically, word similarity in vector space reflects similar contexts of words in the corpus. Depending on preprocessing, unigram or n-gram models can be trained. Word2vec includes two algorithms. CBOW predicts a given word from the window of surrounding words, while skip-gram (SG) predicts surrounding words from the current word. Secondly, FastText [7] is based on the skip-gram model, however, in contrast to word2vec, it makes use of sub-word information, and represents words as bag of character n-grams.

Hyperparameter tuning has a large impact on the performance of embedding models [15]. In the evaluations, we compare the results for different parameters settings for both datasets, details on those settings are found in Sect. 4.1.

3.4   Implementation

As mentioned, two teams worked independently on dataset translation and model training, which leads to the provision of two independent GitHub repositories[1][2]. The repositories can be used to reproduce the results, and to evaluate alternative methods – based on the book translation used. All library requirements, usage, and evaluation, and most importantly the datasets, are found in the repositories. For model creation and evaluation the popular Gensim library is used [6]. The implementation contains two main evaluation modules, one for the *analogies* task, and one for *word intrusion*. In the repository, word intrusion is coined *doesn't-match*, as this is the name of the respective Gensim function.

---

[1] https://github.com/DenisRomashov/nlp2018_hp_asoif_rus.

[2] https://github.com/ishutov/nlp2018_hp_asoif_rus.

Third parties can either reuse the provided evaluation scripts on a given embedding model, or use the datasets directly. The dataset format is the same as in word2vec [2] for *analogies*, and for the word intrusion task it is simple the understand, with the 4 words of the task unit, and the intruder marked.

4   Evaluation

4.1   Evaluation Setup

Book Corpora and Dataset Translation. The models analyzed in the evaluations are trained on two popular fantasy novel corpora, "A Song of Ice and Fire" (ASOIF) by GRR Martin, and "Harry Potter" (HP) by JK Rowling.

From ASOIF, we took the first four books; the corpus size is 11.8 MB of plain text, with a total of 10.5M tokens (11.1M before preprocessing). The book series includes a large world with an immense number of characters, whereby about 30–40 main characters exists. Narration is mostly linear and the story is told in first person from the perspective of different main characters.

The HP book series consists of seven books, with a size of 10.7 MB and 9.2M tokens (9.8M before preprocessing). The books tell the story of young Harry Potter and his friends in a world full of magic. The complexity of the world, and the number of characters, is generally lower than in ASOIF.

The basics of dataset translations were already mentioned in Sect. 3.2. Two independent teams worked on the task. In case of ASOIF, both teams based their translation work and also model creation on the same Russian book corpus. For HP an original translation into Russian exists, which is still the most popular one. Later other translations emerged. In order to cover a wider range of corpora, and to investigate the differences between those translations, the teams used two different translations. The exact book versions are listed in the respective GitHub repositories[3][4].

Preprocessing. In principle, we tried to keep corpus preprocessing to a minimum, and did only the following common steps: removal of punctuation symbols (except hyphens), removal of lines that contain no letters (page numbers, etc.), and sentence splitting. However, in comparison to the English original version of datasets and corpus, for Russian we found that term frequencies of the terms in the datasets were significantly lower. A substantial amount of dataset terms even fell below the `min_count` frequency used in model training and was thereby excluded from the models. This can be attributed to the rich morphology of Russian language, and other reasons elaborated in the discussion section (Sect. 5). For this reason, we created a second version of the corpora with lemmatization applied to all tokens of the book series[5]. In the evaluations, we present and compare results of both corpora versions, with and without lemmatization applied.

---

[3] github.com/ishutov/nlp2018_hp_asoif_rus/blob/master/Results.md

[4] github.com/DenisRomashov/nlp2018_hp_asoif_rus/blob/master/RESULTS.md

[5] Using this toolkit: tech.yandex.ru/mystem.

Furthermore, for the creation of n-gram annotated corpora the *word2phrase* tool included in the word2vec toolkit [2] was utilized.

Models and Settings. As mentioned, we train word2vec and FastText models on the book corpora – using the Gensim library. In the upcoming evaluations, we use the following algorithms and settings to train models:

w2v-default: This is a word2vec model trained with the Gensim default settings: 100-dim. vectors, word window size and minimum number of term occurrence are both set to 5, iter (number of epochs): 5, CBOW.

w2v-SG-hs: Defaults, except: 300dim. vectors, 15 iterations, number of negative samples: 0, with hierarchical softmax, with the skip gram method[6].

w2v-SG-hs-w12: like *w2v-SG-hs*, but with a word window of 12 words.

w2v-SG-ns-w12: like *w2v-SG-w12*, with negative sampling set to 15 instead of hierarchical softmax.

w2v-CBOW: like *w2v-SG-hs*, but with the CBOW method instead of skip-gram.

FastText-default: A FastText model trained with the Gensim default settings: those are basically the same default settings as for *w2v-default*, except for FastText-specific parameters.

FastText-SG-hs-w12: Defaults, except: 300dim. vectors, 15 iterations, a word-window of 12 words, number of negative samples: 0, with hierarchical softmax and the skip-gram method[7].

FastText-SG-ns-w12: like *ft-SG-hs-w12*, but with negative sampling (15 samples) instead of hierarchical softmax.

Table 1. Number of tasks and dataset sections (in parentheses) in the Russian language datasets – with the dimensions of task type, book corpus and unigram/n-gram

Book corpus	Task type	Unigram	N-Gram
HP	Analogies	4790 (17)	92 (7)
	Word intrusion	8340 (19)	1920 (7)
ASOIF	Analogies	2848 (8)	192 (2)
	Word intrusion	11180 (13)	2000 (7)

Datasets. In total, we provide eight datasets. This number stems from three datasets dimensions: the task type (analogies and word intrusion, the two book series, and the distinction between unigram and n-gram datasets). Table 1 gives an overview of the number of tasks within the datasets, and also of the number of sections per task. Sections reflect a subtask with specific characteristics and

---

[6] size=300, -negative=0, sg=1, hs=1, iter=15.
[7] size=300, -negative=0, sg=1, hs=1, iter=15, -window=12.

difficulty, for example analogy relations between *husband* and *wife*, or between between a *creature* (individual) and its *species*. Typically, per section, the items on a given side of the relation are members of the same word or named entity category, therefore a distributional language model can be more deeply analyzed for its performance in those subtasks.

In contrast to popular word similarity datasets like WordSim-353 [12] or MEN [13], most of the dataset terms are named entities. Herbelot [3] studies some of the properties of named entities in distributional models. For example, in the unigram word intrusion dataset, only around 7% (ASOIF) and 17% (HP) of terms are *kinds*, the rest are named entities.

4.2   Evaluation Results

In this section, we present and analyze the evaluation results for the presented datasets using word embedding models. We start with an overview of the results of the *analogies* and *word intrusion* tasks, followed by more fine-grained results for the different subtasks of the *analogies* tasks. For the *word intrusion* task, we investigate evaluation results depending on task difficulty, and finally, a summary of results on n-gram datasets is presented. Further details on the results can be found on github (See footnote 3 and 4).

Table 2. Overall *analogies* accuracy of the Russian unigram datasets for both book series. Results are given for models with and without lemmatization of the corpora. Values for English given in parenthesis.

Book series	ASOIF		HP	
Preprocessing	Minimal	Lemmatization	Minimal	Lemmatization
w2v-default	0.57 (8.15)	2.35 (–)	00.42 (6.88)	1.75 (–)
w2v-SG-hs	17.61 (28.44)	24.40 (–)	13.40 (25.11)	23.00 (–)
w2v-SG-hs-w12	**24.56** (37.11)	**32.66** (–)	**20.34** (30.00)	**28.95** (–)
w2v-SG-ns-w12	21.58 (29.32)	20.97 (–)	13.17 (20.84)	12.99 (–)
w2v-w12-CBOW	0.57 (2.67)	1.07 (–)	0.68 (7.22)	2.62 (–)
FastText-default	0.42 (1.33)	2.31 (–)	0.08 (0.87)	0.42 (–)
FastText-SG-hs-w12	11.04 (29.81)	21.58 (–)	8.57 (25.46)	19.30 (–)
FastText-SG-ns-w12	0.99 (14.64)	0.8 (–)	3.77 (14.23)	4.13 (–)

Table 2 provides an overview of results for the *analogies* task. It includes the results for the two book series ASOIF and HP. We distinguish two types of input corpora, namely with and without the application of lemmatization ("minimal" preprocessing vs. "lemmatization"). Embedding models were trained on the corpora with the settings described in Sect. 4.1. Furthermore, the evaluation scores for English language corpora and datasets are given for comparison.

The results in Table 2 indicate that models trained with the skip-gram algorithm clearly outperform CBOW for analogy relations. Another important fact is that for Russian language lemmatization of the corpus tokens before training has a strong and consistent positive impact on results. However, the numbers for Russian stay below the numbers for English. Both the performance impact of lemmatization, and the differences between English may partly be the result of differences in corpus term frequency. This intuition will be investigated and discussed in Sect. 5.

Table 3. Overall *word intrusion* accuracy (in percent) for the Russian unigram datasets for both book series. Results are given for models with and without lemmatization of the corpora. Values for English given in parenthesis.

Book series	ASOIF		HP	
Preprocessing	Minimal	Lemmatization	Minimal	Lemmatization
w2v-default	62.03 (86.53)	64.83 (–)	34.69 (64.83)	53.59 (–)
w2v-SG-hs	65.93 (77.9)	**73.30** (–)	55.99 (73.3)	60.44 (–)
w2v-SG-hs-w12	67.11 (74.86)	68.89 (–)	**61.09** (68.69)	59.87 (–)
w2v-SG-ns-w12	**68.09** (75.15)	67.1 (–)	58.43 (74.43)	57.01 (–)
w2v-w12-CBOW	57.35 (75.61)	61.28 (–)	42.39 (61.28)	48.73 (–)
FastText-default	61.59 (73.82)	56.56 (–)	41.92 (56.56)	46.50 (–)
FastText-SG-hs-w12	66.82 (75.99)	70.20 (–)	60.99 (70.2)	60.27 (–)
FastText-SG-ns-w12	67.81 (75.38)	68.41 (–)	59.13 (76.41)	**61.54** (–)
Stock embeddings	27.3 (–)	– (–)	25.36 (–)	– (–)
Random baseline	25.00	25.00	25.00	25.00

Table 3 gives the results for the *word intrusion* tasks. Again, we distinguish between preprocessing with and without lemmatization, and between the results of different models trained on the two book series. For the word intrusion task, the differences observed between skip-gram and CBOW, and regarding lemmatization, are smaller as compared to *analogies* results in Table 2; however, the tendencies still exist. For comparison, we also applied pretrained FastText models ("Stock Embeddings") trained on Wikipedia[8] to the task. As expected, those models perform very poorly, only slightly over the random baseline.

All datasets are split into various sections, which reflect specific relation types, for example *child-father* or *houses-and-their-seats*. Those relations have certain characteristics, such as involving person names, location entities, or other, which allow a fine-grained analysis and comparison of embedding models and their performance. Table 4 shows some selected sections from the ASOIF analogies dataset. The performance varies strongly over the different subtasks, but the data indicates, that models that do well in total, are also more suitable on the individual tasks (Table 5).

---

[8] https://github.com/facebookresearch/fastText/blob/master/pretrained-vectors.md.

Table 4. ASOIF Analogies Russian dataset: Accuracy of different word embedding models on selected analogies task sections, and total accuracy. In parenthesis, values from the English language dataset are given for comparison.

Task section	first-lastname	husband-wife	loc-type	houses-seats	Total
Number of tasks:	2368	30	168	30	2848
w2v-default	1.93 (8.78)	0.0 (6.67)	5.0 (4.76)	10.0 (20.0)	2.35 (8.15)
w2v-SG-hs	26.98 (32.01)	15.0 (10.0)	7.5 (11.9)	26.67 (40.0)	24.4 (28.44)
w2v-SG-hs-w12	**36.24** (42.36)	**20.0** (10.0)	6.25 (19.64)	**30.0** (33.33)	**32.66** (37.71)
w2v-SG-ns-w12	32.62 (40.62)	5.0 (6.67)	**12.5** (22.62)	26.67 (40.0)	29.62 (36.41)
w2v-w12-CBOW	0.69 (1.27)	5.0 (6.67)	2.5 (11.9)	6.67 (30.0)	1.07 (2.67)
FastText-default	2.33 (1.06)	5.0 (3.33)	0.0 (3.57)	3.33 (3.33)	2.31 (1.33)
FastText-SG-hs-w12	23.86 (34.04)	15.0 (6.67)	6.25 (13.69)	20.0 (46.67)	21.58 (29.81)
FastText-SG-ns-w12	27.33 (35.09)	15.0 (3.33)	5.0 (11.9)	13.33 (26.67)	24.4 (30.44)

Table 5. Accuracy results with regards to task difficulty – Russian ASOIF word intrusion dataset (unigrams), trained on a lemmatized corpus.

Task difficulty	1 (hard)	2 (med-hard)	3 (medium)	4 (easy)	AVG
Number of tasks:	2795	2795	2795	2795	11180
w2v-default	61.82	**72.9**	70.16	91.74	**74.17**
w2v-SG-hs	46.27	67.72	73.38	85.33	68.18
w2v-SG-hs-w12	39.18	67.73	**75.67**	85.83	67.1
w2v-SG-ns-w12	57.53	70.98	74.35	90.84	73.43
FastText-default	**61.86**	65.62	66.26	**96.85**	72.65
FastText-SG-hs-w12	46.76	67.59	73.38	87.48	68.8
FastText-SG-ns-w12	39.28	66.87	73.56	86.37	66.52

The word intrusion datasets were created with the idea of four task difficulty levels. The *hard* level includes near misses; on the *medium-hard* level, the outlier still has some semantic relation to the target terms, and is of the same word (or NE) category. On the *medium* level outliers are of the same word category, but have little semantic relatedness to the target term. And finally, in the *easy* category, the terms have no specific relation to the target terms. As an example, if the target terms are Karstark Greyjoy Lannister, ie. names of *houses*, then a *hard* intruder might be Theon, who is a person from one of the houses. Bronn will be a *med-hard* intruder, also a person, not from those houses. Winterfell (a location name) will be in the *medium* category, and raven in the easy one.

Very interestingly, models using the CBOW algorithm (such as *w2v-default* and *FastText-default*) provide very good results on the hardest task category with over 60% of correct intruders selected. On the other hand, SG-based models only show 39%-46%. For the easiest category, *FastText-default* excels with almost 97% accuracy. In general word embeddings, esp. when trained on small datasets and using cosine similarity for the task, struggle to single out terms in the *hard* category by a specific (minor) characteristic of the target terms. This will be further discussed in Sect. 5.

N-Gram Results. As mentioned, in addition to the four unigram dataset, we created complementary n-gram datasets for the two book series and the two task types. In correspondence with the n-gram detection method used [2], in the datasets n-grams are words connected by the underscore symbol. Many of the terms are person or location names such as *Forbidden_Forest* or *Maester_Aemon*. For reasons of brevity, we will not include result tables here (see GitHub for details). The general tendency is that n-gram results are below unigram results in the *analogies* task, for word intrusion results are comparable with around 70% accuracy (depending on model settings). In comparison with word2vec, FastText-based models perform better on n-gram than unigram tasks, this can be explained by the capability of FastText to leverage subword-information within n-grams.

5   Discussion

In general, there is quite a big difference in performance between Russian and English datasets and models, when the same (minimal) preprocessing is being applied to the corpora. For example, in Table 2 the best ASOIF performance for Russian (with minimal preprocessing) is 24.56%, but 37.11% for English, and for HP the values are 20.34% for Russian, and 30.00% for English. In the case of *word intrusion*, the same pattern repeats: 68.09% for Russian ASOIF, and 86.53% for English, and finally, 61.09% for the Russian HP dataset versus 74.43% (English).

Our first intuition was, that the rich morphology of the Russian language, where also proper nouns have grammatical inflections by case, might reduce the frequency of dataset words in the corpora. Sahlgren and Lenci [15] show the impact of term frequency on task accuracy in the general domain. Subsequent analysis shows, that eg. for the HP dataset, the average term frequency of dataset terms is 410 for the English terms and English book corpus, while for the Russian it is only 249. We then decided to apply lemmatization to the Russian corpora, which helped to raise Russian average term frequency to 397. Also the evaluation results improved overall, as seen in the tables in Sect. 4.2. However, despite the positive effects, lemmatization also introduces a source of errors. For some dataset terms, the frequency even becomes lower; after lemmatization, the number of Russian dataset terms that are below the *min_count* threshold to be introduced into the word embedding models rises. An example of such problem cases is the word "Fluffy" from HP, which was translated as "Пушок". But then the lemmatizer wrongly changed it to "пушка" (a gun), so that "Пушок" disappeared from the trained models.

A number of other difficulties and reasons for the lower performance on the Russian datasets emerged: a) When comparing the output of the two translation teams, we observed words that have many meaningful translations to Russian, thereby lowering term frequency. For example, "intelligence" can be translated as "ум", "интеллект", "осознание", "остроумие" and so on. b) The translit-eration of English words into Russian is not always clear, and we have found in the analysis that even within the same book corpora (translations) it is not always consistent, even more so between different translators. For example, there is no [æ] phonetic sound in Russian, so it can be transliterated to multiple let-ters: a, e, э. c) In Russian, the ё letter is often replaced with e. If this happens inconsistently, it impacts term frequencies.

Another interesting aspect is the performance of the models on various difficulty levels in the word intrusion tasks. If difficulty is low, then common word embedding models already work very well, in our experiments with an accuracy up to 97%. However, in the *hard* category terms are very similar in their overall semantics and context, but the target terms possess one character-istic that the intruder lacks. Cosine similarity just looks at overall vicinity in vector space, with a success rate of ca. 40–60%. There has been some work on named entities and the distinction between *individuals* (and their properties) and *kinds* within distributional models [3,4,16], but there is still much room on how to tackle such issues in a general way.

6   Conclusions

In this work, we present Russian language datasets in the digital humanities domain for the evaluation of distributional semantics models. The datasets cover two basic task types, *analogy* relations and *word intrusion* for two well-known fantasy novel book series. The provided baseline evaluations with word2vec and FastText models show that models for the Russian versions of the corpora and datasets offer lower accuracy than for the English originals. The contributions of the work include: a) the translation to Russian and provision (on GitHub) of eight datasets in the digital humanities domain, b) providing baseline evalu-ations and comparisons for various settings of popular word embedding models, c) studying the effects of preprocessing (esp. lemmatization) on performance, d) analyzing the reasons for differences between Russian and English language evaluations, most notably term frequency and issues arising from translation.

Acknowledgments. This work was supported by the Government of the Russian Federation (Grant 074-U01) through the ITMO Fellowship and Professorship Program.

References

1. Harris, Z.: Distributional structure. Word **10**(2–3), 146–162 (1954)
2. Mikolov, T., Chen, K., Corrado, G., Dean, J.: Efficient estimation of word repre-sentations in vector space. arXiv preprint arXiv:1301.3781 (2013)
3. Herbelot, A.: Mr. Darcy and Mr. Toad, gentlemen: distributional names and their kinds. In: Proceedings of the 11th International Conference on Computational Semantics, pp. 151–161 (2015)

4. Boleda, G., Padó, S., Gupta, A.: Instances and concepts in distributional space. In: 2017 EACL, Valencia, Spain, 3–7 April 2017, Volume 2: Short Papers, pp. 79–85 (2017)
5. Wohlgenannt, G., Chernyak, E., Ilvovsky, D., Barinova, A., Mouromtsev, D.: Relation extraction datasets in the digital humanities domain and their evaluation with word embeddings. In: CICLING 2018. Volume upcoming of Lecture Notes in Computer Science., Hanoi, Vietnam, Springer (2018). upcoming
6. Řehůřek, R., Sojka, P.: Software framework for topic modelling with large corpora. In: Proceedings of the LREC 2010 Workshop on New Challenges for NLP Frameworks, Valletta, Malta, ELRA, pp. 45–50 (2010)
7. Bojanowski, P., Grave, E., Joulin, A., Mikolov, T.: Enriching word vectors with subword information. arXiv preprint arXiv:1607.04606 (2016)
8. Ghannay, S., Favre, B., Estève, Y., Camelin, N.: Word embedding evaluation and combination. In: Proceedings of the LREC 2016, Paris, France, ELRA (2016)
9. Pennington, J., Socher, R., Manning, C.D.: Glove: global vectors for word representation. In: Empirical Methods in Natural Language Processing (EMNLP), pp. 1532–1543 (2014)
10. Kutuzov, A., Kuzmenko, E.: WebVectors: a toolkit for building web interfaces for vector semantic models. In: Ignatov, D.I., et al. (eds.) AIST 2016. CCIS, vol. 661, pp. 155–161. Springer, Cham (2017). https://doi.org/10.1007/978-3-319-52920-2_15
11. Panchenko, A., Loukachevitch, N.V., Ustalov, D., Paperno, D., Meyer, C., Konstantinova, N.: RUSSE: the first workshop on Russian semantic similarity. CoRR abs/1803.05820 (2015)
12. Finkelstein, L., et al.: Placing search in context: the concept revisited. In: Proceedings of the 10th International Conference of the World Wide Web, pp. 406–414. ACM (2001)
13. Bruni, E., Tran, N.K., Baroni, M.: Multimodal distributional semantics. J. Artif. Int. Res. **49**(1), 1–47 (2014)
14. Gladkova, A., Drozd, A., Matsuoka, S.: Analogy-based detection of morphological and semantic relations with word embeddings: what works and what doesn't. In: SRW@HLT-NAACL, pp. 8–15. ACL (2016)
15. Sahlgren, M., Lenci, A.: The effects of data size and frequency range on distributional semantic models. In: EMNLP, pp. 975–980. ACL (2016)
16. Boleda, G., Padó, S., Pham, N.T., Baroni, M.: Living a discrete life in a continuous world: reference in cross-modal entity tracking. In: IWCS 2017–12th International Conference on Computational Semantics – Short papers (2017)
17. Elson, D.K., Dames, N., McKeown, K.R.: Extracting social networks from literary fiction. In: Proceedings of the 48th Annual Meeting of the ACL (ACL 2010), pp. 138–147. Association for Computational Linguistics, Stroudsburg, PA, USA (2010)
18. Jayannavar, P., Agarwal, A., Ju, M., Rambow, O.: Validating literary theories using automatic social network extraction. In: CLFL@ NAACL-HLT, pp. 32–41 (2015)
19. Vala, H., Jurgens, D., Piper, A., Ruths, D.: Mr. Bennet, his coachman, and the archbishop walk into a bar but only one of them gets recognized: on the difficulty of detecting characters in literary texts. In: Màrquez, L.E.A., ed.: EMNLP, pp. 769–774. ACL (2015)

Towards the Automatic Processing of Language Registers: Semi-supervisedly Built Corpus and Classifier for French

Gwénolé Lecorvé[1(✉)], Hugo Ayats[1], Benoît Fournier[1], Jade Mekki[1,3], Jonathan Chevelu[1], Delphine Battistelli[3], and Nicolas Béchet[2]

[1] Univ Rennes, CNRS, IRISA/6, rue de Kerampont, Lannion, France
{gwenole.lecorve,hugo.ayats,benoit.fournier,
jade.mekki,jonathan.chevelu}@irisa.fr
[2] Université de Bretagne Sud, CNRS, IRISA/Campus Tohannic, Vannes, France
[3] Université Paris Nanterre, CNRS, MoDyCo/av. République, Nanterre, France
delphine.battistelli@parisnanterre.fr

Abstract. Language registers are a strongly perceptible characteristic of texts and speeches. However, they are still poorly studied in natural language processing. In this paper, we present a semi-supervised approach which jointly builds a corpus of texts labeled in registers and an associated classifier. This approach relies on a small initial seed of expert data. After massively retrieving web pages, it iteratively alternates the training of an intermediate classifier and the annotation of new texts to augment the labeled corpus. The approach is applied to the casual, neutral, and formal registers, leading to a 750 M word corpus and a final neural classifier with an acceptable performance.

Keywords: Language registers · Labeled data · Annotated corpus

1 Introduction

The language registers provide a lot of information about a communicator and the relationship with the recipients of her/his messages. Their automatic processing could show whether two persons are friends or are in a hierarchical relation, or give hints about someone's educational level. Modeling language registers would also benefit in natural language generation by enabling to modulate the style of artificial discourses. However, language registers are still poorly studied in natural language processing (NLP), particularly because of the lack of large training data. To overcome this problem, this paper presents a semi-supervised, self-training, approach to build a text corpus labeled in language registers.

The proposed approach relies on a small set of manually labeled data and a massive collection of automatically collected unlabeled web pages. Text segments extracted from these web pages are iteratively labeled using a classifier—a neural network—trained on the labeled data. For a given iteration, text segments that

© Springer Nature Switzerland AG 2023
A. Gelbukh (Ed.): CICLing 2019, LNCS 13451, pp. 480–492, 2023.
https://doi.org/10.1007/978-3-031-24337-0_34

are classified with a high confidence are added to the training data, and a new classifier is then trained for a next iteration. Through this process, we expect to label as many segments as possible, provided that the classifier accuracy remains good enough when augmenting the training data. In practice, this process is applied on a set of 400, 000 web pages, and results in a corpus of about 750 million words labeled in casual, neutral and formal register. Alongside, when testing on 2 different test sets, the final classifier performs accuracies of 87% and 61%. The set of descriptors used includes 46 characteristics of various natures (lexical, morphological, syntactical . . .) questions of a preliminary expert analysis.

This paper is a first step in the process of modelling language registers in NLP. As such, the will of the authors is to report about first experiments, popularize the issue of language registers, and provide a baseline for future improvements and tasks that are more elaborate. Especially, the presented semi-supervised process is not new and could be improved. Likewise, the associated classifier could also probably benefit from various sophistications.

In this paper, Sect. 2 presents a state of the art about language registers, while Sect. 3 details the semi-supervised approach. Then, Sects. 4, 5, and 6 describe the collected data, the classifier training, and the resulting corpus.

2   State of the Art and Positioning

The notion of register refers to the way in which linguistic productions are evaluated and categorized within the same linguistic community [1, 2]. A register is characterized by multiple specific features (more or less complex terms, word order, verb tenses, length of sentences, etc.) and can be compared to others, sometimes with in a given ordering (e.g., formal, literary, neutral, casual, slang. . .). Such a partitioning depends on the angle from which linguistic communities are observed, for instance, the influence of the communication media [3] or the degree of specialization [4, 5]. Its granularity may also vary [6–8]. In this work, we consider 3 registers: casual, neutral and formal. This choice is primarily motivated by pragmatism, as this division is relatively consensual and unambiguous for manual labeling, while not prohibiting possible refinements in the future. The neutral register involves a minimal set of assumptions about any specific knowledge of the message recipient, and is therefore based on the grammar and vocabulary of the language, with no rare constructions and terms. On the contrary, the formal register assumes the recipient to have a high proficiency, whereas the casual one allows voluntary or faulty deviations of the linguistic norms. In this paper, we do not focus on the sociolinguistics, but seek to enable NLP on language registers by building a large labeled corpus.

Registers got very little attention in NLP. [9,10] proposed to classify documents as formal or informal. In [11], the authors train a regression model to predict a level of formality of sentences. In these papers, features are derived from a linguistic analysis. Although these features are a good basis for our work, they are designed for English and do not apply for French. Moreover, they work on few data (about 1 K documents) whereas we expect to build a large corpus (>100 K documents). More generally, the study of language register shares similarities with authorship attribution [12,13] and the analysis of new media like blogs, SMSs, tweets, etc. [14–18], where research is backed by the release of reference corpora. Such a corpus does not exist for the language registers.

Automatic style processing methods are all based on a set of relevant features derived from the texts to be processed. Due to its historical importance, author attribution work can list a wide range of features. As indicated by [12], an author's preferences or writing choices are reflected at several levels of language. The most obvious—and most studied—is the lexical level, e.g. through the length of words and sentences in a text, the richness of its vocabulary or frequencies of words and word n-grams [19,20]. In this respect, it is generally accepted in the community that grammatical words (prepositions, articles, auxiliaries, modal verbs, etc.) are of significant interest while the others (nouns, adjectives, etc..) should be avoided for style processing [21,22], according to a principle of orthogonality between style and meaning. This principle emphasizes the need to abstract some elements of meaning, otherwise the analysis risking to be be biased by the text's topic. Nevertheless, semantics can prove to be useful, for example through the frequencies of synonyms and hypernyms, or the functional relations between propositions (clarification of a proposition by another, opposition, etc.) [22,23]. Moreover, whatever their meaning, some specific words explicitly testifies to the fact that the text is of a specific style [24], especially in the case of language registers. Syntactically, the use of descriptors derived from morphosyntactic and syntactic analyzes is very widely used to characterize the style [21,25,26]. Finally, other work has been interested in graphical information by considering n-gram of characters, types of graphemes (letter, number, punctuation, capital letters, etc.) or information compression measures [21,27,28]. In our work, a preliminary linguistic study was conducted in this sense, leading to a set of descriptors for the 3 registers considered, as detailed in the description of the trained classifier in Sect. 4. Before that, next section introduces the overall semi-supervised joint construction of the corpus and classifier.

3   Proposed Approach

The semi-supervised process used to build the labeled corpus is schematized in Fig. 1. This process follows a self-training approach where seed data is augmented with automatically labeled texts. This approach has been experimented in various other NLP tasks [29,30]. In our approach, the corpus used for data augmentation is collected from the web. Queries are submitted to a search engine. These queries are derived from two specialized lexicons, one for the casual register, and the other for the formal one. Then, the collected texts are filtered using a neural network classifier to extract the most relevant ones for each of the 3 considered registers. Since the classifier requires labeled training data and data labeling requires a classifier, the approach is iterative. That is, a first classifier is initially trained on a small initial manually annotated seed. This first classifier makes it possible to select texts whose predicted register is considered as reliable. These texts are added to those already labeled, and a new iteration starts. In the end, this process results in a set of categorized texts and a classifier.

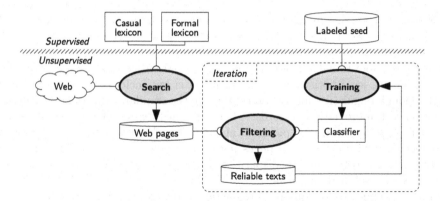

Fig. 1. Overview of the semi-supervised process.

Note that the use of the Internet is not an originality of our work since many similar examples exist in literature, for example [31] (although our collection process is not iterative here) or [32] for the collection of thematic pages. Then, self-training approaches is known to potentially degrade along iterations, due to classification mistakes, i.e., the augmented data progressively moves away from the original target task. An important objective in our work was to make sure that classification accuracy would not drop along the iterations.

The considered classes are *casual*, *neutral*, and *formal*. The assumption is also made that some texts do not belong to any of the three registers, either because they are badly formed (foreign language, SMS style, non-natural text, etc.) or because the register is not homogeneous (e.g. user comments). Our condition on classification reliability makes it possible to model this.

4   Data

In practice, the lexicons on which the collection of web pages is based are words and expressions automatically retrieved from a backup of the French version of Wiktionary. For a given register, only the unambiguous words belonging to a register are considered, that is to say the terms having all their meanings annotated as belonging to the same register. Precisely, the terms annotated as slang, casual, popular and vulgar were grouped within the casual lexicon and those categorized as literary and formal within the formal lexicon, each thus totaling respectively 6,000 and 500 entries. Equal numbers of queries are built for each register by randomly combining selected elements of the associated lexicon. Queries are empirically bound from 2 to 6 words in order to ensure a non-zero number of results and a minimal relevance for the returned pages. Web requests are made using the Bing API. In total, 12 K requests are submitted, each limited to a maximum of 50 hits. Online dictionaries were excluded at the time of the request in order to only retrieve pages where the searched terms are in context, and not isolated in a definition or an example. 76% of the queries returned at least one hit and 49% reached the maximum limit, be it for casual or formal queries. This is results in a collection of 400 K web pages.

The textual content of the web pages is extracted automatically thanks to a dedicated tool that looks for the central textual part of the page. It excludes titles, menus, legal notices, announcements, etc. but includes comments if they have enough linguistic content and conform to the standard editorial style (punctuation, not abbreviation of words. . .). The cleaned texts were segmented on the paragraph boundaries into pieces of about 5, 000 characters to avoid a lack of homogeneity within long web pages (eg forums) and not to introduce training biases related to text length disparities. Furthermore, non-French textual segments were excluded, resulting in 825 K segments, representing 750 M words. While all web pages are supposed to contained register-specific terms from their query, segmentation also enables introducing segments where none is present.

The seed collection of hand-tagged texts gathers 435 (about 440 K words) segments from novels, journals[1] and web pages[2]. Segments were jointly labeled by 2 qualified annotators and are balanced over the 3 registers. The seed texts have been selected such that there is no ambiguity about their register. As such, they can be regarded as stereotypical. Examples are given in the appendix. This seed is divided into training (40%, i.e., 174 segments), development (20%) and test (40%) sets. In addition, a second test set of 139 segments is randomly sampled from the collected corpus of web segments. These segments were labeled as follows: 27 as casual (19%), 69 as neutral (50%), 38 as formal (27%), and 5 as none (4%) because ambivalent[3]. This second set represents a more realistic situation since our seed has been designed to be unambiguous. The distribution over the registers and the presence of the "none" class illustrate this difference.

Text segments are described by 46 global features derived from related work in French linguistics [33–36]. The exhaustive list is given in Table 1. These features are relative frequencies of various linguistic phenomena covering lexical, phonetic, morphosyntactic and syntactic aspects. We address a few remark regarding these features. First, it can be noticed that no lexicon exists for the neutral register. Then, some words may be ambiguous regarding their membership in a registry. Thus, two feature variants are considered for register-specific words frequencies. The first weights the frequency of a word by the number of acceptations identified as belonging to the given register divided by the total number of its acceptations. The other variant is stricter. It only counts a word if all its acceptations are identified as belonging to the register. The case of the phrases or expressions does not require this duality because they are generally less ambiguous. Finally, most of the features denote well-known phenomena highlighting deviations to the norm of the language, for instance through mappings of non-written usages (especially speech) into the written language or syntac-

---

[1] Among which: *Kiffe kiffe demain* (Faïza Guène), *Albertine disparue* (Marcel Proust), *Les Mohicans de Paris* (Alexandre Dumas), The Bridge-Builders (Rudyard Kipling), *Les misérables* (Victor Hugo), and archives from the newspaper *L'Humanité*.

[2] These web pages do not come from the automatically collected set.

[3] Often because of mixed narrative and active parts.

Table 1. Features used by the classifier.

Lexicon
– Casual words weighted by their number of acceptations as casual: 7 828 items
– Formal words weighted by their number of acceptations as formal : 565 items
– Purely casual words (all acceptations are casual) : 3 075 items
– Purely formal words (all acceptations are formal) : 166 items
– Casual phrases : 3 453 items
– Formal phrases : 143 items
– Animal names : 78 items
– Onomatopoeia (e.g., *"ah"*, *"pff"*...) : 125 items
– SMS terms (e.g., *"slt"*, *"lol"*, *"tkt"*...) : 540 items
– Lexical and syntactic anglicisms
– Unknown words
– Word *"ça"* ("it"/"this")
– Word *"ce"* ("it"/"this")
– Word *"cela"* ("it"/"this")
– Word *"des fois"* ("sometimes")
– Word *"là"* ("there"/"here")
– Word *"parfois"* ("sometimes")

Phonetics
– Vowel elision (*"m'dame"*, *"p'tit"*...)
– Elision of 'r' (*"vot'"*, *"céleb'"*...)
– Written liaisons *"z"* (*"les zanimaux"*)

Morphology
– Repeating syllables (*"baba"*, *"dodo"*...)
– Repeating vowels (*"saluuuut"*)
– Word endings in *"-asse"*
– Word endings in *"-iotte"*
– Word endings in *"-o"*
– Word endings in *"-ou"*
– Word endings in *"-ouze"*

Morphosyntax
– All verb tenses and modes
– All types of subject pronouns
– Verb groups (French peculiarity)

Syntax
– Double possessive form (e.g., *"son manteau à lui"*, "his coat belonging to him")
– Structure *"c'est ... qui"* ("it's ... who/which")
– Use of *"est-ce que"* (specific French interrogative form)
– Conjunction *"et"* ("and")
– Shortened negative forms without *"ne"* (e.g., "il vient pas")
– Other, uncategorized, syntactic irregularities

tic mistakes. The lexical richness is also part of this deviation since there is an infinity to diverge from the norm. Hence, the casual register is rich.

In practice, feature were extracted using dedicated dictionaries, orthographic and grammatical analyzes (LanguageTool), and *ad hoc* scripts.

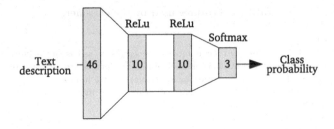

Fig. 2. Architecture of the neural network classifier.

5   Classifier Training

As illustrated in Fig. 2, the classifier is a multilayer neural network, fed by the 46 global features of a segment, predicting probabilities for each register (softmax layer of size 3). 2 dense hidden layers of size 10 are considered, respectively with leaky ReLU[4] and *tanh* activation function. No extensive tuning on the development set has been performed on this architecture but the use of a simple architecture and global features is voluntary since the seed data is small[5]. The reliability of a prediction is directly given by the majority class probability, and a threshold is applied to decide whether to validate the segment's label or not.

The experiments were conducted in Python using Keras and TensorFlow. Apart from learning the first model on the seed, successive classifiers are trained by batch of 100 instances over 20 epochs using the optimization algorithm *rmsprop* and the mean absolute error as loss function. At each iteration, the newly selected segments among the web data are injected into the training set for 80% and the development set for the rest. The test set is never modified in order to measure the progress of the classifier throughout the process.

The Fig. 3 shows the accuracy evolution through iterations on the test set and the manually labeled sample of the collected segments. Different selection thresholds are reported, ranging from 0.8 to 1 (i.e., the classifier is sure of its predictions). These high values are justified by the high accuracy of 87% on the seed. Overall, the classifier is rather stable despite the insertion of new data, regardless of the dataset, showing that the inserted data is relevant. Still, the results on the labeled sample are lower than those on the test set, although still better than a random or naive classification. This is not an overfitting on the seed since this data is completely diluted once the training corpus is augmented with the selected web segments. Instead, we think that this discrepancy is because the automatically collected data is less clean and less stereotypical than the seed. Hence, their register is more difficult to predict. Regarding data selection, the strictest threshold value, 1, leads to deteriorate the results, while thresholds 0.9 and 0.99 produce the best results.

---

[4] Parameter α set to 0.1.

[5] Especially, word-based models, e.g., RNNs, could not reasonably be applied here due to this initially limited amount of data. However, such models will be studied in the future using the final corpus.

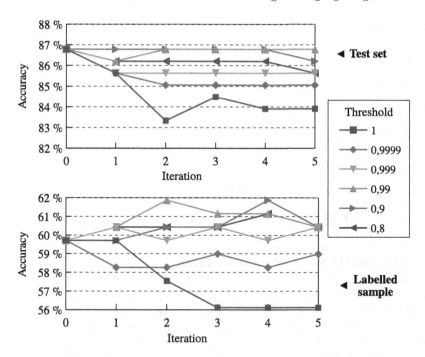

Fig. 3. Evolution of accuracy on the test set (top) and the labeled sample (bottom).

Table 2. Recall, precision, and F-measure for each register after 5 iterations with a threshold set to 0.99) on the test set and the labeled sample.

	Test set			Labeled sample		
	Casual	Neutral	Formal	Casual	Neutral	Formal
Recall	.90	.78	.93	.53	.72	.45
Precision	.84	.90	.87	.52	.64	.61
F-measure	.87	.83	.90	.52	.68	.52

Table 2 shows the recall, precision, and F-measure at the end of the process for the threshold 0.99, on the test set and on the labeled sample. On the test set, it appears that the results are relatively homogeneous between registers, the lowest F-measure being for the neutral register because of a lower recall. Conversely, the results on the labeled sample—which are worse as previously highly—show that the neutral register is the one best recognized as the casual and formal registers present weak F-measures. For the first one, difficulties seem to be global with precision and recall just greater than 0.5. For the second, the weak results come from a high proportion of false negatives (low recall). Therefore, while the results are encouraging, they also call for further improvements. Especially, attention should concentrate on maximizing precision, i.e., avoiding false positives, because they tend to distort the convergence of the semi-supervised process.

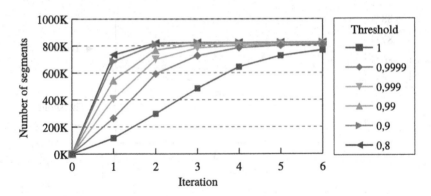

Fig. 4. Size of the labeled corpus for each iteration.

6   Automatically Labeled Corpus

Figure 4 illustrates the size evolution of the labeled corpus for different selection thresholds. First, it appears that all or almost all the collected segments end up the process with a label. Given the already mentioned complexity of the data (noise, ambiguity), this again urges on further developments on false positives and a stopping criterion. Then, it appears that the process is quick, converging in a few iterations. For example, 89% labels are validated at the end of the first pass for a selection threshold of 0.8.

At the end of the process for the threshold of 0.99, texts labeled as casual come for 68% from casual queries and, therefore, for 32% from formal queries. These ratios are 47/53% and 37/63% for the neutral and formal registers. Hence, the lexicons are appropriate to initiate the process since they do not fully confine the collected pages in the register of their original query. However, a manual analysis shows that a considerable number of texts classified as casual but coming from formal queries (and vice versa) should not be. Hence, some phenomena are still poorly understood and the method should be refined. For instance, it is likely that the model excessively trusts some features, especially register-specific terms. One solution may be to introduce a dropout mechanism when training the neural network.

Finally, as detailed by Fig. 5, a deeper analysis of the corpus evolution shows that the class distribution automatically evolves. Starting from the initially balanced setting of the seed, the proportions of the casual, neutral, and formal registers in the final step corpus are 28%, 48%, and 24%—consistently what is observed on the labeled sample of the web segments. 2 examples of automatically labeled segments are given in Table 3.

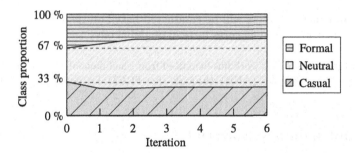

Fig. 5. Evolution of class distribution in the labeled corpus (percentage on the number of texts, threshold = 0.99).

Table 3. Excerpts from 2 texts automatically labeled as neutral and formal.

Neutral	Formal
Oui, Monsieur Adrien Richard, si vous aimez mieux, le directeur de l'usine, mais nous, nous ne l'appelons que Monsieur Adrien, parce qu'on a été à l'école ensemble et qu'il nous appelle aussi par notre prénom.	D'ailleurs, nous retrouvons la même distinction dédaigneuse à l'égard des professionnels et de leur " vil salaire " qui ne les empêche pas de mourir " ès hôpitaux ", chez le docte Muret.
Yes, Mr. Adrien Richard, if you like better, the director of the factory, but we only call him Mr. Adrien because we went to school together and he also calls us our first name.	*Moreover, we find the same disdainful distinction with regard to professionals and their "vile salary" which does not prevent them from dying "in hospitals", at the doctor Muret's.*

7 Conclusion

In this paper, we have presented a semi-supervised process that jointly builds a text corpus labeled in language registers and an associated classifier. Based on a large set of text segments and a few initial expert resources, the result of this approach is a corpus of 825 K textual segments representing a total of about 750 M words. The classifier achieves a good accuracy of 87% on the test set, and more modest results on a manually labeled subset of the collected segments. These results seem to demonstrate the validity of the approach, while also highlighting the need for refinements and for a less stereotypical seed.

Among the lines of future work, questions about classification uncertainty and about the model's over-confidence will be dealt in priority. The use of scaled memberships (instead of binary ones), of an "undetermined" label, and of dropout or cross validation during training should help in these perspectives. An increased attention to false positives (e.g., within the objective function of the neural network) should also be paid. Moreover, the selection criterion for new segments could be improved, e.g., by combining the classifier's output probability with the probability to make an error. Finally, in the long term, advanced studies of language registers will be conducted based on the labeled corpus (discriminative features, local features instead of global ones, sequence models, etc.).

Aknowledgements. This work has benefited from the financial support of the French National Research Agency (ANR) through the TREMoLo project (ANR-16-CE23-0019).

A   Supplementary Material

The following supplementary material can be downloaded at ftp://ftp.cicling.org/in/CICLing-2019/CICLing_58.zip:

- Exhaustive list of features.
- CSV data (seed and web segments).
- Examples of raw seed and automatically labeled web texts.

References

1. Ure, J.: Introduction: approaches to the study of register range. Int. J. Sociol. Lang. **1982** (1982)
2. Biber, D., Conrad, S.: Register, Genre, and Style. Cambridge University Press, Cambridge (2009)
3. Charaudeau, P.: Le discours d'information médiatique: la construction du miroir social. Nathan (1997)
4. Borzeix, A., Fraenkel, B.: Langage et travail (communication, cognition, action). CNRS éd (2005)
5. Moirand, S.: Les discours de la presse quotidienne. Observer, analyser, comprendre. Puf (2007)
6. Sanders, C.: Sociosituational Variation. Cambridge University Press, Cambridge (1993)
7. Biber, D., Finegan, E.: Sociolinguistic Perspectives on Register. Oxford University Press on Demand, Oxford (1994)
8. Gadet, F.: Niveaux de langue et variation intrinsèque. Palimpsestes **10** (1996)
9. Peterson, K., Hohensee, M., Xia, F.: Email formality in the workplace: a case study on the enron corpus. In: Proceedings of the Workshop on Languages in Social Media (2011)
10. Pavlick, E., Tetreault, J.: An empirical analysis of formality in online communication. Trans. Assoc. Comput. Linguist. **4** (2016)
11. Sheikha, F.A., Inkpen, D.: Automatic classification of documents by formality. In: IEEE International Conference on Natural Language Processing and Knowledge Engineering (NLP-KE) (2010)
12. Stamatatos, E.: A survey of modern authorship attribution methods. J. Assoc. Inf. Sci. Technol. **60** (2009)
13. Iqbal, F., Binsalleeh, H., Fung, B.C., Debbabi, M.: A unified data mining solution for authorship analysis in anonymous textual communications. Inf. Sci. **231** (2013)
14. Schler, J., Koppel, M., Argamon, S., Pennebaker, J.W.: Effects of age and gender on blogging. In: Proceedings of the AAAI Spring Symposium: Computational Approaches to Analyzing Weblogs, vol. 6 (2006)

15. Kobus, C., Yvon, F., Damnati, G.: Normalizing SMS: are two metaphors better than one? In: Proceedings of COLING (2008)
16. Gianfortoni, P., Adamson, D., Rosé, C.P.: Modeling of stylistic variation in social media with stretchy patterns. In: Proceedings of the Workshop on Algorithms and Resources for Modelling of Dialects and Language Varieties (2011)
17. Eisenstein, J.: What to do about bad language on the internet. In: Proceedings of HLT-NAACL (2013)
18. Cougnon, L.A., Fairon, C.: SMS Communication: A Linguistic Approach, vol. 61. John Benjamins Publishing Company, Amsterdam (2014)
19. De Vel, O., Anderson, A., Corney, M., Mohay, G.: Mining e-mail content for author identification forensics. ACM Sigmod Rec. **30** (2001)
20. Sanderson, C., Guenter, S.: Short text authorship attribution via sequence kernels, markov chains and author unmasking: an investigation. In: Proceedings of EMNLP (2006)
21. Koppel, M., Schler, J.: Exploiting stylistic idiosyncrasies for authorship attribution. In: Proceedings of IJCAI Workshop on Computational Approaches to Style Analysis and Synthesis, vol. 69 (2003)
22. Argamon, S., Whitelaw, C., Chase, P., Hota, S.R., Garg, N., Levitan, S.: Stylistic text classification using functional lexical features. J. Assoc. Inf. Sci. Technol. **58** (2007)
23. McCarthy, P.M., Lewis, G.A., Dufty, D.F., McNamara, D.S.: Analyzing writing styles with Coh-Metrix. In: Proceedings of the FLAIRS Conference (2006)
24. Tambouratzis, G., Markantonatou, S., Hairetakis, N., Vassiliou, M., Carayannis, G., Tambouratzis, D.: Discriminating the registers and styles in the modern greek language-part 2: extending the feature vector to optimize author discrimination. Lit. Linguist. Comput. **19** (2004)
25. Hirst, G., Feiguina, O.: Bigrams of syntactic labels for authorship discrimination of short texts. Literar. Linguist. Comput. **22** (2007)
26. Sidorov, G., Velasquez, F., Stamatatos, E., Gelbukh, A., Chanona-Hernández, L.: Syntactic n-grams as machine learning features for natural language processing. Expert Syst. Appl. **41** (2014)
27. Marton, Y., Wu, N., Hellerstein, L.: On compression-based text classification. In: Proceedings of the European Conference on Information Retrieval (ECIR), vol. 3408 (2005)
28. Escalante, H.J., Solorio, T., Montes-y Gómez, M.: Local histograms of character N-grams for authorship attribution. In: Proceedings of HTL-ACL (2011)
29. McClosky, D., Charniak, E., Johnson, M.: Effective self-training for parsing. In: Proceedings of HLT-NAACL (2006)
30. He, Y., Zhou, D.: Self-training from labeled features for sentiment analysis. Inf. Process. Manag. **47** (2011)
31. Baroni, M., Bernardini, S.: BootCaT: bootstrapping corpora and terms from the web. In: Proceedings of LREC (2004)
32. Lecorvé, G., Gravier, G., Sébillot, P.: On the use of web resources and natural language processing techniques to improve automatic speech recognition systems. In: Proceedings of LREC (2008)
33. Gadet, F.: La variation, plus qu'une écume. Langue française (1997)
34. Gadet, F.: Is there a French theory of variation? Int. J. Sociol. Lang. **165** (2003)
35. Bilger, M., Cappeau, P.: L'oral ou la multiplication des styles. Langage et société (2004)

36. Ilmola, M.: Les registres familier, populaire et vulgaire dans le canard enchaîné et charlie hebdo: étude comparative. Master's thesis, University of Jyväskylä, Finland (2012)

Machine Translation

Evaluating Terminology Translation in MT

Rejwanul Haque[1]([✉]), Mohammed Hasanuzzaman[2], and Andy Way[3]

[1] Department of Computing, South East Technological University, Carlow, Ireland
{rejwanul.haque,mohammed.hasanuzzaman,andy.way}@adaptcentre.ie
[2] School of Computing, Munster Technological University, Cork, Ireland
[3] School of Computing, Dublin City University, Glasnevin, Dublin 9, Dublin, Ireland

Abstract. Terminology translation plays a crucial role in domain-specific machine translation (MT). Preservation of domain knowledge from source to target is arguably the most concerning factor for clients in translation industry, especially for critical domains such as medical, transportation, military, legal and aerospace. Evaluation of terminology translation, despite its huge importance in the translation industry, has been a less examined area in MT research. Term translation quality in MT is usually measured with domain experts, either in academia or industry. To the best of our knowledge, as of yet there is no publicly available solution to automatically evaluate terminology translation in MT. In particular, manual intervention is often needed to evaluate terminology translation in MT, which, by nature, is a time-consuming and highly expensive task. In fact, this is unimaginable in an industrial setting where customised MT systems are often needed to be updated for many reasons (e.g. availability of new training data or leading MT techniques). Hence, there is a genuine need to have a faster and less expensive solution to this problem, which could aid the end-users to instantly identify term translation problems in MT. In this study, we propose an automatic evaluation metric, *TermEval*, for evaluating terminology translation in MT. To the best of our knowledge, there is no gold-standard dataset available for measuring terminology translation quality in MT. In the absence of gold-standard evaluation test set, we semi-automatically create a gold-standard dataset from English–Hindi judicial domain parallel corpus.

We trained state-of-the-art phrase-based SMT (PB-SMT) and neural MT (NMT) models on two translation directions: English-to-Hindi and Hindi-to-English, and use TermEval to evaluate their performance on terminology translation over the created gold-standard test set. In order to measure the correlation between TermEval scores and human judgments, translations of each source terms (of the gold-standard test set) is validated with human evaluator. High correlation between TermEval and human judgements manifests the effectiveness of the proposed terminology translation evaluation metric. We also carry out comprehensive manual evaluation on terminology translation and present our observations.

Keywords: Terminology translation · Machine translation · Neural machine translation

© Springer Nature Switzerland AG 2023
A. Gelbukh (Ed.): CICLing 2019, LNCS 13451, pp. 495–520, 2023.
https://doi.org/10.1007/978-3-031-24337-0_35

1   Introduction

Terms are productive in nature, and new terms are being created all the time. A term could have multiple meanings depending on the context in which it appears. For example, words "terminal" ('a bus terminal' or 'terminal disease' or 'computer terminal') and "play" ('play music' or 'plug and play' or 'play football' or 'a play') could have very different meanings depending on the context in which they appear. A polysemous term (e.g. terminal) could have many translation equivalents in a target language. For example, the English word 'charge' has more than twenty target equivalents in Hindi (e.g. 'dam' for 'value', 'bhar' for 'load', 'bojh' for 'burden'). While encountering a judicial document, translation of "charge" has to be a particular Hindi word: 'aarop'. The target translation could lose its meaning if one does not take the term translation and domain knowledge into account. So the preservation of domain knowledge from source to target is pivotal in any translation workflow (TW), and this is one of the customers' primary concerns in the translation industry. Naturally, translation service providers (TSPs) who use MT in production expect translations to be consistent with the relevant context and the domain in question. However, evaluation of terminology translation has been one of the least explored arena in MT research. No standard automatic MT evaluation metric (e.g. BLEU [36]) can provide much information on how good or bad a MT system is in translating domain-specific expressions. To the best of our knowledge, as of now no one has proposed any effective way to automatically evaluate terminology translation in MT. In industrial TWs, in general, TSPs hire manual experts relating to the concerned domain for identifying term translation problems in MT. Nevertheless, such human evaluation process is normally laborious, expensive and time-consuming task. Moreover, in an industrial setting, retraining of customer-specific MT engine from scratch is carried out quite often when a reasonable size of new training data pertaining to the domain and styles of that on which that MT system was built or a new state-of-the-art MT technique is available. In industry, carrying out human evaluation on term translation from scratch each time when a MT system is updated is an unimaginable task in a commercial context. This is an acute problem in industrial TW and renowned TSPs want it to be solved for their own interest. A suitable solution to the problem of terminology translation evaluation would certainly aid the MT users who want to quickly assess theirs MT systems in the matter of the domain-specific term translation, and be a blessing for the translation industry.

In this work, we propose an automatic evaluation metric, TermEval, to quickly assess terminology translation quality in automatic translation. With this, we aim to provide the MT users a solution to the problem of the terminology translation evaluation. The proposed automatic evaluation metric TermEval, we believe, would ease the problem of those MT users who often need to carry out evaluation on terminology translation. Since there is no publicly available gold-standard for term translation evaluation, we create a gold-standard evaluation test set from a legal domain data (i.e. judicial proceedings) following a semi-automatic terminology annotation strategy. We use our inhouse bilingual term annotation tool, *TermMarker* (cf. Sect. 4). In short, TermMarker marks source and target terms in either side of a test set, incorporate lexical and inflectional variations of the terms relevant to the context in which they appear, with exploiting the automatic terminology extraction technique of [18, 19]. The annotation technique needs little manual intervention to validate the term tagging and mapping, provided a rather noisy automatic bilingual terminology in annotation interface. In an industrial set-up, TSPs would view this method as an ideal and onetime solution to the one that we pointed out above since the annotation scheme is to be a cheaper and faster exercise and will result a reusable gold-standard in measuring term translation quality.

PB-SMT [31], a predictive modeling approach to MT, was the main paradigm in MT research for more than two decades. NMT [4,9,26,48,51], an emerging prospect for MT research, is an approach to automatic translation in which a large neural network (NN) is trained by deep learning techniques. Over the last six years, there has been incremental progress in the field of NMT to the point where some researchers are claiming to have parity with human translation [20]. Nowadays, NMT, despite being a relatively new addition to this field of research, is regarded as a preferred alternative to previous mainstream methods and represents a new state-of-the-art in MT research. We develop competitive PB-SMT and NMT systems with a less examined and low-resource language pair, English–Hindi. Hindi is a morphologically rich and highly inflected Indian language. Our first investigation is from a less inflected language to a highly inflected language (i.e. English-to-Hindi), and the second one is the other way round (i.e. Hindi-to-English). With this, we compare term translation in PB-SMT and NMT with a difficult translation pair involving two morphologically divergent languages. We use TermEval to evaluate MT system's performance on terminology translation on gold-standard test set. To check how TermEval correlates with the human judgements, the translation of each source term (of the gold-standard test set) is validated with human evaluators. We found that TermEval represents a promising metric for automatic evaluation of terminology translation in MT, with showing very high correlation with the human judgements. In addition, we demonstrate a comparative study on term translation in PB-SMT and NMT in two set-ups: automatic and manual. To summarize, our main contributions in this paper are as follows:

1. We semi-automatically create a gold-standard evaluation test set for evaluating terminology translation in MT. We demonstrate various linguistic issues and challenges in relation to the annotation scheme.
2. To the best of our knowledge, we are the first to propose an automatic evaluation metric for evaluating terminology translation in MT, namely TermEval. We test TermEval and found it be an excellent in quality while measuring term translation errors in MT.
3. We compare PB-SMT and NMT in terminology translation on two translation directions: English-to-Hindi and Hindi-to-English. We found that NMT is less error-prone than PB-SMT with regard to the domain-specific terminology translation.

The remainder of the paper is organised as follows. In Sect. 2, we discuss related work. Section 3 describes our MT systems used in our experiments. In Sect. 4, we present how we created gold-standard dataset and examine challenges in relation to the termbank creation process. In Sect. 5, we report our evaluation plan and experimental results. Section 6 describes discussion and analysis while Sect. 7 concludes, and provides avenues for further work.

2   Related Work

2.1   Terminology Annotation

Annotation techniques have been widely studied in many areas of natural language processing (NLP). To the best of our knowledge, no one has explored the area in relation to the term annotation in corpus barring [38] who investigated term extraction, tagging and mapping techniques for under-resourced languages. They mainly present methods for term extraction, term tagging in documents, and bilingual term mapping from comparable corpora for four under-resourced languages: Croatian, Latvian, Lithuanian, and Romanian. The paper primary focused on acquiring bilingual terms from comparable Web crawled narrow domain corpora similar to the study of [18, 19] who automatically create bilingual termbank from parallel corpus.

In our work, we select a test set from a (judicial) domain parallel corpus, and semi-automatically annotate the test set sentences by marking source and target terms in either side of that and incorporating lexical and inflectional variations of the terms relevant to the context in which they appear. For annotation we took support from a rather noisy bilingual terminology that was automatically created from juridical corpus. For automatic bilingual term extraction we followed the approach of [18, 19].

2.2   Term Translation Evaluation Method

As far as measuring terminology translation quality in MT is concerned, researchers or end-users (e.g. translation industry) generally carry out manual evaluation with domain experts. Farajian et al. [13] proposed an automatic terminology translation evaluation metric which computes the proportion of terms in the reference set that are correctly translated by the MT system. This metric looks for source and target terms in the reference set and translated documents, given a termbank. There is a potential problem with this evaluation method. There could be a possible instance where a source term from the input sentence is incorrectly translated into the target translation, and the reference translation of the source term spuriously appears in the translation of a different input sentence. In such cases, the above evaluation method would make hit counts which are likely to be incorrect. In addition to the above problem, there are two more issues that [13] cannot address, which we discuss in Sect. 4.4 where we highlight the problem with the translation of ambiguous terms and in Sect. 4.2 where we point out the consideration of the lexical and inflectional variations for a reference term, given the context in which the reference translation appears.

In order to evaluate the quality of the bilingual terms in MT, Arčan et al. [2] manually created a terminology gold-standard for the IT domain. They hired annotators with a linguistic background to mark all domain-specific terms in the monolingual GNOME and KDE corpora [49]. Then, the annotators manually created a bilingual pair of two domain-specific terms found in a source and target sentence, one being the translation of the other. This process resulted in the identification of 874 domain-specific bilingual terms in the two datasets [1]. The end goal (i.e., evaluating the quality of term translation in MT) of their manual annotation task was identical to that of this study. However, our annotation task is a semi-automatic process that helps create a terminology gold-standard more quickly. In this work, we intent to ease this problem with proposing an automatic term translation evaluation metric. In short, the annotation task takes support from a bilingual terminology that is automatically created from a bilingual domain corpus. For automatic bilingual term extraction, we followed the approach of [18,19]. In this context, an obvious challenge in relation to the term annotation task is that there is a lack of a clear definition of terms (i.e., what entities can be labelled as terms [37]). While it is beyond the scope of this article to discuss this matter, the various challenges relating to terminology annotation, translation and evaluation will be presented in more detail.

2.3 PB-SMT Versus NMT: Terminology Translation & Evaluation

Since the introduction of NMT to MT community, researchers investigate to what extents and in what aspects NMT are better (or worse) than PB-SMT. In a quantitative comparative evaluation, [25] report performance of PB-SMT and NMT across fifteen language pairs and thirty translation directions on the United Nations Parallel Corpus v1.0 [56] and show that for all translation directions, NMT is either on par with or surpasses PB-SMT. We refer interested reader [50] for more works in this direction. In a nutshell, most of the studies in this direction show that the NMT systems provide better translation quality (e.g. more fluent, less lexical, reordering and morphological errors) than the PB-SMT systems.

We now turn our particular attention to the papers that looked into terminology translation in NMT. Burchardt et al. [7] conduct a linguistically driven fine-grained evaluation to compare rule-based, phrase-based and neural MT engines for English–German based on a test-suite for MT quality, confirming the findings of previous studies. In their German-to-English translation task, PB-SMT, despite reaching the lowest average score, is the best-performing system on named-entities and terminology translation. However, when tested on reverse translation direction (i.e. English-to-German), a commercial NMT engine becomes the winner as long as the term translation is concerned. In a similar experimental set-up, Macketanz et al. [34] report that their PB-SMT system outperforms NMT system on terminology translation on both in-domain (IT domain) and general domain test suites in an English-to-German translation task. Specia et al. [46] carried out an error annotation process using the Multidimensional Quality Metrics error annotation framework [33] on MT PE environment. The list of errors is divided into three main categories: accuracy, fluency and terminology. According to the annotation results, terminology-related errors are found more in NMT translations than in PB-SMT translations in English-to-German task (139 vs 82), and other way round in English-to-Latvian task (31 vs 34). Beyer et al. [5], from their manual evaluation procedure, report that PB-SMT outperforms NMT on term translation, which they speculate could be because their technical termbank was part of the training data used for building their PB-SMT system. Špela [53] conducts an automatic and small-scale human evaluation on terminology translation quality of Google Translator NMT model [54] compared to its earlier PB-SMT model for Slovene–English language pair and in the specialised domain of karstology. The evaluation result of Slovene-to-English task confirms NMT is slightly better than PB-SMT in terminology translation, while the opposite direction (i.e. English-to-Slovene task) shows a reversed picture with PB-SMT outperforming NMT. Špela [53] carries out a little bit qualitative analysis, with counting terms that are dropped in target translations and detailing instances where MT systems often failed to preserve the domain knowledge. More recently, we investigated domain term translation in PB-SMT and NMT on English–Hindi translation tasks [16,17] and found that the NMT systems commit fewer lexical, reordering and morphological errors than the PB-SMT systems. The opposite picture is observed in the case of term omission in translation, with NMT omitting more terms in translation than PB-SMT.

As far as terminology translation quality evaluation in PB-SMT alone is concerned, given its relatively a longer history, there has been many papers that has investigated this problem. For an example, [23] investigated term translation in a PB-SMT task and observed that more than 10% of high-frequency terms are incorrectly translated by their PB-SMT decoder, although the system's BLEU [36] score is quite high, i.e. 63.0 BLEU. One common thing in the papers [5,7,23,34,46,53], above is that they generally carried out subjective evaluation in order to measure terminology translation quality in MT, which, as mentioned earlier, is a time-consuming and expensive task. In this paper, we propose an automatic evaluation strategy that automatically determines terminology translation errors in MT. We observe terminology translation quality in PB-SMT and NMT on two translation directions: English-to-Hindi and Hindi-to-English, and two set-ups: automatic and subjective evaluations.

3   MT Systems

3.1   PB-SMT System

For building our PB-SMT systems we used the Moses toolkit [30]. We used 5-gram LM trained with modified Kneser-Ney smoothing using KenLM toolkit [22]. For LM training we combine a large monolingual corpus with the target-side of the parallel training corpus. Additionally, we trained a neural LM with NPLM toolkit[1] [52] on the target-side of parallel training corpus alone. Our PB-SMT log-linear features include: (a) 4 translational features (forward and backward phrase and lexical probabilities), (b) 8 lexicalised reordering probabilities (*wbe-mslr-bidirectional-fe-allff*), (c) 2 5-gram LM probabilities (Kneser-Ney and NPLM), (d) 5 OSM features [12], and (e) word-count and distortion penalties. In our experiments word alignment models are trained using GIZA++ toolkit[2] [35], phrases are extracted following *grow-diag-final-and* algorithm of [31], Kneser-Ney smoothing is applied at phrase scoring, and a smoothing constant (0.8u) is used for training lexicalized reordering models. The weights of the parameters are optimized using the margin infused relaxed algorithm [8] on the development set. For decoding the cube-pruning algorithm [24] is applied, with a distortion limit of 12. We call the English-to-Hindi and Hindi-to-English PB-SMT systems EHPS and HEPS, respectively.

3.2   NMT System

For building our NMT systems we used the MarianNMT [25] toolkit. The NMT systems are Google transformer models [51]. In our experiments we followed the

---

[1] https://www.isi.edu/natural-language/software/nplm/.

[2] http://www.statmt.org/moses/giza/GIZA++.html.

recommended best set-up from [51]. The tokens of the training, evaluation and validation sets are segmented into sub-word units using the Byte-Pair Encoding (BPE) technique [14], which was proposed by [43]. Since English and Hindi are written in Roman and Devanagari scripts and have no overlapping characters, the BPE is applied individually on the source and target languages. We performed 32,000 join operations. Our training set-up is detailed below. We consider size of encoder and decoder layers 6. As in [51], we employ residual connection around layers [21], followed by layer normalisation [3]. The target embeddings and output embeddings are tied in output layer [41]. Dropout [15] between layers is set to 0.10 . We use mini-batches of size 64 for update. The models are trained with Adam optimizer [27], with learning-rate set to 0.0003 and reshuffling the training corpora for each epoch. As in [51], we also use the learning rate warm-up strategy for Adam. The validation on development set is performed using three cost functions: cross-entropy, perplexity and BLEU. The early stopping criteria is based on cross-entropy, however, the final NMT system is selected as per highest BLEU scores on the validation set. The beam size for search is set to 12.

Initially, we use parallel training corpus to build our English-to-Hindi and Hindi-to-English baseline transformer models. We translate monolingual sentences (cf. Table 1) with the baseline models and create source synthetic sentences [42]. Then, we append this synthetic training data to the parallel training data, and retrain the baseline models. We make our final NMT model with ensembles of 4 models that are sampled from the training run. We call our final English-to-Hindi and Hindi-to-English NMT systems EHNS and HENS, respectively.

3.3 Data Used

For experimentation we used the IIT Bombay English-Hindi parallel corpus[3] [32] that is compiled from a variety of existing sources, e.g. OPUS[4] [49]. That is why the parallel corpus is a mixture of various domains. For building additional language models (LMs) for Hindi and English we use the HindEnCorp monolingual corpus [6] and monolingual corpus from various sources (e.g. the European Parliamentary proceedings [29]) from the OPUS project, respectively. Corpus statistics are shown in Table 1. We selected 2,000 sentences (test set) for the evaluation of the MT systems and 996 sentences (development set) for validation from the Judicial parallel corpus (cf. Table 1) which is a juridical domain corpus (i.e. proceedings of legal judgments). The MT systems were built with the training set shown in Table 1 that includes the remaining sentences of the Judicial parallel corpus. In order to perform tokenisation for English and Hindi, we used the standard tokenisation tool[5] of the Moses toolkit.

3.4 PB-SMT Versus NMT

In this section, we present the comparative performance of the PB-SMT and NMT systems in terms of automatic evaluation metrics: BLEU, METEOR [11], TER [45], chrF3

---

[3] http://www.cfilt.iitb.ac.in/iitb_parallel/.

[4] http://opus.lingfil.uu.se/.

[5] https://github.com/moses-smt/mosesdecoder/blob/master/scripts/tokenizer/tokenizer.perl.

Table 1. Corpus statistics. English (EN), Hindi (HI).

English-Hindi parallel corpus			
	Sentences	Words (EN)	Words (HI)
Training set	1,243,024	17,485,320	18,744,496
(Vocabulary)		180,807	309,879
Judicial	7,374	179,503	193,729
Development set	996	19,868	20,634
Test set	2,000	39,627	41,249
Monolingual Corpus		Sentences	Words
Used for PB-SMT Language Model			
English		11M	222M
Hindi		10.4M	199M
Used for NMT Back Translation			
English		1M	20.2M
Hindi		903K	14.2M

[39] and BEER [47]. The BLEU, METEOR and TER are standard metrics which are widely used by the MT community. The chrF3 and BEER metrics, which are character-level n-gram precision-based measures, are proved to have high correlation with human evaluation. Note that TER is an error metric, which means lower values indicate better translation quality. We report evaluation results in Table 2.

Table 2. Performance of PB-SMT and NMT systems on automatic evaluation metrics.

	BLEU	METEOR	TER	chrF3	BEER
EHPS	28.8	30.2	53.4	0.5247	0.5972
EHNS	36.6	33.5	46.3	0.5854	0.6326
HEPS	34.1	36.6	50.0	0.5910	0.6401
HENS	39.9	38.5	42.0	0.6226	0.6644

As can be seen from Table 2, EHPS and EHNS produce reasonable BLEU scores (28.8 BLEU and 36.6 BLEU) on the test set given a difficult translation pair. These BLEU scores, in fact, underestimate the translation quality, given the relatively free word order in Hindi, providing just single reference translation set for evaluation. Many TSPs consider 30.0 BLEU score as a benchmarking value and use those MT systems in their TW that produce BLEU scores above the benchmarking value. For example, [44] successfully used an English-to-Latvian MT system with a similar BLEU score

(35.0 BLEU) in SDL Trados CAT tool.[6] In this perspective, EHPS is just below par and EHNS is well above the benchmarking value.

As far as the Hindi-to-English translation task is concerned, HEPS and HENS produce moderate BLEU scores (34.1 BLEU and 39.9 BLEU) on the test set. As expected, MT quality from the morphologically-rich to morphologically-poor language improves. The differences in BLEU scores of PB-SMT and NMT systems in both the English-to-Hindi and Hindi-to-English translation tasks are statistically significant [28]. This trend is observed with the remaining evaluation metrics.

4   Creating Gold-Standard Evaluation Set

This section describes a technique that semi-automatically creates a gold-standard evaluation test data for evaluating terminology translation in MT. Our proposed automatic evaluation metric, TermEval, is based on the gold-standard test set. To exemplify, we present the semi-automatic test data creation technique for the English–Hindi language pair below. For evaluating term translation with our MT systems (cf. Sect. 3) we use the test set (cf. Table 1) that contains 2,000 sentence-pairs and is from judicial domain. We annotated the test set by marking term-pairs on the source- and target-sides of the test set. The annotation process is accomplished with our own bilingual term annotation tool, TermMarker. It is an user-friendly GUI developed with PyQT5.[7] We have made TermMarker publicly available to the research community via a software repository.[8] The annotation process starts with displaying a source–target sentence-pair from the test set at TermMarker's interface. If there is a source term present in the source sentence, its translation equivalent (i.e. target term) is found in the target sentence, then the source–target term-pair is marked. The annotation process is simple, which is carried out manually. The annotator, who is a native Hindi evaluator with excellent English skills, is instructed to mark those words as terms that belong to legal or judicial domains. The annotator was also instructed to mark those sentence-pairs from the test set that contains errors (e.g. mistranslations, spelling mistakes) in either source or target sentences. The annotator reported 75 erroneous sentence-pairs which we discard from the test set. In addition to this, there are 655 sentence-pairs of the test set that do not contain any terms. We call the set of remaining 1,270 sentence-pairs *gold-testset*. Each sentence-pair of gold-testset contains at least one aligned source-target term-pair. We have made the gold-testset publicly available to the research community.[9]

4.1   Annotation Suggestions from Bilingual Terminology

TermMarker supports annotation suggestions from an external terminology, if supplied. We recommend this option for faster annotation, although this is optional. For example, in our case, while manually annotating bilingual terms in the judicial domain test set

---

[6] https://en.wikipedia.org/wiki/SDL_Trados_Studio.

[7] https://en.wikipedia.org/wiki/PyQt.

[8] https://github.com/rejwanul-adapt/TermMarker.

[9] https://github.com/rejwanul-adapt/EnHiTerminologyData.

we took support from a rather noisy bilingual terminology that was automatically created from the Judicial corpus (cf. Table 1). For automatic bilingual term extraction we followed the benchmark approach of [18,19] which is regarded as the state-of-the-art terminology extraction technique and works well even on as few as 5,000 parallel segments. The user chooses one of the three options for an annotation suggestion: accept, skip and reject. The rejected suggestion is excluded from the bilingual terminology to make sure it never appears as the annotation suggestion in future. The newly marked term-pair is included in the bilingual termbank, which are to be used in annotation process.

Table 3. Statistics of occurrences of terms in gold-testset.

	Number of Source–Target Term-pairs	3,064
English	Terms with LIVs	2,057
	LIVs/Term	5.2
Hindi	Terms with LIVs	2,709
	LIVs/Term	8.4

In Table 3, we show statistics of occurrences of terms in the gold-standard evaluation set (i.e. gold-testset). We found 3,064 English terms and their target equivalents (3,064 Hindi terms) in source- and target-sides of gold-testset, respectively, i.e. the number of aligned English–Hindi term-pairs in gold-testset is 3,064. We observed presence of the nested terms (i.e. overlapping terms) in gold-testset, e.g. 'oral testimony' and 'testimony', 'pending litigation' and 'litigation', 'attesting witness' and 'witness'. In nested terms, we call a higher-gram overlapping term (e.g. 'oral testimony') a *superterm*, and a lower-gram overlapping term (e.g. 'testimony') a *subterm*. A nested term may have more than one subterm, but it has to have only one superterm. TermMarker allows us to annotate both subterms and superterms.

4.2 Variations of Term

A term could have more than one domain-specific translation equivalent. The number of translation equivalents for a source term could vary from language to language depending on the morphological nature of the target language. For example, translation of the English word 'affidavit' has multiple target equivalents (lexical and inflectional variations) in Hindi if the translation domain is legal or juridical: 'shapath patr', 'halaphanaama', 'halaphanaame', 'halaphanaamo'. The term 'shapath patr' is the lexical variation of Hindi term 'halaphanaama'. The base form 'halaphanaama' could have many inflectional variations (e.g. 'halaphanaame', 'halaphanaamo') given the sentence's syntactic and morphological profile (e.g. gender, case). In similar contexts, translation of the English preposition 'of' has multiple variations (postpositions) ('ka', 'ke') in Hindi. For this, an English term 'thumb impression' may have many translations in Hindi, e.g. 'angoothe **ka** nishaan' and 'angoothe **ke** nishaan', where 'angoothe' means 'thumb' and 'nishaan' means 'impression'.

For each term we check whether the term has any additional variations (lexical or inflectional) pertaining to the juridical domain and relevant to the context of the sentence. If this is the case, we include the relevant variations as the alternatives of the term. The idea is to create termbank as exhaustive as possible. In Table 3, we report the number of English and Hindi terms for which we added lexical and inflectional variations (LIVs), and the average number of variations per such term. As expected, both the numbers are higher in Hindi than English.

During annotation, the user can manually add relevant variations for a term through TermMarker's interface. However, we again exploit the method of [18, 19] for obtaining variation suggestions for a term. The automatically extracted bilingual terminology of [18, 19] comes with the four highest-weighted target terms for a source term. If the user accepts an annotation suggestion (source–target term-pair) from the bilingual terminology, the remaining three target terms are considered as the variation suggestions of the target term. Like the case of an annotation suggestion above, the user chooses one of the three options for a variation suggestion: accept, skip and reject. The rejected variation is excluded from the bilingual terminology to make sure it never appears as a variation suggestion in future. The newly added variation is included in the bilingual terminology for future use. Note that TermMarker has also an option to conduct annotation in both ways (source-to-target and target-to-source) at the same time. For this, the user can optionally include a target-to-source bilingual terminology.

4.3   Consistency in Annotation

As pointed out in the sections above, new term-pairs and variations of terms are often added to the terminology at the time of annotation. This may cause inconsistency in annotation since new term-pairs or variations could be omitted for annotation in the preceding sentences that have already been annotated. In order to eliminate the inconsistency problem, TermMarker includes a *checkup module* that traces annotation history, mainly with storing rejected and skipped items. The checkup module notifies the human annotator when any of the preceding sentences has to be annotated for a newly included term-pair or variation of a term to the termbank.

On completion of the annotation process, a set of randomly selected 100 sentence-pairs from gold-testset were further annotated for agreement analysis. Inter-annotator agreement was computed using Cohen's kappa [10] at word-level. This means for a multi-word term we consider number of words in it for this calculation. For each word we count an agreement whenever both annotators agree that it is a term (or part of term) or non-term entity. We found the kappa coefficient to be very high (i.e. 0.95) for the annotation task. This indicates that our terminology annotation is to be excellent in quality.

4.4   Ambiguity in Terminology Translation

One can argue that term annotation can be accomplished automatically if a bilingual terminology is available for the target domain. If we automatically annotate a test set with a given terminology, the automatic annotation process will likely to introduce noise into the test set. As an example, the translation of an English legal term 'case' is 'mamla' in

Hindi. The translation of the word 'case' could be the same Hindi word (i.e. 'mamla') even if the context is not legal. A legal term 'right' can appear in a legal/juridical text with a completely different context (e.g. fracture of right elbow). The automatic annotation process will ignore the contexts in which these words ('case', 'right', 'charge') belong and incorrectly mark these ambiguous words as terms. The automatic annotation process will introduce even more noise while adding variations for a term from a termbank or other similar sources (e.g. dictionary) for the same reason.

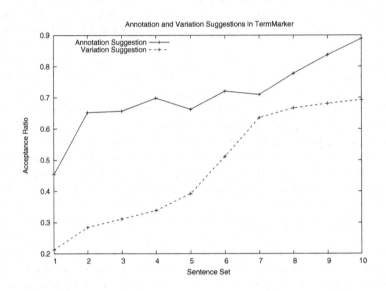

Fig. 1. Curves for acceptance ratio of suggestions.

4.5   Measuring Performance of TermMarker

We tested whether the use of annotation and variation suggestions from bilingual terminology [18, 19] makes the annotation process more faster. For this, we divided our judicial domain test set (2,000 segments, cf. Table 1) into 10 sets, each of which contains 200 segment-pairs. Note that we started our annotation process on set 1 and ended on set 10. We count the number of total annotation and variation suggestions as well as the number of accepted annotation and variation suggestions over each set. We plot the ratio of these (e.g. accepted annotation suggestions/total annotation suggestions) against the segment set number in Fig. 1. The x-axis and y-axis of Fig. 1 represent the segment set number and acceptance ratio, respectively. We see from Fig. 1 that both curves (annotation and variation suggestions) move upward over time, i.e. the acceptance rate of suggestions increases over time. This is because anomalous entries are rejected from the terminology and new valid entries are added into the terminology all the time, which makes the annotation process iteratively faster.

5   Evaluating Terminology Translation in MT

5.1   Automatic Evaluation: TermEval

The proposed evaluation metric starts evaluation process with forming a tuple with each source sentence from the test set, its translation (i.e. hypothesis), the list of source terms appearing in the source sentence, their reference translations (i.e. terms), and LIVs of the reference terms. First, we look for the reference terms (or the LIVs of the reference terms) in the hypothesis.[10] If a reference term (or one of its LIVs) is found in the hypothesis, this indicates that the MT system has correctly translated the source term into the target language. If this is not the case, there is likely to be an error in that term translation. At the end of the iteration (over all tuples), we obtain the TermEval score. The evaluation method includes an additional procedure that takes nested overlapping terms and multiple identical reference terms (or variations) into account. More formally, TermEval is calculated provided the test and translation sets using (1):

$$
\text{TermEval} = \frac{\sum_{n=1}^{N} \sum_{s=1}^{S} \sum_{v=1}^{V} \begin{cases} 1 & \text{if } R_v \in \text{Hyp}_n; \ break; \\ 0 & \text{otherwise} \end{cases}}{\text{NT}}
\tag{1}
$$

where: N : number of sentences in the test set
S : number of source terms in the nth source sentence
V : number of reference translations (including LIVs) for s^{th} the source term
R_v : v^{th} reference term for s^{th} the source term
Hyp_n : translation of nth source input sentence
NT : total number of terms in the test set

5.2   Term Translation Accuracy with TermEval

In this section we obtain evaluation results to evaluate the performance of the MT models on terminology translation using gold-testset that includes 3,064 source–target term-pairs. The terminology translation quality is measured using the TermEval metric (cf. (1)), and the scores are reported in Table 4. For clarity, we also report the total number of correct translation (CT) in the table, which, in fact, is the numerator of the right side of (1). As can be seen from the table, EHPS and EHNS correctly translate 2,610 and 2,680 English terms (out of total 3,064 terms), respectively, into Hindi, resulting the

---

[10] Since Moses can supply word-to-word alignments with its output (i.e. translation) from the phrase table (if any), one can exploit this information to trace target translation of a source term in the output. However, there are few potential problems with the alignment information, e.g. there could be null or erroneous alignments. Note that, at the time of this work, the transformer models of MarianNMT could not supply word-alignments (i.e. attention weights). In fact, our intention is to make our proposed evaluation method as generic as possible so that it can be applied to the output of any MT system (e.g. an online commercial MT engine). This led us to abandon such dependency.

Table 4. PB-SMT vs NMT: terminology translation accuracy in TermEval.

	CT	TermEval
EHPS	2,610	0.852
EHNS	2,680	0.875
HEPS	2,554	0.834
HENS	2,540	0.829

TermEval scores of 0.852 and 0.875, respectively. We use approximate randomization [55] to test the statistical significance of the difference between two systems. We found that the difference between the scores is statistically significant. On the other direction (i.e. from Hindi-to-English), HEPS and HENS correctly translate 2,554 and 2,540 English terms (out of total 3,064 terms), respectively, into Hindi, resulting TermEval of 0.834 and 0.829, respectively. Unlike the above, we found that the difference between the scores are not statistically significant.

Table 5. Number of seen LIVs in translations.

	LIVs	%
EHPS	703	26.0
EHNS	665	24.5
HEPS	241	11.7
HENS	245	11.9

Proportion of LIVs Seen in Translation. We counted the number of instances where a source term is correctly translated into target translation and the translation-equivalent of that term is one of the variations of the reference term; the percentage with respect to the total number of reference terms which includes variations (cf. Table 3) in the gold-testset is shown in Table 5. We see from the table that these numbers are much higher in the English-to-Hindi PB-SMT task (26.0% and 24.5%) compared to the Hindi-to-English task (11.7% and 11.9%). Hindi is a morphologically rich and highly inflected language, and we see from Table 1 that the training set vocabulary size is much higher in Hindi compared to English. This could be the reason why the numbers are much higher in the English-to-Hindi task.

5.3 Manual Evaluation Method

This section presents our manual evaluation plan. As mentioned above, the sentences of gold-testset were translated with the English-to-Hindi and Hindi-to-English MT systems (cf. Sect. 3). Translations of the source terms of gold-testset were manually validated and classified into two categories: error and correct. This was accomplished with the human evaluators. The manual evaluation was carried out with a GUI that randomly displays a source sentence and its reference translation from gold-testset, and the automatic translation by one of the MT systems. For each source term the GUI highlights the source term and the corresponding reference term from source and reference sentences, respectively, and displays the LIVs of the reference term, if any. The GUI lists the correct and incorrect categories. The evaluator, a native Hindi speaker with the excellent English and Hindi skills, is instructed to follow the following criterion for evaluating the translation of a source term: (a) judge correctness/incorrectness of the translation of the source term in hypothesis and label it with an appropriate category listed on GUI, (b) do not need to judge whole translation instead look at the local context in which both source term and its translation belong to, and (c) take the syntactic and morphological properties of the source term and its translation into account. v The manual classification process was completed for all MT system types. We randomly selected additional 100 source terms for further classification. The idea is to measure agreement in manual classification. We considered the binary categories (correct or incorrect term translation) in calculation, i.e. we count an agreement whenever both evaluators agree that it is correct (or incorrect) term translation, with agreement by chance = 1/2. As far as the agreements on classification of terminology translations with the four MT systems are concerned, we found that the kappa coefficient for the binary classes ranges from 0.97 to 1.0. It is believed that a kappa coefficient between 0.6–0.8 represents substantial agreement, with anything above 0.8 indicating perfect agreement. In this sense, our manual term translation classification quality can be labeled as excellent.

Table 6. PB-SMT vs NMT: terminology translation accuracy (Manual). CT: number of correct translation, ACC: accuracy.

	CT	ACC
EHPS	2,761	0.901
EHNS	2,811	0.917
HEPS	2,668	0.870
HENS	2,711	0.884

5.4 Measuring Term Translation Accuracy from Manual Classification Results

Given the manual classification results above, we obtain the performance of our MT systems on terminology translation. We measure term translation accuracy given the number of correct term translations and the total number of terms in gold-testset, i.e. 3,064. As in above, we report the total number of correct translations (CT) in Table 6. For comparison, we also report accuracy (ACC) (i.e. this is measured as the fraction of the correct term translations over the total number of term translations) in Table 6. As can be seen from the table, EHPS and EHNS correctly translate 2,761 and 2,811 English terms (out of total 3,064 terms), respectively, into Hindi, resulting ACC of 0.901 and

0.917, respectively. We use approximate randomization to test the statistical significance of the difference between two systems. We found that the difference between the scores is statistically significant. In the case of the Hindi-to-English task, we see that HEPS and HENS correctly translate 2,668 and 2,711 English terms (out of total 3,064 terms), respectively, into Hindi, resulting ACC of 0.870 and 0.884, respectively. As in above, the difference between the scores is statistically significant.

Table 7. Number of overlaps in correct and incorrect translation in PB-SMT and NMT

	PB-SMT	NMT	PB-SMT ∩ NMT
English-to-Hindi Task			
Correct	2,761	2,811	2614
Incorrect	303	253	86
Hindi-to-English Task			
Correct	2,668	2,711	2,483
Incorrect	396	353	115

5.5 Overlapping Correct and Incorrect Term Translations

Toral and Sánchez-Cartagena [50] carried out a comparative analysis with measuring overlap between an output by NMT and another by PB-SMT systems. Likewise, we report the number of pairwise overlap, i.e. number of instances in which NMT and PB-SMT have identical manual term classification outcomes (correct). The last columns of Table 7 show the numbers of pairwise overlap. The number of overlapping instances under the incorrect type are 86 and 115 that are nearly one third of the total numbers of error committed by the PB-SMT and NMT systems alone, indicating majority of the errors in PB-SMT are complementary with those in NMT. This finding on terminology translation is corroborated with that of [40] who finds complementarity with the issues relating to the translations of NMT and PB-SMT.

5.6 Validating TermEval

This section presents how we measure the correlation of human evaluation and automatic evaluation in the matter of terminology translation in MT.

Contingency Table. Using the TermEval metric we obtained the number of correct (and incorrect) term translations by the MT systems, which were reported in Table 4. With the human evaluation we obtain the actual number of correct (and incorrect) term translations by the MT systems on the test set, which were shown in Table 6. Therefore, given the automatic and human evaluation results, it is straightforward for us to evaluate our proposed evaluation metric, TermEval. Hence, given the automatic and manual evaluation results, we create contingency tables for both the English-to-Hindi and

Table 8. Contingency tables.

English-to-Hindi Task

PB-SMT			NMT		
	2,610	454		2,680	384
2,761	2,602	159	2,811	2,677	134
303	8	295	253	3	250

Hindi-to-English Task

PB-SMT			NMT		
	2,554	510		2,540	524
2,668	2,554	114	2,711	2,540	171
396	0	396	353	0	353

Hindi-to-English tasks and show them is Table 8. As can be seen from Table 8, there are two contingency tables for each task, left side tables are for the PB-SMT tasks and the right side tables are for the NMT tasks. The first row and column of each table represent automatic and manual classes, respectively. Each row or column shows two numbers, denoting the correct and incorrect term translations by the MT systems. Thus, the numbers from the manual classes are distributed over the automatic classes, and the other way round. For an example, manual evaluator labels 2,761 term translations (out of 3,064 total source terms) as correct translations in the English-to-Hindi PB-SMT task, and these 2,761 term translations belong to 2 categories (correct: 2,602, and incorrect: 159) as per the automatic evaluation.

Measuring Accuracy. Give the contingency tables (cf. Table 8), we measure accuracy of TermEval in the English–Hindi translation task. For measuring accuracy we make use of three widely used metrics: precision, recall and F1. We report the scores in Table 9. As can be seen from Table 9, we obtain roughly similar scores in four translation tasks (i.e. ranging from F1 of 0.967 to F1 of 0.978), and generally very high precision and slightly low recall scores in all tasks. TermEval represents a promising metric as per the scores in Table 9. We provide our in-depth insights on the results from Table 9 below (cf. Sect. 6).

6   Discussion and Analysis

In this section, first we discuss the scenario in which TermEval labeled those term translations as correct, which, are, in fact, incorrect as per the human evaluation results (false positives, cf. Table 8). Then, we discuss the reverse scenario in which TermEval labeled those term translations as incorrect, which, are, in fact, correct as per the human evaluation results (false negatives, cf. Table 8).

Table 9. Accuracy of proposed automatic term translation evaluation method: precision, recall and F1 metrics.

	EHPS	EHNS
P	0.997	0.999
R	0.942	0.953
F1	0.968	0.975
	EHPS	EHNS
P	1.0	1.0
R	0.957	0.937
F1	0.978	0.967

6.1 False Positives

As can be seen from fifth row of Table 8, there are 8 and 3 false-positives in the English-to-Hindi PB-SMT and NMT tasks, respectively. This indicates that in each case TermEval labels the term translation as correct, because the corresponding reference term (or one of its LIVs) is found in hypothesis, although the manual evaluator labeled that term translation as incorrect. We verify these cases in translations with the corresponding reference terms and their LIVs. We found that in 8 cases out of 11 cases the LIV lists contain incorrect inflectional variations for the reference terms. In each of these cases, an inflectional variation of the reference term misfits in the context of the reference translation when used in the place of the corresponding reference term. These incidents can be viewed as annotation errors as these erroneous inflectional variations for the reference terms were included in gold-testset at the time of its creation. For the 3 remaining cases we found that the English-to-Hindi PB-SMT system made correct lexical choice for the source terms, although the meanings of theirs target-equivalents in the respective translation are different to those of the source terms. This can be viewed as a cross-lingual disambiguation problem. For an example, one of the three source terms is 'victim' (reference translation 'shikaar') and the English-to-Hindi PB-SMT system makes a correct lexical choice ('shikaar') for 'victim', although the meaning of 'shikaar' is completely different in the target translation, i.e. here, its meaning is equivalent to English 'hunt'. It would be a challenging task for any evaluation metric as was the case for TermEval to correctly recognise such term translation errors. We keep this topic as a subject of future work.

6.2 False Negatives

We see from Table 8 that the number of false negatives (e.g. 159 in the English-to-Hindi PB-SMT task) across all MT tasks are much higher than that of false positives. This is, in fact, responsible for the slightly worse recall scores (cf. Table 9). We point out below why TermEval failed to label such term translations (e.g. 159 terms in the English-to-Hindi PB-SMT task) as correct despite the fact those are correct translations as per the human evaluation results.

Reordering Issue. Here, first we highlight the word ordering issue in term translation. For an example, a Hindi source term "khand nyaay peeth ke nirnay" (English reference term: "division bench judgment") is correctly translated into the following English translation by the Hindi-to-English NMT system: "it shall also be relevant to refer to article 45–48 of the *judgment of the division bench*". Nevertheless, TermEval implements a simple word matching module that, essentially, failed to capture such word ordering at target translation. In Table 10, we report the number of instances where TermEval failed to distinguish those term translation in the PB-SMT and NMT tasks that contains all words of the reference term (or one of its LIVs) but in a order different to the reference term (or one of its LIVs). As can be seen from Table 10, these numbers are slightly high when the target language is English. In order to automatically capture a term translation whose word-order is different to that of the reference term (or one of its LIVs), we need to incorporate language-specific or lexicalised reordering rules into TermEval, which we intend to investigate in future.

Table 10. False negatives in the PB-SMT and NMT tasks when TermEval failed to distinguish reordered correct term translation.

False Negative [Due to Term Reordering]			
	PB-SMT	NMT	PB-SMT ∩ NMT
English-to-Hindi	4	7	4
Hindi-to-English	13	11	4

Inflectional Issue. We start the discussion with an example from the Hindi-to-English translation task. There is a source Hindi term 'abhikathan', its reference term is 'allegation', and a portion of the reference translation is 'an allegation made by the respondent ...'. The LIV list of the reference term includes two lexical variations for 'allegation': 'accusation' and 'complaint'. A portion of the translation produced by the Hindi-to-English NMT system is 'it was *alleged* by the respondent ...', where we see the Hindi term 'abhikathan' is translated into 'alleged', which is a correct translation-equivalent of Hindi legal term 'abhikathan'. TermEval failed to label the above term translation as correct due to the fact that its morpholgical form is different to that of the reference term. Here, we show one more example. Consider a source Hindi sentence 'sbachaav mein koee bhee *gavaah* kee kisee bhee apeel karanevaale ne jaanch nahee kee gaee hai' and the English reference translation 'no *witness* in defence has been examined by either of the appellants.' Here, 'gavaah' is a Hindi term and its English equivalent is 'witness'. The translation of the source sentence by the Hindi-to-English NMT system is 'no appeal has been examined by any of the *witnesses* in defence'. Here, the translation of the Hindi term 'gavaah' is 'witnesses' which is correct as per the context of the target translation. Again, TermEval failed to trace this term translation. In Table 11, we report number of instances (i.e. false negatives) in the English–Hindi PB-SMT and NMT tasks, where TermEval failed to label term translations as correct due to the above reason. In Table 11, we see a mix bag of results, i.e. such instances are more seen in PB-SMT when target is Hindi and the other way round (i.e. more seen in NMT) when target is English.

Table 11. False Negatives in the PB-SMT and NMT tasks when TermEval fails to capture a correct term translation whose morphological form is different to the reference term (or one of its LIVs).

False Negative [Due to Inflectional Issue]			
	PB-SMT	NMT	PB-SMT ∩ NMT
English-to-Hindi	112	87	31
Hindi-to-English	75	107	48

We recall the rule that we defined while forming LIV list for a reference term from Sect. 4.2. *We considered only those inflectional variations for a reference term that would be grammatically relevant to the context of the reference translation in which they would appear.* In practice, translation of a source sentence can be generated in numerous ways. It is possible that a particular inflectional variation of a reference term could be grammatically relevant to the context of the target translation, which, when replaces the reference term in the reference translation, may (syntactically) misfit in the context of the reference translation. This is the reason why the term annotator, at the time of annotation, did not consider such inflectional variation for a reference term. These cases are likely to be seen more with the morphologically-rich languages, e.g. Hindi. However, in our case, we see from Table 11 that this has happened with both Hindi and English. This is to be a challenging problem for any intelligent evaluation metric to address.

In this context, we mention the standard MT evaluation metric METEOR [11] that, to an extent, tackles the above problems (reordering and inflectional issues) with two special modules (i.e. paraphrase and stem matching modules). In future, we intend to investigate this problem in the light of terminology translation in MT.

Table 12. False Negatives in the PB-SMT and NMT translation tasks when TermEval fails to capture correct term translations for various reasons.

False Negative [Due to Miscellaneous Reasons]			
	PB-SMT	NMT	PB-SMT ∩ NMT
English-to-Hindi	8	4	–
Hindi-to-English	2	20	–

Miscellaneous Reasons. In Table 12, we report the number of false negatives in PB-SMT and NMT tasks when TermEval fails to capture a correct term translation for various reasons. There are mainly four reasons:

- Reordering and inflectional issue: this can be viewed as the combination the above two types: 'reordering and inflectional issues'. In short, the translation of a source term contains all words of the reference term (or one of its LIVs) but in a order different to the reference term (or one of its LIVs) (i.e. 'reordering issue' in above) and one or more words of that translation include inflectional morphemes which are different to those of the words of the reference term (or one of the LIVs) (i.e. 'inflectional issue').
- Term transliteration: translation-equivalent of a source term is the transliteration of the source term itself. We observed that this happened only when the target language is Hindi. In practice, many English terms (transliterated form) are often used in Hindi text (e.g. 'tariff orders', 'exchange control manual').
- Term co-refereed: translation-equivalent of a source term is not found in hypothesis, however, it is rightly co-refereed in target translation.
- Semantically coherence translation: translation-equivalent of a source term is not seen in hypothesis, however, its meaning is rightly transferred into target. For an example, consider the source Hindi sentence "sabhee apeelakartaon ne aparaadh sveekaar nahin kiya aur muqadama chalaaye jaane kee maang kee" and reference sentence "all the appellants pleaded not guilty to the charge and claimed to be tried". Here, 'aparaadh sveekaar nahin' is a Hindi term and its English translation is 'pleaded not guilty'. The Hindi-to-English NMT system produces the following English translation "all the appellants did not accept the crime and sought to run the suit." for the source sentence. In this example, we see the meaning of source term 'aparaadh sveekaar nahin' is preserved at target translation.

Table 13. False negatives in the PB-SMT and NMT tasks when TermEval fails to capture term translations as correct due to the absence of the appropriate LIVs in gold-testset.

False Negative [Due to Missing LIVs]			
	PB-SMT	NMT	PB-SMT ∩ NMT
English-to-Hindi	33	36	10
Hindi-to-English	24	33	5

Missing LIVs. In few cases, we found that a source term is correctly translated into target, but the translation is neither the reference term nor any of its LIVs. These can be viewed as annotation mistake since annotators missed to add relevant LIVs for the reference term into the gold-testset. In Table 13, we report number of false negatives in PB-SMT and NMT tasks when TermEval fails to capture correct term translations due to this reason.

7 Conclusion

In this study, we proposed an automatic evaluation metric for evaluating terminology translation in MT, TermEval. Due to the unavailability of the gold-standard for evaluating terminology translation in MT, we adopted a technique that semi-automatically created a gold-standard test set from English–Hindi judicial domain parallel corpus. We found that TermEval represents a promising metric for the automatic evaluation of terminology translation in MT, while showing very high correlation with the human judgements. We examined why the automatic evaluation technique failed to distinguish term translation in few cases, and identified reasons (e.g. reordering and inflectional issues in term translation) for such aberration.

We also demonstrated our observations on term translation in state-of-the-art PB-SMT and NMT in two evaluation settings: automatic and manual. In manual evaluation, we found that NMT is less error-prone than PB-SMT (0.901 versus 0.917 and 0.870 versus 0.884 ACC (accuracy) scores in English-to-Hindi and Hindi-to-English translation tasks, respectively; differences in scores are statistically significant). We also found that the majority of the term translation errors by the PB-SMT systems are complementary with those by the NMT systems.

In this work, we also created a gold-standard test set that can be regarded as an important language resource in MT research. The gold-standard test set can also be used for the evaluation of a related natural language processing tasks, e.g. terminology extraction. This can also serve itself as a test-suite for automatic monolingual and bilingual term annotation tasks. We demonstrated various linguistic issues and challenges while creating our gold-standard data set, which could provide insights for such annotation scheme.

In future, we aim to make our gold-standard evaluation test set as exhaustive as possible with adding missing LIVs and correcting erroneous LIVs (cf. Sect. 5.6) of the reference terms. We also intend to incorporate lexical rules in our automatic term evaluation metric, which can help raise its accuracy. We plan to test our evaluation technique with different language pairs and domains.

Acknowledgments. The ADAPT Centre for Digital Content Technology is funded under the Science Foundation Ireland (SFI) Research Centres Programme (Grant No. 13/RC/2106) and is co-funded under the European Regional Development Fund.

References

1. BitterCorpus. https://hlt-mt.fbk.eu/technologies/bittercorpus. Accessed 28 Aug 2019
2. Arčan, M., Turchi, M., Tonelli, S., Buitelaar, P.: Enhancing statistical machine translation with bilingual terminology in a cat environment. In: Proceedings of the 11th Biennial Conference of the Association for Machine Translation in the Americas, pp. 54–68 (2014)
3. Ba, J.L., Kiros, J.R., Hinton, G.E.: Layer normalization. CoRR abs/1607.06450 (2016). https://arxiv.org/abs/1607.06450
4. Bahdanau, D., Cho, K., Bengio, Y.: Neural machine translation by jointly learning to align and translate. In: Proceedings of the 3rd International Conference on Learning Representations, pp. 1–15. San Diego, CA (2015)

5. Beyer, A.M., Macketanz, V., Burchardt, A., Williams, P.: Can out-of-the-box NMT beat a domain-trained Moses on technical data? In: Proceedings of EAMT User Studies and Project/Product Descriptions, pp. 41–46. Prague, Czech Republic (2017)
6. Bojar, O., et al.: Hindencorp - Hindi-English and Hindi-only corpus for machine translation. In: Proceedings of the Ninth International Conference on Language Resources and Evaluation, LREC, pp. 3550–3555 (2014)
7. Burchardt, A., Macketanz, V., Dehdari, J., Heigold, G., Peter, J.T., Williams, P.: A linguistic evaluation of rule-based, phrase-based, and neural MT engines. Prague Bull. Math. Linguist. **108**(1), 159–170 (2017)
8. Cherry, C., Foster, G.: Batch tuning strategies for statistical machine translation. In: Proceedings of the 2012 Conference of the North American Chapter of the Association for Computational Linguistics: Human Language Technologies, pp. 427–436. Association for Computational Linguistics, Montréal, Canada (2012)
9. Cho, K., van Merriënboer, B., Gülçehre, Ç., Bougares, F., Schwenk, H., Bengio, Y.: Learning phrase representations using RNN encoder-decoder for statistical machine translation. In: Proceedings of the 2014 Conference on Empirical Methods in Natural Language Processing (EMNLP), pp. 1724–1734. Doha, Qatar, October 2014
10. Cohen, J.: A coefficient of agreement for nominal scales. Educ. Psychol. Meas. **20**(1), 37–46 (1960)
11. Denkowski, M., Lavie, A.: Meteor 1.3: automatic metric for reliable optimization and evaluation of machine translation systems. In: Proceedings of the Sixth Workshop on Statistical Machine Translation, pp. 85–91. Association for Computational Linguistics, Edinburgh, Scotland, July 2011
12. Durrani, N., Schmid, H., Fraser, A.: A joint sequence translation model with integrated reordering. In: Proceedings of the 49th Annual Meeting of the Association for Computational Linguistics: Human Language Technologies, pp. 1045–1054. Association for Computational Linguistics, Portland, Oregon, USA, June 2011
13. Farajian, M.A., Bertoldi, N., Negri, M., Turchi, M., Federico, M.: Evaluation of terminology translation in instance-based neural MT adaptation. In: Proceedings of the 21st Annual Conference of the European Association for Machine Translation, pp. 149–158. Alicante, Spain (2018)
14. Gage, P.: A new algorithm for data compression. C Users J. **12**(2), 23–38 (1994)
15. Gal, Y., Ghahramani, Z.: A theoretically grounded application of dropout in recurrent neural networks. CoRR abs/1512.05287 (2016). https://arxiv.org/abs/1512.05287
16. Haque, R., Hasanuzzaman, M., Way, A.: Investigating terminology translation in statistical and neural machine translation: a case study on English-to-Hindi and Hindi-to-English. In: Proceedings of RANLP 2019: Recent Advances in Natural Language Processing, pp. 437–446. Varna, Bulgaria (2019)
17. Haque, R., Hasanuzzaman, M., Way, A.: Analysing terminology translation errors in statistical and neural machine translation. Mach. Transl. **34**(2), 149–195 (2020)
18. Haque, R., Penkale, S., Way, A.: Bilingual termbank creation via log-likelihood comparison and phrase-based statistical machine translation. In: Proceedings of the 4th International Workshop on Computational Terminology (Computerm), pp. 42–51. Dublin, Ireland (2014)
19. Haque, R., Penkale, S., Way, A.: TermFinder: log-likelihood comparison and phrase-based statistical machine translation models for bilingual terminology extraction. Lang. Resour. Eval. **52**(2), 365–400 (2018). https://doi.org/10.1007/s10579-018-9412-4
20. Hassan, H., et al.: Achieving human parity on automatic Chinese to English news translation, March 2018. ArXiv e-prints
21. He, K., Zhang, X., Ren, S., Sun, J.: Deep residual learning for image recognition. CoRR abs/1512.03385 (2015). http://arxiv.org/abs/1512.03385

22. Heafield, K., Pouzyrevsky, I., Clark, J.H., Koehn, P.: Scalable modified kneser-ney language model estimation. In: Proceedings of the 51st Annual Meeting of the Association for Computational Linguistics (Volume 2: Short Papers), pp. 690–696. Association for Computational Linguistics, Sofia, Bulgaria, August 2013

23. Huang, G., Zhang, J., Zhou, Y., Zong, C.: A simple, straightforward and effective model for joint bilingual terms detection and word alignment in SMT. Nat. Lang. Underst. Intell. Appl. ICCPOL/NLPCC 2016 **10102**, 103–115 (2016)

24. Huang, L., Chiang, D.: Forest rescoring: faster decoding with integrated language models. In: Proceedings of the 45th Annual Meeting of the Association of Computational Linguistics, pp. 144–151. Association for Computational Linguistics, Prague, Czech Republic, June 2007

25. Junczys-Dowmunt, M., Dwojak, T., Hoang, H.: Is neural machine translation ready for deployment? A case study on 30 translation directions. ArXiv e-prints (2016)

26. Kalchbrenner, N., Blunsom, P.: Recurrent continuous translation models. In: Proceedings of the 2013 Conference on Empirical Methods in Natural Language Processing (EMNLP), pp. 1700–1709. Seattle, WA, October 2013

27. Kingma, D.P., Ba, J.: Adam: a method for stochastic optimization. CoRR abs/1412.6980 (2014). http://arxiv.org/abs/1412.6980

28. Koehn, P.: Statistical significance tests for machine translation evaluation. In: Lin, D., Wu, D. (eds.) Proceedings of the 2004 Conference on Empirical Methods in Natural Language Processing (EMNLP), pp. 388–395. Association for Computational Linguistics, Barcelona, Spain, July 2004. http://acl.ldc.upenn.edu/acl2004/emnlp/pdf/Koehn.pdf

29. Koehn, P.: Europarl: a parallel corpus for statistical machine translation. In: Proceedings of MT Summit X: The Tenth Machine Translation Summit, pp. 79–86. Phuket, Thailand (2005)

30. Koehn, P., et al.: Moses: open source toolkit for statistical machine translation. In: ACL 2007, Proceedings of the Interactive Poster and Demonstration Sessions, pp. 177–180. Prague, Czech Republic (2007)

31. Koehn, P., Och, F.J., Marcu, D.: Statistical phrase-based translation. In: HLT-NAACL 2003: Conference Combining Human Language Technology Conference Series and the North American Chapter of the Association for Computational Linguistics Conference Series, pp. 48–54. Edmonton, AB (2003)

32. Kunchukuttan, A., Mehta, P., Bhattacharyya, P.: The IIT Bombay English-Hindi parallel corpus. CoRR 1710.02855 (2017). https://arxiv.org/abs/1710.02855

33. Lommel, A.R., Uszkoreit, H., Burchardt, A.: Multidimensional quality metrics (MQM): a framework for declaring and describing translation quality metrics. Tradumática: tecnologies de la traducció (12), 455–463 (2014)

34. Macketanz, V., Avramidis, E., Burchardt, A., Helcl, J., Srivastava, A.: Machine translation: phrase-based, rule-based and neural approaches with linguistic evaluation. Cybern. Inf. Technol. **17**(2), 28–43 (2017). https://content.sciendo.com/view/journals/pralin/108/1/article-p159.xml

35. Och, F.J., Ney, H.: A systematic comparison of various statistical alignment models. Comput. Linguist. **29**(1), 19–51 (2003)

36. Papineni, K., Roukos, S., Ward, T., Zhu, W.J.: Bleu: a method for automatic evaluation of machine translation. In: ACL-2002: 40th Annual Meeting of the Association for Computational Linguistics, pp. 311–318. ACL, Philadelphia, PA (2002)

37. Pazienza, M.T., Pennacchiotti, M., Zanzotto, F.M.: Terminology extraction: an analysis of linguistic and statistical approaches. In: Sirmakessis, S. (ed.) Knowledge Mining, vol. 185, pp. 255–279. Springer, Berlin, Heidelberg (2005). https://doi.org/10.1007/3-540-32394-5_20

38. Pinnis, M., Ljubešić, N., Ştefănescu, D., Skadiņa, I., Tadić, M., Gornostay, T.: Term extraction, tagging, and mapping tools for under-resourced languages. In: Proceedings of the 10th

Conference on Terminology and Knowledge Engineering (TKE 2012), pp. 193–208. Madrid, Spain (2012)

39. Popović, M.: chrF: character n-gram f-score for automatic MT evaluation. In: Proceedings of the Tenth Workshop on Statistical Machine Translation, pp. 392–395. Association for Computational Linguistics, Lisbon, Portugal, September 2015

40. Popović, M.: Comparing language related issues for NMT and PBMT between German and English. Prague Bull. Math. Linguist. **108**(1), 209–220 (2017)

41. Press, O., Wolf, L.: Using the output embedding to improve language models. CoRR abs/1608.05859 (2016). http://arxiv.org/abs/1608.05859

42. Sennrich, R., Haddow, B., Birch, A.: Improving neural machine translation models with monolingual data. CoRR abs/1511.06709 (2015). http://arxiv.org/abs/1511.06709

43. Sennrich, R., Haddow, B., Birch, A.: Neural machine translation of rare words with subword units. In: Proceedings of the 54th Annual Meeting of the Association for Computational Linguistics (Volume 1: Long Papers), pp. 1715–1725. Association for Computational Linguistics, Berlin, Germany, August 2016

44. Skadiņš, R., Puriņš, M., Skadina, I., Vasiļjevs, A.: Evaluation of SMT in localization to under-resourced inflected language. In: Proceedings of the 15th International Conference of the European Association for Machine Translation (EAMT 2011), pp. 35–40. Leuven, Belgium (2011)

45. Snover, M., Dorr, B., Schwartz, R., Micciulla, L., Makhoul, J.: A study of translation edit rate with targeted human annotation. In: In Proceedings of the 7th Biennial Conference of the Association for Machine Translation in the Americas (AMTA-2006), pp. 223–231. Cambridge, Massachusetts (2006)

46. Specia, L., et al.: Translation quality and productivity: a study on rich morphology languages. In: Proceedings of MT Summit XVI, the 16th Machine Translation Summit, pp. 55–71. Asia-Pacific Association for Machine Translation, Nagoya, Japan (2017)

47. Stanojević, M., Sima'an, K.: Beer: better evaluation as ranking. In: Proceedings of the Ninth Workshop on Statistical Machine Translation, pp. 414–419. Association for Computational Linguistics, Baltimore, Maryland, USA, June 2014

48. Sutskever, I., Vinyals, O., Le, Q.V.: Sequence to sequence learning with neural networks. In: Proceedings of the 27th International Conference on Neural Information Processing Systems, pp. 3104–3112. NIPS 2014, Montreal, Canada (2014)

49. Tiedemann, J.: Parallel data, tools and interfaces in OPUS. In: Proceedings of the 8th International Conference on Language Resources and Evaluation (LREC'2012), pp. 2214–2218. Istanbul, Turkey (2012)

50. Toral, A., Sánchez-Cartagena, V.M.: A multifaceted evaluation of neural versus phrase-based machine translation for 9 language directions. CoRR abs/1701.02901 (2017). http://arxiv.org/abs/1701.02901

51. Vaswani, A., et al.: Attention is all you need. CoRR abs/1706.03762 (2017). http://arxiv.org/abs/1706.03762

52. Vaswani, A., Zhao, Y., Fossum, V., Chiang, D.: Decoding with large-scale neural language models improves translation. In: Proceedings of the 2013 Conference on Empirical Methods in Natural Language Processing, pp. 1387–1392. Association for Computational Linguistics, Seattle, Washington, USA, October 2013

53. Vintar, V.: Terminology translation accuracy in statistical versus neural MT: an evaluation for the English-Slovene language pair. In: Du, J., Arcan, M., Liu, Q., Isahara, H. (eds.) Proceedings of the LREC 2018 Workshop MLP-MomenT: The Second Workshop on Multi-Language Processing in a Globalising World and The First Workshop on Multilingualism at the intersection of Knowledge Bases and Machine Translation, pp. 34–37. European Language Resources Association (ELRA), Miyazaki, Japan, May 2018

54. Wu, Y., et al.: Google's neural machine translation system: Bridging the gap between human and machine translation. CoRR abs/1609.08144 (2016). http://arxiv.org/abs/1609.08144

55. Yeh, A.: More accurate tests for the statistical significance of result differences. In: Proceedings of the 18th Conference on Computational Linguistics - Volume 2, COLING 2000, pp. 947–953. Saarbrücken, Germany (2000)

56. Ziemski, M., Junczys-Dowmunt, M., Pouliquen, B.: The united nations parallel corpus v1.0. In: Proceedings of the Tenth International Conference on Language Resources and Evaluation (LREC 2016). European Language Resources Association (ELRA), Portorož, Slovenia (2016)

Detecting Machine-Translated Paragraphs by Matching Similar Words

Hoang-Quoc Nguyen-Son[(✉)], Tran Phuong Thao, Seira Hidano,
and Shinsaku Kiyomoto

KDDI Research Inc., Saitama, Japan
{ho-nguyen,th-tran,se-hidano,kiyomoto}@kddi-research.jp

Abstract. Machine-translated text plays an important role in modern life by smoothing communication from various communities using different languages. However, unnatural translation may lead to misunderstanding, a detector is thus needed to avoid the unfortunate mistakes. While a previous method measured the naturalness of continuous words using a N-gram language model, another method matched noncontinuous words across sentences but this method ignores such words in an individual sentence. We have developed a method matching similar words throughout the paragraph and estimating the paragraph-level coherence, that can identify machine-translated text. Experiment evaluates on 2000 English human-generated and 2000 English machine-translated paragraphs from German showing that the coherence-based method achieves high performance (accuracy = 87.0%; equal error rate = 13.0%). It is efficiently better than previous methods (best accuracy = 72.4%; equal error rate = 29.7%). Similar experiments on Dutch and Japanese obtain 89.2% and 97.9% accuracy, respectively. The results demonstrate the persistence of the proposed method in various languages with different resource levels.

Keywords: Machine translation · Human-created paragraph · Coherence · Similar word matching

1 Introduction

Machine translation is the most vital assistance in communication between two persons comprehending different languages, so renowned international companies such as Facebook and Google integrate translators into text content including blogs, web-pages, and comments. While translation is increasingly developed for rich resource languages, especially in European community having a strong connection in economy and culture; the low resource languages such as Arabic, Pashto, and Dari are also initially investigated to prevent potential risks from criminal actions[1], for example, terrorism and kidnapping.

Although a machine preserves meaning in a translated text, the use of 'strange' words reduces readability. Figure 1 illustrates different quality of orig-

---

[1] https://www.pri.org/stories/2011-04-26/machine-translation-military.

© Springer Nature Switzerland AG 2023
A. Gelbukh (Ed.): CICLing 2019, LNCS 13451, pp. 521–532, 2023.
https://doi.org/10.1007/978-3-031-24337-0_36

Human-created paragraph p_H	"The third *idea that we have* is **instant** feedback. With **instant** feedback, the computer **grades** exercises. *I mean,* how else do you teach 150,000 students? **Your** computer *is* **grading** all *the* **exercises**. *And* we've all **submitted** homeworks, and *your grades* come back two weeks *later, you've* forgotten *all* about it. *I* *don't* *think* I've still **received** some of my **homeworks** *from* my *undergraduate days. ....*"
Machine-translated paragraph p_M	"The third **concept** is **called immediate** feedback. With **immediate** feedback, the computer **rates** *the* exercises. How else do you teach 150 000 students? **The** computer **evaluates** all **tasks**. We've all **done** homework and forgotten about it *during the* two-week *correction period*. I still do not **have** some of my first **chores** back...."

Fig. 1. Coherence of human-created vs machine-generated paragraph.

inal and translation[2] despite same content in each. A machine can correctly generate grammatical text, but the selection of vague words may result in misunderstanding, especially in the last sentence of the figure. The confusing can be reduced by recognizing and notifying translated text to readers.

Many methods have been published different approaches to detecting translation text. These approaches can be categorized by core techniques: parsing tree, N-gram model, word distribution, and word similarity. The first approach extracted distinguishable features from *parsing trees* [3,6], but such trees are only parsed from an individual sentence. To overcome this problem, other methods [1,2,8] based on *N-gram language model* extract such features from nearby words in both inside and outside a sentence. The limitation of this model is that meaningful features are only given from few nearby words, common in three. Other work [5,9] analyzes the histogram of *word distribution* from a massive amount of words, particularly suitable for document level. A recent method [10], the closest one to this paper, estimates text coherence by mutually matching words across pairwise sentences using *word similarity*. This method, however, ignores the connection in such words within a sentence.

The coherence of a human-generated paragraph is often higher than that of a machine-translated one. In Fig. 1, for instance, the high coherence makes the human-generated text more comprehensible. We analyze coherence by highlighting the difference of word usage in italic. A machine commonly uses quite different words, emphasized in bold, that affect the preservation of intrinsic meaning. For example, in the third sentence, *"tasks"* replaces *"exercises"* easily leading to misunderstanding. Moreover, the translation misses some subordinate words that are marked in underline in the man-made paragraph. According to Volanskey et al. [12], such words significantly improve text comprehension. Thus, the missing of the certain words, especially in the last sentence, makes the translated text more confusable.

---

[2] https://www.ted.com/talks/anant_agarwal_why_massively_open_online_courses_still _matter/transcript.

In this paper, we have proposed a method for matching similar words in a paragraph with maximum similarity. The similarity is then used to estimate coherence that can determine whether a paragraph is translated by a machine or created by a human.

We collected TED talk[3] transcripts written by native speakers and chose only transcripts aligned with both German and English. While English represents for human-generated text, the German is translated into English by Google to produce machine-generated one. The best translator, Google, can create not only the highest quality translation but also the most difficult to distinguish, as demonstrated in Aharoni et al.'s work [1]. We then randomly select 2000 paragraph pairs for conducting experiments. The results show that the coherence-based method accomplishes superior accuracy 87.0% and low equal error rate 13.0%. It surpasses previous methods with the best accuracy 72.4% and equal error rate 29.7%.

The coherence-based method also reaches the highest performances on Dutch and Japanese with similar experiments. It demonstrates the persistence of the proposed method in various languages. While Dutch has competitive results with German, Japanese even obtains impressive accuracy 97.9% and mere equal error rate 1.9%. It indicates lack of coherence on the low resource language. Based on this finding, translators can enhance text coherence by enriching linguistic resources.

In the rest of this paper, Sect. 2 outlines the main previous methods of machine-translated text detection. Section 3 describes a step-by-step guide to extract coherence features. Experiments on these features and comparison of the coherence-based with previous methods on various languages are shown and analyzed in Sect. 4. Finally, Sect. 5 summaries some main key points and mentions future work.

2 Related Work

Since machine-translated text detection is an important task of natural language processing, many researchers have involved suggesting useful solutions. The previous solutions can be grouped by the core usage including parsing tree, N-gram model, word distribution, and word similarity. Some main methods of each group are summarized in below.

2.1 Parsing Tree

In this approach, researchers aimed to extract detectable features from a parsing tree for use in machine-translated text identification. For example, Chae and Nenova [3] claimed that parsing of machine text is commonly simpler than that of human one. The authors indicated that a simple parsing often contains short main constituents, that is noun, verb, and adjective phrases. Following the

---

[3] https://www.ted.com/.

intuition, the authors extracted meaningful features, such as parsing tree depth, phrase type proportion, average phrase length, phrase type rate, and phrase length, before using them to distinguish computer- with human-generated text.

Li et al. [6] inherited several above features including parsing tree depth and phrase type proportion. In addition, they investigated that the structure of human parsing is more balancing than that of machine one. They thus suggested some useful features: the ratio of right- compared to left-branching nodes, the number of left-branching nodes for noun phrases. The main limitation of parsing-based methods is that they just generate a parsing tree for an individual sentence. An integrated tree cannot be built for larger scope of multiple sentences such as paragraph or document.

2.2   *N*-gram Model

To overcome the limitation of parsing-based approach, Arase and Zhou recommended another method [2] based on fluency estimation. They mainly used *N*-gram language model to estimate the fluency of continuous words. The restriction of this model is that it efficiently examines only on few continuous words, common in three. The authors reduced the deficiency by using sequential pattern mining to measure the fluency of non-continuous words. In-fluent patterns in human text are mined, such as *"not only * but also," "more * than,"* that contrast with that in machine-generated text, for example, *"after * after the," "and also * and."* There are two other reasonable combinations also aim to diminish the restriction of *N*-gram model. The first combination [8] extracted the specific noise words often used by a human, that is misspelled and reduction words, or by a machine, namely untranslated words. This combination, however, is only efficient in online social network in which contains a substantial number of such noises. The second combination [1] focused on functional words abundantly occurring in machine-translated text. Additional features in the three combinations achieve non-high performances but these features effectively improve the overall performances when they are integrated with the original *N*-gram model.

2.3   Word Distribution

Another approach recognizes machine-generated text by analyzing a histogram of word distribution. For example, Labbé and Labbé suggested an inter-textual metric for estimating the similarity of word distributions [5]. This metric is perfectly used for classifying artificial and real papers with accuracy up to 100%, but Nguyen-Son et al. [9] indicated that the inter-textual metric is just suitable for paper detection and they developed another method for translation detection also based on word distribution. This method pointed out that a word distribution of human text is closer with a Zipfian distribution than that of machine one. They also offered some valuable features to support the word distribution, that is specific phrases (e.g., idiom, cliché, ancient, dialect, phrasal verb) and co-reference resolution. The restriction of distribution-based methods is that they are only stable with a large number of words. However, the deficiency is

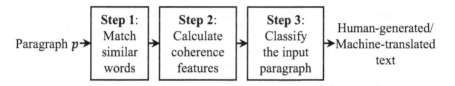

Fig. 2. Proposed schema for computer-generated paragraph detection.

revealed on homologous texts that refers to same sources such as paraphrasing and translation because such text mostly contains a same set of words.

2.4 Word Similarity

The closest method with our work was suggested by Nguyen-Son et al. [10]. They matched similar words in pairwise sentences of a paragraph. In two sentences, each word is only matched with another word at most so that total similarity of matching is maximum. We extend this idea by matching similar words in both internal and external sentences throughout a paragraph, so a word can be used as a bridge of other words in the text.

3   Proposed Method

3.1   Overview

The proposed schema distinguishes between machine-translated and human-generated paragraphs in three steps shown in Fig. 2.

- **Step 1** (*Match similar words*): Each word is matched with other words in the input paragraph p. The similarity of matched pairs is measured by Euclidean distance. The maximum similarity is distributed into disparity groups based on part of speech of the matched words.
- **Step 2** (*Calculate coherence features*): Mean and variance metrics are calculated for all similarity in each group. These metrics are used as features to estimate the coherence of p.
- **Step 3** (*Classify the input paragraph*): The coherence features are used to determine whether the input p is created by a human or is translated by a machine.

The three-step is presented in detail and demonstrated by the human-created paragraph p_H and the machine-translated one p_M in Fig. 1.

3.2   Detail

Matching Similar Words (Step 1). Words in the input paragraph p are separated and labeled with parts of speech (POS) using Stanford tagger [7]. A

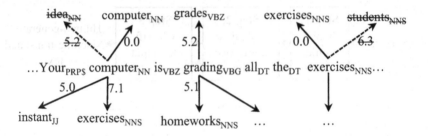

Fig. 3. Matching main words in human-generated paragraph p_H.

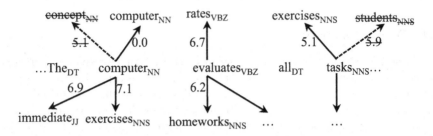

Fig. 4. Matching main words in machine-translated paragraph p_M.

word is then matched to other words, and each similarity of two matched words is measured. The similarity is estimated by the distance of two vectors on a word embedding. We use a common word embedding, GloVe [11], that is trained from Wikipedia 2014 merging with Gigaword 5 and produces 400 K vectors with 300 dimensions in each. The Euclidean is chosen here due to much wider distance comparing with Cosine. In Euclidean space, the distance of vectors is larger, so the difference of words is clearer. In there, the higher similarity of two words indicates the lower value of the distance. Some of matched pairs are plotted in Fig. 3 and Fig. 4 for the human paragraph p_H and the machine one p_M, respectively.

If a word is matched with other words having the same POSs, then the minimum distance is preserved. In Fig. 3, for example, a singular noun "computer"(NN) can be matched with two singular nouns, namely "idea" and "computer," having the similarities 5.2 and 0.0, respectively. The lower distance "computer-computer"(0.0) is chosen while the other matching is eliminated that is marked in strike-through.

As shown in the two figures, the similarity of a human-generated text tends lower than that of a machine-translated one. It demonstrates the high coherence of human-created passages, on the other hand, the use of low-coherent words reduces the overall coherence of the machine-translated text. For example, a pair "evaluates-rates" (6.7) in p_M causes to slightly drop the coherence when it is compared to a higher coherence pair "grading-grades" (5.2) in p_H. The difference also affects to other matching such as "grading-

1035 POS pairs

...	...	...	...	...	...	...	
...	0.0	7.1	5.0	5.2	5.1	0.0	...
...	NN-NN	NN-NNS	NN-JJ	VBG-VBZ	VBG-NNS	NNS-NNS	...

Fig. 5. Distributing similarities to part of speech (POS) groups.

homeworks" versus "*evaluates-homeworks.*" Similar cases occur in other pairs, for instance, { "*exercises-exercises,*" "*computer-instance*"} versus { "*exercises-tasks,*" "*computer-immediate*"} in human versus machine text, correspondingly. It is easy to confuse readers who possibly understand the meaning in various ways.

Calculating Coherence Features (Step 2). The similarity of the remaining pairs is distributed to certain groups based on their POSs. For example, while "*computer-computer*" (0.0) in p_H is allocated to NN-NN group, "*computer-exercises*" (7.1) is delivered to NN-NNS as shown in Fig. 5. The number of groups equals 1035 created from a list of 45 separate POSs.

Means and variances are calculated in each group for estimating the text coherence. 2070 values including 1035 means and 1035 variances are used to detect machine-translated paragraphs in the next step.

Classifying the Input Paragraphs (Step 3). The statistical values represent as coherence features to determine whether the input p is a computer- or a human-generated paragraph. These features are examined on three common machine learning classifiers that were chosen in state-of-the-art methods including linear classification [4], support vector machine (SVM) optimized by stochastic gradient descent (SGD), SVM optimized by sequential minimal optimization (SMO). The SVM (SMO) reaches the best performance, so it is selected as the final classifier.

4 Evaluation

4.1 Dataset

We collected 3088 English and 2253 German transcripts, that are composed by native speakers in TED talks and posted from June 2009 to November 2018. We then choose the transcripts existing in both English and German and aligned in paragraph-by-paragraph. While the rest English text is considered as a human creation, the German is translated into English by Google for generating machine-generated paragraphs. Finally, we randomly selected 2000 aligned pairs to conduct experiments. Each paragraph contains 14.4 sentences in average.

Table 1. Comparison with previous methods on accuracy (ACC) and equal error rate (ERR) metrics. The underline describes for the best classifiers, which are selected in the previous methods. The best values of each work are emphasized in bold, and the highest performance among them is highlighted by red.

Method		LINEAR		SGD(SVM)		SMO(SVM)	
		ACC	EER	ACC	EER	ACC	EER
Word distribution and coreref [9]		66.5%	33.4%	<u>66.6%</u>	**33.3%**	66.9%	33.3%
Parsing tree [6]		**67.9%**	<u>33.4%</u>	67.0%	34.4%	67.6%	**32.8%**
N-gram and functional words [1]		**69.5%**	**30.5%**	67.0%	32.9%	<u>69.3%</u>	<u>30.8%</u>
Word similarity [10]		**72.4%**	**29.7%**	69.6%	31.1%	70.9%	30.8%
Cosine	Mean	83.8%	15.5%	80.0%	23.5%	**84.6%**	**15.6%**
	Variance	67.8%	32.5%	70.3%	31.1%	**72.8%**	**27.3%**
	Combination	83.9%	17.0%	81.1%	21.3%	**85.4%**	**14.6%**
Euclidean	Mean	83.2%	19.0%	83.3%	18.5%	**85.6%**	**14.0%**
	Variance	75.7%	21.4%	76.3%	25.3%	**79.5%**	**20.7%**
	Combination	84.1%	16.8%	84.8%	15.4%	87.0%	13.0%

4.2   Comparison

The dataset is evaluated by 10-fold cross-validation on three common machine learning algorithms including linear classification (LINEAR), support vector machine (SVM) optimized by stochastic gradient descent SVM (SGD), or by sequential optimization SVM (SMO). These classifiers reached best performances on previous machine translation detection methods. Since F-measure and accuracy (ACC) are analogous to results, the only ACC is shown in this experiment. We also calculate equal error rate (EER) to test the persistence of each classifier. The coherence-based method using mean, variance, and their combination in both Cosine and Euclidean distance is compared with four previous methods on the same task. While three methods based on word distribution with coreference resolution (coreref) [9], N-gram model [1], and word similarity [10] can directly extract features from a paragraph, the other [6] based on parsing tree only obtains such features from an individual sentence. Thus, we adopt this method for a paragraph by calculating average on the features. The results of the comparison are shown in Table 1.

The accuracy in Table 1 is in harmony with EER that demonstrates the high persistence through classifiers. Through all previous methods, the highest performances (in bold) are identical or competitive to best classifiers indicated in italic with the maximum deviation only 0.6%. In these classifiers, the method based on word distribution and coreference resolution (coreref) [9] attained the lowest performance. The main reason is that the distribution is affected by a limited number of words within a paragraph. The parsing-based method [6] slightly improves the performance. However, the parsing can only be built from an individual sentence, so the relationship of words in cross-sentence is ignored. Another method [1] based on N-gram model and functional words are more suitable for paragraphs but this model is just efficient on few consecutive words.

On the other hand, a similarity-based method [10] exploits the extra connections among nonconsecutive words and accomplishes the current state-of-the-art performance (accuracy = 72.4%; EER = 29.7%).

The coherence-based method achieves the superior performances comparing with previous work through all classifiers. The Euclidean distance brings higher results due to the large diversity in measuring word similarity. Although mean is more appropriate to evaluate the text coherence than variance, the later one significantly supports to enhance the overall outcomes. Therefore, the combination archives topmost performances. The best performance (accuracy = 87.0%; EER = 13.0%) is obtained when using SMO(SVM) classifier. This classifier is thus chosen for further experiments on the coherence-based method while the other methods are evaluated on the best classifiers chosen in corresponding papers (underline in Table 1).

4.3   Individual Features

We examine the coherence-based method on each individual feature with the experimental results shown in Table 2. The performances are sorted by combination accuracy for finding most important features contributing to estimate the coherence. The outcomes demonstrate the mutual support between mean and variance in the combinations. In top pairs, the colon indicates the pivot role when it is combined with other POSs. Since the ':' rarely occurs in a machine-translated text, this mark significantly provides for recognizing the artificial translation. Because the mark is often used to explanation, the missing of colon causes reducing the clarity of the translated text comparing with the original version. The translators can upgrade the text coherence by integrating this mark into the translation.

Table 2. Performances of top five POS pairs.

Rank	POS pair	Mean	Variance	Combination
1	TO-:	48.3%	60.5%	72.0%
2	VBP-RB	61.2%	55.7%	68.1%
3	:-WP	60.9%	56.3%	66.6%
4	WRB-:	55.8%	59.1%	66.5%
5	PRP-RB	63.2%	49.8%	66.2%

4.4   Other Languages

We conduct similar experiments with two other languages including Dutch and Japanese. While Dutch is another rich resource like German, the Japanese is a low resource language. 2000 English paragraphs are chosen in each language for

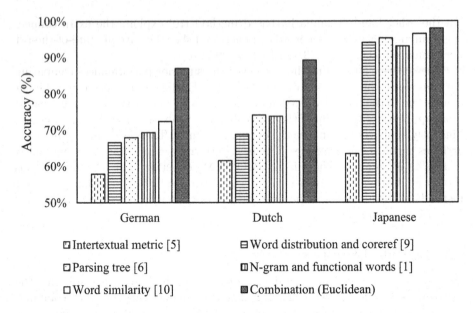

Fig. 6. Evaluation on various languages.

human-generated text. The aligned 2000 Dutch and 2000 Japanese paragraphs are translated into English by Google for producing machine text. The results of the experiments are plotted in Fig. 6. Because equal error rate associates with accuracy, the only accuracy is shown in the chart. We also evaluate on another method [5] based on word distribution that is recommended by Labbé and Labbé. While other methods extract features to run on machine learning classifiers, the distribution-based method suggested an inter-textual distance to measure the distance between two distributions.

In Fig. 6, although the inter-textual metric impressively recognizes an artificial document, it is insufficient to apply for lower granularity such as a paragraph. In other words, the method is degraded by a limited amount of words in a text. Therefore, this work archives almost same performances due to similar word distribution on these languages. On the other hand, other methods measure the text fluency, so the performances are obviously changed between low and rich resource languages. In remaining methods, while the performances on Dutch are similar to German, evaluation on Japanese archives notable improvement. It demonstrates the significant impact of resource level on the fluency measurement. Moreover, the grammar structure is also another important aspect. English uses a common structure SVO, i.e. a subject follows by a verb and an objective, but most of Japanese sentences are SOV. In previous methods, the approach [10] based on word similarity is the most stable and reaches higher performances.

The coherence-based method outperforms other methods in all three languages. While Dutch is similar to German, Japanese clearly improves the accuracy up to 97.9%. It indicates the poor coherence of machine-translated text in

Machine-translated paragraph (from German)	"The third concept is called immediate feedback. With immediate feedback, the computer rates the exercises. How else do you teach 150 000 students? The computer evaluates all tasks. We've all done homework and forgotten about it during the two-week correction period. I still do not have some of my first chores back...."
Machine-translated paragraph (from Japanese)	"The idea of the third eye is instant feedback Feedback with instant feedback Computer gains a practice Does any other way to teach 150,000 students? Computer gets results of all the exercises So at our time I got a problem after submitting the assignment I got back after a couple of weeks and have forgotten everything My problem at my college days has not been returned at all..."

Fig. 7. The machine-translated text from German and Japanese.

the low-resource language. Another translation of the text in Fig. 1 is translated from Japanese as shown in Fig. 7. Comparing to German, the Japanese version is obviously lower quality, so it leads to hard-understand the intrinsic meaning. The finding can be used to justify the quality of a machine translator. Based on that, the translator can improve the text coherence by enriching resources in such languages.

5   Conclusion

We propose a method for identifying machine-translated paragraph using coherence features. Each word is matched to other words through a paragraph with the maximum similarity. The similarity represents for text coherence and is used to distinguish human-created with machine-translated paragraphs. Experiments on German show that the coherence-based method archives superior performance (accuracy = 87.0%; equal error rate = 13.0%) when it is compared with other state-of-the-art methods. Similar experiments on Dutch obtain equivalent results while evaluation on Japanese reaches superb accuracy 97.9% and mere equal error rate 1.9%. The results indicate that text coherence is affected by a resource level. The coherence can also be used to measure the quality of a machine translator.

The current work focuses on classifying human-written and machine-translated text. In future work, we aim to produce and evaluate man-made and artificial translation. We also target on estimating the coherence on a website for referring readability. Moreover, a deep learning network can be used to enhance the coherence measurement.

References

1. Aharoni, R., Koppel, M., Goldberg, Y.: Automatic detection of machine translated text and translation quality estimation. In: Proceedings of the 52nd Annual Meeting of the Association for Computational Linguistics (ACL), pp. 289–295. Association for Computational Linguistics (2014)

2. Arase, Y., Zhou, M.: Machine translation detection from monolingual web-text. In: Proceedings of the 51st Annual Meeting of the Association for Computational Linguistics (ACL), pp. 1597–1607. Association for Computational Linguistics (2013)
3. Chae, J., Nenkova, A.: Predicting the fluency of text with shallow structural features: case studies of machine translation and human-written text. In: Proceedings of the 12th Conference of the European Chapter of the Association for Computational Linguistics (EACL), pp. 139–147. Association for Computational Linguistics (2009)
4. Fan, R.E., Chang, K.W., Hsieh, C.J., Wang, X.R., Lin, C.J.: LIBLINEAR: a library for large linear classification. J. Mach. Learn. Res. **9**, 1871–1874 (2008)
5. Labbé, C., Labbé, D.: Duplicate and fake publications in the scientific literature: how many SCIgen papers in computer science? Scientometrics **94**(1), 379–396 (2013)
6. Li, Y., Wang, R., Zhao, H.: A machine learning method to distinguish machine translation from human translation. In: Proceedings of the 29th Pacific Asia Conference on Language, Information and Computation (PACLIC), pp. 354–360 (2015)
7. Manning, C.D., Surdeanu, M., Bauer, J., Finkel, J., Bethard, S.J., McClosky, D.: The stanford CoreNLP natural language processing toolkit. In: Proceedings 52nd Annual Meeting of the Association for Computational Linguistics (ACL): System Demonstrations, pp. 55–60. Association for Computational Linguistics (2014)
8. Nguyen-Son, H.-Q., Echizen, I.: Detecting computer-generated text using fluency and noise features. In: Hasida, K., Pa, W.P. (eds.) PACLING 2017. CCIS, vol. 781, pp. 288–300. Springer, Singapore (2018). https://doi.org/10.1007/978-981-10-8438-6_23
9. Nguyen-Son, H.Q., Tieu, N.D.T., Nguyen, H.H., Yamagishi, J., Echizen, I.: Identifying computer-generated text using statistical analysis. In: Proceedings of Asia-Pacific Signal and Information Processing Association Annual Summit and Conference (APSIPA ASC), pp. 1504–1511. IEEE (2017)
10. Nguyen-Son, H.Q., Tieu, N.D.T., Nguyen, H.H., Yamagishi, J., Echizen, I.: Identifying computer-translated paragraphs using coherence features. In: Proceedings of the 32nd Pacific Asia Conference on Language, Information and Computation (PACLIC) (2018)
11. Pennington, J., Socher, R., Manning, C.D.: Glove: global vectors for word representation. In: Proceedings of the Conference on Empirical Methods in Natural Language Processing (EMNLP), pp. 1532–1543 (2014)
12. Volansky, V., Ordan, N., Wintner, S.: On the features of translationese. Digit. Scholarsh. Humanit. **30**(1), 98–118 (2013)

Improving Low-Resource NMT with Parser Generated Syntactic Phrases

Kamal Kumar Gupta[⊠], Sukanta Sen, Asif Ekbal, and Pushpak Bhattacharyya

Department of Computer Science and Engineering,
Indian Institute of Technology Patna, Patna, India
{kamal.pcs17,sukanta.pcs15,asif,pb}@iitp.ac.in

Abstract. Recently, neural machine translation (NMT) has become highly successful achieving state-of-the-art results on many resource-rich language pairs. However, it fails when there is a lack of sufficiently large amount of parallel corpora for a domain and/or language pair. In this paper, we propose an effective method for NMT under a low-resource scenario. The model operates by augmenting the original training data with the examples extracted from the parse trees of the target-side sentences. It provides important evidences to the model as these phrases are relatively smaller and linguistically correct. Our experiment on the benchmark WMT14 dataset shows an improvement of 3.28 BLEU and 3.41 METEOR score for Hindi to English translation. Evaluation on the same language pair with relatively much smaller datasets of judicial and health domains also show the similar trends with significant performance improvement in terms of BLEU (15.63 for judicial and 15.97 for health) and METEOR (14.30 for judicial and 15.93 for health).

Keywords: Neural machine translation · Low resource machine translation · Low resource NMT

1 Introduction

Neural machine translation (NMT) has recently attracted a lot of attention to the translation community due to its promising results on several language pairs [4] and rapid adoption in the deployment services [6, 10, 22]. The advantages of an NMT system over the Statistical Machine Translation (SMT) are the followings: an entire machine translation (MT) system can be implemented with a single end-to-end architecture; it is better than SMT at generating the fluent outputs [14]. However, NMT requires a huge parallel corpus and the absence of which makes outputs suffer from adequacy.

Efficiency of any NMT model [14] greatly depends on the size of the parallel corpus. It performs well with a very large size of training data, but performs poorly when there is a scarcity of such a large corpus. In addition to that, SMT models are known to perform better when we do not have a sufficiently large amount of parallel data. However, [14] has shown that NMT based method makes a huge jump in BLEU score as we increase the training data size while SMT improves the BLEU score with a fixed rate.

Building NMT models under a low-resource scenario is a great challenge to the researchers. We consider the scenarios to be low-resource when we do not have a sufficient amount of parallel data for a certain language pair or do not have a sufficient

© Springer Nature Switzerland AG 2023
A. Gelbukh (Ed.): CICLing 2019, LNCS 13451, pp. 533–544, 2023.
https://doi.org/10.1007/978-3-031-24337-0_37

amount of parallel data for a particular domain. In our case, we develop an NMT system for Hindi-English language pair which has relatively less parallel data as compared to some European language pairs. Our domains are judicial and health for which we do not have sufficient amount of parallel data, especially for language pair like Hindi-English. Translating documents related to health and judiciary are very crucial in a multilingual country like India. In general, the health information (electronic medical records, health tips available in social media etc.) are available in English and making these available in Hindi would be useful to the common people.

For both health and judicial domains, we have a dearth of parallel data, and we treat this situation as low-resource scenario. In this paper, we propose an approach for NMT that can effectively work for the purpose of improving low-resource NMT without using any additional monolingual data. With the help of a constituency parser, we extract the phrases from the parsed trees of the target sentences. Though the quality of these phrases depends on the robustness of the parser, we can consider these phrases[1] useful for providing important evidences during training. First, we extract noun, verb and prepositional phrases and then we *i.* back-translate these phrases to generate the source-target parallel phrases; and *ii.* make identical copies of these phrases in the source side to obtain the copied parallel phrases. These (translated and copied) parallel phrases are added with the original training data as training examples. Our method is inspired by the idea of adding additional monolingual data through back-translation [21] in order to reduce data sparsity. Our method does not add additional monolingual data because for some pair of languages or for the particular domains, sufficient data may not be present and adding out-of-domain data (to the parallel corpus) may create ambiguity.

Target side monolingual data helps to improve fluency. In encoder-decoder NMT architecture, decoder is indeed a Recurrent Neural Network (RNN) language model. So, it is important that target side data must be accurate in fluency. For our Hindi→English translation system extracting phrases from Hindi (source) and adding them to the English (target) side using back-translation and copied technique will affect the fluency because the augmented synthetic English phrases may not be very fluent. That is why we use phrases from English (target) side only. Our experiments with Hindi→English language for judicial, health domain and WMT14 dataset show impressive performance gains due to the inclusion of these phrases. The key points of our proposed work as follows:

– We augment the original training data using syntactic phrases extracted from the original training data with the help of a constituency parser. This augmented data is used for training of an NMT system.
– We empirically show that our proposed approach of augmenting training data by syntactic phrase pairs, which uses no additional monolingual data, can improve the performance significantly over the baseline NMT model developed with only the original training data.

The remainder of the paper is organized as follows: in Sect. 2, we present the related works briefly. Section 3 describes in details our proposed approach. Section 4 and 5 provide the details of the datasets used and the experimental setup, respectively.

---

[1] Linguistically more accurate as the lengths are short.

We show the results and analysis in Sect. 6. Finally, in Sect. 7 we conclude our work with future work road-maps.

2   Related Works

Neural Machine Translation (NMT) requires a significantly large-scale parallel data for training. Many language pairs, especially under the low resource scenario, do not have this abundance of information. Hence researchers are currently focusing on exploring methods that could be effective in a low-resource scenario. The use of monolingual target language data with the available parallel corpus is one such method that researchers have attempted in recent times. [9] trained a language model using monolingual target language words and integrate it with NMT system trained on a low-resource language pair. They showed improvement in translation for Turkish-English language pair. [21] introduced the back-translation method in which instead of training language model on monolingual target data, they translated the monolingual target data into source language and use this synthetic parallel data along with the original parallel data to train our NMT models. Inspired by the back-translation method, [23] proposed a method by adding the translated monolingual source-side data to the target-side, and create the synthetic parallel data. [7] made an identical copy of monolingual target data at source side to make copied parallel data which they used along with the original training data for training NMT models.

Our approach is different from the above mentioned approaches as these make use of additional monolingual data with the original bilingual corpus while, in our case, we do not use any additional monolingual data with the original training corpus. Our approach aims at extracting additional information from the original target side training sentences, itself, in the form of syntactic phrases and use it with bilingual corpus to train the NMT model.

3   Proposed Method

Our proposed method is based on the standard attentional encoder-decoder approach [1], and augments the original training data with additional synthetic examples. This method of augmenting data does not force any changes in the NMT architecture. In our method, we are not just dividing the target sentence into small phrases, rather we extract NP, VP and PP phrases in a way so that we get the whole sentence in the form of small to large sequences. Adding instances to the training data in such a way helps the model to learn the sentences from smaller to larger sequences as mentioned in the example of Sect. 3.1. This, in turn, helps to preserve adequacy as well as fluency. For generating the synthetic training examples, our proposed model leverages the information obtained from parsing and back translation. We extract the phrases by parsing the target side sentences, and generate their equivalents in the source sides. Thereafter, we combine these with the original parallel data to use it to train the NMT model for source→target. Figure 1 gives an idea about the steps to be followed in this approach.

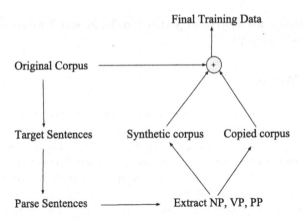

Fig. 1. Overall architecture. '+' means append.

3.1   Phrase Extraction from Parse Tree

We consider the target sentences from original training corpus, and parse it using a constituency parser. We perform experiments for the Hindi→English, and for parsing our target side (English) sentences, we use the Stanford parser[2]. Suppose in Hindi-English training data a source sentences is "इस अधिनियम को इंपीरियल लाइब्रेरी का नाम राष्ट्रीय पुस्तकालय में बदलने के लिए पारित किया गया था।" *is adhiniyam ko impeeriyal laibreree ka naam raashtreey pustakaalay mein badalane ke lie paarit kiya* and its aligned target sentences is *"This act was passed to change the name of the Imperial Library to National Library."* After getting the parse tree of a target sentence, we extract the NP, VP and PP from the parse trees. For example, from the parse tree, as shown in Fig. 2, we extract the syntactic phrases given in Table 1.

We collect the phrases in such a way that the phrases of the same category (e.g., NP, PP or VP) comes in the order of a small sequence to the large sequence. It means we just do not extract the single largest phrase from NP, VP and PP, rather we extract all the possible constituent phrases (in case of nested phrases). For example, first we obtain a NP "the Imperial Library" and then when we get the larger noun phrase "the name of the Imperial Library", we also consider its subset (i.e. constituent) NP as the potential candidate. We can see that all these phrases are linguistically sound.

3.2   Phrase Based SMT for Target→Source

We need a target to source translation system for generating synthetic source phrases by back translating the target phrases (c.f. Sect. 3.3). After extracting phrases from the target sentences of the original training corpus, we need their aligned parallel source translations. For this translation, we prefer PB-SMT [15] over NMT [1] because PB-SMT performs better in case of a small parallel corpus. For judicial domain, we have

---

[2] https://nlp.stanford.edu/software/lex-parser.shtml.

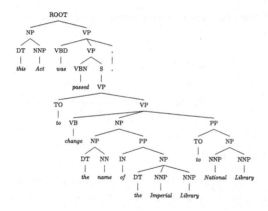

Fig. 2. Parsed tree

Table 1. English phrases from parse tree with their Hindi translation

Phrase in English language	Phrase in Hindi language
National Library [NP]	राष्ट्रीय पुस्तकालय
to National Library [PP]	राष्ट्रीय पुस्तकालय के लिए
the Imperial Library [NP]	इंपीरियल पुस्तकालय
of the Imperial Library [PP]	इंपीरियल पुस्तकालय के
the name [NP]	नाम
the name of the Imperial Library [NP]	इंपीरियल पुस्तकालय का नाम
change the name of the Imperial Library to National Library [VP]	के नाम परिवर्तन करने के लिए राष्ट्रीय पुस्तकालय इंपीरियल पुस्तकालय
to change the name of the Imperial Library to National Library [VP]	के नाम में परिवर्तन करने के लिए राष्ट्रीय पुस्तकालय इंपीरियल पुस्तकालय
passed to change the name of the Imperial Library to National Library [VP]	के नाम में परिवर्तन करने के लिए पारित इंपीरियल पुस्तकालय के लिए राष्ट्रीय पुस्तकालय
was passed to change the name of the Imperial Library to National Library [VP]	के लिए किया गया था इंपीरियल पुस्तकालय का नाम को राष्ट्रीय पुस्तकालय
this Act [NP]	यह अधिनियम

only 5000 parallel sentences, whereas, for the health domain, we have 23,000 parallel sentences. Table 2 shows the statistics of the extracted phrases from the target sentences.

3.3 Synthetic Parallel Corpus Using Back-Translation

We take the target phrases obtained through parsing the target sentences of the training corpus (c.f. Sect. 3.1) and translate them into source language using a phrase-based SMT system (descried in Sect. 3.2) to obtain the source-target parallel phrases. It has been shown in the literature that back-translated parallel corpus, when added to the original parallel corpus, helps in improving the performance of the system even though it may contain incorrect source translation [21].

Table 2. Number of English phrases generated from English sentences

Dataset	Sentences	Phrases
WMT14	263,654	595,969
Judicial	5,000	81,308
Health	23,000	260,487

3.4 Copied Parallel Corpus

[20] have used additional target side monolingual corpus with dummy source sentences. Following the work of [7], we take the additional target side monolingual data and convert it into a parallel corpus by making each source sentence identical to its target counterpart. We refer to this parallel corpus as the copied corpus. The decoder in encoder-decoder is essentially the RNN language model that also conditioned on source context. However, even though the source and target sentences are the same, NMT model performs better at predicting the next output word given the current output. Suppose, a target phrase is *the Imperial Library*, then using this method at training time, we feed *the Imperial Library* to the encoder and try to predict *the Imperial Library* at the decoder.

3.5 NMT Training with Synthetic and Copied Corpus

After obtaining the synthetic parallel data and copied parallel data, we mix these two with the original parallel corpus. It makes our final training data bigger. Statistics regarding the size of the available parallel corpus, number of extracted phrases and the size of synthetic and copied parallel corpus is described in Sect. 4. We take the system trained using the original training corpus as a baseline. Apart from the original training corpus, we create three new parallel corpora by adding (i) only synthetic data; (ii) only copied corpus; and (iii) both synthetic and copied data with the original corpus. We shuffle these augmented corpora. Now we have four kinds of parallel corpora for the same language pair. We use these to train an attention based NMT model [1].

4 Datasets

We perform all our experiments for Hindi to English using the parallel corpora from WMT14 [3,5] and two other domains: *judicial* and *health*.

Table 2 shows the number of phrases generated by the parse trees. These phrases are used to create synthetic and copied data used as additional inputs to the training. Table 4 shows the number of sentences and the number of tokens present in the training data before and after adding synthetic and copied phrases. Size of development set for WMT14, Judicial and Health dataset is 520, 1000 and 1000 respectively. Size of test set for WMT14, Judicial and Health dataset is 2507, 1561 and 1000 respectively.

Table 3. Vocabulary size of each dataset

Dataset	System	Hindi	English
WMT14	*Baseline, Baseline+BT*	112, 344	104, 016
	Baseline+Copied, Baseline+BT+Copied	216, 360	104, 016
Judicial	*Baseline, Baseline+BT*	8, 357	9, 324
	Baseline+Copied, Baseline+BT+Copied	17, 681	9, 324
Health	*Baseline, Baseline+BT*	19, 996	17, 255
	Baseline+Copied, Baseline+BT+Copied	37, 251	17, 255

5   Experimental Setup

We train two types of SMT models: one is English→Hindi (used for back translation) and anther is Hindi→English (used for checking the performance of phrase-based
SMT on original training data). For both, we use the Moses [13] toolkit for training.
We tokenize and true-case the sentences and remove the sentences with words more
than 80 in the preprocessing step. We build 4-gram language model with modified
Kneser-Ney smoothing [12] using IRSTLM [8]. For word alignment, we use GIZA++
[17] with grow-diag-final-and heuristics. For other parameters we use the default set-
tings of Moses. The trained systems are tuned using Minimum Error Rate Training
(MERT) [16]. For getting synthetic parallel data using back translation, we use the
English→Hindi PB-SMT system.

All the NMT models we train are based on attention-based encoder-decoder [1] approach. We train the models at word-level using the Nematus[3] toolkit [19] with the following settings: hidden layer size 512, word embedding dimension 256, max sentence
length 80, batch size 40, and learning rate 0.0001. We use the default early-stopping[4]
criteria of the Nematus. We use the Adam [11] optimizer.

We consider all the vocabulary words. The vocabulary size for each model is shown
in Table 3. For decoding, we use beam width as 5. For other parameters of the Nematus,
the default values are used.

---

[3] https://github.com/EdinburghNLP/nematus.
[4] It is based on BLEU score with patience value = 10.

Table 4. Evaluation results with BLEU and METEOR scores of different Hindi→English systems. #TrainingExamples, #SourceTokens and #TargetTokens columns show the training data size (*Increased dataset size is original training samples augmented with syntactic phrases*).

Domain	System	#TrainingExample	#SourceTokens	#TargetTokens	BLEU	METEOR
WMT14	PB-SMT	263,654	3,330,273	3,033,689	8.24	22.63
	Baseline	263,654	3,330,273	3,033,689	7.52	15.39
	Baseline+Copied	859,623	9,663,437	9,366,853	**7.73**	**15.20**
	Baseline+BT	859,623	9,606,960	9,366,853	**10.80**	**18.80**
	Baseline+BT+Copied	1,455,592	15,940,124	15,700,017	**10.41**	**18.26**
Judicial	PB-SMT	5,000	129,971	121,430	23.13	29.94
	Baseline	5,000	129,971	121,430	1.87	6.47
	Baseline+Copied	86,308	627,578	619,020	**8.15**	**12.53**
	Baseline+BT	86,308	661,291	619,020	**13.81**	**18.09**
	Baseline+BT+Copied	167,616	1,158,898	1,116,627	**17.50**	**20.77**
Health	PB-SMT	23,000	418,853	391,943	20.02	30.10
	Baseline	23,000	418,853	391,943	2.19	8.90
	Baseline+Copied	283,487	1,670,053	1,643,143	**13.04**	**19.74**
	Baseline+BT	283,487	1,716,070	1,643,143	**16.53**	**23.39**
	Baseline+BT+Copied	543,974	2,967,270	2,894,343	**18.16**	**24.83**

6   Results and Analysis

We evaluate our proposed approach using BLEU [18] and METEOR [2] metrics. We summarize the results in Table 4. Some translation outputs produced using our proposed approach are shown in Table 5. Apart from the standard phrase-based SMT system (*PB-SMT*), we train the following NMT systems:

(i) *Baseline*: trained on the original training corpus only.
(ii) *Baseline+Copied*: trained on the original training corpus mixed with copied data.
(iii) *Baseline+BT*: trained on the original training corpus mixed with back-translated data.
(iv) *Baseline+BT+Copied*: trained on the original training corpus mixed with the back-translated and copied data.

BLEU scores for NMT baselines (*Baseline*) are very poor (7.52, 1.87 and 2.19) for *WMT14, judicial* and *health* data due to the size of training corpus. Table 4 shows the improvements in terms of BLEU and METEOR scores after adding the phrases in the baseline corpus. We observe that adding phrases using back translation (i.e. (*Baseline+BT*)) and back translation with copied data (i.e. (*Baseline+BT+copied*)) yield higher scores. Outputs produced by *Baseline+copied* are improved in terms of fluency over the baseline (*Baseline*). However, it lacks in adequacy because of the missing translations for some words. Adequacy increases when we use synthetic data with the original training data in *Baseline+BT* as it translates the missing terms that *Baseline+copied* fails to translate. But the lack of proper sequence of phrases reduces the fluency. Further, by adding the copied data to *Baseline+BT* increases the fluency and maintains adequacy of the outputs generated by the *Baseline+BT+copied*.

Table 5. Some example outputs produced by different proposed systems. *We can observe that Baseline+BT+copied gives better results close to reference.*

	Output #1 (Judicial domain)
Source	यह शिकायतकर्ता द्वारा दायर अभियोजन से मुक्त होने का हकदार था .
Transliteration	yah shikaayatakarta dvaara daayar abhiyojan se mukt hone ka hakadaar tha .
Gloss	it complainant by filed prosecution from exempt be entitled was .
Reference	it was entitled to be exempt from prosecution filed by the complainant .
PB-SMT	it filed by the complainant of free from prosecution was entitled .
Baseline	The petitioner has been filed by the petitioner in this case .
Baseline+Copied	It is filed by the complainant passed by the prosecution .
Baseline+BT	It was entitled to the prosecution filed by the complainant .
Baseline+BT+copied	It was entitled to be free from the prosecution filed by the complainant .
	Output #2 (Health domain)
Source	अमेरिका के एक नामचीन प्रोफेसर के अनुसार एंटीएजिंग क्रीमों के इस्तेमाल से त्वचा विषैले पदार्थों के सम्पर्क में आ सकती है और सूरज की रोशनी से होनेवाले नुकसान की सम्भावना भी बढ़ जाती है ।
Transliteration	amerika ke ek naamacheen prophesar ke anusaar enteeejing kreemon ke istemaal se tvacha vishaile padaarthon ke sampark mein aa sakatee hai aur sooraj kee roshanee se honevaale nukasaan kee sambhaavana bhee badh jaatee hai .
Gloss	America of famous professor according anti-ageing creams of use skin poisonous substances of contact into come can and sun of light from harm of possibility also increase is .
Reference	According to a renowned American Professor, the skin may come in contact with harmful substances by using anti-ageing creams and the chance of skin damage from the sunlight increases .
PB-SMT	according to a famous professor of America skin by the use of anti - ageing creams toxic substances can come in contact with the and the possibility of the harm because of also increases .
Baseline	According to Dr . . . : This disease is used in the form of which there is a lot of benefit by which there is a lot of benefit in the skin and the patient gets destroyed .
Baseline+Copied	According to a famous appendix of America by using skin the skin of the skin becomes strong and the possibility of fear occurring due to the light there is also increase .
Baseline+BT	According to a famous professor of America skin can come in contact with ageing creams and the possibility of harm because of the light of Sun .
Baseline+BT+copied	According to a famous professor of America with the use of ageing creams and skin may come into contact with poisonous rashes and the possibility of the harm because of the accumulation of sun also increases .

From the outputs #1 shown in Table 5, we can see that *Baseline* system translates the phrase *"that the petitioner has been filed by the petitioner"* wrongly, and generates some phrases (e.g., *the petitioner*) multiple times. Though fluency is improved by *Baseline+Copied*, some words (e.g., मुक्त -*Exempt*, हकदार -*Entitled*) are not translated. Further, the *Baseline+BT* system translates some missing words, but still lacks in ordering of the phrases and also drops some phrases (*be free from*). The *Baseline+BT+copied* system translates all the missing words and phrases and improves the fluency by maintaining the phrase order. Similar observation regarding the improvement in translation quality and behavior of our models can be seen in Output #2 where *Baseline+copied* is better at fluency compared to the *Baseline*. However, it drops the translation of few terms like *'professor'*, *'creams'* etc. and repeats the terms like *"the skin"* unnecessarily. The *Baseline+BT* improves adequacy by translating previously missing words (not all) but lacks in maintaining proper phrase ordering. *Baseline+BT+copied* maintains adequacy and fluency substantially except translating a word as *'ageing'* instead of *'anti-ageing'*. One more important observation should be made here that in both the examples PB-SMT translates the words correctly but badly fails in fluency because of the wrong ordering of phrases but because of its end to end nature NMT preserves the fluency and with sufficient amount of training data it improves the translation quality in context of adequacy too.

We have done significance tests and the results are significant with 95% confidence level (with $p = 0$ which is <0.05) and 99% confidence level (with $p = 0$ which is <0.01). Analysis shows that performance improvement in our proposed model is statistically significant over the baselines.

7   Conclusion

Training an NMT system requires large training set, which is not easily available for many languages. In this work, we proposed a technique to improve the translation quality of a NMT systems under the low-resource scenario by injecting syntactic phrases extracted from the parse trees of target-side training data. We extract the noun, verb and prepositional phrases from target sentences of the training data, and perform back translation to generate phrases for the source side. We use these synthetic phrase pairs as additional training data. We empirically showed that our method of augmenting original training data, without using any additional monolingual data, can improve the baseline NMT system for Hindi→English translation in several domains. In future, we will investigate the effectiveness of this approach for the other low-resource Indian languages and domains. We will extract syntactic phrases using parser from both the source and target sides to analyze its impact on the translation quality.

Acknowledgement. We gratefully acknowledge TDIL, MeitY who supported this research work under development of the project "Hindi to English machine translation for judicial domain".

References

1. Bahdanau, D., Cho, K., Bengio, Y.: Neural machine translation by jointly learning to align and translate. In: International Conference on Learning Representation (ICLR) (2015)
2. Banerjee, S., Lavie, A.: METEOR: an automatic metric for MT evaluation with improved correlation with human judgments. In: Proceedings of the ACL Workshop on Intrinsic and Extrinsic Evaluation Measures for Machine Translation and/or Summarization, pp. 65–72. Association for Computational Linguistics (2005). http://www.aclweb.org/anthology/W05-0909

3. Bojar, O., et al.: Findings of the 2014 workshop on statistical machine translation. In: Proceedings of the Ninth Workshop on Statistical Machine Translation, pp. 12–58 (2014)
4. Bojar, O., et al.: Findings of the 2016 conference on machine translation. In: ACL 2016 First Conference on Machine Translation (WMT16), pp. 131–198. The Association for Computational Linguistics (2016)
5. Bojar, O., et al.: Hindencorp-Hindi-English and Hindi-only corpus for machine translation. In: LREC, pp. 3550–3555 (2014)
6. Crego, J., et al.: Systran's pure neural machine translation systems. arXiv preprint arXiv:1610.05540 (2016)
7. Currey, A., Barone, A.V.M., Heafield, K.: Copied monolingual data improves low-resource neural machine translation. In: Proceedings of the Second Conference on Machine Translation, pp. 148–156 (2017)
8. Federico, M., Bertoldi, N., Cettolo, M.: IRSTLM: an open source toolkit for handling large scale language models. In: Ninth Annual Conference of the International Speech Communication Association (2008)
9. Gulcehre, C., et al.: On using monolingual corpora in neural machine translation. arXiv preprint arXiv:1503.03535 (2015)
10. Junczys-Dowmunt, M., Dwojak, T., Hoang, H.: Is neural machine translation ready for deployment? A case study on 30 translation directions. In: In Proceedings of the International Workshop on Spoken Language Translation (IWSLT) (2016)
11. Kingma, D.P., Ba, J.: Adam: a method for stochastic optimization. In: International Conference on Learning Representation (ICLR) (2015)
12. Kneser, R., Ney, H.: Improved backing-off for m-gram language modeling. In: Acoustics, Speech, and Signal Processing, 1995. ICASSP-95, 1995 International Conference on, vol. 1, pp. 181–184. IEEE (1995)
13. Koehn, P., et al.: Moses: open source toolkit for statistical machine translation. In: Proceedings of the 45th Annual Meeting of the ACL on Interactive Poster and Demonstration Sessions, pp. 177–180. Association for Computational Linguistics (2007)
14. Koehn, P., Knowles, R.: Six challenges for neural machine translation. In: Proceedings of the First Workshop on Neural Machine Translation, pp. 28–39. Association for Computational Linguistics, Vancouver, August 2017. http://www.aclweb.org/anthology/W17-3204
15. Koehn, P., Och, F.J., Marcu, D.: Statistical phrase-based translation. In: Proceedings of the 2003 Conference of the North American Chapter of the Association for Computational Linguistics on Human Language Technology, vol. 1, pp. 48–54. Association for Computational Linguistics (2003)
16. Och, F.J.: Minimum error rate training in statistical machine translation. In: Proceedings of the 41st Annual Meeting on Association for Computational Linguistics, vol. 1, pp. 160–167. Association for Computational Linguistics (2003)
17. Och, F.J., Ney, H.: A systematic comparison of various statistical alignment models. Comput. Linguist. **29**(1), 19–51 (2003)
18. Papineni, K., Roukos, S., Ward, T., Zhu, W.J.: BLEU: a method for automatic evaluation of machine translation. In: Proceedings of the 40th annual meeting on association for computational linguistics, pp. 311–318. Philadelphia, Pennsylvania (2002)
19. Sennrich, R., et al.: Nematus: a toolkit for neural machine translation. arXiv preprint arXiv:1703.04357 (2017)
20. Sennrich, R., Haddow, B., Birch, A.: Improving neural machine translation models with monolingual data. arXiv preprint arXiv:1511.06709 (2015)
21. Sennrich, R., Haddow, B., Birch, A.: Improving neural machine translation models with monolingual data. In: Proceedings of the 54th Annual Meeting of the Association for Computational Linguistics, ACL 2016, 7–12 August 2016, Berlin, Germany (2016)

22. Wu, Y., et al.: Google's neural machine translation system: bridging the gap between human and machine translation. CoRR abs/1609.08144 (2016). http://arxiv.org/abs/1609.08144
23. Zhang, J., Zong, C.: Exploiting source-side monolingual data in neural machine translation. In: Proceedings of the 2016 Conference on Empirical Methods in Natural Language Processing, pp. 1535–1545 (2016)

How Much Does Tokenization Affect Neural Machine Translation?

Miguel Domingo[1]($\boxtimes$), Mercedes García-Martínez[2], Alexandre Helle[2],
Francisco Casacuberta[1], and Manuel Herranz[2]

[1] Pattern Recognition and Human Language Technology Research Center,
Universitat Politècnica de València, Camino de Vera s/n, 46022 Valencia, Spain
{midobal,fcn}@prhlt.upv.es
[2] Pangeanic/B.I Europa PangeaMT Technologies Division, Valencia, Spain
{m.garcia,a.helle,m.herranz}@pangeanic.com

Abstract. Tokenization or segmentation is a wide concept that covers simple processes such as separating punctuation from words, or more sophisticated processes such as applying morphological knowledge. Neural Machine Translation (NMT) requires a limited-size vocabulary for computational cost and enough examples to estimate word embeddings. Separating punctuation and splitting tokens into words or subwords has proven to be helpful to reduce vocabulary and increase the number of examples of each word, improving the translation quality. Tokenization is more challenging when dealing with languages with no separator between words. In order to assess the impact of the tokenization in the quality of the final translation on NMT, we experimented on five tokenizers over ten language pairs. We reached the conclusion that the tokenization significantly affects the final translation quality and that the best tokenizer differs for different language pairs.

Keywords: Machine translation · Tokenization · Recurrent neural network

1 Introduction

Segmentation is an essential process that has been extensively studied in literature [3,4,13,14]. It covers simple processes such as separating punctuation from words (tokenization), splitting words in subparts based on their frequency or more sophisticated processes such as applying morphological knowledge. In this work, we use tokenization referring to separating punctuation and splitting tokens into words or subwords.

Tokenizing words has proven to be helpful to reduce vocabulary and increase the number of examples of each word. It is extremely important for languages in which there is no separation between words and, therefore, a single token corresponds to more than one word. The way in which tokens are split can greatly change the meaning of the sentence. For example, the Japanese word 警 means *admonish*, and 察 means *observe*. However, together they form the word *police* (警察) . Therefore, a correct tokenization can help to improve translation quality.

In this study, we aim to find the impact of tokenization on the quality of the final translation. To do so, we experimented with five tokenizers over ten language pairs.

© Springer Nature Switzerland AG 2023
A. Gelbukh (Ed.): CICLing 2019, LNCS 13451, pp. 545–554, 2023.
https://doi.org/10.1007/978-3-031-24337-0_38

To the best of our knowledge, this is the first work in which an exhaustive comparison between tokenizers has been run for NMT. We include tokenizers based on morphology that could guide the splitting of the words [17].

Some previous works include studying the effect of word-level preprocessing for Arabic on Statistical Machine Translation (SMT). A comparison of several segmenters for Chinese on SMT was done by Zhao et al. [24]. Huck et al. [6] compared morphological segmenters for German in NMT. Finally, Kudo [11] compared their statistical word segmenter with other well-known Japanese morphological segmenters, reaching the conclusions that statistical segmenters worked better than morphological ones.

Our main contributions are as follows:

- First study of tokenizers for neural machine translation.
- Experimentation with five different tokenizers over ten language pairs.

The rest of this document is structured as follows: Sect. 2 introduces the neural machine translation system used in this work. After that, in Sect. 3, we present the tokenizers applied for comparison purposes. Then, in Sect. 4, we describe the experimental framework, whose results are presented and discussed in Sect. 5. Section 6 shows some translation examples of the results. Finally, in Sect. 7, conclusions are drawn.

2   Neural Machine Translation

Given a source sentence $x_1^J = x_1, \ldots, x_J$ of length J, NMT aims to find the best translated sentence $\hat{y}_1^{\hat{I}} = \hat{y}_1, \ldots, \hat{y}_{\hat{I}}$ of length $\hat{I}$:

$$\hat{y}_1^{\hat{I}} = \underset{I, y_1^I}{\arg\max}\, Pr(y_1^I \mid x_1^J) \tag{1}$$

where the conditional translation probability is modelled as:

$$Pr(y_1^I \mid x_1^J) = \prod_{i=1}^{I} Pr(y_i \mid y_1^{i-1}, x_1^J) \tag{2}$$

NMT frequently relies on a Recurrent Neural Network (RNN) encoder-decoder framework. The source sentence is projected into a distributed representation at the encoding step. Then, the decoder generates, at the decoding step, its translation word by word [21].

The input of the system is a word sequence in the source language. Each word is projected linearly to a fixed-size real-valued vector through an embedding matrix. Then, these word embeddings are fed into a bidirectional [18] Long Short-Term Memory (LSTM) [5] network. As a result, a sequence of annotations is produced by concatenating the hidden states from the forward and backward layers.

An attention mechanism [1] allows the decoder to focus on parts of the input sequence, computing a weighted mean of annotated sequences. A soft alignment model computes these weights, weighting each annotation with the previous decoding state.

Another LSTM network is used for the decoder. This network is conditioned by the representation computed by the attention model and the last generated word. Finally, a distribution over the target language vocabulary is computed by the deep output layer [16].

The model is trained by applying stochastic gradient descent jointly to maximize the log-likelihood over a bilingual parallel corpus. At decoding time, the model approximates the most likely target sentence with beam-search [21].

3 Tokenizers

In this section, we present the tokenizers we employed in order to assess their impact on the quality of the final translation.

SentencePiece[1]: an unsupervised text tokenizer and detokenizer mainly for Neural Network-based text generation systems where the vocabulary size is predetermined prior to the neural model training. It can be used for any language, but its models need to be trained for each of them. To do so, we used the unigram [12] mode and a vocabulary size of 32000 over each corpora's training partition. Figure 1a shows an example of tokenizing a sentence using *SentencePiece*.

Mecab[2]: an open source morphological analysis engine for Japanese, based on conditional random fields. It extracts morphological and syntactical information from sentences and splits tokens into words. Figure 1b shows an example of tokenizing a sentence using *Mecab*.

Stanford Word Segmenter [22]: a Chinese word segmenter based on conditional random fields. Using a set of morphological and character reduplication features, it is able to split Chinese tokens into words. In this work, we use the toolkit's CTB scheme. Figure 1c shows an example of tokenizing a sentence using *Stanford Word Segmenter*.

OpenNMT tokenizer [8]: the tokenizer included with the *OpenNMT* toolkit. It normalizes characters (e.g., quotes Unicode variants) and separates punctuation from words. It can be used with any language. Figure 1d shows an example of tokenizing a sentence using *OpenNMT tokenizer*.

Moses tokenizer [10]: the tokenizer included with the *Moses* toolkit. It separates punctuation from word—preserving special tokens such as URL or dates—and normalizes characters (e.g., quotes Unicode variants). It can be used with any language. Figure 1e shows an example of tokenizing a sentence using *Moses tokenizer*.

4 Experimental Framework

In this section, we describe the corpora, systems and metrics used in order to asses our proposal.

---

[1] https://github.com/google/sentencepiece.

[2] http://taku910.github.io/mecab/.

Original: *In a browser window (Internet Explorer or Firefox) browse to www.dellconnect.com.*
Segmented: *In ‿a ‿browser ‿window ‿(Internet ‿Explorer ‿or ‿Firefox) ‿browse ‿to ‿www . dell connect . com .*

(a) Example of a sentence tokenized using *SentencePiece*. ‿ indicates the start of a word in the original sentence. The tokenization has split punctuation and transformed the url into several words.

Original: ブラウザウィンドウ*(Internet Explorer*また
は*Firefox)*で、*www.dellconnect.com*にアクセスします。
Segmented: ブラウザウィンドウ (*Internet Explorer* または *Firefox*) で 、 *www . dellconnect . com* に アクセス し ます 。

(b) Example of a sentence tokenized using *Mecab*. In the original sentence, the only spaces were written to separate foreign words (*Internet Explorer*). The tokenization has added spaces between Japanese words, split the punctuation and transformed the url into several words.

Original: 到 *http://www.kace.com/trial*，然后"下 *K1000* 用版"，将的 *OVF* （放虚化格
式）文件下到 *vSphere* 系。
Segmented: 到 *http : //www.kace.com/trial* ， 然后 " 下 *K1000* 用版 " ， 将 的 *OVF* （放 虚 化 格 式） 文件 下 到 *vSphere* 系 。

(c) Example of a sentence tokenized using *Stanford Word Segmenter*. The original sentence only contained spaces to separate foreign words (e.g., *vSphere*). The tokenization has added spaces between the Chinese words, split the punctuation, and separated the *http:* from the url.

Original: *In a browser window (Internet Explorer or Firefox) browse to www.dellconnect.com.*
Segmented: *In a browser window (Internet Explorer or Firefox) browse to www . dellconnect . com .*

(d) Example of a sentence tokenized using *OpenNMT tokenizer*. The tokenization has split punctuation and transformed the url into several words.

Original: *In a browser window (Internet Explorer or Firefox) browse to www.dellconnect.com.*
Segmented: *In a browser window (Internet Explorer or Firefox) browse to www.dellconnect.com.*

(e) Example of a sentence tokenized using *Moses tokenizer*. The tokenization has split the punctuation, without modifying the url.

Fig. 1. Examples of segmenting sentences with each word segmenter.

4.1 Corpora

The corpora selected for our experimental session was extracted from translation memories from the translation industry. The files are the result of professional translation tasks demanded by real clients. The general domain is technical (see Table 1 for the specific content of each language pair), which is harder for NMT than other general domains such as news. Unlike in other domains, in technical domains certain words correspond to specific terms and have a different translation to their most frequent one: e.g., *rear arm* translates into German as *hinterer Arm*. However, in this domain, it should be translated as *hinterer Querlenker*. In order to increase language diversity, we selected the following language-pairs: Japanese–English, Russian–English, Chinese–English, German–English, and Arabic–English. Table 2 shows the corpora statistics.

Table 1. Specific domains for each language pair. *Ja* stands for Japanese, *En* for English, *Ru* for Russian, *Zh* for Chinese, *De* for German and *Ar* for Arabic.

Specific domain	Language				
	Ja–En	Ru–En	Zh–En	De–En	Ar–En
Computer software - instructions for use		X			X
Medical equipment and supplies	X	X	X	X	X
Consumer electronics	X		X	X	X
Industrial electronics		X		X	
Stores and retail distribution	X	X	X		
Healthcare		X			

The training dataset is composed of around three million sentences in the German–English language pair and around half a million sentences in the rest of the language pairs. Development and test datasets are composed of two thousand sentences for all the language pairs.

4.2 Systems

NMT systems were trained with *OpenNMT* [8]. We used LSTM units taking into account the findings in [2]. The size of the LSTM units and word embeddings were set to 1024. We used Adam [7] with a learning rate of 0.0002 [23], a beam size of 6 and a batch size of 20. We reduced the vocabulary using Byte Pair Encoding (BPE) [19], training the models with a joint vocabulary of 32000 BPE units. Finally, the corpora were lowercased and, later, recased using *OpenNMT*'s tools.

4.3 Evaluation Metrics

We made use of the following well-known metrics to assess our proposal:

Table 2. Corpora statistics. *Ja* stands for Japanese, *En* for English, *Ru* for Russian, *Zh* for Chinese, *De* for German and *Ar* for Arabic. *Tokens*BPE and *Vocabulary*BPE are the number of tokens and size of the vocabulary after applying BPE to the corpora. K stands for thousand and M for millions.

Partition	Type	Language				
		Ja–En	Ru–En	Zh–En	De–En	Ar–En
Train	Sentences	532.0 K	496.0 K	460.8 K	2.9 M	557.0 K
	Tokens	10.0/7.3 M	7.6/7.4 M	6.7/6.4 M	35.9/39.4 M	7.3/7.8 M
	Vocabulary	41.5/111.6 K	180.9/133.3 K	82.8/102.6 K	1.1 M/615.7 K	115.5/61.8 K
	Tokens$_{BPE}$	10.5/8.3 M	9.8/9.5 M	7.5/7.4 M	49.8/49.0 M	8.4/8.7 M
	Vocabulary$_{BPE}$	16.0/17.1 K	24.8/11.6 K	22.0/16.6 K	25.6/22.3 K	21.6/10.7 K
Development	Sentences	2000	2000	2000	2000	2000
	Tokens	39.0/27.6 K	34.0/32.2 K	27.8/27.8 K	42.4/45.4 K	21.1/21.7 K
	Vocabulary	2.3/3.4 K	7.6/5.4 K	2.7/3.8 K	6.2/4.4 K	3.6/2.9 K
	Tokens$_{BPE}$	42.1/31.3 K	41.2/38.5 K	29.5/31.2 K	53.7/51.0 K	23.3/24.2 K
	Vocabulary$_{BPE}$	1.9/2.5 K	6.5/3.7 K	2.5/2.9 K	4.9/3.6 K	3.4/2.1 K
Test	Sentences	2000	2000	2000	2000	2740
	Tokens	18.4/26.8 K	28.6/28.3 K	48.7/30.5 K	41.7/44.6 K	22.1/23.3 K
	Vocabulary	3.5/3.9 K	7.3/5.1 K	9.2/3.8 K	6.0/4.3 K	3.2/2.6 K
	Tokens$_{BPE}$	39.5/30.2 K	98.7/94.4 K	32.9/35.6 K	83.9/82.8 K	34.4/32.9 K
	Vocabulary$_{BPE}$	1.8/2.7 K	8.0/5.4 K	2.7/3.0 K	8.3/6.8 K	4.1/2.3 K

BiLingual Evaluation Understudy (BLEU) [15]: corresponds to the geometric average of the modified n-gram precision. It is multiplied by a brevity factor to penalize short sentences.

Translation Error Rate (TER) [20]: number of word edit operations (insertion, substitution, deletion, and swapping), normalized by the number of words in the final translation.

Confidence intervals ($p = 0.05$) are computed for all metrics by means of bootstrap resampling [9].

5 Results

In this section, we present the results of the experiments conducted in order to assess the impact of the tokenizer on the translation quality. Table 3 shows the experimental results.

For the Ja–En experiment, the best results were yielded by *Moses tokenizer* and *Mecab*. It must be taken into account that in both experiments, the English side of the corpus was segmented with *Moses tokenizer*, this means that the segmentation of the target side has a greater impact on the translation quality. Overall, there is a quality improvement of around 4 points in terms of BLEU and 3 points in terms of TER with respect to the tokenizer which yielded the second best results.

For En–Ja, the best results were yielded by *Mecab*, representing a significant improvement (around 12 points in terms of BLEU and 15 points in terms of TER) with respect to the tokenizer which yielded the second best results. Most likely, this is due to *Mecab* being developed specifically to segment Japanese.

Table 3. Experimental results comparing the translation quality produced by using the different tokenizers. In the columns *Mecab* and *Stanford*, *Moses tokenizer* was used for segmenting the English part of the corpora since both *Mecab* and *Stanford Word Segmenter* only work for Japanese and Chinese respectively. Best results are denoted in bold.

Language	SentencePiece		OpenNMT tokenizer		Moses tokenizer		Mecab		Stanford	
	BLEU	TER	BLEU	TER	BLEU	TER	BLEU	TER	BLEU	TER
Ja–En	32.0 ± 1.3	51.1 ± 1.5	29.1 ± 1.4	54.7 ± 1.4	**36.3 ± 1.4**	**47.5 ± 1.3**	36.0 ± 1.5	48.6 ± 1.4	–	–
En–Ja	26.5 ± 1.4	62.5 ± 1.9	25.0 ± 4.4	89.9 ± 4.1	33.6 ± 2.3	61.0 ± 2.5	**45.8 ± 1.3**	**43.7 ± 1.3**	–	–
Ru–En	12.9 ± 0.9	72.7 ± 1.1	11.9 ± 0.9	74.9 ± 1.3	**15.3 ± 1.0**	**68.6 ± 1.2**	–	–	–	–
En–Ru	12.2 ± 0.8	75.0 ± 1.0	11.3 ± 0.9	77.3 ± 1.1	**16.3 ± 1.2**	**70.4 ± 1.6**	–	–	–	–
Zh–En	20.5 ± 1.1	64.8 ± 1.2	23.1 ± 1.3	64.8 ± 1.3	**27.5 ± 1.3**	59.8 ± 1.2	–	–	26.0 ± 1.3	**59.3 ± 1.2**
En–Zh	17.1 ± 1.2	71.2 ± 1.2	10.4 ± 3.9	101.1 ± 3.1	21.4 ± 2.0	65.8 ± 1.7	–	–	**29.9 ± 1.2**	**55.6 ± 1.2**
De–En	21.4 ± 0.8	67.8 ± 2.1	29.6 ± 0.9	54.2 ± 0.9	**30.3 ± 0.9**	**52.8 ± 0.9**	–	–	–	–
En–De	16.1 ± 0.7	76.4 ± 2.3	22.5 ± 0.9	65.0 ± 1.5	**23.6 ± 0.9**	**62.9 ± 1.0**	–	–	–	–
Ar–En	**17.9 ± 0.8**	66.9 ± 1.3	14.8 ± 0.8	71.3 ± 1.1	**19.1 ± 0.9**	**65.4 ± 1.9**	–	–	–	–
En–Ar	10.1 ± 0.6	75.3 ± 1.3	9.2 ± 0.6	77.2 ± 0.9	**12.4 ± 0.7**	**69.8 ± 0.9**	–	–	–	–

For Ru–En and En–Ru, *Moses tokenizer* yielded the best results (with improvements of around 2 to 4 points in terms of BLEU and 5 points in terms of TER). It is worth noting that, in both cases, *SentencePiece* and *OpenNMT tokenizer* yielded similar results.

The Chinese experiments behaved similarly to the Japanese experiments: *Moses tokenizer* and *Stanford Word Segmenter* (the specific Chinese word tokenizer, which included using *Moses tokenizer* for segmenting the English part of the corpus) achieved the best results when translating to English (yielding an improvement of around 7 points in terms of BLEU and 5 points in terms of TER), and *Stanford Word Segmenter* achieved the best results when translating to Chinese (yielding an improvement of around 8 points in terms of BLEU and 20 points in terms of TER).

For the German experiments, the best results were yielded by both *OpenNMT tokenizer* and *Moses tokenizer*, representing an improvement of around 7 to 9 points in terms of BLEU and 14 to 17 points in terms of TER. It is worth noting how, despite being the largest corpora, *SentencePiece*—which learns how to segment from the corpora's training data—yielded the worst results. As a future study, we should evaluate the relation between the size of the corpora and the quality yielded by *SentencePiece*.

Finally, Arabic behaved similarly to Russian, with *Moses tokenizer* yielding the best results for both Ar–En and En–Ar (representing improvements of around 2 to 4 points in terms of BLEU and 4 to 6 points in terms of TER). However, *SentencePiece* performed similar to *Moses tokenizer* when translating to English. When translating to Arabic, both *SentencePiece* and *OpenNMT tokenizer* yielded similar results.

Overall, *Moses tokenizer* yielded the best results for German, Russian and Arabic experiments. When using specialized morphologically oriented tokenizers, the system using *Mecab* obtained the best results for Japanese experiments; and *Stanford Word Segmenter* for Chinese experiments. Additionally, *OpenNMT tokenizer* and *SentencePiece* yielded the worst translation quality in all experiments. An explanation for these poor results is that *OpenNMT tokenizer* is fairly simple: it only separates punctuation symbols from words. However, this is not the case for *SentencePiece*. We think that using *SentencePiece* in a bigger training dataset in order to better learn the segmentation could help to improve their results. Nonetheless, as mentioned before, we have to corroborate this in a future work.

Table 4. English to German translation examples comparing *SentencePiece*, *OpenNMT tokenizer* and *Moses tokenizer*. First line corresponds to the source sentence in English, second line to the German reference and third, forth and fifth lines to the translations generated using *SentencePiece*, *OpenNMT tokenizer* and *Moses tokenizer* respectively to segment the corpora. Correct translations hypothesis are denoted in bold, and incorrect translations are denoted in italic.

Example 1	
Source	Revalidation of single-pilot single-engine class ratings
Reference	Verlängerung von klassenberechtigungen für einmotorige flugzeuge mit einem piloten
SentencePiece	**verlängerung** *der einzelantriebsklasse einmotorischer motorklasse*
OpenNMT tokenizer	*zur* **validierung** *der einmotorik-einzelmaschine* **mit** *einzelantrieb*
Moses tokenizer	**verlängerung von klassenberechtigungen für einmotorige flugzeuge mit einem piloten**
Example 2	
Source	Cold drawing of wire
Reference	Herstellung von kaltgezogenem draht
SentencePiece	*kalt zeichnung des drahtes*
OpenNMT tokenizer	*kaltbildzeichnung*
Moses tokenizer	**herstellung von kaltgezogenem draht**

6   Qualitative Analysis

We obtained a better performance using *Moses tokenizer* than *OpenNMT tokenizer* and *SentencePiece*. In order to qualitatively analyze this performance, Table 4 shows a couple of examples of translation outputs generated using *SentencePiece*, *OpenNMT tokenizer* and *Moses tokenizer* for segmenting the corpora.

The first example clearly shows a better performance when using *Moses tokenizer* rather than *SentencePiece*. The translation output from the system trained using *Moses tokenizer* for segmenting matches the reference. However, the output translations of the systems using *OpenNMT tokenizer* and *SentencePiece* are wrong. Translation segmented with *OpenNMT tokenizer* contains many repetitions and lacks sense. Additionally, translation segmented by *SentencePiece* has problems repeating some words in the translation (e.g., *motor*) and missing some translation words (e.g., the translation of *pilot*).

The system's behavior using *Moses tokenizer* in the second example is similar: its translation matches the reference. By contrast, the systems using *SentencePiece* and *OpenNMT tokenizer* translated wrongly. The system using *SentencePiece* translated all the words from the source but its translation is not grammatically correct. A correct translation could be *kalte Zeichnung des Drahtes*. Lastly, *OpenNMT tokenizer*'s performance is the worst in this case: the translation of its system ignored the word *wire*.

Therefore, we observed that, despite sharing the same data and model architecture, the behavior of the systems' translation changed as a result of using a different tokenizer.

7 Conclusions

In this study, we tested different tokenizers to evaluate their impact on the quality of the final translation. We experimented using 10 language pairs and arrived to the conclusion that tokenization has a great impact on the translation quality, achieving gains of up to 12 points of BLEU and 15 points of TER.

Additionally, we observed that there was not a single best tokenizer. Each one produced the best results for certain language pairs. Although, in some cases, those best results overlapped with the ones yielded by other tokenizers. Moreover, we have seen different behaviors depending on the language pair direction. The system using *SentencePiece* obtained the best results for Ar–En, but not for En–Ar translation.

As a future work, we would like to evaluate the relation between the size of the corpora and the quality yielded by *SentencePiece*—which uses each language's training corpora to learn how to segment. It would also be interesting to compare more segmentation strategies such as separating by characters or fixed n-grams. Finally, we would like to confirm that repeating these experiments on some of the general domain training data used for these languages achieves similar effects.

Acknowledgments. The research leading to these results has received funding from the Centro para el Desarrollo Tecnológico Industrial (CDTI) and the European Union through Programa Operativo de Crecimiento Inteligente (EXPEDIENT: IDI-20170964). We gratefully acknowledge the support of NVIDIA Corporation with the donation of a GPU used for part of this research.

References

1. Bahdanau, D., Cho, K., Bengio, Y.: Neural machine translation by jointly learning to align and translate. arXiv preprint arXiv:1409.0473 (2015)
2. Britz, D., Goldie, A., Luong, T., Le, Q.: Massive exploration of neural machine translation architectures. arXiv preprint arXiv:1703.03906 (2017)
3. Dyer, C.: Using a maximum entropy model to build segmentation lattices for MT. In: Proceedings of the Annual Conference of the North American Chapter of the Association for Computational Linguistics, pp. 406–414 (2009)
4. Goldwater, S., McClosky, D.: Improving statistical MT through morphological analysis. In: Proceedings of the Conference on Human Language Technology and Empirical Methods in Natural Language Processing, pp. 676–683 (2005)
5. Hochreiter, S., Schmidhuber, J.: Long short-term memory. Neural Comput. 9(8), 1735–1780 (1997)
6. Huck, M., Riess, S., Fraser, A.: Target-side word segmentation strategies for neural machine translation. In: Proceedings of the Conference on Machine Translation, pp. 56–67 (2017)
7. Kingma, D.P., Ba, J.: Adam: a method for stochastic optimization. arXiv preprint arXiv:1412.6980 (2014)
8. Klein, G., Kim, Y., Deng, Y., Senellart, J., Rush, A.M.: OpenNMT: open-source toolkit for neural machine translation. arXiv preprint arXiv:1701.02810 (2017)
9. Koehn, P.: Statistical significance tests for machine translation evaluation. In: Proceedings of the Conference on Empirical Methods in Natural Language Processing, pp. 388–395 (2004)
10. Koehn, P., et al.: Moses: open source toolkit for statistical machine translation. In: Proceedings of the Annual Meeting of the Association for Computational Linguistics, pp. 177–180 (2007)

11. Kudo, T.: Sentencepiece experiments (2018). https://github.com/google/sentencepiece/blob/master/doc/experiments.md
12. Kudo, T.: Subword regularization: improving neural network translation models with multiple subword candidates. In: Proceedings of the Annual Meeting of the Association for Computational Linguistics, pp. 66–75 (2018)
13. Nguyen, T., Vogel, S., Smith, N.A.: Nonparametric word segmentation for machine translation. In: Proceedings of the International Conference on Computational Linguistics, pp. 815–823 (2010)
14. Nießen, S., Ney, H.: Statistical machine translation with scarce resources using morphosyntactic information. Comput. Linguist. **30**(2), 181–204 (2004)
15. Papineni, K., Roukos, S., Ward, T., Zhu, W.J.: BLEU: a method for automatic evaluation of machine translation. In: Proceedings of the Annual Meeting of the Association for Computational Linguistics, pp. 311–318 (2002)
16. Pascanu, R., Gulcehre, C., Cho, K., Bengio, Y.: How to construct deep recurrent neural networks. arXiv preprint arXiv:1312.6026 (2013)
17. Pinnis, M., Krišlauks, R., Deksne, D., Miks, T.: Neural machine translation for morphologically rich languages with improved sub-word units and synthetic data. In: Proceedings of the International Conference on Text, Speech, and Dialogue, pp. 237–245 (2017)
18. Schuster, M., Paliwal, K.K.: Bidirectional recurrent neural networks. IEEE Trans. Signal Process. **45**(11), 2673–2681 (1997)
19. Sennrich, R., Haddow, B., Birch, A.: Neural machine translation of rare words with subword units. In: Proceedings of the Annual Meeting of the Association for Computational Linguistics, pp. 1715–1725 (2016)
20. Snover, M., Dorr, B., Schwartz, R., Micciulla, L., Makhoul, J.: A study of translation edit rate with targeted human annotation. In: Proceedings of the Association for Machine Translation in the Americas, pp. 223–231 (2006)
21. Sutskever, I., Vinyals, O., Le, Q.V.: Sequence to sequence learning with neural networks. In: Proceedings of the Advances in Neural Information Processing Systems, pp. 3104–3112 (2014)
22. Tseng, H., Chang, P., Andrew, G., Jurafsky, D., Manning, C.: A conditional random field word segmenter. In: Proceedings of the Special Interest Group of the Association for Computational Linguistics Workshop on Chinese Language Processing, pp. 168–171 (2005)
23. Wu, Y., et al.: Google's neural machine translation system: bridging the gap between human and machine translation. arXiv preprint arXiv:1609.08144 (2016)
24. Zhao, H., Utiyama, M., Sumita, E., Lu, B.L.: An empirical study on word segmentation for Chinese machine translation. In: Proceedings of the Computational Linguistics and Intelligent Text Processing, pp. 248–263 (2013)

Take Help from Elder Brother: Old to Modern English NMT with Phrase Pair Feedback

Sukanta Sen[1]([✉]), Mohammed Hasanuzzaman[2], Asif Ekbal[1],
Pushpak Bhattacharyya[1], and Andy Way[2]

[1] Department of Computer Science and Engineering,
Indian Institute of Technology Patna, Bihta, India
{sukanta.pcs15,asif,pb}@iitp.ac.in
[2] ADAPT Centre, School of Computing, Dublin City University, Dublin, Ireland
hasanuzzaman.im@gmail.com, andy.way@adaptcentre.ie

Abstract. Due to the ever-changing nature of the human language and the variations in writing style, age-old texts in one language may be incomprehensible to a modern reader. In order to make these texts familiar to the modern reader, we need to rewrite them manually. But this is not always feasible if the volume of texts is very large. In this paper, we present this rewriting task as a neural machine translation (NMT) problem. We propose an effective approach for training NMT system using a tiny parallel corpus comprising of only 2.7k parallel sentences. We inject parallel phrase pairs extracted using Statistical Machine Translation (SMT) as additional training examples to NMT. We choose publicly available old-modern English parallel texts for our experiments. Evaluation results show that our proposed approach outperforms the baseline NMT system by more than 18 BLEU points without using any additional training data.

Keywords: Neural machine translation · Low resource NMT · Phrase extraction · Old-modern English

1 Introduction

Human languages are constantly evolving and changing over time to reflect socio-cultural changes, fit current conventions, mores, expressions, and needs. This change in a language often requires rewriting the old texts for the modern readers in the same language. In line with global trends, old texts are increasingly available in the forms that computer can process. These ever expanding records (e.g. historical records, scanned books, academic papers, large-scale corpora, maps etc.)—either digitally born or reconstructed through digitization pipelines—are too big to be rewritten manually. Humanities researchers including historians have a keen interest in computational approaches to process and study digitized old texts for research, writing, and dissemination of knowledge.

© Springer Nature Switzerland AG 2023
A. Gelbukh (Ed.): CICLing 2019, LNCS 13451, pp. 555–566, 2023.
https://doi.org/10.1007/978-3-031-24337-0_39

In this paper, we pose this rewriting task as a problem of machine translation and use Neural Machine Translation (NMT) as a framework to solve. The NMT [3,6,11,16,25] has recently drawn significant attention to the researchers due to its encouraging performance on publicly available benchmark datasets [5] and rapid adoption in production systems [8,15,26]. The key points of NMT are: it generates fluent outputs and it can be implemented as a single end-to-end neural system unlike long-dominant phrase-based SMT [20] which combines many sub-modules. However, the task is not a trivial one. It requires a huge amount of parallel data (often in the range of millions) for building a good NMT system. In the absence of sufficient amount of data, a model learns poorly because of low-counts of source-target units, and makes NMT suffer from the adequacy problem. This is also true in case of an old-modern parallel texts, where we have a dearth of parallel corpus.

In this work, we propose an NMT system using a very small parallel corpus (approximately 2.7 k parallel sentences for old-modern English). We extract phrase pairs from the training data with the help of phrase-based SMT [20] training, and augment the original training corpus by adding these phrase pairs as parallel sentences. By phrase, we do not necessarily mean any linguistic phrase – it is rather a consecutive sequence of words. We evaluate our approach using the BLEU [21], METEOR [4] and TER [24] metrics against the baseline model. The baseline is trained using the original training data only. Our experiments show that our proposed approach attains significant performance gain over the baseline model. We also compare our approach with the well established back-translation [23] method under different settings. We summarize the contributions of our current work as follows:

- We propose an effective NMT approach with feedback from SMT phrases for translating old-to-modern English texts.
- We empirically show that when we do not have enough parallel sentences, parallel phrases, extracted from the training data, can bring significant improvement.
- We also find that our approach can further improve a model trained using back-translation based method.

2   Related Works

Archer et al. [1] presented a summary of previously attempted manual and semi-automatic methods to map historical spelling variants to modern equivalents in order to use it for several natural language processing (NLP) applications.

Domingo et al. [9] used SMT to modernize historical documents. To the best of our knowledge, there exists no significant attempt which aims to automatically translate such old English text into the modern English using NMT.

However, our research is not specific to the old-modern English. Our approach is also applicable to the low-resource scenarios. For training an NMT system, we require a large amount of parallel corpus which is not readily available for every languages and domains. In order to address this issue, however, there have been few attempts to build the NMT systems for low-resource language pairs [10,12,23,27] by incorporating huge monolingual corpus in both the source and target sides.

Sennrich et al. [23] have incorporated monolingual data on the target side to investigate two methods of filling the source side of the monolingual data. In the first method, they have used a dummy source sentence for every target sentence and in the second method, they used a synthetic source sentence obtained via back-translation. They claimed that the second method is more effective. However, if there is not enough parallel data, quality of the back-translation is again a problem.

Zhang and Zong [27] explored the effect of incorporating large-scale source-side monolingual data in NMT in different ways. In the first approach, inspired by [23], they first built a baseline system and then obtained parallel synthetic data by translating the monolingual data. This parallel data along with the original data is used for training an attention-based encoder-decoder NMT system. The second method used the multi-task learning framework to generate the target translation.

Arthur et al. [2] have proposed a model to incorporate translation lexicons through calculating lexical predictive probability and adding this probability to the input of the softmax. Zoph et al. [28] applied transfer learning for low-resourced NMT. Although there have been attempts to expand the training data through back-translated monolingual corpus, the effect of adding source-target phrases into the training data is less explored. We hypothesize that augmenting phrase pairs into training data may be useful for generating adequate translations.

3   Proposed Method

We focus on a low-resource scenario where we have a very small parallel data only. To deal with this situation, we add the phrase pairs extracted from the training corpus as feedback to the NMT framework during training. The proposed method uses a state-of-the-art attention-based encoder-decoder NMT architecture [3]. Here, we briefly describe the architecture first and then present the details of the proposed method.

3.1   Overview of NMT

The goal of NMT is to translate a sequence of source words into a sequence of target words with the help of a large neural network. The basic architecture of an NMT uses two recurrent neural networks, one is called encoder and other is known as decoder. The encoder converts the source sentence into a dense fixed-length vector and then the decoder generates target sentence from that vector. But the main drawback of this encoder-decoder approach is that it fails drastically as length of the input sentence grows. The encoder-decoder approach assumes that the encoder can encode the whole sentence into a fixed length vector, which is not realistic, specifically for the longer sentences. To mitigate this drawback, Bahdanau et al. [3] came up with an idea which focused on the whole input sentence while generating the outputs.

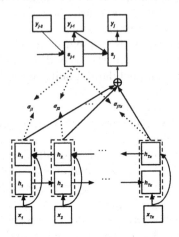

Fig. 1. Attention-based encoder-decoder architecture

Formally, given a sequence of source words x $(= x_1, x_2, x_3, ..., x_n)$ and the previously translated $i - 1$ words y $(= y_1, y_2, y_3, ..., y_{i-1})$, the probability of the ith translated word y_i is calculated as:

$$p(y_i|s_{i-1}, y_{i-1}, c_i) = softmax(W_o t_i) \tag{1}$$

where t_i, the input to the *softmax* is computed as:

$$t_i = tanh(W_s s_i + W_e y_{i-1} + W_c c_i) \tag{2}$$

where W_s, W_e, W_c, W_o are the model parameters. The hidden state s_i in the decoder at time step i is computed as:

$$s_i = g(s_{i-1}, y_{i-1}, c_i) \tag{3}$$

Here g is a nonlinear transformation function, which is usually a Long short-term memory (LSTM) [14] or a gated recurrent unit (GRU) [6], and c_i is the context vector at time step i, which is calculated as a weighted sum of the input annotations h_j:

$$c_i = \sum_{j=1}^{Tx} \alpha_{ij} h_j \tag{4}$$

where Tx is the length of the source sequence and h_j is the encoder hidden state at jth time step and computed using a nonlinear transformation function as:

$$h_j = f(h_{j-1}, x_j) \tag{5}$$

The normalized weight α_{ij} for h_j is calculated as:

$$\alpha_{ij} = \frac{\exp(e_{ij})}{\sum_{k=1}^{Tx} \exp(e_{ik})} \tag{6}$$

$$e_{ij} = V_a^T tanh(U_a s_{i-1} + W_a h_j) \tag{7}$$

where V_a, U_a and W_a are the trainable parameters. All of the parameters in the NMT model are optimized to maximize the following conditional log-likelihood of the N parallel sentences

$$\ell(\theta) = \frac{1}{N} \sum_{n=1}^{N} \sum_{j=1}^{Ty} logp(y_j|s_j, y_{j-1}, c_j) \tag{8}$$

where Ty is the length of the target sequence.

3.2 Phrase Augmentation

Our proposed method uses the state-of-the-art attention-based encoder-decoder [3] NMT architecture. Apart from feeding sentence pairs into the attention-based encoder-decoder, we also feed phrase pairs as training examples. This gives an illusion of having a larger corpus. In order to do this, we first extract parallel phrases from the parallel corpus and then add these parallel phrases in the training set. We use the Moses [19] SMT system. We train a source-target phrase-based SMT [20] and extract all the phrase pairs from the phrase table. However, all of these parallel phrases are not sound, i.e. there can be some wrong source-target phrases [20]. We set different conditions while choosing the phrases from the phrase table. Assuming every source phrase e is aligned to a set of target phrase $F = (f_1, f_2, ..., f_n)$, we consider all or some target phrases for a source phrase. Note that n may vary for each source phrase. We select three sets of phrase pairs from the phrase table.

(i) For first set: we select all the parallel phrases from the phrase table.
(ii) For second set: we select phrase pairs (e, f_t) with $P(f_t|e) \geq 0.5$
(iii) For third set: we select phrase pairs (e, f_t) with $P(f_t|e) = 1$

where f_t is a target phrase for a source phrase e. Since the number of phrase pairs is larger than the number of original parallel sentences, to maintain a fair ratio between them, we use the following formula for combining phrase pairs with the original training set.

$$\begin{aligned} Augmented\ Corpus = N \times Original\ corpus \\ + Extracted\ phrase\ pairs \end{aligned} \tag{9}$$

We combine the extracted set (of parallel phrases) with N times of the original corpus, where, N is calculated as

$$N = \frac{Number\ of\ extracted\ phrase\ pairs}{Number\ of\ original\ parallel\ sentences}$$

Otherwise, training set will contain mostly the phrases and since the phrases are smaller in length, they may make the model biased towards the phrase length.

4   Data Sets

We use publicly available old-modern parallel text[1]. As old English texts, we use publicly available *The Homilies of the Anglo-Saxon Church*(See footnote 1) by Ælfric of Eynsham (c.950–c.1010) who was a prolific author in old English, and its translation by Benjamin Thorpe (c.1782–c.1870) as modern English texts. We call it OE-ME corpus. The OE-ME corpus is tiny in size and it has 720 parallel paragraphs divided into 40 sections. Most of the parallel segments have equal number of parallel sentences which help in aligning the parallel sentences. For experiments, we randomly split it into three sets: *train*, *test*, and *dev* containing 2,716, 500, and 500 sentences, respectively. For tokenization, we use *tokenizer.perl* of the Moses SMT system. We observe that OE has a larger vocabulary than ME, however, ME has more tokens than OE in all *train*, *test*, and *dev* sets. This also gives an intuition that even though they belong to the same language, they are not linguistically same at all. We also use the openly available English Bible corpus[2] [7] for back-translation [23] into old English. Since our original training corpus comprises of religious texts, we choose the Bible corpus. It has approximately 31 k modern English sentences.

Details of the datasets are presented in Table 1.

Table 1. Number of sentences for different data sets

	train	dev	test	Bible
# Sentence	2,716	500	500	31,102

5   Experimental Setup

We use Nematus [22] for training the NMT models. All our NMT models are based on attention-based encoder-decoder approach and trained at word level. We set embedding size as 128, hidden size as 256 and learning rate as 0.001. The models are trained with mini-batch size of 40 and we restrict the maximum sentence length to 80 words. We consider all vocabulary words (vocabulary size for each system has been shown in Table 2). We use the Adam optimizer [17] for optimizing the models. The training stops on meeting the early-stopping criteria[3]. We save a model on every 2,500 updates. For testing, we set beam size as 3

---

[1] https://en.wikisource.org/wiki/The_Homilies_of_the_Anglo-Saxon_Church.
[2] https://github.com/christos-c/bible-corpus/blob/master/bibles/English.xml.
[3] We use early stopping based on BLEU measure with early-stopping patience value 10. All the models run for 110–140 k (approx.) updates before early-stopping.

and select the model that produced the highest BLEU score on the development set for scoring the test set. For other parameters, default values of Nematus were used. We use the Moses [19] for training the phrase-based SMT (*PBSMT*) and as well as for phrase extraction. For training, we keep the following settings in the Moses: grow-diag-final-and heuristics for word alignment, msd-bidirectional-fe for reordering model, and 4-gram language model (LM) with modified Kneser-Ney smoothing [18] using KenLM [13]. However, we note that the order of LM does not affect the phrase table. We train the following different types of NMT systems:

(i) *Baseline-NMT*: The NMT model is trained using only original parallel corpus.

(ii) *BackTrans*: The NMT model is trained using original corpus along with the back-translated parallel sentences (of size 10 k, 20 k, and 31 k). Back-translated corpus (BT) is generated (translating the Bible corpus from modern to old English) using a SMT system trained on the original training corpus.

(iii) *Phrase-Augmented*: These systems are trained to improve the *Baseline-NMT* and *BackTrans* systems by injecting the phrase pairs extracted from the respective training data. We train three types of *Phrase-Augmented* systems (see Table 2):
  – *Type-A*: For these systems, the training data is augmented using phrases extracted from original training data only.
  – *Type-B*: For these systems, the training data is augmented using phrases extracted from original training data and back-translated data together.
  – *Type-C*: For these systems, we add back-translated data with the original data.

6   Results and Analysis

Table 2 summarizes the results of different systems. We can see that SMT is still the best choice in the absence of sufficient parallel corpus. *However, our motivation is not to beat the SMT system, but to develop an appropriate NMT based system in the absence of sufficiently large parallel corpus.* It is well established that SMT performs better in the absence of sufficient amount of corpus. In contrast, NMT is better at fluency and being an end-to-end system, it is easy adoptable and scalable.

From Table 2, it is evident that the baseline NMT system *Baseline-NMT* is outperformed by all of our proposed systems. The third system ($B_N + Phrase_{Org.}$) obtains the best performance among all the three proposed systems (with phrase extracted from original corpus only, i.e. *Type-A*). Though back-translation ($B_N + BT$) performs better than our third system $B_N + Phrase_{Org.}$ by a very small (0.34 BLEU) margin, it is to be noted that it requires significantly large amount of external data compared to the original training data. Our proposed system ($B_N + Phrase_{Org.}$), however, outperforms $B_N + 10$ k BT and $B_N + 20$ k BT by 6.86 and 4.53 BLEU points, respectively.

Table 2. Scores of different systems in terms of BLUE, METEOR and TER. p is the probability of a phrase pair. Org:original training data. $PHR_{Org.}$: phrase pairs extracted from *Org*. *BT*: Parallel corpus obtained through back-translation. $PHR_{Org+BT,p}$: phrase pairs, having probability p, extracted from *Org*. and *BT*. **Data Size** column shows the training data size using the Eq. 9. **Vocab** column shows vocabulary sizes for old and modern English

System		Data size	Vocab		BLEU	METEOR	TER
			Old	Mod			
	PBSMT	2,716	8,878	5,102	39.95	36.96	37.99
	Baseline-NMT (B_N)	2,716			**10.03**	**15.95**	**90.06**
Back-Trans	B_N+10k BT	12,716	15,067	10,948	21.90	24.95	64.66
	B_N+20k BT	22,716	17,341	13,162	24.23	25.83	58.15
	B_N+BT	33,818	19,083	14,859	**29.10**	**30.70**	**52.48**
	Proposed Approach						
Type-A	B_N+$PHR_{Org,p=1.0}$	341,659	8,878	5,102	20.83	25.84	84.66
	B_N+$PHR_{Org,p\geq0.5}$	385,015			25.41	27.79	69.42
	B_N+PHR_{Org}	485,739			**28.76**	**28.56**	**56.37**
Type-B	B_N+$PHR_{Org+BT,p=1.0}$	4,850,270	19,080	14,855	25.30	26.50	63.33
	B_N+$PHR_{Org+BT,p\geq0.5}$	5,094,325			27.93	28.76	60.58
	B_N+$PHR_{Org.+BT}$	6,068,402			25.17	26.95	68.76
Type-C	B_N+BT+PHR_{Org}	512,613	19,083	14,859	**32.35**	**31.88**	**50.36**
	B_N+BT+PHR_{Org+BT}	6,073,265			**29.37**	**30.31**	**56.02**

The improvements are also consistent with respect to the other evaluation metrics METEOR and TER as well. Evaluation suggests that our proposed model can provide a promising solution when we do not have a sufficient amount of parallel corpus and a large amount of monolingual corpus. We also note that when we select phrase pairs with probabilities $p \geq 0.5$ and $p = 1$ (for *Type-A*) we lose many phrase pairs (with probabilities $p \leq 0.5$ and $p < 1$, respectively). We can see that considering all phrases extracted from original training data results the best performance (B_N+$Phrase_{Org.}$). However, when we consider all phrases, there will be some wrong parallel phrases as well but the result shows that they do not affect the overall BLEU score.

We also experiment using phrases extracted from the original training data *Org* and back translated (*BT*) data combined. We observe performance improvement when these are used for training the model (c.f. performance of B_N+ $PHR_{Org+BT,p=1.0}$ and B_N + $PHR_{Org+BT,p\geq0.5}$ improve over B_N+ $PHR_{Org,p=1.0}$ and B_N + $PHR_{Org,p\geq0.5}$, respectively). However, B_N+ PHR_{Org+BT} does not improve compared to B_N+ PHR_{Org} because of the following reason: here we consider all the phrases and a significant number of phrases are wrong as back-translated parallel corpus has many wrong source sentences.

Table 3. Output examples of different NMT systems. *Output 1 is from B_N+ Phrase$_{Org}$ and Output 2 is from $B_N+ BT+Phrase_{Org}$*

Old → *Modern*	
Source	Uton nu gehyran be ōan Halgan Gaste, hwæt he sý.
Reference	Let us now hear concerning the Holy Ghost, what he is.
Baseline-NMT	Let now say as much as a Holy Ghost speak Blessed much a Holy Ghost speak, let the Holy Ghost how he might clear himself, what he might manifestly promised us.
Output 1	Let us now be prophesied of the Holy Ghost, what he might not.
Output 2	Let us now hear concerning the Holy Ghost, what he might overcome them

Our experiments (*Type-C*) show that *BackTrans* can be further improved by augmenting phrases. We note that systems B_N+PHR_{Org+BT} and $B_N+BT+PHR_{Org+BT}$ perform poorer than their preceding counterparts, i.e. systems B_N+PHR_{Org} and $B_N+BT+PHR_{Org}$, respectively. This might have happened due to a number of incorrect phrases that are being added to original training set as a result of phrase extraction from $Org+BT$. This approach does not generate perfect phrases because BT itself contains many wrong source sentences.

To conclude the analysis, we observe that when we add phrases from the original training data only, we can select all the phrases. However, when we extract phrases from the training and back-translated data combined, we need to ignore many phrases because back-translated data produces many wrong phrases.

Table 3 shows few outputs produced by the proposed systems. We observe that the output of the baseline system is not adequate and many parts of the output are repetitive. For example the phrases "*Holy Ghost*" and "*he might*" are output multiple times. This is because the baseline model (trained on only original training set) does not learn the mapping between theses phrases as the corpus is very small. In contrast, our proposed system generates better translation.

7   Conclusion

In this paper, we proposed a method to train an NMT system that performs very effectively on a tiny parallel corpus. We extracted phrase pairs from the original training corpus and used them to augment the training data. This gives an illusion of having more training examples. We choose publicly available old-modern English parallel corpus which comprises only 2.7 k sentences, and posed the rewriting task of old English to modern English as NMT task. We experimentally showed that our approach can significantly improve an NMT system when have we very little amount of training data. We followed standard attention-based encoder-decoder network and our approach improved the baseline old-to-modern English NMT system by a margin of up to 18 BLEU points.

Back-translation has been shown to improve the baseline significantly in many previous NMT tasks. However, quality of back-translation highly depends on the size of original training data. We showed that in a extremely low-resource scenario like old-modern English translation, our approach further improves an NMT system build on top of back-translation.

Our approach augments the original training set and is not specific to any NMT architecture. Thus it can be used in any NMT setting. In this work, we used attention-based encoder-decoder architecture for a specific problem of old-to-modern English translation. In future, we would like apply our approach to various NMT architectures for many low-resource language pairs. In addition to that, we would like to see if the extracted phrases improve an NMT system when the original training corpus is sufficiently large.

Acknowledgments. Asif Ekbal acknowledges Young Faculty Research Fellowship (YFRF), supported by Visvesvaraya PhD scheme for Electronics and IT, Ministry of Electronics and Information Technology (MeitY), Government of India, being implemented by Digital India Corporation (formerly Media Lab Asia). Mohammed Hasanuzzaman and Andy Way would like to acknowledge ADAPT Centre for Digital Content Technology, funded under the SFI Research Centres Programme (Grant 13/RC/2106) and is co-funded under the European Regional Development Fund.

References

1. Archer, D., Kytö, M., Baron, A., Rayson, P.: Guidelines for normalising early modern English corpora: decisions and justifications. ICAME J. **39**(1), 5–24 (2015)
2. Arthur, P., Neubig, G., Nakamura, S.: Incorporating discrete translation lexicons into neural machine translation. In: Proceedings of the 2016 Conference on Empirical Methods in Natural Language Processing, pp. 1557–1567 (2016)
3. Bahdanau, D., Cho, K., Bengio, Y.: Neural machine translation by jointly learning to align and translate. In: International Conference on Learning Representation (ICLR) (2015)
4. Banerjee, S., Lavie, A.: METEOR: an automatic metric for MT evaluation with improved correlation with human judgments. In: Proceedings of the ACL Workshop on Intrinsic and Extrinsic Evaluation Measures for Machine Translation and/or Summarization, pp. 65–72. Association for Computational Linguistics (2005)
5. Bojar, O., et al.: Findings of the 2016 conference on machine translation. In: ACL 2016 First Conference on Machine Translation (WMT16), pp. 131–198. The Association for Computational Linguistics (2016)
6. Cho, K., Van Merriënboer, B., Bahdanau, D., Bengio, Y.: On the properties of neural machine translation: encoder-decoder approaches. In: Proceedings of SSST-8, Eighth Workshop on Syntax, Semantics and Structure in Statistical Translation, pp. 103–111 (2014)
7. Christodouloupoulos, Christos, Steedman, Mark: A massively parallel corpus: the Bible in 100 languages. Lang. Resour. Eval. **49**(2), 375–395 (2014). https://doi.org/10.1007/s10579-014-9287-y
8. Crego, J., et al.: Systran's pure neural machine translation systems. arXiv preprint arXiv:1610.05540 (2016)

9. Domingo, M., Chinea-Rios, M., Casacuberta, F.: Historical documents moderniza-
tion. Prague Bull. Math. Linguist. **108**(1), 295–306 (2017)
10. Fadaee, M., Bisazza, A., Monz, C.: Data augmentation for low-resource neural
machine translation. In: Proceedings of the 55th Annual Meeting of the Association
for Computational Linguistics, vol. 2, pp. 567–573. Association for Computational
Linguistics, Vancouver, Canada (2017)
11. Forcada, Mikel L.., Neco, Ramón P..: Recursive hetero-associative memories for
translation. In: Mira, José, Moreno-Díaz, Roberto, Cabestany, Joan (eds.) IWANN
1997. LNCS, vol. 1240, pp. 453–462. Springer, Heidelberg (1997). https://doi.org/
10.1007/BFb0032504
12. Gulcehre, C., Firat, O., Xu, K., Cho, K., Bengio, Y.: On integrating a language
model into neural machine translation. Comput. Speech Lang. **45**, 137–148 (2017)
13. Heafield, K.: KenLM: faster and smaller language model queries. In: Proceedings of
the Sixth Workshop on Statistical Machine Translation, pp. 187–197. Association
for Computational Linguistics (2011)
14. Hochreiter, S., Schmidhuber, J.: Long short-term memory. Neural Comput. **9**(8),
1735–1780 (1997)
15. Junczys-Dowmunt, M., Dwojak, T., Hoang, H.: Is neural machine translation ready
for deployment? A case study on 30 translation directions. In: Proceedings of the
International Workshop on Spoken Language Translation (IWSLT) (2016)
16. Kalchbrenner, N., Blunsom, P.: Recurrent continuous translation models. In: Proceedings of the 2013 Conference on Empirical Methods in Natural Language Processing, pp. 1700–1709 (2013)
17. Kingma, D.P., Ba, J.: Adam: a method for stochastic optimization. In: International Conference on Learning Representation (ICLR) (2015)
18. Kneser, R., Ney, H.: Improved backing-off for m-gram language modeling. In: 1995
International Conference on Acoustics, Speech, and Signal Processing (ICASSP
1995), vol. 1, pp. 181–184. IEEE (1995)
19. Koehn, P., et al.: Moses: open source toolkit for statistical machine translation.
In: Proceedings of the 45th Annual Meeting of the ACL on Interactive Poster and
Demonstration Sessions, pp. 177–180. Association for Computational Linguistics
(2007)
20. Koehn, P., Och, F.J., Marcu, D.: Statistical phrase-based translation. In: Proceedings of the 2003 Conference of the North American Chapter of the Association
for Computational Linguistics on Human Language Technology, vol. 1, pp. 48–54.
Association for Computational Linguistics (2003)
21. Papineni, K., Roukos, S., Ward, T., Zhu, W.J.: BLEU: a method for automatic
evaluation of machine translation. In: Proceedings of the 40th Annual Meeting on
Association for Computational Linguistics, pp. 311–318. Philadelphia, Pennsylvania (2002)
22. Sennrich, R., et al.: Nematus: a toolkit for neural machine translation. arXiv
preprint arXiv:1703.04357 (2017)
23. Sennrich, R., Haddow, B., Birch, A.: Improving neural machine translation models
with monolingual data. In: Proceedings of the 54th Annual Meeting of the Association for Computational Linguistics, ACL 2016, 7–12 August 2016, Berlin, Germany
(2016)
24. Snover, M., Dorr, B., Schwartz, R., Micciulla, L., Makhoul, J.: A study of translation edit rate with targeted human annotation. In: Proceedings of Association for
Machine Translation in the Americas, vol. 200, pp. 223–231 (2006)

25. Sutskever, I., Vinyals, O., Le, Q.V.: Sequence to sequence learning with neural networks. In: Advances in Neural Information Processing Systems, pp. 3104–3112 (2014)
26. Wu, Y., et al.: Google's neural machine translation system: bridging the gap between human and machine translation. arXiv preprint arXiv:1609.08144 (2016)
27. Zhang, J., Zong, C.: Exploiting source-side monolingual data in neural machine translation. In: Proceedings of the 2016 Conference on Empirical Methods in Natural Language Processing, pp. 1535–1545 (2016)
28. Zoph, B., Yuret, D., May, J., Knight, K.: Transfer learning for low-resource neural machine translation. In: Proceedings of the 2016 Conference on Empirical Methods in Natural Language Processing, pp. 1568–1575. Association for Computational Linguistics, Austin, Texas (2016)

Adaptation of Machine Translation Models with Back-Translated Data Using Transductive Data Selection Methods

Alberto Poncelas[(⊠)] [ID], Gideon Maillette de Buy Wenniger[ID], and Andy Way[ID]

ADAPT Centre, School of Computing, Dublin City University, Dublin, Ireland
{alberto.poncelas,gemdbw}@gmail.com, andy.way@adaptcentre.ie

Abstract. Data selection has proven its merit for improving Neural Machine Translation (NMT), when applied to authentic data. But the benefit of using synthetic data in NMT training, produced by the popular back-translation technique, raises the question if data selection could also be useful for synthetic data?

In this work we use Infrequent n-gram Recovery (INR) and Feature Decay Algorithms (FDA), two transductive data selection methods to obtain subsets of sentences from synthetic data. These methods ensure that selected sentences share n-grams with the test set so the NMT model can be adapted to translate it.

Performing data selection on back-translated data creates new challenges as the source-side may contain noise originated by the model used in the back-translation. Hence, finding n-grams present in the test set become more difficult. Despite that, in our work we show that adapting a model with a selection of synthetic data is an useful approach.

Keywords: Data selection · Back-translation · Synthetic data · Neural machine translation · Feature decay algorithms · Infrequent n-gram recovery

1 Introduction

Neural Machine Translation (NMT) models tend to perform better with larger amounts of data. However, a smaller model trained with data in the same domain as the document to be translated (test set) may perform better than a bigger general-domain model.

Data selection algorithms can be applied as a technique to obtain data of a particular domain. Generally speaking, these methods start from a large set of sentences, and from this set select a subset of sentences that are closer to the domain of interest than other sentences in the large set. Among these methods, Transductive Algorithms (TA) perform the selection by using the test set as seed and retrieving those sentences that are relatively closer to this seed than others. Models built using the output of TA also perform better than general-domain models [1,2].

© Springer Nature Switzerland AG 2023
A. Gelbukh (Ed.): CICLing 2019, LNCS 13451, pp. 567–579, 2023.
https://doi.org/10.1007/978-3-031-24337-0_40

Alternatively, a general-domain model can also be adapted to a certain domain by applying the technique known as *fine-tuning* [3–5]. This consists of training the last epochs of an NMT model (built with out-domain data) using a smaller but in-domain set of sentences.

Unfortunately, additional data that are closer to the test set are not always available. The work of [6] showed that the inclusion of back-translated data can boost the performance of NMT models. Since then, adding synthetic data for training Machine Translation (MT) models has become more popular.

In this work we want to investigate whether it is useful to apply TA to synthetic data selection, in order to retrieve artificial sentences closer to the test set. We study the performance of TA on the task of synthetic data selection, applied in two different configurations (see Fig. 1):

1. Batch processing: The first approach involves back-translating a monolingual set of sentences completely and then selecting sentences from synthetic parallel set. The selection criteria of TA are based on the overlap of *n*-grams of the test set (the seed) with those in the source-side of the parallel set. For this reason, the performance of TA may be worse on back-translated data as the *n*-grams, which have been artificially generated, may be unnatural in terms of word-order.

2. Online processing: This involves selecting the necessary monolingual, target-side, sentences and afterwards back-translating the selected set. The advantage of the online process is that it is not necessary to back-translate the complete data set before selecting data. Nevertheless, as the selection is performed in monolingual target-language we cannot use the test set (which is in

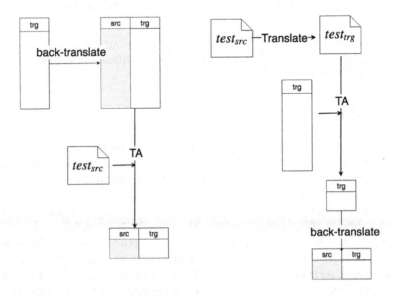

Fig. 1. Pipeline of the batch (left) and online (right) processing to obtain TA-selected synthetic data.

the source-side language) as seed. To solve this, we can proceed as described in the work of [7] and translate the test set using a generic-domain NMT model. Then, this translated text can be used as seed.

2   Related Work

2.1   Transductive Data Selection Algorithms

In this section we describe the algorithms used in the paper, which belong to the family of transductive [8] data selection methods. Such methods select the most relevant sentences for the test set using the (source-side) test set itself. The methods score each sentence s in the candidate data U (the set of sentences that have not been yet selected), and then the sentence with the highest score is added to selected pool L, which is initially empty. Note that this process is done iteratively as the scores (which depend on U and L) are updated after a sentence has been selected.

Infrequent n-gram Recovery (INR): In the work of [9,10] they propose extracting sentences containing n-grams (present in the test set) that are considered infrequent. Therefore, words such as stop words are ignored. The sentences in the candidate data U are scored according to Eq. (1):

$$score(s, U) = \sum_{ngr \in S_{test}} max(0, t - C_{S_I + L}(ngr)) \tag{1}$$

where t is the threshold that indicates whether an n-gram is frequent or not. If the count of the n-gram ngr ($C_{S_I + L}(ngr)$) in the selected pool L (and an in-domain set S_I used for initialization) exceeds the value of t then it will not contribute to the score of the sentence.

Feature Decay Algorithms (FDA): Feature Decay Algorithms [11] selects data by promoting sentences containing many n-grams from the test set, but penalizing those n-grams that have been selected several times. Each n-gram ngr is assigned an initial score, then each time a sentence containing ngr is selected the score of ngr is decreased. The default scoring function is defined as in Eq. (2):

$$score(s, L) = \frac{\sum_{ngr \in S_{test}} 0.5^{C_L(ngr)}}{length(s)} \tag{2}$$

Observe that the more occurrences of ngr are in the selected pool L ($C_L(ngr)$) the less it contributes towards the scoring of the sentence s.

2.2   Using Approximated Target Side

The methods presented in Sect. 2.1 use the test set as seed in order to retrieve sentences. However, a similar approach can be executed by using an approximated translation of the test set (approximated target side) as seed [7]. This seed can be generated by another MT model.

The output of a TA, such as INR or FDA, can be represented as a sequence of sentences $TA_{src} = (s_1^{(src)}, s_2^{(src)}, s_3^{(src)}, ...s_N^{(src)})$ of N sentences. We use the subscript src to indicate that the seed is a text in the source language. However, we can first translate the test set using a generic NMT model and execute the TA using the translation as a seed. The output of this execution could also be represented as a sequence of sentences $TA_{trg} = (s_1^{(trg)}, s_2^{(trg)}, s_3^{(trg)}, ...s_N^{(trg)})$

The two outputs, TA_{src} and TA_{trg}, can be combined as a new sequence of N sentences as in Eq. (3)

$$TA = (s_1^{(src)}, ...s_{N*\alpha}^{(src)}, s_1^{(trg)}, ...s_{N*(1-\alpha)}^{(trg)}) \qquad (3)$$

where the top sentences from each output are concatenated. The value of $\alpha \in [0, 1]$ represents the proportion of data that are selected from TA_{src} and TA_{trg}.

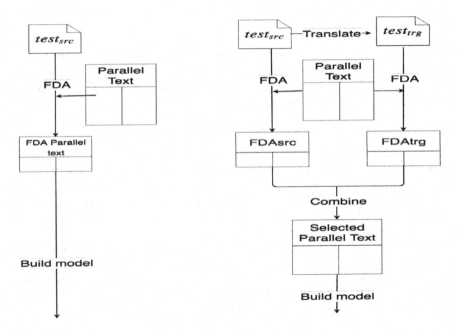

Fig. 2. Pipeline of the traditional usage of FDA (left) and pipeline of our proposal, using the target-side (right) [7].

Figure 2 (right) shows the pipeline that we followed to build the mixture of the outputs using both seeds. Although the data obtained from TA_{trg} are not always useful for adapting an MT model for the test set, mixing the data selected using the test set and the approximated target side can lead to improvements [7].

3   Fine-Tuning Models with Synthetic Data

The work of [6] showed that NMT models can be improved by adding synthetic training data. In their work they use monolingual sentences in the target language and translate them into the source-language with an NMT model. This creates a parallel corpus in which the source side has been artificially generated and the target side is human-produced data (and hence, the fluency of the translation will not be affected). Models built with back-translated data alone (or mixed with back-translated data) can have a performance comparable to those built with real data [12].

In this work we want to explore the performance of NMT models when fine-tuned with TA-selected synthetic data so they are adapted to a given test set. We are interested in exploring three main Research Questions (RQ):

- RQ1: **Does a model adapted with TA-selected back-translated data achieve improvements over the non-adapted model?**
  The strength of performing the fine-tuning technique is to adapt a model with data in the same domain as the document to be translated. Although TA can retrieve relevant data, we do not know the performance when executed using synthetic data. The artificially-generated sentences may contain unusual n-grams, so the overlap with the test set is lower. This prevents TA from retrieving relevant sentences.
- RQ2: **Does a model adapted with TA-selected back-translated data perform better than a model adapted with TA-selected authentic data?**
  Suppose that using synthetic data for adaptation leads to improvements, we also want to compare the performance to that of a model adapted with TA-retrieved authentic data. The quality of the back-translated (source) data, in terms of being an exact translation of the target, is expected to be lower than that of the source-side in the corresponding authentic sentence pairs (which were after all created by human translators). However, the authentic data have already been used to build the model to be adapted, whereas the selected artificial (source) sentences is a set of newly generated data, which may add useful new information not present in the original authentic data set. For this reason, the selected synthetic data might add more value to training the model and may also improve generalization. Therefore, fine-tuning with selected back-translated data may yield larger performance gains than fine-tuning with (repeated) authentic sentences.
- RQ3: **Is it preferable to follow the batch or the online processing?**
  As both processing (batch and online) retrieve different subsets of data, we want to study the performance of the models when they are adapted with a mixture of both outputs. The strategy we follow to combine the outputs is to concatenate them in different proportion in a similar way (using different sizes of α) as explained in Sect. 2.2.

4   Experiments

4.1   Experimental Settings

We build German-to-English models with the parallel data provided in the WMT 2015 [13] (*training data*). All data sets are tokenized and truecased. We also apply Byte Pair Encoding (BPE) [14] with 89500 merge operations. The synthetic data are built by translating the target-side (English) into the source language (German). We use an NMT model built with 1M randomly-selected sentences.

The NMT models are built using OpenNMT-py[1] [15] with the default parameter values: 2-layer LSTM with 500 hidden units, vocabulary size of 50000 words for each language.

All the models built are evaluated on two test sets using BLEU [16], TER [17] and METEOR [18] evaluation metrics. These metrics provide an estimation of the quality of the translation compared to a human-translated reference. The two test sets used to evaluate the models are: (i) *NEWS test set* provided in WMT 2015 News Translation Task; and (ii) *BIO test set*, the Cochrane[2] dataset from the WMT 2017 biomedical translation shared task [19].

In each table, we mark in bold the scores that are better than the baseline, and if they constitute a statistically significant improvement (at level $p = 0.01$) we mark them with an asterisk. This was computed with multeval [20] using bootstrap resampling [21].

4.2   Model Adaptation with Subsets of Data

The general-domain model used in this work as baseline is an NMT model trained with the complete training dataset for 13 epochs. The result of the model can be seen in Table 1

Table 1. Results of the general-domain model evaluated in the NEWS test set and BIO test set.

	NEWS	BIO
BLEU	0.2634	0.3314
TER	0.5441	0.4679
METEOR	0.3009	0.3457

The experiments carried out consist of using INR and FDA to select different sizes of data: 100 K, 200 K and 500 K sentence pairs. In INR method, a low value of t causes the method to be more strict and retrieve less sentences. We use the larger value so the execution does not exceed 48 h (i.e. $t = 80$ for NEWS test

---

[1] https://github.com/OpenNMT/OpenNMT-py.
[2] http://www.himl.eu/test-sets.

Table 2. Results of the models built with different sizes of INR_{src} and INR_{trg} using authentic data.

		Baseline	$\alpha = 1$	$\alpha = 0.75$	$\alpha = 0.50$	$\alpha = 0.25$	$\alpha = 0$
NEWS							
100 K	BLEU	0.2634	**0.2649**	**0.2659**	**0.2664***	**0.2655**	**0.2659***
	TER	0.5441	**0.5419**	**0.5408***	**0.5417***	**0.5413**	**0.5430***
	METEOR	0.3009	**0.3021***	**0.3030***	**0.3037***	**0.3033***	**0.3034***
200 K	BLEU	0.2634	**0.2644**	**0.2661***	**0.2666***	**0.2655**	**0.2649**
	TER	0.5441	**0.5435**	**0.5410***	**0.5406***	**0.5413***	**0.5437***
	METEOR	0.3009	**0.3012**	**0.3025***	**0.3028***	**0.3029***	**0.3027***
BIO							
100 K	BLEU	0.3314	**0.3352***	**0.3346**	**0.3347**	**0.3370***	**0.3339**
	TER	0.4679	**0.4592***	**0.4631**	**0.462**	**0.4591***	**0.4605***
	METEOR	0.3457	**0.3477**	**0.3478**	**0.3463**	**0.3488***	**0.3475**
200 K	BLEU	0.3314	**0.3388***	**0.3362***	**0.3403***	**0.3386***	**0.3343**
	TER	0.4679	**0.459***	**0.4589***	**0.457***	**0.4563***	**0.4590***
	METEOR	0.3457	**0.3494***	**0.3477**	**0.3502***	**0.3489***	**0.3495***

set and $t = 640$ for BIO test set). However, the amount of sentences retrieved are below 500 K, so in the experiments we only evaluate the models adapted with 100 K and 200 K INR-selected sentences. The sentences retrieved are used to adapt the general-domain model. In particular, we adapt the 12th epoch of the model by fine-tuning it with the selected data.

In Table 2 and Table 3 we show the performance of the models when fine-tuned with different sizes of selected authentic data. In the tables we also indicate the proportions of data selected using the test set or the approximated target side as seed.

As we can see, the performance of the adapted models are higher than that of the general-domain model (Table 1). In addition, using a mixture of TA_{src} and TA_{trg} (columns $\alpha = 0.75$, $\alpha = 0.50$ and $\alpha = 0.25$) can achieve a higher performance than TA_{src} or TA_{trg} alone.

In our experiments we follow the same procedure using synthetic data in order to perform comparisons among the general-domain model, models adapted with authentic data, and models adapted with synthetic data.

5   Results

The results of the models adapted with synthetic data are shown in Table 4 (INR method) and Table 5 (FDA method). In order to answer RQ1, we include in the first column, as baseline, the performance of the 13th epoch of the general-domain model (Table 1). We mark in bold those scores that indicate a better

Table 3. Results of the models built with different sizes of FDA_{src} and FDA_{trg} using authentic data.

		Baseline	$\alpha = 1$	$\alpha = 0.75$	$\alpha = 0.50$	$\alpha = 0.25$	$\alpha = 0$
NEWS							
100 K	BLEU	0.2634	**0.2649**	**0.2665***	**0.2642***	**0.2643**	0.2633
	TER	0.5441	**0.5421**	**0.5412***	**0.5413***	**0.5416***	**0.5416***
	METEOR	0.3009	**0.3021***	**0.3027***	**0.3022***	0.3019	0.3020
200 K	BLEU	0.2634	**0.2655**	**0.2665***	**0.2651**	**0.2652**	**0.2654***
	TER	0.5441	**0.5417***	**0.5412***	**0.5413***	**0.5421***	**0.5404***
	METEOR	0.3009	**0.3024***	**0.3027***	**0.3025***	**0.3025***	**0.3027***
500 K	BLEU	0.2634	**0.264***	**0.2658***	**0.2671***	**0.2654**	0.2650
	TER	0.5441	0.5447	**0.5414***	**0.5412***	**0.5415***	**0.5404***
	METEOR	0.3009	**0.3010***	**0.3028***	**0.3028***	**0.3024***	**0.3028***
BIO							
100 K	BLEU	0.3314	**0.3368***	**0.3377***	**0.3391***	**0.339***	0.3331
	TER	0.4679	**0.4597***	**0.4611***	**0.4599***	**0.4597***	0.4649
	METEOR	0.3457	**0.3471**	**0.3473**	**0.3476**	**0.3485**	0.3463
200 K	BLEU	0.3314	**0.3396***	0.3414*	**0.3375***	**0.3391***	**0.3370***
	TER	0.4679	**0.4564***	**0.459***	**0.4574***	**0.4596***	**0.4572***
	METEOR	0.3457	**0.3501***	**0.3503***	**0.3491***	**0.3484***	**0.3496***
500 K	BLEU	0.3314	**0.3375***	**0.3406***	**0.3358***	**0.3354***	0.3336
	TER	0.4679	**0.4592***	**0.4552***	**0.4593***	**0.4574***	0.4617
	METEOR	0.3457	**0.3492***	**0.3496***	**0.3485**	**0.3494***	**0.3485***

performance than the baseline and add an asterisk if they are statistically significant at level p = 0.01.

In the tables we observe that adapted models with artificial data tend to perform better on NEWS test set than BIO test set (e.g. BLEU scores are only higher in the NEWS test set). This manifests that the domain of the model used for back-translating plays an important role. In our experiments the above model is closer to the news domain because it was built using a sample of the authentic training data.

METEOR scores of adapted models are higher than those of the general-domain model for both test sets, and in many cases the improvements are statistical significant (with p = 0.001). In contrast, TER scores are lower than the baseline. This may be caused by the synonym or conjugation chosen by the adapted model. For example, the sentence "auch Schulen" is translated by the general-domain model as "schools too" (the same as in the reference), but adapted model produced "also schools".

Table 4. Results of the models built with different sizes of INR_{src} and INR_{trg} using back-translated data.

		Baseline	$\alpha = 1$	$\alpha = 0.75$	$\alpha = 0.50$	$\alpha = 0.25$	$\alpha = 0$
NEWS							
100 K	BLEU	0.2634	**0.2664**	**0.267**	**0.2671**	**0.2679***	**0.2675***
	TER	0.5441	0.5492	0.5496	0.55	0.5496	0.5513
	METEOR	0.3009	**0.3058***	**0.3062***	**0.3063***	**0.3067***	**0.3061***
200 K	BLEU	0.2634	**0.2666**	**0.2673***	**0.2678***	**0.2673***	**0.2672***
	TER	0.5441	0.5485	0.5486	0.5478	0.5481	0.5481
	METEOR	0.3009	**0.3064***	**0.3061***	**0.3068***	**0.3066***	**0.3068***
BIO							
100 K	BLEU	0.3314	0.324	0.327	0.3263	0.3269	0.3251
	TER	0.4679	0.4762	0.4747	0.4753	0.4751	0.4764
	METEOR	0.3457	**0.3486**	**0.3490**	**0.3502***	**0.351***	**0.3489**
200 K	BLEU	0.3314	0.3241	0.3255	0.3255	0.3254	0.3251
	TER	0.4679	0.4782	0.4755	0.4732	0.4742	0.4745
	METEOR	0.3457	**0.3487**	**0.3501***	**0.3508***	**0.3509***	**0.3505***

5.1 Model Adaptation with Synthetic Data

In our experiments, the back-translated data used for the adaptation are new data unseen by the model (the authentic data used to adapt the models presented in Tables 2 and 3 are subsets of the same data used to build the general-domain model). The outcomes observed in the experiments show that adapting the models with synthetic data does not achieve as good results as adapting them with authentic data (which answers the RQ2). If we compare cell-wise (i.e. same value of α and same size of selected sentences) Tables 2 and 4 or Tables 3 and 5 we see slight improvements for the BLEU and METEOR scores for the news test set (NEWS subtables). However, none of these are statistically significant at $p = 0.01$.

As mentioned previously, the sentences produced by the model used for back-translation may contain mistakes such as word-ordering, incorrect translations etc. which reduces the potential sentences that TA can retrieve. For example, in our experiments we find the following sentence in the NEWS test set "Auf der Hüpfburg beim Burggartenfest war am Sonnabend einiges los." (according to the reference "Something is happening on the bouncy castle at the Burggartenfest.") contains the word "Hüpfburg" ("bouncy castle") which is used by TA to retrieve sentences. There are 18 occurrences of this word in the authentic data set. However, in the synthetic data there are no instances of this word. Instead, the back-translated counterparts of sentences containing "Hüpfburg" include words such as "bouncer" (copied from the English side) or "bounmit" (a word that does not exist). Nevertheless, in some cases back-translated sentences may be closer to literal translation than those found in the authentic

Table 5. Results of the models built with different sizes of FDA_{src} and FDA_{trg} using back-translated data.

		Baseline	$\alpha = 1$	$\alpha = 0.75$	$\alpha = 0.50$	$\alpha = 0.25$	$\alpha = 0$
NEWS							
100 K	BLEU	0.2634	**0.2639**	**0.2654**	**0.264**	**0.2655**	**0.2672***
	TER	0.5441	0.5525	0.5509	0.5522	0.5511	0.5493
	METEOR	0.3009	**0.305***	**0.3054***	**0.3051***	**0.3055***	**0.3062***
200 K	BLEU	0.2634	**0.2655**	**0.2658**	**0.2663**	**0.2666**	**0.2679***
	TER	0.5441	0.5497	0.5512	0.5504	0.5493	0.5484
	METEOR	0.3009	**0.3051***	**0.3053***	**0.306***	**0.3055***	**0.3063***
500 K	BLEU	0.2634	**0.2662**	**0.2674***	**0.2668**	**0.2679***	**0.2664**
	TER	0.5441	0.5483	0.5494	0.5501	0.5488	0.5489
	METEOR	0.3009	**0.3061***	**0.3068***	**0.3062***	**0.3068***	**0.3062***
BIO							
100 K	BLEU	0.3314	0.3228	0.3248	0.3238	0.3254	0.3262
	TER	0.4679	0.4755	0.475	0.4751	0.4742	0.4744
	METEOR	0.3457	**0.349**	**0.3488**	**0.3497***	**0.3521***	**0.3500***
200 K	BLEU	0.3314	0.3214	0.3245	0.3258	0.3255	0.3241
	TER	0.4679	0.478	0.4743	0.4737	0.4751	0.4749
	METEOR	0.3457	**0.3487**	**0.3495**	**0.3501***	**0.349**	**0.3482**
500 K	BLEU	0.3314	0.3215	0.3223	0.3229	0.3241	0.3226
	TER	0.4679	0.4842	0.4843	0.4817	0.4813	0.4811
	METEOR	0.3457	**0.3478**	**0.3488**	**0.3486**	**0.3491**	**0.349**

set [7,22]. For example, in the authentic data set we find the sentence-pair ⟨"er ist verheiratet und hat zwei Kinder.", "since then, he has had a long career on stage, in film and on television. He has also established himself as a singer and an author in recent years."⟩ which do not convey the same meaning. However, the machine-produced source-side is "seitdem hat er eine lange Karriere auf der Bühne, im Film und im Fernsehen absolviert und hat sich auch als Sängerin und Autor in den letzten Jahren etabliert" which is closer in meaning to the target-side sentence. Another example is the pair ⟨"10 %!", "one tenth!"⟩. Although, they have the same meaning, in the back-translated counterpart the source-side sentence is "ein Zehntel!", which is a literal translation.

5.2 Batch and Online Processing

In order to answer RQ3 we need to compare columns $\alpha = 1$ (batch processing, i.e. extract from back-translated data using the test set) and $\alpha = 0$ (online processing, i.e. extract from authentic data using the approximated set). In Table 4 and Table 5 we see that in our experiments following the online process the results tend to be better.

Using an approximated target side as seed is risky, as it can be of low quality. For example, the sentence "Das Buch wurde neu für 48$ verkauft." ("The book was selling for $48 new.") is translated as "The book was sold for 48$." by the general-domain model. As we can see, the word "new" is omitted in the translation. This means that the TA will not consider the word "new" when selecting sentences.

Despite that, we find that the generated target-side seed may contain n-grams that better represent the context of the input document. For example, the sentence in the test set "Ich liebe es, in einem Probenraum zu sein." is translated, according to the reference, as "I love being in a rehearsal room.". The model adapted with $100\,$K sentences from FDA_{src} ($\alpha = 1$) generates the translation "I love to be in a sample room.", whereas the model adapted with FDA_{trg} ($\alpha = 0$) produces a sentence that conveys the same meaning to the reference: "I love to be in a rehearsal room.".

We observe that the occurrences of "Proben" (due to BPE, the word is splitted as "Proben@@ raum") are translated as "sample" or "rehersal" depending on the context. The fact that in the approximated target side the word has been accurately translated as "rehearsal room" induces FDA_{trg} to select more sentences that include the term "rehearsal". In contrast, FDA_{src} retrieves sentences based on the word "Proben" in the seed (as it is present in the test set). However, in the training data this word has been artificially produced and it replaces words such as "Messwasser" ("water sample") or "Musterproduktion" ("sample production").

6   Conclusion and Future Work

In this paper we have analyzed various use-cases of synthetic data for adapting a general-domain model. We have seen that using a TA it is possible to obtain sentences from synthetic data that can improve the model, even if the sentences used for adaptation are an artificial version of the same sentences used to construct the general model.

In addition, we have seen that performing the adaptation online, extracting just the necessary monolingual target-language sentences (using an approximated translation of the test set as seed) and back-translating them afterwards, is a reasonable approach that can even perform better than selecting directly from synthetic sentences.

In the future, we want to further extend this research and explore the effects on the performance of combining both authentic and synthetic data or the use of forward-translation [23]. In addition, we are interested in exploring whether the results observed in this paper are the same when using other language pairs or other configurations of INR and FDA [24,25].

Acknowledgements. This research has been supported by the ADAPT Centre for Digital Content Technology which is funded under the SFI Research Centres Programme (Grant 13/RC/2106) and is co-funded under the European Regional Development Fund.

This work has also received funding from the European Union's Horizon 2020 research and innovation programme under the Marie Skłodowska-Curie grant agreement No 713567.

References

1. Poncelas, A., de Buy Wenniger, G.M., Way, A.: Feature decay algorithms for neural machine translation. In: Proceedings of the 21st Annual Conference of the European Association for Machine Translation, pp. 239–248. Alacant, Spain (2018)

2. Silva, C.C., Liu, C.H., Poncelas, A., Way, A.: Extracting in-domain training corpora for neural machine translation using data selection methods. In: Proceedings of the Third Conference on Machine Translation: Research Papers, pp. 224–231, Brussels, Belgium (2018)

3. Luong, M.T., Manning, C.D.: Stanford neural machine translation systems for spoken language domains. In: Proceedings of the International Workshop on Spoken Language Translation, pp. 76–79. Da Nang, Vietnam (2015)

4. Freitag, M., Al-Onaizan, Y.: Fast domain adaptation for neural machine translation. arXiv preprint arXiv:1612.06897 (2016)

5. van der Wees, M., Bisazza, A., Monz, C.: Dynamic data selection for neural machine translation. In: Proceedings of the 2017 Conference on Empirical Methods in Natural Language Processing, pp. 1400–1410. Copenhagen, Denmark (2017)

6. Sennrich, R., Haddow, B., Birch, A.: Improving neural machine translation models with monolingual data. In: Proceedings of the 54th Annual Meeting of the Association for Computational Linguistics, vol. 1, pp. 86–96. Berlin, Germany (2016)

7. Poncelas, A., de Buy Wenniger, G.M., Way, A.: Data selection with feature decay algorithms using an approximated target side. In: 15th International Workshop on Spoken Language Translation, pp. 173–180. Bruges, Belgium (2018)

8. Vapnik, V.N.: Statistical Learning Theory. Wiley-Interscience (1998)

9. Parcheta, Z., Sanchis-Trilles, G., Casacuberta, F.: Data selection for NMT using infrequent n-gram recovery. In: Proceedings of the 21st Annual Conference of the European Association for Machine Translation, pp. 219–227. Alacant, Spain (2018)

10. Gascó, G., Rocha, M.A., Sanchis-Trilles, G., Andrés-Ferrer, J., Casacuberta, F.: Does more data always yield better translations? In: Proceedings of the 13th Conference of the European Chapter of the Association for Computational Linguistics, pp. 152–161. Avignon, France (2012)

11. Biçici, E., Yuret, D.: Instance selection for machine translation using feature decay algorithms. In: Proceedings of the Sixth Workshop on Statistical Machine Translation, pp. 272–283. Edinburgh, Scotland (2011)

12. Poncelas, A., Shterionov, D., Way, A., de Buy Wenniger, G.M., Passban, P.: Investigating backtranslation in neural machine translation. In: 21st Annual Conference of the European Association for Machine Translation, pp. 249–258. Alacant, Spain (2018)

13. Bojar, O., et al.: Findings of the 2015 workshop on statistical machine translation. In: Proceedings of the Tenth Workshop on Statistical Machine Translation, pp. 1–46. Lisboa, Portugal (2015)

14. Sennrich, R., Haddow, B., Birch, A.: Neural machine translation of rare words with subword units. In: Proceedings of the 54th Annual Meeting of the Association for Computational Linguistics, vol. 1, pp. 1715–1725. Berlin, Germany (2016)

15. Klein, G., Kim, Y., Deng, Y., Senellart, J., Rush, A.M.: OpenNMT: open-source toolkit for neural machine translation. In: Proceedings of the 55th Annual Meeting of the Association for Computational Linguistics-System Demonstrations, pp. 67–72. Vancouver, Canada (2017)
16. Papineni, K., Roukos, S., Ward, T., Zhu, W.J.: BLEU: a method for automatic evaluation of machine translation. In: Proceedings of 40th Annual Meeting of the Association for Computational Linguistics, pp. 311–318. Philadelphia, Pennsylvania, USA (2002)
17. Snover, M., Dorr, B., Schwartz, R., Micciulla, L., Makhoul, J.: A study of translation edit rate with targeted human annotation. In: Proceedings of the 7th Conference of the Association for Machine Translation in the Americas, pp. 223–231. Cambridge, Massachusetts, USA (2006)
18. Banerjee, S., Lavie, A.: METEOR: an automatic metric for MT evaluation with improved correlation with human judgments. In: Proceedings of the ACL Workshop on Intrinsic and Extrinsic Evaluation Measures for Machine Translation and/or Summarization, pp. 65–72. Ann Arbor, Michigan (2005)
19. Yepes, A.J., et al.: Findings of the WMT 2017 biomedical translation shared task. In: Proceedings of the Second Conference on Machine Translation, pp. 234–247 (2017)
20. Clark, J.H., Dyer, C., Lavie, A., Smith, N.A.: Better hypothesis testing for statistical machine translation: controlling for optimizer instability. In: Proceedings of the 49th Annual Meeting of the Association for Computational Linguistics: Human Language Technologies, vol. 2, pp. 176–181. Portland, Oregon (2011)
21. Koehn, P.: Statistical significance tests for machine translation evaluation. In: Proceedings of the 2004 Conference on Empirical Methods in Natural Language Processing, pp. 388–395. Barcelona, Spain (2004)
22. Poncelas, A., Way, A., Sarasola, K.: The ADAPT system description for the IWSLT 2018 Basque to English translation task. In: 15th International Workshop on Spoken Language Translation, pp. 76–82. Bruges, Belgium (2018)
23. Chinea-Rios, M., Peris, A., Casacuberta, F.: Adapting neural machine translation with parallel synthetic data. In: Proceedings of the Second Conference on Machine Translation, pp. 138–147. Copenhagen, Denmark (2017)
24. Poncelas, A., Way, A., Toral, A.: Extending feature decay algorithms using alignment entropy. In: Quesada, J.F., Martín Mateos, F.J., López-Soto, T. (eds.) FETLT 2016. LNCS (LNAI), vol. 10341, pp. 170–182. Springer, Cham (2017). https://doi.org/10.1007/978-3-319-69365-1_14
25. Poncelas, A., de Buy Wenniger, G.M., Way, A.: Applying n-gram alignment entropy to improve feature decay algorithms. Prague Bull. Math. Linguist. **108**, 245–256 (2017)

Morphology, Syntax, Parsing

Automatic Detection of Parallel Sentences from Comparable Biomedical Texts

Rémi Cardon[1,2]([⊠]) and Natalia Grabar[1,2]

[1] CNRS, UMR 8163, 59000 Lille, France
remi.cardon@univ-lille.fr
[2] Univ. Lille, UMR 8163 - STL - Savoirs Textes Langage, 59000 Lille, France

Abstract. Parallel sentences provide semantically similar information which can vary on a given dimension, such as language or register. Parallel sentences with register variation (like expert and non-expert documents) can be exploited for the automatic text simplification. The aim of automatic text simplification is to better access and understand a given information. In the biomedical field, simplification may permit patients to understand medical and health texts. Yet, there is currently no such available resources. We propose to exploit comparable corpora which are distinguished by their registers (specialized and simplified versions) to detect and align parallel sentences. These corpora are in French and are related to the biomedical area. Our purpose is to state whether a given pair of specialized and simplified sentences is to be aligned or not. Manually created reference data show 0.76 inter-annotator agreement. We treat this task as binary classification (alignment/non-alignment). We perform experiments on balanced and imbalanced data. The results on balanced data reach up to 0.96 F-Measure. On imbalanced data, the results are lower but remain competitive when using classification models train on balanced data. Besides, among the three datasets exploited (semantic equivalence and inclusions), the detection of equivalence pairs is more efficient.

Keywords: Parallel sentences · Text alignment · Biomedical texts

1 Introduction

Parallel sentences provide semantically similar information which can vary on a given dimension. Typically, parallel sentences are collected in two languages and correspond to mutual translations. In the general language, the Europarl [1] corpus provides such sentences in several pairs of languages. Yet, the dimension on which the parallelism is positioned can come from other levels, such as expert and non-expert register of language. The following pair of sentences (first in expert and second in non-expert languages) illustrates this:

- *Drugs that inhibit the peristalsis are contraindicated in that situation*
- *In that case, do not take drugs intended for blocking or slowing down the intestinal transit*

© Springer Nature Switzerland AG 2023
A. Gelbukh (Ed.): CICLing 2019, LNCS 13451, pp. 583–594, 2023.
https://doi.org/10.1007/978-3-031-24337-0_41

Indeed, pairs of parallel sentences provide useful information on lexicon used, syntactic structures, stylistic features, etc., as well as the correspondences between the languages or registers. Hence, pairs built from different languages are widely used in machine translation, while pairs differentiated by the register of language can be used for the text simplification. The purpose of text simplification is to provide simplified versions of texts, in order to remove or replace difficult words or information. Simplification can be concerned with different linguistic aspects, such as lexicon, syntax, semantics, pragmatics and even document structure.

Automatic text simplification can be used as a preprocessing step for NLP applications or for producing suitable versions of texts for humans. In this second case, simplified documents are typically created for children [2], for people with low literacy or foreigners [3], for people with mental or neurodegenerative disorders [4], or for laypeople who face specialized documents [5]. Our work is related to the creation of simplified medical documents for laypeople, such as patients and their relatives. It has indeed been noticed that medical and health documents contain information that is difficult to understand by patients and their relatives, mainly because of the presence of technical and specialized terms and notions. This situation has a negative effect on the healthcare process [6–8]. Hence, helping patients to better understand medical and health information is an important issue, which motivates our work.

In order to perform biomedical text simplification, we propose to collect parallel sentences, which align difficult and simple information, as they provide crucial and necessary indicators for automatic systems for the text simplification. Indeed, such pairs of sentences contain cues on transformations which are suitable for the simplification, such as lexical substitutes and syntactic modifications. Yet, this kind of resources is seldom available, especially in languages other than English. As a matter of fact, it is easier to access comparable corpora: they cover the same topics but are differentiated by their registers (documents created for medical professionals and documents created for patients). More precisely, we can exploit an existing monolingual comparable corpus with medical documents in French [9]. The purpose of our work is to detect and align parallel sentences from this comparable corpus. We also propose to test what is the impact of imbalance on categorization results: imbalance of categories is indeed the natural characteristics in textual data.

The existing work on searching parallel sentences in monolingual comparable corpora indicates that the main difficulty is that such sentences may show low lexical overlap but be nevertheless parallel. Recently, this task gained in popularity in general-language domain thanks to the semantic text similarity (STS) initiative. Dedicated *SemEval* competitions have been proposed for several years [10–12]. The objective, for a given pair of sentences, is to predict whether they are semantically similar and to assign a similarity score going from 0 (independent semantics) to 5 (semantic equivalence). This task is usually explored in general-language corpora. Among the exploited methods, we can notice:

- lexicon-based methods which rely on similarity of subwords or words from the processed texts or on machine translation [13]. The features exploited can

be: lexical overlap, sentence length, string edition distance, numbers, named entities, the longest common substring [14–18];
- knowledge-based methods which exploit external resources, such as Word-Net [19] or PPDB [20]. The features exploited can be: overlap with external resources, distance between the synsets, intersection of synsets, semantic similarity of resource graphs, presence of synonyms, hyperonyms or antonyms [21–23];
- syntax-based methods which exploit the syntactic modelling of sentences. The features often exploited are: syntactic categories, syntactic overlap, syntactic dependencies and constituents, predicat-argument relations, edition distance between syntactic trees [24–27];
- corpus-based methods which exploit distributional methods, latent semantic analysis (LSA), topics modelling, word embeddings, etc. [28–33].

Yet, there is no work on detection and alignment of parallel sentences in specialized areas, like biomedicine. Our work is positioned in this area.

In what follows, we first present the linguistic material used, and the methods proposed. We then present and discuss the results obtained, and conclude with directions of future work.

2   Method

We use the CLEAR comparable medical corpus [9] available online[1] which contains three comparable sub-corpora in French. Documents within these sub-corpora are contrasted by the degree of technicality of the information they contain with typically specialized and simplified versions of a given text. These corpora cover three genres: drug information, summaries of scientific articles, and encyclopedia articles. We also exploit a reference dataset with sentences manually aligned by two annotators.

2.1   Comparable Corpora

Table 1 indicates the size of the comparable corpus in French: number of documents, number of words (occurrences and lemmas) in specialized and simpli-

Table 1. Size of the three source corpora. Column headers: number of documents, total of occurrences (specialized and simple), total of unique lemmas (specialized and simple)

corpus	# docs	# occ_{sp}	# occ_{simpl}	# $lemmas_{sp}$	# $lemmas_{simpl}$
Drugs	11,800 × 2	52,313,126	33,682,889	43,515	25,725
Scient.	3,815 × 2	2,840,003	1,515,051	11,558	7,567
Encyc.	575 × 2	2,293,078	197,672	19,287	3,117

---

[1] http://natalia.grabar.free.fr/resources.php#clear.

Table 2. Size of the reference data with consensual alignment of sentences. Column headers: number of documents, sentences and word occurrences for each subset, alignment rate

corpus	# doc.	Specialized				Simplified				Alignment rate (%)	
		source		aligned		source		aligned			
		# sent.	it # occ.	# pairs.	# occ.	# sent.	# occ.	# pairs.	# occ.	sp.	simp.
Drugs	12 × 2	4,416	44,709	502	5,751	2,736	27,820	502	10,398	18	11
Scient.	13 × 2	553	8,854	112	3,166	263	4,688	112	3,306	20	43
Encyc.	14 × 2	2,494	36,002	49	1,100	238	2,659	49	853	2	21

fied versions. This information is detailed for each sub-corpus: drug information (*Drugs*), summaries of scientific articles (*Scient.*), and encyclopedia articles (*Encyc.*).

The *Drug* corpus contains drug information such as provided to health professionals and patients. Indeed, two distinct sets of documents exist, each of which contains common and specific information. This corpus is built from the public drug database[2] of the French Health ministry. Specialized versions of documents provide more word occurrences than simplified versions.

The *Scientific* corpus contains summaries of meta-reviews of high evidence health-related articles, such as proposed by the Cochrane collaboration [34]. These reviews have been first intended for health professionals but recently the collaborators started to create simplified versions of the reviews (*Plain language summary*) so that they can be read and understood by the whole population. This corpus has been built from the online library of the Cochrane collaboration[3]. Here again, specialized version of summaries is larger than the simplified version, although the difference is not very important.

The *Encyclopedia* corpus contains encyclopedia articles from Wikipedia[4] and Vikidia[5]. Wikipedia articles are considered as technical texts while Vikidia articles considered as their simplified versions (they are created for children from 8 to 13 year old). Similarly to the works done in English, we associate Vikidia with Simple Wikipedia[6]. Only articles indexed in the medical portal are exploited in this work. From Table 1, we can see that specialized versions (from Wikipedia) are also longer than simplified versions.

Those three corpora have different degrees of parallelism: Wikipedia and Vikidia articles are written independently from each other, drug information documents are related to the same drugs but the types of information presented for experts and laypeople vary, while simplified summaries from the *scientific* corpus are created starting from the expert summaries.

---

[2] http://base-donnees-publique.medicaments.gouv.fr/.
[3] http://www.cochranelibrary.com/.
[4] https://fr.wikipedia.org.
[5] https://fr.vikidia.org.
[6] http://simple.wikipedia.org.

2.2   Reference Data

The reference data with aligned sentence pairs, which associate technical and simplified contents, are created manually. We have randomly selected 2× 14 *encyclopedia* articles, 2×12 *drug* documents, and 2×13 *scientific* summaries. The sentence alignment is done by two annotators following these guidelines:

1. exclude identical sentences or sentences with only punctuation and stopword difference ;
2. include sentence pairs with morphological variations (e.g. *Ne pas dépasser la posologie recommandée.* and *Ne dépassez pas la posologie recommandée.* – both examples can be translated by *Do not take more than the recommended dose.*);
3. exclude sentence pairs with overlapping semantics, when each sentence brings own content, in addition to the common semantics;
4. include sentence pairs in which one sentence is included in the other, which enables many-to-one matching (e.g. *C'est un organe fait de tissus membraneux et musculaires, d'environ 10 à 15 mm de long, qui pend à la partie moyenne du voile du palais.* and *Elle est constituée d' un tissu membraneux et musculaire. – It is an organ made of membranous and muscular tissues, approximately 10 to 15 mm long, that hangs from the medium part of the soft palate.* and *It is made of a membranous and muscular tissue.*);
5. include sentence pairs with equivalent semantics – other than semantic intersection and inclusion (e.g. *Les médicaments inhibant le péristaltisme sont contre-indiqués dans cette situation.* and *Dans ce cas, ne prenez pas de médicaments destinés à bloquer ou ralentir le transit intestinal. – Drugs that inhibit peristalsis are contraindicated in that situation.* and *In that case, do not take drugs intended for blocking or slowing down the intestinal transit.*).

The judgement on semantic closeness may vary according to the annotators. For this reason, the alignments provided by each annotator undergo consensus discussions. This alignment process provides a set of 663 aligned sentence pairs. The inter-annotator agreement is 0.76 [35]. It is computed within the two sets of sentences proposed for alignment by the two annotators.

Because the three corpora vary in their capacity to provide parallel sentences, we compute their *alignment rate*. The alignment rate for a given corpus is the number of sentences that are part of an aligned pair relative to the total number of sentences. As expected, only a tiny fraction of all possible pairs corresponds to aligned sentences. We can observe that the *scientific* corpus is the most parallel with the highest alignment rate of sentences, while the two other corpora (*drugs* and *encyclopedia*) contain proportionally less parallel sentences. Sentences from simplified documents in the *scientific* and *drugs* corpora are longer than sentences from specialized documents because they often add explanations for technical notions, like in this example: *We considered studies involving bulking agents (a fibre supplement), antispasmodics (smooth muscle relaxants) or antidepressants (drugs used to treat depression that can also change pain perceptions) that used outcome measures including improvement of abdominal pain, global assessment (overall relief of IBS symptoms) or symptom score.* In the *encylopedia* corpus such notions are replaced by simpler words, or removed. Finally, in all corpora, we observe frequent substitutions by synonyms, like *{nutrition, food}*, *{enteral, directly in the stomach}*, or *{hypersensitivity, allergy}*. Notice that with such substitutions, lexical similarity between sentences is reduced.

The documents are pre-processed. They are segmented into sentences using strong punctuation (*i.e.* . *?!;:*). We removed, from each subcorpus, the sentences that are found in at least half of the documents of a given corpus. Those sentences are typically legal notices, section titles, and remainders from the conversion of the HTML versions of the documents. The lines that contain no alphabetic characters have also been removed. That reduces the total number of possible pairs for each document pair approximately from 940,000 to 590,000.

2.3   Automatic Detection and Alignment of Parallel Sentences

Automatic detection and alignment of parallel sentences is the main step of our work. The unity processed is a pair of sentences. The objective is to categorize the pairs of sentences in one of the two categories:

- alignment: the sentences are parallel and can be aligned;
- non-alignment: the sentences are non-parallel and cannot be aligned.

The reference data provide 663 positive examples (parallel sentence pairs). In order to perform the automatic categorization, we also need negative examples, which are obtained by randomly pairing all sentences from all the document pairs and removing the sentence pairs that are already found to be parallel. Approximately, 590,000 non-parallel sentences pairs are created in this way.

For the automatic alignment of parallel sentences, we first use a binary classification model that relies on logistic regression. Our goal is to propose features that can work on textual data in different languages and registers. We use several features which are mainly lexicon-based and corpus-based, so that they can be easily applied to textual data in other corpora, speacialized areas and languages or transposed on them. The features are computed on word forms (occurrences). The features are the following:

1. *Number of common non-stopwords.* This feature permits to compute the basic lexical overlap between specialized and simplified versions of sentences [28]. This feature exploits external knowledge (set of stopwords), which are nevertheless very common linguistic data;
2. *Number of common stopwords.* This feature also exploits external knowledge (set of stopwords). It concentrates on non-lexical content of sentences;
3. *Percentage of words from one sentence included in the other sentence, computed in both directions.* This features represents possible lexical and semantic inclusion relations between the sentences;
4. *Sentence length difference between specialized and simplified sentences.* This feature assumes that simplification may imply stable association with the sentence length;

5. *Average length difference in words between specialized and simplified sentences.* This feature is similar to the previous one but takes into account average difference in sentence length;
6. *Total number of common bigrams and trigrams.* This feature is computed on character ngrams. The assumption is that, at the sub-word level, some sequences of characters may be meaningful for the alignment of sentences if they are shared by them;
7. *Word-based similarity measure exploits three scores (cosine, Dice and Jaccard).* This feature provides a more sophisticated indication on word overlap between two sentences. Weight assigned to each word is set to 1;
8. *Character-based minimal edit distance* [36]. This is a classical acception of edit distance. It takes into account basic edit operations (insertion, deletion and substitution) at the level of characters. The cost of each operation is set to 1;
9. *Word-based minimal edit distance* [36]. This feature is computed with words as units within sentence. It takes into account the same three edit operations with the same cost set to 1. This feature permits to compute the cost of lexical transformation of one sentence into another.

2.4   Experimental Design

The set with manually aligned pairs is divided into three subsets:

- *equivalence*: 238 pairs with equivalent semantics,
- *tech in simp*: 237 pairs with inclusion where the content of technical sentence is fully included in simplified sentence, and simplified sentence provides additional content,
- *simp in tech*: 112 pairs with inclusion where the content of simplified sentence is fully included in technical sentence, and technical sentence provides additional content.

For each subset, we perform two sets of experiments:

1. We train and test the model with balanced data (we randomly select as many non-aligned pairs as aligned pairs), and then we progressively increase the number of non-aligned pairs until we reach a ratio of 3000:1, which is close to the real data.
2. Then, for each ratio, we apply the obtained model to the whole dataset and evaluate the results.

As there is some degree of variability coming with the subset of non-aligned pairs that are randomly selected for the imbalance ratio, every single one of those experiments has been performed fifty times: the results that are presented correspond to the mean values over the fifty runs.

2.5   Evaluation

For evaluating the results, in each experiment we divide the indicated datasets in two parts: two thirds for training and one third for testing. The metrics we use are Recall, Precision and F1 scores. As we are primarily focused on detection of the aligned pairs, we only report scores for that class. Another reason to exclude the negative class and the global score from the observations is that when the data are imbalanced (negative class is growing progressively), misclassifying the positive data has little influence over the global scores, which thus always appear to be high (metrics above 0.99).

3   Presentation and Discussion of Results

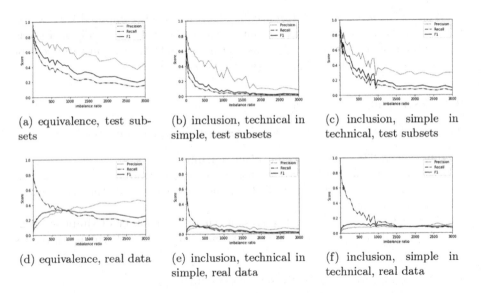

(a) equivalence, test sub-sets

(b) inclusion, technical in simple, test subsets

(c) inclusion, simple in technical, test subsets

(d) equivalence, real data

(e) inclusion, technical in simple, real data

(f) inclusion, simple in technical, real data

Fig. 1. Precision, Recall and F-1 for the various experiments and subsets.

We present the results in Fig. 1: The x axis represents the growing of imbalance (the first position is 1 and corresponds to balanced data), while the y axis represents the values of Precision, Recall and F-measure. The results for three subsets are presented: equivalence (Figs. 2(a) and 2(d)), inclusion of technical sentence in simple sentence (Figs. 2(b) and 2(e)), and inclusion of simple sentence in technical sentence (Figs. 2(c) and 2(f)). Besides, Figs. 2(a), 2(b) and 2(c) present the results obtained by training and testing the model on the same with the same imbalance ratio (first set of experiments described in Sect. 2.4). As for Figs. 2(d), 2(e) and 2(f), they present the results obtained by the models mentioned above that are applied on the whole set of manually annotated data (second set of experiments described in Sect. 2.4).

The most visible conclusion we can draw from those experiments is that equivalent pairs (Figs. 2(a) and 2(d)) are easier to classify than inclusion pairs (the rest of the Figures). Values of both, Precision and Recall, are higher on the equivalence dataset at different imbalance points. For instance, on Fig. 2(a) at the starting point, we obtain 0.96 Precision, 0.93 Recall and 0.94 F-measure. This result is positive because the equivalence dataset usually provides the main and the most complete information on transformations required for the simplification. As for the inclusion relations, at the same point and experimental setting, we obtain 0.90 Precision, 0.89 Recall and 0.89 F-measure on technical in simple inclusion dataset, and 0.92 Precision, 0.93 Recall and 0.92 F-measure on simple in technical inclusion dataset. We assume that the inclusion classification models cover a large variety of situations which do not necessarily correspond to the searched information. We need to design additional filters to make the results more suitable for our purpose.

We can also observe from Fig. 1 that the use of balanced data provides very high results, both for Precision and Recall, which are very close to the reference data (>0.90 performance). This is true for the three subsets tested (equivalence and inclusions). This means that models dealing with balanced data can efficiently detect pairs of sentences with parallel contents in balanced and imbalanced datasets. As expected, when imbalance is introduced in the data, the performance of the models decreases. This means that imbalance introduces additional confusion between sentences that should be aligned and those that should not be aligned. Yet, the imbalance has greater effect on the inclusion datasets, while again the equivalence dataset resists better. We can conclude from these results that, when processing real data, it is more suitable to exploit classification models trained on balanced data. Such models show better discrimination for the detection of sentence with parallel contents.

Another interesting finding is that the values of Precision remain higher than the values of Recall. This is particularly observable with experiments using models trained on balanced data (Figs. 2(a), 2(b) and 2(c)). We assume that these models can efficiently detect the positive pairs of sentences, which makes the Precision to remain high. Yet, with the increasing imbalance, additional confusion is introduced in data and the results.

Overall, we consider that the results obtained are very good when balanced data are processed. Because imbalance is a natural situation in the task we aim, as it can be observed in Table 2, our future work will concentrate in proposing additional filters to remove non-alignable sentences or to exclude pairs of sentences which should not be aligned.

4   Conclusion and Future Work

We proposed to address the task of detection and alignment of parallel sentences from monolingual comparable corpora in French. The comparable dimension is due to the technicality of documents, which contrast technical and simplified versions of documents and sentences. We use the CLEAR corpus related to the biomedical area.

Several experiments are performed. More specifically, we work with three subsets of data (equivalence and inclusions between sentences), and with balanced and imbalanced datasets. On balanced dataset, we reach up to 0.93 F-measure, with a very good balance between Precision and Recall. On imbalanced dataset, the performance of classifiers decreases. Yet, the alignment results remain better when models trained on balanced datasets are exploited.

In future, we plan to exploit the best models generated for enriching the set of parallel sentences. The Recall scores may be the main measure for choosing the best classifier and approach. Specific attention will be paid to the filtering of the imbalanced data in order to remove non-alignable sentences and pairs. Enriching the existing reference dataset will permit to prepare data necessary for the development of simplification methods for the medical documents in French. Other directions for future work are concerned with the exploitation of other features and approaches for the alignment of sentences. As we have seen, the lexical distance between technical and simplified sentences may be high, so the use of word embeddings or the exploitation of external knowledge may be useful to smooth lexical variation.

Acknowledgements. This work was funded by the French National Agency for Research (ANR) as part of the *CLEAR* project (*Communication, Literacy, Education, Accessibility, Readability*), ANR-17-CE19-0016-01.

References

1. Koehn, P.: Europarl: a parallel corpus for statistical machine translation. In: Conference Proceedings: The Tenth Machine Translation Summit, pp. 79–86. Phuket, Thailand, AAMT, AAMT (2005)
2. Vu, T.T., Tran, G.B., Pham, S.B.: Learning to simplify children stories with limited data. In: Nguyen, N.T., Attachoo, B., Trawiński, B., Somboonviwat, K. (eds.) ACIIDS 2014. LNCS (LNAI), vol. 8397, pp. 31–41. Springer, Cham (2014). https://doi.org/10.1007/978-3-319-05476-6_4
3. Paetzold, G.H., Specia, L.: Benchmarking lexical simplification systems. In: LREC, pp. 3074–3080 (2016)
4. Chen, P., Rochford, J., Kennedy, D.N., Djamasbi, S., Fay, P., Scott, W.: Automatic text simplification for people with intellectual disabilities. In: AIST, pp. 1–9 (2016)
5. Leroy, G., Kauchak, D., Mouradi, O.: A user-study measuring the effects of lexical simplification and coherence enhancement on perceived and actual text difficulty. Int. J. Med. Inform. **82**, 717–730 (2013)
6. AMA: Health literacy: report of the council on scientific affairs. Ad hoc committee on health literacy for the council on scientific affairs, American Medical Association. JAMA, **281**, 552–557 (1999)
7. Mcgray, A.: Promoting health literacy. J. Am. Med. Inform. Assoc. **12**, 152–163 (2005)
8. Rudd, E.: Needed action in health literacy. J. Health Psychol. **18**, 1004–10 (2013)
9. Grabar, N., Cardon, R.: CLEAR - Simple corpus for medical French. In: Workshop on Automatic Text Adaption (ATA), pp. 1–11 (2018)
10. Agirre, E., Cer, D., Diab, M., Gonzalez-Agirre, A., Guo, W.: *SEM 2013 shared task: semantic textual similarity. In: *SEM, pp. 32–43 (2013)

11. Agirre, E., et al.: SemEval-2015 task 2: semantic textual similarity, English, Spanish and pilot on interpretability. SemEval **2015**, 252–263 (2015)
12. Agirre, E., et al.: SemEval-2016 task 1: semantic textual similarity, monolingual and cross-lingual evaluation. SemEval **2016**, 497–511 (2016)
13. Madnani, N., Tetreault, J., Chodorow, M.: Re-examining machine translation metrics for paraphrase identification. In: NAACL-HLT, pp. 182–190 (2012)
14. Clough, P., Gaizauskas, R., Piao, S.S., Wilks, Y.: METER: Measuring text reuse. In: ACL, pp. 152–159 (2002)
15. Zhang, Y., Patrick, J.: Paraphrase identification by text canonicalization. In: Australasian Language Technology Workshop, pp. 160–166 (2005)
16. Qiu, L., Kan, M.Y., Chua, T.S.: Paraphrase recognition via dissimilarity significance classification. In: Empirical Methods in Natural Language Processing, pp. 18–26. Sydney, Australia (2006)
17. Nelken, R., Shieber, S.M.: Towards robust context-sensitive sentence alignment for monolingual corpora. In: EACL, 161–168 (2006)
18. Zhu, Z., Bernhard, D., Gurevych, I.: A monolingual tree-based translation model for sentence simplification. COLING **2010**, 1353–1361 (2010)
19. Miller, G.A., Beckwith, R., Fellbaum, C., Gross, D., Miller, K.: Introduction to wordnet: An on-line lexical database. Technical report, WordNet (1993)
20. Ganitkevitch, J., Van Durme, B., Callison-Burch, C.: PPDB: the paraphrase database. In: NAACL-HLT, pp. 758–764 (2013)
21. Mihalcea, R., Corley, C., Strapparava, C.: Corpus-based and knowledge-based measures of text semantic similarity. In: AAAI, pp. 1–6 (2006)
22. Fernando, S., Stevenson, M.: A semantic similarity approach to paraphrase detection. In: Comp Ling UK, pp. 1–7 (2008)
23. Lai, A., Hockenmaier, J.: Illinois-LH: a denotational and distributional approach to semantics. In: Workshop on Semantic Evaluation (SemEval 2014), pp. 239–334. Dublin, Ireland (2014)
24. Wan, S., Dras, M., Dale, R., Paris, C.: Using dependency-based features to take the "para-farce" out of paraphrase. In: Australasian Language Technology Workshop, pp. 131–138 (2006)
25. Severyn, A., Nicosia, M., Moschitti, A.: Learning semantic textual similarity with structural representations. In: Annual Meeting of the Association for Computational Linguistics, pp. 714–718 (2013)
26. Tai, K.S., Socher, R., Manning, C.D.: Improved semantic representations from tree-structured long short-term memory networks. In: Annual Meeting of the Association for Computational Linguistics, pp. 1556–1566. Beijing, China (2015)
27. Tsubaki, M., Duh, K., Shimbo, M., Matsumoto, Y.: Non-linear similarity learning for compositionality. In: AAAI Conference on Artificial Intelligence, pp. 2828–2834 (2016)
28. Barzilay, R., Elhadad, N.: Sentence alignment for monolingual comparable corpora. In: EMNLP, pp. 25–32 (2003)
29. Guo, W., Diab, M.: Modeling sentences in the latent space. In: ACL, pp. 864–872 (2012)
30. Zhao, J., Zhu, T.T., Lan, M.: ECNU: one stone two birds: ensemble of heterogenous measures for semantic relatedness and textual entailment. In: Workshop on Semantic Evaluation (SemEval 2014), pp. 271–277 (2014)
31. Kiros, R., et al.: Skip-thought vectors. In: Neural Information Processing Systems (NIPS), pp. 3294–3302 (2015)
32. He, H., Gimpel, K., Lin, J.: Multi-perspective sentence similarity modeling with convolutional neural networks. In: EMNLP, pp. 1576–1586. Lisbon, Portugal (2015)

33. Mueller, J., Thyagarajan, A.: Siamese recurrent architectures for learning sentence similarity. In: AAAI Conference on Artificial Intelligence, pp. 2786–2792 (2016)
34. Sackett, D.L., Rosenberg, W.M.C., MuirGray, J.A., Haynes, R.B., Richardson, W.S.: Evidence based medicine: what it is and what it isn't. BMJ **312**, 71–2 (1996)
35. Cohen, J.: A coefficient of agreement for nominal scales. Educ. Psychol. Measur. **20**, 37–46 (1960)
36. Levenshtein, V.I.: Binary codes capable of correcting deletions, insertions and reversals. In: Soviet Physics. Doklady, p. 707 (1966)

MorphBen: A Neural Morphological Analyzer for Bengali Language

Ayan Das$^{(\boxtimes)}$ and Sudeshna Sarkar

Department of Computer Science and Engineering, Indian Institute of Technology Kharagpur, Kharagpur 721302, WB, India
{ayan.das,sudeshna}@cse.iitkgp.ernet.in

Abstract. Rule-based systems based on two-level morphology for tagging the morphological features of a word work quite well for Bengali language and are able to predict all possible morphological derivations for standard forms of words whose roots occur in the dictionary. However many words have multiple morphological derivations and the correct morphological derivation depends upon the context of the word. Non-dictionary words are also very frequent. Machine learning based methods have been used for predicting the values of morphological features of a word which take into account the context of the word. Although the machine learning systems to some extent can disambiguate the cases related to the words with multiple possible values, these systems needs to be improved to make more efficient use of the character-level information. Character-level information is particularly important for analysis of out-of-vocabulary (OOV) words which are not seen in the training data. We propose a method which makes use of both the context of the word as well as makes efficient use of the constituent characters of the words in order to develop a high quality morphological analyzer for Bengali. In this work we show that using character-level information along with the contextual information improves the performance of the morphological analyzer both for the OOV words and in predicting the correct analyses for the instances of the words that can have multiple morphological derivations.

Keywords: Morphological analyzer · Bengali language · Text analysis

1 Introduction

Morphological analysis is a very important natural language processing (NLP) task. Good quality morphological analyzers are crucial for more complex NLP tasks such as syntax analysis, semantic role labelling, machine translation etc. for highly inflected languages such as Bengali. Given annotated data, supervised machine learning techniques and rule-based systems have been used to develop highly accurate morphological analyzers in a language.

Our objective is to develop a good quality a morphological analyzer for Bengali language. Bengali belongs to the Indo-European language family. It is the

A. Gelbukh (Ed.): CICLing 2019, LNCS 13451, pp. 595–607, 2023.
https://doi.org/10.1007/978-3-031-24337-0_42

official language of Bangladesh and the second most spoken language after Hindi in India. Bengali is a very morphologically rich language. In Bengali the average number of surface forms corresponding to a root form is high as compared to other languages such as Hindi or English.

Bengali language shows both fusional as well as agglutinative inflectional properties. Besides having a large number of inflectional wordforms, more than one suffixes can be inflected with a word; e.g., in the word *chhelegulike*, the suffixes *-guli* (plural) and *-ke* (to) are used to convey the sense *to a group of boys*. In Table 1 we present some examples of inflectional forms of nouns and verbs. At the same time the syntactic forms does not always uniquely determine the morphological features. A word form can have several morphological derivations depending upon its context. In Table 2 we present some examples of such words which have multiple morphological derivations. These characteristics of Bengali language make its computational processing extremely challenging. It is necessary to capture the character-level information for predicting the morphological derivations of the words.

Table 1. Examples of inflected wordforms of different root forms corresponding to different parts of speech

Root form	POS	Inflected wordforms
khA (eat)	Verb	*khAchchhi, khAba, kheyechi, kheyechhilAm,* *khAo/khA, kheyechho, kheyechhe,kheyechhilo ,kheyechhen , kheyechhilen*
thAk (stay)	Verb	*thAkchhi, thAkbo, thekechhi, thekechhilam, thAko/thAk,* *thekechho, thekechhe, thekechhilo ,thekechhen, thekechhilen*
Ami (I)	Pronoun	*AmAr, AmAder, AmAke, AmArTA*
chhele (boys)	Noun	*chheleTA, chheleTAr, chheleder, chheleguli, chhelera, chheleTAke, chhelegulike*

In Table 2 we present some examples of some wordforms that can be derived from different rootforms depending on context.

Table 2. Example of wordforms that can be derived from multiple rootforms

Wordform	Morphological derivations
kare	root=*kar*, Verb, Person=3, Tense=Present
	root=*kare*, Affix
	root=*kar*, Noun, Case=Locative, Animacy=Inanimate
pAn	root=*pA*, Verb, Honorific=Hon, Person=3, Tense=Present
	root=*pAn*, Noun
	root=*pAn*, Noun, Case=Accusative,Animacy=Inanimate, Number=Singular
mAtAl	root=*mAtal*, Adjective
	root=*mAtAl*, Noun
	root=*mAtA* Verb, Person=3, Tense=Past

The rule-based system based on two-level morphology for tagging the morphological features proposed by [4] for Bengali language performs quite well in predicting the all possible morphological derivations of the words whose roots are present in its dictionary. However, contextual information is necessary to obtain the correct morphological derivation of an instance of a word that can have multiple morphological derivations depending on context. Also the system has to guess the morphological derivations for the OOV words.

The machine learning based approaches [5,8,12,13] for morphological analysis that have shown state-of-the-art performance over several languages takes into account the context of the word being analyzed. These are neural systems that use the representations of the words in a sentence to encode the sentence using a bidirectional recurrent neural networks. The representations of the words in these encoded sentences contain contextual information and hence to some extent helps in disambiguating the morphological derivation of an instance of a word when it can have multiple derivations. The input to these systems are the sentence-level and and the character-level encodings of the words. However these systems do not make use of the character-level information fully. But it is necessary to capture direct character-level information for particularly predicting the morphological derivations of the OOV words or the sub-word informations that are helpful for finding the morphological derivations.

We do not approach the task of morphological analysis as a sequence tagging task where the entire morphological derivation of a word is treated as single tag [5,8,12]. Instead we aim to predict the values corresponding to each morphological feature category (Case, Tense, Number, Person etc.) of a word and bundle up the individual predicted values corresponding to each feature category to form the morphological derivation of the word [13]. In this work, we propose a neural model for predicting the morphological derivations that makes use of both the contextual information and the character-level information of the words. We show that our system makes more efficient use of the character-level information derived from the constituent characters of a words. We also show that the contextual information along with the character-level representation significantly improves performance of the morphological analyzer in case of the OOV words and in deriving the correct derivation of the instances of the words with multiple possible morphological derivations.

For our work we have developed a dataset of 2776 sentences (40191 tokens) tagged using Universal Dependencies (UD) (universaldependencies.org) tagset. The UD provides a universal collection of PoS, morphological tagset [11], dependency relations [10], and, common annotation guidelines to facilitate consistent annotation of similar syntactic constructs across languages and also allows language-specific extensions. A morphological analyzer trained on data annotated using UD tags shall be helpful for the development of cross-lingual and multi-lingual systems for Bengali language.

The rest of the paper is organized as follows. In Sect. 2 we discuss some of the state-of-the-models for morphological analysis and the approaches for morphological analysis in Bengali language reported in the literature. In Sect. 3

we discuss in details our neural model for morphological analysis. In Sect. 4 we discuss in details the dataset used for our experiments, the settings for our experiments and detailed analysis of the results and the Sect. 5 concludes the paper.

2 Related Work

In this section, we discuss some of the most recent approaches to morphological analysis and the work on morphological analysis in Bengali language that has been reported in the literature.

2.1 Recent Models for Morphological Analysis

In this section we discuss some of the recent models for morphological analysis.

Morphological Anlyzer: The system reported by [12] in CONLL 2018 shared task on Universal Dependencies gave the highest average morphological features accuracy over all the languages among the other submitted system. They have used the architecture proposed by [5] where they have used a two-layer Bi-LSTM structure. In the first layer, a sentence is encoded both at character-level and the word-level using two separate BiLSTMs. The BiLSTM layer at the second level encodes the sentence using the representation of a word derived from the outputs of the first-level Bi-LSTMs. The representation of a word input to the BiLSTM at the second level consists of the character-level encoding and word-level encodings in the form of hidden states of the first-layer BiLSTMs corresponding to the word. These systems treats the entire morphological derivation of a word as a single tags that boils down the task of morphological analysis to a sequence tagging task.

[13] proposed an encoder-decoder based model for morphological analysis where they used a character-level Bi-LSTM to encode each word. They used these encoded word representations as input to a BiLSTM to obtain an encoding of a sentence where each word is finally represented by the hidden state of the BiLSTM corresponding to the word. They have explored a number of decoders to predict the various morphological features from the encodings.

Morphological Analysis in Bengali Language: [4] proposed a rule-based two-layer morphological analyzer for Bengali language. [2] gave a detailed discussion on the rules for morphological analysis of the nouns and verbs in Bengali language. [3] proposed a rule-based approach to identify correct morphological analysis of each Bengali word-form by cancelling (or pruning) all analyses which show contextual incompatibility in case a word contains multiple morphological alternatives. In this approach they achieved an accuracy of 89.24% and were able to disambiguate a single set of morph features for 96.35% words. [7] proposed a morphanalyzer where they predict only the root form of a word. In this work they have used gold-standard part-of-speech as additional information.

3 Our Neural Model for Morphological Analysis

Although the rule-based systems [4] perform quite well for the wordforms whose roots are present in its dictionary, these systems do not take into consideration the context of the words and hence performs poorly for the words that can have multiple morphological derivations depending on context. Also these systems predicts the morphological derivations of the OOV words by guessing and thus performs poorly for the OOV words also.

In the system proposed by [13] the final prediction is done from the representation of a word encoded at the sentence-level and hence lacks sufficient character-level information which hinders its performance on OOV words. We call the system **Context**.

In this section, we describe our model for morphological analysis where we use both the contextual information as well as the character-level information to predict the morphological derivation of a word. We name our system **ContextChar**.

Figure 1 shows the block diagram of our proposed architecture where each block represents a component of our system.

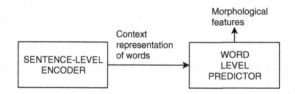

Fig. 1. Block diagram of our system for morphological analysis

Our system consists of two components: a sentence-level encoder and a morphological analyzer. The sentence level encoder generates the contextual representation of a word using a BiLSTM using pre-trained word embeddings and character-level encodings of the words. The morphological analyzer takes the contextual representation of the word as input and uses it as an additional input along with the character embeddings of a word to predict its morphological features.

3.1 Sentence Level Encoder

The sentence level encoder consists of a 3-layer Bi-LSTM which encodes a sentence as a sequence of words. Each word is represented by a concatenation of a pre-trained embedding and a character-level representation. The character-level representation of a word is obtained by encoding the word using a CNN which takes the embeddings of the constituent characters of the word as input. We also carried out an experiment where we used BiLSTM based character-level word representation along with the pre-trained word embeddings [13]. The results comparable but slightly worse. Hence, we report on the result for CNN based representation only. The concatenation of the hidden states of the forward and backward hidden states corresponding to a word in the sentence is passed to the word-based predictor as contextual representation of a word.

From the sentence level encodings of the words we also predict the PoS tags of the words by using a CRF model. We pass the contextual representation of the word as input to a CRF layer to predict the part of speech of the word.

Figure 2 shows the steps for encoding the character-level representations of the word *Ami* using CNN (Fig. 2a) and the sentence level encoder model (Fig. 2b).

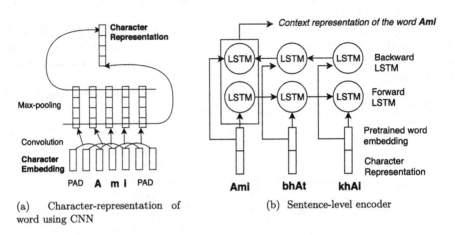

(a)   Character-representation of word using CNN

(b)   Sentence-level encoder

Fig. 2. Figure 2a shows the generation of character-level representaion of a word using CNN and 2b shows the sentence encoder model using BiLSTM. The system is run on the example sentence *Ami bhAt khAi* (*I eat rice*). The sentence is encoded by the BiLSTM using the representations of the words. The input to the BiLSTM for the word is the concatenation of its pre-trained word embedding and its character-level representation.

3.2   Word Level Predictor

Our morphological analyzer model consists of a CNN which takes as input the representations of the constituent characters of the word as input. Each character is represented by the concatenation of its pre-trained embedding and the *contextual representation* of the corresponding word from the sentence-level encoder. Figure 3 shows our morphanalyzer model. Each character is represented by the concatenation of its embeddings and the context-level representation of the word obtained from the sentence-level encoder. The context-level representation of the word is the concatenation of the forward and backward BiLSTM of the sentence-level encoder corresponding to the word in the sentence. In order to predict the values corresponding to the different morphological categories we apply a MLP layer with ReLU activation function followed by a softmax layer. For each morphological category we use a separate MLP layer.

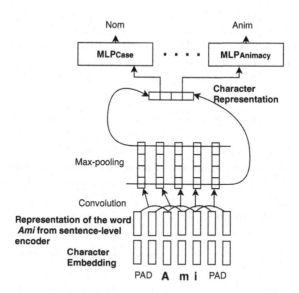

Fig. 3. The neural architecture for generation of representation of a word using CNN. Each character of the word being encoded is represented by the concatenation of its pretrained embedding and the contextual representation of the word from the sentence-level encoder.

4 Dataset, Experiments and Results

In this section we describe in details the experiments and the dataset used.

4.1 Data

For our experiments we have used a manually annotated Bengali treebank which consists of 2776 sentences (40191 tokens) annotated using UD PoS, morphology and dependency relations. For training we have randomly chosen a subset of 2331 sentences and tested our system on 345 sentences. We used the remaining 100 sentences as development set. In the test data, 15.5% of the words are OOV.

For pre-trained word and character embeddings we have used the fastText [6] word embeddings pre-trained on Wikipedia.

In Table 3 we present the morphological feature categories and their corresponding values. Corresponding to each morphological feature we present the PoS groups associated with it. The groups are *noun* group which includes nouns, pronouns, proper nouns, and the verb group include the verbs and auxiliary verbs. We also show the parts of speech groups for which the morphological features are applicable.

The data is also tagged with UD PoS tags. Our PoS tagset consists of 24 UD PoS tags. The PoS tags are as follows; DET, VERB:VINF, PART, X, NOUN, AFFIX, PUNCT, NUM, CCONJ, ADV, NOUN:VN, VERB:VB, VERB:VC, ADP, VERB:VBI, PROPN, VERB:VBN, AUX, INTJ, ADJ, SYM, PRON, SCONJ. As seen from the tagset, we have classified the verbs into 5 finer categories instead of using separate finer PoS tags.

Table 3. Morph features and their types.

Morph features	PoS categories	Feature values
Case	noun, adposition, verbal noun	Accusative, Nominative, Oblique, Locative, Ablative, Instrumental, Dative, Genitive
Polarity	verb, adverb	Negative
Animacy	noun	Inanimate, Animate
Mood	verb	Indicative, Imperative, Inter
Definiteness	noun, determiner	Definite
Number	noun	Singular, Unspecified_Number, Plural
Causative	verb	Yes
Person	pronoun, verb	1, 2, 3
Tense	verb	Past, Future, Present
Aspect	verb	Imperfect, Perfect
Honorific	noun, verb	Honorific, Unspecified_Honorific
Proper	noun	True
Emphatic	all	Emphatic
Prontype	pronoun	Relative, Interrogative, Total, Demonstrative, Personal, Indefinite, Reflexive

4.2   Implementation Details

We implemented our system using Tensorflow 1.10 [1]. The word and character embeddings sizes were set to 300 and 100 respectively. We used a dropout rate of 0.5 throughout for the LSTMs. The hidden layer size in the sentence-level encoder was set to 180. For the CNN, we used 100 filters and filter size of 3 which were empirically found to give the best result.

4.3 Morphological Analysis

For tagging the words with values corresponding to different morphological categories we have explored the following word representations.

Baseline Models: We compare our proposed model with the *Context* model proposed by [13]. In this model we got best results with the decoder that predicts the feature values using separate classifiers for each feature category.

In Table 4 we report the results corresponding to the different models.

Table 4. Comparison of overall accuracies of morphological analyzer corresponding to the different morphological categories for different settings.

Model	Case	Polar	Anim	Mood	Per	Caus	Num	Def	Tens	Aspect	Honor	Prop	Emph	Pron
Context	87.5	**99.9**	90.9	98.9	98.1	**99.5**	83.2	99.3	99.2	98.3	87.6	**98.2**	**99.7**	99.1
ContextChar	**90.6**	99.9	**93.1**	**99.1**	**98.5**	99.5	**86.4**	**99.4**	**99.3**	**98.7**	**90.8**	98.0	99.7	**99.2**

We observe that our model performs significantly better than the baseline system in for the *Case, Animacy, Mood, Number, Aspect, Honorific* and *Proper*. We observe that among the morphological feature categories, the models give very high accuracy for *Polarity, Causative, Definiteness, Tense, Emphatic, Person* and *Prontype*.

From the results we observe the following;

- Since the inflections carry significant information about the role of a word in a sentence, the values corresponding to some feature categories can be distinctly identified even for unseen words.
- Jointly using both the context representation and character representation of a word gives the better results in terms of most of the morphological feature categories.

All Tags Accuracy: In Table 5 we report the accuracy of the systems over all the morphological tags. We consider the automatic tagging of a word to be correct if the values under all the feature categories are correct. We classified the words into *nouns, verbs* and *others*. The *nouns* group include the nouns, pronouns, proper nouns and the verbal nouns, the *verbs* group include the finer classes of verbs and auxiliary verbs and the *others* group includes the rest of the parts of speech. We find that the drop in accuracy of the words belonging to the group of nouns is due to the errors in the *Case* and *Number* features. In case of the verbs the errors in *Honorific* feature is one of the most potent reason for drop in accuracy. Besides, we also found that the *Aspect* feature is also quite ambiguous needs an understanding of the meaning of the sentence for disambiguation which results is drop in accuracy for the verbs.

Table 5. Accuracy of morph feature tagger for different parts of speech. The *nouns* group includes *nouns, pronouns, proper nouns* and *verbal nouns*, the *verbs* group includes all the finer classes of verbs, and the rest of the parts of speech are included in the *others* group.

Features	Context	ContextChar
Nouns	46.4	**58.3**
Verbs	63.7	**72.0**
Others	86.1	**90.4**

In Table 6 we report the results corresponding to the morphological features for which significant differences in accuracy is observed under the different categories. The features are as follows; *Case, Animacy, Number* and *Honorific*. Although the performance of the systems are comparable in terms of *Aspect* and *Proper*, we analyze the errors committed by the systems since the accuracies are not very high. We classify the words under two criteria: OOV or non-OOV, and, *single-feature* or *multi-feature*. By *single-feature* we mean that a particular word form can take a single value corresponding to a morphological feature.

Case: In case of OOV words we observe that our model performs particularly better in identifying the *Genitive* and *Accusative* case when the word explicitly contains a distinct marker indicative of these cases e.g. in the words *Abe-dankArIke, debAshiske, Ahmedke* the *-ke* marker is an indicator of *Accusative* case and for the words *hoTeler, italir* the *-r* marker is an indicator of the Genitive marker. Also in case of the words such as *dhare, niye, kare* the words can take either *Locative* case or no case at all depending on whether its PoS is ADP or VERB. We observe that the combined representation better disambiguates these cases than our baseline.

However, the case markers in Bengali language are highly ambiguous e.g., the *null* suffix can both indicate *nominative* as well as *accusative* case. Similarly, *-er* suffix can be used for both *accusative* as well as *genitive* case. This ambiguity is one of the primary reason for high error rate for the *Case* feature.

Animacy: We observe that our model labels the words which have certain suffixes that are specific to the *animate* nouns such as *-der, -ke, -bAbu, -mantrI, -rA* and *inanimate* nouns such as *-e, -te*. It is interesting to note that the character-level representation essentially helps to capture the information regarding the animacy of a word from sub-word information. For example, tags the OOV words *mistri, bhAiyer, kachikAnchArA* as animate which actually have subwords common with *rAjmistrir, bhAike, bhAi, kachikAnchAder* in the training data. Similarly, observations hold for the following words *compAniguli, owArDti, pUr-basthalI* as inanimate which share sub-words with the words *compAni, owArDer, pUrbasthalIte* in the training data. We observe that the words such as *tAr* is always a pronoun but may be both animate or inanimate depending on object it is referring to in a context. Also *pare, bhitare* are inanimate when used as noun but does not have any animate attribute as postposition. These ambiguities are difficult to resolve. However, the combination of context and character-level information resolved some more cases in the data correctly as compared to model using only context information.

Number: We observe that in case of most of the OOV words the baseline system either tags it as an entity whose number cannot be specified (Unspecified number) or abstains from assigning a tag. However, we observe that our system identifies the words with -*TA*, -*ke*, -*er* suffixes as Singular and -*der*, -*ra* as plural. In case of the words such as *ei*, *tate*, *ta* which take the number of the noun they are referring to it is difficult to predict the correct number without sufficient context information. However, the character-level information along with the contextual information shows better performance in terms of the multi-feature words in terms of this feature. We observe significant number of errors occur in case of the words for which the information about number is not explicitly available from the word itself e.g., *pAnchTi chhele*. The word *chhele* can stand for both singular and plural. The word *pAnchTi* (five) indicates that in this context the number of *chhele* is plural.

Honorific: We observe that the majority of the errors under this feature category occur in case of the nouns. This is because the honorificity of a noun usually depends upon the honorificity of its head verb which will require a dependency relation to resolve. However, we observe that in case of the OOV words our systems tens to label the nouns as unspecifiable and does not tag the non-nouns and non-verbs at all. However, the baseline does not allocate any tag to the majority of the OOV words. However, we observe that our system correctly tags some nouns as *honourable* when has some suffixes such as -*ttam*, -*bAbu* and some subwords such as *pratibeshI*, *AtmIya*, *bAngAli* etc. We also note that it correctly tags some verbs as honourable that ends with -*n* such as *tolen*, *kAtAben*. The honorificity of most of the words that are *multi-featured* with respect to the Honorific category depend upon their PoS tag in a context. For example, the words *kare*, *hay*, *pare* can take the PoS tags VERB, AFFIX and ADP depending upon its usage. These words take honorific value only as VERB.

Table 6. Comparison of the accuracies of the morphological analyzer models for different morphological features. The accuracies are reported based on the words that belong to any of the categories: *OOV:* Out of vocabulary words, *non-OOV:* Words seen during training, *single-feature* and *multiple-features*.

Features	OOV		non-OOV		single-feature		multiple-features	
	Context	ContextChar	Context	ContextChar	Context	ContextChar	Context	ContextChar
Case	75.7	**79.1**	90.4	**92.7**	92.8	**94.6**	79.1	**82.9**
Animacy	77.4	**84.1**	92.2	**94.7**	93.7	**95.9**	77.5	**83.8**
Number	69.2	**75.9**	85.5	**89.5**	91.1	**93.3**	65.0	**74.2**
Honorific	74.4	**82.5**	89.9	**92.9**	92.1	**95.1**	77.9	**83.2**

4.4   PoS tagger

From the encoded sentence, the PoS tag of a word is predicted by a conditional random fields [9] (CRF) layer which takes the contextual representations of the words as input. The contextual representation of a word consists of the concatenation of the final hidden layers of the forward and backward LSTMs of the PoS tagger corresponding to the word. The representation is passed through a multilayer perceptron (MLP) with ReLU activation function to reduce the dimension before passing it to the CRF layer.

In Table 7 we report the accuracies of the PoS tagger. We observe that although we use the same neural architecture for predicting the PoS tags, training the PoS tagger with our proposed morphological analysis model results in improvement in performance of PoS taggers, particularly in case of OOV words and the words that can take multiple PoS tags.

Table 7. Accuracy of the PoS tagger.

PoS tagger model	Overall accuracy	OOV accuracy	Words in multiple PoS classes
PoS tagger (morph. from context representation)	89.8	71.6	82.2
PoS tagger (morph. using our model)	92.1	80.1	84.0

5   Conclusions

In this paper we propose a model for morphological analysis in Bengali language which uses both character-level word representation and the context representation from word-level sentence encoding to predict the values corresponding to the different morphological categories. We show that using character-level encoding actually helps to improve the accuracy of prediction of the values corresponding to some feature values based on character patterns within a word. Also when trained with a PoS tagger by parameter sharing it helps to improve the performance of the PoS tagger.

We aim to explore some opportunities of introducing some language specific features to improve the performance of the system and improve the agreement among the morphological feature values in Bengali language.

References

1. Abadi, M., et al.: TensorFlow: large-scale machine learning on heterogeneous systems (2015). https://www.tensorflow.org/. Software available from tensorflow.org
2. Ali, M.N.Y., Al-Mamun, S.M.A., Das, J.K., Nurannabi, A.M.: Morphological analysis of bangla words for universal networking language. In: 2008 Third International Conference on Digital Information Management, pp. 532–537 (Nov 2008). https://doi.org/10.1109/ICDIM.2008.4746734

3. Barik, B., Sarkar, S.: Pattern based pruning of morphological alternatives of bengali wordforms. In: 2014 International Conference on Advances in Computing, Communications and Informatics (ICACCI), pp. 1724–1730 (2014). https://doi.org/10.1109/ICACCI.2014.6968551

4. Bhattacharya, S., Choudhury, M., Sarkar, S., Basu, A.: Inflectional morphology synthesis for bengali noun, pronoun and verb systems. In: In Proceedings of the National Conference on Computer Processing of Bangla NCCPB, pp. 34–43 (2005)

5. Bohnet, B., McDonald, R., Simoes, G., Andor, D., Pitler, E., Maynez, J.: Morphosyntactic tagging with a meta-bilstm model over context sensitive token encodings. arXiv preprint arXiv:1805.08237 (2018)

6. Bojanowski, P., Grave, E., Joulin, A., Mikolov, T.: Enriching word vectors with subword information. Trans. Assoc. Comput. Linguist. 5, 135–146 (2017)

7. Chakrabarty, A., Garain, U.: Benlem (a bengali lemmatizer) and its role in WSD. ACM Trans. Asian Low-Resour. Lang. Inf. Process. 15(3), 12:1–12:18 (Feb 2016). https://doi.org/10.1145/2835494, http://doi.acm.org/10.1145/2835494

8. Dozat, T., Qi, P., Manning, C.D.: Stanford's graph-based neural dependency parser at the conll 2017 shared task. In: Proceedings of the CoNLL 2017 Shared Task: Multilingual Parsing from Raw Text to Universal Dependencies, pp. 20–30 (2017)

9. Lafferty, J., McCallum, A., Pereira, F.C.: Conditional random fields: Probabilistic models for segmenting and labeling sequence data (2001)

10. Nivre, J., et al.: Universal dependencies v1: a multilingual treebank collection. In: Proceedings of the 10th International Conference on Language Resources and Evaluation (LREC 2016), pp. 1659–1666. European Language Resources Association, Portorož, Slovenia (2016)

11. Petrov, S., Das, D., McDonald, R.: A universal part-of-speech tagset. In: Chair, N.C.C., et al., (eds.) Proceedings of the Eight International Conference on Language Resources and Evaluation (LREC 2012). European Language Resources Association (ELRA), Istanbul, Turkey (2012)

12. Smith, A., Bohnet, B., de Lhoneux, M., Nivre, J., Shao, Y., Stymne, S.: 82 treebanks, 34 models: universal dependency parsing with multi-treebank models. In: Proceedings of the CoNLL 2018 Shared Task: Multilingual Parsing from Raw Text to Universal Dependencies, pp. 113–123. Association for Computational Linguistics (2018), http://aclweb.org/anthology/K18-2011

13. Tkachenko, A., Sirts, K.: Modeling composite labels for neural morphological tagging. In: Proceedings of the 22nd Conference on Computational Natural Language Learning, pp. 368–379. Association for Computational Linguistics (2018). http://aclweb.org/anthology/K18-1036

CCG Supertagging Using Morphological and Dependency Syntax Information

Luyện Ngọc Lê[iD] and Yannis Haralambous[(✉)][iD]

IMT Atlantique & UMR CNRS 6285 Lab-STICC, CS 83818,
29238 Brest Cedex 3, France
luyenlengoc@gmail.com, yannis.haralambous@imt-atlantique.fr

Abstract. After presenting a new CCG supertagging algorithm based on morphological and dependency syntax information, we use this algorithm to create a CCG French Tree Bank corpus (20,261 sentences) based on the FTB corpus by Abeillé *et al.* We then use this corpus, as well as the Groningen Tree Bank corpus for the English language, to train a new BiLSTM+CRF neural architecture that uses (a) morphosyntactic input features and (b) feature correlations as input features. We show experimentally that for an inflected language like French, dependency syntax information allows significant improvement of the accuracy of the CCG supertagging task, when using deep learning techniques.

Keywords: CCG supertagging · Dependency syntax · FTB corpus · BiLSTM · CRF

1 Introduction

Combinatory Categorial Grammars (CCG) [36] provide a transparent interface between syntax and underlying semantic representation. They allow access to a deep semantic structure of the sentence and facilitate recovering of non-local dependencies involved in the construction such as coordination, extraction, control, and raising. CCGs have been introduced by Mark Steedman [34,35] as a non-transformational grammatical theory relying on combinatory logic. CCGs are strongly lexicalized in the sense that words are associated with one or more syntactic types, called *lexical categories*. These can be basic (e.g., S, NP, PP) or complex, obtained by using the *functors* / and \ on basic categories (e.g., S/NP, NP\PP etc.). Each lexical category has a specific meaning, for example, NP is the noun or the noun phrase, S\NP represents a verb phrase or an intransitive verb that requires a subject (NP) on its left as argument (cf. Fig. 1 for an example). The process of assigning lexical categories to words is called *supertagging* because, contrarily to POS tags, CCG tags are detailed syntactic structures.

In this paper we first assign CCG labels to words using syntax dependencies and POS tags, and then we build complete CCG derivation trees for sentences in a traditional approach. Afterwards we use this information as input features to

A. Gelbukh (Ed.): CICLing 2019, LNCS 13451, pp. 608–621, 2023.
https://doi.org/10.1007/978-3-031-24337-0_43

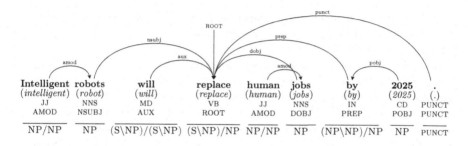

Fig. 1. A sentence with POS tags, dependencies and CCG lexical categories.

train a neural network, in order to improve the accuracy of a CCG supertagging task model.

CCG supertagging plays an important role in parsing systems, as a preliminary step to the build of complete CCG derivation trees. In general, this task can be considered as a sequence labeling problem with input sentence $s_{\text{input}} = (w_1, w_2, \ldots, w_n)$ and the CCG supertags $s_{\text{output}} = (t_1, t_2, \ldots t_n)$ as output. Input features can be words or they can be extracted from words, such as suffix, capitalization property or characters selection [2,23,25,40,41]. We will use morphosyntactic annotations such as lemma, suffix, POS tags and dependency relations [20] to build feature sets. These annotations are extremely useful in order to add additional information about word as well as long-range dependencies in the sentence (Fig. 1). These novel features allow us to improve accuracy of a supertagging neural network. We also consider adding correlations between features as additional input features of the network and examine the results.

In the past few years, Recurrent Neural Networks (RNN) [15] along with its variants such as Long-Short Term Memory (LSTM) [14,17] and GRU [8] have been proven to be effective for many NLP tasks, and especially for sequence labeling such as POS tagging, Named Entity Recognition etc. In the CCG supertagging task, different RNN-based models have been proposed and have obtained high accuracy results. Following this trend, we base our model on the Bi-Directional LSTM (BiLSTM) architecture associated with Conditional Random Fields (CRF) as output layer. Thus, we take advantage of the ability to remember the information of previous and next words in the sentence with the BiLSTM network and increase the ability to learn from the relationship of output labels with CRF.

The main contributions of our work are:

1. the use of morphosyntactic information for the traditional supertagging task;
2. the creation of a CCG Tree Bank for French language using this method;
3. the use of a new neural network architecture based on BiLSTM and CRF for the supertagging task trained on a standard English Tree Bank and on our French Tree Bank.

As we will see, morphosyntactic information and the new neural network architecture improve the task performance significantly in the case of French language, but not significantly in the case of English language.

The remainder of the paper is organized as follows. In the next section we describe our approach of using POS tags and dependency relations in (traditional) CCG supertagging. In Sect. 3, we discuss the state of the art of machine learning methods for CCG supertagging. In Sect. 4 we present our new neural network model architecture. In Sect. 5, we evaluate our method on an English and a French corpus, the latter been developed by ourselves using the methods described in Sect. 1. Finally, we conclude and discuss open questions.

2 From Dependency Syntax to CCG Derivation Tree

In [1], Abeillé, Clément & Kinyon announce a French Tree Bank based on syntax constituents. In [6,7] Candito, Crabbé & Denis convert this Tree Bank into syntax dependencies. We use information from these two corpora to build a CCG Tree Bank, as described in this section.

Dependency parsing consists in building a tree rooted at the head of the sentence (usually the verb), the edges of which, called *dependencies*, connect words and are labeled by syntactic functions, e.g., subject, object, oblique, determiner, attribute etc. Dependency syntax trees are obtained by parsers such as Malt-Parser [30], Stanford Parser [12], MST parser [28], Spacy [19], etc. These tools also provide POS tags of words.

In order to assign CCG lexical categories to words of a sentence, we start by calculating its dependency tree. Then, we process words which have unique lexical categories in the corpus: e.g., nouns have lexical category NP, adjectives have lexical category NP/NP or NP\NP depending whether they are on the left or on the right of the noun, etc. Once we have assigned these unique (or position-dependent, as in adjectives) lexical categories, we move over to verbs.

The main verb of the sentence, which is normally the root of the dependency tree, may have *argument dependencies*, labeled *suj, obj, a_obj, de_obj, p_obj*, i.e., correspondences with subject, direct and indirect object, and/or *adjunct dependencies* labeled *mod, ats*, etc., representing complementary information such as number, time, place, and so on. We assign lexical category S\NP to a main verb having a subject to its left, and then we add a /NP (or a \NP, depending on its position with respect to the verb) for each direct object or indirect object (in the order of words in the sentence).

Our next step is to binarize the dependency tree on the basis of information about dominant sentence structure: In French, most sentences are SVO, as in *"Mon fils (S) achète (V) un cadeau (O)"* (My son buys a gift), or SOV as in *"Il (S) le (O) donnera (V) à sa mère (indirect O)"* (He will give it to his mother). Using this general linguistic property, we can extract and classify the components of the sentence into: subject, direct object, indirect object, verbs, complement phrases.

The algorithm we propose for transforming a dependency tree into a binary tree consists is subdivided into two steps:

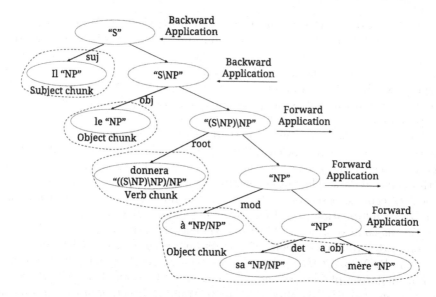

Fig. 2. CCG derivation tree of sentence "He will give it to his mother."

1. we extract chunks from the dependency tree based on syntactic information and dependency labels between words. For example, the subject chunk is obtained by finding a word that has a dependency labeled *suj*, the verb chunk corresponds to the root of the dependency structure, direct or indirect object chunks are obtained as words with links directed to the root verb and having labels *obj* or *p_obj*, etc.;

2. we build a binary tree for each chunk, and then combine the binary trees in inverse order of the dominant sentence structure. For example if SVO is the dominant structure, we start by building the binary tree of the *object* chunk, then combine it with the binary tree of the *verb* chunk, and finally we obtain the binary tree of the *subject* chunk.

In Fig. 2, the reader can see four chunk groups in the dependency tree, displayed as regions of the binarized tree.

By applying this approach to the complete set of 21,550 dependency trees of the FTB corpus, we have obtained a bank of CCG derivation trees for 94,02 % of the sentences. This new corpus is available (together with the code implementing the algorithm described above) as a resource in the frame of the verifiability, reproducibility, and working description policy of CICLING 2019 conference.

3   Machine Learning and Supertagging

One of the first applications of machine learning to CCG supertagging is the development of a statistical parser by Clark & Curran [10]. They proceed in two main steps: supertagging and combining of lexical categories. Their supertagging approach is based on the log-linear model by using the lexical category set in a local five-word context to obtain a distribution. The model's features are words POS tags included in the five-word window, plus the two previously assigned lexical categories (to the left). They applied their method on the CCG Bank corpus [18] with 92.6% of accuracy for words and (only) 36.8% of accuracy for complete sentences (that is the percentage of sentences of which *all* words are tagged correctly).

Like many others supervised methods, CCG supertagging requires a sufficiently large amount of labeled training data to achieve a good result. Mike & Mark [26] have introduced a semi-supervised approach to improve a CCG parser with unlabeled data. They have constructed a model for the prediction of CCG lexical categories, based on vector-space embeddings. Features are words and some other information (e.g., POS tagging, chunking, named-entity recognition, etc.) in the context window. Their experiments used the neural network model of Collobert [11] in association with conditional random fields (CRF) [38].

Using RNN for CCG supertaging has been proven to provide better results with a similar set of features and window size in the work of Xu [42]. However, the conventional RNN is often difficult to train and there still exist problems such as gradient vanishing and exploding, in the layers over long sequences [4,31]. Therefore, LSTM networks—a special variant of RNN which is capable of learning long-term dependencies—were proposed to overcome these RNN limitations. In particular, Bi-directional LSTM network models have been created with the ability to store two-way information, and the majority of literature in the area [2,23,25,40,41] uses this model with different training procedures and achieves high accuracy.

The performance of BiLSTM networks models has been improved by combining them with a CRF model for the sequence labeling task [21,27,36]. Using a BiLSTM-CRF model similar to the one in [21], the authors of [22] have shown the efficiency of CRF by achieving a higher accuracy in CCG supertagging and multi-tagging tasks.

In most of the above works, similarly to many sequence labeling tasks, the model inputs are words and their features are extracted directly from words. However, we claim that lexical categories assignment to words can use morphological and dependency syntax to enrich the feature set. In the following section, we present a neural network model based on BiLSTM-CRF architecture with moprhosyntactic features.

4   Neural Network Model

4.1   Input Features

We will use the following input features for words in sentences:

- the *word* per se (word);
- the *word lemma* (lemma);
- the *POS tag* of the word (postag);

– the *dependency relation* (deprel) of the word with its parent in the dependency tree (and the tag "root" for the head of the dependency tree, which has no parent).

Each one of these features provides predictive information about the CCG supertag label. Therefore, our input sentence will be $s = \{x_1, x_2, ...x_n\}$ where each x_i is a vector of the features $x_i = [\text{word}_i, \text{lemma}_i, \text{postag}_i, \text{deprel}_i]$.

Before describing our model, let us briefly review, in the following section, pre-existant models with which we will compare it.

4.2   Basic Bi-directional LSTM and CRF Models

Unidirectional LSTM Model. As mentioned earlier, the shortcomings of standard RNNs in practice involve gradient vanishing and an explosion problem when dealing with long term dependencies. LSTMs are designed to cope with these gradient problems. Basically, a conventional RNN is defined as follows: the input $x = (x_1, x_2, \ldots, x_T)$ feeds the network, and the network computes the hidden vector sequence $h = (h_1, h_2, \ldots, h_T)$, and the output sequence, $y = (y_1, y_2, \ldots, y_T)$, from $t = 1, \ldots, T$ where T is the number of time steps as in the following formulas:

$$h_t = f(Ux_t + Wh_{t-1} + b_h) \tag{1}$$
$$y_t = g(Vh_t + b_y), \tag{2}$$

where U, W, V denote weight matrices that are computed in training time, b denotes bias vectors and $f(z)$, $g(z)$ are activation functions.

Based on the basic architecture of a RNN, an LSTM layer is formed from a set of memory blocks [16,17]. Each block contains one or more recurrently connected memory cells and three gate units: input, output and forget gate. More specifically, activation computation in a memory cell at time step t is defined by the following formulas:

$$i_t = \sigma(W_{xi}x_t + W_{hi}h_{t-1} + W_{ci}c_{t-1} + b_i) \tag{3}$$
$$f_t = \sigma(W_{xf}x_t + W_{hf}h_{t-1} + W_{cf}c_{t-1} + b_f) \tag{4}$$
$$o_t = \sigma(W_{xo}x_t + W_{ho}h_{t-1} + W_{co}c_{t-1} + b_o) \tag{5}$$
$$c_t = f_t \odot c_{t-1} + i_t \odot \tanh(W_{xc}x_t + W_{hc}h_{t-1} + b_c) \tag{6}$$
$$h_t = o_t \odot \tanh(c_t), \tag{7}$$

where i_t, f_t, o_t, c_t correspond to input gate, forget gate, output gate and cell vectors, σ is the logistic sigmoid function, tanh is the hyperbolic tangent function, W terms denote weight matrices, and b terms denote bias vectors.

Bidirectional LSTM Model. In order to assign a supertag to a word, we need to use the word's information and its relations to the previous and next word in the sentence. Two-way information access from past to future and vice versa gives global information in a sequence. However, the LSTM cell only retrieves information from the past using input and output of the previous LSTM cell. In other words, an LSTM cell does not receive any information from the LSTM cell *following* it. Therefore, a Bi-Directional LSTM (BiLSTM) model has been proposed in [3,33] to overcome this problem, as follows:

$$\text{Bi-LSTM}_{\text{sequence}}(x_{1:n}) = \text{LSTM}_{\text{backward}}(x_{n:1}) \circ \text{LSTM}_{\text{forward}}(x_{1:n}). \quad (8)$$

In general architectures, one may have one forward LSTM layer and one backward LSTM layer for the complete sequence and run them in reverse time. The features of the two layers are concatenated at the level of the output layers. Thus, information from both the past and the future is transmitted to each memory LSTM cell. The hidden state is computed as follows:

$$h_t = f(W_{\overleftarrow{h}}\overleftarrow{h_t} + W_{\overrightarrow{h}}\overrightarrow{h_t}), \quad (9)$$

where $\overleftarrow{h_t}$ is backward hidden sequence, $\overrightarrow{h_t}$ is the forward hidden sequence.

CRF Model. BiLSTM networks are used to build efficient predictive models of the output sequence based on the features of the input sequence. However, they can not consider the correlation between output labels and their neighborhoods. In our case, CCG supertags, by nature, *always* have correlations with the previous or next labels, for example, an output CCG supertag of a word is NP/NP (usually an article), which allows us to predict the fact that the next CCG supertag is NP.

In order to enhance the ability to predict next labels from current label in an output sequence, two approaches can be used:

1. building a tag distribution for each training step and using an heuristic search algorithm to find optimal tag sequences [39];
2. focusing on the context with sentence-level information instead of only word-level information. The leading work of this approach is the CRF model of [24].

We use the second approach in the output layer of our model. The combination of BiLSTM network and CRF network can improve the efficiency of the model by strengthening the relationship between the output labels through the CRF layer, based on the input features through the BiLSTM layer.

4.3 Our Model for Feature Set Enrichment and CCG Supertagging

In the model we propose (see Fig. 3), each input is a set of features: word, lemma, POS tag and dependency relation. These features are vectorized with a fixed size by using the embedding matrix in the embedding layer. In the next layer, the correlation between pairs of features is calculated by combining them. Then, we use a BiLSTM network to memorize and learn the relations with other words in the context of the sentence for these pairs of features. After that, all features are concatenated to become the input of the second BiLSTM network layer. Finally, CCG supertag labels are obtained in the output of the CRF network layer, the input of which is the output of the 2nd output BiLSTM layer.

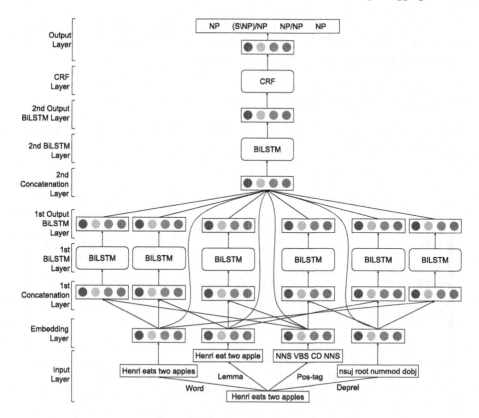

Fig. 3. Architecture of the BiLSTM network with a CRF layer on the outputs.

5   Evaluation

5.1   Dataset and Preprocessing

We use the two different corpora to experiment our model, one in English and one in French. The first corpus is the Groningen Meaning Bank (GMB) corpus [5] which has been built for deeper semantic analysis on a discourse scope. The second one is our CCG Corpus for French which we extracted from the French Tree Bank (FTP) corpus [1] by using the dependency analysis of the sentence (see Sect. 2).

In order to obtain a standard dataset for training process, we extract all sentences with annotations for each word such as lemma, POS tag, dependency relation and CCG label, for each corpus. As the GMB corpus does not contain dependency relations, we have used the Stanford Parser [12] to add it a posteriori.

We compare the structures of the two corpora in Table 1. In particular, there is a difference in the distribution of sentences according to their length (see also Fig. 4). In the GMB corpus, the distribution of the number of short sentences and long sentences is relatively similar. This is quite different in the FTB corpus where there are more short sentences, and where long sentences spread over a wider range. This difference of the distribution rate in the datasets can affect the training process outcomes of the two corpora.

Table 1. Statistics on the two corpora.

Statistic	#Sentences	#Words	#Word tokens	#Lemma tokens	#POS labels	#Deprels	#CCG labels
GMB	23,451	1,037,739	32,073	26,987	43	56	636
FTB	18,724	570,054	28,748	18,762	29	27	73

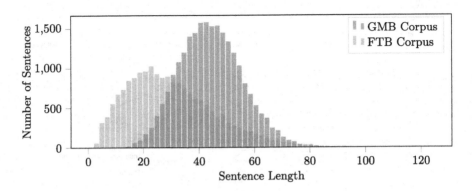

Fig. 4. Histogram of sentence length in the corpora.

5.2 Training Procedure

We implement our neural network by using the Keras deep learning library [9]. The datasets are divided into three sets: training set, validation set and test set with the proportion

$$(0.8 * (\text{training\_set}) + 0.2 * (\text{validation\_set})) * 0.8 + 0.2 * (\text{test\_set}).$$

Validation sets are used to measure performance at each epoch. Final evaluation on the test set is based on the best accuracy results in the validation sets.

Pre-trained Word Embeddings. In order to work with numeric data in the neural network, we use pre-trained word embeddings to transform words or lemmas of the corpora into numeric vectors. More specifically, we use Glove [32] (a 200-dimensional embedding trained on 6 billion words collected from Wikipedia) for the GMP corpus, and Word2vec [29] (the French version by Fauconnier [13], also with 200 dimensions, and trained on 1.6 billion words collected from the web) for the FTB corpus. For out-of-vocabulary words, we assign embeddings by random samples. Based on the distribution by length of sentences (Fig. 4), we assign a fixed length of 120 words to all sentences, so that the input dimension is $[120, 200]$ for each sentence input. Finally, the other features are transformed to numeric vectors by using a one-hot encoding matrix with size depending on their number in the dictionary.

Parameters and Hyperparameters. We fix the number of training examples (batch size) as 32 for each forward or backward propagation. Each training process runs 20 times to evaluate and compare outcomes (epoch). In addition, we have experimented with the number of different hidden states, such as 64, 128, 256, 512 to find a configuration that is optimally consistent with the model. We decided to carry on the experiment with 128 hidden states because this choice optimally balances accuracy and performance.

Optimization Algorithm. Choosing an optimizer is a crucial part of the model building process. Our model uses the Root Mean Square Prop (RMSprop) optimizer [37] which proceeds by keeping an exponentially weighted average of the squares from past gradients. To increase convergence, the learning rate is divided by this average:

$$v_{dw} := \beta v_{dw} + (1 - \beta) \cdot dw^2 \tag{10}$$

$$v_{db} := \beta v_{db} + (1 - \beta) \cdot db^2 \tag{11}$$

$$W := W - \alpha \frac{dw}{\sqrt{v_{dw}} + \epsilon} \tag{12}$$

$$b := b - \alpha \frac{db}{\sqrt{v_{dw}} + \epsilon}, \tag{13}$$

where v_{dw} v_{db} are the exponentially weighted averages from past squares of gradients, dw^2 and db^2 are cost gradient related to the current layer weight, W and b denote weight and bias, α is the learning rate from 0.9 to 0.0001 ($\alpha = 0.01$ is the default setting), β is an hyperparameter to be tuned and ϵ is very small to avoid dividing by zero.

5.3   Experimental Results

In order to evaluate the proposed input features and the model, we conduct two sorts of comparison:

1. we compare the outcomes of different feature sets such as [word], [word, suffix], [word, suffix, cap(italization)], [word, lemma, suffix, cap], [word, lemma, postag, suffix, cap], [word, lemma, postag, deprel, suffix, cap], [lemma, postag, deprel], [lemma, postag], [lemma];
2. we compare the outcomes of different neural network architectures such as BiLSTM [25], standard BiLSTM CRF [21], Double-BiLSTM CRF [22], and ours.

Evaluation on our test set is shown on Table 2 for the French FTB corpus and on Table 3 for the English GMB corpus.

According to our architecture and since we use correlations of features as additional features, our model requires at least two features in the input data. Therefore, we can not produce results on input data with a single feature like [word] or [lemma]. Nevertheless we compare the outcome of other models on our input features, including single input features.

Table 2. 1-best tagging accuracy comparison results on the test set in the French FTB Corpus.

Feature set	BiLSTM	BiLSTM CRF	Double BiLSTM CRF	Our model
word	*78.60*	*78.76*	*77.14*	–
word, suffix	*78.97*	*78.80*	*76.58*	78.90
word, suffix, cap	*78.56*	*78.97*	**75.96**	84.43
word, lemma, suffix, cap	79.16	79.67	78.78	78.49
word, lemma, postag, suffix, cap	81.28	81.84	81.24	81.50
word, lemma, postag, deprel, suffix, cap	83.23	83.95	83.56	84.06
word, lemma, postag, deprel	83.43	83.98	83.70	**85.05**
lemma, postag, deprel	83.00	83.05	83.15	82.40
lemma, postag	80.20	80.37	81.40	80.05
lemma	77.61	77.83	76.66	–

Let us first start with the French corpus. In Table 2 we have displayed methods from the literature in italics: word, suffix and cap(italization) as input features, BiLSTM, BiLSTM+CRF and Double BiLSTM+CRF as architectures. As the reader can see, by applying pre-existing methods we obtain a maximum accuracy of 75.96%. By using our input features with pre-existing architectures we obtain a maximum accuracy of 83.98%. By using our architecture with input features used by others we get an accuracy of 84.43%. Both of these results are significantly better than those in previous works. Finally, by combining our input features with our model we manage to gain another 1% and achieve a topmost accuracy of 85.05%.

It is interesting to notice that, even though the lemma feature carries less information than the word feature (as expected), the combination of lemma and POS tag features provides better results than the word feature, and that these results are systematically increased by 2% when dependency relations are added as well.

The accuracy results for the English GMB corpus are displayed on Table 3. Here differences are less significant, and the results all lie in the 92–94% range, with a single exception: the case of the lemma feature, where we lose about 2% of accuracy. Nevertheless, when we add the POS tag feature to the lemma feature, we get a slightly better result than the word feature (the difference is about 1%). The best result is an accuracy of 94.31%, obtained by our model, but not

Table 3. 1-best tagging accuracy comparison results on the test set in English GMB Corpus.

Feature set	BiLSTM	BiLSTM CRF	Double BiLSTM CRF	Our model
word	*92.83*	*92.49*	*91.16*	–
word, suffix	*93.08*	*92.93*	*91.57*	92.92
word, suffix, cap	***93.30***	*93.20*	*91.48*	**94.31**
word, lemma, suffix, cap	93.33	93.26	91.78	93.38
word, lemma, postag, suffix, cap	93.29	93.02	93.25	92.44
word, lemma, postag, deprel, suffix, cap	93.45	93.18	93.15	92.46
word, lemma, postag, deprel	93.35	93.18	93.24	92.90
lemma, postag, deprel	93.25	93.13	92.98	93.26
lemma, postag	93.21	93.12	92.98	92.95
lemma	90.56	90.13	89.86	–

with our morphosyntactic features but rather with the legacy word, suffix and cap(italization) features.

A general conclusion could be that morphosyntactic information brings a real advantage for neuronal supertagging of French (a language the verbs of which are highly inflected). It would be interesting to test the model with even more inflected languages such as German, Russian or Greek.

6 Conclusion

We have presented a new CCG supertagging task based on morphological and dependency syntax information, which has allowed us to create a CCG version of the French Tree Bank corpus FTB. We used this corpus to train a new BiLSTM+CRF neural architecture that uses new, morphosyntactic, input features as well as feature correlations as separate input features. We have experimentally shown that, at least for an inflected language as French, dependency syntax information is useful for improving the accuracy of the CCG supertagging task when using deep learning techniques.

Both the CCG French Tree Bank corpus we have developed as the code we used for the traditional and for the deep learning supertaggers are available in the frame of the verifiability, reproducibility, and working description policy of the CICLING 2019 conference.

References

1. Abeillé, A., Clément, L., Toussenel, F.: Building a treebank for French. In: Treebanks: Building and Using Parsed Corpora, pp. 165–187. Kluwer (2003)
2. Ambati, B.R., Deoskar, T., Steedman, M.: Shift-reduce CCG parsing using neural network models. In: Proceedings of NAACL 2016, pp. 447–453 (2016)

3. Baldi, P., Brunak, S., Frasconi, P., Pollastri, G., Soda, G.: Bidirectional dynamics for protein secondary structure prediction. In: Sun, R., Giles, C.L. (eds.) Sequence Learning. LNCS (LNAI), vol. 1828, pp. 80–104. Springer, Heidelberg (2000). https://doi.org/10.1007/3-540-44565-X_5

4. Bengio, Y., Frasconi, P., Simard, P.: The problem of learning long-term dependencies in recurrent networks. In: IEEE International Conference on Neural Networks 1993, pp. 1183–1188. IEEE (1993)

5. Bos, J., Basile, V., Evang, K., Venhuizen, N.J., Bjerva, J.: The Groningen meaning bank. In: Ide, N., Pustejovsky, J. (eds.) Handbook of Linguistic Annotation, vol. 2, pp. 463–496. Springer, Dordrecht (2017). https://doi.org/10.1007/978-94-024-0881-2_18

6. Candito, M., Crabbé, B., Denis, P.: Statistical French dependency parsing: treebank conversion and first results. In: Proceedings of LREC 2010, pp. 1840–1847 (2010)

7. Candito, M., Crabbé, B., Denis, P., Guérin, F.: Analyse syntaxique du français: des constituants aux dépendances. In: Proceedings of TALN 2009 (2009)

8. Cho, K., Van Merriënboer, B., Bahdanau, D., Bengio, Y.: On the properties of neural machine translation: encoder-decoder approaches. arXiv:1409.1259

9. Chollet, F.: Deep Learning with Python. Manning Publications (2018)

10. Clark, S., Curran, J.R.: Wide-coverage efficient statistical parsing with CCG and log-linear models. Comput. Linguist. **33**(4), 493–552 (2007)

11. Collobert, R.: Deep learning for efficient discriminative parsing. In: Proceedings of the Fourteenth International Conference on Artificial Intelligence and Statistics, pp. 224–232 (2011)

12. De Marneffe, M.C., MacCartney, B., Manning, C.D., et al.: Generating typed dependency parses from phrase structure parses. In: Proceedings of LREC 2006, vol. 6, pp. 449–454 (2006)

13. Fauconnier, J.P.: French word embeddings (2015). http://fauconnier.github.io

14. Gers, F.A., Schmidhuber, J., Cummins, F.: Learning to forget: continual prediction with LSTM. In: Proceedings of ICANN 1999. IET (1999)

15. Goller, C., Kuchler, A.: Learning task-dependent distributed representations by backpropagation through structure. Neural Netw. **1**, 347–352 (1996)

16. Graves, A., Schmidhuber, J.: Framewise phoneme classification with bidirectional LSTM and other neural network architectures. Neural Netw. **18**(5–6), 602–610 (2005)

17. Hochreiter, S., Schmidhuber, J.: Long short-term memory. Neural Comput. **9**(8), 1735–1780 (1997)

18. Hockenmaier, J., Steedman, M.: CCGbank: a corpus of CCG derivations and dependency structures extracted from the Penn Treebank. Comput. Linguist. **33**(3), 355–396 (2007)

19. Honnibal, M., Johnson, M.: An improved non-monotonic transition system for dependency parsing. In: Proceedings of the 2015 Conference on Empirical Methods in Natural Language Processing, pp. 1373–1378 (2015)

20. Honnibal, M., Kummerfeld, J.K., Curran, J.R.: Morphological analysis can improve a CCG parser for English. In: Proceedings of Coling 2010, pp. 445–453 (2010)

21. Huang, Z., Xu, W., Yu, K.: Bidirectional LSTM-CRF models for sequence tagging. arXiv:1508.01991

22. Kadari, R., Zhang, Y., Zhang, W., Liu, T.: CCG supertagging via Bidirectional LSTM-CRF neural architecture. Neurocomputing **283**, 31–37 (2018)

23. Kadari, R., Zhang, Y., Zhang, W., Liu, T.: CCG supertagging with bidirectional long short-term memory networks. Nat. Lang. Eng. **24**(1), 77–90 (2018)

24. Lafferty, J., McCallum, A., Pereira, F.C.: Conditional random fields: probabilistic models for segmenting and labeling sequence data. In: Proceedings of ICML 2001, pp. 282–289 (2001)
25. Lewis, M., Lee, K., Zettlemoyer, L.: LSTM CCG parsing. In: Proceedings of NAACL 2016, pp. 221–231 (2016)
26. Lewis, M., Steedman, M.: Improved CCG parsing with semi-supervised supertagging. Trans. ACL **2**, 327–338 (2014)
27. Ma, X., Hovy, E.: End-to-end sequence labeling via bi-directional LSTM-CNNS-CRF. arXiv:1603.01354
28. McDonald, R., Pereira, F., Ribarov, K., Hajič, J.: Non-projective dependency parsing using spanning tree algorithms. In: Proceedings of EMNLP 2005, pp. 523–530 (2005)
29. Mikolov, T., Sutskever, I., Chen, K., Corrado, G.S., Dean, J.: Distributed representations of words and phrases and their compositionality. In: Proceedings of NIPS 2013, pp. 3111–3119 (2013)
30. Nivre, J., Hall, J., Nilsson, J.: MaltParser: a data-driven parser-generator for dependency parsing. In: Proceedings of LREC 2006, pp. 2216–2219 (2006)
31. Pascanu, R., Mikolov, T., Bengio, Y.: On the difficulty of training recurrent neural networks. In: International Conference on Machine Learning, pp. 1310–1318 (2013)
32. Pennington, J., Socher, R., Manning, C.D.: Glove: global vectors for word representation. In: Proceedings of EMNLP 2014, pp. 1532–1543 (2014)
33. Schuster, M., Paliwal, K.K.: Bidirectional recurrent neural networks. IEEE Trans. Signal Process. **45**(11), 2673–2681 (1997)
34. Steedman, M.: Surface Structure and Interpretation. MIT Press, Cambridge (1996)
35. Steedman, M.: The Syntactic Process. MIT Press, Cambridge (2000)
36. Steedman, M., Baldridge, J.: Combinatory categorial grammar. In: Non-transformational Syntax: Formal and Explicit Models of Grammar, pp. 181–224. Wiley-Blackwell (2011)
37. Tieleman, T., Hinton, G.: Lecture 6.5-rmsprop: divide the gradient by a running average of its recent magnitude. COURSERA: Neural Netw. Mach. Learn. **4**(2), 26–31 (2012)
38. Turian, J., Ratinov, L., Bengio, Y.: Word representations: a simple and general method for semi-supervised learning. In: Proceedings of the 48th Annual Meeting of the Association for Computational Linguistics, pp. 384–394 (2010)
39. Vaswani, A., Bisk, Y., Sagae, K., Musa, R.: Supertagging with LSTMs. In: Proceedings of NAACL 2016, pp. 232–237 (2016)
40. Wu, H., Zhang, J., Zong, C.: A dynamic window neural network for CCG supertagging. In: Proceedings of AAAI 2017, pp. 3337–3343 (2017)
41. Xu, W.: LSTM shift-reduce CCG parsing. In: Proceedings of EMNLP 2016, pp. 1754–1764 (2016)
42. Xu, W., Auli, M., Clark, S.: CCG supertagging with a recurrent neural network. In: Proceedings of the 53rd Annual Meeting of the Association for Computational Linguistics, vol. 2, pp. 250–255 (2015)

Representing Overlaps in Sequence Labeling Tasks with a Novel Tagging Scheme: Bigappy-Unicrossy

Gözde Berk, Berna Erden$^{(\boxtimes)}$, and Tunga Güngör

Department of Computer Engineering, Boğaziçi University,
Bebek, 34342 Istanbul, Turkey
{gozde.berk,berna.erden,gungort}@boun.edu.tr

Abstract. Multiword expression (MWE) identification can be handled by using sequence tagging approach accompanied with stochastic models and variants of IOB tagging scheme. In this paper, we introduce a new tagging scheme called *bigappy-unicrossy* to rise to the challenge of overlapping MWEs. The bigappy-unicrossy tagging scheme is compared with the two other well-known tagging schemes which are IOB2 and gappy 1-level in the verbal multiword expression (VMWE) identification task using bidirectional Long Short-Term Memory model with a Conditional Random Field layer on top (bidirectional LSTM-CRF). Both the bigappy-unicrossy and the gappy 1-level tagging schemes outperform the IOB2 tagging scheme. The bigappy-unicrossy tagging scheme competes with the gappy 1-level tagging scheme. We believe that our tagging scheme will show better performance on corpora with higher frequency of overlapping cases.

Keywords: IOB tagging scheme · Multiword expressions · Gappy 1-level tagging scheme · Bigappy-unicrossy tagging scheme · Long Short-Term Memory

1 Introduction

Multiword expressions (MWEs) are lexical items consisting of more than one word that show some degree of idiomaticity at the lexical, syntactic, semantic, pragmatic, and/or statistical levels. The idiomatic character of MWEs means that the properties of MWEs cannot be derived from their component items. For example, the semantics of *kick the bucket* whose lexical meaning is *to die* is not predictable from its parts [1].

The processing of MWEs is a key issue for natural language processing (NLP) tasks such as parsing and machine translation. In the context of MWEs, the automatic annotation of MWEs in running text is defined as the task of MWE identification. The techniques used in identifying MWEs can be grouped into

---

G. Berk and B. Erden—These authors contributed equally to the work.

© Springer Nature Switzerland AG 2023
A. Gelbukh (Ed.): CICLing 2019, LNCS 13451, pp. 622–635, 2023.
https://doi.org/10.1007/978-3-031-24337-0_44

four main categories: rule-based methods, classifiers, sequence tagging models, and parsing [4].

PARSEME (PARSing and Multi-word Expressions) [11], a research community which is involved in the treatment of MWEs, organizes a shared task on automatic identification of verbal multiword expressions (VMWEs) which is one of the subtypes of MWEs. The community aims to compare and evaluate language-independent VMWE identification systems using a gold standard corpus annotated by PARSEME participants. For Edition 1.1. of the PARSEME Shared Task on automatic identification of VMWEs in 2018, the PARSEME network released annotated corpora for 20 languages. The corpora for each language are sampled from various resources such as news, newswire and articles, not restricted to a specific domain. In PARSEME Shared Task, systems can submit their results in two tracks: open and closed. In the open track, systems are allowed to use additional resources like MWE lexicons, raw corpora, word embeddings and so on.

In recent years, deep neural network architectures have been broadly applied for a wide range of NLP tasks, especially for sequence tagging. The most recent sequence tagging approaches [7,8,10] have delivered the state-of-the-art results for part-of-speech (POS) tagging and Named Entity Recognition (NER). Moreover, the performance of the studies can be evaluated on various languages without requiring feature engineering methods. It is possible to address the identification of MWEs in the same sophisticated ways. However, different from other sequence labeling tasks, the discontinuity and overlapping properties inherent in MWEs necessitate developing a special procedure for the MWE identification task.

In this study, we present a novel tagging scheme called *bigappy-unicrossy* to treat MWE-specific challenges using bidirectional Long Short-Term Memory with a Conditional Random Field (BiLSTM-CRF) neural network architecture proposed by [7]. Additionally, we test our proposed solution on the corpora released in Edition 1.1. of the PARSEME Shared Task for 19 languages. In our experiments, we juxtapose *bigappy-unicrossy* with IOB2 [14] and gappy 1-level [17] tagging schemes. All in all, the difficulties in identification of MWEs can be tackled in some degree by representing all gappy words and crossing boundaries of nested MWEs in sequences.

2   Related Work

In the literature on MWEs, several variations of standard IOB tagging scheme have been evaluated and discussed so far. The IOBES tagging scheme within a supervised approach based on a neural network model is used by [9]. A new tagging scheme named as gappy (discontinuous) 1-level tagging is described by [17] in order to encode gappy MWEs.

In Edition 1.1. of the PARSEME Shared Task, the participated systems exploited several neural architectures. While SHOMA [18], which is the best performing system with respect to the overall macro-average MWE-based F1 score, employs a combination of convolutional network, bidirectional long-short term memory (BiLSTM) network and conditional random field (CRF), Deep-BGT [2] makes use of BiLSTM-CRF with the gappy-1 level tagging scheme in the open track. The two systems use pre-trained word embeddings released by fastText [6]. Mumpitz [5] uses only BiLSTM layer, while GBD-NER [3] adds a graph-based encoding layer to the BiLSTM layer. Besides Deep-BGT, Veyn [19] also presents a recurrent neural network (RNN) model that combines three different tagging schemes.

Similar to Deep-BGT and SHOMA, we decide to explore a neural network architecture and benefit from the availability of fastText word embeddings for several languages. Also, Deep-BGT adopts an advanced tagging scheme to its system, which is different from other competitors, but it does not extend it to all languages. Therefore, we choose to implement the same model which is bidirectional LSTM-CRF for all languages with our new tagging scheme.

3   Corpus

The corpora provided by the PARSEME Shared Task Edition 1.1 consist of 20 languages. We cover 19 languages which are Bulgarian (BG), German (DE), Greek (EL), English (EN), Spanish (ES), Basque (EU), Farsi (FA), French (FR), Hebrew (HE), Hindu (HI), Crotian (HR), Hungarian (HU), Italian (IT), Lithuanian (LT), Polish (PL), Portuguese (PT), Romanian (RO), Slovenian (SL), and Turkish (TR). We do not use the Arabic (AR) corpus because it is not publicly available yet. The corpus of each language is divided into training, test, and development sets. Most of the languages contain more than 2000 VMWEs but EN and LT include less than 1000 VMWEs which are the smallest corpora among the datasets. The language specific statistics are summarized in Table 1. It contains information about number of tokens and VMWEs in each training corpus. Also, percentage of VMWEs and percentage of discontinuous VMWEs in each test corpus are provided.

The corpus is released in cupt format [11], which is publicly available in [12]. The cupt format is the extension of the conllu format[1]. The current format represents each token in a sentence by 11 columns. The first 10 columns specify the rank, token, lemma, part-of-speech, morphological features, and syntactic dependencies, as in the conllu format. The 11th column introduces the VMWE annotation.

The language team members have annotated the corpora following the annotation guidelines prepared by PARSEME [11]. The categories of VMWEs are the followings:

- Universal categories which exist in all participated languages:
  - Light Verb Constructions with two subtypes (LVC.full and LVC.cause)
  - Verbal Idioms (VID)

---

[1] http://universaldependencies.org/format.html.

- Quasi-universal categories which some languages include:
  - Inherently Reflexive Verbs (IRV)
  - Verb-Particle Constructions with two subtypes (VPC.full and VPC.semi)
  - Multi-verb Constructions (MVC)
- Language-specific categories
  - Inherently Adpositional Verbs (IAV)
- Optional experimental category which is added after the annotation process:
  - Inherently Clitic Cerbs (LS.ICV).

Table 1. Language-specific statistics for the corpora.

Languages	# of tokens	# of VMWEs	VMWE %	Discontinuity %
BG	480413	6704	1.40	29
DE	173293	3823	2.21	46
EL	224762	2405	1.07	45
EN	124203	832	0.67	41
ES	182364	2739	1.50	28
EU	157807	3823	2.42	19
FA	61568	3453	5.61	21
FR	528132	5677	1.08	44
HE	369013	2239	0.61	24
HI	35430	1034	2.92	7
HR	89536	2451	2.74	42
HU	156336	7760	4.97	8
IT	430789	4257	0.99	33
LT	208512	812	0.39	40
PL	274318	5152	1.89	30
PT	638002	5536	0.89	43
RO	1015623	5891	0.58	33
SL	280522	3378	1.20	51
TR	376464	7141	1.90	59

4 MWE Identification

According to [4], MWE identification is the process of annotating MWE instances in a given corpus and an automatic annotation system is called MWE tagger. Most of the NLP applications such as parsing, machine translation, etc. are in need of MWE identification [4]. However, MWE identification is not straightforward since it brings about some challenges. These challenges also require specialized metrics to evaluate the success of an MWE tagger.

4.1   Challenges

Challenges for MWE identification are listed as discontinuity, overlaps, ambiguity, and variability [4]. These challenges stem from the nature of MWEs. There may be tokens other than the ones belonging to MWE between the tokens of the MWE. For example, *take seriously* can have other tokens in between to be in the form of *take someone/something seriously*, but the VMWE is still *take seriously* here. Figure 1 shows an example use. Discontinuity varies from language to language [4]. This variability can be seen in Table 1.

Overlaps include different cases such as nesting, shared tokens, and so on [4]. Nesting can be defined as having at least one MWE inside an another MWE. We can give the sentence *I took her decision to move on seriously* in Fig. 1 as an example. Here, we have two VMWEs which are *took seriously* and *move on*. So, *move on* is referred as a nested VMWE. Two or more MWEs can also share one or more tokens. Additionally, there is a special case of overlaps which is called crossing. In this case, some or all tokens of different MWEs are positioned crosswise. The sentence *I made not only changes but also additions* is an example of both shared tokens and crossing. In this example, *not only but also* is an MWE because it is a complex function word. Both *made changes* and *made additions* are VMWEs belonging to LVC.full category. As a result, *made changes* and *made additions* together is an example to shared tokens case. The example to crossing case is *made changes* and *not only but also*. The example sentence also contains nesting due to *made additions* and *not only but also*. Moreover, we can infer that overlaps also contain the challenge of discontinuity.

Ambiguity stems from the fact that the group of tokens can be either an MWE or a non-MWE [4]. In other words, tokens can lose their original meanings or each token can contribute to the sentence with its original meaning. Also, the same MWE can belong to different types of MWE. An example can be given by these two sentences: *He teaches mathematics as well as physics* and *His performance in mathematics is as well as his performance in physics*. Here, *as well as* is ambiguous.

MWEs do not always appear in fixed forms. This flexibility is called variability [4]. Regarding the example in Fig. 1, *take seriously* changes its form and becomes *took seriously*.

4.2   Evaluation Metrics

Evaluation metrics are important to correctly determine the quality of a model in machine learning. However, evaluating a model only with general accuracy score may not explain how much the model resolves the domain-specific problems. Therefore, the PARSEME network [11] defines MWE-specific evaluation metrics which focus on the following challenges:

- Continuity metric is calculated for continuous (EN: *set up* a meeting) and discontinuous (EN: *set me up*) cases.
- Length metric is calculated separately for single-token VMWEs
  (DE: *aufmachen*) and multi-token VMWEs (DE: *macht es auf*).

– In terms of Novelty metric, if a VMWE is annotated at least once in the training corpus, it is accepted as "seen"; otherwise it is considered as "unseen".
– Variability metric is provided for seen VMWEs which are not identical to their original form in the training corpus.

Also, MWE-based score is calculated over fully predicted VMWE sequences, and token-based score is calculated over partial matches. In this paper, we evaluate our system results based on the aforementioned metrics.

5 Tagging Schemes

In this work, we approach the MWE identification task as a sequence labeling problem together with IOB encoding. However, the challenges of discontinuity and overlaps cannot be addressed using the classical IOB tagging schemes [13, 14]. Therefore, we make use of the gappy 1-level tagging scheme [17] but still it is not adequate for overlaps. For this reason, we developed a novel tagging scheme called *bigappy-unicrossy* to rise to the challenge of overlapping MWEs. Furthermore, we accept that MWEs can be both single-token and multi-token. Since there is a difference between the IOB1 [13] and the IOB2 [14] tagging schemes regarding the single-token ones, we applied the IOB2 tagging scheme. We also modified the gappy 1-level tagging scheme because it does not accept single-token MWEs in its original definition.

5.1 IOB1 Tagging Scheme

The IOB tagging scheme was first proposed by [13] which approach text chunking as a tagging problem. In this scheme, the tag set is $\{I, O, B\}$. I represents a token in the chunk, B stands for a token which is the beginning of the chunk spanning more than one token, and O denotes a token outside of any chunk. Therefore, B cannot be used alone without I. In other words, a single-token chunk gets the I tag. The IOB tagging scheme is also called the IOB1 tagging scheme in the literature [16] after the IOB2 tagging scheme is introduced.

5.2 IOB2 Tagging Scheme

The IOB2 tagging scheme is derived from the idea of giving B tag to every initial token of the chunk ignoring the chunk size [14,16]. It has same tag set with IOB1 which is $\{I, O, B\}$. B represents a token in the beginning of the chunk, I is used for a token in the chunk other than the initial token, and O is used for a token outside of any chunk. Hence, the only difference is that each chunk begins with B and B is followed by I if the chunk contains more than one word. In other words, a single-token chunk gets the B tag. We think that the IOB2 tagging scheme is more suitable for MWE identification by taking single-token MWEs into consideration. Hence, this tagging scheme rather than the IOB1 scheme is used in the experiments to create a baseline for comparing the performance of the other tagging schemes.

5.3   Gappy 1-Level Tagging Scheme

Both of the IOB tagging schemes are more suitable for tagging continuous chunks. However, MWEs include not only continuous chunks but also discontinuous chunks. The nature of MWEs can also pose nesting which cannot be represented by the IOB tagging schemes. Therefore, a new tagging scheme which is called the gappy 1-level tagging scheme is introduced by [17].

The tag set of the gappy 1-level tagging scheme is $\{I, O, B, i, o, b\}$. I, O, B tags are similar to the ones in the IOB tagging schemes. I symbolizes a token in the chunk, B represents a token which is at the beginning of the chunk, and O is used for a token outside of the chunk. Since [17] accepts MWEs as chunks containing more than one word, the difference between the IOB1 and IOB2 tagging schemes disappears. So, all B tags are followed by one or more I tags.

i, o, b tags have similar roles as I, O, B tags. The only difference is that the lowercase tags are used for nested chunks. i symbolizes a token in the nested chunk, b represents a token which is at the beginning of the nested chunk, and o is used for a token outside of the nested chunk which is also a gap for the chunk outside. Again, all b tags are followed by one or more i tags. Figure 1 shows an example use of this scheme.

The gappy 1-level tagging scheme accepts only multi-token MWEs. According to our definition of an MWE, we allow single-token MWEs too. Therefore, we modified the gappy 1-level tagging scheme such that it also allows single-token MWEs in the IOB2 tagging scheme fashion by using the B tag for the single-token MWEs that are not nested and the b tag for the nested single-token MWEs. The modified version is used in the experiments and it is referred as *gappy 1-level*.

The gappy 1-level tagging scheme solves the discontinuity and nesting problems partially. The discontinuity problem is solved by the o tag. Nesting problem which is a particular case of overlaps is solved partially because the gappy 1-level tagging scheme accepts only continuous nested MWEs.

5.4   Bigappy-Unicrossy Tagging Scheme

The variants of IOB tagging schemes propose partial solutions to the challenges of MWE identification. The IOB2 tagging scheme only identifies continuous chunks. The gappy 1-level tagging scheme partially solves the discontinuity problem because it does not allow discontinuous nested MWEs. Due to the elimination of discontinuous nested MWEs, nesting is also partially solved. Also, there is no attempt to solve other cases of overlaps such as crossing and shared tokens and so on. For this reason, there is a need of a new tagging scheme. We developed a novel tagging scheme called *bigappy-unicrossy* to represent overlaps in sequence labeling tasks and to solve the discontinuity problem accordingly.

The tag set of the bigappy-unicrossy tagging scheme is $\{I, O, B, i, o, b\}$. B stands for the beginning of the chunk. It is also used for single-token chunks. I represents a token in the chunk. It is used in the case of multi-token chunks where B is followed by one or more I. O is used for a token outside of the chunk.

The bigappy-unicrossy tagging scheme allows two levels of discontinuity, one level of nesting, and one level of crossing. The name *bigappy-unicrossy* is given accordingly. *i, o, b* tags are in charge of nested chunks, chunks with crosswise positioned tokens, and discontinuous chunks.

The *o* tag is used for a token outside of the nested chunk. It is also used for a token that does not belong to the nested chunk but between the tokens of the nested chunk. Tokens that are in between of the chunk but does not belong to the chunk can be called gaps [17]. In other words, *o* is used for all gappy chunks and each gap is tagged with *o*. We treated all the gaps the same. We do not differ the gaps between a chunk or a nested chunk because we assume that the tokens belonging to the gaps have the same role, which is being a token that can be inserted to the chunk without being a part of the chunk.

As an example, consider the sentence *I take her decision to make some changes seriously. take seriously* and *make changes* are VMWEs. The gaps for *take seriously* are *her, decision, to, make, some,* and *changes.* Since *make changes* is a nested VMWE here, the actual gaps are *her, decision, to,* and *some.* The gap for *make changes* is *some. some* has the same role for both of the VMWEs and tagged with *o*. Consequently, bigappy-unicrossy handles two levels of discontinuity: one level for gaps in the outer chunks and one level for gaps in the nested chunks.

Lowercase tags resemble uppercase tags in terms of their roles. *b* stands for the beginning of the nested or crossy chunk. It is also used for single-token ones. *i* represents a token in the nested or crossy chunk. It is used in the case of multi-token chunks where *b* is followed by one or more *i*. *o* is used for a token belonging to a gap of the gappy chunk.

In this tagging scheme, crossing cases and nesting cases are treated in a similar way, because the identification of crossing cases are like nesting cases. Here is our tagging procedure: Firstly, the first token of the chunk (we can call it chunk *X*) is identified if it is the first chunk appeared in the sentence and it is tagged with *B*. If the chunk *X* is multi-token, the remaining part is tagged with *I*. If there is a token belonging to beginning of another chunk (we can call it chunk *Y*) within the chunk *X*, the chunk *Y* can be nested and/or positioned crosswise with chunk *X*. Then, it is tagged with *b*. If the chunk *Y* is multi-token, the remaining part is tagged with *i*. Here, the index of the last token of chunk *Y* can be either smaller or bigger than the index of the last token of chunk *X*. The first case is called nesting and the latter one is called crossing. If the middle tokens of chunk *X* and chunk *Y* are positioned crosswise in the first case, there is also crossing.

The bigappy-unicrossy tagging scheme allows only one level of nesting or crossing because we have only two types of tag sets which are uppercase and lowercase. Here, the *B* and *I* tags belong to chunk *X*. So, they are held until the end of chunk *X*. After the last token of *X*, the *B* and *I* are released. The same rule applies for the *b* and *i* tags. The *b* and *i* tags belong to chunk *Y*. So, they are held until the end of chunk *Y*. After the last token of *Y*, the *b* and *i* are released.

On the other hand, it solves discontinuity problem in two levels. Incorporating more levels is not necessary since such a case is very scarce. Some examples are given in Table 2. Example 3 shows a crossing case. Other examples include nesting cases. Shared tokens are ignored in this tagging scheme. Table 2 shows different ways of eliminating shared tokens. In Example 3, *made additions* is eliminated. In example 4, *made changes* is eliminated. Therefore, it overcomes the challenge of overlaps partially.

Table 2. Examples to the bigappy-unicrossy tagging scheme.

Example 1	I	took	her	decision	to		move	on	seriously	
	O	B	o	o	o		b	i	I	
Example 2	I	take	her	decision	to		make	some	changes	seriously
	O	B	o	o	o		b	o	i	I
Example 3	I	made	not	only	changes	but	also	additions		
	O	B	b	i	I	i	i	O		
Example 4	I	made	not	only	changes	but	also	additions		
	O	B	b	i	o	i	i	I		

6   Model and Experiments

We design a language-independent system based on the bidirectional LSTM-CRF model provided by [7]. Similar to Deep-BGT system [2], we make use of the pre-trained word embeddings provided by fastText [6]. The word embeddings were trained on Common Crawl and Wikipedia. The dimension of word embedding vector is 300. In addition to the word embeddings, we choose the POS and dependency relation (DEPREL) tags that are available in the cupt files as inputs for the system.

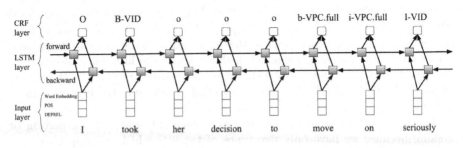

Fig. 1. The bidirectional LSTM-CRF model of Deep-BGT [2].

As shown in Fig. 1, the architecture of the bidirectional LSTM-CRF network composes of three layers. The inputs are fed into the BiLSTM layer. The bidirectional LSTM network processes both past and future features, respectively, in the

forward and backward units whose dimensions are set to 20. The outputs of the
LSTM units pass to the CRF layer, which decodes the VMWE labels. We optimize the parameters of the model for each language with the Nadam optimizer
without exceeding batch size 32 as suggested by [15]. All chosen hyperpameters
of the model are displayed in Table 3. A fixed dropout rate of 0.1 is applied on
all the bidirectional LSTM layers. Since the size of the training data plays a
major role in deep learning models, we add the development set to the training
set if a language has a development set and do not make use of the development
set separately. As we use non-deterministic approach, we run our experiments
five times in order to maintain reproducible and reliable results and take the
average.

Table 3. Model parameters.

Languages	Batch size	# of epochs
BG, FR, HE, LT, PT, RO, TR	32	12
DE, EL, ES, EU, HI, HU	16	15
FA, IT, PL, SL	16	12
EN, HR	8	15

7 Results

Table 4 shows the language-specific results for the IOB2, the gappy 1-level and
the bigappy-unicrossy tagging schemes. MWE-based and token-based F-measure
(F1) are presented for all tagging schemes. The results cover 19 languages. Each
language also has the F1 score of the system which is the best for that language
in the open track of PARSEME shared task Edition 1.1 and it is referred as
shared task. The last row in the table shows the cross-lingual macro-averages
which is calculated by averaging the F1 scores for 19 languages.

According to the MWE-based results, both the bigappy-unicrossy and the
gappy 1-level tagging schemes outperform the IOB2 tagging scheme. The reason behind is that bigappy-unicrossy and gappy 1-level capture discontinuous
VMWEs whereas IOB2 cannot capture them.

On the other hand, there is a slight difference of 0.38 between gappy 1-level
and bigappy-unicrossy. The gappy 1-level tagging scheme is the best in 11 languages while the bigappy-unicrossy tagging scheme is the best in 8 of them. The
results are close in this experiment because the frequency of overlapping cases
is low in the corpora of languages used for the experiment. The corpora only
contains VMWEs. In the case of all other types of MWEs, the overlap frequency
will be higher and the bigappy-unicrossy tagging scheme will show better performance. The bigappy-unicrossy tagging scheme is not only for MWEs. It can be

also used in other sequence labeling tasks. The bigappy-unicrossy tagging scheme can prove itself better in the other domains of NLP or in their combinations.

The F1 scores are close to each other for all the three tagging schemes in terms of the token-based results in Table 4. While the tagging scheme becomes more complex, identification also becomes more complex in the case of deep learning systems.

The experiments reveal that our application of the gappy 1-level and bigappy-unicrossy tagging schemes competes with the MWE-based best shared task results. When the tagging schemes and the best shared task results are compared, it is observed that gappy 1-level surpasses the best shared task results in 4 languages consisting of BG, DE, FR, HI and bigappy-unicrossy surpasses the best shared task results in 5 languages consisting of EL, FA, HR, LT, SL. In the case of token-based results, gappy 1-level is the best in BG, DE, EL and bigappy-unicrossy is the best in FA.

Table 4. The language-specific results for the IOB2, the gappy 1-level, the bigappy-unicrossy tagging schemes and the best PARSEME shared task results in the open track.

Lang.	MWE-based				Token-based			
	IOB2	Gappy 1-level	Bigappy-unicrossy	Shared task	IOB2	Gappy 1-level	Bigappy-unicrossy	Shared task
BG	64.60	67.03	66.89	65.56	67.21	67.72	67.24	66.85
DE	42.62	50.75	49.73	45.53	54.28	55.10	53.31	54.65
EL	52.10	60.54	61.11	58.00	63.73	66.97	65.61	66.79
EN	26.97	31.60	31.73	33.27	30.15	31.19	30.86	34.36
ES	31.07	33.59	35.00	38.39	37.54	38.64	39.76	44.69
EU	69.91	72.62	73.07	77.04	75.38	75.03	76.06	80.21
FA	75.07	79.31	81.37	78.35	82.01	81.33	84.48	82.95
FR	53.92	61.96	58.55	60.88	65.05	64.78	61.57	65.80
HE	24.93	27.45	26.74	38.91	28.56	29.30	28.74	44.02
HI	71.28	73.35	72.54	72.71	74.06	74.78	74.35	75.62
HR	44.62	51.85	52.83	47.84	53.68	54.58	56.18	58.19
HU	70.53	74.83	73.84	85.83	73.90	76.48	76.13	86.73
IT	31.52	38.17	37.58	45.40	40.28	42.73	43.61	55.13
LT	19.15	22.85	24.04	22.86	25.31	22.89	24.49	28.13
PL	58.54	65.87	64.65	63.60	64.78	67.70	66.41	67.23
PT	54.62	61.32	60.21	68.17	62.29	62.91	62.39	73.51
RO	82.34	85.89	84.60	87.18	85.31	86.33	85.19	88.69
SL	45.30	54.06	54.22	52.27	56.30	56.46	57.50	61.55
TR	52.26	55.95	52.93	58.66	58.12	57.52	54.30	61.63
AVG	51.12	56.26	55.88	57.92	57.79	58.55	58.33	62.99

Figure 2 shows the relationship between the percentage of discontinuous VMWEs in the corpora for all languages and the relative success of the bigappy-unicrossy tagging scheme over the IOB2 tagging scheme. Discontinuity percentages are also available in Table 1 which provides the language-specific statistics. The relative success is found by subtracting the MWE-based F1 score of bigappy-unicrossy from that of IOB2. It is seen that there is a correlation to some extent between the discontinuity ratio and the success improvement with the bigappy-unicrossy scheme. This result denotes that the effect of the bigappy-unicrossy tagging scheme increases more on discontinuous MWEs.

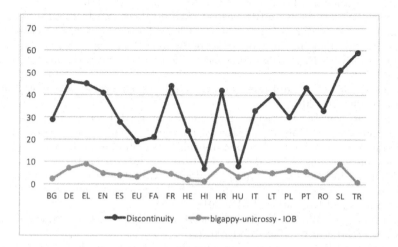

Fig. 2. The discontinuity percentages versus the MWE-based F1 score differences between the bigappy-unicrossy and the IOB tagging schemes.

8   Conclusion

In this study, we described a new tagging scheme called bigappy-unicrossy to address overlaps in sequence labeling tasks. It is attempted to solve the challenges of discontinuity and overlaps that include nesting and crossing.

Additionally, we presented an empirical study that explores the effect of a tagging scheme for VMWE identification on 19 languages by using bidirectional LSTM-CRF network. The code is publicly available[2]. The performance of the bigappy-unicrossy tagging scheme is close to the gappy 1-level tagging scheme and it gets ahead in 8 languages. To conclude, the bigappy-unicrossy tagging scheme is expected to be by far the best scheme on data sets with higher frequency of overlaps and we plan to apply our tagging scheme on such data sets as future work.

---

[2] https://github.com/deep-bgt/Deep-BGT.

Acknowledgements. This research was supported by Boğaziçi University Research Fund Grant Number 14420.

References

1. Baldwin, T., Kim, S.N.: Multiword expressions. Handb. Nat. Lang. Process. **2**, 267–292 (2010)
2. Berk, G., Erden, B., Güngör, T.: Deep-BGT at PARSEME shared task 2018: bidirectional LSTM-CRF model for verbal multiword expression identification. In: Proceedings of the Joint Workshop on Linguistic Annotation, Multiword Expressions and Constructions (LAW-MWE-CxG-2018), pp. 248–253 (2018)
3. Boroş, T., Burtica, R.: GBD-NER at PARSEME shared task 2018: multi-word expression detection using bidirectional long-short-term memory networks and graph based decoding. In: Proceedings of the Joint Workshop on Linguistic Annotation, Multiword Expressions and Constructions (LAW-MWE-CxG-2018), pp. 254–260 (2018)
4. Constant, M., et al.: Multiword expression processing: a survey. Comput. Linguist. **43**(4), 837–892 (2017)
5. Ehren, R., Lichte, T., Samih, Y.: Mumpitz at PARSEME shared task 2018: a bidirectional LSTM for the identification of verbal multiword expressions. In: Proceedings of the Joint Workshop on Linguistic Annotation, Multiword Expressions and Constructions (LAW-MWE-CxG-2018), pp. 261–267 (2018)
6. Grave, E., Bojanowski, P., Gupta, P., Joulin, A., Mikolov, T.: Learning word vectors for 157 languages. In: Proceedings of the International Conference on Language Resources and Evaluation (LREC 2018) (2018)
7. Huang, Z., Xu, W., Yu, K.: Bidirectional LSTM-CRF models for sequence tagging. arXiv preprint arXiv:1508.01991 (2015)
8. Lample, G., Ballesteros, M., Subramanian, S., Kawakami, K., Dyer, C.: Neural architectures for named entity recognition. arXiv preprint arXiv:1603.01360 (2016)
9. Legrand, J., Collobert, R.: Phrase representations for multiword expressions. In: Proceedings of the 12th Workshop on Multiword Expressions, pp. 67–71. Association for Computational Linguistics (2016). https://doi.org/10.18653/v1/W16-1810, http://aclweb.org/anthology/W16-1810
10. Ma, X., Hovy, E.: End-to-end sequence labeling via bi-directional LSTM-CNNs-CRF. arXiv preprint arXiv:1603.01354 (2016)
11. Ramisch, C., et al.: Edition 1.1 of the PARSEME shared task on automatic identification of verbal multiword expressions. In: Proceedings of the Joint Workshop on Linguistic Annotation, Multiword Expressions and Constructions (LAW-MWE-CxG 2018). Association for Computational Linguistics, Santa Fe (2018)
12. Ramisch, C., et al.: Annotated corpora and tools of the PARSEME shared task on automatic identification of verbal multiword expressions (edition 1.1) (2018). http://hdl.handle.net/11372/LRT-2842. LINDAT/CLARIN digital library at the Institute of Formal and Applied Linguistics (ÚFAL), Faculty of Mathematics and Physics, Charles University
13. Ramshaw, L., Marcus, M.: Text chunking using transformation-based learning. In: Third Workshop on Very Large Corpora (1995). http://aclweb.org/anthology/W95-0107
14. Ratnaparkhi, A.: Maximum entropy models for natural language ambiguity resolution. Ph.D. thesis, University of Pennsylvania, Philadelphia, PA, USA (1998). aAI9840230

15. Reimers, N., Gurevych, I.: Reporting score distributions makes a difference: performance study of LSTM-networks for sequence tagging. arXiv preprint arXiv:1707.09861 (2017)
16. Sang, E.F.T.K., Veenstra, J.: Representing text chunks. In: Proceedings of the Ninth Conference on European Chapter of the Association for Computational Linguistics, EACL 1999, pp. 173–179. Association for Computational Linguistics, Stroudsburg (1999). https://doi.org/10.3115/977035.977059
17. Schneider, N., Danchik, E., Dyer, C., Smith, N.A.: Discriminative lexical semantic segmentation with gaps: running the MWE gamut. Trans. Assoc. Comput. Linguist. **2**, 193–206 (2014)
18. Taslimipoor, S., Rohanian, O.: SHOMA at parseme shared task on automatic identification of VMWEs: neural multiword expression tagging with high generalisation. arXiv preprint arXiv:1809.03056 (2018)
19. Zampieri, N., Scholivet, M., Ramisch, C., Favre, B.: Veyn at PARSEME shared task 2018: recurrent neural networks for VMWE identification. In: Proceedings of the Joint Workshop on Linguistic Annotation, Multiword Expressions and Constructions (LAW-MWE-CxG-2018), pp. 290–296 (2018)

*Paris is Rain. or It is raining in Paris?: Detecting Overgeneralization of Be-verb in Learner English

Ryo Nagata[1,2](✉) ⓘ, Koki Washio[3] ⓘ, and Hokuto Ototake[4] ⓘ

[1] Konan University, 8-9-1 Okamoto, Higashinada, Kobe, Hyogo 658-8501, Japan
nagata-cicling@ml.hyogo-u.ac.jp
[2] Japan Science and Technology Agency, PRESTO, 4-1-8 Honcho, Kawaguchi, Saitama 332-0012, Japan
[3] The University of Tokyo, 3-8-1 Komaba, Meguroku, Tokyo 153-8902, Japan
kokiwashio@g.ecc.u-tokyo.ac.jp
[4] Fukuoka University, 19-1 Nanakuma 8-Chome, Fukuoka 814-0180, Japan
ototake@fukuoka-u.ac.jp

Abstract. This paper addresses the detection of overgeneralization of be-verb found in learner English. It is an error where the subject and complement are not semantically equivalent in a be-verb sentence as in *Paris is rain. This type of error often appears in the writing of learners whose native language has a be-verb equivalent that has usages other than those which English be-verb does. This paper presents a method for detecting overgeneralization of be-verb by predicting through word embeddings whether a given subject and complement pair is semantically equivalent or not. It also presents an effective and efficient way of determining the hyperparameters in the method. Experiments show that the present method outperforms four baseline methods based on corpus statistics and WordNet ontology despite the fact that it is a rather simple method. Looking into the detection results reveals the performance limitations of the proposed method.

Keywords: Grammatical error detection/correction · Overgeneralization of be-verb · Learner English

1 Introduction

Although grammatical error detection/correction has made tremendous progress with the advent of neural network-based methods, there still exist errors to which researchers have paid far less attention. **Overgeneralization of be-verb**, as in *Paris is rain.*, is one of the typical examples; as far as we know, there has been no work on its detection. It is an error where be-verb links the subject and complement which are not semantically equivalent[1]; in the example sentence

---

[1] The precise definition is introduced in Sect. 2.

© Springer Nature Switzerland AG 2023
A. Gelbukh (Ed.): CICLing 2019, LNCS 13451, pp. 636–647, 2023.
https://doi.org/10.1007/978-3-031-24337-0_45

above, the subject *Paris* is NOT *rain*, which violates the basic English rule of *be*-verb (or strictly copula *be*), that subject-complement [11] should be semantically equivalent to its subject.

Overgeneralization of *be*-verb often appears in learner English. This is especially true for learners whose native language has a *be*-verb equivalent that has usages other than English *be*-verb does. A typical example is Japanese. For instance, the following expressions are valid in the corresponding Japanese expressions: **Airplanes are danger.* (correctly *Airplanes are dangerous.*), **The meeting is five.* (correctly, *The meeting is scheduled at five.* or *The meeting is at five.*), and **Paris is rain* (correctly, *It is raining in Paris.*)[2]. In other words, overgeneralization of *be*-verb is (at least partly) ascribed to language transfer or mother tongue interference. This suggests that it will likely be beneficial to such groups of learners to give them feedback explaining how *be*-verb functions in English and why such expressions as **Paris is rain.* and **Airplanes are danger.* are not correct. To achieve it, one has to recognize overgeneralization of *be*-verb in learner English in the first place, distinguishing it from the other error types.

Unfortunately, however, previous error detection/correction methods would not suit this application. Previous error-specific methods, which typically rely on an unannotated native corpus, solve error detection/correction as a classification problem as in article and preposition error detection/correction methods [4,6] (i.e., selecting the correct article or preposition). This way of detection/correction does not apply well to this type of error; it is not a problem to select the correct *be*-verb, but to determine whether a given subject and complement pair is semantically equivalent. It is not trivial at all how to detect this type of error as a classification problem, relying solely on an unannotated native corpus. Another typical way is to predict directly whether a given word or phrase is correct or not as found in the method [6]. However, it would not be practical at all considering the fact that it requires a learner corpus annotated with overgeneralization of *be*-verb, which are rare at present; as far as we know, such publicly available data do not exist. Machine translation-based methods [7] and neural network-based methods [2,14] would probably be capable of detecting/correcting part of overgeneralization of *be*-verb, but not of distinguishing it from the other error types because they detect/correct multiple error types simultaneously. It is crucial to detect it as overgeneralization of *be*-verb in order to realize such feedback as mentioned above. Besides, it is often the case that overgeneralization of *be*-verb requires the rewrite of the whole structure as in **Paris is rain.* $\rightarrow$ *It is raining in Paris.*, which would be difficult for all the methods above.

In view of this background, this paper presents a method for detecting overgeneralization of *be*-verb. It uses word embedding vectors (simply, word embeddings) to predict whether a given subject and complement pair is semantically equivalent or not, which plays the central role in the error detection procedure.

---

[2] Admittedly, such expressions as *I am coffee.* can also be correct in English. However, they are used in limited contexts and situations, and the usage rarely appears in the writings of learners of English (e.g., essay writing).

They have been shown to be effective in detecting historical meaning changes [3] and differences in meaning between loan words and their originals [12]. These results imply that they will also likely be effective in the present task. The present method detects the triple of *be*-verb, its subject, and its complement as overgeneralization of *be*-verb if a given pair is predicted not to be semantically equivalent.

The contributions of this paper are four-fold: (i) it presents the first-ever method that detects overgeneralization of *be*-verb; (ii) it also presents a method for determining the hyperparameters of the method efficiently and effectively; (iii) the resulting method significantly outperforms four baselines based on corpus statistics and WordNet [9] ontology; (iv) it investigates detection results to show the performance limitations empirically and theoretically.

2 *Be*-verb Sentence and Overgeneralization of *be*-verb

In this paper, the *be*-verb sentence, which is the target of error detection, is defined as follows:

be-verb sentence: sentence consisting of S, V, C where S is a noun phrase (NP) that is the subject of the sentence, V is the *be*-verb, and C is also an NP that is its subject-complement [11].

Hereafter, subject-complement will be referred to just as complement.

Under this definition, the *be*-verb sentence has the following two basic usages [11]:

(a) Identification:
  e.g., Kevin is my brother.
(b) Characterization:
  e.g., Dwight is an honest man.

The usages (a) and (b) are roughly summarized as the rule that the subject and complement in a *be*-verb sentence should be semantically equivalent. For example, *Kevin* is a *brother* and *Dwight* is a *man*. Hereafter, the rule will be referred to as **subject complement equivalence**.

Learners of English often violate this rule as in the examples we have already seen in Sect. 1. This is especially true for the writer whose native language has a *be*-verb equivalent that has usages other than the above two. A typical example is Japanese; the three erroneous examples are all valid in Japanese. This type of error is defined as **overgeneralization of *be*-verb** in this paper. Part of the reasons why it occurs is that the other usages in the native language are negatively transferred into English. Considering this, it would be useful to explain to this group of learners why the usage is erroneous and how English *be*-verb functions.

Note that only characterization attributes normally allow reversal of subject and complement without affecting the semantic relation. Learners might also violate this. However, this paper excludes this type of error from the detection target. We will discuss this problem again in Sect. 5.

3 Proposed Method

3.1 Detection Procedure

The procedure of the proposed method is as follows:

Step (1) Input
Step (2) Subject complement pair extraction
Step (3) Subject complement equivalence check
Step (4) Error detection
Step (5) Postprocessing
Step (6) Output

In Step (1), each sentence is read from the detection target text. In Step (2), its dependency structure is obtained by using a parser. Then, its subject complement pairs are extracted from the parse (if any); only head nouns are extracted. In Step (3), the extracted pairs are examined as to whether they are semantically equivalent or not based on word embeddings. The details are described in Subsect. 3.2. In Step (4), it is determined whether they are correct or not; if they are predicted to be semantically equivalent in Step (3), then they are judged to be correct; otherwise, erroneous (overgeneralization of *be*-verb). In Step (5), as a postprocessing step, those that contain one of the following words are filtered out to achieve a better error detection performance: *it, they, this, these, that, those, thing, things.* These are the words that can refer to a wide variety of things, and thus it would be hard to predict whether a given pair is semantically equivalent or not. Filtered-out pairs are always judged to be correct. Finally, in Step (6), the detection result is output, either 0 (correct) or 1 (erroneous). Alternatively, the triple (subject, *be*-verb, and complement) are marked in the target sentence when the result is erroneous.

3.2 Subject Complement Equivalence Check

Figure 1 shows the big picture of subject complement equivalence check. As already mentioned, subject complement equivalence check is done based on word embeddings. They are learned in advance from a large native corpus. Before learning, all words are put into lowercase to decrease the vocabulary size.

To formalize the subject complement equivalence check procedure, let s and c be a subject and its corresponding complement in a *be*-verb sentence, respectively. Also, let v_s and v_c be their corresponding word embeddings (i.e., vectors), respectively, and $\cos(v_s, v_c)$ be the cosine similarity between the two.

Then, the check is done by the following function with a threshold θ:

$$f(v_s, v_c) = \begin{cases} 1 & (\cos(v_s, v_c) > \theta \\ 0 & (\text{otherwise}) \end{cases} \tag{1}$$

where 1 and 0 denote that the pair is semantically equivalent and not, respectively (Subsect. 3.3 will shortly describe how to determine the threshold θ).

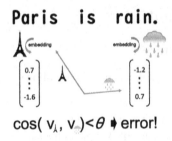

Fig. 1. Illustration of subject complement equivalence check.

Equation (1) can be interpreted as follows. The similarity between subject and complement is measured by the cosine of the corresponding two vectors. Then, the check whether or not the subject complement pair is semantically equivalent is approximated to be the problem of determining whether or not the pair is similar enough in terms of the angle between the two vectors as illustrated in Fig. 1. This may seem a too crude approximation, but it works well in practice as shown in Sect. 4. For the moment, take as an example the erroneous sentences *Paris is rain.*, *Airplanes are danger*, and *The meeting is five*. Their corresponding cosine similarities are 0.05, 0.08, and −0.04, respectively, which shows that their vectors are almost orthogonal[3]. This observation agrees well with our intuition that each subject has little or nothing to do with its corresponding complement. Unlike these erroneous pairs, the values become much larger for correct expressions such as *Paris is a city.* (0.24), *Airplanes are a machine.* (0.21), *The meeting is a gathering (of someone).* (0.57).

3.3   How to Determine Hyperparameters

Performance of the proposed method greatly depends on the values of its hyperparameters including the threshold[4] θ in Eq. (1) and those for word embeddings such as the dimension of the vector and the window size. Ideally, it would be best to determine them with a development set, that is, a learner corpus with which overgeneralization of *be*-verb errors are manually annotated. Unfortunately, however, it would not be possible considering that there exists no such learner corpus at present.

To overcome this problem, the proposed method automatically generates pseudo-training data from a native corpus as shown in Fig. 2. To achieve this, it first extracts subject complement pairs from a native corpus just as in Step (2) described in Subsect. 3.1 ((1) *Extraction* in Fig. 2). It discards the pairs whose

---

[3] The cosine similarities were calculated by the word embeddings with the window size of 10 and the dimension of 200 whose details are described in Sect. 4.

[4] Strictly, the threshold θ is rather a parameter of the method than a hyperparameter. However, it will be referred to as a hyperparameter for the simplicity of explanation in this paper.

subject or complement does not appear[5] in a learner corpus in order to create a learner corpus-like training data set ((2) *Filtering* in Fig. 2). It would be safe to say that (almost) all these pairs are free from overgeneralization of *be*-verb because they are from a native corpus. In other words, they can be regarded as correct instances consisting of subjects and complements that likely appear in learner English. It then generates pseudo-erroneous pairs from them by sampling out their subjects and complements independently ((3) *Random Sampling* in Fig. 2). Namely, it randomly chooses a subject from one of them and a complement from another to make a pseudo-erroneous pair. An exception is that it excludes those already found in the correct instance set to avoid including possibly correct pairs. Finally, it merges correct and pseudo-erroneous instances into a pseudo-training data set ((4) *Merging* in Fig. 2).

One thing we should take care of is that we have to determine the ratio of pseudo-erroneous instances to correct ones, which corresponds to the error ratio of overgeneralization of *be*-verb in learner English. This paper assumes that it is empirically given considering the fact that a learner corpus annotated with overgeneralization of *be*-verb is not publicly available.

The proposed method uses the resulting pseudo-training data just as a standard development set. In other words, it applies Step (3)–(6) in Subsect. 3.1 to them to estimate its performance with an arbitrary set of values for the hyperparameter; it selects the setting that maximizes performance (*F*-measure, for example).

Here, it should be emphasized that the present development set is a pseudo-one (i.e., automatically generated). For this reason, the estimated best setting may not perform well on a learner corpus. Even if it were a real one, the resulting setting may not, suffering from other problems such as overfitting.

To reduce the problem, the proposed method takes a vote of the detection results obtained through word embeddings with different settings (different window sizes and different dimensions, for example). For this, it learns a number of them from a native corpus. It determines the threshold θ for each of them by the same method as above. Finally, it takes a vote of their detection results (whether correct or not) to determine whether a given pair is really correct or not. This way of detection will likely reduce the influence from the problems above.

4 Evaluation

We chose the Konan-JIEM Learner Corpus fifth edition [10] (KJ) as our detection target. We manually annotated it with overgeneralization of *be*-verb. We first extracted *be*-verb sentences, and in turn subject complement pairs from it by using the LexicalizedParser of Stanford Parser Ver.3.5.0[6]. As a result, we

---

[5] Note that occurrences other than as a subject or a complement are considered when the subject and complement in question are checked whether they appear in learner English.

[6] https://nlp.stanford.edu/software/lex-parser.shtml. Sentences longer than 50 tokens are excluded from parsing.

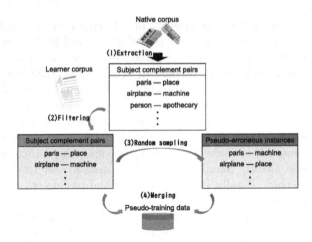

Fig. 2. Procedure for generating pseudo-training data set.

obtained 294 subject complement pairs. The first and second authors independently annotated them with *correct* or *erroneous*. After that, they discussed and solved disagreements, which identified 83 errors.

We obtained word embeddings[7] from news.en-00001-of-00100 to news.en-00099-of-00100 of one Billion Word Language Model Benchmark[8]. We used 15 sets of them with different settings of their hyperparameters (the combinations of the dimensions ranging over 200, 400, 800 and the window sizes ranging over 5, 10, 15, 20, 25). We did not include words appearing less than five times in the word embeddings; we did not apply the proposed method to subject complement pairs containing one of these words (i.e., they were always regarded as correct when they appeared in a subject complement pair in KJ). The other hyperparameters were fixed as described in the footnote.

We used the same native corpus to determine the values of the threshold θ in Eq. (1). We applied the method described in Subsect. 3.3 to it to obtain a pseudo-training set. We set the ratio of pseudo-erroneous instances to the whole training data to 28% (=83/294), which equals the error rate in KJ. This resulted in a pseudo-training set consisting of 155,402 correct instances and 60,434 erroneous instances. We selected the value of the threshold θ that maximizes F-measure on it, ranging over $0 \leq \theta \leq 1$ with an interval of 0.01. Note that because we used the prior knowledge about the error rate in the target corpus, the evaluation was not strictly done by a blind test.

For comparison, we implemented four baseline methods. The first one was simply based on co-occurrence of the subject complement pair in question. Namely, it detected as errors those that did not appear in the native corpus above. The second one was based on Pointwise Mutual Information (PMI) between a given

---

[7] We used the word2vec software (https://github.com/tmikolov/word2vec) with the options: -negative 25 -sample 1e-4 -iter 15 -cbow 1 -min-count 5.

[8] http://www.statmt.org/lm-benchmark/.

Table 1. Detection performance.

Method	Accuracy	Recall	Precision	$F_{1.0}$
Proposed (given best hyperparameter)	0.786	0.614	0.622	0.618
Proposed (estimated best hyperparameter, voting)	**0.762**	0.578	**0.578**	**0.578**
Proposed (estimated best hyperparameter)	0.643	0.639	0.414	0.502
Majority class	0.713	—	—	—
Co-occurrence	0.673	0.530	0.436	0.478
Lin's similarity (given best hyperparameter)	0.599	0.735	0.389	0.509
PMI (given best hyperparameter)	0.480	0.759	0.321	0.451
Is-a recognition (given best hyperparameter)	0.411	**0.816**	0.282	0.419

subject complement pair. It detected as errors those whose PMI was smaller than a threshold. We set it to the one that maximized F-measure on KJ to show the upper bound of its performance. We defined the probability of co-occurrence in PMI as that of subject complement pairs in the *be*-verb sentences in the native corpus. We also defined the probability of single word occurrence as the unigram probability of each word in the native corpus. We estimated all probabilities by Laplace Estimator. The third one was based on the Lin's similarity [8] calculated from WordNet. Similar to the second one, it detected as errors those whose similarity was less than a threshold, which we determined the same way as in the PMI-based method. The fourth one was an adaptation of the neural network-based method [13] for recognizing semantic relations between given word pairs. We trained it[9] so that it can predict whether a given word pair has the *is-a* relation or not; we detected it as an error if a given pair was predicted not to have the *is-a* relation. We used the same corpora and the same parser as in the proposed method to implement the four baselines. Also, we excluded from error detection subject complement pairs that appeared less than five times in KJ (they were always predicted to be correct as in the proposed method). Note that the evaluation for the PMI-based and Lin's similarity-based methods were also not strictly done by a blind test because their thresholds were optimized on KJ.

To evaluate performance, we used accuracy, which was defined as the number of instances whose subject complement equivalence was correctly predicted divided by the number of subject complement pairs. We also used recall, precision, and F-measure.

Table 1 shows the results. For comparison, Table 1 includes the performances of the proposed method with the hyperparameter setting optimized on KJ (threshold $\theta = 0.10$; window size: 10 words; and dimension: 200), that with the best setting estimated from the pseudo-training data ($\theta = 0.10$; window size: 5 words; and dimension: 800) without voting, and another baseline that always predicts as correct; they are denoted as *Proposed (given best hyperparameter)*, *Proposed (estimated best hyperparameter)*, and *Majority class* in Table 1, respectively.

---

[9] BLESS [1], which contains the information about semantic relations, was used as the training data for recognizing semantic relations.

Table 1 reveals that the co-occurrence-based method is a strong baseline, which outperforms the other two baselines that exploit other sources of information (corpus statistics and WordNet). Besides, the thresholds in the latter two are optimized on KJ. Nevertheless, the former achieves a better accuracy. This implies that subject complement co-occurrences obtained from a large native corpus is a good source of evidence to tell that a given pair is correct. At the same time, its accuracy is still low even compared to the majority class baseline. This also implies that it suffers from the data sparseness problem even when it uses such a large native corpus as the one Billion Word Language Model Benchmark. This probably applies to the PMI-based method, too. In addition, PMI is suitable for measuring how correlated a pair is but not to predict whether it is semantically equivalent or not. Contrary to our expectation, the Lin's similarity-based method does not perform well either even though it is based on a kind of semantic knowledge (WordNet). Note that the similarity is not defined for pairs containing a proper noun or a pronoun (except I) because they are not included in the synsets of WordNet. This partly explains why it does not perform well. Similarly, the method based on *is-a* recognition does not work well either. Part of the reasons is ascribed to the miss-match between the target language (learner English) and the semantic relation corpus *BLESS* (native English).

In contrast, the proposed method with voting outperforms even the best-performing baseline (co-occurrence) both in accuracy and F-measure; the difference in accuracy is statistically significant at a significance level of 0.05 (McNemar's test, $p = 0.015$). Importantly, its accuracy and precision are especially high compared to the three baselines. This property of the proposed method is particularly preferable in applications to language learning assistance that put more emphasis on precision over recall.

Given the best setting of the hyperparameters, the proposed method improves further, achieving an accuracy of 0.786 and an F-measure of 0.618. In contrast, its performance degrades when it relies solely on the best estimated setting from the pseudo-training data set, suggesting that the estimation can be unreliable in some cases. The proposed method with voting successfully overcomes the problem, achieving much better performance.

The evaluation results are summarized as follows. The word embedding-based method is effective in predicting whether or not a given subject complement pair is semantically equivalent and in turn in detecting overgeneralization of *be*-verb despite the fact that it is a rather simple method requiring no manually-annotated learner corpus. All it does is use the information about the error rate of overgeneralization of *be*-verb in the target text. At the same time, it is crucial to set the hyperparameters properly. The voting method aptly avoids selecting just one setting.

5   Discussion

We investigated the detection results of the proposed method with the given best hyperparameter setting. It gave a cosine similarity of below zero to 20 subject

Table 2. Error typology and its recall.

Error type	Recall
Whole structure rewrite	0.47 (9/19)
e.g., *Japan was a winter.* → *It was winter in Japan.*	
Change of subject/complement to another noun	0.50 (9/18)
e.g., *My job is an acceptance.* → *My job is a receptionist.*	
Change of *be*-verb to another verb	0.50 (5/10)
e.g., *I can be fun.* → *I can have fun.*	
Change of complement to participle	0.88 (7/8)
e.g., *I was warry about it.* → *I was worried about it.*	
Change of complement to adjective	1.0 (7/7)
e.g., *Airplanes are danger.* → *Airplanes are dangerous.*	
Addition of preposition after *be*-verb	0.50 (3/6)
e.g., *The story is basketball.* → *The story was about basketball.*	
Reversal of subject and complement	0.00 (0/1)
e.g., *Fruits are bananas.* → *Bananas are fruits.*	
Other	0.57 (8/14)
TOTAL	0.58 (48/83)

complement pairs out of 294, which were accordingly detected as overgeneralization of *be*-verb. Only 60% (12 instances) of them were actually erroneous. At first sight, this seemed that it was not a good measure for subject complement equivalence check. Looking into the pairs, however, revealed that five out of eight false positives were in proper nouns and that the accuracy on common nouns was much higher (83%=10/12) than on them (0.25%=2/8). We observed high accuracy at the other end of the cosine similarity, too; 48 pairs received a cosine similarity of more than 0.3, which were determined as correct. Most of them (41) were actually correct use, achieving an accuracy of 85%. These results show that the proposed method is effective in detecting overgeneralization of *be*-verb at least in common nouns.

Proper nouns were found to be problematic in the whole data set. One possible reason for this is that proper nouns from the writer's native language (such as Japanese names and places in the present case) were less frequent than other English common nouns in the native corpus used to learn word embeddings. Consequently, their word embeddings were likely to be less reliable. Also, some had coincidently the exact same spelling as an English proper noun. Examples were *Kobe* (a Japanese city) and *Himeji Oden* (a kind of Japanese local food). In the native corpus used as training data for word embeddings, they both often appeared as a person name (*Kobe Bryant* and *Greg Oden*, both basketball players); we actually sampled 100 instances of *Kobe* out of it and recognized 87% of them as a person (mostly, *Kobe Bryant*). This explains well why false positives occurred in these proper nouns in KJ as in *Kobe is a nice place*. They appeared seven times in total, four of which had a detection failure, suggesting that proper nouns require special care in order to achieve better performance.

For better understanding of its detection tendency, we classified the 83 errors into subcategories according to the treatment they require to become a valid English sentence. As a result, we recognized seven types; Table 2 shows them with their corresponding recall. It shows that the proposed method is capable of detecting all types of error, except **Reversal of subject and complement**, to some extent.

Let us finally discuss the performance limitations of the proposed method. One of the major limitations is that it cannot distinguish between senses in homographs and even polysemes. In other words, a noun, a homograph/polyseme or not, is represented by one word embedding vector. This property of word embeddings leads to false positives and negatives as found in the detection failures in the proper nouns described above. It may mitigate this problem to encode the information on the entire NPs of the head nouns (subject and complement) in question. It can be achieved, for example, by taking the averages of their word embeddings or encoding them into LSTM [5]. It would be interesting to see how well such encoded vectors work on this problem.

Another performance limitation is that it does not at all target the reversal of subject and complement errors as in *Fruits are bananas*. Obviously, it gives the same value of the cosine similarity for a subject complement pair and its reversal. Accordingly, it always predicts both to be correct or incorrect, failing to detect this type of error. This is one of the performance limitations of the proposed method in theory. At the same time, this type of error is relatively infrequent even in learner English as shown in Table 2. This is probably because it is not a problem of language use but a problem of logic. Therefore, after a certain age, language learners are expected to have much less trouble with it; what poses the difficulty for them is the interference from their native language when the *be*-verb equivalent has usages other than those which English *be*-verb does. Considering this, the limitation will likely be less problematic in practice.

6   Conclusions

This paper addressed the problem of the detection of overgeneralization of *be*-verb. The presented method solved it as a problem of predicting the subject complement equivalence through word embeddings. Evaluation showed that it outperformed the three baselines exploiting corpus statistics and the Word-Net ontology. Detailed investigations of the results revealed its performance limitations.

Acknowledgments. This work was supported by Japan Science and Technology Agency (JST), PRESTO Grant Number JPMJPR1758, Japan

References

1. Baroni, M., Lenci, A.: How we BLESSed distributional semantic evaluation. In: Proceedings of the GEMS 2011 Workshop on GEometrical Models of Natural Language Semantics, pp. 1–10. Association for Computational Linguistics (2011). http://aclweb.org/anthology/W11-2501
2. Chollampatt, S., Taghipour, K., Ng, H.T.: Neural network translation models for grammatical error correction. In: Proceedings of 25th International Joint Conference on Artificial Intelligence, pp. 2768–2774 (2016). http://www.ijcai.org/Proceedings/16/Papers/393.pdf
3. Hamilton, W.L., Leskovec, J., Jurafsky, D.: Diachronic word embeddings reveal statistical laws of semantic change. In: Proceeings of 54th Annual Meeting of the Association for Computational Linguistics (Volume 1: Long Papers), pp. 1489–1501 (2016). http://www.aclweb.org/anthology/P16-1141
4. Han, N.R., Chodorow, M., Leacock, C.: Detecting errors in English article usage by non-native speakers. Nat. Lang. Eng. **12**(2), 115–129 (2006)
5. Hochreiter, S., Schmidhuber, J.: Long short-term memory. Neural Comput. **9**(8), 1735–1780 (1997). https://doi.org/10.1162/neco.1997.9.8.1735
6. Izumi, E., Uchimoto, K., Saiga, T., Supnithi, T., Isahara, H.: Automatic error detection in the Japanese learners' English spoken data. In: Proceedings of 41th Annual Meeting of the Association for Computational Linguistics, pp. 145–148 (2003)
7. Junczys-Dowmunt, M., Grundkiewicz, R.: Phrase-based machine translation is state-of-the-art for automatic grammatical error correction. In: Proceedings of the 2016 Conference on Empirical Methods in Natural Language Processing, pp. 1546–1556 (2016). https://aclweb.org/anthology/D16-1161
8. Lin, D.: An information-theoretic definition of similarity. In: Proceedings of 15th International Conference on Machine Learning, pp. 296–304 (1998). http://dl.acm.org/citation.cfm?id=645527.657297
9. Miller, G.A.: WordNet: a lexical database for English. Commun. ACM **38**(11), 39–41 (1995)
10. Nagata, R., Sakaguchi, K.: Phrase structure annotation and parsing for learner English. In: Proceedings of 54th Annual Meeting of the Association for Computational Linguistics, pp. 1837–1847 (2016). https://doi.org/10.18653/v1/P16-1173
11. Quirk, R., Greenbaum, S., Leech, G., Svartvik, J.: A Comprehensive Grammar of the English Language. Longman, New York (1985)
12. Takamura, H., Nagata, R., Kawasaki, Y.: Analyzing semantic change in Japanese loanwords. In: Proceedings of 15th Conference of the European Chapter of the Association for Computational Linguistics: Volume 1, Long Papers, pp. 1195–1204 (2017). http://www.aclweb.org/anthology/E17-1112
13. Washio, K., Kato, T.: Filling missing paths: modeling co-occurrences of word pairs and dependency paths for recognizing lexical semantic relations. In: Proceedings of 2018 Conference of the North American Chapter of the Association for Computational Linguistics, pp. 1123–1133 (2018). https://doi.org/10.18653/v1/N18-1102, http://aclweb.org/anthology/N18-1102
14. Yannakoudakis, H., Rei, M., Andersen, Ø.E., Yuan, Z.: Neural sequence-labelling models for grammatical error correction. In: Proceedings of the 2017 Conference on Empirical Methods in Natural Language Processing, pp. 2795–2806 (2017). https://www.aclweb.org/anthology/D17-1297

Speeding up Natural Language Parsing by Reusing Partial Results

Michalina Strzyz[ID] and Carlos Gómez-Rodríguez[✉][ID]

FASTPARSE Lab, LyS Research Group, Departamento de Computación,
Universidade da Coruña, CITIC, Campus de Elviña, s/n, 15071 A Coruña, Spain
{michalina.strzyz,carlos.gomez}@udc.es

Abstract. This paper proposes a novel technique that applies case-based reasoning in order to generate templates for reusable parse tree fragments, based on PoS tags of bigrams and trigrams that demonstrate low variability in their syntactic analyses from prior data. The aim of this approach is to improve the speed of dependency parsers by avoiding redundant calculations. This can be resolved by applying the predefined templates that capture results of previous syntactic analyses and directly assigning the stored structure to a new n-gram that matches one of the templates, instead of parsing a similar text fragment again. The study shows that using a heuristic approach to select and reuse the partial results increases parsing speed by reducing the input length to be processed by a parser. The increase in parsing speed comes at some expense of accuracy. Experiments on English show promising results: the input dimension can be reduced by more than 20% at the cost of less than 3 points of Unlabeled Attachment Score.

Keywords: Natural language processing · Parsing · Efficiency

1 Introduction

Current state-of-art parsing algorithms are facing high computational costs, which can be an obstacle when applied to large-scale processing. For example, the BIST parser [7] reports speeds of around 50 sentences per second in modern CPUs, and even the fastest existing parsers, which forgo recurrent neural networks, are limited to a few hundred sentences per second [12]. While these speeds can be acceptable for small-scale processing, they are clearly prohibitive when the intention is to apply natural language parsing at the web scale. Thus, there is a need for approaches that can parse faster, even if this comes at some

This work has received funding from the European Research Council (ERC), under the European Union's Horizon 2020 research and innovation programme (FASTPARSE, grant agreement No 714150), from the TELEPARES-UDC project (FFI2014-51978-C2-2-R) and the ANSWER-ASAP project (TIN2017-85160-C2-1-R) from MINECO, and from Xunta de Galicia (ED431B 2017/01). We gratefully acknowledge NVIDIA Corporation for the donation of a GTX Titan X GPU.

A. Gelbukh (Ed.): CICLing 2019, LNCS 13451, pp. 648–657, 2023.
https://doi.org/10.1007/978-3-031-24337-0_46

accuracy cost, as differences in accuracy above a certain threshold have been shown to be unimportant for some downstream tasks [3].

In this paper we propose a novel approach to improve the speed of existing parsers by avoiding redundant calculations. In particular, we identify fragments of the input for which a syntactic parse is known with high confidence from prior data, so that we can reuse the existing result directly instead of parsing the fragment again. This effectively reduces the length of the input being processed by the parser, which is a major factor determining parsing time.

We test a prototype of the approach where the reusable fragments are bigrams and trigrams, the matching criteria are based on part-of-speech tags, and the parsers to be optimized are linear-time transition-based dependency parsers. Note, however, that the technique is generic enough to be applied to any kind of parsing algorithm (including constituent parsers) and speed improvements are expected to be higher for higher-complexity parsers such as those based on dynamic programming, which have been the traditional target of pruning and optimization techniques [2,13].

The aim of this approach is to significantly improve the parsing speed while keeping an acceptable accuracy level. This involves a trade-off between speed and accuracy, depending on how aggressively the technique is applied: more lenient confidence criteria for reusing fragments will lead to larger reductions in input length and thus faster parsing but at a cost to accuracy. We expect that the accuracy cost can be reduced in the future by using more fine-grained matching criteria (based on forms or lemmas) and augmenting training data so more fragments can be confidently reused.

2 Reuse of Partial Results

A natural language obeys the so-called Zipf's law, which shows how words are distributed with respect to their frequency. Namely, a language has a few high-frequency tokens and many uncommon tokens. This phenomenon is also present when investigating lemmas [1]. Similarly, n-gram phrases fit this distribution [4,5]. Therefore, to some extent, repetitions of identical n-grams are likely to be found in large corpora.

This implies that a parser will probably encounter known n-grams in a text. Since most of the time repetitions of the same n-gram will have identical syntactic structure (e.g. the phrase "the 10 provinces of Canada" can be reasonably expected to always have the same internal structure), we can exploit this by reusing previous analyses for these known n-grams. More generally, this can be extended to similar n-grams ("the 50 provinces of Spain" can be expected to have the same analysis as the phrase above), which can be identified with templates: the two examples above can be captured by a template "the CD provinces of NNP" (where CD and NNP are the part-of-speech tags for numerals and proper nouns), associated with a partial syntactic analysis to be reused. This combination of template matching and result reuse can be seen as an instantiation of case-based reasoning, a cognitively-rooted problem-solving technique based on solving new problems by reusing solutions to similar past problems [6,10].

Given a template like in the example above, we implement case-based reasoning by examining the training set and counting the number of n-grams that match it, as well as the different syntactic analyses that they have been assigned. If the variability of these analyses is higher than a given threshold, then the template will not be useful. However, if matching n-grams exhibit the same syntactic structure the vast majority of the time, then we can create a rule assigning that parse tree fragment to the template. At test time, we will reuse the parse tree fragment whenever we encounter an n-gram matching this rule.

3   Generating the Templates

In this prototype we will use limited training data. Since the frequency of word pairs is higher than multi-word phrases [11] only templates consisting of bigrams or trigrams of PoS tags are considered in this implementation in order to not suffer excessively from sparsity problems. More detailed templates (such as those including lemmas or word forms) would require augmenting the training set.

In order to calculate the level of confidence of a syntactic analysis for a given template, the following patterns for a bigram of PoS tags, as detailed in Fig. 1, are taken into consideration.

To generate a template for a given PoS tag bigram, we count the frequency of each of these head patterns relative to the total frequency of the bigram in the training set. Then, we focus on the most frequent pattern for the bigram. If this pattern has a dependent pointing to a word outside the bigram (Fig. 1b) or multiple unconnected roots (Fig. 1c), then the bigram will be discarded for template generation. The reason is that our reuse approach is based on replacing a sentence fragment (for which the parse is extracted from a rule) with its head word. The parser will operate on this head word only, and the parsed fragment will then be linked with the resulting tree. To do this, reusable fragments must have a single head and no external material depending on their dependents.

If the dominant pattern is eligible according to these criteria, then we will consider it if its relative frequency (confidence) is above a given threshold (head threshold) and, since our parsing is labeled, if the most frequent dependency label involved in the pattern has, in turn, a relative frequency above a second threshold (label threshold).

The confidence of heads and labels for a trigram is calculated analogously. The number of possible patterns of a trigram increases to a total of 19. As can be seen in Fig. 2, a trigram can be a fully connected tree where the dependents have no descendants outside the scope (7 patterns) or where at least one of the dependents points to a word outside the scope (7 patterns)[1]. A trigram can also be not fully connected or not connected at all (5 patterns).

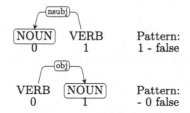

(a) bigram with dependents with no descendants outside the scope

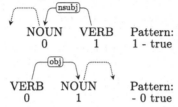

(b) bigram with a dependent pointing to a word outside the scope

NOUN   VERB   Pattern:
0   1   - - false

(c) bigram with missing relation

Fig. 1. Possible patterns for a bigram of PoS tags NOUN VERB and VERB NOUN used for calculating the confidence of a template. Each pattern denotes the index of each word's head ("-" in case of headless word) and whether a dependent is pointing to a word outside the scope of the bigram ("true") or not ("false").

---

[1] In this implementation we only consider projective trees for trigrams.

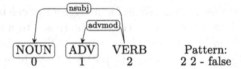

(a) trigram with dependents with no descendants outside the scope

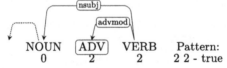

(b) trigram with a dependent pointing to a word outside the scope

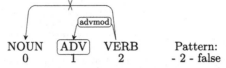

(c) trigram with missing relation with one of its components

Fig. 2. Some of the possible patterns for a trigram of PoS tags NOUN ADV VERB used for calculating the confidence of a template. Each pattern denotes the index of word's head ("-" in case of headless word) and whether a dependent is pointing to a word outside the scope of the trigram ("true") or not ("false").

Similarly to the case of bigrams, only patterns for trigrams highlighted in Fig. 2a that exceed predefined thresholds for confidence will be used as templates. If the input fragment matches a template, all dependents will be removed from the input to be processed by a parser. The patterns from Fig. 2b and Fig. 2c are discarded from being candidates for a template.

4   Experiments

4.1   Data and Evaluation

We conducted our experiments on the English treebank from Universal Dependencies (UD) v2.1 [9] and the results are evaluated with Unlabeled and Labeled Attachment Scores (UAS and LAS). The speeds are measured in tokens/second on CPU[2].

4.2   Model

During training time our model computes a set of rules that surpasses two thresholds: one for the dominant heads and the second for the dominant labels for a

---

[2] Intel Core i7-7700 CPU 4.2 GHz.

given n-gram. In our experiments the thresholds were set manually. Each template contains information about the n-gram's most likely head(s) and label(s) in order to automatically assign a syntactic analysis to the input that matches that template at parsing time.

As an example, Table 1 illustrates some templates that surpass the thresholds in the training set. The thresholds were set to 83% for confidence of head and 83% for confidence of label. This generated 141 unique templates, of which 97 had confidence of 100% for both thresholds but with low frequency.

Table 1. Example of templates generated during training time that surpass the predefined thresholds: 83% for confidence of the dominant head pattern and 83% for confidence of the dominant label pattern for a given n-gram of PoS tags.

Template	Dominant head pattern	Confidence of head pattern (%)	Dominant label pattern	Confidence of label pattern (%)
DET NOUN	1 - false	86.50	det - false	86.49
SCONJ PROPN VERB	2 2 - false	100	mark nsubj - false	100
AUX ADJ	1 - false	93.71	cop - false	93.64
ADV VERB	1 - false	93.17	advmod - false	91.31

The rules are applied to matching bigrams and trigrams on the training set, removing words other than the head.[3] This produces a reduced training set that no longer contains n-grams matching any of the templates. Our parser is then trained on this reduced training set. At parsing time templates are applied to the input, which is reduced in the same way, the parser is then ran on this shorter input and finally the parse tree fragments are attached to the resulting output. We verified experimentally that, as expected, a parser achieves better accuracy combined with our technique when trained on the reduced than on the entire training set.

4.3 Results

In our experiments we run the following dependency parsers on the remaining test set: transition-based BIST Parser [7] that uses bidirectional long short memory (BiLSTM) networks, as well as MaltParser [8] with the arc-eager and Covington transition systems.

Table 2 demonstrates the results after applying templates with threshold variation on the development set. In the experimental setup we use templates consisting of bigrams alone, trigrams alone or combined in order to compare their performance. While the specific parameters to choose depend on the speed-accuracy balance one wants to achieve, we selected the models where the thresholds for the dominant head and label pattern are 83-83 and 87-87 as our "optimal models",

---

[3] In case a bigram and trigram overlap, the n-gram with higher head confidence will be chosen and its dependents will be removed.

considering that they provide a reasonable trade-off between speed and accuracy. Thus, only these optimal thresholds were used afterwards when testing the technique on the development and test sets in order to investigate more in detail and to compare the accuracy and parsing speed. We use the notation $M_z^{x,y}$ where x indicates whether bigrams (2) were used, y trigrams (3) and z level of confidence for head and label pattern respectively.

Table 3 compares the performance of the parsers with and without applying our technique. The results are revealing in several ways. The experiments confirm that the more lenient confidence thresholds result in larger reductions of input length and thus faster parsing, but at the expense of accuracy. This applies to all parsers and indicates that our approach can be generic and applicable to diverse parsing algorithms. Moreover, the results show that it is feasible to reduce input size by more than 20% at the cost of less than 3 points of unlabeled attachment score.

Table 2. Performance of MaltParser with arc-eager transition system and the % of the text reduction after applying templates with different thresholds, on the development set. The total number of words in the development set: 25150 and test set: 25097.

Setup	UAS (%)	LAS (%)	Word Reduction (%)
$M_{90-90}^{2,3}$	84.73	82.15	5.0
$M_{87-87}^{2,3}$	**84.38**	**81.64**	**8.3**
M_{85-85}^{3}	84.15	81.45	8.5
M_{85-85}^{2}	84.01	81.17	11.1
$M_{85-85}^{2,3}$	83.1	80.15	18.2
$M_{83-83}^{2,3}$	**82.95**	**80.03**	**20.7**
$M_{80-80}^{2,3}$	81.51	78.42	22.7
$M_{80-70}^{2,3}$	81.2	77.71	24.2

5   Ongoing Work

One of the main weaknesses of the approach presented above is that the final templates generated during training are only based on adjacent words (i.e., words whose position indexes differ by 1). It does not take into account longer dependency arcs that could be captured by a bigram or trigram after removing the intervening dependents with shorter arcs. This approach can be improved by iteratively finding new templates and recalculating their confidence based on the outcome of applying the preceding template and removing the dependents it captured. In this way, an n-gram would capture a longer arc after intervening words have been removed. In this new approach, the order of applying templates generated during training time is crucial, instead of treating templates as a *bag-of-rules* that match an n-gram from the input.

Table 3. Performance of BIST Parser and MaltParser with the arc-eager and Covington transition systems and after applying templates compared with the baseline. The reported parsing speed (tokens/sec) only refers to the runtime of the dependency parser on the entire data set (baseline) or remaining text (that was passed to the parser after extracting the fragments captured by the templates) excluding the time needed to run the technique which is already negligible and will be optimized in the future versions.

Parser	Data Set	Setup	UAS (%)	LAS (%)	Word Reduction (%)	Tokens/Sec	Speed-up Factor
BIST Parser	dev	baseline	**88.07**	**86.08**	**NA**	1818 ± 51	NA
		$M^{2,3}_{87-87}$	86.94	84.49	8.3	2006 ± 20	1.10x
		$M^{2,3}_{83-83}$	85.20	82.34	20.7	2328 ± 34	1.28x
	test	baseline	**87.64**	**85.65**	**NA**	1860 ± 36	NA
		$M^{2,3}_{87-87}$	86.11	83.63	8.7	2006 ± 83	1.08x
		$M^{2,3}_{83-83}$	84.79	82.21	20.7	2291 ± 39	1.23x
MaltParser arc-eager	dev	baseline	**85.07**	**82.65**	**NA**	17387 ± 711	NA
		$M^{2,3}_{87-87}$	84.38	81.64	8.3	18118 ± 860	1.04x
		$M^{2,3}_{83-83}$	82.95	80.03	20.7	19758 ± 588	1.14x
	test	baseline	**84.58**	**82.00**	**NA**	17748 ± 801	NA
		$M^{2,3}_{87-87}$	83.32	80.44	8.7	18626 ± 459	1.05x
		$M^{2,3}_{83-83}$	82.02	79.13	20.7	19286 ± 889	1.09x
MaltParser Covington	dev	baseline	**83.99**	**81.65**	**NA**	16121 ± 581	NA
		$M^{2,3}_{87-87}$	82.84	80.24	8.3	16500 ± 819	1.02x
		$M^{2,3}_{83-83}$	81.69	78.92	20.7	18210 ± 484	1.13x
	test	baseline	**83.68**	**81.34**	**NA**	16009 ± 1035	NA
		$M^{2,3}_{87-87}$	82.75	80.02	8.7	16561 ± 629	1.03x
		$M^{2,3}_{83-83}$	81.36	78.6	20.7	17395 ± 1407	1.09x

We performed a preliminary experiment where new templates are generated based on the outcome of applying preceding templates and removing dependents. We believe it can be beneficial to localize noun phrases first in the input sentence, because they cover vast part of sentences. We look at the distance between PoS tags in an n-gram where the head is a noun. Some PoS tags show tendency to appear closer to a noun than others. We give priority to templates that capture PoS tags closest connected to a noun. In subsequent steps, we generate templates that show the highest confidence at each iteration, and add them to a list that is applied in order. This technique applied on MaltParser with the arc-eager transition system reduces the dev set by almost 21% with UAS of 82.14 and a LAS of 79.20.

6   Conclusion

We have obtained promising results where the input length can be reduced by more than 20% at the cost of less than 3 points of UAS. We believe that our work can be a starting point for developing templates that in the future can significantly speed up parsing time by avoiding redundant syntactic analyses at the minimal expense of accuracy.

The present study has investigated two approaches. In the first technique, which is the main focus of the paper (Sect. 4.2) templates were treated as a *bag-of-rules* that have to exceed predefined thresholds for the dominant head and label pattern for a given PoS tag n-gram, prioritizing ones with the highest confidence. In the second approach (Sect. 5) more importance is given to the order in which templates should be applied. However, the second technique is still in a preliminary stage, and requires some refinement. Research into solving this problem is already in progress. To further our research we plan to use both PoS tags and lemmas in our templates. The sparsity problem in finding n-grams involving lemmas will be tackled by augmenting training data with parsed sentences.

While we have tested our approach on transition-based dependency parsers, it is worth noting that the technique is generic enough to be applied to practically any kind of parser. Since fragment reuse is implemented as pre and postprocessing step, it works regardless of the inner working of the parser. As the technique reduces the input length received by the parser, speed gains can be expected to be larger on parsers with higher polynomial complexity, like those based on dynamic programming. The same idea would also be applicable to other grammatical representations, for example in constituent parsing, by changing the reusable fragments to the relevant representation (e.g. subtrees of a constituent tree).

Further studies will need to be undertaken in order to show the results of the approach when applied on other kinds of parsers, and on other languages different from English.

References

1. Baroni, M.: Distributions in text. In: Corpus Linguistics: An international handbook, vol. 2, pp. 803–821. Mouton de Gruyter (2009)
2. Bodenstab, N., Dunlop, A., Hall, K., Roark, B.: Beam-width prediction for efficient context-free parsing. In: Proceedings of the 49th Annual Meeting of the Association for Computational Linguistics: Human Language Technologies - Volume 1, HLT 2011, pp. 440–449. Association for Computational Linguistics, Stroudsburg (2011). http://dl.acm.org/citation.cfm?id=2002472.2002529
3. Gómez-Rodríguez, C., Alonso-Alonso, I., Vilares, D.: How important is syntactic parsing accuracy? An empirical evaluation on rule-based sentiment analysis. Artif. Intell. Rev. **52**(3), 2081–2097 (2017). https://doi.org/10.1007/s10462-017-9584-0
4. Ha, L.Q., Hanna, P., Ming, J., Smith, F.: Extending Zipf's law to n-grams for large corpora. Artif. Intell. Rev. **32**(1–4), 101–113 (2009). https://doi.org/10.1007/s10462-009-9135-4
5. Ha, L.Q., Sicilia-Garcia, E.I., Ming, J., Smith, F.J.: Extension of Zipf's law to words and phrases. In: Proceedings of the 19th International Conference on Computational Linguistics-Volume 1, pp. 1–6. Association for Computational Linguistics (2002)
6. Hüllermeier, E.: Case-Based Approximate Reasoning, Theory and Decision Library, vol. 44. Springer, Cham (2007). https://doi.org/10.1007/1-4020-5695-8

7. Kiperwasser, E., Goldberg, Y.: Simple and accurate dependency parsing using bidirectional LSTM feature representations. TACL **4**, 313–327 (2016). https:// transacl.org/ojs/index.php/tacl/article/view/885
8. Nivre, J., Hall, J., Nilsson, J.: Maltparser: a data-driven parser-generator for dependency parsing. In: Proceedings of LREC, vol. 6, pp. 2216–2219 (2006)
9. Nivre, J., et al.: Universal dependencies 2.1 (2017). http://hdl.handle.net/11234/ 1-2515. LINDAT/CLARIN digital library at the Institute of Formal and Applied Linguistics ('UFAL), Faculty of Mathematics and Physics, Charles University
10. Richter, M.M., Aamodt, A.: Case-based reasoning foundations. Knowl. Eng. Rev. **20**(3), 203–207 (2005). https://doi.org/10.1017/S0269888906000695
11. Smith, F., Devine, K.: Storing and retrieving word phrases. Inf. Process. Manage. **21**(3), 215–224 (1985)
12. Straka, M., Straková, J.: Tokenizing, pos tagging, lemmatizing and parsing ud 2.0 with udpipe. In: Proceedings of the CoNLL 2017 Shared Task: Multilingual Parsing from Raw Text to Universal Dependencies, pp. 88–99. Association for Computational Linguistics, Vancouver, August 2017. http://www.aclweb.org/anthology/K/ K17/K17-3009.pdf
13. Vieira, T., Eisner, J.: Learning to prune: exploring the frontier of fast and accurate parsing. Trans. Assoc. Comput. Linguist. **5**, 263–278 (2017). https://transacl.org/ ojs/index.php/tacl/article/view/924

Unmasking Bias in News

Javier Sánchez-Junquera[1][(✉)], Paolo Rosso[1], Manuel Montes-y-Gómez[2], and Simone Paolo Ponzetto[3]

[1] PRHLT Research Center, Universitat Politècnica de València, Valencia, Spain
jjsjunquera@gmail.com, prosso@dsic.upv.es
[2] Instituto Nacional de Astrofísica Óptica y Electrónica, Puebla, Mexico
mmontesg@inaoep.mx
[3] Data and Web Science Group, University of Mannheim, Mannheim, Germany
simone@informatik.uni-mannheim.de

Abstract. We present experiments on detecting hyperpartisanship in news using a 'masking' method that allows us to assess the role of style vs. content for the task at hand. Our results corroborate previous research on this task in that topic related features yield better results than stylistic ones. We additionally show that competitive results can be achieved by simply including higher-length n-grams, which suggests the need to develop more challenging datasets and tasks that address implicit and more subtle forms of bias.

Keywords: Bias in information · Hyperpartisanship and orientation · Masking technique

1 Introduction

Media such as radio, TV channels, and newspapers control which information spreads and how it does it. The aim is often not only to inform readers but also to influence public opinion on specific topics from a hyperpartisan perspective.

Social media, in particular, have become the default channel for many people to access information and express ideas and opinions. The most relevant and positive effect is the democratization of information and knowledge but there are also undesired effects. One of them is that social media foster information bubbles: every user may end up receiving only the information that matches her personal biases, beliefs, tastes and points of view. Because of this, social media are a breeding ground for the propagation of fake news: when a piece of news outrages us or matches our beliefs, we tend to share it without checking its veracity; and, on the other hand, content selection algorithms in social media give credit to this type of popularity because of the click-based economy on which their business are based. Another harmful effect is that the relative anonymity of social networks facilitates the propagation of toxic, hate and exclusion messages. Therefore, social media contribute to the misinformation and polarization of society, as we have recently witnessed in the last presidential elections in USA

A. Gelbukh (Ed.): CICLing 2019, LNCS 13451, pp. 658–667, 2023.
https://doi.org/10.1007/978-3-031-24337-0_47

or the Brexit referendum. Clearly, the polarization of society and its underlying discourses are not limited to social media, but rather reflected also in political dynamics (e.g., like those found in the US Congress [1]): even in this domain, however, social media can provide a useful signal to estimate partisanship [4].

Closely related to the concept of controversy and the "filter bubble effect" is the concept of bias [2], which refers to the presentation of information according to the standpoints or interests of the journalists and the news agencies. Detecting bias is very important to help users to acquire balanced information. Moreover, how a piece of information is reported has the capacity to evoke different sentiments in the audience, which may have large social implications (especially in very controversial topics such as terror attacks and religion issues).

In this paper, we approach this very broad topic by focusing on the problem of detecting hyperpartisan news, namely news written with an extreme manipulation of the reality on the basis of an underlying, typically extreme, ideology. This problem has received little attention in the context of the automatic detection of fake news, despite the potential correlation between them. Seminal work from [5] presents a comparative style analysis of hyperpartisan news, evaluating features such as characters n-grams, stop words, part-of-speech, readability scores, and ratios of quoted words and external links. The results indicate that a topic-based model outperforms a style-based one to separate the left, right and mainstream orientations.

We build upon previous work and use the dataset from [5]: this way we can investigate hyperpartisan-biased news (i.e., extremely one-sided) that have been manually fact-checked by professional journalists from BuzzFeed. The articles originated from 9 well-known political publishers, three each from the mainstream, the hyperpartisan left-wing, and the hyperpartisan right-wing. To detect hyperpartisanship, we apply a masking technique that transforms the original texts in a form where the textual structure is maintained, while letting the learning algorithm focus on the writing style or the topic-related information. This technique makes it possible for us to corroborate previous results that content matters more than style. However, perhaps surprisingly, we are able to achieve the overall best performance by simply using higher-length n-grams than those used in the original work from [5]: this seems to indicate a strong lexical overlap between different sources with the same orientation, which, in turn, calls for more challenging datasets and task formulations to encourage the development of models covering more subtle, i.e., implicit, forms of bias.

The rest of the paper is structured as follows. In Sect. 2 we describe our method to hyperpartisan news detection based on masking. Section 3 presents details on the dataset, experimental results and a discussion of our results. Finally, Sect. 4 concludes with some directions for future work.

2   Investigating Masking for Hyperpartisanship Detection

The masking technique that we propose here for the hyperpartisan news detection task has been applied to text clustering [3], authorship attribution [7], and

Table 1. Examples of masking style-related information or topic-related information.

Original text	Masking topic-related words	Masking style-related words
Officers **went after** Christopher **Few after** watching **an** argument **between him and his** girlfriend outside **a** bar **just before the** 2015 shooting	* went after * Few after * an * between him and his * a * just before the # *	Officers * * Christopher * * watching * argument * * * * girlfriend outside * bar * * * 2015 shooting

Table 2. Statistics of the original dataset and its subset used in this paper.

	Left-wing	Mainstream	Right-wing	Σ
Original data [5]	256	826	545	1627
Cleaned data	252	787	516	1555

recently to deception detection [6] with encouraging results. The main idea of the proposed method is to transform the original texts to a form where the textual structure, related to a general style (or topic), is maintained while content-related (or style-related) words are masked. To this end, all the occurrences (in both training and test corpora) of non-desired terms are replaced by symbols.

Let W_k be the set of the k most frequent words, we mask all the occurrences of a word $w \in W_k$ if we want to learn a *topic-related model*, or we mask all $w \notin W_k$ if we want to learn a *style-based model*. Whatever the case, the way in which we mask the terms in this work is called *Distorted View with Single Asterisks* and consists in replacing w with a single asterisk or a single # symbol if the term is a word or a number, respectively. For further masking methods, refer to [7].

Table 1 shows a fragment of an original text and the result of masking style-related information or topic-related information. With the former we obtain distorted texts that allow for learning a *topic-based model*; on the other hand, with the latter, it is possible to learn a *style-based model*. One of the options to choose the terms to be masked or maintained without masking is to take the most frequent words of the target language [7]. In the original text from the table, we highlight some of the more frequent words in English.

3   Experiments

We used the BuzzedFeed-Webis Fake News Corpus 2016 collected by [5] whose articles were labeled with respect to three political orientations: mainstream, left-wing, and right-wing (see Table 2). Each article was taken from one of 9 publishers known as hyperpartisan left/right or mainstream in a period close to the US presidential elections of 2016. Therefore, the content of all the articles is related to the same topic.

During initial data analysis and prototyping we identified a variety of issues with the original dataset: we cleaned the data excluding articles with empty or bogus texts, e.g. *'The document has moved here'* (23 and 14 articles respectively). Additionally, we removed duplicates (33) and files with the same text but inconsistent labels (2). As a result, we obtained a new dataset with 1555 articles out of 1627.[1] Following the settings of [5], we balance the training set using random duplicate oversampling.

3.1 Masking Content vs. Style in Hyperpartisan News

In this section, we reported the results of the masking technique from two different perspectives. In one setting, we masked *topic-related information* in order to maintain the predominant writing style used in each orientation. We call this approach a *style-based model*. With that intention we selected the k most frequent words from the target language, and then we transformed the texts by masking the occurrences of the rest of the words. In another setting, we masked *style-related information* to allow the system to focus only on the topic-related differences between the orientations. We call this a *topic-based model*. For this, we masked the k most frequent words and maintained intact the rest.

After the text transformation by the masking process in both the training and test sets, we represented the documents with character n-grams and compared the results obtained with the *style-based* and the *topic-related models*.

3.2 Experimental Setup

Text Transformation: We evaluated different values of k for extracting the k most frequent words from English[2]. For the comparison of the results obtained by each model with the ones of the state-of-the-art, we only showed the results fixing $k = 500$.

Text Representation: We used a standard bag-of-words representation with *tf* weighting and extract character n-grams with a frequency lower than 50.

Classifier: We compared the results obtained with Naïve Bayes (NB), Support Vector Machine (SVM) and Random Forest (RF); for the three classifiers we used the versions implemented in *sklearn* with the parameters set by default.

Evaluation: We performed 3-fold cross-validation with the same configuration used in [5]. Therefore, each fold comprised one publisher from each orientation (the classifiers did not learn a publisher's style). We used macro F_1 as the evaluation measure since the test set is unbalanced with respect to the three classes. In order to compare our results with those reported in [5], we also used accuracy, precision, and recall.

Baseline: Our baseline method is based on the same text representation with the character n-grams features, but without masking any word.

---

[1] The dataset is available at https://github.com/jjsjunquera/UnmaskingBiasInNews.
[2] We use the BNC corpus (https://www.kilgarriff.co.uk/bnc-readme.html) for the extraction of the most frequent words as in [7].

Table 3. Results of the proposed masking technique ($k = 500$ and $n = 5$) applied to mask topic-related information or style-related information. NB: Naive Bayes; RF: Random Forest; SVM: Support Vector Machine. The last two rows show the results obtained by applying the system from [5][5] to our cleaned dataset (Sect. 3).

Masking Method	Classifier	Macro F_1	Accuracy	Precision			Recall			F_1		
				left	right	main	left	right	main	left	right	main
Baseline model	NB	0.52	0.56	0.28	0.57	0.81	**0.49**	0.58	0.56	0.35	0.57	0.66
	RF	0.56	0.62	0.28	0.61	0.80	0.36	0.72	0.63	0.32	0.66	0.70
	SVM	**0.70**	**0.77**	**0.55**	**0.75**	**0.84**	0.42	**0.79**	**0.87**	**0.47**	**0.77**	**0.85**
Style-based model	NB	0.47	0.52	0.20	0.51	0.73	0.28	0.65	0.49	0.23	0.57	0.59
	RF	0.46	0.53	0.24	0.58	0.64	0.36	0.34	0.73	0.29	0.43	0.68
	SVM	0.57	0.66	0.33	0.66	0.75	0.26	0.61	0.84	0.29	0.62	0.79
Topic-based model	NB	0.54	0.60	0.26	0.63	0.74	0.36	0.62	0.65	0.29	0.62	0.69
	RF	0.53	0.55	0.27	0.64	0.71	0.44	0.60	0.58	0.33	0.61	0.64
	SVM	0.66	0.74	0.48	0.73	0.81	0.38	0.78	0.82	0.42	0.75	0.82
System from [5] (applied to our cleaned dataset)												
Style	RF	0.61	0.63	0.29	0.62	0.71	0.16	0.62	0.80	0.20	0.61	0.74
Topic	RF	0.63	0.65	0.27	0.65	0.72	0.15	0.62	0.84	0.19	0.63	0.77

3.3   Results and Discussion

Table 3 shows the results of the proposed method. We compare with [5] against their topic and style-based methods. In order to compare our results with those reported in [5], we report the same measures the authors used. We also include the macro F_1 score because of the unbalance test set. For these experiments we extract the character 5-grams from the transformed texts, taking into account that as more narrow is the domain more sense has the use of longer n-grams. We follow the steps of [7] and set k = 500 for this comparison results.

Similar to [5], the topic-based model achieves better results than the style-related model. However, the differences between the results of the two evaluated approaches are much higher (0.66 vs. 0.57 according to Macro F_1) than those shown in [5]. The highest scores were consistently achieved using the SVM classifier and masking the style-related information (i.e., the topic-related model). This could be due to the fact that all the articles are about the same political event in a very limited period of time. In line with what was already pointed out in [5], the left-wing orientation is harder to predict, possibly because this class is represented with fewer examples in the dataset.

Another reason why our masking approach achieves better results could be that we use a higher length of character n-grams. In fact, comparing the results of [5] against our baseline model, it is possible to note that even without masking any word, the classifier obtains better results. This suggests that the good results are due to the length of the character n-grams rather than the use of the masking technique.

Robustness of the approach to different values of k and n. With the goals of: (i) understanding the robustness of the approach to different parameter values; and to see if (ii) it is possible to overcome the $F_1 = 0.70$ from the baseline model, we vary the values of k and n and evaluate the macro F_1 using SVM.

Figure 1 shows the results of the variation of $k \in \{100, 200, ..., 5000\}$. When $k > 5000$, we clearly can see that the topic-related model, in which the k most frequent terms are masked, is decreasing the performance. This could be explained by the fact that relevant topic-related terms start to be masked too. However, a different behavior is seen in the style-related model, in which we tried to maintain only the style-related words without masking them. In this model, the higher is k the better is the performance. This confirms that for the used dataset, taking into account only style-related information is not good, and observing also topic-related information benefits the classification. When k tends to the vocabulary size, the style-related model tends to behave like the baseline model, which we already saw in Table 3 that achieves the best results.

From this experiment, we conclude that: (i) the topic-related model is less sensitive than the style-related model when $k < 500$, i.e. the k most frequent terms are style-related ones; and (ii) when we vary the value of k, both models achieve worse results than our baseline.

On the other hand, the results of extracting character 5-grams are higher than extracting smaller n-grams, as can be seen in Fig. 2. These results confirm that perhaps the performance of our approach overcomes the models proposed in [5] because of the length of the n-grams[6].

Relevant features. Table 4 shows the features with the highest weights from the SVM (we use `scikit-learn`'s method to collect feature weights). It is possible to note that the mention of *cnn* was learned as a discriminative feature when the news from that publisher were used in the training (in the topic-based model). However, this feature is infrequent in the test set where no news from CNN publisher was included (Table 5).

Features like *donal* and *onal* are related to Donald Trump, while *illar* and *llary* refer to Hillary Clinton. Each of these names is more frequent in one of the hyperpartisan orientation, and none of them occurs frequently in the mainstream orientation. On the other hand, the relevant features from the style-based model involve function words that are frequent in the three classes (e.g., *out, you, and, of*) even if the combination between function words and other characters can lightly differ in different orientations.

---

[6] In [5] the authors used $n \in [1, 3]$.

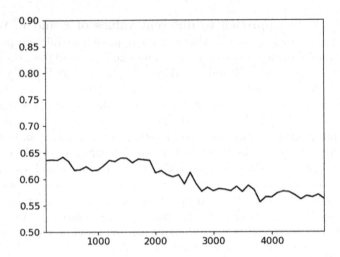

(a) Varying k values and masking the most frequent words: topic-based model.

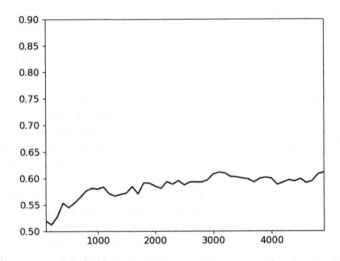

(b) Varying k values and maintaining without masking the most frequent words: style-based-model.

Fig. 1. Macro F_1 results of the proposed masking technique. We set n = 5 for comparing results of different values of k.

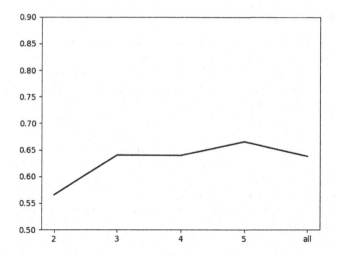

(a) Varying n values and masking the 500 most frequent words.

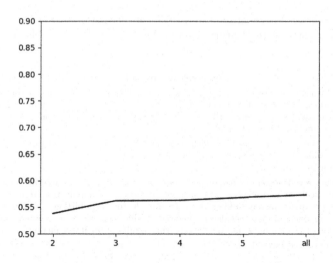

(b) Varying n values and maintaining without masking the 500 most frequent words.

Fig. 2. Macro F_1 results of the proposed masking technique.

Table 4. Most relevant features to each class in each model.

Baseline model			Style-based model			Topic-based model		
left	main	right	left	main	right	left	main	right
_imag	_cnn_	e_are	but_*	n_thi	y_**_	ant_*	_cnn_	hilla
_that	said_	lary_	out_w	s_*_s	out_a	_imag	cs_*_	_*_da
e_tru	_said	_your	t_**_	_how_	as_to	lies_	ics_*	als_*
e_don	y_con	n_pla	you_h	at_he	o_you	_*_ex	sday_	_*_le
_here	ry_co	e_thi	t_and	m_*_t	ell_*	etty_	ed_be	_dail
s_of_	_cnn_	s_to_	_is_a	*_*_u	and_n	donal	_cnn_	_*_te
e_tru	n_ame	_your	h_*_a	e_#_*	hat_w	n_*_c	day_*	*_ame
for_h	said_	illar	_of_#	and_*	*_#_#	onald	cs_*_	_*_am
donal	_said	hilla	or_hi	**_*_	_it_t	ying_	ics_*	illar
racis	ore_t	llary	for_h	t_the	e_of_	thing	*_*_e	llary
_here	said_	and_s	**_*_	and_*	o_you	_*_*	ed_be	_*_le
_kill	story	_hill	_in_o	_*_tw	n_it_	e_*_	y_con	_*_ri
_that	_said	_let_	hat_*	*_two	and_n	n_*_	tory_	_hill
_trum	tory_	_comm	f_**_	s_**_	_all_	t_*_	story	_bomb
trump	ed_be	lary_	*_so_	*_onl	f_*_w	_imag	d_bel	*_*_r

Table 5. Fragments of original texts and their transformation by masking the k most frequent terms. Some of the features from Table 4 using the topic-related model are highlighted.

	Topic-related model
left	(...)which his son pretty much confirmed in a foolish statement. The content of those tax returns has been the subject of much speculation, but given Trumps long history of tax evasion and political bribery, it doesnt take much imagination to assume hes committing some kind of fraud
	* * son pretty * confirmed * foolish statement * content * * tax returns * * * subject * * speculation * * Trump * * * tax evasion * * bribery * doesn * * imagination * assume * committing * * * fraud
main	Obama proved beyond a shadow of a doubt in 2011 when he released his long-form birth certificate (...) CNN and Fox News cut away at points in the presentation. Networks spent the day talking about Trump's history as a birther (...) Before Friday, the campaign's most recent deception came Wednesday when campaign advisers told reporters that Trump would not be releasing results of his latest medical exam
	Obama proved beyond shadow * doubt * 2011 * * released * ** birth certificate (...) CNN * Fox News cut * *points * * presentation Networks spent * * talking * Trump * * birther (...) * Friday * campaign * recent deception * Wednesday * campaign advisers told reporters * Trump * * * releasing results * * latest medical exam
right	The email, which was dated March 17, 2008, and shared with POLITICO, reads: Jim, on Kenya your person in the field might look into the impact there of Obama's public comments about his father. I'm told by State Dept officials that Obama publicly derided his father on (...) Blumenthal, a longtime confidant of both Bill and Hillary Clinton, emerged as a frequent correspondent in the former secretary of (...)
	* email * * dated March 17 2008 * shared * POLITICO reads Jim * Kenya * * * * field * * * * impact * * Obama * comments * * * told * * Dept officials * Obama publicly derided * * (...) Blumenthal longtime confidant * * Bill * Hillary Clinton emerged * frequent correspondent * * former secretary * (...)

4   Conclusions

In this paper we presented initial experiments on the task of hyperpartisan news detection: for this, we explored the use of masking techniques to boost the performance of a lexicalized classifier. Our results corroborate previous research on the importance of content features to detect extreme content: masking, in addition, shows the benefits of reducing data sparsity for this task comparing our results with the state of the art. We evaluated different values of the parameters and see that finally our baseline model, in which we extract character 5-grams without applying any masking process, achieves the better results. As future work we plan to explore more complex learning architectures (e.g., representation learning of masked texts), as well as the application and adaptation of unsupervised methods for detecting ideological positioning from political texts 'in the wild' for the online news domain.

Acknowledgments. The work of Paolo Rosso was partially funded by the Spanish MICINN under the research project MISMIS-FAKEnHATE on Misinformation and Miscommunication in social media: FAKE news and HATE speech (PGC2018-096212-B-C31).

References

1. Andris, C., Lee, D., Hamilton, M.J., Martino, M., Gunning, C.E., Selden, J.A.: The rise of partisanship and super-cooperators in the US house of representatives. PLoS ONE **10**(4), e0123507 (2015)
2. Baeza-Yates, R.: Bias on the web. Commun. ACM **61**(6), 54–61 (2018). https://doi.org/10.1145/3209581
3. Granados, A., Cebrian, M., Camacho, D., de Borja Rodriguez, F.: Reducing the loss of information through annealing text distortion. IEEE Trans. Knowl. Data Eng. **23**(7), 1090–1102 (2011)
4. Hemphill, L., Culotta, A., Heston, M.: # polarscores: measuring partisanship using social media content. J. Inf. Tech. Polit. **13**(4), 365–377 (2016)
5. Potthast, M., Kiesel, J., Reinartz, K., Bevendorff, J., Stein, B.: A stylometric inquiry into hyperpartisan and fake news. In: Proceedings of the 56th Annual Meeting of the Association for Computational Linguistics (Volume 1: Long Papers), pp. 231–240 (2018)
6. Sánchez-Junquera, J.: Adaptación de dominio para la detección automática de textos engañosos. Master's thesis, Instituto Nacional de Astrofísica, Óptica y Electrónica (2018). http://inaoe.repositorioinstitucional.mx/jspui/handle/1009/1470. In Spanish
7. Stamatatos, E.: Authorship attribution using text distortion. In: Proceedings of the 15th Conference of the European Chapter of the Association for Computational Linguistics: Volume 1, Long Papers, vol. 1, pp. 1138–1149 (2017)

Author Index

Printed in the United States
by Baker & Taylor Publisher Services